FUNDAMENTALS OF GENETICS

OF RELATED INTEREST
FROM THE BENJAMIN/CUMMINGS SERIES IN THE LIFE SCIENCES

FUNDAMENTALS OF GENETICS

Second Edition

Peter J. Russell

REED COLLEGE

An imprint of Addison Wesley Longman, Inc.

San Francisco • Reading, Massachusetts • New York • Harlow, England
Don Mills, Ontario • Sydney • Mexico City • Madrid • Amsterdam

Executive Editor: Erin Mulligan
Sponsoring Editor: Lynn Cox
Project Editor: Evelyn Dahlgren
Developmental Editor: Becky Strehlow
Publishing Assistant: Kristina Roth
Managing Editor: Laura Kenney
Senior Art Supervisor: Donna Kalal
Illustrators: Imagineering; J/B Woolsey Associates, Inc.
Copyeditor: Alan Titche
Cover Designer: Yvo Riezebos
Prepress Supervisor: Vivian McDougal
Compositor: Pre-Press Company, Inc.
Marketing Manager: Gay Meixel

On the cover: Computer-generated model of Cro protein bound to DNA.
©Ken Eward/Science Source/Photo Researchers, Inc.

Library of Congress Cataloging-in-Publication Data
Russell, Peter J.
    Fundamentals of genetics/Peter J. Russell—2nd ed.
        p.   cm.
    Includes index.
    ISBN 0-321-03626-3
    1. Genetics.   I. Title
OH430.R8685   1999   99-43584
576.5—dc21   CIP

1 2 3 4 5 6 7 8 9 10—RNV—03 02 01 00 99

Benjamin/Cummings, an imprint of Addison Wesley Longman, Inc.
1301 Sansome Street
San Francisco, CA 94111

BRIEF CONTENTS

DETAILED CONTENTS

14

CLONING AND MANIPULATION OF DNA   297

15

REGULATION OF GENE EXPRESSION IN BACTERIA AND BACTERIOPHAGES   334

16

EUKARYOTIC GENE REGULATION 357

22

QUANTITATIVE GENETICS 507

PREFACE

OVERVIEW OF THE TEXT

It is now more than 40 years since the mysteries of DNA were first unraveled, and in that time genetics has become one of the most exciting and ground-breaking of the sciences. A continual flood of discoveries not only expands our understanding about heredity but also affects our daily lives in areas ranging from disease therapy to courtroom evidence.

Fundamentals of Genetics, Second Edition, reflects this dynamic nature of the field. It has been extensively updated to include many significant advances made since the last edition, particularly those advances resulting from the use of molecular experimental approaches. *Fundamentals of Genetics*, Second Edition, is a shorter and less complex version of my *Genetics* textbook. *Fundamentals of Genetics*, Second Edition, has been crafted to present a full coverage of genetics in a careful and quality manner, but without overwhelming the student with details and complex explanations. This text shares with my *Genetics* text an "experimental" approach in which a solid treatment is given of many of the research experiments that contributed to our knowledge of the subject. However, *Fundamentals of Genetics*, Second Edition, is not simply a "cut-and-paste" version of *Genetics*. Rather, this text was written using the more comprehensive text as a starting point for the overall treatment of the subject, with great attention paid to reducing the complexity of the discussions and making the text easier to read. Thus, *Fundamentals of Genetics*, Second Edition, is an ideally suitable text for students who have a limited background in biology and chemistry. The length of this new text is much shorter than more comprehensive texts, making it ideal not only for one-semester courses in genetics but also for one-quarter courses.

Fundamentals of Genetics, Second Edition, retains the overall approach, organization, and pedagogical features (such as Principal Points, Keynotes, Summaries, and Analytical Approaches for Solving Genetics Problems) that made the previous edition a valued learning tool for students of genetics. Problem solving remains a major feature of the book, and this edition includes more and updated end-of-chapter Questions and Problems. It also retains a format that allows instructors to use the chapters out of sequence to accommodate various teaching approaches.

The second edition includes the following improvements:

- Human examples continue to be used extensively throughout the text, and they have been updated particularly with regard to new molecular understandings of various human genetic diseases.
- All the molecular aspects of genetics are updated, so that the book continues to reflect our current understanding of genes at the molecular level. New material is presented, for example, on the structure of eukaryotic chromosomes (Chapter 10), the control of the cell cycle (Chapter 11), promoter elements and initiation of transcription in eukaryotes (Chapter 12), 3' end generation and polyadenylation in eukaryotes (Chapter 13), alternative splicing in the production of mRNAs (Chapter 16), modern molecular screens for the isolation of mutants (Chapter 18), and investigation of genetic relationships by mitochondrial DNA analysis (Chapter 20).
- The coverage of mapping genes in human chromosomes (Chapter 5) now involves a modern presentation of constructing physical and genetic maps of the human genome.
- The coverage of recombinant DNA technology has been updated and expanded to reflect the rapid developments in that area. New material is presented, for example, on the use of PCR (polymerase chain reaction) in the analysis of diseases and the current status of the Human Genome Project.
- The coverage of developmental genetics has been updated to include current information on *Drosophila* development and on the relationship between homeotic genes of *Drosophila* and mammals. The experiment demonstrating that the nucleus of an adult sheep cell could direct the development of a normal lamb from an enucleated egg cell is described.
- A new chapter, "Genetics of Cancer" (Chapter 17), discusses the relationship of the cell cycle to cancer, mutational models for cancer, genes and

cancer, the multistep nature of cancer, and chemicals and radiation as carcinogens. Cancer is of obvious importance to the human population, and this chapter imparts important knowledge to students about the nature of the disease and the various roles that genes play in its development.

- Approximately 160 new questions and problems have been added throughout the book.

ORGANIZATION AND COVERAGE

The three major areas of genetics—transmission genetics, molecular genetics, and population and quantitative genetics—are covered in 22 chapters. Chapter 1 is an introductory chapter designed to summarize the main branches of genetics, to explain what geneticists do and what their areas of research encompass, and to discuss cellular reproduction in eukaryotes, in particular mitosis and meiosis. The next six chapters deal with transmission of the genetic material. Chapters 2 and 3 present the basic principles of genetics in relation to Mendel's laws. Chapter 2 is focused on Mendel's contributions to our understanding of the principles of heredity, while Chapter 3 covers the experimental evidence for the relationship between genes and chromosomes, and methods of sex determination. Mendelian genetics in humans is introduced in Chapter 2 with a focus on pedigree analysis and autosomal traits. The topic is continued in Chapter 3 with respect to sex-linked genes.

The exceptions to and extensions of Mendelian analysis (such as the existence of multiple alleles, the modification of dominance relationships, gene interactions and modified Mendelian ratios, essential genes and lethal alleles, and the relationship between genotype and phenotype) are described in Chapter 4. In Chapter 5 genetic mapping in eukaryotes is presented. We describe how the order of and distance between the genes on eukaryotic chromosomes are determined in genetic experiments designed to quantify the crossovers that occur during meiosis. Specialized topics discussed are the procedures for constructing genetic and physical maps in humans, and tetrad analysis. In Chapter 6, we discuss the ways of mapping genes in bacteriophages and in bacteria, which take advantage of the processes of transformation, conjugation, and transduction. Fine structure analysis of bacteriophage genes concludes this chapter. With the understanding of the relationship between genes and chromosomes obtained from Chapters 2 through 6, chromosomal mutations—changes in normal chromosome structure or chromosome number—are discussed in Chapter 7. Chromosomal mutations in eukaryotes and human disease syndromes that result from chromosomal mutations, including triplet repeat mutations, are emphasized.

Chapters 8 through 14 comprise the "molecular core" of *Fundamentals of Genetics,* Second Edition, detailing the current level of our knowledge about the molecular aspects of genetics. In Chapter 8 we examine some aspects of gene function, such as the genetic control of the structure and function of proteins and enzymes and the role of genes in directing and controlling biochemical pathways. A number of examples of human genetic diseases that result from enzyme deficiencies are described to reinforce the concepts. The discussion of gene function in Chapter 8 enables students to understand the important concept that genes specify proteins and enzymes, setting them up for the following chapters in which gene structure and expression are discussed.

In Chapter 9 we cover the structure of DNA, presenting the classical experiments that revealed DNA to be genetic material and that established the double helix model as the structure of DNA. The details of DNA structure and organization in prokaryotic and eukaryotic chromosomes are set out in Chapter 10. We cover DNA replication in prokaryotes and eukaryotes and recombination between DNA molecules in Chapter 11.

After thoroughly explaining the nature of the gene and its relationship to chromosome structure, in Chapter 12 we discuss the first step in the expression of a gene—transcription. First, we describe the general process of transcription and then present the currently understood details of the transcription of messenger RNA, transfer RNA, and ribosomal RNA genes, and the production of mature mRNA by processing of the initial transcript. In Chapter 13 we describe the structure of proteins, the evidence for the nature of the genetic code, and a detailed description of our current knowledge of translation in both prokaryotes and eukaryotes. In Chapter 14 we discuss recombinant DNA technology and other molecular techniques that are now essential tools of most areas of modern genetics. There are descriptions of the use of recombinant DNA technology to clone and characterize genes and to manipulate DNA, followed by a discussion of the applications of recombinant DNA technology in the analysis of biological processes, the diagnosis of human diseases, the isolation of human genes, the Human Genome Project (with the goal of mapping and sequencing the complete genomes of humans and other selected organisms), forensics (DNA typing), gene therapy, the development of commercial products, and the genetic engineering of plants.

The next two chapters focus on regulation of gene expression in prokaryotes (Chapter 15) and eukaryotes (Chapter 16). In Chapter 15 we discuss the operon as a unit of gene regulation, the current molecular details in the regulation of gene expression in bacterial operons, and regulation of genes in bacteriophages. In Chapter 16 we explain how eukaryotic gene expression is regulated, stressing molecular changes that accompany gene regulation, short-term gene regulation in simple and complex eukaryotes, gene regulation in development and differentiation, and immunogenetics.

Next, in Chapter 17—the new chapter on the genetics of cancer—we discuss the relationship of the cell cycle to cancer, and the various types of genes that, when mutated, play a role in the development of cancer. We also discuss the fact that cancer usually requires a number of independent mutational events in order for it to develop. Lastly, we consider the induction of cancer by chemicals and radiation (carcinogens).

In Chapters 18 and 19 we describe some of the ways in which genetic material can change or be changed. Chapter 18 covers the processes of gene mutation, the procedures that screen for potential mutagens and carcinogens (the Ames test), some of the mechanisms that repair damage to DNA, and some of the procedures that are used to screen for particular types of mutants. Chapter 19 presents the structures and movements of transposable genetic elements in prokaryotes and eukaryotes.

In Chapter 20 we address the organization and genetics of extranuclear genomes of mitochondria and chloroplasts. We cover the current molecular information about the organization of genes within the extranuclear genomes and the classical genetic experiments that are used to study the inheritance of extranuclear genes. New to this chapter are discussions of RNA editing and genomic imprinting.

In Chapters 21 and 22 we describe the genetics of populations and quantitative genetics, respectively. In Chapter 21, "Population Genetics," we present the basic principles in population genetics, extending our studies of heredity from the individual organism to a population of organisms. In Chapter 22, "Quantitative Genetics," we consider the heredity of traits in groups of individuals that are determined by many genes simultaneously. In this chapter we also discuss heritability: the relative extent to which a characteristic is determined by genes or by the environment. Both Chapters 21 and 22 include discussions of the application of molecular tools to these areas of genetics. Chapter 21, for example, includes a section on measuring genetic variation with RFLPs and DNA sequencing as well as a discussion of molecular evolution.

PEDAGOGICAL FEATURES

Because the field of genetics is complex, making the study of it potentially difficult, we have incorporated a number of special pedagogical features to assist students and to enhance their understanding and appreciation of genetic principles. These features have proved to be very effective in the previous edition of this text:

- Each chapter opens with an outline of its contents and a section called "Principal Points." Principal Points are short summaries that alert students to the key concepts they will encounter in the material to come.
- Throughout each chapter, strategically placed "Keynote" summaries emphasize important ideas and critical points.
- Important terms and concepts—highlighted in bold—are clearly defined where they are introduced in the text. For easy reference, they are also compiled in a glossary at the back of the book.
- Some chapters include boxes covering special topics related to chapter coverage. Some of these boxed topics are: "Genetic Terminology" (Chapter 2); "Hardy, Weinberg, and the History of Their Contribution to Population Genetics" (Chapter 21); and "Analysis of Genetic Variation with Protein Electrophoresis" (Chapter 21).
- Chapter summaries close each chapter, further reinforcing the major points that have been discussed.
- With the exception of the introductory Chapter 1, all chapters conclude with a section entitled "Analytical Approaches for Solving Genetics Problems." Genetics principles have always been best taught with a problem-solving approach. However, beginning students often do not acquire the necessary experience with basic concepts that would enable them to attack assigned problems methodically. In the "analytical approaches" sections (pioneered in earlier editions of this text), typical genetics problems are talked through in step-by-step detail to help students understand how to tackle a genetics problem by applying fundamental principles.
- The problem sets that close the chapters include approximately 460 questions and problems designed to give students further practice in solving genetics problems. The problems for each chapter represent a range of topics and difficulty. The answers to questions indicated by an asterisk (*)—approximately one-half of the questions—can be found at the back of the book, and answers to all questions are available in a separate supplement, the *Study Guide and Solutions Manual*.

- Comprehensive and up-to-date suggested readings for each chapter are listed at the back of the book.
- Special care has been taken to provide the most useful index—extensive, accurate, and well cross-referenced.

SUPPLEMENTS

For the Student

A *Study Guide and Solutions Manual* to accompany this text has been prepared by Bruce Chase of the University of Nebraska. In addition to detailed solutions for all the problems in the text, the guide contains the following features for each chapter: a review of key terms, symbols, and concepts; a "Thinking Analytically" section, which provides guidance and tips on solving problems and avoiding common pitfalls; and additional questions for practice and review.

Genetics: Practice Problems and Solutions by Joseph Chinnici and David Matthes provides more than 400 problems of various types, including short-answer, conceptual, challenge, and collaborative problems.

BiologyLabs On-Line: Genetics Version allows students to get instant results for experiments that would take longer in the real world. The Genetics Version provides access to five genetics labs: FlyLab (formerly Virtual FlyLab), TranslationLab, HemoglobinLab, PedigreeLab, and EvolutionLab. **http:// biologylab.awlonline.com**

The *Fundamentals of Genetics Companion Web Site* allows instructors to offer on-line quizzing, create syllabi, conduct threaded discussion groups, and administer on-line content.

For the Instructor

The printed *Test Bank*, prepared by Holly Ahern of Adirondack Community College, includes over 800 multiple-choice, fill-in-the-blanks, and short-answer questions. The entire set of questions is also available in Windows and Macintosh test-generating software.

A set of 125 full-color transparencies—featuring large, clear labels for easy viewing—complements the text.

An art CD-ROM contains animations of key concepts and most of the line art from the text, saved as high-resolution files and appropriate for use in popular presentation programs, including PowerPoint.

The full complement of supplemental teaching materials is available to instructors who adopt the text.

ACKNOWLEDGMENTS

Making an invaluable contribution to this text are John and Bette Woolsey of J/B Woolsey Associates, Inc., who designed and generated much of the art for *Genetics* and *Fundamentals of Genetics*, First Edition. I also thank the artists at Imagineering for assistance in art production and art supervisor Donna Kalal for coordinating the art program for this edition.

I would also like to thank all of the reviewers involved in this edition:

Anna G. Brownell, Chapman University
Bruce J. Cochrane, University of South Florida
Joyce Cuff, Thiel College
J. Roger Eagan, Adirondack Community College
William F. Ettinger, Gonzaga University
Ann Evans, University of New Mexico, Albuquerque
Nancy H. Ferguson, Clemson University
Rodney S. Foth, University of Dubuque
Anne Gerber, University of North Dakota, Grand Forks
Michael A. Goldman, San Francisco State University
Kerry L. Hull, Bishop's University, Canada
Christina King-Smith, St. Joseph's University
Charles C. Lambert, California State University, Fullerton
Ron W. Leavitt, Brigham Young University
James C. Lin, Northwestern State University
Paul F. Lurquin, Washington State University, Pullman
Peter Luykx, University of Miami
Clint Magill, Texas A & M University
Katharine Mixter Mayne, Birmingham-Southern College
Michael P. McCann, St. Joseph's University
Florian Muckenthaler, Bridgewater State University
C. Neal McReynolds, Blue Mountain College
Mary Anne Nelson, University of New Mexico, Albuquerque
Gail R. Patt, Boston University
Vickie M. Peck, University of New Mexico, Albuquerque
Gregory Podgorski, Utah State University, Logan
DeLois Mitchell Powell, University of Maryland, Eastern Shore/Highline Community College
John Ringo, University of Maine, Orono
Albert D. Robinson, State University of New York, Potsdam
R. R. Schalles, Kansas State University, Manhattan
Robert Sharrock, Montana State University, Bozeman
Peter J. Wejksnora, University of Milwaukee
D. A. Whited, North Dakota State University

For this edition, I would also like to thank Bruce Chase (University of Nebraska, Omaha), who provided excellent new end-of-chapter problems; thanks also to accuracy checker Elizabeth Peake.

I am grateful to the literary executor of the late Sir Ronald A. Fisher, F.R.S., to Dr. Frank Yates, F.R.S., and to Longman Group Ltd. London, for permission to reprint Table IV from their book *Statistical Table for Biological, Agricultural and Medical Research* (Sixth Edition, 1974).

Finally, I wish to thank the publishing professionals who helped to make *Fundamentals of Genetics,* Second Edition, a physical reality. In particular, I thank Evelyn Dahlgren, project editor; Becky Strehlow, developmental editor; Kristina Roth, publishing assistant; Donna Kalal, senior art supervisor; Laura Kenney, managing editor; and Vivian McDougal, prepress supervisor. I also thank Barbara Littlewood for her continued efforts in maintaining an excellent, comprehensive index for this book.

Peter J. Russell

CHAPTER *1*

GENETICS: AN INTRODUCTION

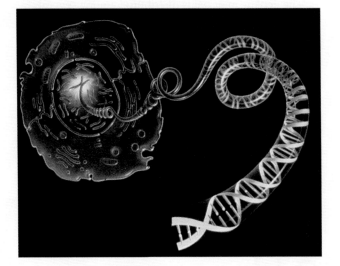

PRINCIPAL POINTS

~ Genetics is often divided into three main branches: transmission genetics, which deals with the transmission of genes from generation to generation; molecular genetics, which deals with the structure and function of genes at the molecular level; and population genetics, which deals with the distribution and behavior of genes in populations. Geneticists use the methods of science to investigate all aspects of genes. Depending upon whether the goal is obtaining a fundamental understanding of genetic phenomena, or overcoming problems in society or exploiting discoveries, genetics research is considered to be basic or applied, respectively.

~ Eukaryotes are organisms with cells in which the genetic material is located in a membrane-bound nucleus. The genetic material is distributed among several linear chromosomes.

~ Prokaryotes, the simplest cellular organisms, lack a membrane-bound nucleus. All bacteria are prokaryotes. The genetic material is DNA.

~ Diploid eukaryotic cells have two haploid sets of chromosomes, one set coming from each parent. The members of a pair of chromosomes are called homologous chromosomes. Haploid eukaryotic cells have only one set of chromosomes.

~ Mitosis is the process of nuclear division in eukaryotic cells represented by M in the cell cycle (G_1, S, G_2, and M). Mitosis results in the production of daughter nuclei that contain identical chromosome numbers and are genetically identical to one another and to the parent nucleus from which they arose. Prior to mitosis, the chromosomes replicate, producing duplicates of each. Mitosis usually is followed by cytokinesis, the actual division of the cell into two genetically identical progeny cells.

~ Meiosis occurs in all sexually reproducing eukaryotes. It is a process in which a specialized diploid cell (or cell nucleus) with two sets of chromosomes is transformed through one round of DNA replication and two rounds of nuclear division into four haploid cells (or four nuclei), each with one set of chromosomes. In the first of two divisions, pairing, crossing-over, and synapsis of homologous chromosomes occur. The meiotic process, in combination with fusion, results in the conservation of the number of chromosomes from generation to generation.

~ Meiosis generates genetic variability (a) through the processes by which maternal and paternal chromosomes are reassorted in progeny nuclei and (b) by crossing-over between members of a homologous pair of chromosomes.

*W*elcome to the study of **genetics**, the science of heredity. Genetics is primarily concerned with understanding biological properties that are transmitted from parent to offspring. The subject matter of genetics includes heredity; the molecular nature of the genetic material; how genes—the determinants of the characteristics of organisms—control life functions; and the distribution and behavior of genes in populations.

Genetics is central to biology because gene activity underlies all life processes, from cell structure and function to reproduction. Learning what genes are, how genes are transmitted from generation to generation, how genes are expressed, and how gene expression is regulated is the focus of this book. Genetics is expanding so rapidly, though, that it is simply not possible to describe everything we know about it between these covers. The important principles and concepts are presented carefully and thoroughly;

readers who want to go further can consult the suggested reading at the end of the text.

We assume that you have at least an awareness of the general features of genetics that were presented in your introductory biology course. This chapter presents a brief introduction to genetics and geneticists to put your studies in context relative to other areas of biology. It also discusses the transmission of chromosomes from cell division to cell division, and from generation to generation, by the processes of mitosis and meiosis, respectively. This knowledge will help you understand your subsequent study of genes and the results of genetic crosses.

THE BRANCHES OF GENETICS

Geneticists—scientists who study the processes of genetics—often divide genetics into three main branches. **Transmission genetics** (sometimes called classical genetics) is the branch dealing with the transmission of genes and genetic traits from generation to generation, and how genes recombine. Analyzing the pattern of transmission of traits in a human pedigree or in crosses of experimental organisms is an example of a transmission genetics study. **Molecular genetics** is the branch dealing with the molecular structure and function of genes. Analyzing the molecular events involved in the expression of genes is an example of a molecular genetics study. Finally, **population genetics** is the branch dealing (usually in mathematical terms) with the distribution and behavior of genes in populations. Analyzing the frequency of a disease-causing gene in the human population is an example of a population genetics study. Although these branches help us think about genes from different perspectives, there are no sharp boundaries between them. Increasingly, for example, population geneticists analyze molecular data to determine gene frequencies in large groups. Historically, transmission genetics developed first, followed by population genetics and then molecular genetics.

Through the products they encode, genes influence all aspects of an organism's life. Understanding transmission genetics and population genetics will help you understand such things as population biology, ecology, evolution, and animal behavior. Similarly, understanding molecular genetics is useful when you study such topics as neurobiology, cell biology, developmental biology, animal physiology, plant physiology, and immunology.

GENETICISTS AND GENETICS RESEARCH

The material presented in this book is the result of an incredible amount of research done by geneticists working in many areas of biology. In this section we discuss geneticists as scientists, the nature of their research, and the types of organisms geneticists have found to be fruitful objects of research.

What Do Geneticists Do?

Geneticists use the methods of science in their studies. As researchers, geneticists typically use the **hypothetico-deductive method of investigation**. This consists of making *observations*, forming *hypotheses* to explain the observations, making experimental *predictions* based on the hypotheses, and finally *testing* the

predictions. The last step provides new observations, producing a cycle that leads to a refinement of the hypotheses and perhaps, eventually, to the establishment of a theory that attempts to explain a set of observations.

As is true of all other areas of scientific research, the exact path a research project will follow cannot be predicted precisely. In part, the unpredictability of research makes it exciting and motivates the scientists engaged in it. The discoveries that have revolutionized genetics were not planned; they developed out of research in which basic genetic principles were being examined. Barbara McClintock's work on the inheritance of patches of color on corn kernels is an excellent example. After accumulating a large amount of data from genetic crosses, she hypothesized that the appearance of colored patches was the result of the movement (transposition) of a segment of DNA from one place to another in the genome. Only many years later were these DNA segments—called *transposons* or *transposable elements*—isolated and characterized in detail. (A more complete discussion of this discovery and of Barbara McClintock's life is presented in Chapter 19.) We know now that transposons are ubiquitous, playing a role not only in the evolution of species, but also in some human diseases.

What Are Basic and Applied Research?

Genetics research, and research in general, may be either basic or applied. In **basic research**, experiments are done to gain an understanding of fundamental phenomena, whether or not the knowledge gained leads to any immediate applications. In **applied research**, experiments are done with an eye toward overcoming specific problems in society or exploiting discoveries. Basic research was responsible for most of the facts we discuss in this book. For example, we know how the expression of many prokaryotic and eukaryotic genes is regulated as a result of basic research on model organisms such as the bacterium *Escherichia coli* (*E. coli*), the yeast *Saccharomyces cerevisiae*, and the fruit fly *Drosophila melanogaster*. The knowledge obtained from basic research is used largely to fuel more basic research.

Applied research is done with different goals in mind. In agriculture, applied genetics has contributed significantly to improvements in animals bred for food (such as reducing the amount of fat in beef and pork) and in crop plants (such as increasing the amount of protein in soybeans). In medicine, a number of diseases are caused by genetic defects, and great strides are being made in understanding the molecular bases of some of those diseases. Drawing on knowledge gained from basic research, applied genetics research involves, for example, developing rapid diagnostic tests for genetic diseases and producing new pharmaceuticals for treating diseases.

There is no sharp dividing line between basic and applied research. Indeed, in both areas researchers use similar techniques and depend on the accumulated body of information for building hypotheses. For example, **recombinant DNA technology**—procedures that allow molecular biologists to splice a DNA fragment from one organism into DNA from another organism and to clone (make many identical copies of) the new recombinant DNA molecule—has had a profound effect on both basic and applied research. Many biotechnology companies owe their existence to recombinant DNA technology, as they seek to clone and manipulate genes in developing their products. In the area of plant breeding, recombinant DNA technology has made it easier to introduce traits such as disease resistance from noncultivated species into cultivated species. Such crop improvement traditionally was achieved using conventional breeding experiments. Figure 1.1a shows the Flavr Savr tomato, a fruit that has been genetically engineered to spoil more slowly. These tomatoes can be picked later than normal tomatoes, when they have developed more flavor. In animal breeding, recombinant DNA technology is being used in the beef, dairy cattle, and poultry industries, for example, to increase the amount of lean meat, the amount of milk, and the number of eggs. In medicine the results are equally impressive. Recombinant DNA technology is being used in the production of many antibiotics, hormones, and other medically important agents such as clotting factor and human insulin (marketed under the name Humulin; Figure 1.1b), and in the diagnosis and treatment of a number of human genetic diseases. In forensics, *DNA typing* (also called *DNA fingerprinting*) is being used in, for example, paternity cases, criminal cases, and anthropological studies. In short, the science of genetics is currently in an exciting and dramatic growth phase, and there is still much to discover.

KEYNOTE

Genetics is often divided into three main branches: transmission genetics, dealing with the transmission of genes from generation to generation; molecular genetics, dealing with the structure and function of genes at the molecular level, and population genetics, dealing with the distribution and behavior of genes in populations. Geneticists use the methods of science to investigate all aspects of genes. Depending upon whether the goal is obtaining a fundamental understanding of genetic phenomena, or overcoming problems in society or exploiting discoveries, genetics research is considered to be basic or applied, respectively.

~ FIGURE 1.1

Examples of products developed as a result of recombinant DNA technology. (a) The Flavr Savr tomato. This tomato has been genetically engineered so that its rate of spoilage is much slower than usual; as a result, it can be picked at a later stage of ripeness and has more flavor than a normal tomato. (b) Humulin, human insulin for insulin-dependent diabetics.

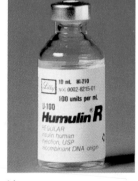

a)   b)

What Organisms Are Suitable for Genetic Experimentation?

Geneticists use various organisms in their research. The principles of heredity were first established in the nineteenth century by Gregor Mendel's experiments with the garden pea. Since Mendel's time, many organisms have been used in genetics experiments. Among the qualities that make an organism well suited for genetic experimentation are the following:

• The genetic history of the organism involved must be well known.

• The organism must have a relatively short life cycle so that a large number of generations occur within a relatively short time. In this way, researchers can obtain data readily over many generations. Fruit flies, for example, produce offspring in 10 to 14 days.

• A mating must produce a large number of offspring.

• The organism should be easy to handle. For example, hundreds of fruit flies can easily be kept in small bottles.

• Most importantly, genetic variation must exist among the individuals in the population so that the inheritance of traits can be studied. The more marked the variations, the easier the genetic analysis.

In the following, some basic organizational features of eukaryotic and prokaryotic cells are presented along with examples of organisms that have been useful historically in genetics research. A number of viruses that infect prokaryotic and eukaryotic organisms have also been useful in genetics research; they will be introduced in later chapters.

EUKARYOTES.   Eukaryotes (meaning "true nucleus") are organisms with cells within which the genetic material (DNA) is located in the **nucleus**, a discrete structure bounded by a nuclear envelope. Eukaryotes can be unicellular or multicellular. Many eukaryotic organisms have been and continue to be used in genetic research. In genetics today, a great deal of genetics research is done with six eukaryotes (Figure 1.2a–f): *Saccharomyces cerevisiae* (a yeast); *Drosophila melanogaster* (fruit fly); *Caenorhabditis elegans* (a nematode worm); *Arabidopsis thaliana* (a small weed of the mustard family); *Mus musculus* (mouse); and *Homo sapiens* (human). Humans are included not because they meet the criteria for an organism well suited for genetic experimentation, but because ultimately we wish to understand as much as we can about human genes and their expression. With this understanding we will be able to combat genetic diseases as well as

gain fundamental knowledge about our species' development and evolution.

Over the years, research with the following seven eukaryotes has also contributed significantly to our understanding of genetics (Figure 1.2g–m): *Neurospora crassa* (orange bread mold); *Tetrahymena* (a protozoan); *Paramecium* (a protozoan); *Chlamydomonas reinhardtii* (a green alga); *Pisum sativum* (garden pea); *Zea mays* (corn); and *Gallus* (chicken). Of these, *Tetrahymena*, *Paramecium*, *Chlamydomonas*, and *Saccharomyces* are unicellular organisms, and the rest are multicellular.

You learned about many features of eukaryotic cells in your introductory biology course. Figure 1.3 shows a generalized higher plant cell and a generalized animal cell. Surrounding the cytoplasm of both plant cells and animal cells is the *plasma membrane*. Plant cells, but not animal cells, have a rigid cell wall outside of the plasma membrane. The nucleus of eukaryotic cells contains DNA complexed with proteins and organized into a number of linear structures called *chromosomes*. The nucleus is separated from the rest of the cell—the cytoplasm and associated organelles—by the double membrane of the nuclear envelope. The membrane is selectively permeable and has pores about 20–80 nm (nm = nanometer = 10^{-9} meter) in diameter that make it possible for materials to move between the nucleus and the cytoplasm. For example, ribosomes, which function in the cytoplasm to synthesize proteins, are assembled within the nucleus in the **nucleolus** and reach the cytoplasm through the pores.

The cytoplasm of eukaryotic cells contains many different materials and organelles. Of special interest to geneticists are the *centrioles*, the *endoplasmic reticulum* (ER), *ribosomes*, *mitochondria*, and *chloroplasts*. Centrioles (also called basal bodies) are found in the cytoplasm of nearly all animal cells (see Figure 1.3b), but not in most plant cells. In animal cells, a pair of centrioles is located at the center of the centrosome, a region of undifferentiated cytoplasm that organizes the spindle fibers that function in mitosis and meiosis, processes discussed later in this chapter.

The endoplasmic reticulum (ER) is a double-membrane system, continuous with the nuclear envelope, that runs throughout the cell. So-called rough ER has ribosomes attached to it, giving it a rough appearance, while smooth ER does not. Ribosomes bound to rough ER synthesize proteins to be secreted by the cell or to be localized in either the cell membrane or particular organelles within the cell. The synthesis of proteins other than those distributed via the ER is performed by ribosomes that are free in the cytoplasm.

Mitochondria (singular: mitochondrion) (see Figure 1.3) are large organelles surrounded by a double membrane; the inner membrane is highly convoluted. Mitochondria play a crucial role in processing

~ FIGURE 1.2

Eukaryotic organisms that have contributed significantly to our knowledge of genetics. (a) *Saccharomyces cerevisiae* (a budding yeast); (b) *Drosophila melanogaster* (fruit fly); (c) *Caenorhabditis elegans* (a nematode); (d) *Arabidopsis thaliana* (Thale cress, a member of the mustard family); (e) *Mus musculus* (mouse); (f) *Homo sapiens* (human); (g) *Neurospora crassa* (orange bread mold); (h) *Tetrahymena* (a protozoan); (i) *Paramecium* (a protozoan); (j) *Chlamydomonas reinhardtii* (a green alga); (k) *Pisum sativum* (a garden pea); (l) *Zea mays* (corn); (m) *Gallus* (chicken).

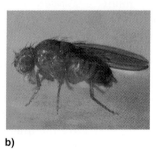

a)

b)

c)

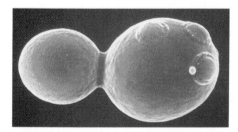

d)

e)

f)

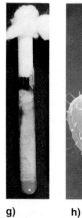

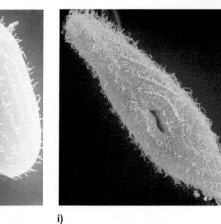

g)

h)

i)

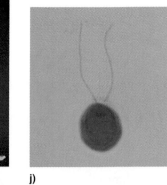

j)

k)

l)

m)

~ FIGURE 1.3

Eukaryotic cells: Cutaway diagrams of (a) a generalized higher plant cell, and (b) a generalized animal cell, showing the main organizational features and the principal organelles in each.

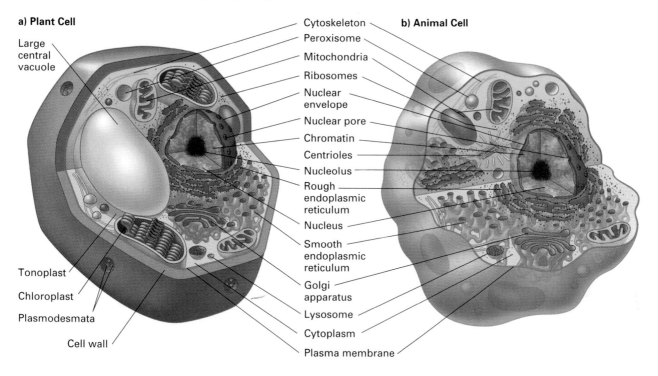

a) Plant Cell

- Large central vacuole
- Tonoplast
- Chloroplast
- Plasmodesmata
- Cell wall

- Cytoskeleton
- Peroxisome
- Mitochondria
- Ribosomes
- Nuclear envelope
- Nuclear pore
- Chromatin
- Centrioles
- Nucleolus
- Rough endoplasmic reticulum
- Nucleus
- Smooth endoplasmic reticulum
- Golgi apparatus
- Lysosome
- Cytoplasm
- Plasma membrane

b) Animal Cell

energy for the cell. They also contain DNA that encodes some of the proteins that function in the mitochondrion, as well as some components of the mitochondrial protein synthesis machinery.

Many plant cells contain chloroplasts, large, triple-membraned, chlorophyll-containing organelles involved in photosynthesis (see Figure 1.3a). Chloroplasts also contain DNA that encodes some of the proteins that function in the chloroplast, as well as some components of the chloroplast protein synthesis machinery.

KEYNOTE

Eukaryotes are organisms that have cells in which the genetic material is located in a membrane-bound nucleus. The genetic material is distributed among several linear chromosomes.

PROKARYOTES. Unlike eukaryotes, **prokaryotes** (meaning "prenuclear") do not have a nuclear envelope surrounding their DNA (Figure 1.4); this is the major distinguishing feature of prokaryotes. Included in the prokaryotes are all the **bacteria**, which are spherical, rod-shaped, or spiral-shaped organisms; most are single-celled, although a few are multicellular and filamentous. The shape of a bacterium is maintained by a rigid cell wall located outside the cell membrane. The bacteria are divided into two distantly related groups, the *eubacteria* and the *archaebacteria.* Eubacteria are the common varieties found in living organisms (naturally or by infection), in soil, and in water. Archaebacteria are often found in much more inhospitable conditions, such as hot springs, salt marshes, methane-rich marshes, or in the ocean depths. Bacteria generally vary in size from about 100 nm in diameter to 10 μm in diameter and 60 μm long; one species, the surgeonfish symbiont *Epulopiscium fishelsoni*, is as large as 60 × 800 μm, fully a million times larger than *E. coli!*

In most cases, the bacteria studied in genetics are eubacteria. The most intensely studied is *E. coli* (Figure 1.5), a rod-shaped bacterium common in human intestines. Studies of *E. coli* have significantly advanced our understanding of the regulation of gene expression and the development of molecular biology. *E. coli* is also used extensively in recombinant DNA experiments.

~ FIGURE 1.4

Cutaway diagram of a generalized prokaryotic cell.

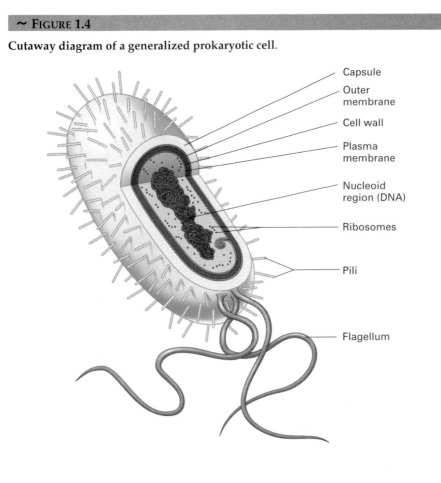

Capsule
Outer membrane
Cell wall
Plasma membrane
Nucleoid region (DNA)
Ribosomes
Pili
Flagellum

~ FIGURE 1.5

Colorized scanning electron micrograph of *Escherichia coli,* a rod-shaped eubacterium common in human intestines.

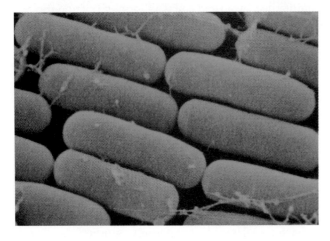

KEYNOTE

Prokaryotes, the simplest cellular organisms, lack a membrane-bound nucleus. All bacteria are prokaryotes. The genetic material is DNA.

CELLULAR REPRODUCTION IN EUKARYOTES: MITOSIS AND MEIOSIS

Since genes are located on chromosomes, the segregation of genes from generation to generation parallels the segregation of chromosomes from generation to generation. To prepare us for our discussion of gene inheritance, we now consider the transmission of chromosomes from cell division to cell division, and from generation to generation, by the processes of *mitosis* and *meiosis*, respectively.

Chromosome Complement of Eukaryotes

The genetic material of eukaryotes is distributed among multiple, linear chromosomes; the number of chromosomes, with rare exceptions, is characteristic of the species. Many eukaryotes have two copies of each type of chromosome in their nuclei, so their chromosome complement is said to be **diploid**, or 2N. Diploid eukaryotes are produced by the fusion of two **gametes**—mature reproductive cells that are specialized for sexual fusion—one from the female parent and one from the male parent. The fusion produces a diploid **zygote**, which then undergoes embryological development. Each gamete has only one set of chromosomes and is said to be **haploid** (N). The complete complement of genetic information in a haploid chromosome set is called the **genome**. Two examples of diploid organisms are humans, with 46 chromosomes (23 pairs), and *Drosophila melanogaster*, with eight chromosomes (four pairs). By contrast, the yeast *Saccharomyces cerevisiae*, with 16 chromosomes, is haploid.

In diploid organisms, the members of a chromosome pair that contain the same genes and that pair at meiosis are called **homologous chromosomes**; each member of a pair is called a **homolog**, and one homolog is inherited from each parent. Chromosomes that contain different genes and that do not pair during meiosis are called **nonhomologous chromosomes**. Figure 1.6 illustrates the chromosomal organization of haploid and diploid organisms.

In animals and in some plants, male and female cells are distinct with respect to their complement of **sex chromosomes**, the chromosomes in many eukaryotic organisms that are represented differently in the two sexes. One sex has a matched pair of sex chromosomes, while the other sex has an unmatched pair of sex chromosomes or a single sex chromosome. For example, human females have two X chromosomes (XX), while human males have one X and one Y (XY). Chromosomes other than sex chromosomes are called **autosomes**.

KEYNOTE

Diploid eukaryotic cells have two haploid sets of chromosomes, one set coming from each parent. The members of a pair of chromosomes, one from each parent, are called homologous chromosomes. Haploid eukaryotic cells have only one set of chromosomes.

Under the microscope, we can see that chromosomes differ in size and morphology (appearance) within and between species. Each chromosome often has a constriction along its length called a **centromere**, which is important for the behavior of the chromosomes during cellular division. The location of the centromere in one of four general positions in the chromosome is useful in classifying eukaryotic chromosomes (Figure 1.7). A **metacentric chromosome** has the centromere in approximately the center, so the chromosome appears to have two approximately equal arms. **Submetacentric chromosomes** have one arm longer than the other; **acrocentric chromosomes** have one arm with a stalk, and often with a

~ **FIGURE 1.6**

Chromosomal organization of haploid and diploid organisms.

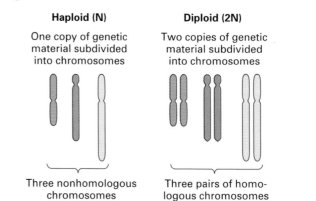

"bulb" (called a *satellite*) on it; and **telocentric chromosomes** have only one arm because the centromere is at the end. Chromosomes also vary in relative size. Chromosomes of mice, for example, are all similar in length, whereas those of humans have a wide range of relative lengths. Chromosome length and centromere position are constant for each chromosome and help in identifying individual chromosomes.

Asexual and Sexual Reproduction

Eukaryotes can reproduce by asexual or sexual reproduction. In **asexual reproduction** a new individual develops from either a single cell or from a group of cells (vegetative reproduction) in the absence of any sexual process. Asexual reproduction is found in both unicellular (single-celled) and multicellular eukaryotes. Single-celled eukaryotes (such as yeast) grow, double their genetic material, and generate two progeny cells, each of which contains an exact copy of the genetic material found in the parental cell.

Sexual reproduction is the fusion of two haploid gametes to produce a single diploid cell called a

~ **FIGURE 1.7**

General classification of eukaryotic chromosomes as metacentric, submetacentric, acrocentric, and telocentric types, based on the position of the centromere.

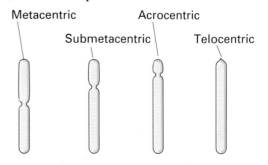

zygote, from which a new multicellular individual develops. Sexual reproduction involves the alternation of diploid and haploid phases. Importantly, sexual reproduction achieves genetic recombination; that is, it generates gene combinations in the offspring that are distinct from those in the parents. With the exception of self-fertilizing organisms (such as many plants), the two gametes are from different parents. Note that genetic recombination takes place during the production of gametes (or spores in plants). Figure 1.8 shows the cycle of growth and sexual reproduction in animals; here the haploid gametes are sperm and eggs. Note that the number of chromosomes keeps constant from generation to generation.

In general, sexually reproducing organisms have two sorts of cells: *somatic (body) cells* and *germ (sex) cells*. Somatic cells are haploid or diploid, depending on the eukaryote; the somatic cells of simpler eukaryotes such as yeast are often haploid, while those of more complex eukaryotes typically are diploid. The products of meiosis, which are always haploid, may be gametes (animals) or spores (plants). In contrast, all somatic cells reproduce by a process called mitosis, which we explore next. Figure 1.9 illustrates the differences in chromosome complement between asexual and sexual reproduction.

Mitosis

In both unicellular and multicellular eukaryotes, cellular reproduction is a cyclical process of growth, **mitosis** (*nuclear division* or *karyokinesis*), and (usually) **cell division** (*cytokinesis*). The cycle of growth, mitosis, and cell division is called the **cell cycle**. In proliferating somatic cells, the cell cycle consists of two phases: the mitotic (or division) phase (M) and an interphase between divisions (Figure 1.10). Interphase consists of three stages: G_1 (gap 1), S, and G_2 (gap 2). During G_1 (the presynthesis stage), the cell prepares for DNA and chromosome replication, which take place in the S stage. In G_2 (the postsynthesis stage), the cell prepares for cell division, or the M phase. Put another way, the precise replication of chromosomes takes place in interphase and then mitosis occurs, resulting in the distribution of a complete chromosome set to each of two progeny nuclei.

The relative time spent in each of the four stages of mitosis varies greatly among cell types. In a given organism, variation in the length of the cell cycle depends primarily on the duration of G_1; the duration of S plus G_2 plus M is approximately the same in all cell types. For example, some cancer cells and early fetal cells of humans spend minutes in G_1, while some differentiated adult cells (such as nerve cells) spend

∼ FIGURE 1.8

Cycle of growth and reproduction in sexually reproducing animals. Sexual reproduction involves the alternation of diploid and haploid phases.

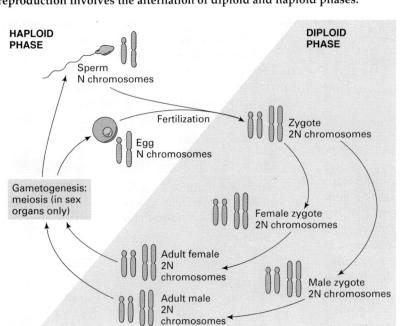

~ FIGURE 1.9

The differences in chromosome complement of organisms undergoing asexual and sexual reproduction.

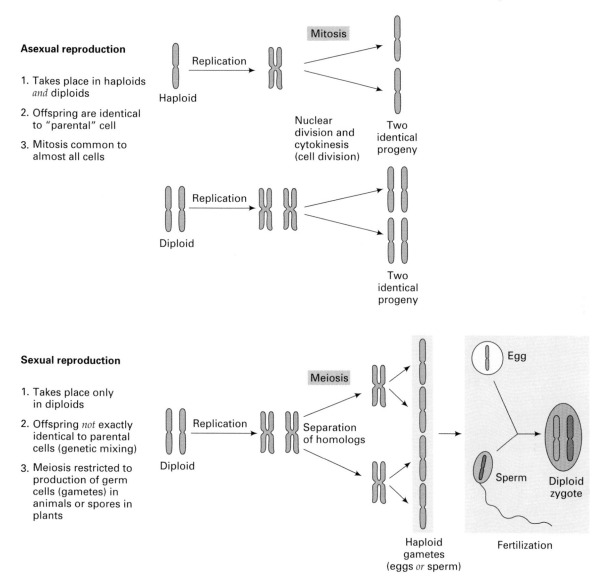

Asexual reproduction

1. Takes place in haploids *and* diploids

2. Offspring are identical to "parental" cell

3. Mitosis common to almost all cells

Sexual reproduction

1. Takes place only in diploids

2. Offspring *not* exactly identical to parental cells (genetic mixing)

3. Meiosis restricted to production of germ cells (gametes) in animals or spores in plants

years in G_1. Finally, some cells exit the cell cycle from G_1 and enter a quiescent, nondividing state called G_0 (G zero).

During interphase, the individual chromosomes are elongated and are difficult to see under the light microscope. The DNA of each chromosome is replicated in the S phase, giving two exact copies, called **sister chromatids**, which are held together by the replicated but unseparated centromeres. (Because the centromeres have not separated, only one centromere structure is visible under the microscope.) More precisely, a **chromatid** is one of the two visibly distinct, longitudinal subunits of all replicated chromosomes

that becomes visible between early prophase and metaphase of mitosis. Later, when the centromeres separate, the sister chromatids become known as **daughter chromosomes**.

Mitosis occurs in both haploid and diploid cells. It is a continuous process, but for purposes of discussion it is usually divided into four cytologically distinguishable stages called *prophase, metaphase, anaphase,* and *telophase.* Figure 1.11 shows the four stages in simplified diagrams. The photographs in Figure 1.12 show the typical chromosome morphology in interphase and in the four stages of mitosis in animal (whitefish early embryo) cells.

~ FIGURE 1.10

Eukaryotic cell cycle. This cycle assumes a period of 24 hours although great variation exists among cell types and organisms.

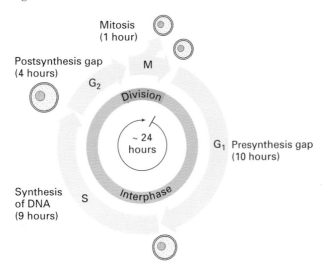

METAPHASE. Metaphase (see Figures 1.11 and 1.12c) begins when the nuclear envelope has completely disappeared. During metaphase the kinetochore microtubules orient the chromosomes so that their centromeres become aligned in one plane halfway between the two spindle poles, with the long axes of the chromosomes at 90 degrees to the spindle axis. The plane where the chromosomes become aligned is called the **metaphase plate**.

Figure 1.14 shows electron micrographs of human chromosomes in metaphase. Note the highly condensed state of the sister chromatids. The micrograph in Figure 1.14c shows a human chromosome from which much of the protein has been removed. In the center of the chromosome is a dense framework of protein called a *scaffold*, which still has the form of the chromosome. The scaffold is surrounded by a halo of DNA that has uncoiled and spread outward.

ANAPHASE. Anaphase (see Figures 1.11 and 1.12d) begins when the joined centromeres of sister chromatids separate, giving rise to two daughter chromosomes. Once the paired kinetochores on each chromosome separate, the sister chromatid pairs undergo **disjunction** (separation), and the daughter chromosomes (the former sister chromatids) move toward the poles. In anaphase the daughter chromosomes are pulled toward the opposite poles of the cell by the shortening kinetochore microtubules. As they are pulled the chromosomes have characteristic shapes related to the location of the centromere along the chromosome's length. For example, a metacentric chromosome will be V-shaped as the two roughly equal-length chromosome arms trail the centromere in its migration toward the pole. Similarly, a submetacentric chromosome will have a J shape with a long and short arm. Cytokinesis (cell division) usually begins in the latter stages of anaphase.

PROPHASE. At the beginning of **prophase** (see Figures 1.11 and 1.12b) the chromatids are very elongated. In preparation for mitosis they begin to coil tightly so that they appear shorter and fatter under the microscope. By late prophase each chromosome, which was duplicated during the preceding S phase of interphase, can be seen to consist of two sister chromatids.

Many mitotic events depend upon the *mitotic spindle* (spindle apparatus), a structure consisting of fibers composed of microtubules made of special proteins called *tubulins*. The mitotic spindle assembles outside of the nucleus during prophase. In most animal cells, the centrioles (see Figure 1.3b) are the focal points for spindle assembly; higher-plant cells usually lack centrioles, but they do have a mitotic spindle. Centrioles are arranged in pairs and, prior to the S phase, the cell's centriole pair replicates. Then, during mitosis, each new centriole pair becomes the focus for a radial array of microtubules called the *aster*. Early in prophase the two asters are next to one another and close to the nuclear envelope; by late prophase the two asters have moved far apart along the outside of the nucleus and are spanned by the microtubular spindle fibers.

Near the end of prophase, a substage called *prometaphase* occurs in which the nucleolus or nucleoli cease to be discrete areas within the nucleus and the nuclear envelope breaks down, allowing the spindle to enter the nuclear area. Specialized structures called **kinetochores** form on each face of the centromere of each chromosome and become attached to special microtubules called *kinetochore microtubules* (Figure 1.13).

TELOPHASE. During **telophase** (see Figures 1.11 and 1.12e) the migration of daughter chromosomes to the two poles is completed, and the two sets of daughter chromosomes are assembled into two groups at opposite ends of the cell. The chromosomes begin to uncoil and assume the elongated state characteristic of interphase. A nuclear envelope forms around each chromosome group, the spindle microtubules disappear, and the nucleolus or nucleoli re-form. At this point, nuclear division is complete: the cell has two nuclei.

CYTOKINESIS. *Cytokinesis* refers to division of the cytoplasm; usually it follows the nuclear division stage of mitosis and is completed by the end of telophase. Cytokinesis compartmentalizes the two

~ FIGURE 1.11

Interphase and mitosis in an animal cell.

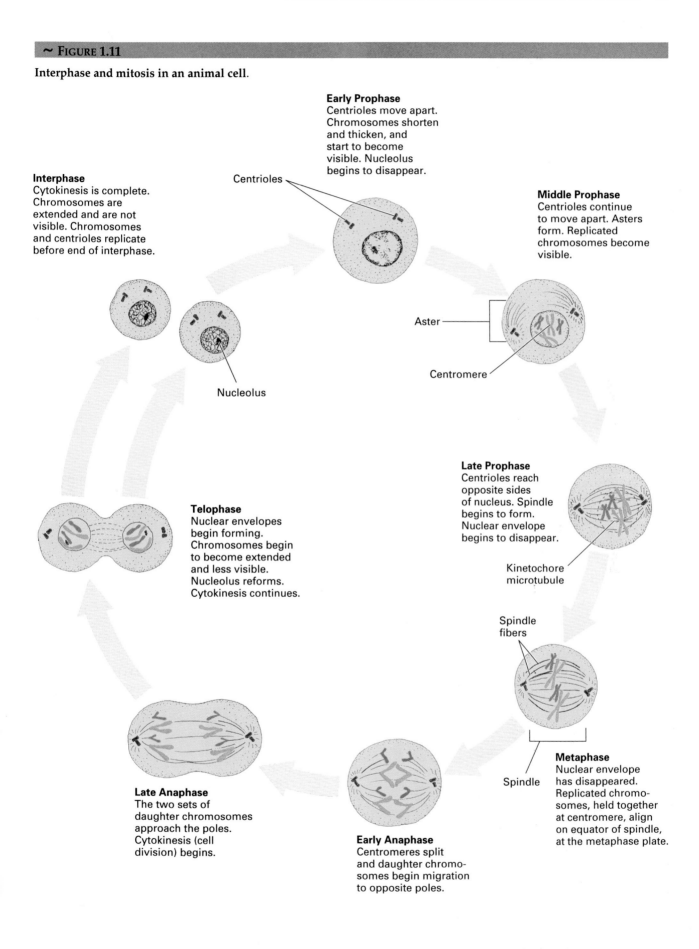

Early Prophase
Centrioles move apart. Chromosomes shorten and thicken, and start to become visible. Nucleolus begins to disappear.

Centrioles

Interphase
Cytokinesis is complete. Chromosomes are extended and are not visible. Chromosomes and centrioles replicate before end of interphase.

Middle Prophase
Centrioles continue to move apart. Asters form. Replicated chromosomes become visible.

Aster

Centromere

Nucleolus

Late Prophase
Centrioles reach opposite sides of nucleus. Spindle begins to form. Nuclear envelope begins to disappear.

Kinetochore microtubule

Spindle fibers

Telophase
Nuclear envelopes begin forming. Chromosomes begin to become extended and less visible. Nucleolus reforms. Cytokinesis continues.

Metaphase
Nuclear envelope has disappeared. Replicated chromosomes, held together at centromere, align on equator of spindle, at the metaphase plate.

Spindle

Late Anaphase
The two sets of daughter chromosomes approach the poles. Cytokinesis (cell division) begins.

Early Anaphase
Centromeres split and daughter chromosomes begin migration to opposite poles.

~ FIGURE 1.12

Interphase and the stages of mitosis in whitefish early embryo cells. (a) Interphase; (b) Late prophase; (c) Metaphase; (d) Early anaphase; (e) Telophase.

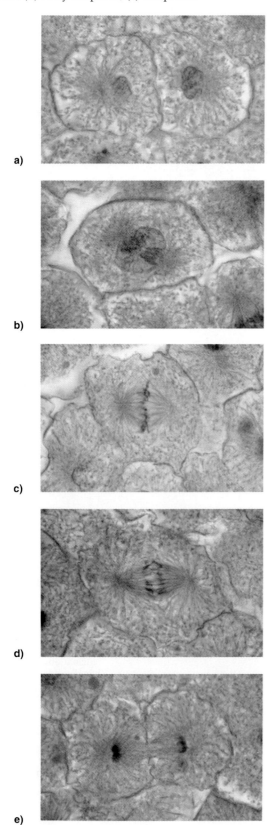

a)

b)

c)

d)

e)

~ FIGURE 1.13

Kinetochores and kinetochore microtubules.

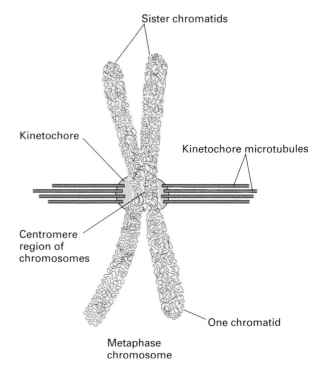

new nuclei into separate daughter cells, completing the mitosis/cell division process (Figure 1.15). In animal cells, cytokinesis proceeds with the formation of a constriction in the middle of the cell; the constriction continues until two daughter cells are produced (see Figure 1.15a). Most plant cells, by contrast, do not divide by the formation of a constriction. Instead, a new cell membrane and cell wall are assembled between the two new nuclei to form a *cell plate* (see Figure 1.15b). Cell wall material coats each side of the cell plate, and the result is two progeny cells.

GENE SEGREGATION IN MITOSIS. Mitosis maintains a constant amount of genetic material and a constant set of genes from cell generation to cell generation. Mitosis is a highly ordered process in which one copy of each duplicated chromosome segregates into each of the two daughter cells. Thus, for a haploid (N) cell, chromosome duplication produces a cell in which each chromosome has doubled its content. Mitosis then results in two progeny haploid cells, each with one complete set of chromosomes (= a genome). For a diploid (2N) cell, which has two sets of chromosomes (= two genomes), chromosome duplication produces a cell in which each chromosome set has doubled its content. Mitosis then results in two genetically identical progeny diploid cells, each with two sets of chromosomes (= two genomes). As a result, an equal distribution

~ FIGURE 1.14

Human metaphase chromosome. (a) Transmission electron micrograph of an intact chromosome; (b) Colorized scanning electron micrograph of an intact chromosome; (c) Transmission electron micrograph of an intact chromosome from which much of the protein has been removed; the DNA filaments have uncoiled.

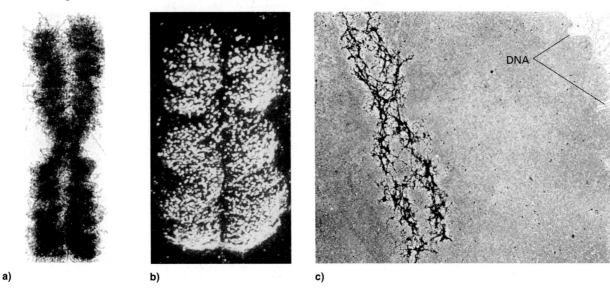

a)   b)   c)

of genetic material occurs, and no genetic material is lost.

KEYNOTE

Mitosis is the process of nuclear division in eukaryotes. It is one part of the cell cycle (G_1, S, G_2, and M) and it results in the production of daughter nuclei that contain identical chromosome numbers and that are genetically identical to one another and to the parent nucleus from which they arose. Prior to mitosis, the chromosomes duplicate. Mitosis is usually followed by cytokinesis. Both haploid and diploid cells proliferate by mitosis.

Meiosis

Meiosis is the two successive divisions of a *diploid* nucleus following only one DNA replication (chromosome duplication) cycle. That original diploid nucleus contains one haploid set of chromosomes from the mother and one set from the father. Meiosis occurs only at a special point in an organism's life cycle. In animals it results in the formation of haploid gametes (**gametogenesis**). Prior to meiosis, homologous chromosomes replicate; during meiosis they pair and then undergo two divisions—meiosis I and meio-

sis II—each consisting of a series of stages (Figure 1.16). Meiosis I results in reduction in the number of chromosomes from diploid to haploid (reductional division), and meiosis II results in separation of the sister chromatids. As a consequence, each of the four nuclei resulting from the two meiotic divisions receives one chromosome of each chromosome set

~ FIGURE 1.15

Cytokinesis (cell division). (a) Diagram of cytokinesis in an animal cell; (b) Diagram of cytokinesis in a plant cell.

a) **Animal cell**   b) **Plant cell**

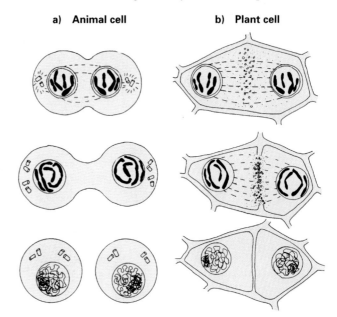

~ **FIGURE 1.16**

The stages of meiosis in an animal cell.

Early prophase I

Chromosomes, already replicated, become visible.

Middle prophase I

Homologous chromosomes shorten and thicken. The chromosomes synapse and crossing-over occurs.

Late prophase I

Results of crossing-over become visible as chiasmata. Nuclear envelope begins to disappear. Spindle apparatus begins to form.

Metaphase I

Assembly of spindle is completed. Each chromosome pair (bivalent) aligns across the equatorial plane of the spindle.

Telophase II

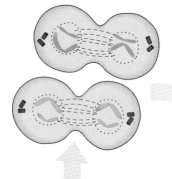

4 gametes

Anaphase I

Homologous chromosome pairs (dyads) separate and migrate toward opposite poles.

Anaphase II

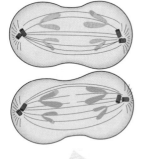

Telophase I

Chromosomes (each with two sister chromatids) complete migration to the poles and new nuclear envelopes may form. (Other sorting patterns are possible.)

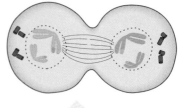

Metaphase II

Prophase II

Cytokinesis

In most species, cytokinesis occurs to produce two cells. Chromosomes do not replicate before meiosis II.

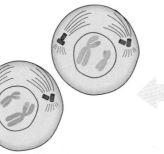

(that is, one complete haploid genome). In most cases the divisions are accompanied by cytokinesis, so the result of meiosis of a single diploid cell is four haploid cells.

MEIOSIS I: THE FIRST MEIOTIC DIVISION.   Meiosis I, in which the chromosome number is reduced from diploid to haploid, consists of four stages: *prophase I, metaphase I, anaphase I,* and *telophase I.*

Prophase I.   As prophase I begins, the chromosomes have already duplicated (see Figure 1.16). Prophase I is divided into a number of substages. Except for the behavior of homologous pairs of chromosomes and crossing-over, prophase I of meiosis is very similar to prophase of mitosis.

In **leptonema** (early prophase I—the leptotene stage) the chromosomes have begun to coil. Once a cell enters leptonema, it is committed to the meiotic process. A key event in leptonema is *pairing* of homologous chromosomes, the loose alignment of homologous regions of the chromosomes. Then a most significant event begins: **crossing-over,** the reciprocal physical exchange of chromosome segments at corresponding positions along pairs of homologous chromosomes. If there are genetic differences between the homologs, crossing-over can produce new gene combinations in a chromatid. There is usually no loss or addition of genetic material to either chromosome, because crossing-over involves reciprocal exchanges. A chromosome that emerges from meiosis with a combination of genes that differs from the combination with which it started is called a **recombinant chromosome.** Therefore, crossing-over is a mechanism that can give rise to **genetic recombination.**

In **zygonema** (early/mid-prophase I—the zygotene stage) a key event is **synapsis,** the formation of a very tight association of homologous chromosomes brought about by the formation along the length of the chromatids of a zipperlike structure called the **synaptonemal complex.** Because of the replication that occurred earlier, each synapsed set of homologous chromosomes consists of four chromatids and is referred to as a **bivalent** or a *tetrad.* The chromosomes are maximally condensed before the synaptonemal complex forms.

Following zygonema is **pachynema** (midprophase I—the pachytene stage). At the end of pachynema the synaptonemal complex is disassembled, and the chromosomes have started to elongate.

In **diplonema** (mid/late prophase I—the diplotene stage) the chromosomes begin to move apart. The result of crossing-over becomes visible during diplonema as a cross-shaped structure called a **chiasma** (plural, *chiasmata*; Figure 1.17). Since all four

~ **FIGURE 1.17**

Appearance of chiasmata, the visible evidence of crossing-over, in diplonema.

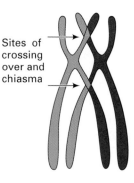

Sites of crossing over and chiasma

chromatids may be involved in crossing-over events along the length of the homologs, the chiasma pattern at this stage may be very complex.

Diplonema is followed rapidly in most organisms by the remaining stages of meiosis. However, in many animals the oocytes (egg cells) can remain in diplonema for very long periods. In human females, for example, oocytes go through meiosis I up to diplonema by the seventh month of fetal development and then remain arrested in this stage for many years. At the onset of puberty and until menopause, one oocyte per menstrual cycle completes meiosis I and is ovulated. If the oocyte is fertilized by a sperm as it passes down the fallopian tube, it quickly completes meiosis II and, by fusion with haploid sperm, a functional zygote is produced.

In **diakinesis** (late prophase), the nucleolus and nuclear envelope break down. Simultaneously, the spindle is assembled. The chromosomes can be most easily counted at this stage of meiosis.

The pairing, crossing-over, and synapsis phenomena described for prophase I apply to homologous chromosomes—namely, the autosomes. Even though the sex chromosomes are not homologous chromosomes, the mammalian Y chromosome has a small terminal portion called the *pseudoautosomal region* that is homologous to a portion of the X chromosome. In meiosis in males, crossing-over between the pseudoautosomal regions of the X and Y chromosomes is required for correct segregation of X and Y chromosomes as meiosis proceeds.

Metaphase I.   By the beginning of metaphase I (see Figure 1.16), the nuclear envelope has completely broken down and the bivalents become aligned on the equatorial plane of the cell. The spindle is completely formed now, and the microtubules are attached to the kinetochores of the homologs. Note particularly that the *pairs* of homologs (the bivalents) are found at the metaphase plate. In contrast, in mitosis of most organisms, replicated homologous chromosomes (sister chromatid pairs) align *independently* at the metaphase plate.

Anaphase I. In anaphase I (see Figure 1.16) the chromosomes in each bivalent separate, so the chromosomes of each homologous pair disjoin and migrate toward opposite poles, the areas in which new nuclei will form. (At this stage, each of the separated chromosomes is called a *dyad*.) This migration assumes that (1) maternally derived and paternally derived centromeres segregate randomly to each pole (except for the parts of chromosomes exchanged during the crossing-over process) and (2) at each pole there is a haploid complement of replicated centromeres with associated chromosomes. *At this time, the segregated sister chromatid pairs remain attached at their respective centromeres. In other words, a key difference between mitosis and meiosis I is that sister chromatids remain joined after metaphase in meiosis I, whereas in mitosis they separate.*

Telophase I. In telophase I (see Figure 1.16) the dyads complete their migration to opposite poles of the cell,

and (in most cases) new nuclear envelopes form around each haploid grouping. In most species, cytokinesis follows, producing two haploid cells. Thus, meiosis I, which begins with a diploid cell that contains one maternally derived and one paternally derived set of chromosomes, ends with two nuclei, each of which is haploid and contains one mixed-parental set of dyads. After cytokinesis, each of the two progeny cells has a nucleus with a haploid set of dyads.

MEIOSIS II: THE SECOND MEIOTIC DIVISION.
The second meiotic division is very similar to a mitotic division (see Figure 1.16).

In **prophase II** the chromosomes condense.

In **metaphase II**, each of the two daughter cells organizes a spindle apparatus that attaches to the centromeres that still connect the sister chromatids. The centromeres line up on the equator of the second-division spindles.

During **anaphase II** the centromeres split and the chromatids are pulled to the opposite poles of the spindle. One sister chromatid of each pair goes to one pole while the other goes to the opposite pole. The

~ **FIGURE 1.18**

Comparison of mitosis and meiosis in a diploid cell.

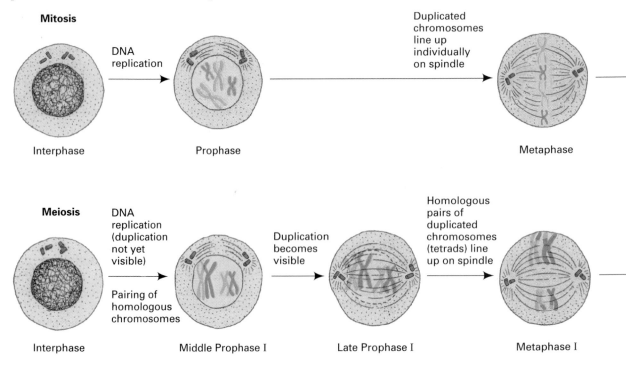

separated chromatids are considered chromosomes in their own right.

In the last stage, **telophase II**, a nuclear envelope forms around each set of chromosomes, and cytokinesis takes place. After telophase II, the chromosomes become more elongated and are no longer visible under the light microscope.

The end products of the two meiotic divisions are four haploid cells (gametes in animals) from one original diploid cell (see Figure 1.16). Each of the four progeny cells has one chromosome from each homologous pair of chromosomes. Moreover, these chromosomes are not exact copies of the original chromosomes because of crossing-over. Figure 1.18 compares mitosis and meiosis.

GENE SEGREGATION IN MEIOSIS.   Meiosis has three significant results:

1. Meiosis generates haploid cells with half the number of chromosomes found in the diploid cell that entered the process because two division cycles follow only one cycle of DNA replication (S period). Fusion of haploid nuclei re-

stores the diploid number. Therefore, through a cycle of meiosis and fusion, the chromosome number is maintained in sexually reproducing organisms.

2. In metaphase I, each maternally derived and paternally derived chromosome has an equal chance of aligning on one or the other side of the equatorial metaphase plate. As a result, each nucleus generated by meiosis will have some combination of maternal and paternal chromosomes.

The number of possible chromosome combinations in the haploid nuclei resulting from meiosis is large, especially when the number of chromosomes in an organism is large. Consider a hypothetical organism with two pairs of chromosomes in a diploid cell entering meiosis. Figure 1.19 shows the two possible combinations of maternal and paternal chromosomes that can occur at the metaphase plate. Understanding this concept is useful when we consider the segregation of genes in Chapter 2.

The general formula for the number of possible chromosome arrangements at the metaphase

~ **FIGURE 1.18 continued**

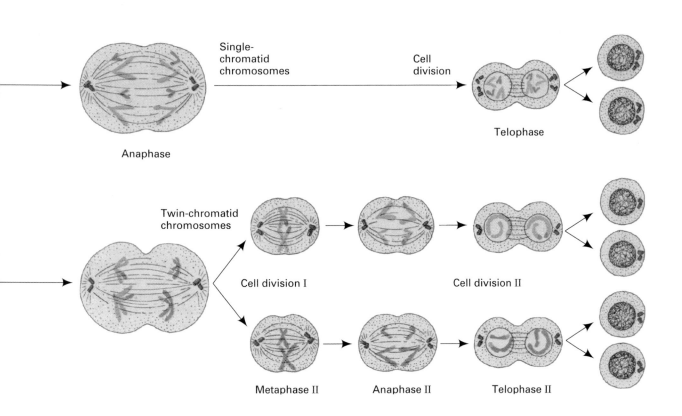

Anaphase

Single-chromatid chromosomes

Cell division

Telophase

Twin-chromatid chromosomes

Cell division I

Cell division II

Metaphase II

Anaphase II

Telophase II

~ FIGURE 1.19

The two possible arrangements of two pairs of homologous chromosomes on the metaphase plate of the first meiotic division. Paternal chromosomes are shown in yellow and green; maternal chromosomes in purple and tan.

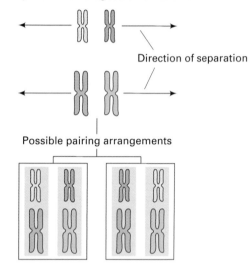

Two pairs of homologous chromosomes

Direction of separation

Possible pairing arrangements

plate in meiosis is 2^{n-1}, where n is the number of chromosome pairs. Similarly, the general formula for the number of possible chromosome combinations in the nuclei resulting from meiosis is 2^n. In *Drosophila*, which has four pairs of chromosomes, the number of possible combinations in nuclei resulting from meiosis is 2^3, or 8; in humans, which have 23 chromosome pairs, over 4 million combinations are possible. Therefore, because there are many gene differences between the maternally derived and paternally derived chromosomes, the nuclei produced by meiosis will be genetically quite different from the parental cell and from one another.

3. The crossing-over between maternal and paternal chromatid pairs during meiosis I generates still more variation in the final combinations. Crossing-over occurs during every meiosis, and because the sites of crossing-over vary from one meiosis to another, the number of different kinds of progeny nuclei produced by this process is extremely large. Given the genetic features of meiosis, this process is of critical importance for understanding the behavior of genes.

The events that occur in meiosis are the bases for the segregation and independent assortment of genes according to Mendel's laws, discussed in Chapter 2.

KEYNOTE

Meiosis occurs in all sexually reproducing eukaryotes. It is a process by which a specialized diploid (2N) cell or cell nucleus with two sets of chromosomes is transformed, through one round of chromosome replication and two rounds of nuclear division, into four haploid (N) cells or nuclei, each with one set of chromosomes. In the first of two divisions, pairing, crossing-over, and synapsis of homologous chromosomes occur. The meiotic process, in combination with fertilization, results in the conservation of the number of chromosomes from generation to generation. It also generates genetic variability through the various ways in which maternal and paternal chromosomes are combined in the progeny nuclei and by crossing-over (the reciprocal physical exchange of chromosome segments at corresponding positions along pairs of homologous chromosomes).

LOCATIONS OF MEIOSIS IN THE LIFE CYCLE. Next we discuss the locations of meiosis in animals and plants.

Meiosis in Animals. Most multicellular animals are diploid through most of their life cycles. In such animals, meiosis produces haploid gametes, which produce a diploid zygote when their nuclei fuse in fertilization. The zygote then divides by mitosis to produce the new diploid organism. Thus the gametes are the only haploid stages of the life cycle. Gametes are only formed in specialized cells. In males the gamete is the sperm, produced through a process called **spermatogenesis**; in females the gamete is the egg, produced by **oogenesis**. Spermatogenesis and oogenesis are illustrated in Figure 1.20.

In male animals the **sperm cells** (**spermatozoa**) are produced within the testes. The testes contain the primordial germ cells (*primary spermatogonia*), which, via mitosis, produce *secondary spermatogonia*. Spermatogonia transform into *primary spermatocytes* (*meiocytes*), each of which undergoes meiosis I and gives rise to two *secondary spermatocytes*. Each secondary spermatocyte undergoes meiosis II. The results of these two divisions are four haploid *spermatids* that eventually differentiate into the male gametes, the spermatozoa.

In female animals the ovary contains the primordial germ cells (*primary oogonia*), which, by mitosis, give rise to *secondary oogonia*. These cells transform

~ FIGURE 1.20

Spermatogenesis and oogenesis in an animal cell.

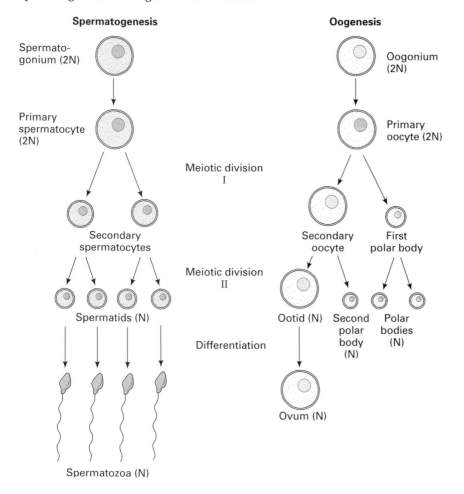

into *primary oocytes*, which grow until the end of oogenesis. The diploid primary oocyte goes through meiosis I and unequal cytokinesis to give two cells: a large one called the **secondary oocyte** and a very small one called the *first polar body*. In meiosis II the secondary oocyte produces two haploid cells. One is a very small cell called a *second polar body;* the other is a large cell that rapidly matures into the mature egg cell, or **ovum**. The first polar body may or may not divide during meiosis II. The polar bodies have no function in most species and degenerate; only the ovum is a viable gamete. (In many animals, the cell that is actually fertilized is the secondary oocyte; however, nuclear fusion must await completion of meiosis by that oocyte.) Thus in the female animal, only one mature gamete (the ovum) is produced by meiosis of a diploid cell. In humans, all oocytes are formed in the fetus, and one oocyte completes meiosis I each month in the adult female, but does not progress any further unless stimulated to do so by fertilization by a sperm.

Meiosis in Plants. The life cycle of sexually reproducing plants typically has two phases: the **gametophyte** or haploid stage, in which gametes are produced, and the **sporophyte** or diploid stage, in which haploid spores are produced by meiosis.

In angiosperms (the flowering plants), the flower is the structure in which sexual reproduction occurs. Figure 1.21 shows a generalized flower containing both male and female reproductive organs, the **stamens** and **pistils**, respectively. Each stamen consists of a single stalk, the filament, on the top of which is an anther. From the anther are released the pollen grains, which are immature male gametophytes (gamete-producing phase). The pistil, which contains the female gametophytes, typically consists of the stigma, a sticky surface specialized to receive the

~ FIGURE 1.21

Generalized structure of a flower.

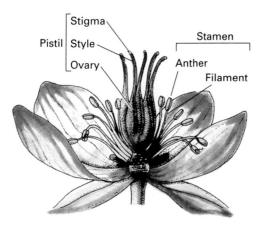

pollen; the style, a thin stalk down which a pollen tube grows from the adhered pollen grain; and, at the base of the structure, the ovary, within which are the ovules. Each ovule encloses a female gametophyte (the *embryo sac*) containing a single egg cell. When the egg cell is fertilized, the ovule develops into a seed.

Among living organisms, only plants produce gametes from special bodies called gametophytes. Thus plant life cycles have two distinct reproductive

~ FIGURE 1.22

Alternation of gametophyte and sporophyte generations in flowering plants.

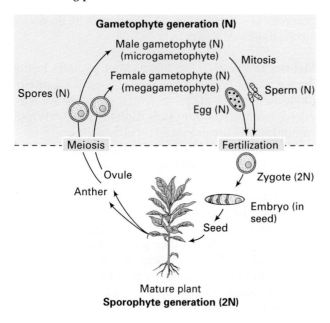

phases, called the **alternation of generations** (Figure 1.22). Meiosis and fertilization constitute the transitions between these stages. The haploid *gametophyte generation* begins with spores that are produced by meiosis. In flowering plants the spores are the cells that ultimately become pollen and embryo sac. Fertilization initiates the diploid *sporophyte generation*, which produces the specialized haploid cells called spores, completing the cycle.

SUMMARY

This chapter introduced the subject of genetics briefly. Genetics is often divided into three main branches: transmission genetics, molecular genetics, and population genetics. Geneticists investigate all aspects of genes and, depending upon whether the goal is to obtain fundamental knowledge, or to solve problems in society or exploit discoveries, genetics research is considered to be basic or applied, respectively.

This chapter also introduced some of the organisms that have been fruitful subjects of genetic studies. As the simplest cellular organisms, prokaryotes lack a membrane-bound nucleus and have a single chromosome located in a nucleoid region. Eukaryotic organisms may be single-celled or multicellular and have several chromosomes located within a membrane-bound nucleus.

Eukaryotic cells contain either one or two sets of chromosomes: the former are haploid cells and the latter are diploid cells. Both haploid and diploid cells proliferate by mitosis. Mitosis involves one round of DNA replication followed by one round of nuclear division, often accompanied by cell division. Thus, mitosis results in the production of daughter nuclei that contain identical chromosome numbers and that are genetically identical to one another and to the parent nucleus from which they arose.

In all sexually reproducing organisms, meiosis occurs at a particular stage in the life cycle. Meiosis is the process by which a diploid cell (never a haploid cell) or cell nucleus undergoes one round of DNA replication and two rounds of nuclear division to produce four specialized haploid cells or nuclei. The products of meiosis are gametes or meiospores. Unlike mitosis, meiosis generates genetic variability in two main ways: (1) through the various ways in which maternal and paternal chromosomes are combined in progeny nuclei; and (2) through crossing-over between maternally derived and paternally derived homologs to produce recombinant chromosomes with some maternal and some paternal genes.

QUESTIONS AND PROBLEMS

Solutions to Questions and Problems marked with an asterisk are found at the end of this text, starting on page S-1. Solutions to all problems are found in the Student Study Guide and Solutions Manual that accompanies this text.

In 1.1 through 1.3, select the correct answer.

*1.1 Interphase is a period corresponding to the cell cycle phases of
a. mitosis.
b. S.
c. $G_1 + S + G_2$.
d. $G_1 + S + G_2 + M$.

1.2 Chromatids joined together by a centromere are called
a. sister chromatids.
b. homologs.
c. alleles.
d. bivalents (tetrads).

*1.3 Mitosis and meiosis always differ in regard to the presence of
a. chromatids.
b. homologs.
c. bivalents.
d. centromeres.
e. spindles.

1.4 State whether each of the following statements is true or false. Explain your choice.
a. The chromosomes in a somatic cell of any organism are all morphologically alike.
b. During mitosis the chromosomes divide and the resulting sister chromatids separate at anaphase, ending up in two nuclei, each of which has the same number of chromosomes as the parental cell.
c. At zygonema, any chromosome may synapse with any other chromosome in the same cell.

1.5 For each mitotic event described in the following table, write the name of the event in the blank provided in front of the description. Then put the events in the correct order (sequence); start by placing a 1 next to the description of interphase, and continue through 6, which should correspond to the last event in the sequence.

NAME OF EVENT		ORDER OF EVENT
_____	The cytoplasm divides and the cell contents are separated into two separate cells.	_____
_____	Chromosomes become aligned along the equatorial plane of the cell.	_____
_____	Chromosome replication occurs.	_____
_____	The migration of the daughter chromosomes to the two poles is complete.	_____
_____	Replicated chromosomes begin to condense and become visible under the microscope.	_____
_____	Sister chromatids begin to separate and migrate toward opposite poles of the cell.	_____

*1.6 Decide whether the answer to these statements is *yes* or *no*. Then explain the reasons for your decision.
a. Can meiosis occur in haploid species?
b. Can meiosis occur in a haploid individual?

1.7 Select the correct answer. The general life cycle of a eukaryotic organism has the sequence
a. 1N → meiosis → 2N → fertilization → 1N.
b. 2N → meiosis → 1N → fertilization → 2N.
c. 1N → mitosis → 2N → fertilization → 1N.
d. 2N → mitosis → 1N → fertilization → 2N.

*1.8 Which statement is true?
a. Gametes are 2N; zygotes are 1N.
b. Gametes and zygotes are 2N.
c. The number of chromosomes can be the same in gamete cells and in somatic cells.
d. The zygotic and the somatic chromosome numbers cannot be the same.
e. Haploid organisms have haploid zygotes.

1.9 All of the following happen in prophase I of meiosis *except*
a. chromosome condensation.
b. pairing of homologs.
c. chiasma formation.
d. terminalization.
e. segregation.

*1.10 Give the name of the stage of mitosis or meiosis at which each of the following events occurs:

a. Chromosomes are located in a plane at the center of the spindle.
b. The chromosomes move away from the spindle equator to the poles.

1.11 Given the diploid, meiotic mother cell shown below, diagram the chromosomes as they would appear
a. in late pachynema;
b. in a nucleus at prophase of the second meiotic division;
c. in the first polar body resulting from oogenesis in an animal.

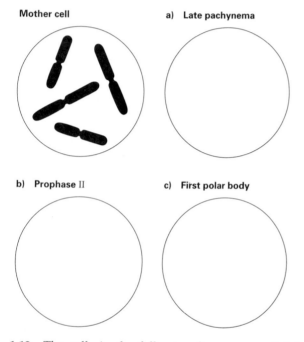

1.12 The cells in the following figure were all taken from the same individual (a mammal). Identify the cell division events occurring in each cell, and explain your reasoning. What is the sex of the individual? What is the diploid chromosome number?

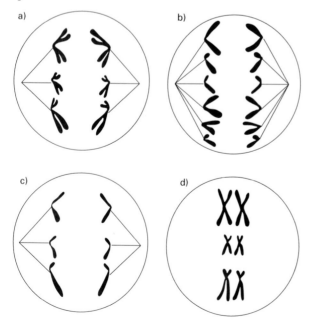

1.13 Does mitosis or meiosis have greater significance in the study of heredity? Explain your choice.

*__1.14__ Consider a diploid organism that has three pairs of chromosomes. Assume that the organism receives chromosomes A, B, and C from the female parent and A′, B′, and C′ from the male parent. To answer the following questions, assume that no crossing-over occurs.
a. What proportion of the gametes of this organism would be expected to contain all the chromosomes of maternal origin?
b. What proportion of the gametes would be expected to contain some chromosomes from both maternal and paternal origin?

*__1.15__ Normal diploid cells of a theoretical mammal are examined cytologically at the mitotic metaphase stage for their chromosome complement. One short chromosome, two medium-length chromosomes, and three long chromosomes are present. Explain how the cells might have such a set of chromosomes.

1.16 Is the following statement true or false? Explain your decision. "Meiotic chromosomes may be seen after appropriate staining in nuclei from rapidly dividing skin cells."

*__1.17__ Is the following statement true or false? Explain your decision. "All of the sperm from one human male are genetically identical."

1.18 The horse has a diploid set of 64 chromosomes, and a donkey has a diploid set of 62 chromosomes. Mules (viable but usually sterile progeny) are produced when a male donkey is mated to a female horse. How many chromosomes will a mule cell contain?

*__1.19__ The red fox has 17 pairs of large, long chromosomes. The arctic fox has 26 pairs of shorter, smaller chromosomes.
a. What do you expect to be the chromosome number in somatic tissues of a hybrid between these two foxes?
b. The first meiotic division in the hybrid fox shows a mixture of paired and single chromosomes. Why do you suppose this occurs? Can you suggest a possible relationship between this fact and the observed sterility of the hybrid?

*__1.20__ At the time of synapsis preceding the reduction division in meiosis, the homologous chromosomes align in pairs, and one member of each pair passes to each of the daughter nuclei. In an animal with five pairs of chromosomes, assume that chromosomes 1, 2, 3, 4, and 5 have come from the father, and 1′, 2′, 3′, 4′, and 5′ have come from the mother. Assuming no crossing-over, in what proportion of the gametes of this animal will all the paternal chromosomes be present together?

CHAPTER 2

MENDELIAN GENETICS

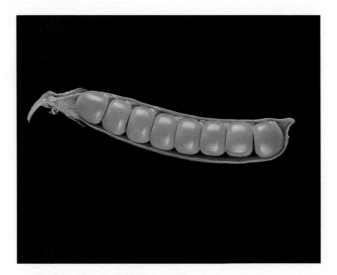

PRINCIPAL POINTS

~ The genotype is the genetic makeup of an organism, whereas the phenotype is the observable properties (structural and functional) of an organism that are produced by the interaction between its genotype and the environment.

~ Genes provide the potential for the development of characteristics; this potential is affected by interactions with other genes and with the environment.

~ Mendel's first law, the principle of segregation, states that the two members of a gene pair segregate from each other in the formation of gametes.

~ To determine an unknown genotype (usually in an individual expressing the dominant phenotype), a cross

is made between that individual and a homozygous recessive individual. This cross is called a testcross.

~ Mendel's second law, the principle of independent assortment, states that members of different gene pairs are transmitted independently of one another during the production of gametes.

~ Mendelian principles apply to all eukaryotes. The study of the inheritance of genetic traits in humans is complicated by the fact that controlled crosses cannot be done. Instead, human geneticists examine genetic traits by pedigree analysis—that is, by following the occurrence of a trait in family trees in which the trait is segregating.

---

*B*y simple observation you can see that a lot of variation exists among individuals of a given species. Among dogs, for example, are many breeds, including Bernese mountain dogs, Dalmatians, pointers, dachshunds, Pomeranians, and so on. Even though all dogs belong to the same species, *Canis familiaris*, each breed has characteristic sizes, shapes, colors, and behaviors. Similarly, individual humans vary in eye color, height, skin color, and hair color, even though all humans belong to the species *Homo sapiens*. The differences among individuals within and among species are mainly the result of differences in the DNA sequences of their genes, because it is the genes that determine the structure, function, and development of the cell and organism.

The understanding of how genes are transmitted from parent to offspring began with the work of Gregor Johann Mendel (1822–1884), an Augustinian monk. The goal of this chapter is for you to learn the basic principles of the transmission of genes through an examination of Mendel's work. Be aware that even though Mendel analyzed the segregation of hereditary traits, he did not know that genes control the traits, or that genes are located in chromosomes, or even that chromosomes existed.

GENOTYPE AND PHENOTYPE

Before we begin our study of Mendel's work, we must distinguish between the nature of the genetic material and the physical characteristics that result from the expression of genes.

The characteristics of an individual that are transmitted from one generation to another are sometimes called **hereditary traits** (Mendel called them **characters**). These traits are under the control of DNA segments called **genes** (Mendel called them *factors*). The genetic constitution of an organism is called its **genotype**, and the **phenotype** is the observable properties (structural and functional) of an organism produced by the interaction between its genotype and the environment.

~ **FIGURE 2.1**

Influences on the physical manifestation (phenotype) of the genetic blueprint (genotype): interactions with other genes and their products (such as hormones) and with the environment (such as nutrition).

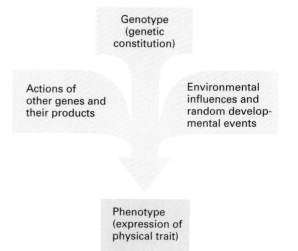

Genes provide only the potential for developing a particular phenotypic characteristic. The extent to which that potential is realized depends both on interactions with other genes and their products and on environmental influences and random developmental events (Figure 2.1). A person's height, for example, is controlled by many genes, the expression of which can be significantly affected by internal and external environmental influences such as the effects of hormones during puberty (an internal environmental influence) and nutrition (an external environmental influence). In other words, genes are a starting point for determining the structure and function of an organism, and the route to the mature phenotypic state is highly complex and involves many interacting biochemical pathways.

KEYNOTE

The genotype is the genetic constitution of an organism; the phenotype is the observable properties (structural and functional) of an organism produced by the interaction between its genotype and the environment. The genes provide the potential for the development of characteristics; this potential is often affected by interactions with other genes and with the environment. Thus, individuals with the same genotype can have different phenotypes, and individuals with the same phenotypes may have different genotypes.

MENDEL'S EXPERIMENTAL DESIGN

The work of Gregor Johann Mendel (Figure 2.2) is considered the foundation of modern genetics. In 1843 he was admitted to the Augustinian Monastery in Brno (now Brünn, Czech Republic). In 1854 he began a series of breeding experiments with the garden pea *Pisum sativum* to learn something about the mechanisms of heredity. Likely as a result of his creativity, Mendel discovered some fundamental principles of genetics.

From the results of crossbreeding pea plants with different characteristics such as height, flower color, and seed shape, Mendel developed a simple theory to explain the transmission of hereditary characteristics or traits from generation to generation. (Mendel had no knowledge of mitosis and meiosis, so he did not know, as we now know, that genes segregate according to chromosome behavior.) Although Mendel reported his conclusions in 1865, their significance

~ FIGURE 2.2

Gregor Johann Mendel, founder of the science of genetics.

was not fully realized until the late 1800s and early 1900s.

Mendel's experimental approach was effective because he made simple interpretations of the ratios of the types of progeny he obtained from his crosses, and because he then carried out direct and convincing experiments to test his hypotheses. In his initial breeding experiments he took the simplest approach of studying the inheritance of one trait at a time. (This is how you should work genetics problems.) He made carefully controlled matings (crosses) between strains of peas that had obvious differences in heritable traits and, most importantly, he kept very careful records of the outcome of the crosses and the number of each type of pea produced. The numerical data he obtained enabled him to do a rigorous analysis of the hereditary transmission of characteristics.

Generally, genetic crosses are done as follows. Two diploid individuals differing in phenotype are allowed to produce haploid gametes by meiosis. Fusion of male and female gametes produces zygotes from which the diploid progeny individuals are generated. The phenotypes of the offspring are analyzed to provide clues to the heredity of those phenotypes.

Mendel did all his significant genetic experiments with the garden pea. The garden pea was a good choice because it fits many of the criteria that make an organism suitable for use in genetic experiments (see Chapter 1): It is easy to grow, bears flowers and fruit in the same year a seed is planted, and produces a large number of seeds. When the seeds are planted, they germinate to produce new pea plants.

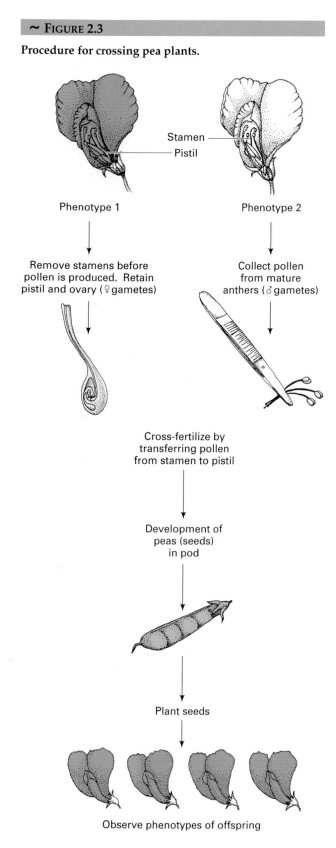

~ Figure 2.3

Procedure for crossing pea plants.

Stamen
Pistil

Phenotype 1

Phenotype 2

Remove stamens before
pollen is produced. Retain
pistil and ovary (♀ gametes)

Collect pollen
from mature
anthers (♂ gametes)

Cross-fertilize by
transferring pollen
from stamen to pistil

Development of
peas (seeds)
in pod

Plant seeds

Observe phenotypes of offspring

Figure 2.3, which presents the procedure for cross-ing pea plants, begins with a cross section of a flower of the garden pea, showing the stamens (male re-productive organs) and the pistils (female reproduc-tive organs). The pea normally reproduces by **self-fertilization**; that is, the anthers at the ends of the stamen produce pollen, which lands on the pistil with-in the same flower and fertilizes the plant. This pro-cess is also called **selfing**. Fortunately, we can prevent self-fertilization of the pea by removing the stamens from a developing flower bud before their anthers produce any mature pollen. Then pollen taken from the stamens of another flower can be dusted onto the pistil of the emasculated one to pollinate it.

Cross-fertilization, or simply **cross**, is the term used for the fusion of male gametes (in this case, pollen) from one individual and female gametes (eggs) from another. Once cross-fertilization has oc-curred, the zygote develops in the seeds (peas), which are then planted. The phenotypes of the plants that grow from the seeds are then analyzed.

Mendel obtained 34 strains of pea plants that dif-fered in a number of traits. He allowed each strain to self-fertilize for many generations to be sure that the traits he wanted to study were inherited. This prelimi-nary work ensured that he worked only with pea strains in which the trait under investigation remained unchanged from parent to offspring for many genera-tions. Such strains are called **true-breeding** or **pure-breeding strains**.

Next, Mendel selected seven traits to study in breeding experiments. Each trait (or character) had two easily distinguishable, alternative appearances (phe-notypes). Mendel studied the following character pairs (Figure 2.4):

1. flower and seed coat color (grey versus white seed coats, and purple versus white flowers; note that a single gene controls these particular color properties of both seed coats and flowers)

2. seed color (yellow versus green)

3. seed shape (smooth versus wrinkled)

4. pod color (green versus yellow)

5. pod shape (inflated versus pinched)

6. stem height (tall versus short)

7. flower position (axial versus terminal)

MONOHYBRID CROSSES AND MENDEL'S PRINCIPLE OF SEGREGATION

Before we discuss Mendel's experiments, we must be clear on the terminology used in breeding experi-

~ FIGURE 2.4

Seven character pairs in the garden pea that Mendel studied in his breeding experiments.

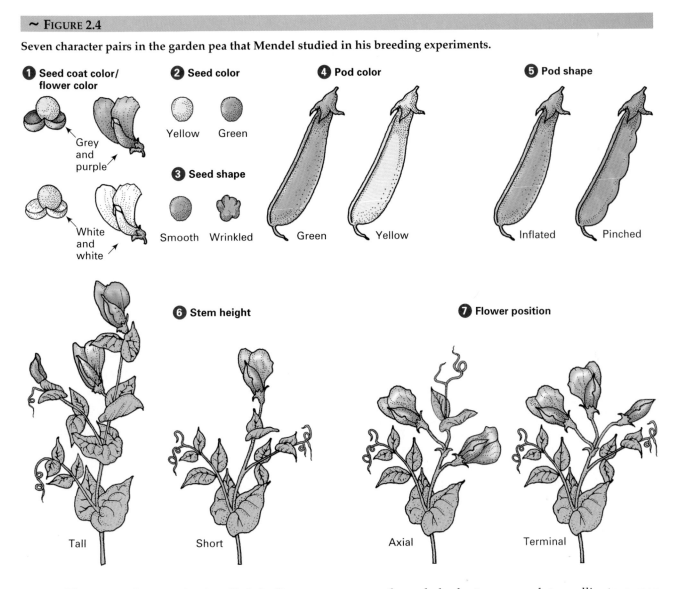

1 Seed coat color/flower color — Grey and purple, White and white

2 Seed color — Yellow, Green

3 Seed shape — Smooth, Wrinkled

4 Pod color — Green, Yellow

5 Pod shape — Inflated, Pinched

6 Stem height — Tall, Short

7 Flower position — Axial, Terminal

ments. The parental generation is called the **P generation**. The progeny of the P mating is called the **first filial generation**, or **F_1**. The subsequent generation produced by breeding together the F_1 offspring is the **F_2 generation**. Interbreeding the offspring of each generation results in F_3, F_4, F_5 generations, and so on.

Mendel first performed crosses between true-breeding strains of peas that differed in a single trait. Such crosses are called **monohybrid crosses**. For example, when he pollinated pea plants that gave rise only to smooth seeds[1] with pollen from a true-breeding variety that produced only wrinkled seeds, the outcome was all smooth seeds (Figure 2.5). The same result was obtained when the parental types were reversed; that is, when the pollen from a smooth-seeded plant was used to pollinate a pea plant that gave wrinkled seeds, the outcome was all smooth seeds. Matings that are done both ways—smooth female [♀] × wrinkled male [♂] and wrinkled female [♀] × smooth male [♂]—are called **reciprocal crosses.** Conventionally, the female is given first in crosses of plants. If the results of reciprocal crosses are the same, it means that the trait does not depend on the sex of the organism.

The significant point of this cross is that all the F_1 progeny seeds of the smooth × wrinkled reciprocal crosses were smooth; that is, they exactly resembled only one of the parents in this character rather than being a blend of both parental phenotypes. The finding that all offspring of true-breeding parents are alike is sometimes referred to as the *principle of uniformity in F_1.*

Next, Mendel planted the seeds and allowed the F_1 plants to self-fertilize to produce the F_2 seed. Both smooth and wrinkled seeds appeared in the F_2 generation, and both types could be found within the same

[1]Seeds are the diploid progeny of sexual reproduction. If a phenotype concerns the seed itself, the results of the cross can be seen directly by looking at the seeds. If a phenotype concerns a part of the mature plant, such as flower color, then the seeds must be germinated and grown to maturity before that phenotype can be seen.

~ FIGURE 2.5

Results of one of Mendel's breeding crosses. In the parental generation he crossed a true-breeding pea strain that produced smooth seeds with one that produced wrinkled seeds. All the F₁ progeny seeds were smooth.

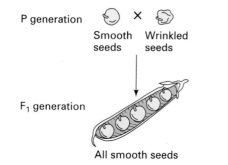

pod. Typical of his analytical approach to the experiments, Mendel counted the number of seeds of each type. He found that 5,474 were smooth and 1,850 were wrinkled (Figure 2.6). The calculated ratio of smooth seeds to wrinkled seeds was 2.96:1, which is very close to a 3:1 ratio.

Mendel observed that although the F₁ resembled only one of the parents in their phenotype, they did not breed true, a fact that distinguishes the F₁ from the parent they resembled. Moreover, the F₁ could produce some F₂ progeny with the parental phenotype that had disappeared in the F₁. But how can a trait present in the P generation disappear in the F₁ and then reappear in the F₂? Mendel concluded that the alternative traits in the cross—smoothness or wrinkledness of the seeds—were determined by **particulate factors**. He reasoned that these factors, which were transmitted from parents to progeny through the gametes, carried hereditary information. We now know these factors by another name: *genes*.

Since Mendel was examining a pair of traits (wrinkled/smooth), each factor was considered to exist in alternative forms (which we now call **alleles**), each of which specified one of the traits. For the gene controlling pea seed shape, there is one form (allele) that results in the production of a smooth seed and another allele that results in a wrinkled seed.

Mendel reasoned further that a true-breeding strain of peas must contain a pair of identical factors. Since the F₂ exhibited both traits while the F₁ exhibited only one of those traits, then each F₁ individual must have contained both factors, one for each of the alternative traits. In other words, crossing two different true-breeding strains brings together in the F₁ one factor from each strain: the eggs contain one factor from one strain and the pollen grains contain one factor from the other strain. Further, since only one of the traits was seen in the F₁ generation, the expression of the "missing" trait must somehow have been

masked by the visible trait; this masking is called *dominance*. For the smooth × wrinkled cross, the F₁ seeds were all smooth. Thus, the allele for smoothness is masking or **dominant** to the allele for wrinkledness. Conversely, wrinkled is said to be **recessive** to smooth because the factor for wrinkled is masked.

A simple way to visualize the crosses is to use symbols for the alleles, as Mendel did. For the smooth × wrinkled cross we can give the symbol *S* to the allele for smoothness and the symbol *s* to the allele for wrinkledness. The letter used is based on the dominant phenotype, and the convention in this case is that the dominant allele is given the capital letter and the recessive allele the small letter. (This convention was used for many years, particularly in plant genetics. Now it is more conventional to base the letter assignment on the recessive phenotype. We will use this convention later.)

Using these symbols, the genotype of the parental plant grown from the smooth seeds is *SS*, and that of the wrinkled parent is *ss*. True-breeding individuals that contain two copies of the same specific allele of a particular gene are said to be **homozygous** for that

~ FIGURE 2.6

The F₂ progeny of the cross shown in Figure 2.5. When the plants grown from the F₁ seeds were self-pollinated, both smooth and wrinkled F₂ progeny seeds were produced. Commonly, both seed types were found in the same pod. In his experiments Mendel counted 5,474 smooth and 1,850 wrinkled F₂ progeny seeds for a ratio of 2.96:1.

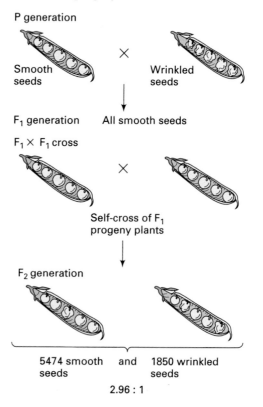

~ FIGURE 2.7

Dominant and recessive alleles of a gene for seed shape in peas.

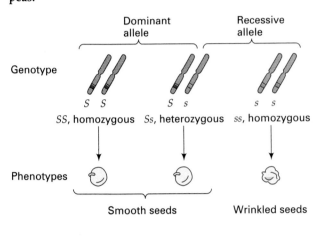

gene (Figure 2.7). When plants produce gametes by meiosis (see Chapter 1), each gamete contains only one copy of the gene (one allele); the plants from smooth seeds produce S-bearing gametes, and the plants from wrinkled seeds produce s-bearing gametes. When the gametes fuse during fertilization, the resulting zygote has one S allele and one s allele, a genotype of Ss. Plants that have two different alleles of a particular gene are said to be **heterozygous**. Because of the dominance of the smooth S allele, Ss plants produce smooth seeds (see Figure 2.7).

Figure 2.8 diagrams the smooth × wrinkled cross using genetic symbols; the production of the F_1 is shown in Figure 2.8a, and that of the F_2 in Figure 2.8b. (In Figures 2.7 and 2.8, the genes are shown on chromosomes. Keep in mind that the segregation of genes from generation to generation follows the behavior of chromosomes.) The true-breeding, smooth-seeded parent has the genotype SS, and the true-breeding, wrinkle-seeded parent has the genotype ss. Since each parent is true breeding and diploid (that is, has two

~ FIGURE 2.8

The same cross as in Figures 2.5 and 2.6, using genetic symbols to illustrate the principle of segregation of Mendelian factors. (a) Production of the F_1 generation; (b) Production of the F_2 generation.

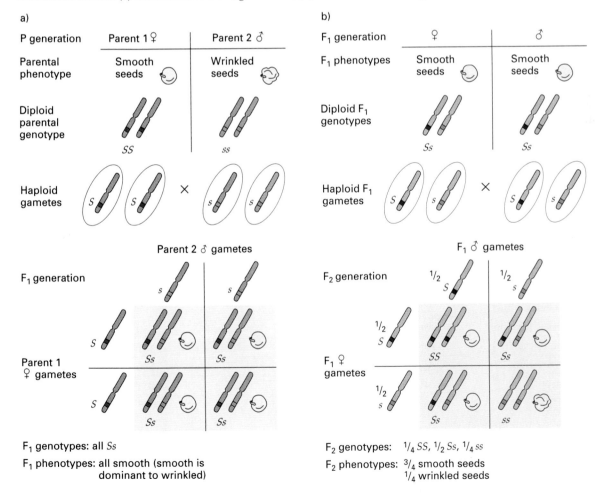

sets of chromosomes), each must contain two copies of the same allele. All the F_1 plants produce smooth seeds, and all are Ss heterozygotes.

The plants grown from the F_1 seeds differ from the smooth parent in that they produce equal numbers of two types of gametes: S-bearing gametes and s-bearing gametes. All the possible fusions of F_1 gametes are shown in the matrix in Figure 2.8b, called a **Punnett square** after its originator, R. Punnett. These fusions give rise to the zygotes that produce the F_2 generation.

In the F_2 generation, three types of genotypes are produced: SS, Ss, and ss. As a result of the random fusing of gametes, the relative proportion of these zygotes is 1:2:1, respectively. However, since the S factor is dominant to the s factor, both the SS and Ss seeds are smooth, and the F_2 generation seeds show a phenotypic ratio of 3 smooth : 1 wrinkled.

Mendel also analyzed the behavior of the six other pairs of traits. Qualitatively and quantitatively, the same results were obtained (Table 2.1). From the seven sets of crosses he made the following general conclusions about his data:

1. The results of reciprocal crosses were always the same.

2. All F_1 progeny resembled one of the parental strains, indicating the dominance of one allele over the other.

3. In the F_2 generation, the parental trait that had disappeared in the F_1 generation reappeared. Further, the trait seen in the F_1 was always found

in the F_2 at about three times the frequency of the other trait.

The Principle of Segregation

From the sort of data we have discussed, Mendel proposed what has become known as his **first law**, the **principle of segregation**: *Recessive characters, which are masked in the F_1 from a cross between two true-breeding strains, reappear in a specific proportion in the F_2.* In modern terms this means that *the two members of a gene pair (alleles) segregate (separate) from each other during the formation of gametes.* As a result, half the gametes carry one allele, and the other half carry the other allele. In other words, each gamete carries only a single allele of each gene. The progeny are produced by the random combination of gametes from the two parents.

In proposing the principle of segregation, Mendel had differentiated between the factors (genes) that determined the traits (the genotype) and the traits themselves (the phenotype). From a modern perspective we know that genes are on chromosomes, and that the specific location of a gene on a chromosome is called its **locus** (or **gene locus**; plural *loci*). Further, Mendel's first law means that at the gene level the members of a pair of alleles segregate during meiosis, and that each offspring receives only one allele from each parent. Thus, **gene segregation** parallels the separation of homologous pairs of chromosomes at anaphase I in meiosis (see Chapter 1).

~ TABLE 2.1

Mendel's Results in Crosses Between Plants Differing in One of Seven Characters

CHARACTER[a]	F_1	F_2 (NUMBER)			F_2 (RATIO)
		DOMINANT	RECESSIVE	TOTAL	DOMINANT: RECESSIVE
Seeds: smooth versus wrinkled	All smooth	5,474	1,850	7,324	2.96:1
Seeds: yellow versus green	All yellow	6,022	2,001	8,023	3.01:1
Seed coats: grey versus white [b] Flowers: purple versus white	All grey All purple }	705	224	929	3.15:1
Flowers: axial versus terminal	All axial	651	207	858	3.14:1
Pods: inflated versus pinched	All inflated	882	299	1,181	2.95:1
Pods: green versus yellow	All green	428	152	580	2.82:1
Stem: tall versus short	All tall	787	277	1,064	2.84:1
Total or average		**14,949**	**5,010**	**19,959**	**2.98:1**

[a] The dominant trait is always written first.

[b] A single gene controls both the seed coat and the flower color trait.

BOX 2.1

GENETICS TERMINOLOGY

Alleles: alternative forms of a gene. For example, *S* and *s* alleles represent the smoothness and wrinkledness of the pea seed. (Like gene symbols, allele symbols are <u>underlined</u> or *italicized*.)

Cross: a mating between two individuals, leading to the fusion of gametes.

Diploid: a eukaryotic cell or organism with two homologous sets of chromosomes.

Gamete: a mature reproductive cell that is specialized for sexual fusion. Each gamete is haploid and fuses with a cell of similar origin but of opposite sex to produce a diploid zygote.

Gene (Mendelian factor): the determinant of a characteristic of an organism. Gene symbols are <u>underlined</u> or *italicized*.

Genotype: the genetic constitution of an organism. A diploid organism in which both alleles are the same at a given gene locus is said to be **homozygous** for that allele. Homozygotes produce only one gametic type with respect to that locus. For example, true-breeding smooth-seeded peas have the genotype *SS*, and true-breeding wrinkle-seeded peas have the genotype *ss*; both are homozygous. The smooth parent is **homozygous dominant**; the wrinkled parent is **homozygous recessive**.

Diploid organisms that have two different alleles at a specific gene locus are said to be **heterozygous**. Thus, F_1 hybrid plants from the cross of *SS* and *ss* parents have one *S* allele and one *s* allele. Individuals heterozygous for two allelic forms of a gene produce two kinds of gametes (*S* and *s*).

Haploid: a cell or an individual with one copy of each chromosome.

Locus (gene locus; plural = loci): the specific place on a chromosome where a gene is located.

Phenotype: the physical manifestation of a genetic trait that results from a specific genotype and its interaction with the environment. In our example, the *S* allele was dominant to the *s* allele, so in the heterozygous condition the seed is smooth. Therefore, both the homozygous dominant *SS* and the heterozygous *Ss* seeds have the same phenotype (smooth), even though they differ in genotype.

Zygote: the cell produced by the fusion of male and female gametes.

Box 2.1 presents a summary of the genetics concepts and terms we have discussed so far in this chapter. A thorough familiarity with these terms is essential to your study of genetics.

KEYNOTE

Mendel's first law, the principle of segregation, states that the two members of a gene pair (alleles) segregate (separate) from each other in the formation of gametes; half the gametes carry one allele, and the other half carry the other allele.

Representing Crosses with a Branch Diagram

The use of a Punnett square to represent the pairing of all possible gamete types from the two parents in a monohybrid cross (shown in Figure 2.8) is a relatively simple way to predict the relative frequencies of genotypes and phenotypes in the next generation. In this section, however, we examine an alternative method, one which you are encouraged to master: the branch or fork diagram. (Box 2.2 discusses some elementary principles of probability that will help you understand this approach.) In order to use the branch diagram approach, it is necessary to know the dominance/recessiveness relationship of the allele pair so that the progeny phenotypic classes can be determined. Figure 2.9 illustrates the application of branch diagram analysis of the F_1 selfing of the smooth $\times$ wrinkled cross diagrammed in Figure 2.8.

The F_1 seeds from the cross in Figure 2.8 have the genotype *Ss*. In meiosis we expect half of the gametes to be *S* and half to be *s* (see Figure 2.9). Thus, 1/2 is the predicted frequency of each of these two types. But just as tossing a coin many times does not always give exactly half heads and half tails, the two gametes may not be produced in an exactly 1:1 ratio. However, the more chances (tosses), the closer the observed frequency will come to the predicted frequency.

From the rules of probability, we can predict the expected frequencies of the three possible genotypes

BOX 2.2

ELEMENTARY PRINCIPLES OF PROBABILITY

A **probability** is the ratio of the number of times a particular event is expected to occur to the number of trials during which the event could have happened. For example, the probability of picking a heart from a deck of 52 cards, 13 of which are hearts, is $p(\text{heart}) = 13/52 = 1/4$. That is, we would expect, on the average, to pick a heart from a deck of cards once in every four trials.

Probabilities and the *laws of chance* are involved in the transmission of genes. As a simple example, let us consider a couple and the chance that their child will be a boy (or a girl). Assume that an exactly equal number of boys and girls are born (which is not precisely true, but we can assume it to be so for the sake of discussion). The probability that the child will be a boy is 1/2, or 0.5. Similarly, the probability that the child will be a girl is 1/2.

Now, a rule of probability can be introduced: the **product rule**. The product rule states that *the probability of two independent events occurring simultaneously is the product of each of their individual probabilities.* Thus, the probability that both children in families with two children will be girls is 1/4. That is, the probability of the first child being a girl is 1/2, the probability of the second being a girl is also 1/2, and by the product rule the probability of both children being girls is $1/2 \times 1/2 = 1/4$. Similarly, the probability of having three boys in a row is $1/2 \times 1/2 \times 1/2 = 1/8$.

Another rule of probability, the **sum rule**, states that *the probability of either one of two independent (mutually exclusive) events occurring is the sum of their individual probabilities.* For example, if two dice are thrown, what is the probability of getting two sixes or two ones? The individual probabilities are calculated as follows: The probability of getting two sixes is found by using the product rule. The probability of getting one six, $p(\text{one six})$, is 1/6, since there are six faces to a die. Therefore, the probability of getting two sixes, $p(\text{two sixes})$, when two dice are thrown is $1/6 \times 1/6 = 1/36$. Similarly, $p(\text{two ones}) = 1/36$. To roll two sixes *or* two ones involves independent events, so the sum rule is used. The sum of the individual probabilities is $1/36 + 1/36 = 2/36 = 1/18$. To return to our family example, the probability of having two boys or two girls is $1/4 + 1/4 = 1/2$.

～ FIGURE 2.9

Using the branch diagram approach to calculate the ratios of phenotypes in the F₂ generation of the cross in Figure 2.8.

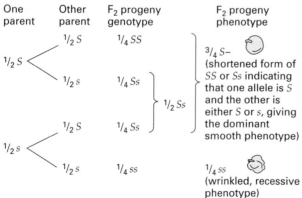

in the F₂ generation. To produce an *SS* plant, an *S* egg must pair with an *S* pollen grain. The frequency of *S* eggs in the population of eggs is 1/2, and the frequency of *S* pollen grains in the pollen population is 1/2. Therefore, the expected proportion of *SS* smooth plants in the F₂ is $1/2 \times 1/2 = 1/4$. Similarly, the expected proportion of *ss* wrinkled progeny in the F₂ is $1/2 \times 1/2 = 1/4$.

What about the *Ss* progeny? Again, the frequency of *S* in one gametic type is 1/2, and the frequency of *s* in the other gametic type is also 1/2. However, there are two ways in which *Ss* progeny can be obtained. The first involves the fusion of an *S* egg with *s* pollen, and the second is a fusion of an *s* egg with *S* pollen. Using the product rule (see Box 2.2), the probability of each of these events occurring is $1/2 \times 1/2 = 1/4$. By using the sum rule (see Box 2.2), the probability of *one or the other occurring* is the sum of the individual probabilities, or $1/4 + 1/4 = 1/2$.

The prediction, then, is that one-fourth of the F₂ progeny will be *SS*, half will be *Ss*, and one-fourth will be *ss*, exactly as was found with the Punnett square method shown in Figure 2.8. Either method—the

Punnett square or the branch diagram—may be used with any cross, but as crosses become more complicated, the Punnett square method becomes cumbersome.

Confirming the Principle of Segregation: The Use of Testcrosses

When formulating his principle of segregation, Mendel did a number of tests to ensure the correctness of his results. He continued the self-fertilizations at each generation up to the F_6 and found that in every generation both the dominant and recessive characters were found. He concluded, therefore, that the principle of segregation was valid no matter how many generations were involved.

Another important test concerned the F_2 plants. As shown in Figure 2.8, a ratio of 1:2:1 occurs for the genotypes SS, Ss, and ss for the smooth × wrinkled example. Phenotypically, the ratio of smooth to wrinkled is 3:1. At the time of Mendel's experiments, the presence of segregating factors that were responsible for the smooth and wrinkled phenotypes was only an hypothesis. To test his factor hypothesis, Mendel allowed the F_2 plants to self-pollinate. As he expected, the plants produced from wrinkled seeds bred true, supporting his conclusion that they were pure for the s factor (gene).

Selfing the plants derived from the F_2 smooth seeds produced two different types of progeny. *One-third* of the smooth F_2 seeds produced all smooth-seeded progeny, whereas the other *two-thirds* produced both smooth and wrinkled seeds in each pod in a ratio of 3 smooth : 1 wrinkled (Figure 2.10). For the plants that produced both seed types in the progeny, the actual ratio of smooth:wrinkled seeds was 3:1; that is, the same ratio as seen for the F_2 progeny. These results completely support the prin-

ciple of segregation of genes. The random combination of gametes that form the zygotes of the original F_2 produces two genotypes that give rise to the smooth phenotype (see Figures 2.8 and 2.9); the relative proportion of the two genotypes SS and Ss is 1:2. The SS seeds give rise to true-breeding plants, whereas the Ss seeds give rise to plants that behave exactly like the F_1 plants when they are self-pollinated in that they produce a 3:1 ratio of smooth:wrinkled progeny. *Mendel explained these results by proposing that each plant had two factors, while each gamete had only one. He also proposed that the random combination of the gametes generated the progeny in the proportions he found. Mendel obtained the same results in all seven sets of crosses.*

The self-fertilization test of the F_2 progeny proved a useful way of confirming the genotype of a plant with a given phenotype. A more common test to ascertain the genotype of an organism is to perform a **testcross**, a cross of an individual of unknown genotype (usually expressing the dominant phenotype) with a homozygous recessive individual in order to determine the unknown genotype.

Consider again the cross shown in Figure 2.8. We can predict the outcome of a testcross of the F_2 progeny showing the dominant, smooth-seed phenotype. If the F_2 individuals are homozygous SS, then the result of a testcross with an ss plant will be all smooth seeds. As Figure 2.11a shows, the Parent 1 smooth SS plants produce only S gametes. Parent 2 is homozygous recessive wrinkled, ss, so it produces only s gametes. Therefore, all zygotes are Ss, and all the resulting seeds have the smooth phenotype. In actual practice, then, if a plant with a dominant trait is testcrossed, and only the dominant phenotype is seen among the progeny, then the plant must have been homozygous for the dominant allele. In contrast, heterozygous Ss F_2 plants testcrossed with a homozygous ss plant give a 1:1 ratio of dominant to recessive phenotypes. As Figure 2.11b shows, the Parent 1 smooth Ss produces both S and s gametes in equal proportion, while the homozygous ss Parent 2 produces only s gametes. As a result, half the progeny of the testcross are Ss heterozygotes and have a smooth phenotype because of the dominance of the S allele, and the other half are ss homozygotes and have a wrinkled phenotype. In actual practice, then, if a plant with a dominant trait is testcrossed, and the progeny exhibit a 1:1 ratio of dominant to recessive phenotypes, then the plant must have been heterozygous.

In summary, testcrosses of the F_2 progeny from Mendel's crosses that showed the dominant phenotype resulted in a 1:2 ratio of homozygous dominant : heterozygous genotypes in the F_2 progeny.

~ FIGURE 2.10

Determining the genotypes of the F_2 smooth progeny of Figure 2.8 by selfing the plants grown from the smooth seeds.

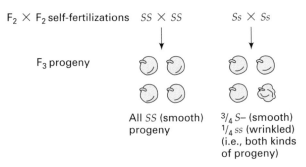

$F_2 \times F_2$ self-fertilizations $SS \times SS$ $Ss \times Ss$

F_3 progeny

All SS (smooth) progeny

$3/4$ $S-$ (smooth)
$1/4$ ss (wrinkled)
(i.e., both kinds of progeny)

~ FIGURE 2.11

Determining the genotypes of the F$_2$ generation smooth seeds (Parent 1) of Figure 2.8 by testcrossing plants grown from the seed with a homozygous recessive wrinkled (ss) strain (Parent 2). (a) If Parent 1 is *SS*, then all progeny seeds are smooth; (b) If Parent 1 is *Ss*, then 1/2 of the progeny seeds are smooth and 1/2 are wrinkled.

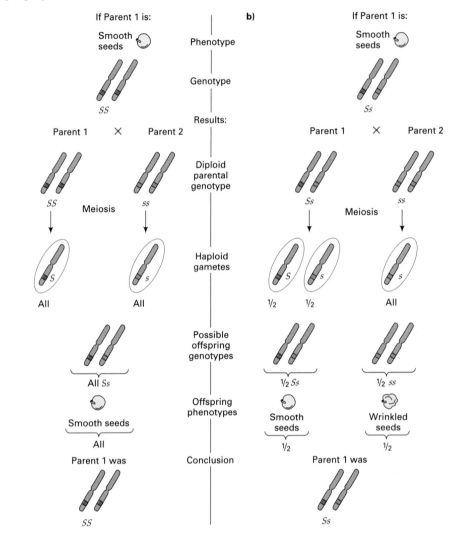

That is, when crossed with the homozygous recessive, one-third of the F$_2$ progeny with the dominant phenotype gave rise only to progeny with the dominant phenotype and were therefore homozygous for the dominant allele. The other two-thirds of the F$_2$ progeny with the dominant phenotype produced progeny with a 1:1 ratio of dominant phenotype to recessive phenotype, and therefore were heterozygous.

KEYNOTE

A testcross is a cross of an individual of unknown genotype (usually expressing the dominant phenotype) with a known homozygous recessive individual in order to determine the unknown genotype. The phenotypes of the progeny of the testcross indicate the genotype of the individual tested. If all the progeny show the dominant phenotype, then the individual was homozygous dominant. If the ratio of progeny with dominant and recessive phenotypes is approximately 1:1, then the unknown individual was heterozygous.

DIHYBRID CROSSES AND THE MENDELIAN PRINCIPLE OF INDEPENDENT ASSORTMENT

The Principle of Independent Assortment

Mendel also analyzed a number of crosses in which two pairs of traits were simultaneously involved. In each case he obtained the same results. From these experiments he proposed his **second law**, the **principle of independent assortment**, which states that *the factors for different traits assort independently of one another.* In modern terms, this means that *genes on different chromosomes behave independently in the production of gametes.*

Consider an example involving smooth (*S*)/wrinkled (*s*) and yellow (*Y*)/green (*y*) seed traits (yellow is dominant to green). When Mendel made crosses between true-breeding smooth-yellow plants (*SS YY*) and wrinkled-green plants (*ss yy*), he got the results shown in Figure 2.12. All the F_1 seeds from this cross were smooth and yellow, as the results of the monohybrid crosses predicted. As Figure 2.12a shows, the smooth-yellow parent produces only *S Y* gametes, which give rise to *Ss Yy* zygotes upon fusion with the *s y* gametes from the wrinkled-green parent. Because of the dominance of the smooth and the yellow traits, all F_1 seeds are smooth and yellow.

The F_1 are heterozygous for two pairs of alleles at two different loci. Such individuals are called dihybrids, and a cross between two of these dihybrids of the same type is called a **dihybrid cross**.

When Mendel self-pollinated the dihybrid F_1 plants to give rise to the F_2 generation (Figure 2.12b), he considered two possible outcomes. One was that the genes for the traits from the original parents would be transmitted together to the progeny. In this case, a phenotypic ratio of 3:1 smooth-yellow: wrinkled-green would be predicted.

The other possibility was that the traits would be inherited independently of one another. In this case, the dihybrid F_1 would produce four types of gametes: *S Y*, *S y*, *s Y*, and *s y*. Given the independence of the two pairs of genes, each gametic type is predicted to occur with equal frequency. In $F_1 \times F_1$ crosses, the four types of gametes would be expected to fuse randomly in all possible combinations to give rise to the zygotes and, hence, the progeny seeds. All the possible gametic fusions are represented in the Punnett square in Figure 2.12b. In a dihybrid cross there are 16 possible gametic fusions. The result is nine different genotypes but, because of dominance, only four phenotypes are predicted:

1 *SS YY*, 2 *Ss YY*, 2 *SS Yy*, 4 *Ss Yy* = 9 smooth-yellow

1 *SS yy*, 2 *Ss yy* = 3 smooth-green

1 *ss YY*, 2 *ss Yy* = 3 wrinkled-yellow

1 *ss yy* = 1 wrinkled-green

According to the rules of probability, if pairs of characters are inherited independently in a dihybrid cross, then the F_2 from an $F_1 \times F_1$ cross will give a 9:3:3:1 ratio of the four possible phenotypic classes. This ratio is the result of the independent assortment of the two gene pairs into the gametes and of the random fusion of those gametes.

This prediction was met in all the dihybrid crosses Mendel performed. In every case the F_2 ratio was close to 9:3:3:1. For our example he counted 315 smooth-yellow, 108 smooth-green, 101 wrinkled-yellow, and 32 wrinkled-green seeds—very close to the predicted ratio. To Mendel this result meant that the factors (genes) determining the specific, different character pairs he was analyzing were transmitted independently. This means that he rejected the possibility that the two traits were inherited together.

~ FIGURE 2.12a

The principle of independent assortment in a dihybrid cross. This cross, actually done by Mendel, involves the smooth/wrinkled and yellow/green character pairs of the garden pea. (a) Production of the F_1 generation.

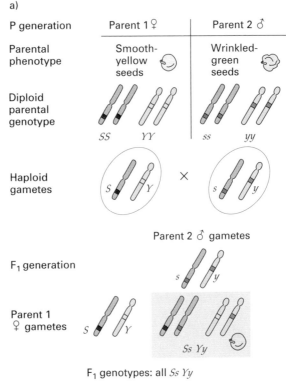

F$_1$ genotypes: all *Ss Yy*

F$_1$ phenotypes: all smooth-yellow seeds

~ FIGURE 2.12b

(b) The F_2 genotypes and 9:3:3:1 phenotypic ratio of smooth-yellow:smooth-green:wrinkled-yellow:wrinkled-green, derived by using the Punnett square. (Note that, compared with previous figures of this kind, only one box is shown in the F_1, instead of four. This is because only one class of gametes exists for Parent 2 and only one class for Parent 1. Previously we showed two gametes from each parent, even though those gametes were identical.)

b)

F₂ genotypes:

$1/16$ ($SS\ YY$) + $2/16$ ($Ss\ YY$) + $2/16$ ($SS\ Yy$) + $4/16$ ($Ss\ Yy$) = $9/16$ smooth-yellow seeds
$1/16$ ($SS\ yy$) + $2/16$ ($Ss\ yy$) = $3/16$ smooth-green seeds
$1/16$ ($ss\ YY$) + $2/16$ ($ss\ Yy$) = $3/16$ wrinkled-yellow seeds
$1/16$ ($ss\ yy$) = $1/16$ wrinkled-green seeds

Branch Diagram of Dihybrid Crosses

Rather than using a Punnett square, it is easier to get into the habit of calculating the expected ratios of phenotypic or genotypic classes by using a branch diagram to apply the laws of probability to the traits one at a time. *With practice you should be able to calculate the probabilities of outcomes of various crosses just by using the laws of probability without any need for drawing out the branch diagram.* Diligently working problems really helps to hone this skill.

Using the same example, in which the two gene pairs assort independently into the gametes, we consider each gene pair in turn. Earlier we saw that an F_1 self of an Ss heterozygote gave rise to progeny of which three-fourths were smooth and one-fourth were wrinkled. Genotypically, the former class had at least one dominant S allele; that is, they were SS or Ss. A convenient way to signify this situation is to use a dash to indicate an allele that has no effect on the phenotype. Thus $S-$ means that phenotypically the seeds are smooth and genotypically they are either SS or Ss.

Now consider the F_2 produced from a selfing of Yy heterozygotes; again, a 3:1 ratio is seen, with three-fourths of the seeds being yellow and one-fourth being green. Since this segregation occurs independently of the segregation of the smooth/wrinkled pair, we can consider all possible combinations of the phenotypic classes in the dihybrid cross. For example, the expected proportion of F_2 seeds that are smooth and yellow is the product of the probability that an F_2 seed will be smooth and the probability that it will be yellow, or $3/4 \times 3/4 = 9/16$. Similarly, the expected proportion of F_2 progeny that are wrinkled and yellow is $3/4 \times 1/4 = 3/16$. Extending this calculation to all possible phenotypes, as shown in Figure 2.13, we obtain the ratio of 9 $S-Y-$ (smooth, yellow) : 3 $S-yy$ (smooth, green) :3 $ss Y-$ (wrinkled, yellow) : 1 $ss yy$ (wrinkled, green).

The testcross can be used to check the genotypes of F_1 progeny and F_2 progeny from a dihybrid cross. In our example, the F_1 is a double heterozygote, $Ss Yy$. This F_1 produces four types of gametes in equal proportions: $S Y$, $S y$, $s Y$, and $s y$ (see Figure 2.12b). In a testcross with a doubly homozygous recessive plant—in this case $ss yy$—the phenotypic ratio of the progeny is a direct reflection of the ratio of gametic types produced by the F_1 parent. In a testcross like this one, then, there will be a 1:1:1:1 ratio in the offspring of $Ss Yy:Ss yy:ss Yy:ss yy$ genotypes, which means a 1:1:1:1 ratio of smooth-yellow: smooth-green:wrinkled-yellow:wrinkled-green phenotypes. This 1:1:1:1 phenotypic ratio is diagnostic of testcrosses in which the "unknown" parent is a double heterozygote.

In the F_2 of a dihybrid cross there are nine different genotypic classes but only four phenotypic classes. The genotypes can be ascertained by testcrossing, as we have shown. Table 2.2 lists the expected ratios of progeny phenotypes from such testcrosses. No two patterns are the same, so here the testcross is truly a diagnostic approach to confirm genotypes.

Trihybrid Crosses

Mendel also confirmed his laws for three characters segregating in other crosses. Such crosses are called **trihybrid crosses**. Here the proportions of F_2 genotypes and phenotypes are predicted with precisely the same logic used before—by considering each character pair independently. Figure 2.14 shows a branch diagram derivation of the F_2 phenotypic classes for a trihybrid cross. The independently assorting character pairs in the cross are smooth versus wrinkled seeds, yellow versus green seeds, and purple versus white flowers. There are 64 combinations of

~ **FIGURE 2.13**

Using the branch diagram approach to calculate the F_2 phenotypic ratio of the cross in Figure 2.12.

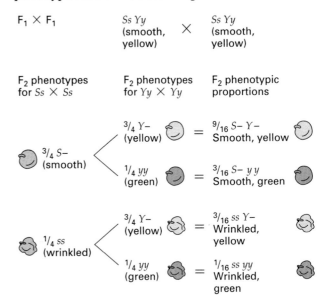

~ TABLE 2.2

Proportions of Phenotypic Classes Expected from
Testcrosses of Strains with Various Genotypes for Two
Gene Pairs

	PROPORTION OF PHENOTYPIC CLASSES			
TESTCROSSES	A– B–	A– bb	aa B–	aa bb
AA BB × aa bb	1	0	0	0
Aa BB × aa bb	1/2	0	1/2	0
AA Bb × aa bb	1/2	1/2	0	0
Aa Bb × aa bb	1/4	1/4	1/4	1/4
AA bb × aa bb	0	1	0	0
Aa bb × aa bb	0	1/2	0	1/2
aa BB × aa bb	0	0	1	0
aa Bb × aa bb	0	0	1/2	1/2
aa bb × aa bb	0	0	0	1

eight maternal and eight paternal gametes. Combination of these gametes gives rise to 27 different genotypes and eight different phenotypes in the F_2 generation. The phenotypic ratio in the F_2 is 27:9:9:9:3:3:3:1.

Now that we have considered enough examples, we can make some generalizations about phenotypic and genotypic classes. In each example discussed, the F_1 is heterozygous for each gene involved in the cross, and the F_2 is generated by selfing (when possible) or by allowing the F_1 progeny to interbreed. In monohybrid crosses there are two phenotypic classes in the F_2, in dihybrid crosses there are four, and in trihybrid crosses there are eight. The general rule is that there are 2^n phenotypic classes in the F_2, where n is the number of independently assorting, heterozygous gene pairs (Table 2.3). (This rule holds *only* when a

~ TABLE 2.3

Number of Phenotypic and Genotypic Classes Expected
from Self-Crosses of Heterozygotes in Which All Genes
Show Complete Dominance

NUMBER OF SEGREGATING GENE PAIRS	NUMBER OF PHENOTYPIC CLASSES	NUMBER OF GENOTYPIC CLASSES
1[a]	2	3
2	4	9
3	8	27
4	16	81
n	2^n	3^n

[a] For example, from Aa × Aa, two phenotypic classes are expected, with genotypic classes of AA, Aa, and aa.

true dominant-recessive relation holds for each of the gene pairs.) Furthermore, we saw that there are three genotypic classes in the F_2 of monohybrid crosses, nine in dihybrid crosses, and 27 in trihybrid crosses. A simple rule is that the number of genotypic classes is 3^n, where n is the number of independently assorting, heterozygous gene pairs (see Table 2.3).

Incidentally, the phenotypic rule (2^n) can also be used to predict the number of classes that will come from a multiple heterozygous F_1 used in a testcross. Here the number of genotypes in the next generation will be the same as the number of phenotypes. For example, from Aa Bb × aa bb there are four progeny genotypes (2^n, where n is 2)—Aa Bb, Aa bb, aa Bb, and aa bb—and four phenotypes: both dominant phenotypes, the A dominant phenotype and b recessive phenotype, the a recessive phenotype and B dominant phenotype, and both recessive phenotypes.

THE "REDISCOVERY" OF MENDEL'S PRINCIPLES

Mendel published his treatise on heredity in 1866 in *Verhandlungen des Naturforschenden Vereines* in Brünn, but it received little attention from the scientific community at the time. At about the turn of the twentieth century, three researchers working independently on breeding experiments came to the same conclusions as had Mendel. The three men were Carl Correns, Hugo de Vries, and Erich von Tschermak. Correns concentrated mostly on maize (corn) and peas, de Vries worked with a number of different plant species, and von Tschermak studied peas.

The first demonstration that Mendelism applied to animals came in 1902 from the work of William Bateson, who experimented with fowl. Bateson also coined the terms *genetics, zygote,* F_1, F_2, and **allelomorph** (literally, "alternative form," meaning one of an array of different forms of a gene), which other researchers shortened to *allele*. The term *gene* as a replacement for Mendelian factor was introduced by W. L. Johannsen in 1909.

STATISTICAL ANALYSIS OF GENETIC DATA: THE CHI-SQUARE TEST

Data from genetic crosses are quantitative. A geneticist typically uses statistical analysis to interpret a set of data from crossing experiments in order to understand the significance of any deviation of observed

~ FIGURE 2.14

Branch diagram derivation of the relative frequencies of the eight phenotypic classes in the F₂ of a trihybrid cross.

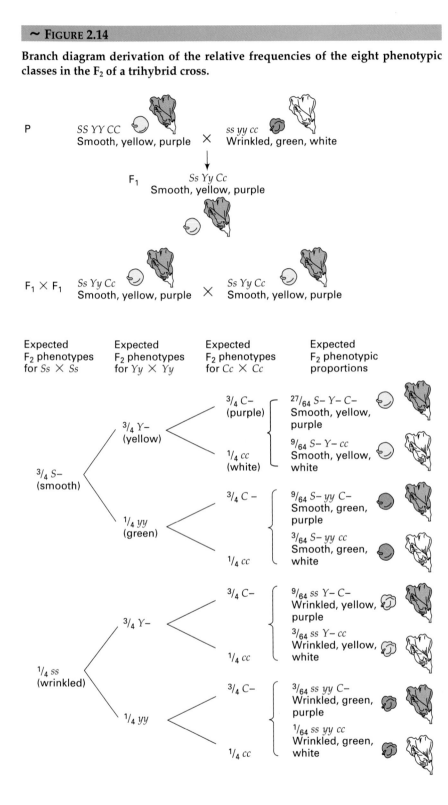

results from the results predicted by the hypothesis being tested. The observed phenotypic ratios among progeny rarely exactly match expected or predicted ratios due to chance factors inherent in biological phenomena. The hypothesis is presented as a **null hypothesis**, which states that there is no real differ-

ence between the observed data and the predicted data. Because every data set will show some difference from what was predicted, appropriate statistical analysis is used to determine whether or not the difference is due to chance. If it is due to chance, then the null hypothesis should be rejected, and a new hypothesis must be developed to explain the data. A relatively simple statistical analysis used to test null hypotheses is called the **chi-square (χ^2) test**, which is essentially a *goodness-of-fit test*. In the genetic crosses we have examined so far, the progeny seemed to fit particular ratios (such as 1:1, 3:1, and 9:3:3:1), and this is where a null hypothesis can be posed and where the chi-square test can tell us if the data are consistent with that hypothesis.

To illustrate the use of the chi-square test, we will analyze theoretical progeny data from a testcross of a smooth-yellow double heterozygote (*Ss Yy*) with a wrinkled-green homozygote (*ss yy*) (see p. 39 and Table 2.2). (Additional applications of the chi-square test are given in Chapter 5.) The progeny data are:

	154 smooth, yellow
	124 smooth, green
	144 wrinkled, yellow
	146 wrinkled, green
Total	568

We hypothesize that the two genes assort independently and use the chi-square test to test the hypothesis, as shown in Table 2.4.

If the two genes assort independently, then a testcross should give a 1:1:1:1 ratio of the four phenotypic classes. First, in column 1 we list the four phenotypes expected in the progeny of the cross. Then we list the observed (*o*) numbers for each phenotype, using actual numbers and not percentages or proportions (column 2). Next, we calculate the expected (*e*) number for each phenotypic class, given the total number of progeny (568) and the hypothesis under evaluation (in this case 1:1:1:1). Thus, in column 3 we list $1/4 \times 568 = 142$, and so on. Now we subtract the expected number (*e*) from the observed number (*o*)

for each class to find differences, called the deviation value (*d*). The sum of the *d* values is always zero (column 4).

In column 5 the deviation squared (d^2) is computed by multiplying each deviation value in column 4 by itself. In column 6 the deviation squared is then divided by the expected number (*e*). The chi-square value, χ^2 (item 7 in the table), is the total of all the values in column 6. The more the observed data deviate from the data expected on the basis of the hypothesis being tested, the higher χ^2 will be. In our example, χ^2 is 3.43. The general formula is

$$\chi^2 = \Sigma \frac{d^2}{e}, \text{ where } \Sigma \text{ means "sum" and } d^2 = (o - e)^2$$

The last value in the table, item 8, is the degrees of freedom (df) for the set of data. The degrees of freedom in a test involving *n* classes are usually equal to $n - 1$.

The χ^2 value and the degrees of freedom are next used to determine the probability (*P*) that the deviation of the observed values from the expected values is due to chance. The *P* value for a set of data is obtained from tables of χ^2 values for various degrees of freedom. Table 2.5 is part of a table of chi-square probabilities. For our example—$\chi^2 = 3.43$ with 3 degrees of freedom—the *P* value is between 0.30 and 0.50. This is interpreted to mean that, with the hypothesis being true, in 30 to 50 out of 100 trials (that is, 30 to 50 percent of the time) we could expect χ^2 values of this magnitude or greater due to chance. We can reasonably regard this deviation as simply being a sampling, or chance, error. We must be cautious how we use this result, however, because a result like this does not tell us that the hypothesis is *correct*; it only indicates that the experimental data provide no statistically compelling argument against the hypothesis.

As a general rule, if the probability of obtaining the observed χ^2 values is greater than 5 in 100 (5 percent of the time: $P > 0.05$), then the deviation of expected from observed is not considered statistically significant, and the hypothesis being tested is not thrown out.

Suppose that in another chi-square analysis of a different set of data we obtained $\chi^2 = 15.85$ with 3 degrees of freedom. By looking up the value in Table 2.5, we see that the *P* value is less than 0.01 and greater than 0.001 ($0.001 < P < 0.01$). Thus, from 0.1 to 1 times out of 100 (0.1 to 1 percent of the time) we could expect χ^2 values of this magnitude or greater due to chance with the hypothesis being true. That this *P* value is less than 0.05 indicates that the results are not statistically consistent with the 1:1:1:1 hypothesis being tested because of the poor fit.

~ TABLE 2.4

Chi-Square Test Example

(1)	(2)	(3)	(4)	(5)	(6)
PHENOTYPES	OBSERVED NUMBER (*o*)	EXPECTED NUMBER (*e*)	*d* (= *o* − *e*)	d^2	d^2/e
Smooth, yellow	154	142	+12	144	1.01
Smooth, green	124	142	−18	324	2.28
Wrinkled, yellow	144	142	+2	4	0.03
Wrinkled, green	146	142	+4	16	0.11
Total	568	568	0		3.43

(7) $\chi^2 = 3.43$ (8) Degrees of freedom (df) = 3

~ TABLE 2.5

Chi-Square Probabilities

	PROBABILITIES									
df	0.95	0.90	0.70	0.50	0.30	0.20	0.10	0.05	0.01	0.001
1	0.004	0.016	0.15	0.46	1.07	1.64	2.71	3.84	6.64	10.83
2	0.10	0.21	0.71	1.39	2.41	3.22	4.61	5.99	9.21	13.82
3	0.35	0.58	1.42	2.37	3.67	4.64	6.25	7.82	11.35	16.27
4	0.71	1.06	2.20	3.36	4.88	5.99	7.78	9.49	13.28	18.47
5	1.15	1.61	3.00	4.35	6.06	7.29	9.24	11.07	15.09	20.52
6	1.64	2.20	3.83	5.35	7.23	8.56	10.65	12.59	16.81	22.46
7	2.17	2.83	4.67	6.35	8.38	9.80	12.02	14.07	18.48	24.32
8	2.73	3.49	5.53	7.34	9.52	11.03	13.36	15.51	20.09	26.13
9	3.33	4.17	6.39	8.34	10.66	12.24	14.68	16.92	21.67	27.88
10	3.94	4.87	7.27	9.34	11.78	13.44	15.99	18.31	23.21	29.59
11	4.58	5.58	8.15	10.34	12.90	14.63	17.28	19.68	24.73	31.26
12	5.23	6.30	9.03	11.34	14.01	15.81	18.55	21.03	26.22	32.91
13	5.89	7.04	9.93	12.34	15.12	16.99	19.81	22.36	27.69	34.53
14	6.57	7.79	10.82	13.34	16.22	18.15	21.06	23.69	29.14	36.12
15	7.26	8.55	11.72	14.34	17.32	19.31	22.31	25.00	30.58	37.70
20	10.85	12.44	16.27	19.34	22.78	25.04	28.41	31.41	37.57	45.32
25	14.61	16.47	20.87	24.34	28.17	30.68	34.38	37.65	44.31	52.62
30	18.49	20.60	25.51	29.34	33.53	36.25	40.26	43.77	50.89	59.70
50	34.76	37.69	44.31	49.34	54.72	58.16	63.17	67.51	76.15	86.66

←————————— | —————————→
Fail to reject | Reject
at 0.05 level

Source: From Table IV in *Statistical Tables for Biological, Agricultural, and Medical Research* by Fisher and Yates, 6th ed., 1974. Reprinted by permission of Addison Wesley Longman Ltd.

MENDELIAN GENETICS IN HUMANS

After the rediscovery of Mendel's laws, geneticists found that the inheritance of genes follows the same principles in all sexually reproducing eukaryotes, including humans. In this section we explore some of the methods used to determine the mechanism of hereditary transmission in humans, and we learn about some inherited human traits.

Pedigree Analysis

The study of human genetics is complicated because controlled matings of humans are impractical. This means that geneticists must study crosses that happen by *chance* rather than by *design*. The inheritance patterns of human traits are usually identified by examining the way the trait occurs in the family trees of individuals who clearly exhibit the trait. Such a study of a family tree, called **pedigree analysis**, involves carefully assembling phenotypic records of the family over several generations. The "affected" individual through whom the pedigree is discovered is called the **proband** (**propositus** if a male, **proposita** if a female).

One of the modern applications of pedigree analysis is **genetic counseling**. A geneticist makes predictions about the probabilities of particular traits (deleterious or not) occurring among the children of a couple. In most cases the couple comes to the counselor because there is some possibility that an undesirable heritable trait exists in one or both families. Pedigree analysis is most useful for traits that are the result of a single gene difference.

Pedigree analysis has its own set of symbols; Figure 2.15 summarizes the basic symbols that will be

~ FIGURE 2.15

Symbols used in human pedigree analysis.

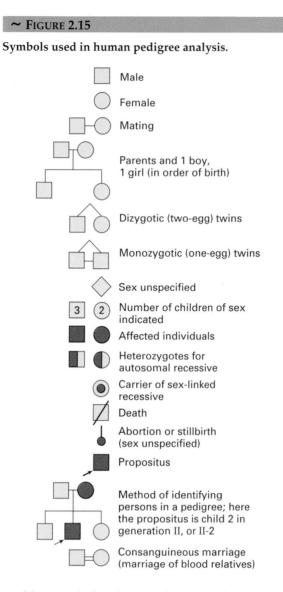

□	Male
○	Female
□—○	Mating
	Parents and 1 boy, 1 girl (in order of birth)
	Dizygotic (two-egg) twins
	Monozygotic (one-egg) twins
◇	Sex unspecified
③ ②	Number of children of sex indicated
■ ●	Affected individuals
	Heterozygotes for autosomal recessive
⊙	Carrier of sex-linked recessive
⧄	Death
	Abortion or stillbirth (sex unspecified)
■	Propositus
	Method of identifying persons in a pedigree; here the propositus is child 2 in generation II, or II-2
	Consanguineous marriage (marriage of blood relatives)

used here and elsewhere in the text. (The terms *autosomal* and *sex-linked* used in the figure are explained in Chapter 3; they are included here for completeness.) Figure 2.16 presents a hypothetical pedigree to show how the symbols are assigned to the family tree.

The trait presented in Figure 2.16 is determined by a recessive mutant allele *a*. (Note that recessive mutant alleles may be rare or common in a population.) Generations are numbered with Roman numerals, while individuals are numbered with Arabic numerals; this makes it easy to refer to particular people in the pedigree. The trait in this pedigree results from homozygosity for the allele, brought about by cousins mating. Since cousins share a fair proportion of their genes, a number of alleles will become homozygous in their offspring; in this case, one mutant recessive allele became homozygous and resulted in an identifiable genetic trait.

Gene symbols are included in this pedigree to show the deductive reasoning possible with such analysis; normally, such symbols are not present, and the researcher would have to analyze the pedigrees without that information. For example, the following reasoning could take place: The trait appears first in generation IV. Since neither parent (the two cousins) had the trait, but they produced two children with the trait (IV-2 and IV-4), the simplest hypothesis is that the trait is caused by a recessive mutant allele. Thus, IV-2 and IV-4 would both have the genotype *aa*, and their parents (III-5 and III-6) must both have the genotype *Aa*. All other individuals who did not have the trait must have at least one *A* allele; that is, they must be *A–* (either *AA* or *Aa*). Since III-5 and III-6 are both heterozygotes, then at least one of each of their parents must have carried an *a* allele. Further, since the trait appeared only after cousins had children, the simplest assumption is that the *a* allele was inherited from individuals with bloodlines shared by III-5 and III-6. This means that II-2 and II-4 are likely both *Aa*, and that one of I-1 and I-2 is *Aa* (perhaps both, unless the allele is rare).

Examples of Human Genetic Traits

RECESSIVE TRAITS. A large number of human traits are known to be caused by homozygosity for mutant alleles that are recessive to the normal allele. Such recessive mutant alleles produce mutant phenotypes because of a *loss of function* or modified function of the gene product resulting from the mutation involved.

Many serious abnormalities or diseases result from homozygosity for recessive mutant alleles. Two individuals expressing the recessive trait of albinism (deficient pigmentation) are shown in Figure 2.17a,

~ FIGURE 2.16

A human pedigree, illustrating the use of pedigree symbols.

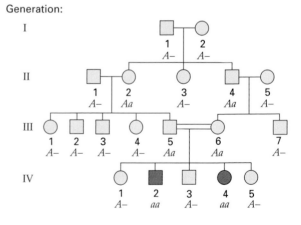

Generation:

~ FIGURE 2.17

Albinism. (a) Two individuals with albinism: the blues musicians Johnny (left) and Edgar Winter (right). (b) A pedigree showing the transmission of the autosomal recessive trait of albinism.

a)

b) Pedigree

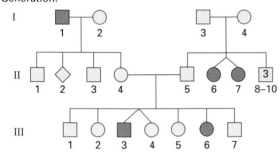

and a pedigree for this trait is shown in Figure 2.17b. Individuals with albinism do not produce the pigment melanin, which protects the skin from harmful ultraviolet radiation. As a consequence, their skin and eyes are very sensitive to sunlight. Frequencies of recessive mutant alleles are usually higher than frequencies of dominant mutant alleles because heterozygotes for the recessive mutant allele are not at a significant selective disadvantage. Nonetheless, individuals homozygous for recessive mutant alleles are usually rare. In the United States approximately 1 in 17,000 of the white population and 1 in 28,000 of the African-American population have albinism. Among the Irish, about 1 in 10,000 have albinism.

The following are some general characteristics of recessive inheritance for a relatively rare trait:

1. Most affected individuals have two normal parents, both of whom are heterozygous. The trait appears in the F_1 since a quarter of the progeny are expected to be homozygous for the recessive allele. If the trait is rare or relatively rare, an individual expressing the trait is likely to mate with a homozygous normal individual; thus, the next generation is represented by heterozygotes who do not express the trait. In other words, recessive traits often "skip" generations. In the pedigree in Figure 2.17b, for example, II-6 and II-7 must both be *aa*, and this means both parents (I-3 and I-4) must be *Aa* heterozygotes. I-1 is also *aa* and so II-4 must be *Aa*. Since II-4 and II-5 produce some *aa* children, II-5 also must be *Aa*.

2. Matings between two normal heterozygotes should produce an approximately 3:1 ratio of normal progeny to progeny exhibiting the recessive trait. However, in the analysis of human populations (families), it is difficult to obtain a large enough sample of individuals to make the data statistically significant. This is especially true if a biochemical test is necessary to confirm the presence of the trait because, in such cases, only the living members of a family can be surveyed. For recessive genes that have less deleterious effects, the allele can reach significant frequencies in the population. An example of such a recessive trait is attached earlobes. The population contains significant numbers of heterozygotes and homozygous recessives for this trait. As a result, there is a strong possibility of *Aa* × *aa* matings, and half the progeny will have the trait.

3. When both parents are affected, all their progeny will usually exhibit the trait.

Other examples of human recessive traits are cystic fibrosis (a lethal disease affecting pancreatic, lung, and digestive functions), sickle-cell anemia (a disease resulting from defective hemoglobin), and phenylketonuria (PKU, a disease in which the amino acid phenylalanine cannot be metabolized).

DOMINANT TRAITS. There are many known dominant human traits. Dominant mutant alleles may produce mutant phenotypes because of a *gain of function* of the gene product resulting from the mutation involved. Figure 2.18a illustrates one such trait, called woolly hair, in which an individual's hair is very tightly kinked, is very brittle, and breaks before it can grow very long. The best examples of pedigrees for this trait come from Norwegian families; one of these pedigrees is presented in Figure 2.18b. Since it is a fairly rare trait and since not all children of an affected parent show the trait, most woolly-haired individuals are probably heterozygous for the dominant allele involved.

Dominant mutant alleles are expressed in a heterozygote when they are in combination with what is usually called the **wild-type allele**—the allele that

~   FIGURE 2.18

Woolly hair. (a) Members of a Norwegian family, some of whom exhibit the trait of woolly hair; (b) Part of a pedigree showing the transmission of the autosomal dominant trait of woolly hair.

a)

b)   Generation:

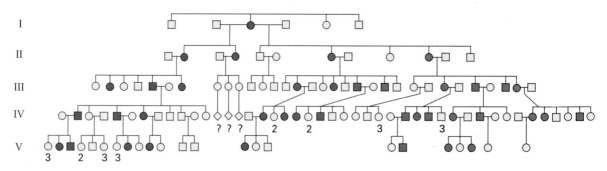

predominates (is present in the highest frequency) in the population found in the "wild." Because many dominant mutant alleles that give rise to recognizable traits are rare, it is extremely unusual to find individuals homozygous for the dominant allele. Thus, an affected person in a pedigree is likely to be a heterozygote, and most pairings that involve the mutant allele are between a heterozygote and a homozygous recessive (wild type). Most dominant mutant genes that are clinically significant (that is, cause medical problems) fall into this category.

The following are some general characteristics of a dominant trait (refer to Figure 2.18b):

1. Every affected person in the pedigree must have at least one affected parent.

2. The trait usually does not skip generations.

3. An affected heterozygous individual will, on average, transmit the mutant gene to half of his or her progeny. If the dominant mutant allele is designated A, and its wild-type allele is a, then most crosses will be $Aa \times aa$. From basic Mendelian principles, half the progeny will be aa (wild type) and the other half will be Aa and show the trait.

Other examples of human dominant traits are achondroplasia (dwarfism resulting from defects in long-bone growth), brachydactyly (malformed hands with short fingers), and Marfan syndrome (connective tissue defects potentially causing death by aortic rupture).

KEYNOTE

Mendelian principles apply to humans, as well as to peas and all other eukaryotes. The study of the human inheritance of genetic traits is complicated because no controlled crosses can be done. Instead, human geneticists analyze genetic traits by pedigree analysis, that is, by examining the occurrences of a trait in family trees of individuals who clearly exhibit the trait. Many recessively inherited and dominantly inherited genetic traits have been identified as a result of pedigree analysis.

SUMMARY

In this chapter we discussed fundamental principles of gene segregation and gene assortment. Genes are DNA segments that control the biological characteristics transmitted from one generation to another, that is, the hereditary traits. An organism's genetic constitution is its genotype, while the physical manifestation of a genetic trait is its phenotype. An organism's genes only provide the potential for the development of that organism's characteristics. That potential is influenced during development by interactions with other genes and with the environment. Thus, individuals with the same genotype can have different phenotypes, and individuals with the same phenotype may have different genotypes.

The first person to obtain some understanding of the principles of heredity—that is, the inheritance of certain traits—was Gregor Mendel. From his breeding experiments with garden peas, Mendel proposed two basic principles of genetics. In modern terms, the principle of segregation states that the two members of a single gene pair (the alleles) segregate from each other in the formation of gametes. For each gene with two alleles, half of the gametes carry one allele, and the other half carry the other allele. The principle of independent assortment, proposed on the basis of experiments involving more than one gene, states that genes for different traits behave independently in the production of gametes. Both principles are recognized by characteristic phenotypic ratios—called Mendelian ratios—in particular crosses. For the principle of segregation, in a monohybrid cross between two true-breeding parents, one exhibiting a dominant phenotype and the other a recessive phenotype, the F_2 phenotypic ratio will be 3:1 for the dominant:recessive phenotypes. For the principle of independent assortment, in a dihybrid cross, the F_2 phenotypic ratio will be 9:3:3:1 for the four phenotypic classes.

The gene segregation patterns can be studied more definitively by determining the genotypes for each phenotypic class. This is done using a testcross, in which an individual of unknown genotype is crossed with a homozygous recessive individual to determine the unknown genotype. For example, in a monohybrid cross resulting in an F_2 3:1 phenotypic ratio, the dominant class can be shown to consist of 1 homozygous dominant : 2 heterozygotes by using a testcross.

Geneticists have found that Mendelian principles of gene segregation apply to all eukaryotes, including humans. The study of the inheritance of genetic traits in humans is complicated because no controlled crosses can be done. Instead, human geneticists analyze genetic traits by pedigree analysis, that is, by examining the occurrences of the trait in family trees of individuals who clearly exhibit the trait. Many recessively inherited and dominantly inherited genetic traits have been identified by pedigree analysis.

ANALYTICAL APPROACHES FOR SOLVING GENETICS PROBLEMS

The most practical way to reinforce Mendelian principles is to solve genetics problems. In this and all following chapters we will discuss how to approach genetics problems by presenting examples of such problems and discussing their answers. These problems use familiar and unfamiliar examples and pose questions designed to get you to think analytically.

Q2.1 A purple-flowered pea plant is crossed with a white-flowered pea plant. All the F_1 plants produced purple flowers. When the F_1 plants are allowed to self-pollinate, 401 of the F_2s have purple flowers and 131 have white flowers. What are the genotypes of the parental and F_1 generation plants?

A2.1 The ratio of plant phenotypes in the F_2 is very close to the 3:1 ratio expected of a monohybrid cross. More specifically, this ratio is expected to result from an $F_1 \times F_1$ cross in which both are heterozygous for a specific gene pair. In addition, since the two parents differed in phenotype and only one phenotypic class appeared in the F_1, it is likely that both parental plants were true breeding. Further, since the F_1 phenotype exactly resembled one of the parental phenotypes, we can say that purple is dominant to white flowers. Assigning the symbol P to the gene that determines purpleness of flowers and the symbol p to the alternative form of the gene that determines whiteness, we can write the genotypes:

P generation: PP, for the purple-flowered plant
pp, for the white-flowered plant

F_1 generation: Pp, which, because of dominance, is purple-flowered

We could further deduce that the F_2 plants have an approximately 1:2:1 ratio of $PP:Pp:pp$ by performing testcrosses.

Q2.2 Consider three gene pairs Aa, Bb, and Cc, each of which affects a different character. In each case the capital letter signifies the dominant allele, and the small letter the recessive allele. These three gene pairs assort independently of each other. Calculate the probability of obtaining:

a. an $Aa\ BB\ Cc$ zygote from a cross of individuals that are $Aa\ Bb\ Cc \times Aa\ Bb\ Cc$;

b. an $Aa\ BB\ cc$ zygote from a cross of individuals that are $aa\ BB\ cc \times AA\ bb\ CC$;

c. an $A\ B\ C$ phenotype (that is, having the dominant phenotypes for each of the three genes) from a cross of individuals that are $Aa\ Bb\ CC \times Aa\ Bb\ cc$;

d. an $a\ b\ c$ phenotype (that is, having the recessive phenotypes for each of the three genes) from a cross of individuals that are $Aa\ Bb\ Cc \times aa\ Bb\ cc$.

A2.2 We must break down the question into simple parts to apply basic Mendelian principles. The key is that the genes assort independently, so we must multiply the probabilities of the individual occurrences to obtain the answers.

a. First, we must consider the Aa gene pair. The cross is $Aa \times Aa$, so the probability of the zygote being Aa is 2/4 since the expected distribution of genotypes is 1 $AA : 2\ Aa : 1\ aa$. Then the probability of BB from $Bb \times Bb$ is 1/4, and that of Cc from $Cc \times Cc$ is 2/4, following the same sort of logic. Using the product rule (see Box 2.2), the probability of an $Aa\ BB\ Cc$ zygote is 1/2 $\times$ 1/4 $\times$ 1/2 = 1/16.

b. Similar logic is needed here, although we must be sure of the genotypes of the parental types since they differ from one gene pair to another. For the Aa pair the probability of getting Aa from $AA \times aa$ has to be 1. Next, the probability of getting BB from $BB \times bb$ is 0, so on these grounds alone we cannot get the zygote asked for from the cross given.

c. This question and the next ask for the probability of getting a particular phenotype, so we must start

thinking about dominance. Again, we consider each character pair in turn. The probability of an A phenotype from $Aa \times Aa$ is 3/4, from basic Mendelian principles. Similarly, the probability of a B phenotype from $Bb \times Bb$ is 3/4. Lastly, the probability of a C phenotype from $CC \times cc$ is 1. Overall, the probability of an $A\ B\ C$ phenotype is 3/4 $\times$ 3/4 $\times$ 1 = 9/16.

d. The probability of an $a\ b\ c$ phenotype from $Aa\ Bb\ Cc \times aa\ Bb\ cc$ is 1/2 $\times$ 1/4 $\times$ 1/2 = 1/16.

Q2.3 In chickens, the white plumage of the leghorn breed is dominant over colored plumage, feathered shanks are dominant over clean shanks, and pea comb is dominant over single comb. Each of the gene pairs segregates independently. If a homozygous white, feathered, pea-combed chicken is crossed with a homozygous colored, clean, single-comb chicken, and the F_1s are allowed to interbreed, what proportion of the birds in the F_2 will produce only white, feathered, pea-combed progeny if mated to colored, clean-shanked, single-combed birds?

A2.3 This example is typical of a question that presents the unfamiliar in an attempt to get at the familiar. The best approach to such questions is to reduce them to their simplest parts and, whenever possible, to assign gene symbols for each character. We are told which character is dominant for each of the three gene pairs, so we can use W for white and w for colored, F for feathered and f for clean shanks, and P for pea comb and p for single comb. The cross involves true-breeding strains and can be written as follows:

P generation: $WW\ FF\ PP \times ww\ ff\ pp$

F_1 generation: $Ww\ Ff\ Pp$

Now the question asks the proportion of the birds in the F_2 that will produce only white, feathered, pea-combed progeny if mated to colored, clean-shanked, single-combed birds. The latter are homozygous recessive for all three genes: that is, $ww\ ff\ pp$, as in the parental generation. For the requested result, the F_2 birds must be white, feathered, and pea-combed, and they must be homozygous for the dominant alleles of the respective genes in order to produce only progeny with the dominant phenotype. What we are seeking, then, is the proportion of the F_2 chickens that are $WW\ FF\ PP$ in genotype. We know that each gene pair segregates independently, so the answer can be calculated by using simple probability rules. We consider each gene pair in turn. For the white/colored case, the $F_1 \times F_1$ is $Ww \times Ww$, and we know from Mendelian principles that the relative proportion of F_2 genotypes will be 1 $WW : 2\ Ww : 1\ ww$. Therefore, the proportion of the F_2s that will be WW is 1/4. The same relationship holds for the other two pairs of genes. Since the segregation of the three gene pairs is independent, we must multiply the probabilities of each occurrence to calculate the probability for *WW FF PP* individuals. The answer is $1/4 \times 1/4 \times 1/4 = 1/64$.

QUESTIONS AND PROBLEMS

*2.1 In tomatoes, red fruit color is dominant to yellow. Suppose a tomato plant homozygous for red is crossed with one homozygous for yellow. Determine the appearance of (a) the F_1; (b) the F_2; (c) the offspring of a cross of the F_1 back to the red parent; (d) the offspring of a cross of the F_1 back to the yellow parent.

2.2 In maize, a dominant allele *A* is necessary for seed color, as opposed to colorless (*a*). Another gene has a recessive allele *wx* that results in waxy starch, as opposed to normal starch (*Wx*). The two genes segregate independently. Give phenotypes and relative frequencies for offspring resulting when a plant of genetic constitution *Aa WxWx* is testcrossed.

*2.3 F_2 plants segregate 3/4 colored : 1/4 colorless. If a colored plant is picked at random and selfed, what is the probability that both colored and colorless plants will be seen among a large number of its progeny?

*2.4 In guinea pigs, rough coat (*R*) is dominant over smooth coat (*r*). A rough-coated guinea pig is bred to a smooth one, giving eight rough and seven smooth progeny in the F_1.
a. What are the genotypes of the parents and their offspring?
b. If one of the rough F_1 animals is mated to its rough parent, what progeny would you expect?

2.5 In cattle, the polled (hornless) condition (*P*) is dominant over the horned (*p*) phenotype. A particular polled bull is bred to three cows. Cow A, which is horned, produces a horned calf; a polled cow B produces a horned calf; and horned cow C produces a polled calf. What are the genotypes of the bull and the three cows, and what phenotypic ratios do you expect in the offspring of these three matings?

*2.6 In the Jimsonweed, purple flowers are dominant to white. Self-fertilization of a particular purple-flowered Jimsonweed produces 28 purple-flowered and 10 white-flowered progeny. What proportion of the purple-flowered progeny will breed true?

*2.7 Two black female mice are crossed with the same brown male. In a number of litters, female X produced 9 blacks and 7 browns, and female Y produced 14 blacks. What is the mechanism of inheritance of black and brown coat color in mice? What are the genotypes of the parents?

2.8 Bean plants may have different symptoms when infected with a virus. Some show local lesions that do not seriously harm the plant; others show general systemic infection. The following genetic analysis was made:

P   local lesions × systemic lesions

F_1 all local lesions

F_2 785 local lesions : 269 systemic lesions

What is probably the genetic basis of this difference in beans? Assign gene symbols to all the genotypes occurring in the genetic analysis. Design a testcross to verify your assumptions.

2.9 A normal *Drosophila* has both brown and scarlet pigment granules in the eyes, which appear red as a result. Brown (*bw*) is a recessive allele on chromosome 2 that, in the homozygous condition, results in the absence of scarlet granules (so that the eyes appear brown). Scarlet (*st*) is a recessive on chromosome 3 that, when homozygous, results in scarlet eyes due to the absence of brown pigment. Any fly homozygous for recessive brown and recessive scarlet alleles produces no eye pigment and has white eyes. The following results were obtained from crosses:

P   brown-eyed fly × scarlet-eyed fly

F_1 red eyes (both brown and scarlet pigment present)

F_2 9/16 red : 3/16 scarlet : 3/16 brown : 1/16 white

a. Assign genotypes to the P and F_1 generations.
b. Design a testcross to verify the F_1 genotype, and predict the results.

*2.10 Grey seed color (*G*) in garden peas is dominant to white seed color (*g*). In the following crosses, the indicated parents with known phenotypes but unknown genotypes produced the listed progeny. Give the possible genotype(s) of each female parent based on the segregation data.

| PARENTS | PROGENY | | FEMALE PARENT |
FEMALE × MALE	GREY	WHITE	GENOTYPE
grey × white	81	82	?
grey × grey	118	39	?
grey × white	74	0	?
grey × grey	90	0	?

*2.11 Fur color in the babbit, a furry little animal and popular pet, is determined by a pair of alleles, *B* and *b*. *BB* and *Bb* babbits are black, while *bb* babbits are white.

A farmer wants to breed babbits for sale. True-breeding white (*bb*) female babbits breed poorly. The farmer purchases a pair of black babbits, and these mate and produce six black and two white offspring. The farmer immediately sells his white babbits, and then he comes to consult you for a breeding strategy to produce more white babbits.

a. If he performed random crosses between pairs of F$_1$ black babbits, what proportion of the F$_2$ progeny would be white?

b. If he crossed an F$_1$ male to the parental female, what is the probability that this cross will produce white progeny?

c. What would be the farmer's best strategy to maximize the production of white babbits?

2.12 In the Jimsonweed, purple flower (*P*) is dominant to white (*p*), and spiny pods (*S*) are dominant to smooth (*s*). In a cross between a Jimsonweed homozygous for white flowers and spiny pods and one homozygous for purple flowers and smooth pods, determine the phenotype of (a) the F$_1$; (b) the F$_2$; (c) the progeny of a cross of the F$_1$ back to the white, spiny parent; (d) the progeny of a cross of the F$_1$ back to the purple, smooth parent.

*2.13 Cleopatra is normally a very refined cat. When she finds even a small amount of catnip, however, she purrs madly, rolls around in the catnip, becomes exceedingly playful, and appears intoxicated. Cleopatra and Antony, who walks past catnip with an air of indifference, have produced five kittens who respond to catnip just like Cleopatra. When the kittens mature, two of them mate and produce four kittens that respond to catnip and one that does not. When another of Cleopatra's daughters mates with Augustus (a nonrelative), who behaves just like Antony, three catnip-sensitive and two catnip-insensitive kittens are produced. Propose an hypothesis for the inheritance of catnip sensitivity that explains these data.

2.14 Using the information in Problem 2.12, what progeny would you expect from the following Jimsonweed crosses? You are encouraged to use the branch diagram approach.

a. *PP ss × pp SS*   d. *Pp Ss × Pp Ss*
b. *Pp SS × pp ss*   e. *Pp Ss × Pp ss*
c. *Pp Ss × Pp SS*   f. *Pp Ss × pp ss*

*2.15 In summer squash, white fruit (*W*) is dominant over yellow (*w*), and disk-shaped fruit (*D*) is dominant over sphere-shaped fruit (*d*). The following problems give the appearances of the parents and their progeny. Determine the genotypes of the parents in each case.

a. White, disk × yellow, sphere gives 1/2 white, disk and 1/2 white, sphere.

b. White, sphere × white, sphere gives 3/4 white, sphere and 1/4 yellow, sphere.

c. Yellow, disk × white, sphere gives all white, disk progeny.

d. White, disk × yellow, sphere gives 1/4 white, disk; 1/4 white, sphere; 1/4 yellow, disk; and 1/4 yellow, sphere.

e. White, disk × white, sphere gives 3/8 white, disk; 3/8 white, sphere; 1/8 yellow, disk; and 1/8 yellow, sphere.

*2.16 Genes *a*, *b*, and *c* assort independently and are recessive to their respective alleles *A*, *B*, and *C*. Two triply heterozygous (*Aa Bb Cc*) individuals are crossed.

a. What is the probability that a given offspring will be phenotypically *A B C*, that is, will exhibit all three dominant traits?

b. What is the probability that a given offspring will be genotypically homozygous for all three dominant alleles?

2.17 In garden peas, tall stem (*T*) is dominant over short stem (*t*), green pods (*G*) are dominant over yellow pods (*g*), and smooth seeds (*S*) are dominant over wrinkled seeds (*s*). Suppose a homozygous short, green, wrinkled pea plant is crossed with a homozygous tall, yellow, smooth one.

a. What will be the appearance of the F$_1$?

b. If the F$_1$s are interbred, what will be the appearance of the F$_2$?

c. What will be the appearance of the offspring of a cross of the F$_1$ back to its short, green, wrinkled parent?

d. What will be the appearance of the offspring of a cross of the F$_1$ back to its tall, yellow, smooth parent?

2.18 *C/c*, *O/o*, and *I/i* are three independently segregating pairs of alleles in chickens. *C* and *O* are dominant alleles, both of which are necessary for pigmentation. *I* is a dominant inhibitor of pigmentation. Individuals of genotype *cc*, or *oo*, or *Ii*, or *II* are white, regardless of what other genes they possess.

Assume that White Leghorns are *CC OO II*, White Wyandottes are *cc OO ii*, and White Silkies are *CC oo ii*. What types of offspring (white or pigmented) are possible, and what is the probability of each, from the following crosses?

a. White Silkie × White Wyandotte

b. White Leghorn × White Wyandotte

c. (Wyandotte–Silkie F$_1$) × White Silkie

2.19 Two homozygous strains of corn are hybridized. They are distinguished by six different pairs of genes, all of which assort independently and produce an independent phenotypic effect. The F$_1$ hybrid is selfed to give an F$_2$.

a. What is the number of possible genotypes in the F$_2$?

b. How many of these genotypes will be homozygous at all six gene loci?

c. If all gene pairs act in a dominant-recessive fashion, what proportion of the F$_2$ will be homozygous for all dominants?

d. What proportion of the F_2 will show all dominant phenotypes?

*2.20 The coat color of mice is controlled by several genes. The agouti pattern, characterized by a yellow band of pigment near the tip of the hairs, is produced by the dominant allele A; homozygous aa mice do not have the band and are nonagouti. The dominant allele B determines black hairs, and the recessive allele b determines brown. Homozygous $c^h c^h$ individuals allow pigments to be deposited only at the extremities (e.g., feet, nose, and ears) in a pattern called Himalayan. The genotype $C–$ allows pigment to be distributed over the entire body.

a. If a true-breeding black mouse is crossed with a true-breeding brown, agouti, Himalayan mouse, what will be the phenotypes of the F_1 and F_2?

b. What proportion of the black agouti F_2 will be of genotype $Aa\ BB\ C^{ch}$?

c. What proportion of the Himalayan mice in the F_2 are expected to show brown pigment?

d. What proportion of all agoutis in the F_2 are expected to show black pigment?

2.21 In cocker spaniels, solid coat color is dominant over spotted coat. Suppose a true-breeding, solid-colored dog is crossed with a spotted dog, and the F_1 dogs are interbred.

a. What is the probability that the first puppy born will have a spotted coat?

b. What is the probability that if four puppies are born, all of them will have solid coats?

2.22 In the F_2 of his cross of red-flowered × white-flowered *Pisum*, Mendel obtained 705 plants with red flowers and 224 with white.

a. Is this result consistent with his hypothesis of factor segregation, from which a 3:1 ratio would be predicted?

b. In how many similar experiments would a deviation as great as or greater than this one be expected? (Calculate χ^2 and obtain the approximate value of P from Table 2.5.)

2.23 In tomatoes, cut leaf and potato leaf are alternative characters, with cut (C) dominant to potato (c). Purple stem and green stem are another pair of alternative characters, with purple (P) dominant to green (p). A true-breeding cut, green tomato plant is crossed with a true-breeding potato, purple plant, and the F_1 plants are allowed to interbreed. The 320 F_2 plants were phenotypically 189 cut, purple; 67 cut, green; 50 potato, purple; and 14 potato, green. Propose an hypothesis to explain the data, and use the χ^2 test to test the hypothesis.

*2.24 The simple case of just two mating types (male and female) is by no means the only sexual system known. The ciliated protozoan *Paramecium bursaria* has a system of four mating types, controlled by two genes (A and B). Each gene has a dominant and a recessive allele.

The four mating types are expressed according to the following scheme:

GENOTYPE	MATING TYPE
AA BB	*A*
Aa BB	*A*
AA Bb	*A*
Aa Bb	*A*
AA bb	*D*
Aa bb	*D*
aa BB	*B*
aa Bb	*B*
aa bb	*C*

It is clear, therefore, that some of the mating types result from more than one possible genotype. We have four strains of known mating type—"A," "B," "C," and "D"—but unknown genotype. The following crosses were made with the indicated results:

MATING TYPE OF PROGENY

CROSS	A	B	C	D
"A" × "B"	24	21	14	18
"A" × "C"	56	76	55	41
"A" × "D"	44	11	19	33
"B" × "C"	0	40	38	0
"B" × "D"	6	8	14	10
"C" × "D"	0	0	45	45

Assign genotypes to "A," "B," "C," and "D."

*2.25 In bees, males (drones) develop from unfertilized eggs and are haploid. Females (workers and queens) are diploid and come from fertilized eggs. W (black eyes) is dominant over w (white eyes). Workers of genotype RR or Rr use wax to seal crevices in the hive; rr workers use resin instead. A $Ww\ Rr$ queen founds a colony after being fertilized by a black-eyed drone bearing the r allele.

a. What will be the appearance and behavior of workers in the new hive, and what are their relative frequencies?

b. Give the genotypes of male offspring, with relative frequencies.

c. Fertilization normally takes place in the air during a "nuptial flight," and any bee unable to fly would effectively be rendered sterile. Suppose a recessive mutation, c, occurs spontaneously in a sperm that fertilizes a normal egg, and that the effect of the mutant gene is to cripple the wings of any adult not bearing the normal allele C. The fertilized egg develops into a normal queen named Madonna. What is the probability that wingless males will be *found in a hive* founded two generations later by one of Madonna's granddaughters?

d. By one of Madonna's great-great-granddaughters?

*2.26 Consider the following pedigree, in which the allele responsible for the trait (*a*) is recessive to the normal allele (*A*).

a. What is the genotype of the mother?
b. What is the genotype of the father?
c. What are the genotypes of the children?
d. Given the mechanism of inheritance involved, does the ratio of children with the trait to children without the trait match what would be expected?

Generation

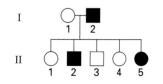

2.27 For the following pedigrees A and B, indicate whether the trait involved in each case could be (a) recessive or (b) dominant. Explain your answer.

Pedigree A

Generation

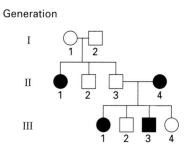

Pedigree B

Generation

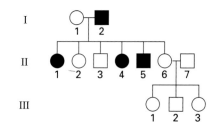

CHAPTER *3*

CHROMOSOMAL BASIS OF INHERITANCE, SEX LINKAGE, AND SEX DETERMINATION

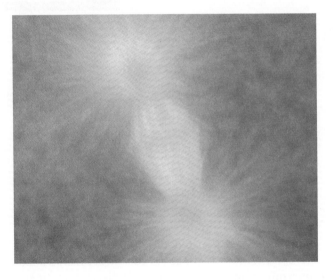

Principal Points

~ The chromosome theory of inheritance states that genes are located on chromosomes.

~ A sex chromosome is a chromosome in eukaryotic organisms that is physically different in the two sexes. In organisms with sex chromosomes, one sex is homogametic and the other is heterogametic.

~ Sex linkage is the physical association of genes with the sex chromosomes of eukaryotes. Such genes are referred to as sex-linked genes.

~ The correlation between gene segregation patterns and the patterns of chromosome behavior in meiosis supports the chromosome theory of inheritance.

~ In many eukaryotic organisms, sex determination is related to the sex chromosomes. In humans and other mammals, for example, the presence of a Y chromosome specifies maleness, while its absence results in female-ness. Several other sex-determination mechanisms are known in eukaryotes.

~ In humans, the allele responsible for a trait can be inherited in one of five main ways: autosomal recessive, autosomal dominant, X-linked recessive, X-linked dominant, or Y-linked.

~ Since Mendel, many exceptions to his rules have been discovered.

---

*O*n Mendel's foundation, early geneticists began to build genetic hypotheses that could be tested by appropriate crosses, and they began to investigate the nature of Mendelian factors. We now know that Mendelian factors are genes and that genes are located on chromosomes; this is called the *chromosome theory of inheritance.* In this chapter we examine the behavior of genes and chromosomes. We then consider the evidence for the chromosome theory of inheritance. In so doing we will learn about the segregation of genes located on the sex chromosomes. Next, we will learn about various mechanisms of sex determination, and lastly we will discuss sex-linked traits in humans. The goal of this chapter, then, is for you to learn how to think about gene segregation in terms of chromosome inheritance patterns.

located on chromosomes. In this section we consider some of the evidence cytologists and geneticists obtained to support this theory.

KEYNOTE

The chromosome theory of inheritance states that genes are located on chromosomes. The first clear formulation of the chromosome theory was made in 1902 by Sutton and by Boveri, who independently recognized that the transmission of chromosomes from one generation to the next closely paralleled the pattern of transmission of genes from one generation to the next.

Chromosome Theory of Inheritance

Around the end of the nineteenth century and the beginning of the twentieth century, cytologists had established that within a given species the total number of chromosomes is constant in all cells, while the chromosome number varies considerably among species (Table 3.1). In 1902, Walter Sutton and Theodor Boveri independently recognized that the transmission of chromosomes from one generation to the next closely paralleled the pattern of transmission of genes from one generation to the next. To explain this correlation, they proposed the **chromosome theory of inheritance**. This theory states that the genes are

Sex Chromosomes

The support for the chromosome theory of inheritance came from experiments that related the hereditary behavior of particular genes to the transmission of the sex chromosomes, the chromosomes in many eukaryotic organisms that are physically different in the two sexes. For many animals the sex chromosome composition of an individual is directly related to the sex of the individual. The sex chromosomes typically are designated the **X chromosome** and the **Y chromosome**. In humans and the fruit fly *Drosophila melanogaster*, for example, the female has two X chromosomes (she is XX with respect to the sex chromosomes), while the male has one X chromosome and

~ TABLE 3.1

Chromosome Number in Various Organisms[a]

ORGANISM	TOTAL CHROMOSOME NUMBER
Human	46
Chimpanzee	48
Dog	78
Cat	72
Mouse	40
Horse	64
Chicken	78
Toad	36
Goldfish	94
Starfish	36
Fruit fly (*Drosophila melanogaster*)	8
Mosquito	6
Australian ant (*Myrecia pilosula*)	♂ 1, ♀ 2
Nematode	♂ 11, ♀ 12
Neurospora (haploid)	7
Sphagnum moss (haploid)	23
Field horsetail	216
Giant sequoia	22
Tobacco	48
Cotton	52
Potato	48
Tomato	24
Bread wheat	42
Baker's yeast (*Saccharomyces cerevisiae*)	34

[a]Except as noted, all chromosome numbers are for diploid cells.

~ FIGURE 3.1

Drosophila melanogaster (fruit fly), an organism used extensively in genetics experiments. (a) Female (left) and male (right). Top: Adult flies; Bottom: Drawings of ventral abdominal surface to show differences in genitalia; (b) Chromosomes of *Drosophila melanogaster* diagrammed to show their morphological differences. A female (left) has four pairs of chromosomes in her somatic cells, including a pair of X chromosomes. The only difference in the male is an XY pair of sex chromosomes instead of two Xs.

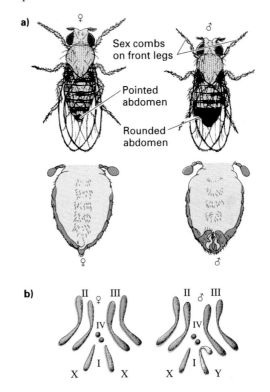

the male produces both X-bearing and Y-bearing gametes. Random fusion of these gametes produces an F_1 generation with 1/2 XX (female) and 1/2 XY (male) flies.

KEYNOTE

A sex chromosome is a chromosome in eukaryotic organisms that is physically different in the two sexes. In many organisms, one sex possesses a pair of identical chromosomes (the X chromosomes). The opposite sex possesses a pair of visibly different chromosomes: one is an X chromosome, and the other, structurally and functionally different, is called the Y chromosome. Commonly, the XX sex is female, and the XY sex is male. The XX and XY sexes are called the homogametic and heterogametic sexes, respectively.

one Y chromosome (he is XY). Figure 3.1a shows male and female *Drosophila*, and Figure 3.1b shows the chromosome sets of the two sexes. Because the male produces two kinds of gametes with respect to sex chromosomes (X or Y), and because the female produces only one type of gamete (X), the male is called the **heterogametic sex** and the female is called the **homogametic sex**. In *Drosophila* the X and Y chromosomes are similar in size, but their shapes are different. (Note: in some organisms the male is homogametic and the female is heterogametic.)

The pattern of transmission of X and Y chromosomes from generation to generation is straightforward (Figure 3.2). In this figure the X is represented by a straight structure much like a slash mark, and the Y by a similar structure topped by a hook to the right. The female produces only X-bearing gametes, while

~ FIGURE 3.2

Inheritance pattern of X and Y chromosomes in organisms where the female is XX and the male is XY.

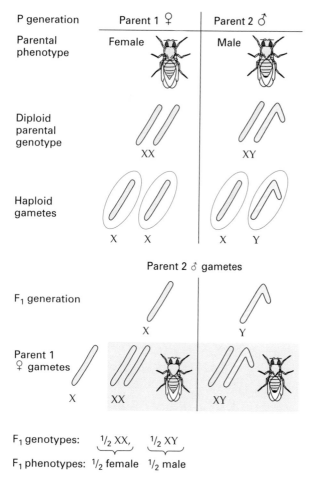

P generation

Parental phenotype

Diploid parental genotype

Haploid gametes

F_1 generation

Parent 1 ♀ gametes

F_1 genotypes: ½ XX, ½ XY

F_1 phenotypes: ½ female ½ male

Sex Linkage

Evidence to support the chromosome theory of heredity came in 1910 when Thomas Hunt Morgan reported the results of genetics experiments with *Drosophila*.

Morgan found in one of his true-breeding stocks a male fly that had white eyes instead of the brick-red eyes characteristic of the **wild type**. The term *wild type* refers to a strain, organism, or gene that is most prevalent in the "wild" population of the organism with respect to genotype and phenotype. For example, a *Drosophila* strain with all wild-type genes has brick-red eyes. Variants of a wild-type strain arise from mutational changes of the wild-type alleles that produce **mutant alleles**; the result is strains with mutant characteristics. Mutant alleles may be recessive or dominant to the wild-type allele; for example, the mutant allele that causes white eyes in *Drosophila* is recessive to the wild-type (red-eye) allele.

Morgan crossed the white-eyed male with a red-eyed female from the same stock and found that all

the F_1 flies were red-eyed. He concluded that the white-eyed trait was recessive. Next, he allowed the F_1 progeny to interbreed and counted the phenotypic classes in the F_2 generation; there were 3,470 red-eyed and 782 white-eyed flies. The number of individuals with the recessive phenotype was too small to fit the Mendelian 3:1 ratio. (Later, he determined that the lower-than-expected number of flies with the recessive phenotype was the result of lower viability of white-eyed flies.) In addition, *Morgan noticed that all the white-eyed flies were male.*

Figure 3.3 diagrams the crosses. The *Drosophila* gene symbolism used is different from the symbolism we used for Mendel's crosses and is described in Box 3.1 (p. 59). You should understand the *Drosophila* gene symbolism before proceeding with this discussion. As we proceed, note that the mother-son inheritance pattern presented in Figure 3.3 is the result of the segregation of genes located on a sex chromosome.

Morgan proposed that the gene for the eye color variant is located on the X chromosome. The condition of X-linked genes in males is said to be **hemizygous** since the gene is present only once in the organism because there is no homologous gene on the Y. For example, the white-eyed *Drosophila* males have an X chromosome with a white allele and no other allele of that gene in their genomes; these males are hemizygous for the white allele. Since the white allele of the gene is recessive, the original white-eyed male must have had the recessive allele for white eyes (designated *w*; see Box 3.1) on his X chromosome. The red-eyed female came from a true-breeding strain, so both of her X chromosomes must have carried the dominant allele for red eyes, w^+ ("w plus").

The F_1 flies are produced in the following way (see Figure 3.3a): The males receive their only X chromosome from their mother and hence have the w^+ allele and are red-eyed. The F_1 females receive a dominant w^+ allele from their mother and a recessive *w* allele from their father, so they are also red-eyed.

In the F_2 produced by interbreeding the F_1 flies, the males that received an X chromosome with the *w* allele from their mother are white-eyed; those that received an X chromosome with the w^+ allele are red-eyed (see Figure 3.3b). The gene transmission shown in this cross—from a male parent to a female offspring ("child") to a male grandchild—is called **crisscross inheritance**.

Morgan also crossed a true-breeding white-eyed female (homozygous for the *w* allele) with a red-eyed male (hemizygous for the w^+ allele) (Figure 3.4). This cross is the *reciprocal cross* of Morgan's first cross—white male × red female—shown in Figure 3.3. All the F_1 females receive a w^+-bearing X from their father and a *w*-bearing X from their mother (see

~ FIGURE 3.3

The X-linked inheritance of red eyes and white eyes in *Drosophila melanogaster*. The symbols w and w^+ indicate the white- and red-eyed alleles, respectively. (a) A red-eyed female is crossed with a white-eyed male; (b) The F_1 flies are interbred to produce the F_2s.

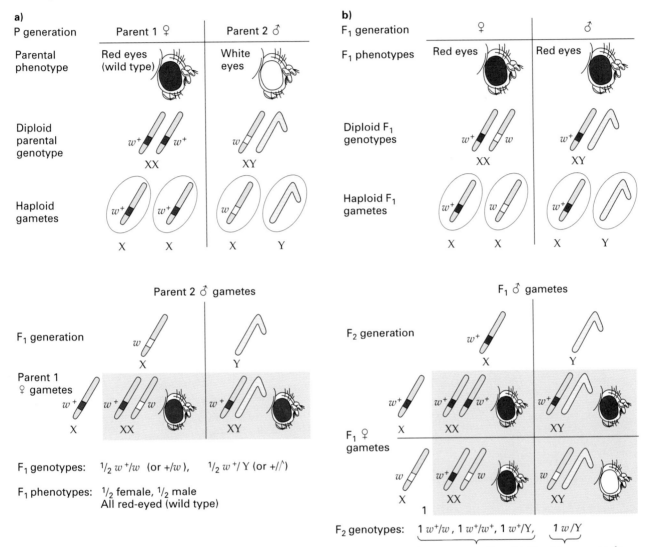

Figure 3.4a). Consequently, they are heterozygous w^+/w and have red eyes. All the F_1 males receive a w-bearing X from their mother and a Y from their father, and so they have white eyes (see Figure 3.4a). This result is distinct from that of the cross in Figure 3.3.

Interbreeding of the F_1 flies (see Figure 3.4b) involves a wY male and a w^+/w female, giving approximately equal numbers of male and female red- and white-eyed flies in the F_2. This ratio differs from the results obtained in the first cross, where an approximately 3:1 ratio of red-eyed:white-eyed flies was obtained and where none of the females and approximately half the males exhibited the white-

eyed phenotype. The difference in phenotypic ratios in the two sets of crosses reflects the transmission patterns of sex chromosomes and the genes they contain.

Morgan's crosses of *Drosophila* involved eye color characteristics that we now know are coded for by a gene found on the X chromosome. These characteristics and the genes that give rise to them are referred to as **sex-linked**—or, more correctly, as **X-linked**—because the gene locus is part of the X chromosome. *X-linked inheritance* is the term used for the pattern of hereditary transmission of X-linked genes. When the results of reciprocal crosses

~ FIGURE 3.4

Reciprocal cross of that shown in Figure 3.3. (a) A homozygous white-eyed female is crossed with a red-eyed (wild-type) male; (b) The F_1 flies are interbred to produce the F_2s. The results of this cross differ from those in Figure 3.3 because of the way sex chromosomes segregate in crosses.

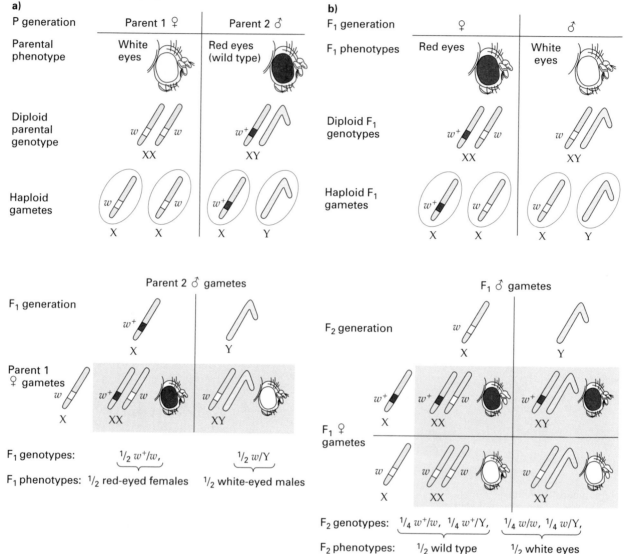

KEYNOTE

Sex linkage is the linkage of genes with the sex chromosomes of eukaryotes. Such genes, as well as the phenotypic characteristics these genes control, are referred to as sex-linked. Those genes only on the X chromosome are called X-linked. Morgan's pioneering work with the inheritance of sex-linked genes of *Drosophila* strongly supported, but did not prove, the chromosome theory of inheritance.

istics may well be involved. By comparison, the results of reciprocal crosses are *always* the same when they involve genes located on the autosomes. Most significantly, Morgan's results strongly supported the hypothesis that genes were located on chromosomes. Morgan found many other examples of genes on the X chromosome in *Drosophila* and in other organisms, thereby showing that his observations were not confined to a single species.

Nondisjunction of X Chromosomes

are not the same, and different ratios are seen for the two sexes of the offspring, sex-linked character-

Proof for the chromosome theory of inheritance came from the work of Morgan's student, Calvin Bridges.

BOX 3.1

GENETIC SYMBOLS REVISITED

Unfortunately, no single system of gene symbols is used by geneticists; the gene symbols used for *Drosophila* are different from those used for peas in Chapter 2. The *Drosophila* symbolism is commonly, but not exclusively, used in genetics today. In this system the symbol + indicates a wild-type allele of a gene. A lowercase letter designates mutant alleles of a gene that are *recessive* to the wild-type allele, and an uppercase letter is used for alleles that are *dominant* to the wild-type allele. *The letters are chosen on the basis of the phenotype of the organism expressing the mutant allele.* For example, a variant strain of *Drosophila* has bright orange eyes instead of the usual brick red. The mutant allele involved is recessive to the wild-type brick-red allele, and because the bright orange eye color is close to vermilion in tint, the allele is designated *v* and is called the vermilion allele. The wild-type allele of *v* is v^+, but when there is no chance of confusing it with other genes in the cross, it is often shortened to +. In the "Mendelian" terminology used up to now, the recessive mutant allele would be *v*, and its wild-type allele would be *V*.

A conventional way to represent the chromosomes (instead of the way we have been using in the figures) is to use a slash (/). Thus, v^+/v or +/*v* indicates that there are two homologous chromosomes, one with the wild-type allele (v^+ or +) and the other with the recessive allele (*v*). The Y chromosome is usually symbolized as a Y or as a bent slash (⁄). Thus, Morgan's cross of a true-breeding red-eyed female fly with a white-eyed male could be written $w^+/w^+ \times w/Y$ or $+/+ \times w/\!\!\!/$.

The same rules apply when the alleles involved are dominant to the wild-type allele. For instance, some *Drosophila* mutants, called *Curly*, have wings that curl up at the end rather than the normal straight wings. The symbol for this mutant allele is *Cy*, and the wild-type allele is Cy^+, or + in the shorthand version. Thus, a heterozygote would be Cy^+/Cy or +/*Cy*.

In the rest of the book, both the *A/a* ("Mendelian") and a^+/a (*Drosophila*) symbolisms will be used, so it is important that you be able to work with both. Since it is easier to verbalize the "Mendelian" symbols ("big *A*, small *a*"), many of our examples will follow that symbolism, even though the *Drosophila* symbolism in many ways is more informative. That is, with the *Drosophila* system, the wild-type and mutant alleles are readily apparent since the wild-type allele is indicated by a +. The "Mendelian" system is commonly used in animal and plant breeding. A good reason for this is that after many years (sometimes centuries) of breeding, it is no longer apparent what the "normal" (= wild-type) gene is.

Morgan's work showed that from a cross of a white-eyed female (*w/w*) with a red-eyed male (w^+/Y), all the F$_1$ males should be white-eyed and all the females should be red-eyed. Bridges found rare exceptions to this result: about 1 in 2,000 of the F$_1$ flies from such a cross are either white-eyed *females* or red-eyed *males*.

To explain these exceptional flies, Bridges hypothesized that a problem had occurred with chromosome segregation in meiosis. Normally, homologous chromosomes (in meiosis I) or sister chromatids (in meiosis II or mitosis) move to opposite poles at anaphase (see pp. 18, 19); when this fails to take place, chromosome **nondisjunction** results. Nondisjunction can involve either autosomes or the sex chromosomes. For the crosses Bridges analyzed, occasionally the two X chromosomes failed to separate, so eggs were produced either with two X chromosomes or with no X chromosomes instead of the usual one. This particular type of nondisjunction is called **X chromosome nondisjunction**. Normal disjunction of the X chromosomes is illustrated in Figure 3.5a, and nondisjunction of the X chromosomes in meiosis I and meiosis II is shown in Figures 3.5b and 3.5c, respectively.

How can nondisjunction of the X chromosomes explain the exceptional flies in Bridges's cross? When nondisjunction occurs in the *w/w* female (Figure 3.6), two classes of exceptional eggs result with equal (and low) frequency: those with two X chromosomes and those with no X chromosomes. The XY male is w^+/Y and produces equal numbers of w^+- and Y-bearing sperm. When these eggs are fertilized by the two types of sperm, the result is four types of zygotes. Two types usually do not survive: the YO class, which has a Y chromosome but no X chromosome, and the triplo-X (XXX) class, which has three X chromosomes and the genotype $w/w/w^+$ (see Figure 3.6).

~ FIGURE 3.5

Nondisjunction in meiosis involving the X chromosome. (Nondisjunction of autosomal chromosomes and of all chromosomes in mitosis occurs in the same way.) (a) Normal X chromosome segregation in meiosis; (b) Nondisjunction of X chromosomes in meiosis I; (c) Nondisjunction of X chromosomes in meiosis II.

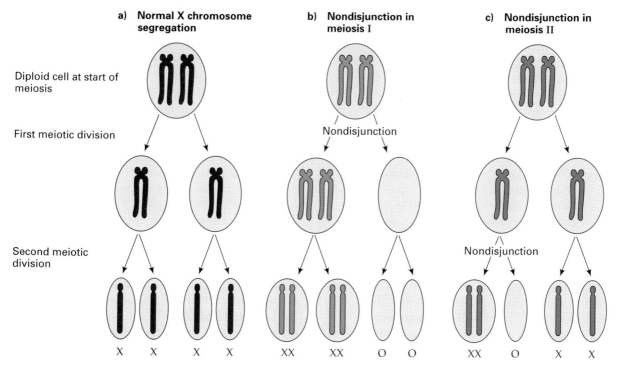

The surviving classes are the red-eyed XO males (in *Drosophila* the XO pattern produces a sterile male), with no Y chromosome and a w^+ allele on the X, and the white-eyed XXY females (in *Drosophila*, XXY produces a fertile female), with a w allele on each X. The males are red-eyed because they received their X chromosome from their fathers, and the females have white eyes because their two X chromosomes came from their mothers. This result is unusual, since sons normally get their X from their mothers, and daughters get one X from each parent.

In sum, gene segregation patterns parallel the patterns of chromosome behavior in meiosis. In Figure 3.7 this parallel is illustrated for a diploid cell with two homologous pairs of chromosomes. The cell is genotypically *Aa Bb*, with the *A/a* gene pair on one chromosome and the *B/b* gene pair on the other chromosome. As the figure shows, the two homologous pairs of chromosomes align on the metaphase plate independently, giving rise to two different segregation patterns for the two gene pairs. Since each of the two alignments, and hence segregation patterns, is equally likely, meiosis results in cells with equal frequencies of the genotypes *A B*, *a b*, *A b*, and *a B*. Genotypes *A B* and *a b* result from one chromosome alignment, and genotypes *A b* and *a B* result from the other alignment. In terms of Mendel's laws, we can see how the principle of segregation applies to the segregation pattern of one homologous pair of chromosomes and the associated gene pair, while the principle of independent assortment applies to the segregation pattern of both homologous pairs of chromosomes.

KEYNOTE

Bridges observed an unexpected inheritance pattern of an X-linked mutant gene in *Drosophila*. He correlated this pattern directly with a rare event during meiosis, called nondisjunction, in which members of a homologous pair of chromosomes do not segregate to the opposite poles. The correlation between gene segregation patterns and the patterns of chromosome behavior in meiosis proved the chromosome theory of inheritance.

SEX DETERMINATION

In this section we discuss some of the mechanisms of sex determination. In *genotypic sex determination systems,* sex is governed by the genotype of the zygote or spores; in *environmental sex determination systems,* sex

~ FIGURE 3.6

Rare primary nondisjunction during meiosis in a white-eyed female *Drosophila melanogaster,* **and results of a cross with a normal red-eyed male.** Note the decreased viability of XXX and YO progeny.

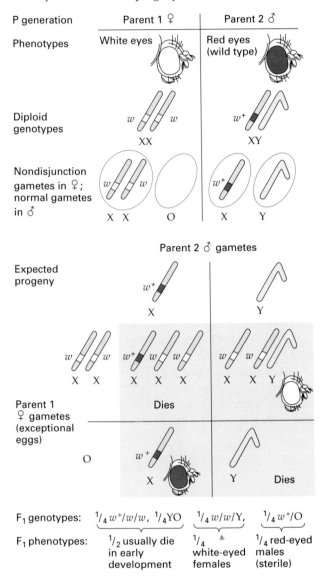

F₁ genotypes: $\frac{1}{4}w^+/w/w$, $\frac{1}{4}$YO | $\frac{1}{4}w/w/$Y, | $\frac{1}{4}w^+/$O

F₁ phenotypes: $\frac{1}{2}$ usually die in early development | $\frac{1}{4}$ white-eyed females | $\frac{1}{4}$ red-eyed males (sterile)

is governed by internal and external environmental conditions.

Genotypic Sex Determination Systems

Genotypic sex determination, in which the sex chromosomes play a decisive role in the inheritance and determination of sex, may occur in one of two ways. In the **Y chromosome mechanism of sex determination** (seen in humans), the Y chromosome determines the sex of an individual. Individuals with a Y chromosome are genetically male, while individuals without a Y chromosome are genetically female. In the

~ FIGURE 3.7

The parallel behavior between Mendelian genes and chromosomes in meiosis. This hypothetical *Aa Bb* diploid cell contains a homologous pair of metacentric chromosomes, which carry the *A/a* gene pair, and a homologous pair of telocentric chromosomes, which carry the *B/b* gene pair. The independent alignment of the two homologous pairs of chromosomes at metaphase I results in equal frequencies of the four meiotic products, *A B, a b, A b,* and *a B,* illustrating Mendel's principle of independent assortment.

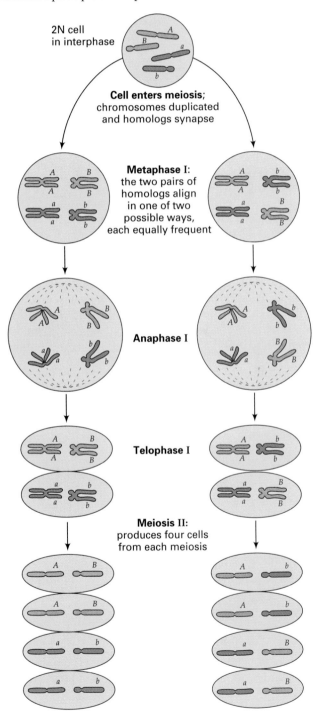

X chromosome–autosome balance system (seen in *Drosophila* and the nematode *Caenorhabditis elegans*), sex is determined by the *ratio between the number of X chromosomes and the number of sets of autosomes*. In this system the Y chromosome has no effect on sex determination, but it is required for male fertility.

SEX DETERMINATION IN MAMMALS. In humans and other placental mammals, sex is determined by the Y chromosome mechanism of sex determination. In the absence of a Y chromosome, the gonads develop as ovaries.

Evidence for the Y Chromosome Mechanism of Sex Determination. Early evidence for the Y chromosome mechanism of sex determination in mammals came from studies in which nondisjunction in meiosis produced an abnormal sex chromosome complement. While sex determination in these cases follows directly from the sex chromosomes present, those individuals with unusual chromosome complements have many unusual characteristics.

Nondisjunction, for example, can produce XO individuals. In humans, XO individuals with the normal two sets of autosomes are female and sterile, and they exhibit **Turner syndrome**. A Turner syndrome individual is shown in Figure 3.8a, and her set of metaphase chromosomes is shown in Figure 3.8b. (A complete set of metaphase chromosomes in a cell is called its **karyotype**; see Chapter 10.) Such individuals have only one sex chromosome—an X chromosome. These females have a genomic complement

designated as 45,X, indicating that they have a total of 45 chromosomes (sex chromosomes plus autosomes), in contrast to the normal 46, and that the sex chromosome complement consists of one X chromosome.

Turner syndrome individuals occur with a frequency of 1 in every 10,000 females born. Up to 99 percent of all 45,X embryos die before birth. Surviving Turner syndrome individuals have few noticeable major defects until puberty, when they fail to develop secondary sexual characteristics. They tend to be shorter than average, and they have weblike necks, poorly developed breasts, and immature internal sexual organs. They have a reduced ability to interpret spatial relationships, and they are usually infertile. All of these defects in XO individuals indicate that two X chromosomes are needed for normal development in females.

Nondisjunction can also result in the generation of XXY humans, who are male and have **Klinefelter syndrome** (Figure 3.9). About 1 in 1,000 males born have Klinefelter syndrome. These 47,XXY males have underdeveloped testes and are often taller than the average male. Some degree of breast development is seen in about 50 percent of affected individuals, and some show subnormal intelligence. Individuals with similar phenotypes are also found with higher numbers of X and/or Y chromosomes—for example, 48,XXXY, 49,XXXXY and 48,XXYY. The defects in Klinefelter individuals indicate that one X and one Y chromosome are needed for normal development in males.

Some individuals have one X and two Y chromosomes; they have *XYY syndrome*. These 47,XYY indi-

---

~ FIGURE 3.8

Turner syndrome (XO). (a) Individual; (b) Karyotype.

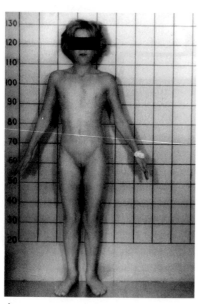

a)

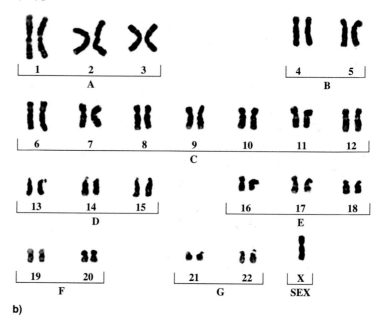

b)

viduals are male because of the Y. The XYY karyotype results from nondisjunction of the Y chromosome in meiosis. About 1 in 1,000 males born have XYY syndrome. They tend to be taller than average, and occasionally there are adverse effects on fertility.

About 1 in 1,000 females born have three X chromosomes instead of the normal two. These 47,XXX (triplo-X) females are mostly completely normal, although they are slightly less fertile and a small number have less than average intelligence.

Table 3.2 summarizes the consequences of exceptional X and Y chromosomes in humans. In every case the normal two sets of autosomes are associated with the sex chromosomes. The Barr bodies mentioned in the table are discussed next.

Dosage Compensation Mechanism for Extra X Chromosomes. Mammals can tolerate abnormalities in the number of sex chromosomes quite well, whereas, with rare exceptions, mammals with an unusual number of autosomes usually die. A **dosage compensation** mechanism in mammals compensates for X chromosomes in excess of one. The somatic-cell nuclei of normal XX females contain a highly condensed mass of chromatin—named the **Barr body** after its discoverer, Murray Barr—not found in the nuclei of normal XY male cells. That is, somatic cells of XX individuals have one Barr body, and somatic cells of XY individuals have no Barr bodies (Figure 3.10 and Table 3.2). In 1961 Mary Lyon and Lillian

~ TABLE 3.2

Consequences of Various Numbers of X and Y Chromosome Abnormalities in Humans, Showing Role of the Y in Sex Determination

CHROMOSOME CONSTITUTION[a]	DESIGNATION OF INDIVIDUAL	EXPECTED NUMBER OF BARR BODIES
46,XX	Normal ♀	1
46,XY	Normal ♂	0
45,X	Turner syndrome ♀	0
47,XXX	Triplo-X ♀	2
47,XXY	Klinefelter syndrome ♂	1
48,XXXY	Klinefelter syndrome ♂	2
48,XXYY	Klinefelter syndrome ♂	1
47,XYY	XYY syndrome ♂	0

[a]The first number indicates the total number of chromosomes in the nucleus, and the Xs and Ys indicate the sex chromosome complement.

Russell expanded this concept into what is now called the *Lyon hypothesis*, which proposed the following:

1. The Barr body is a highly condensed and (mostly) genetically inactive X chromosome (it has become "lyonized" in a process called **lyonization**).

~ FIGURE 3.9

Klinefelter syndrome (XXY). (a) Individual; (b) Karyotype.

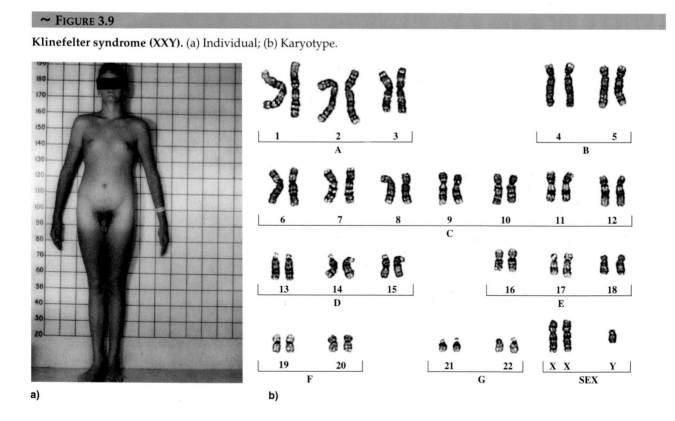

a)

b)

~ **FIGURE 3.10**

Barr bodies. (a) Nuclei of normal human female cells (XX), showing Barr bodies (indicated by arrows); (b) Nuclei of normal human male cells (XY), showing no Barr bodies.

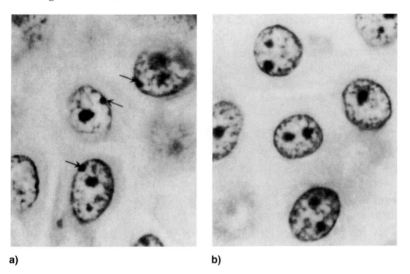

a)   b)

2. The X chromosome inactivated is randomly chosen from the maternal and paternal X chromosomes in a process that is independent from cell to cell. (Once a maternal or paternal X chromosome is inactivated in a cell, all descendants of that cell inherit the inactivation pattern.)

X-chromosome inactivation occurs at about the sixteenth day following fertilization. Because of X-chromosome inactivation, mammalian females heterozygous for X-linked traits are effectively genetic mosaics; that is, some cells show the phenotypes of one X chromosome, while the other cells show the phenotypes of the other X chromosome.

When lyonization operates in cells with extra X chromosomes, all but one of the X chromosomes typically become inactivated to produce Barr bodies. Thus, a general formula for the number of Barr bodies is the number of X chromosomes minus one. This dosage compensation process therefore minimizes the effects of multiple copies of X-linked genes. In normal individuals, lyonization results in one active X chromosome per cell in males and females alike. The number of Barr bodies associated with the normal and abnormal human X chromosome constitutions we have discussed are listed in Table 3.2. Normal 46,XX females have one Barr body, while normal 46,XY males have no Barr bodies. Turner syndrome females (45,X) have no Barr bodies; triplo-X (47,XXX) females have two Barr bodies; 47,XXY Klinefelter syndrome males have one Barr body; and 47,XYY syndrome males have no Barr bodies.

At the molecular level, the process of X inactivation begins in a region of the X chromosome called

the *X-inactivation center* (*XIC*) and spreads in both directions. A gene in the *XIC* region called *XIST* (for X inactive specific transcripts) plays a role in this X inactivation. *XIST* is expressed from the inactive X rather than from the active X, which is the opposite of the expression pattern of other X-linked genes. The *XIST* gene encodes a 15 kb (1 kb = 1,000 bases; thus, a 15,000-base) RNA molecule. When X inactivation begins the nucleus contains a marked increase in the amount of the *XIST* RNA. That RNA coats the X to be inactivated (from which it was transcribed) and silences most of the genes.

The Gene on the Y Chromosome That Determines Maleness. Since the Y chromosome determines maleness in placental mammals, the Y uniquely carries an important gene or genes that sets the switch toward male sexual differentiation. This product is called **testis-determining factor**, and the corresponding hypothesized gene is the testis-determining factor gene. Testis-determining factor causes the tissue that will become gonads to differentiate into testes instead of ovaries. This is the central event in sex determination of many mammals; all other differences between the sexes are secondary effects resulting from hormone action or from the action of factors produced by the gonads. Therefore, sex determination is equivalent to testis determination.

The testis-determining factor gene was found by studying so-called *sex reversal* individuals, that is, males who are XX (instead of XY) and females who are XY (instead of XX). In the XX males, a small fragment from near the tip of the small arm of the Y chromosome had broken off during the production of gametes and become attached to one of the X chromosomes. A number of the XY females had deletions of the same region of the Y chromosome. These findings suggested that the testis-determining factor gene is in that small segment of the Y, and as a result the human *SRY* gene and its mouse counterpart, *Sry*, were identified and cloned. One piece of evidence for why *SRY* and *Sry* are considered to be the testis-determining factor genes of humans and mice, respectively, comes from an experiment in which a DNA clone containing the *Sry* gene was introduced into XX mouse embryos by microinjection. Organisms or cells that have extra genes introduced into their genomes by genetic manipulation are called **transgenic organisms** or **transgenic cells**, and the introduced gene is called a **transgene**. The mice produced from these transgenic embryos were males with normal testis differentiation and subsequent normal male secondary sexual development. Thus, *Sry* alone is capable of causing a full phenotypic sex reversal in an XX transgenic mouse.

SEX DETERMINATION IN DROSOPHILA.

Drosophila melanogaster has four pairs of chromosomes: one pair of sex chromosomes and three pairs of autosomes. In this organism, the homogametic sex is the female (XX) and the heterogametic sex is the male (XY). However, an XXY fly is female and an XO fly is male, indicating that the sex of the fly is not the consequence of the presence or absence of a Y chromosome. In fact, the sex of the fly is determined by the ratio of the number of X chromosomes (X) to the number of *sets* of autosomes (A). Since *Drosophila* is diploid, a wild-type fly has two sets of autosomes, although abnormal numbers of sets can be produced as a result of nondisjunction. *Drosophila* exemplifies the *X chromosome–autosome balance system of sex determination.*

Table 3.3 presents some chromosome complements and the sex of the resulting flies. A normal female has two Xs and two sets of autosomes; the X:A ratio is 1.00. A normal male has a ratio of 0.50. If the X:A ratio is greater than or equal to 1.00, the fly is female; if the X:A ratio is less than or equal to 0.50, the fly is male. If the ratio is between 0.50 and 1.00, the fly is neither male nor female; it is an intersex. Intersex flies are variable in appearance, generally having complex mixtures of male and female attributes for the internal sex organs and external genitalia. Such flies are sterile.

SEX DETERMINATION IN CAENORHABDITIS.

Sex determination in the nematode *Caenorhabditis elegans* also occurs by the X chromosome–autosome balance system. *C. elegans* has become a popular tool for developmental geneticists because an adult has only

~ TABLE 3.3

Sex Balance Theory of Sex Determination in *Drosophila melanogaster*

SEX CHROMOSOME COMPLEMENT	AUTOSOME COMPLEMENT (A)	X:A RATIO[a]	SEX OF FLIES
XX	AA	1.00	♀
XY	AA	0.50	♂
XXX	AA	1.50	Metafemale (sterile)
XXY	AA	1.00	♀
XXX	AAAA	0.75	Intersex (sterile)
XX	AAA	0.67	Intersex (sterile)
X	AA	0.50	♂ (sterile)

[a]If the X chromosome–autosome ratio is greater than or equal to 1.00 (X:A ≥ 1.00), the fly will be a female. If the X chromosome–autosome ratio is less than or equal to 0.50 (X:A ≤ 0.50), the fly will be male. Between these two ratios, the fly will be an intersex.

about a thousand cells, and the lineages of every one of those cells has been carefully defined from egg to adult. *C. elegans* has two sexual types: hermaphrodites and males. Most individuals are **hermaphroditic**; that is, they have both sex organs, an ovary and two testes. They make sperm when they are larvae and store those sperm as development continues. In adults, the ovary produces eggs that are fertilized by the stored sperm as the eggs migrate to the uterus. Self-fertilization in this way almost always produces more hermaphrodites. However, 0.2 percent of the time males are produced from self-fertilization. These males can fertilize hermaphrodites if the two mate, and such matings result in about equal numbers of hermaphrodite and male progeny because the sperm from males has a competitive advantage over the sperm stored in the hermaphrodite. Genetically, hermaphrodites are XX and males are XO; that is, an X chromosome–autosome ratio of 1.00 results in hermaphrodites, and a ratio of 0.50 results in males.

SEX CHROMOSOMES IN OTHER ORGANISMS.

In birds, butterflies, moths, and some fish, the sex chromosome composition is the opposite of that in mammals. The male is the homogametic sex and the female is the heterogametic sex. To prevent confusion with the X and Y chromosome convention, we designate the sex chromosomes in these organisms as Z and W: the males are ZZ and the females are ZW. Genes on the Z chromosome behave just like X-linked genes, except that hemizygosity is found only in females. All the daughters of a male homozygous for a Z-linked recessive gene express the recessive trait, and so on.

Plants exhibit a variety of arrangements of sex organs. Some species (the ginkgo, for example) have plants of separate sexes, with male plants producing flowers that contain only stamens and female plants producing flowers that contain only pistils. These species are called **dioecious** ("two houses"). Other species have both male and female sex organs on the same plant; such plants are said to be **monoecious** ("one house"). If both sex organs are in the same flower, as in the rose and the buttercup, the flower is said to be a *perfect flower*. If the male and female sex organs are in different flowers on the same plant, as in corn, the flower is said to be an *imperfect flower*.

Some dioecious plants have sex chromosomes that differ between the sexes, and a large proportion of these plants have an X-Y system. Such plants typically have an X chromosome–autosome balance system of sex determination like that in *Drosophila*. However, we see many other sex determination systems in dioecious plants.

Many species, particularly eukaryotic microorganisms, do not have sex chromosomes but instead rely on a *genic system* for the determination of sex. In this system the sexes are specified by simple allelic differences at a small number of gene loci. For example, the yeast *Saccharomyces cerevisiae* is a haploid eukaryote that has two "sexes"—**a** and α—referred to as **mating types**. The mating types have the same morphologies, but crosses can occur only between individuals of opposite type. These mating types are controlled by the *MAT***a** and *MAT*α alleles, respectively, of a single gene.

Environmental Sex Determination Systems

In **environmental sex determination**, environmental factors (such as temperature) play a major role in determining the sex of progeny. In certain turtles, for example, eggs incubated above 32°C produce females, eggs incubated below 28°C produce males, and eggs incubated between these temperatures produce a mixture of males and females. No universal system for the temperature control of sex exists, however. Snapping turtle eggs produce females *either* at 20°C or below *or* at 30°C or above, and produce predominantly males at temperatures in between.

It is important to realize that, while environmental factors trigger the particular sexual development pathway in these systems, the pathways themselves are under genetic control. Environmental sex determination mechanisms are much rarer than the genotypic mechanisms we have discussed up to now.

KEYNOTE

Many eukaryotic organisms have sex chromosomes that are represented differentially in the two sexes; in humans and most other mammals, the male is XY and the female is XX. In other eukaryotes with sex chromosomes, the male is ZZ and the female is ZW. Sex determination commonly is related to the sex chromosomes. For humans and many other mammals, for instance, the presence of the Y chromosome confers maleness, and its absence results in femaleness. *Drosophila* and *Caen-orhabditis* have an X chromosome–autosome balance system of sex determination: the sex of the individual is related to the ratio of the number of X chromosomes to the number of sets of autosomes. Several other sex-determining systems are known in eukaryotes, including genic systems, found particularly in simple eukaryotes, and environmental systems.

ANALYSIS OF SEX-LINKED TRAITS IN HUMANS

In Chapter 2 we introduced the analysis of recessive and dominant traits in humans; those traits were not sex-linked but instead were the result of alleles carried on autosomes. In this section we discuss examples of the analysis of X-linked and Y-linked traits in humans.

For the analysis of all pedigrees, whether the trait is autosomal or X-linked, collecting reliable human pedigree data is a difficult task. For example, often one has to rely on a family's recollections. Also, there may not be enough affected people to enable a clear determination of the mechanism of inheritance involved, especially when the trait is rare and the family is small. Further, the expression of a trait may vary, resulting in some individuals erroneously being classified as normal. Lastly, because the same mutant phenotype could result from mutations in more than one gene, it is possible that different pedigrees will indicate, correctly, that different mechanisms of inheritance are involved for the "same" trait.

X-Linked Recessive Inheritance

A trait due to a recessive mutant allele carried on the X chromosome is called an **X-linked recessive trait**. At least 100 human traits are known for which the gene has been traced to the X chromosome. Most of the traits involve X-linked recessive alleles. The best

~ FIGURE 3.11

X-linked recessive inheritance. (a) Painting of Queen Victoria as a young woman. (b) Pedigree of Queen Victoria (III-2) and her descendants, showing the inheritance of hemophilia. (Refer to Figure 2.15, p. 44, for an explanation of symbols used in pedigrees. In the pedigree shown here, marriage partners that were normal with respect to the trait may have been omitted to save space.) Since Queen Victoria was heterozygous for the sex-linked recessive hemophilia allele but no cases occured in her ancestors, the trait may have arisen as a mutation in one of her parents' germ cells (the cells that give rise to the gametes).

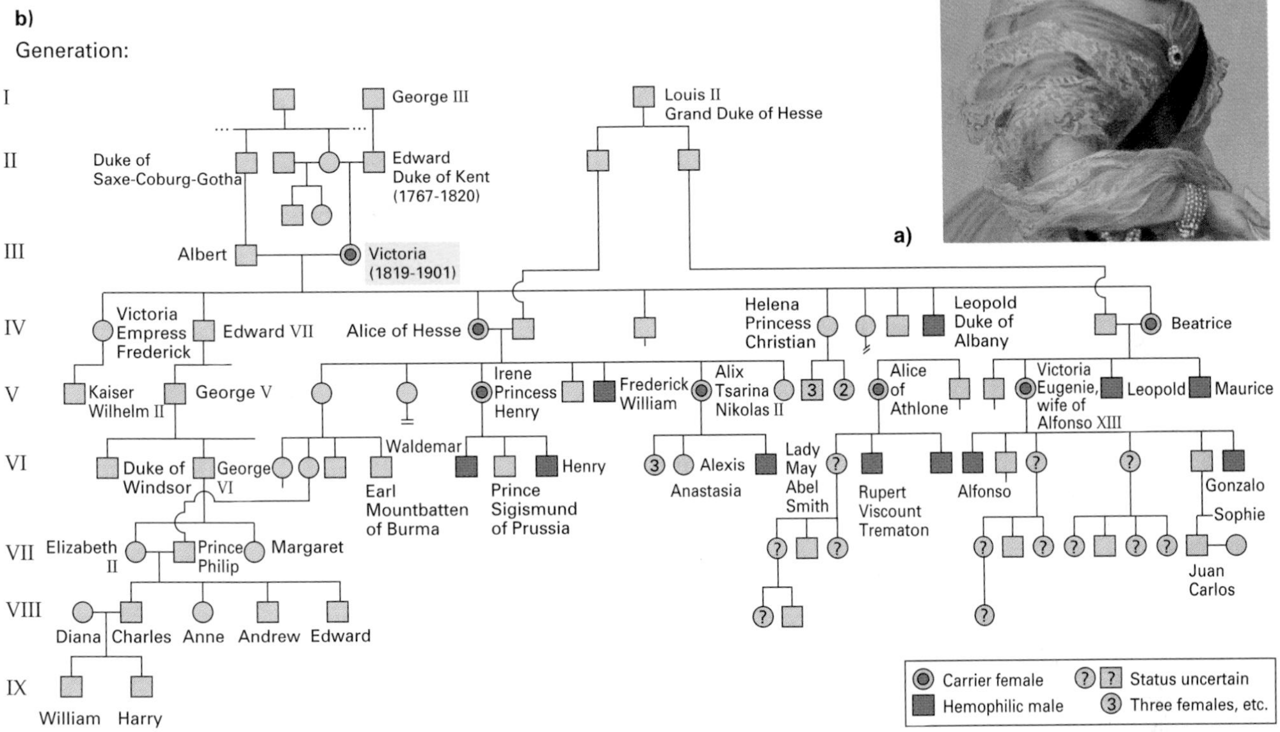

known X-linked recessive pedigree is that of hemophilia A in Queen Victoria's family (Figure 3.11). Hemophilia is a serious ailment in which the blood lacks a clotting factor, so a cut or even a bruise can be fatal to a hemophiliac. In Queen Victoria's pedigree the first instance of hemophilia was in one of her sons. Since she passed the mutant allele on to some of her other children (carrier daughters), she must have been a carrier (heterozygous) herself. It is thought that the mutation occurred on an X chromosome in the germ cells of one of her parents.

In X-linked recessive traits, females usually must be homozygous for the recessive allele in order to express the mutant trait. The trait is expressed in males who possess but one copy of the mutant allele on the X chromosome. Therefore, affected males normally transmit the mutant gene to all their daughters but to none of their sons. The instance of father-to-son inheritance of a rare trait in a pedigree tends to rule out X-linked recessive inheritance.

Other characteristics of X-linked recessive inheritance are the following (refer to Figure 3.11):

1. For X-linked recessive mutant alleles, many more males than females should exhibit the trait, because of the different number of X chromosomes in the two sexes.

2. All sons of an affected (homozygous mutant) mother should show the trait, since males receive their only X chromosome from their mothers.

3. The sons of heterozygous (carrier) mothers should show an approximately 1:1 ratio of normal individuals to individuals expressing the trait; that is, $a^+/a \times a^+/Y$ gives half a^+/Y and half a/Y sons.

4. From a mating of a carrier female with a normal male, all daughters will be normal, but half will be carriers; that is, $a^+/a \times a^+/Y$ gives half a^+/a^+ and half a^+/a females. In turn, half the sons of these carrier females will exhibit the trait.

5. A male expressing the trait, when mated with a homozygous normal female, will produce all normal children, but all the female progeny will be carriers; that is, $a^+/a^+ \times a/Y$ gives a^+/a females and a^+/Y (normal) males.

Other examples of human X-linked recessive traits are Duchenne muscular dystrophy (progressive muscle degeneration that shortens life) and two forms of color blindness.

X-Linked Dominant Inheritance

A trait due to a dominant mutant allele carried on the X chromosome is called an **X-linked dominant trait**. Only a few X-linked dominant traits have been identified.

One example of an X-linked dominant trait—faulty tooth enamel and dental discoloration—is shown in Figure 3.12a; a pedigree for this trait is shown in Figure 3.12b. Note that all the daughters and none of the sons of an affected father (III-1) are affected, and that heterozygous mothers (IV-3) transmit the trait to half their sons and half their daughters. Other X-linked dominant mutant traits are webbing to the tips of the toes in a family in South Dakota (studied in the 1930s) and a severe bleeding anomaly called constitutional thrombopathy. In the latter (also studied in the 1930s), bleeding is not due to the absence of a clotting factor (as in hemophilia) but instead to interference with the formation of blood platelets, which are needed for blood clotting.

X-linked dominant traits follow the same sort of inheritance rules as the X-linked recessives, except that heterozygous females express the trait. In general, X-linked dominant traits tend to be milder in females than in males. Also, since females have twice the number of X chromosomes as males, X-linked dominant traits are more frequent in females than in

---

~ FIGURE 3.12

X-linked dominant inheritance. (a) The teeth of a person with the X-linked dominant trait of faulty enamel; (b) A pedigree showing the transmission of the faulty enamel trait. This pedigree illustrates a shorthand convention that omits parents who do not exhibit the trait. Thus, it is a given that the female in generation I paired with a male who did not exhibit the trait.

a)

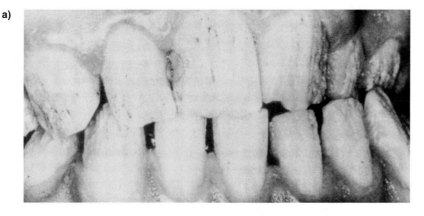

b) Pedigree

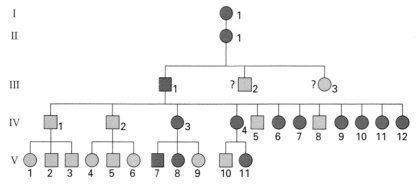

males. If the trait is rare, females with the trait are likely to be heterozygous. These females will pass the trait on to 1/2 of their male progeny and to 1/2 of their female progeny. Males with an X-linked dominant trait will pass the trait on to all of their daughters and to none of their sons.

Y-Linked Inheritance

A trait due to a mutant gene that is carried on the Y chromosome but has no counterpart on the X is called a **Y-linked**, or **holandric** ("wholly male"), **trait**. Such traits should be easily recognizable, since every son of an affected male should have the trait, and no females should ever express it. Several traits with Y-linked inheritance have been suggested. In most cases the genetic evidence for such inheritance is poor or nonexistent. A number of genes on the Y chromosome have been identified, however, including the *SRY* gene for testis-determining factor mentioned earlier.

KEYNOTE

Controlled matings cannot be made in humans, so studying the inheritance of genes in humans must rely on pedigree analysis. The data obtained from pedigree analysis enable geneticists to make judgments, with varying degrees of confidence, about whether a mutant gene is inherited as an autosomal recessive, an autosomal dominant, an X-linked recessive, an X-linked dominant, or a Y-linked allele.

SUMMARY

Chromosome Theory of Inheritance

In Chapter 2, genes were considered abstract entities that control hereditary characteristics. Before 1902, genes were called Mendelian factors. Cytologists working in the nineteenth century accumulated information about cell structure and cell division. The fields of genetics and cytology came together in 1902 when Sutton and Boveri independently hypothesized that genes are on chromosomes, "bodies" within the cell nucleus. The chromosome theory of heredity states that the patterns of chromosome transmission from one generation to the next closely parallel the patterns of gene transmission from one generation to the next.

Support for the chromosome theory of inheritance came from experiments that related the heredi-

tary behavior of particular genes to the transmission of the sex chromosome. The sex chromosome in eukaryotic organisms is the chromosome that is represented differently in the two sexes. In most organisms with sex chromosomes, the female has two X chromosomes, while the male has one X and one Y chromosome. The Y chromosome is structurally and genetically different from the X chromosome. The association of genes with the sex chromosomes of eukaryotes is called sex linkage. Such genes, and the phenotypes they control, are called sex-linked.

Sex Determination Mechanisms

In many cases, sex determination is related to the sex chromosomes. In humans, for example, the presence of a Y chromosome specifies maleness, while its absence results in femaleness. The recent identification and characterization of a gene on the Y chromosome—*SRY*/*Sry*, the product of which directs male sexual differentiation—opens the way to a detailed analysis of the genetic control of sex determination in mammals. Several other sex determination systems are known in eukaryotes, including X chromosome–autosome balance systems (in which sex is determined by the ratio of the number of X chromosomes to the number of sets of autosomes), genic systems (in which a simple allelic difference determines the sex of an individual), and environmental systems (in which sex is determined by environmental cues).

Analysis of Sex-Linked Traits in Humans

In Chapter 2 we discussed the fact that the inheritance patterns of traits in humans are usually studied by charting the family trees of individuals exhibiting the trait, a method called pedigree analysis. In this chapter we considered examples of X- and Y-linked human traits to illustrate the features of those mechanisms of inheritance in pedigrees. Collecting reliable human pedigree data is a difficult task. In many cases the accuracy of record keeping within the families involved is open to question. Also, particularly with small families, there may not be enough affected people to allow an unambiguous determination of the inheritance mechanism. Moreover, the degree to which a trait is expressed may vary, so some individuals may be erroneously classified as normal. It is also possible for the same mutant phenotype to be produced by mutations in different genes, and therefore different pedigrees may correctly indicate different mechanisms of inheritance of the "same" trait.

Analytical Approaches for Solving Genetics Problems

The concepts introduced in this chapter may be reinforced by solving genetics problems similar to those introduced in Chapter 2. When sex linkage is involved, remember that one sex has two kinds of sex chromosomes, whereas the other sex has only one; this feature alters the inheritance patterns slightly. Most of the problems presented in this section center on interpreting data and predicting the outcome of particular crosses.

Q3.1 A female from a true-breeding strain of *Drosophila* with vermilion-colored eyes is crossed with a male from a true-breeding wild-type, red-eyed strain. All the F_1 males have vermilion-colored eyes, and all the females have wild-type red eyes. What conclusions can you draw about the mechanism of inheritance of the vermilion trait, and how could you test them?

A3.1 The observation is the classic one that suggests a sex-linked trait is involved. Since none of the F_1 daughters have the trait and all the F_1 males do, the trait is presumably X-linked recessive. The results fit this hypothesis, because the F_1 males receive the X chromosome with the *v* gene from their homozygous *v/v* mother. Furthermore, the F_1 females are v^+/v since they receive a v^+-bearing X chromosome from the wild-type male parent and a *v*-bearing X chromosome from the female parent. If the trait were autosomal recessive, all the F_1 flies would have had wild-type eyes. If it were autosomal dominant, both the F_1 males and females would have had vermilion-colored eyes. If the trait were X-linked dominant, all the F_1 flies would have had vermilion eyes.

The easiest way to verify this hypothesis is to let the F_1 flies interbreed. This cross is v^+/v female $\times$ *v*/Y male, and the expectation is that there will be a 1:1 ratio of wild-type:vermilion eyes in both sexes in the F_2. That is, half the females are v^+/v and half are *v/v*; half the males are v^+/Y and half are *v/*Y. This ratio is certainly not the 3:1 ratio that would result from an $F_1 \times F_1$ cross for an autosomal gene.

Q3.2 In humans, hemophilia A or B is caused by an X-linked recessive gene. A woman who is a nonbleeder had a father who was a hemophiliac. She marries a nonbleeder, and they plan to have children. Calculate the probability of hemophilia in the female and male offspring.

A3.2 Since hemophilia is an X-linked trait, and since her father was a hemophiliac, the woman must be heterozygous for this recessive gene. If we assign the symbol *h* to this recessive mutation, and h^+ to the wild-type (nonbleeder) allele, she must be h^+/h. The man she marries is normal with regard to blood clotting and hence must be

hemizygous for h^+—that is, h^+/Y. All their daughters receive an X chromosome from the father, and so each must have an h^+ gene. In fact, half the daughters are h^+/h^+ and the other half are h^+/h. Since the wild-type allele is dominant, none of the daughters are hemophiliacs. However, all the sons of the marriage receive their X chromosome from their mother. Therefore, they have a probability of 1/2 that they will receive the chromosome carrying the *h* allele, which means they will be hemophiliacs. Thus, the probability of hemophilia among daughters of this marriage is 0; among sons it is 1/2.

Q3.3 Tribbles are hypothetical animals that have an X-Y sex determination mechanism like that of humans. The trait bald (*b*) is X-linked and recessive to furry (b^+), and the trait long leg (*l*) is autosomal and recessive to short leg (l^+). If you make reciprocal crosses between true-breeding bald, long-legged tribbles and true-breeding furry, short-legged tribbles, do you expect a 9:3:3:1 ratio in the F_2 of either or both of these crosses? Explain your answer.

A3.3 This question focuses on the fundamentals of X-chromosome and autosome segregation during a genetic cross, and it tests whether or not you have grasped the principles involved in gene segregation. Figure 3.A diagrams the two crosses involved, and we can discuss the answer by referring to it.

First, consider the cross of a wild-type female tribble (b^+/b^+, l^+/l^+) with a male double-mutant tribble (*b*/Y, *l/l*). Part (a) of the figure diagrams this cross. These F_1s are all normal—that is, they have furry bodies and short legs—because for the autosomal character both sexes are heterozygous, and for the X-linked character the female is heterozygous and the male is hemizygous for the b^+ allele donated by the normal mother. For the production of the F_2 progeny, the best approach is to treat the X-linked and autosomal traits separately. For the X-linked trait, random combination of the gametes produced gives a 1:1:1:1 genotypic ratio of b^+/b^+ (furry female) : b^+/b (furry female) : b^+/Y (furry male) : *b*/Y (bald male) progeny. Categorizing by phenotypes, 1/2 of the progeny are furry females, 1/4 are furry males, and 1/4 are bald males. For the autosomal leg trait, the $F_1 \times F_1$ is a cross of two heterozygotes, so we expect a 3:1 phenotypic ratio of short-legged:long-legged tribbles in the F_2. Since autosome segregation is independent of the inheritance of the X chromosome, we can multiply the probabilities of the occurrence of the X-linked and autosomal traits to calculate their relative frequencies. The calculations are presented at the bottom of part (a) of the figure.

FIGURE 3.A

a) Furry, short-legged (wild-type) ♀ ×
bald, long-legged ♂

P generation

$b^+/b^+\ l^+/l^+$ ♀ (furry, short) × $b/\wedge\ l/l$ ♂ (bald, long)

F$_1$ generation

$b^+/b\ l^+/l$ ♀ (furry, short) × $b^+/\wedge\ l^+/l$ ♂ (furry, short)

F$_2$ generation

Sex-linked phenotypes and genotypes	Autosomal phenotypes and genotypes

Genotypic results:

$\frac{1}{2}\ b^+$ ($\frac{1}{2}\ b^+/b^+$, $\frac{1}{2}\ b^+/b$; furry) ♀ $\begin{cases} \frac{3}{4}\ l^+\ (l^+/l^+ \text{ and } l^+/l;\ \text{short}) \\ \frac{1}{4}\ l\ (l/l\ ;\ \text{long}) \end{cases}$

$\frac{1}{4}\ b^+$ ($b^+/\wedge$; furry) ♂ $\begin{cases} \frac{3}{4}\ l^+\ (\text{short}) \\ \frac{1}{4}\ l\ (\text{long}) \end{cases}$

$\frac{1}{4}\ b$ ($b/\wedge$; bald) ♂ $\begin{cases} \frac{3}{4}\ l^+\ (\text{short}) \\ \frac{1}{4}\ l\ (\text{long}) \end{cases}$

Phenotypic ratios:

	Furry-short		Furry-long		Bald-short		Bald-long
	$b^+\,l^+$		$b^+\,l$		$b\,l^+$		$b\,l$
♀	6	:	2	:	0	:	0
♂	3	:	1	:	3	:	1
Total	9	:	3	:	3	:	1

b) Bald, long-legged ♀ ×
furry, short-legged (wild-type) ♂

P generation

$b/b\ l/l$ ♀ (bald, long) × $b^+/\wedge\ l^+/l^+$ ♂ (furry, short)

F$_1$ generation

$b^+/b\ l^+/l$ ♀ (furry, short) × $b/\wedge\ l^+/l$ ♂ (bald, short)

F$_2$ generation

Sex-linked phenotypes and genotypes	Autosomal phenotypes and genotypes

Genotypic results:

$\frac{1}{4}\ b^+$ (b^+/b^+; furry) ♀ $\begin{cases} \frac{3}{4}\ l^+\ (l^+/l^+ \text{ and } l^+/l;\ \text{short}) \\ \frac{1}{4}\ l\ (l/l\ ;\ \text{long}) \end{cases}$

$\frac{1}{4}\ b$ (b/b; bald) ♀ $\begin{cases} \frac{3}{4}\ l^+\ (\text{short}) \\ \frac{1}{4}\ l\ (\text{long}) \end{cases}$

$\frac{1}{4}\ b^+$ ($b^+/\wedge$; furry) ♂ $\begin{cases} \frac{3}{4}\ l^+\ (\text{short}) \\ \frac{1}{4}\ l\ (\text{long}) \end{cases}$

$\frac{1}{4}\ b$ ($b/\wedge$; bald) ♂ $\begin{cases} \frac{3}{4}\ l^+\ (\text{short}) \\ \frac{1}{4}\ l\ (\text{long}) \end{cases}$

Phenotypic ratios:

	Furry-short		Furry-long		Bald-short		Bald-long
	$b^+\,l^+$		$b^+\,l$		$b\,l^+$		$b\,l$
♀	3	:	1	:	3	:	1
♂	3	:	1	:	3	:	1
Total	6	:	2	:	6	:	2

The first cross, then, has a 9:3:3:1 ratio of the four possible phenotypes in the F$_2$. However, note that the ratio in each sex is not 9:3:3:1, owing to the inheritance pattern of the X chromosome. This result contrasts markedly with the pattern of two autosomal genes segregating independently, where the 9:3:3:1 ratio is found for both sexes.

The second cross (a reciprocal cross) is diagrammed in part (b) of the figure. Since the parental female in this cross is homozygous for the sex-linked trait, all the F$_1$ males are bald. Genotypically, the F$_1$ males and females differ from those in the first cross with respect to the sex chromosome but are just the same with respect to the autosome. Again, considering the X chromosome first as we go to the F$_2$, we find a 1:1:1:1 genotypic ratio of furry females : bald females : furry males : bald males. In this case, then, half of both males and females are furry, and half are bald, in contrast to the results of the first cross, in which no bald females were produced in the F$_2$. For the autosomal trait we expect a 3:1 ratio of short:long in

the F$_2$, as before. Putting the two traits together, we get the calculations presented in part (b) of the figure. (Note: We use the total 6:2:6:2 here rather than 3:1:3:1 because the numbers add to 16, as does $9 + 3 + 3 + 1$.) So in this case we do not get a 9:3:3:1 ratio; moreover, the ratio is the same in both sexes.

This question has forced us to think through the segregation of two types of chromosomes and has shown that we must be careful about predicting the outcomes of crosses in which sex chromosomes are involved. Nonetheless, the basic principles for this analysis are the same as those used before: reduce the questions to their basic parts and then put the puzzle together step by step.

QUESTIONS AND PROBLEMS

*3.1 In a male *Homo sapiens,* from which grandparent could each sex chromosome have been derived? (Indicate yes or no for each option.)

	MOTHER'S		FATHER'S	
	MOTHER	FATHER	MOTHER	FATHER
X chromosome	___	___	___	___
Y chromosome	___	___	___	___

3.2 In *Drosophila,* white eyes are a sex-linked character. The mutant allele for white eyes (w) is recessive to the wild-type allele for brick-red eye color (w^+). A white-eyed female is crossed with a red-eyed male. An F_1 female from this cross is mated with her father, and an F_1 male is mated with his mother. What will be the eye color of the offspring of these last two crosses?

*3.3 One form of color blindness (c) in humans is caused by a sex-linked recessive mutant gene. A woman with normal color vision (c^+) and whose father was color-blind marries a man of normal vision whose father was also color-blind. What proportion of their offspring will be color-blind? (Give your answer separately for males and females.)

3.4 In humans, red-green color blindness is due to an X-linked recessive gene. A color-blind daughter is born to a woman with normal color vision and a father who is color-blind. What is the mother's genotype with respect to the alleles concerned?

*3.5 In humans, red-green color blindness is recessive and X-linked, while albinism is recessive and autosomal. What types of children can be produced as the result of marriages between two homozygous parents: a normal-visioned albino woman and a color-blind, normally pigmented man?

*3.6 In *Drosophila,* vestigial (partially formed) wings (vg) are recessive to normal long wings (vg^+), and the gene for this trait is autosomal. The gene for the white-eye trait is on the X chromosome. Suppose a homozygous white-eyed, long-winged female fly is crossed with a homozygous red-eyed, vestigial-winged male.
a. What will be the appearance of the F_1?
b. What will be the appearance of the F_2?
c. What will be the appearance of the offspring of a cross of the F_1 back to each parent?

3.7 In *Drosophila,* two red-eyed, long-winged flies are bred together and produce the offspring given in the following table:

	FEMALES	MALES
red-eyed, long-winged	3/4	3/8
red-eyed, vestigial-winged	1/4	1/8
white-eyed, long-winged	—	3/8
white-eyed, vestigial-winged	—	1/8

What are the genotypes of the parents?

3.8 In poultry, a dominant sex-linked gene (B) produces barred feathers, and the recessive allele (b), when homozygous, produces nonbarred feathers. Suppose a nonbarred cock is crossed with a barred hen.
a. What will be the appearance of the F_1 birds?
b. If an F_1 female is mated with her father, what will be the appearance of the offspring?
c. If an F_1 male is mated with his mother, what will be the appearance of the offspring?

*3.9 A man (A) suffering from defective tooth enamel, which results in brown-colored teeth, marries a normal woman. All their daughters have brown teeth, but the sons are normal. The sons of man A marry normal women, and all their children are normal. The daughters of man A marry normal men, and 50 percent of their children have brown teeth. Explain these facts.

3.10 In humans, differences in the ability to taste phenylthiourea are due to a pair of autosomal alleles. Inability to taste is recessive to ability to taste. A child who is a nontaster is born to a couple who can both taste the substance. What is the probability that their next child will be a taster?

*3.11 Cystic fibrosis is inherited as an autosomal recessive. Two parents without cystic fibrosis have two children with cystic fibrosis and three children without. The parents come to you for genetic counseling.
a. What is the numerical probability that their next child will have cystic fibrosis?
b. Their unaffected children are concerned about being heterozygous. What is the numerical probability that a given unaffected child in the family is heterozygous?

3.12 Huntington disease is a human disease inherited as a Mendelian autosomal dominant. The disease results in choreic (uncontrolled) movements, progressive mental deterioration, and eventually death. The disease affects the carriers of the trait any time between 15 and 65 years of age. The American folk singer Woody Guthrie died of Huntington's disease, as did one of his parents. Marjorie Mazia, Woody's wife, had no history of this disease in her family. The Guthries had three children. What is the probability that a particular Guthrie child will die of Huntington's disease?

3.13 Suppose gene A is on the X chromosome, and genes B, C, and D are on three different autosomes. Thus, $A-$ signifies the dominant phenotype in the male or female. An equivalent situation holds for $B-$, $C-$, and $D-$. The cross $AA\ BB\ CC\ DD$ female $\times aY\ bb\ cc\ dd$ male is made.
a. What is the probability of obtaining an $A-$ individual in the F_1?
b. What is the probability of obtaining an a male in the F_1?
c. What is the probability of obtaining an $A-\ B-\ C-\ D-$ female in the F_1?
d. How many different F_1 genotypes will there be?
e. What proportion of F_2s will be heterozygous for the four genes?

f. Determine the probabilities of obtaining each of the following types in the F$_2$: (1) *A– bb CC dd* (female); (2) *aY BB Cc Dd* (male); (3) *AY bb CC dd* (male); (4) *aa bb Cc Dd* (female).

3.14 A Turner syndrome individual would be expected to have how many Barr bodies in the majority of cells?
a. 0   **c.** 2
b. 1   **d.** 3

3.15 An XXY Klinefelter syndrome individual would be expected to have how many Barr bodies in the majority of cells?
a. 0   **c.** 2
b. 1   **d.** 3

*3.16 In human genetics, the pedigree is used for analysis of inheritance patterns. Females are represented by a circle, and males by a square. The following figure presents three, two-generation family pedigrees for a trait in humans. Normal individuals are represented by unshaded symbols, and people with the trait by shaded symbols. For each pedigree (A, B, and C), state, by answering yes or no in the appropriate blank space, whether transmission of the trait can be accounted for on the basis of each of the listed simple modes of inheritance:

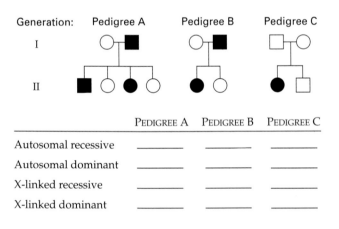

	PEDIGREE A	PEDIGREE B	PEDIGREE C
Autosomal recessive			
Autosomal dominant			
X-linked recessive			
X-linked dominant			

3.17 Looking at the following pedigree, in which shaded symbols represent a "trait," which of the progeny eliminate X-linked recessiveness as a mode of inheritance for the trait?
a. I-1 and I-2
b. II-4
c. II-5
d. II-2 and II-4

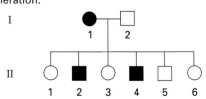

*3.18 When constructing human pedigrees, geneticists often refer to particular persons by a number. The generations are labeled by roman numerals, and the individuals in each generation by Arabic numerals. For example, in the pedigree in the following figure, the female with the asterisk is I-2. Use this means to designate specific individuals in the pedigree. Determine the probable inheritance mode for the trait shown in the affected individuals (the shaded symbols) by answering the following questions. Assume the condition is caused by a single gene.

Generation:

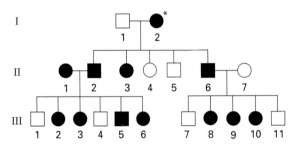

a. Y-linked inheritance can be excluded at a glance. What two other mechanisms of inheritance can be definitely excluded? Why can these be excluded?
b. Of the remaining mechanisms of inheritance, which is the most likely? Why?

3.19 A three-generation pedigree for a particular human trait is shown in the following figure:

Generation:

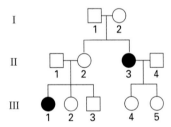

a. What is the mechanism of inheritance for the trait?
b. Which persons in the pedigree are known to be heterozygous for the trait?
c. What is the probability that III-2 is a carrier (heterozygous)?
d. If III-3 and III-4 marry, what is the probability that their first child will have the trait?

3.20 For each of the more complex pedigrees shown in Figure 3.B, determine the probable mechanism of inheritance: autosomal recessive, autosomal dominant, X-linked recessive, X-linked dominant, or Y-linked.

In 3.21 through 3.25, select the correct answer.

*3.21 If a rare genetic disease is inherited on the basis of an X-linked dominant gene, one would expect to find which of the following?

FIGURE 3.B

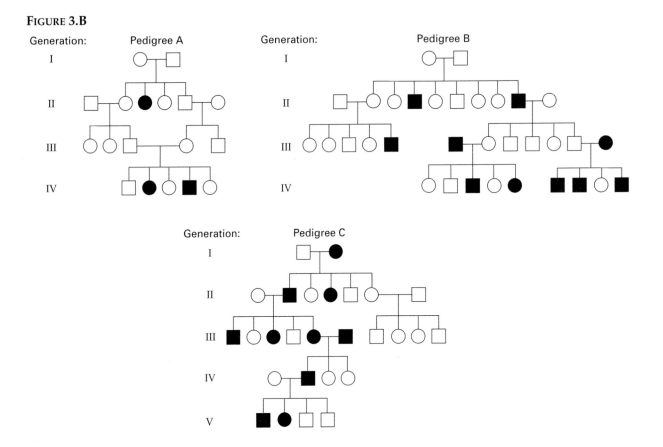

a. Affected fathers have 100 percent affected sons.
b. Affected mothers have 100 percent affected daughters.
c. Affected fathers have 100 percent affected daughters.
d. Affected mothers have 100 percent affected sons.

3.22   If a genetic disease is inherited on the basis of an autosomal dominant gene, one would expect to find which of the following?
a. Affected fathers have only affected children.
b. Affected mothers never have affected sons.
c. If both parents are affected, all of their offspring have the disease.
d. If a child has the disease, one of his or her grandparents also had the disease.

*3.23   If a genetic disease is inherited as an autosomal recessive, one would expect to find which of the following?
a. Two affected individuals never have an unaffected child.
b. Two affected individuals have affected male offspring but no affected female children.
c. If a child has the disease, one of his or her grandparents will have had it.

d. In a marriage between an affected individual and an unaffected one, all the children are unaffected.

3.24   Which of the following statements is *not* true for a disease that is inherited as a rare X-linked dominant trait?
a. All daughters of an affected male will inherit the disease.
b. Sons will inherit the disease only if their mothers have the disease.
c. Both affected males and affected females will pass the trait to half the children.
d. Daughters will inherit the disease only if their father has the disease.

*3.25   Women who were known to be carriers of the X-linked, recessive, hemophilia gene were studied to determine the amount of time required for the blood clotting reaction. It was found that the time required for clotting was extremely variable from individual to individual. The values obtained ranged from normal clotting time at one extreme, to clinical hemophilia at the other extreme. What is the most probable explanation for these findings?

CHAPTER *4*

EXTENSIONS OF MENDELIAN GENETIC ANALYSIS

PRINCIPAL POINTS

~ More than two allelic forms of a gene can exist. However, any given diploid individual can possess only two different alleles of a given gene.

~ With complete dominance, the same phenotype results whether the dominant allele is heterozygous or homozygous. In incomplete dominance, the phenotype of the heterozygote is intermediate between those of the two homozygotes, whereas in codominance the heterozygote exhibits the phenotypes of both homozygotes.

~ In many cases, different genes interact to determine phenotypic characteristics. In epistasis, for example, modified Mendelian ratios occur because of gene interactions: the phenotypic expression of one gene depends on the genotype of another gene locus.

~ Alleles of certain genes may be fatal to the individual. The existence of such lethal alleles of a gene indicates that the product usually produced by the nonlethal allele is essential for the function of the organism; the gene is called an essential gene.

~ Penetrance is the frequency within a population in which an allele manifests itself phenotypically.

Expressivity is the kind or degree of phenotypic manifestation of a gene or genotype in a particular individual.

~ The zygote's genetic constitution only specifies the organism's potential to develop and function. As the organism develops and differentiates, many things can influence gene expression. One such influence is the organism's environment, both internal and external. Examples of the internal environment include age and sex; examples of the external environment include nutrition, light, chemicals, temperature, and infectious agents.

~ Variation in most of the genetic traits considered in the earlier discussion of Mendelian principles is determined predominantly by differences in genotype; that is, phenotypic differences result from genotypic differences. For many traits, however, the phenotypes are influenced by both genes and the environment. The unfortunate debate over the relative contribution of genes and environment to the phenotype—the nature versus nurture controversy—oversimplifies the totality of interaction between genes and genetic effects.

Mendel's principles apply to all eukaryotic organisms and form the foundation for predicting the outcome of crosses in which segregation and independent assortment might occur. As more and more geneticists did experiments, though, they found exceptions to and extensions of Mendel's principles. Several of these cases are discussed in this chapter with the goal of gaining a broader knowledge of genetic analysis, particularly in terms of how genes relate to the phenotypes of an organism.

MULTIPLE ALLELES

So far in our genetic analyses we have considered only pairs of alleles that control characters, such as smooth versus wrinkled seeds in peas, and red versus white eyes in *Drosophila*. The allele that predominates in populations of the organism in the wild is the wild-type allele, and the alternative allele is the mutant allele. In a population of individuals, however, a given gene may have several alleles (one wild type

and the rest mutant), not just two. Such genes are said to have **multiple alleles**, and the alleles are said to constitute a *multiple allelic series* (Figure 4.1). Although a gene may have multiple alleles in a given population of individuals, a *single diploid individual can have*

~ **FIGURE 4.1**

Allelic forms of a gene.

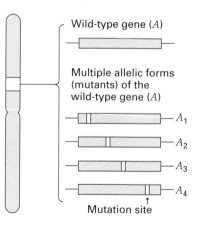

Wild-type gene (A)

Multiple allelic forms (mutants) of the wild-type gene (A)

A_1

A_2

A_3

A_4

Mutation site

only a maximum of two of these alleles, one on each of the two homologous chromosomes carrying the gene locus.

ABO Blood Groups

An example of multiple alleles of a gene is found in the human ABO blood group series. Since certain ABO blood groups are incompatible, these alleles are of particular importance when blood transfusions are done. (There are many blood group series other than ABO; these also can cause problems in blood transfusions and need to be checked through the process of cross-matching.)

Four blood group phenotypes occur in the ABO system: O, A, B, and AB; Table 4.1 lists their possible genotypes. The six genotypes that give rise to the four phenotypes represent various combinations of three ABO blood group alleles: I^A, I^B, and i. People homozygous for the recessive i allele are of blood group O. Both I^A and I^B are dominant to i. Thus, individuals of blood group A are either I^A/I^A or I^A/i, and those of blood group B are either I^B/I^B or I^B/i. Heterozygous I^A/I^B individuals are of blood group AB—that is, essentially of both blood groups A and B simultaneously (see the discussion of codominance in this chapter).

The genetics of this system follows Mendelian principles. An individual who expresses blood group O, for example, must be i/i in genotype. The parents of this person could both be O ($i/i \times i/i$), both could be A ($I^A/i \times I^A/i$, to produce one-fourth i/i progeny), both could be B ($I^B/i \times I^B/i$), or one could be A and one could be B ($I^A/i \times I^B/i$). Simply put, each parent must be either homozygous i or heterozygous, with i as one of the two alleles.

Blood-typing (determining an individual's blood group) and the analysis of blood group inheritance are sometimes used in cases of disputed paternity or maternity or when babies are inadvertently switched in a hospital. In such cases, genetic data cannot prove the identity of the parent. Genetic analysis on the basis of blood group can only be used to show that an individual is *not* the parent of a particular child; for example, a child of phenotype AB (genotype I^A/I^B) could not be the child of a parent of phenotype O (genotype i/i). (Note: In most states, blood-type data alone are usually not sufficient for a legal decision concerning paternity or maternity. In addition, DNA fingerprinting typically is used; see Chapter 14.)

When giving blood transfusions, the blood groups of donors and recipients must be carefully matched because the blood group alleles specify molecules, called *cellular antigens*, on the surfaces of red blood cells. An **antigen**—<u>anti</u>body <u>gen</u>erating substance—is any molecule that is recognized as foreign by an individual. Specific protein molecules called **antibodies** recognize and bind to such antigens and target them for destruction. A given individual has a large number of antigens on cells, many of which would be foreign to another individual—hence the concern over the blood type in blood transfusions. The antigens are usually not recognized as foreign by the individual expressing them (that is, autoimmune diseases are an exception).

The I^A allele specifies the A antigen, the I^B allele specifies the B antigen, and the i allele specifies neither A nor B antigens. People with blood group A (I^A/I^A or I^A/i) have A antigens on their red blood cells, people of the B blood group (I^B/I^B or I^B/i) have B antigens on their red blood cells, people of the AB blood group (I^A/I^B) have both A and B antigens on their red blood cells, and people of the O blood group (i/i) have neither A nor B antigens on their red blood cells. For reasons that are not fully understood, individuals have antibodies in their serum against the blood type antigens they do not express on their own red blood cells. That is, individuals of blood group A have anti-B antibodies, individuals of blood group B have anti-A antibodies, individuals of blood group AB have neither anti-A nor anti-B antibodies, and individuals of blood group O have both anti-A and anti-B antibodies. These antibodies are important clinically. If, for instance, a person with anti-B antibodies (a person of blood group A) receives red blood cells displaying the B antigen (group B blood), then the antibodies will agglutinate (clump) the cells. Since clumped cells cannot move through the fine capillaries, agglutination may lead to organ failure and possibly death. Agglutination of the red blood cells is seen whenever a blood type antibody interacts with the antigen for which it is specific. The antigen-antibody relationships of the ABO blood group system are shown in Figure 4.2.

What transfusions are safe, then, between people with different blood groups in the ABO system?

~ TABLE 4.1

ABO Blood Groups in Humans, Determined by the Alleles I^A, I^B, and i

PHENOTYPE (BLOOD GROUP)	GENOTYPE
O	i/i
A	I^A/I^A or I^A/i
B	I^B/I^B or I^B/i
AB	I^A/I^B

~ **FIGURE 4.2**

Antigen-antibody reactions that characterize the human ABO blood groups. Blood serum from each of the four blood groups was mixed with blood cells from the four types, in all possible combinations. In some cases, such as a mix of B serum with A cells, the cells become clumped.

1. Individuals of blood group A have anti-B antibodies, so they cannot receive blood from individuals of blood groups B or AB. They can receive blood from individuals of blood groups A or O.

2. Those of blood group B have anti-A antibodies, so they cannot receive blood from individuals of blood groups A or AB. They can receive blood from individuals of blood groups B or O.

3. AB individuals have neither anti-A nor anti-B antibodies, so they can receive blood from individuals of any blood group. That is, AB individuals are *universal acceptors*.

4. O individuals have both anti-A and anti-B antibodies, so they can receive blood only from individuals of blood group O. Note that people of blood group O blood can be used as donor blood for any recipient since it causes no reaction. Thus, blood group O individuals are called *universal donors*.

Drosophila Eye Color

Another example of multiple alleles concerns the white (w) locus of *Drosophila*. Recall from Chapter 3 that the w^+ allele results in wild-type brick-red eyes, and that the recessive w allele, when homozygous or hemizygous, results in white eyes. There are actually more than 100 mutant alleles at the white locus. With w^+ symbolizing the wild-type (brick-red) allele of the white-eye gene, w is the recessive mutant white-eye allele, and w^e is the recessive mutant eosin (reddish-orange) allele.

The white locus alleles have different, characteristic eye colors ranging from white to near-wild type when the alleles are homozygous or hemizygous. Eye color phenotype is related to the amount of pigment deposited in the eye cells. Table 4.2 lists the relative amounts of pigment present in wild-type females and in females homozygous for different mutant alleles of the white locus. As expected, the original mutant allele w has the least amount of pigment. Between white and wild, there is a wide range of pigment amounts. The phenotypic expressions of different alleles of the same gene reflect the varying extent to which the biological activity of the protein encoded by the white gene (its *gene product*) has been altered. Various combinations of wild-type and mutant alleles can occur in flies based on the parents involved, but a particular individual can have no more than two different alleles.

~ TABLE 4.2

Eye Pigment Quantification for *Drosophila* White Alleles

GENOTYPES	RELATIVE AMOUNT OF TOTAL PIGMENT
w^+/w^+ (wild type)	1.0000
w/w (white)	0.0044
w^t/w^t (tinged)	0.0062
w^a/w^a (apricot)	0.0197
w^{bl}/w^{bl} (blood)	0.0310
w^e/w^e (eosin)	0.0324
w^{ch}/w^{ch} (cherry)	0.0410
w^{a3}/w^{a3} (apricot-3)	0.0632
w^w/w^w (wine)	0.0650
w^{co}/w^{co} (coral)	0.0798
w^{sat}/w^{sat} (satsuma)	0.1404
w^{col}/w^{col} (colored)	0.1636

KEYNOTE

Many allelic forms of a gene can exist in a population. However, any given diploid individual can possess only two different alleles of a given gene. Multiple alleles obey the same rule of transmission as alleles of which there are only two types.

MODIFICATIONS OF DOMINANCE RELATIONSHIPS

Complete dominance is the phenomenon in which one allele is dominant to another, so that the phenotype of the heterozygote is the same as that of the homozygous dominant. With **complete recessiveness**, the recessive allele is phenotypically expressed only when the organism is homozygous. Complete dominance and complete recessiveness are the two extremes of a range of dominance relationships. While all the allelic pairs Mendel studied showed such complete dominance/complete recessiveness relationships, many allelic pairs do not.

Incomplete Dominance

When one allele is not completely dominant to another allele, it is said to show **incomplete**, or **partial**, **dominance**. With incomplete dominance the heterozygote's phenotype is intermediate to those of individuals homozygous for either individual allele involved.

Plumage color in chickens is an example of incomplete dominance. Crosses between a true-breeding black strain (homozygous $C^B C^B$) and a true-breeding white strain (homozygous $C^W C^W$) give F_1 birds with bluish-grey plumage (Figure 4.3a), called Andalusian blues by chicken breeders. (Note that the symbols are designed to give equal weight to the two alleles, since neither dominates the phenotype. In this particular example, the C signifies color, while "B" and "W" indicate black and white, respectively.) An Andalusian blue is not true breeding because it is heterozygous, so in Andalusian × Andalusian crosses (Figure 4.3b) the two alleles will segregate in the offspring and produce black, Andalusian blue, and white fowl in a ratio of 1:2:1. The most efficient way to produce Andalusian blues is to cross black × white, since all progeny of this mating are Andalusian blues.

Codominance

Codominance is a modification of the dominance relationship that is related to incomplete dominance. In codominance the heterozygote exhibits the phenotypes of *both* homozygotes. Codominance differs from incomplete dominance, in which the heterozygote exhibits a phenotype intermediate between the two homozygotes.

The ABO blood series discussed earlier in this chapter provides a good example of codominance. Heterozygous I^A/I^B individuals are of blood group AB because both the A antigen (product of the I^A allele) and the B antigen (product of the I^B allele) are produced. Thus, the I^A and I^B alleles are codominant.

The human M-N blood group system is another example of codominance. For transfusion compatibility, this system is of less clinical importance than the ABO system. In the M-N system, three blood types occur: M, MN, and N. They are determined by the genotypes L^M/L^M, L^M/L^N, and L^N/L^N, respectively. As in the ABO system, the M-N alleles result in the formation of antigens on the red blood cell surface. The heterozygote in this case has both the M and the N antigens and shows the phenotypes of both homozygotes.

Molecular Explanations of Incomplete Dominance and Codominance

What explains incomplete dominance and codominance at the molecular level? A general interpretation

~ FIGURE 4.3

Incomplete dominance in chickens. (a) A cross between a white bird and a black bird produces F₁s of intermediate grey color, called Andalusian blues. (b) The F₂ generation shows the 1:2:1 phenotype ratio characteristic of incomplete dominance.

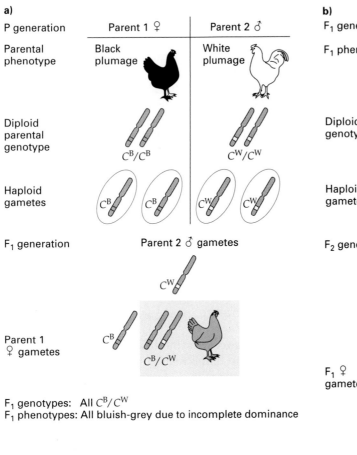

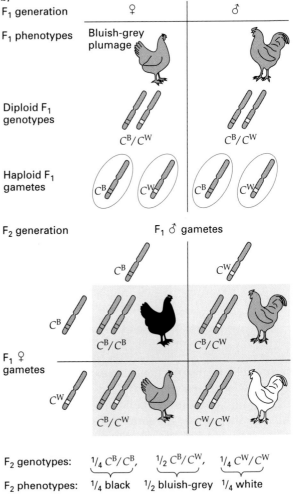

is that in codominance, products result from both alleles in a heterozygote, so that frequently (but not always) both homozygote phenotypes are observed (for example, L^M/L^N individuals express both M and N blood group antigens). In incomplete dominance, only one allele in a heterozygote is expressed to produce a product. A homozygote for the expressed allele, then, has two doses of the gene product, and full phenotypic expression results (for example, black chickens). In a homozygote for the allele that is not expressed, a phenotype characteristic of no gene expression results (white chickens). In a heterozygote, the single allele expressed results in only enough product for an intermediate phenotype (grey chickens).

KEYNOTE

With complete dominance, the same phenotype results whether the dominant allele is heterozygous or homozygous. With complete recessiveness, the allele is phenotypically expressed only when the genotype is homozygous recessive; the recessive allele has no effect on the phenotype of the heterozygote. Complete dominance and complete recessiveness are two extremes between which all transitional degrees of dominance are possible. In incomplete dominance, the phenotype of the heterozygote is intermediate to those of the two homozygotes, while in codominance the heterozygote exhibits the phenotypes of both homozygotes.

GENE INTERACTIONS AND MODIFIED MENDELIAN RATIOS

No gene acts by itself in determining an individual's phenotype; the phenotype is the result of highly complex and integrated patterns of molecular reactions that are under direct gene control. In a number of cases complex interactions between genes can be detected by genetic analysis. We discuss some examples in this section.

Consider two independently assorting gene pairs, each with two alleles: *A* and *a*, and *B* and *b*. The outcome of a cross between individuals, each of whom is doubly heterozygous ($A/a\ B/b \times A/a\ B/b$), will be nine genotypes in the following proportions:

$$1/16\ A/A\ B/B$$
$$2/16\ A/A\ B/b$$
$$1/16\ A/A\ b/b$$
$$2/16\ A/a\ B/B$$
$$4/16\ A/a\ B/b$$
$$2/16\ A/a\ b/b$$
$$1/16\ a/a\ B/B$$
$$2/16\ a/a\ B/b$$
$$1/16\ a/a\ b/b$$

If the phenotypes determined by the two allelic pairs are distinct—for example, smooth versus wrinkled peas, long versus short stems—and there is complete dominance, then we get the familiar dihybrid phenotypic ratio of 9:3:3:1 (see Figure 2.13b). Any alteration in this standard 9:3:3:1 ratio indicates that the phenotype is the product of the interaction of two or more genes.

As we introduced in Chapter 2, the 9:3:3:1 phenotypic ratio can be represented genotypically in a shorthand way as $A/-\ B/-$, $A/-\ b/b$, $a/a\ B/-$, $a/a\ b/b$, respectively. The dash indicates that the phenotype is the same whether the gene is homozygous dominant or heterozygous (for example, $A/-$ means either A/A or A/a). This system cannot be used when incomplete dominance or codominance is involved.

The following sections discuss the main processes that result in modified Mendelian ratios. The first section describes examples of interactions between different genes that control the same general phenotypic attribute. In the second section we discuss examples of interactions of different genes in which an allele of one gene masks the expression of alleles of another gene. This second type of interaction is called

epistasis. In both cases the discussions are confined to dihybrid crosses in which the two pairs of alleles assort independently. In the "real world" there are many more complex examples of gene interactions involving more than two pairs of alleles and/or genes that do not assort independently.

Gene Interactions That Produce New Phenotypes

If the two allelic pairs in a dihybrid cross affect the same phenotypic characteristic, there is a chance for interaction of their gene products to give novel phenotypes, and the result may or may not be modified phenotypic ratios, depending on the particular interaction between the products of the nonallelic genes.

COMB SHAPE IN CHICKENS. In chickens, new comb-shape phenotypes (Figure 4.4) result from interactions between the alleles of two gene loci. Each of the four comb types can be bred true if the alleles involved are homozygous.

Crosses made between true-breeding rose-combed and single-combed varieties show that rose is

~ **FIGURE 4.4**

Four distinct comb shape phenotypes in chickens, resulting from all possible combinations of a dominant and a recessive allele at each of two gene loci. (a) Rose comb ($R/-\ p/p$); (b) Walnut comb ($R/-\ P/-$); (c) Pea comb ($r/r\ P/-$); (d) Single comb ($r/r\ p/p$).

a) Rose comb

b) Walnut comb

c) Pea comb

d) Single comb

completely dominant over single. When the F_1 rose-combed birds are interbred, a ratio of 3 rose : 1 single results in the F_2. Similarly, pea comb is completely dominant over single, with a 3 pea : 1 single ratio in the F_2. When true-breeding rose and pea varieties are crossed, however, the result is new and interesting (Figure 4.5a). Instead of showing either rose or pea combs, all birds in the F_1 show a new comb form, which is called walnut comb because it resembles half a walnut meat.

When the F_1 walnut-combed birds are interbred, not only do walnut-, rose-, and pea-combed birds appear, but so do single-combed birds (Figure 4.5b). These four comb types occur in a ratio of 9 walnut : 3 rose : 3 pea : 1 single. Such a ratio is characteristic in F_2 progeny from two parents each heterozygous for two genes. The doubly dominant class in the F_2 is walnut, while the proportion of the single-combed birds indicates that this class contains both recessive alleles.

The explanation of these results is as follows (see Figure 4.5b): The walnut comb depends on the presence of two dominant alleles, R and P, both located at two independently assorting gene loci. In $R/-$ p/p birds, a rose comb results; in r/r $P/-$ birds, a pea comb results; and in r/r p/p birds, a single comb results. Thus, it is the *interaction* of two dominant alleles, each of which individually produces a different phenotype, that produces a new phenotype.

FRUIT SHAPE IN SUMMER SQUASH (THE 9:6:1 RATIO). Two of the many varieties of summer squash have long fruit and sphere-shaped fruit. The long-fruit varieties are always true breeding. However, in some crosses between different varieties of true-breeding sphere-shaped plants, the F_1 fruit is disk-shaped (Figure 4.6). In such instances, the F_2 fruit are approximately 9/16 disk-shaped, 6/16 sphere-shaped, and 1/16 long-shaped; this is a modification of the typical Mendelian ratio, so two genes are involved. The explanation is as follows: a dominant allele of either gene and homozygosity for the recessive allele of the other gene ($A/-$ b/b or a/a $B/-$) results in the same phenotype—spherical fruit. Two dominant alleles, one of each gene ($A/-$ $B/-$), interact

---

~ FIGURE 4.5

Complete dominance in chickens. The genetic crosses show the interaction of genes for comb shape. (a) The cross of a true-breeding rose-combed bird with a true-breeding pea-combed bird gives all walnut-combed offspring in the F_1. (b) When the F_1 birds are interbred, a 9:3:3:1 ratio of walnut:rose:pea:single occurs in the F_2.

a)
P generation

	Parent 1 ♀	Parent 2 ♂
Parental phenotype	Rose comb	Pea comb
Diploid parental genotype	R/R p/p	r/r P/P
Haploid gametes	R p	r P

Parent 2 ♂ gametes

F_1 generation

r P

Parent 1 ♀ gametes R p

R/r P/p
Walnut

F_1 genotypes: All R/r P/p

F_1 phenotypes: All walnut comb

b)
F_2 generation

F_2 phenotypic ratio for $R/r \times R/r$	F_2 phenotypic ratio for $P/p \times P/p$	Combined F_2 ratios	Expected F_2 phenotypic proportions
$3/4$ $R/-$	$3/4$ $P/-$	$9/16$ $R/-$ $P/-$	$9/16$ walnut
	$1/4$ p/p	$3/16$ $R/-$ p/p	$3/16$ rose
$1/4$ r/r	$3/4$ $P/-$	$3/16$ r/r $P/-$	$3/16$ pea
	$1/4$ p/p	$1/16$ r/r p/p	$1/16$ single

~ **FIGURE 4.6**

Generation of an F$_2$ 9:6:1 ratio for fruit shape in summer squash.

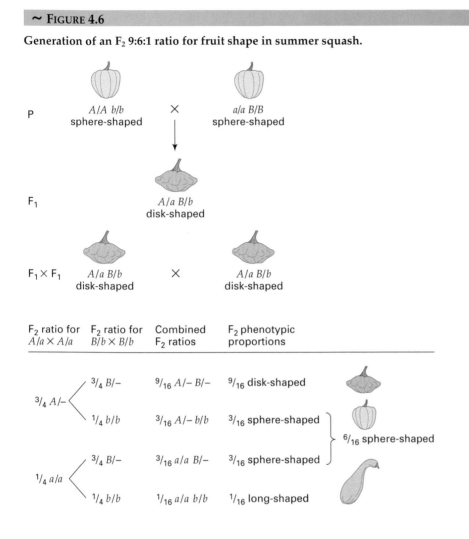

together to produce a new phenotype: disk-shaped fruit. The doubly homozygous recessive (*a/a b/b*) gives a long fruit shape. Thus, the cross of two sphere-shaped squashes here is *A/A b/b* and *a/a B/B*. The F$_1$s are disk-shaped and genotypically *A/a B/b*. The F$_2$ disk-shaped fruits are *A/– B/–*, the spherical fruits are *A/– b/b* or *a/a B/–*, and the long-shaped fruits are *a/a b/b*, giving the 9:6:1 ratio.

Epistasis

Epistasis is a form of gene interaction in which one gene masks the phenotypic expression of another. No new phenotypes are produced by this type of gene interaction. A gene that masks another gene's expression is said to be epistatic, and a gene whose expression is masked is said to be hypostatic. If we think about the F$_2$ genotypes *A/– B/–, A/– b/b, a/a B/–*, and *a/a b/b*, epistasis may be caused by the presence of homozygous recessives of one gene pair, so that *a/a* masks the effect of the *B* allele. Or epistasis may

result from the presence of one dominant allele in a gene pair. For example, the *A* allele might mask the effect of the *B* allele. Epistasis can also occur in both directions between two gene pairs. All these possibilities produce quite a number of modifications of the 9:3:3:1 ratio in a dihybrid cross.

RECESSIVE EPISTASIS: COAT COLOR IN RODENTS (THE 9:3:4 RATIO).   In *recessive epistasis, a/a B/–* and *a/a b/b* individuals have the same phenotype, so the phenotypic ratio in the F$_2$ is 9:3:4 rather than 9:3:3:1. An example is coat color in rodents. Wild mice have a greyish color due to the presence of black and yellow banded hairs in the fur. This coloration, the agouti pattern, aids in camouflage and is found in many wild rodents, including guinea pigs, grey squirrels, and wild mice.

Several other coat colors are seen in domesticated rodents. Albinos, for example, have no pigment in the fur or in the irises of the eyes, so they have a white coat and pink eyes. Albinos are true breeding, and this variation behaves as a complete recessive to any other color. Another variant has black coat color as the result of the absence of the yellow pigment found in the agouti pattern. Black is recessive to agouti.

When true-breeding agouti mice are crossed with albinos, the F_1 progeny are all agouti, and when these F_1 agoutis are interbred, the F_2 progeny consist of approximately 9/16 agouti animals, 3/16 black, and 4/16 albino (Figure 4.7). This pattern occurs because the parents differ in a gene necessary for the development of any color, C, and in a gene for the agouti pattern, A. Phenotypically, $A/-\ C/-$ are agouti, $a/a\ C/-$ are black, and $A/-\ c/c$ and $a/a\ c/c$ are albino, giving a 9:3:4 phenotypic ratio of agouti:black:albino. Thus, this example demonstrates epistasis of c/c over $A/-$. In other words, white hairs are produced in c/c mice, regardless of the genotype at the other locus.

DUPLICATE RECESSIVE EPISTASIS: FLOWER COLOR IN SWEET PEAS (THE 9:7 RATIO).

In the sweet pea, purple flower color is dominant to white and gives a typical 3:1 ratio in the F_2. White-flowered varieties of sweet peas breed true, and crosses between different white varieties usually produce white-flowered progeny. In some cases, however, crosses of two true-breeding white varieties give only purple-flowered F_1 plants. When these F_1 hybrids are self-fertilized, they produce an F_2 consisting of about 9/16 purple-flowered sweet peas and 7/16 white-flowered (Figure 4.8). The 9:7 ratio is a modification of the 9:3:3:1 ratio. Even though they are not all homozygous for the alleles in question, all the F_2 white-flowered plants breed true when self-fertilized. One-ninth of the purple-flowered F_2 plants—the $C/C\ P/P$ genotypes—breed true.

These results may be explained by the interaction of two genes. The 9/16 purple-flowered F_2 plants suggests that colored flowers appear only when two independent dominant alleles are present together, and that the color purple results from some interaction between them. White flower color would then result from homozygosity for the recessive allele of one or both genes. Thus, gene pair C/c specifies whether or not the flower can be colored, and gene pair P/p specifies whether or not purple flower color will result. An interaction of two genes to give

---

~ FIGURE 4.7

Recessive epistasis: generation of an F_2 9:3:4 ratio for coat color in rodents.

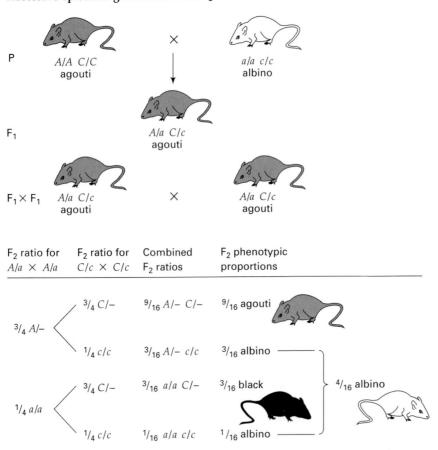

~ **FIGURE 4.8**

Duplicate recessive epistasis: generation of an F₂ 9:7 ratio for flower color in sweet peas.

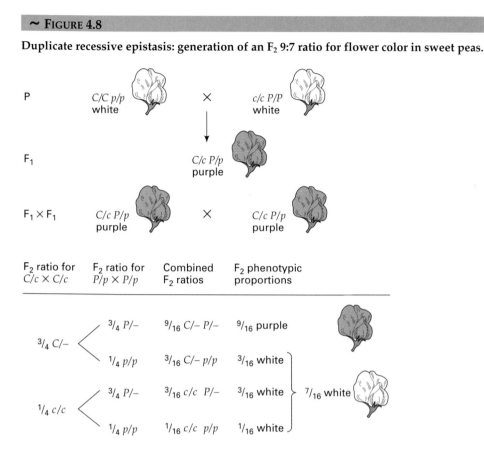

F₂ ratio for $C/c \times C/c$	F₂ ratio for $P/p \times P/p$	Combined F₂ ratios	F₂ phenotypic proportions

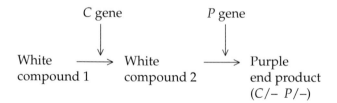

rise to a specific product is a form of epistasis called *duplicate recessive epistasis* or *complementary gene action*.

The following purely theoretical pathway—undoubtedly too simplistic for the actual phenotypes—can be envisioned for the production of the purple pigment:

C gene   P gene

White ──────→ White ──────→ Purple
compound 1   compound 2   end product
                                    (C/– P/–)

In this pathway, a colorless precursor compound (not shown here) is converted in a series of steps to a purple end product. Each step is controlled by the product of a gene. To explain the F₂ ratio, we can propose that gene *C* controls the conversion of white compound 1 to white compound 2, and that gene *P* controls the conversion of white compound 2 to the purple end product. Therefore, homozygosity for the recessive allele of either or both of the *C* and *P* genes will result in a block in the pathway, and only white pigment will accumulate. That is, *C/– p/p, c/c P/–*,

and *c/c p/p* genotypes will all be white. The only plants that will produce purple flowers are those in which both steps of the pathway are completed so that the colored pigment is produced. This situation occurs only in *C/– P/–* plants.

In the cross described (see Figure 4.8), the two white parentals were *C/C p/p* and *c/c P/P*, and the F₁ plants were purple and doubly heterozygous. Interbreeding the F₁ gives a 9:7 ratio of purple:white in the F₂. Recessive epistasis occurs in both directions between two gene pairs. The consequence of this gene interaction is that the same phenotype (white) is exhibited whenever one or the other gene pair is homozygous recessive.

In sum, many types of phenotypic modifications are possible as a result of interactions between the products of different gene pairs. Geneticists detect such interactions when they observe deviations from the expected phenotypic ratios in crosses. We have discussed some examples in which two genes assort independently and in which complete dominance is exhibited in each allelic pair. The ratios we discussed would necessarily be modified further if the genes did not assort independently and/or if incomplete dominance or codominance prevailed. Table 4.3 shows examples of epistatic F₂ phenotypic ratios from an *A/a B/b × A/a B/b* cross.

~ TABLE 4.3

Examples of Epistatic F_2 Phenotypic Ratios from an $A/a\ B/b \times A/a\ B/b$ in Which Complete Dominance Is Shown for Each Gene Pair (From *Science of Genetics*, 6th ed. by George W. Burns and Paul J. Bottino. Copyright © 1989. Reprinted by permission of Prentice-Hall, Inc., Upper Saddle River, NJ.)

		A/A B/B	A/A B/b	A/a B/B	A/a B/b	A/A b/b	A/a b/b	a/a B/B	a/a B/b	a/a b/b
More than four phenotypic classes	A and B both incompletely dominant	1	2	2	4	1	2	1	2	1
	A incompletely B completely dominant	3		6		1	2	3		1
Four phenotypic classes	A and B both completely dominant (classic ratio)	9				3		3		1
Fewer than four phenotypic classes	a/a epistatic to B and b; recessive epistasis	9				3		4		
	A epistatic to B and b; dominant epistasis	12						3		1
	A epistatic to B and b; b/b epistatic to A and a; dominant and recessive epistasis	13ᵃ						3		
	a/a epistatic to B and b; b/b epistatic to A and a; duplicate recessive epistasis	9						7		
	A epistatic to B and b; B epistatic to A and a; duplicate dominant epistasis	15								1
	Duplicate interaction	9				6				1

ᵃThe 13 is composed of the 12 classes immediately above plus the one *a/a b/b* from the last column.

KEYNOTE

In many instances, genes interact to determine phenotypic characteristics. In some cases, interaction between genes results in new phenotypes without modification of typical Mendelian ratios. In another type of gene interaction, called epistasis, interaction between genes causes modifications of Mendelian ratios because one gene interferes with the phenotypic expression of another gene (or genes). The phenotype is controlled largely by the former gene and not the latter when both genes occur together in the genotype. The analysis of epistasis is complicated further when one or both allelic pairs involve incomplete dominance or codominance, or when genes do not assort independently.

ESSENTIAL GENES AND LETHAL ALLELES

For a few years after the rediscovery of Mendel's principles, geneticists believed that mutations only changed the appearance of a living organism, but then they discovered that a mutant allele could cause death. In a sense this mutation is still a change in phenotype, with the new phenotype being lethality. An allele that results in the death of an organism is called a **lethal allele**, and the gene involved is called an essential gene. **Essential genes** are genes that, when mutated, can result in a lethal phenotype. If the mutation is due to a **dominant lethal allele**, both homozygotes and heterozygotes for that allele will show the lethal phenotype. If the mutation is due to a **recessive lethal allele**, only homozygotes for that allele will have the lethal phenotype.

An example of an essential gene is the gene for yellow body color in mice. The yellow variety never breeds true. When yellows are bred to nonyellows, the progeny show an approximately 1:1 ratio of yellow:nonyellow mice. (The nonyellow color depends on other coat color genes.) This ratio is expected from the mating of a heterozygote with a recessive, suggesting that yellow mice are heterozygous. When the yellow heterozygotes are interbred, a phenotypic ratio of about 2 yellow : 1 nonyellow is observed instead of the predicted 3:1 ratio.

It turns out that yellow homozygotes are aborted; in other words, the yellow allele has a *dominant* effect with regard to coat color, but it acts as a *recessive* allele with respect to the lethality phenotype, since only homozygotes die. We now know that homozygotes for the yellow allele die at the embryo stage.

The yellow allele is an allele at the agouti locus (*a*) and has been given the symbol A^Y. The yellow × yellow cross is shown in Figure 4.9. Genotypically, the cross is $A^Y/A \times A^Y/A$. We expect a genotypic ratio of $1/4\ A^Y/A^Y:2/4\ A^Y/A:1/4\ A/A$ among the progeny. The $1/4\ A^Y/A^Y$ mice die before birth, giving a birth ratio of $2/3\ A^Y/A$ (yellow) : $1/3\ A/A$ (nonyellow). Because the A^Y allele causes lethality in the homozygous state, it is called a recessive lethal. Characteristically, when two heterozygotes are crossed, recessive lethal alleles are recognized by a 2:1 ratio of progeny types.

Essential genes are found in all diploid organisms. In humans there are many known recessive lethal alleles. One example is *Tay-Sachs disease* (see p. 187). Homozygotes appear normal at birth, but before about one year of age they show symptoms of central nervous system deterioration. Progressive mental retardation, blindness, and loss of neuromuscular control follow. Afflicted children usually die at three to four years of age. The genetic defect in Tay-Sachs results in an enzyme deficiency that prevents proper nerve function.

There are X-linked lethal mutations as well as autosomal lethal mutations, and dominant lethal mutations as well as recessive lethals. In humans, for example, the genetic disease hemophilia is caused by an X-linked recessive allele. Untreated, hemophilia is lethal. Dominant lethals exert their effect in heterozygotes, resulting in the death of the organism usually at conception or at a fairly young age. Dominant lethals cannot be studied genetically unless death occurs after the organism has reached reproductive age. For example, the symptoms of the

~ **FIGURE 4.9**

Inheritance of a lethal gene A^Y in mice. A mating of two yellow mice gives 1/4 nonyellow (black) mice, 1/2 yellow mice, and 1/4 dead embryos. The viable yellow mice are heterozygous A^Y/A, and the dead individuals are homozygous A^Y/A^Y.

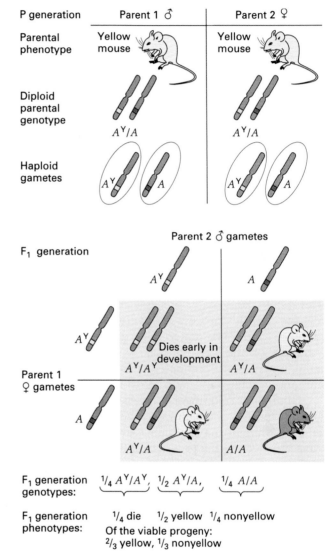

autosomal dominant trait *Huntington disease*—involuntary movements and progressive central nervous system degeneration—may not begin until affected individuals reach their early 30s; as a result, parents may unknowingly pass on the gene to their offspring. Death usually occurs when the afflicted persons are in their 40s or 50s. The well-known American folk singer Woody Guthrie died from Huntington disease.

KEYNOTE

An allele that is fatal to the individual is called a lethal allele. Recessive lethal and dominant lethal alleles exist, and they can be sex-linked or autosomal. The existence of lethal alleles of a gene indicates that the gene's normal product is essential for the function of the organism; the gene is called an essential gene.

THE ENVIRONMENT AND GENE EXPRESSION

The *development* of a multicellular organism from a zygote is a process of *regulated growth and differentiation* that results from the interaction of the organism's genome with the environment, both the internal cellular environment and the external environment. Development is a tightly controlled, programmed series of phenotypic changes that, under normal environmental conditions, is essentially irreversible. Four major processes interact to constitute the complex process of development: (1) replication of the genetic material, (2) growth, (3) differentiation of the various cell types, and (4) the arrangement of differentiated cells into defined tissues and organs.

Think of development as a series of intertwined, complex biochemical pathways. The internal or external environment may influence any of these pathways by affecting the products of the genes controlling the pathways. This phenomenon is most readily studied in experimental organisms where the genotype is unequivocally known. The extent to which the gene manifests its effects under varying environmental conditions can then be seen. We consider some examples in the following section.

Penetrance and Expressivity

In some cases, not all individuals with a particular genotype show the expected phenotype. The frequency with which a dominant or homozygous recessive gene manifests itself in individuals in a population is called the **penetrance** of the gene. Penetrance depends on both the genotype and the environment. Penetrance is complete (100 percent) when all the homozygous recessives show one phenotype, when all the homozygous dominants show another phenotype, and when all the heterozygotes are alike. For example, if all individuals carrying a dominant mutant allele show the mutant phenotype, the allele is completely penetrant. Many genes show complete penetrance; the seven gene pairs in Mendel's experiments and the alleles in the human ABO blood group system are examples.

If less than 100 percent of the individuals with a particular genotype exhibit the phenotype expected, penetrance is incomplete. For example, an organism may be genotypically $A/-$ or a/a but may not display the phenotype typically associated with that genotype. If, say, 80 percent of the individuals carrying a particular gene show the corresponding phenotype, we say that there is 80 percent penetrance.

In humans, for example, brachydactyly, an autosomal dominant trait that causes shortened and malformed fingers, shows 50 to 80 percent penetrance. Individuals with neurofibromatosis, an autosomal dominant trait, develop tumorlike growths (neurofibromas) over the body. This genetic disease also shows 50 to 80 percent penetrance.

Genes may influence a phenotype to different degrees. **Expressivity** refers to the degree to which a penetrant gene or genotype is phenotypically expressed in an individual. Expressivity may be described either qualitatively or quantitatively; for example, expressivity may be strong, intermediate, or slight. Like penetrance, expressivity depends on both the genotype and the environment, and it may be constant or variable. An example of variation in expressivity is found in the human condition called osteogenesis imperfecta. The three main features of this disease are blueness of the sclerae (the whites of the eyes), very fragile bones, and deafness. Osteogenesis imperfecta is inherited as an autosomal dominant with almost 100 percent penetrance. However, the trait shows variable expressivity: A person with the gene may have any one or any combination of the three traits. Moreover, the fragility of the bones for those who exhibit this condition is also very variable.

Variable expressivity is also seen in neurofibromatosis (Figure 4.10). In its mildest form, the disease causes individuals to have only a few pigmented areas on the skin (called *café-au-lait spots* because they are the color of coffee with milk). In more severe cases, one or more other symptoms may be seen, including neurofibromas of various sizes; high blood pressure; speech impediments; headaches; large head; short stature; tumors of the eye, brain, or spinal cord; and curvature of the spine. Therefore, we see that in medical genetics it is important to recognize that a gene may vary widely in its expression, a qualification that makes the task of genetic counseling that much more difficult.

~ FIGURE 4.10

Variable expressivity in individuals with neurofibromatosis. Top: Café-au-lait spot. Middle: Café-au-lait spot and freckling. Bottom: Large number of cutaneous neurofibromas (tumorlike growths).

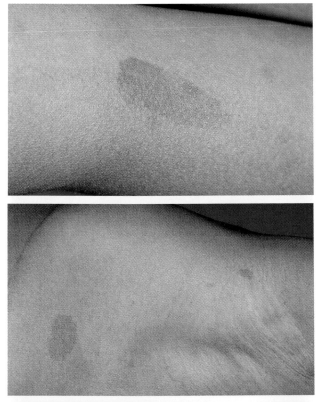

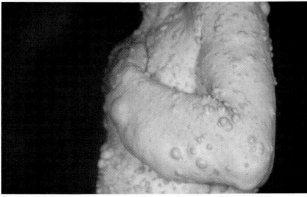

KEYNOTE

Penetrance is the frequency within the population with which a dominant or homozygous recessive allele manifests itself in the phenotype of an individual. Expressivity is the type or degree of phenotypic manifestation of a penetrant allele or genotype in a particular individual.

Effects of the Environment

The phenotypic expression of a gene depends on a number of factors, including the influences of the environment. Next we consider some examples of environmental influences on gene expression.

AGE OF ONSET. The age of the organism creates internal environmental changes that can affect gene function. All genes do not function continually; instead, over time, programmed activation and deactivation of genes occurs as the organism develops and functions. Numerous age-dependent genetic traits occur in humans; pattern baldness (the bald spot creeping forward from the crown of the head) appears in males between 20 and 30 years of age, and Duchenne severe muscular dystrophy appears in children between 2 and 5 years of age. In most cases the nature of the age dependency is not understood.

SEX. The expression of particular genes may be influenced by the sex of the individual. In the case of sex-linked genes, as mentioned earlier, differences in the phenotypes of the two sexes are related to different complements of genes on the sex chromosomes. However, in some cases genes that are on autosomes affect a particular character that appears in one sex but not the other. Traits of this kind are called **sex-limited traits**.

Examples of sex-limited traits in animals are milk yield in dairy cattle (the genes involved obviously operate in females but not in males), the appearance of horns in certain species of sheep (males with genes for horns have horns, and females with genes for horns do not have horns), and the ability to produce eggs or sperm. An example in humans is the distribution of facial hair.

A slightly different situation is found in **sex-influenced traits**, which, like sex-limited traits, are often controlled by autosomal genes. Such traits appear in both sexes, but either the frequency of occurrence in the two sexes is different or the relationship between genotype and phenotype is different.

Pattern baldness is an example of a sex-influenced trait in humans.[1] Pattern baldness is controlled by an autosomal gene that acts as a dominant in males and as a recessive (or at least it is expressed at lower levels) in females. Pattern

[1]Baldness is not a straightforward trait to study. One reason is that there is variable expressivity in the baldness phenotype: baldness may appear first in the crown or on the forehead, and the degree of baldness varies from minimal to extreme. Moreover, a number of genes affect the presence of hair on the head, including the pattern baldness gene, and the final phenotype is the result of the interaction between the environment and the particular set of baldness genes present.

baldness begins in women much later in life than in men because of hormonal influences.

Other human examples of sex-influenced traits are cleft lip and palate (incomplete fusion of the upper lip and palate), in which there is a 2:1 ratio of the trait in males:females; clubfoot (2:1 ratio); gout (8:1 ratio); rheumatoid arthritis (1:3 ratio); osteoporosis (1:3 ratio); and systemic lupus erythematosus (an autoimmune disease—1:9 ratio).

TEMPERATURE.   Temperature may have a significant effect on gene expression, as fur color in Himalayan rabbits illustrates. Certain genotypes of this white rabbit cause dark fur to develop at the ears, nose, and paws, where the local surface temperature is lower (Figure 4.11a). (A similar situation applies to Siamese cats.) However, when a rabbit is reared at a temperature above 30°C, all its fur, including that of the ears, nose, and paws, is white (Figure 4.11b). If a rabbit is raised at 25°C, the typical Himalayan phenotype results (Figure 4.11c). Lastly, if a rabbit is raised at 25°C while part of its body is artificially cooled to a temperature below 25°C, the rabbit develops the Himalayan coat phenotype and exhibits an additional patch of dark fur on the cooled area (Figure 4.11d).

CHEMICALS.   Certain chemicals can have significant effects on an organism, as the following two examples show.

Phenylketonuria.   The human disease phenylketonuria (PKU) is an autosomal recessive trait. In individuals homozygous for the recessive allele, various symptoms appear, most notably mental retardation at an early age. The cause of the phenotypes is a defective gene product used in a biochemical pathway for the metabolism of phenylalanine, an amino acid. (An amino acid is a building block of a protein.) The diet determines how severe the symptoms of PKU will be. Problem foods include protein containing phenylalanine, such as the protein of mother's milk. PKU can be diagnosed soon after birth by a simple blood test and then treated by restricting the amount of phenylalanine in the diet. Testing for PKU after birth is mandatory in the United States. PKU is also discussed in Chapter 8.

Phenocopies Induced by Chemicals.   When a developing embryo is exposed to certain drugs, chemicals, and viruses, these agents may produce a *phenocopy* (phenotypic copy). A **phenocopy** is defined as a nonhereditary phenotypic modification, caused by special environmental conditions, that mimics a similar phenotype caused by a known gene mutation. In other words, although the individual expresses a mutant phenotype, the genotype is *normal*. There are many

~ **FIGURE 4.11**

Effect of temperature on gene expression. (a) Himalayan rabbit; (b) White extremities result when Himalayan rabbits are reared at above 30°C; (c) Normal Himalayan pattern when reared at 25°C; (d) Normal Himalayan pattern when rabbit reared at 25°C, with a dark patch on the side where the flank was cooled to below 25°C.

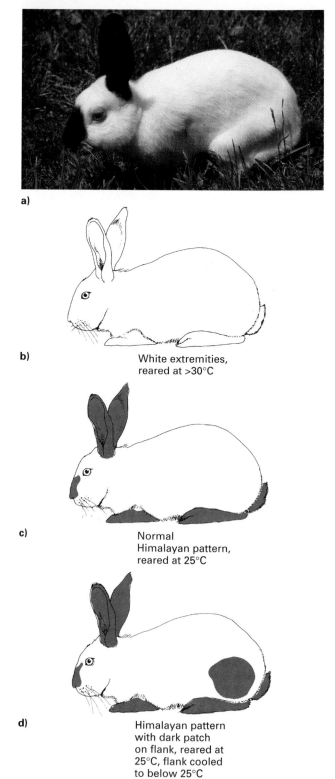

a)

b)   White extremities, reared at >30°C

c)   Normal Himalayan pattern, reared at 25°C

d)   Himalayan pattern with dark patch on flank, reared at 25°C, flank cooled to below 25°C

examples of phenocopies, and, in some instances, studies of phenocopies have provided useful information about the actual molecular defects caused by the phenocopies' mutant counterparts.

In humans, cataracts, deafness, and heart defects are sometimes produced when an individual is homozygous for rare recessive alleles; these disorders may also result if the mother is infected with rubella (German measles) virus during the first 12 weeks of pregnancy. Another human trait for which there is a phenocopy is *phocomelia*, a suppression of the development of the long bones of the limbs, which is caused by a rare dominant allele with variable expressivity. Between 1959 and 1961, similar phenotypes were produced by the sedative thalidomide when it was taken by expectant mothers between the 35th and 50th days of gestation. The drug was removed from the market when its devastating effects were discovered.

KEYNOTE

The phenotypic expression of a gene depends on several factors, including its dominance relationships, the genetic constitution of the rest of the genome (for example, the presence of epistatic or modifier genes), and the influences of the environment. In some cases, special environmental conditions can cause a phenocopy, a nonhereditary and phenotypic modification that mimics a similar phenotype caused by a gene mutation.

Nature Versus Nurture

We are left with the nature–nurture question; that is, what are the relative contributions of genes and environment to the phenotype? (Nature–nurture is discussed more in Chapter 22.) Up to this point, variation in most of the traits we have examined has been determined largely by differences in genotype; that is, phenotypic differences have reflected genetic differences. However, we have already seen that the phenotypes of many traits are influenced by both genes and environment. Let us consider the nature-nurture issue in the context of some human examples.

Human height, or stature, is definitely influenced by genes. On the average, tall parents tend to have tall offspring, and short parents tend to have short offspring. But the environment also plays a role in determining height. For example, human height has increased about 1 inch per generation over the past 100 years as a result of better diets and improved health care.

For a trait such as height, genes set certain limits (or specify potential) for the phenotype. The phenotype an individual develops within these limits depends on the environment. The range of potential phenotypes that a single genotype could develop if exposed to a range of environmental conditions is termed the **norm of reaction**. For some genotypes, the norm of reaction is small; that is, the phenotype produced by a genotype is nearly the same in different environments. For other genotypes, the norm of reaction is large, and the phenotype produced by the genotype varies greatly in different environments.

Many human behavioral traits are the result of interaction between genes and the external environment. One example is alcoholism, which is a major health problem in the United States—about 10 million Americans are problem drinkers, and 6 million are severely addicted to alcohol. Numerous studies have shown that alcoholism is influenced by genes. For instance, sons of alcoholic fathers who are separated from their biological parents at birth and adopted into a family with nonalcoholic parents are four times more likely to become alcoholic than sons adopted at birth whose biological fathers were not alcoholic. However, no gene forces a person to drink alcohol. That is, one cannot become an alcoholic unless one is exposed to an environment in which alcohol is available and drinking is encouraged. What genes do is make certain people more or less susceptible to alcohol abuse; they increase or decrease the risk of developing alcoholism. How genes influence our susceptibility to alcohol abuse is not yet clear. They may affect the way we metabolize alcohol, which might affect how much we drink. Or, genes may influence certain of our personality traits, which makes us more or less likely to drink heavily. The important point is that a behavioral trait such as alcoholism may be influenced by genes, but the genes alone do not produce the phenotype.

Nowhere has the role of genes and environment been more controversial than in the study of human intelligence. In the past, people tended to think of human intelligence as *either* genetically preprogrammed *or* produced entirely by the environment. The clash of these opposing views was termed the nature–nurture controversy. Today, geneticists recognize that neither of these extreme views is correct; human intelligence is the product of both genes and environment.

That genes influence human intelligence is clearly evidenced by genetic conditions that produce mental retardation, such as PKU (see p. 90) and Down syndrome (see Chapter 7, pp. 168–170). Numerous studies also indicate that genes influence differences in IQ among nonretarded people. (IQ, or Intelligence Quotient, is a standardized measure of mental age compared to chronological age; it is relatively stable over time.) For example, adoption studies show that the IQ of adopted children is closer to that of their biological parents than to the IQ of their adoptive parents.

However, IQ is also influenced by environment. Identical twins frequently differ in IQ, which can only be explained by environmental differences. Family size, diet, and culture are environmental factors known to affect IQ. Thus, IQ results from the interaction of genes and environment. Consequently, if two people (other than identical twins) differ in IQ, it is impossible to attribute that difference solely to either genes or environment, because both interact in determining the phenotype. So, although we cannot change our genes, we can alter the environment and thus affect a phenotypic trait like intelligence.

KEYNOTE

Variation in most of the genetic traits we considered in our discussion of Mendelian principles is determined predominantly by differences in genotype; that is, phenotypic differences resulted from genotypic differences. For many traits, however, the phenotypes are influenced by both genes and the environment. The debate over the relative contribution of genes and environment to phenotype has been termed the nature versus nurture controversy.

SUMMARY

A variety of exceptions to and extensions of Mendel's principles were discussed in this chapter. These include the following:

1. *Multiple alleles*: A gene may have many allelic forms in a population, and these alleles are called multiple alleles. A diploid individual can have only two different alleles of a given set of multiple alleles.

2. *Modified dominance relationships*: In complete dominance, the same phenotype results whether an allele is heterozygous or homozygous. In in-

complete dominance, the phenotype of the heterozygote is intermediate between those of the two homozygotes. In codominance, the heterozygote shows the phenotypes of both homozygotes.

3. *Gene interactions and modified Mendelian ratios*: In many cases, genes do not function independently in determining phenotypic characteristics. In epistasis, modified Mendelian ratios occur because of interactions of nonallelic genes: the phenotypic expression of one gene depends on the genotype of another gene locus. In other interactions a new phenotype is produced.

4. *Essential genes and lethal alleles*: Alleles of certain genes result in the lack of production of a necessary functional gene product, and this gives rise to a lethal phenotype. Such lethal alleles may be recessive or dominant. The existence of lethal alleles of a gene indicates that the normal product of the gene is essential for the organism.

5. *Penetrance and expressivity*: Penetrance is the condition in which not all individuals who are known to have a particular allele show the phenotype specified by that allele. That is, penetrance is the frequency with which a dominant or homozygous recessive allele manifests itself in individuals in the population. The related phenomenon of expressivity describes the degree to which a penetrant gene or genotype is phenotypically expressed in an individual. Both penetrance and expressivity depend on the genotype and the external environment.

6. *Dual influence of genes and environment on phenotype*: An organism's potential to develop and function is specified by the zygote's genetic constitution. As an organism develops and differentiates, gene expression is influenced by a number of factors, including dominance relationships, the genetic constitution of the rest of the genome, and the influences of the internal and external environment. That is, the phenotypes of many traits are influenced by both genes and environment. The debate over the relative contributions of genes and environment to the phenotype has been termed the nature versus nurture controversy.

ANALYTICAL APPROACHES FOR SOLVING GENETICS PROBLEMS

Q4.1 In snapdragons, red flower color (C^R) is incompletely dominant to white flower color (C^W); the heterozygote has pink flowers. Also, normal broad leaves

(L^B) are incompletely dominant to narrow, grasslike leaves (L^N); the heterozygote has an intermediate leaf breadth. If a red-flowered, narrow-leaved snapdragon is crossed with a white-flowered, broad-leaved one, what will be the phenotypes of the F₁ and F₂ generations, and what will be the frequencies of the different classes?

A4.1 This basic question on gene segregation includes the issue of incomplete dominance. In the case of incomplete dominance, remember that the genotype can be directly determined from the phenotype. Therefore, we do not need to ask whether or not a strain is true breeding because all phenotypes have a different (and therefore known) genotype.

The best approach here is to assign genotypes to the parental snapdragons. Let $C^R/C^R\ L^N/L^N$ represent the red, narrow plant, and $C^W/C^W\ L^B/L^B$ represent the white, broad plant. The F₁ plants from this cross will all be double heterozygotes, $C^R/C^W\ L^B/L^N$. Owing to the incomplete dominance, these plants are pink-flowered and have leaves of intermediate breadth. Interbreeding the F₁s gives the F₂ generation, but it does not have the usual 9:3:3:1 ratio. Instead, there is a different phenotype for each genotype. These genotypes and phenotypes and their relative frequencies are shown in Figure 4.A.

Q4.2 In snapdragons, red flower color is incompletely dominant to white, with the heterozygote being pink; normal flowers are completely dominant to peloric-shaped ones; and tallness is completely dominant to dwarfness. The three gene pairs segregate independently. If a homozygous red, tall, normal-flowered plant is crossed with a homozygous white, dwarf, peloric-flowered one, what proportion of the F₂ will resemble the F₁ in appearance?

A4.2 Let us assign symbols: C^R = red and C^W = white; N = normal flowers and n = peloric; T = tall and t = dwarf. The initial cross, then, becomes $C^R/C^R\ T/T\ N/N \times C^W/C^W\ t/t\ n/n$. From this cross we see that all the F₁ plants are triple heterozygotes with the genotype $C^R/C^W\ T/t\ N/n$ and with the phenotype pink, tall,

FIGURE 4.A

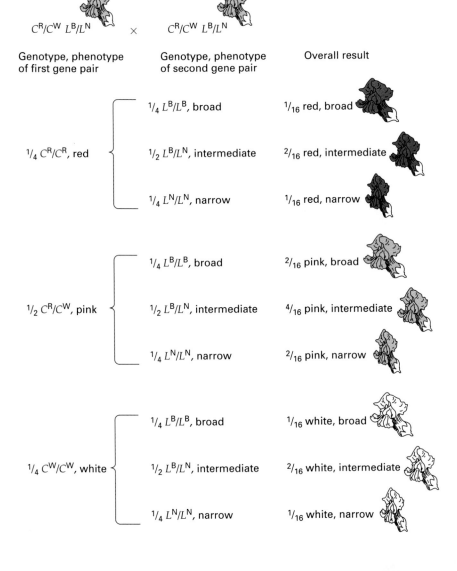

normal-flowered. Interbreeding the F_1 generation will produce 27 different genotypes in the F_2; this answer follows from the rule that the number of genotypes is 3^n, where n is the number of heterozygous gene pairs involved in the $F_1 \times F_1$ cross (see Chapter 2).

Here we are asked specifically for the proportion of F_2 progeny that resemble the F_1 in appearance. We can calculate this proportion directly without needing to display all the possible genotypes and then grouping the progeny in classes according to phenotype. First, we calculate the frequency of pink-flowered plants in the F_2; then we determine the proportion of these plants that have the other two attributes. From a $C^R/C^W \times C^R/C^W$ cross we calculate that half of the progeny will be heterozygous C^R/C^W and therefore pink. Next, we determine the proportion of F_2 plants that are phenotypically like the F_1 with respect to height (tall). Either T/T or T/t plants will be tall, and so 3/4 of the F_2 will be tall. Similarly, 3/4 of the F_2 plants will be normal-flowered like the F_1s. To obtain the probability of all three of these phenotypes occurring together (pink, tall, normal), we must multiply the individual probabilities since the gene pairs segregate independently. The answer is $1/2 \times 3/4 \times 3/4$, or 9/32.

Q4.3

a. An $F_1 \times F_1$ self gives a 9:7 phenotypic ratio in the F_2. What phenotypic ratio would you expect if you test-crossed the F_1?

b. Answer the same question for an $F_1 \times F_1$ cross that gives a 9:3:4 ratio.

c. Answer the same question for a 15:1 ratio.

A4.3 This question deals with epistatic effects. In answering the question, we must consider the interaction between the different genotypes in order to proceed with the testcross. Let us set up the general genotypes that we will deal with throughout. The simplest are allelic pairs a^+ and a, and b^+ and b, where the wild-type alleles are completely dominant to the other member of the pair.

a. A 9:7 ratio in the F_2 implies that both members of the F_1 are double heterozygotes and that epistasis is involved. Essentially, any genotype with a homozygous recessive condition has the same phenotype, so the 3, 3, and 1 parts of a 9:3:3:1 ratio are phenotypically combined into one class. In terms of genotype, 9/16 are $a^+/- b^+/-$ types, and the other 7/16 are $a^+/- b/b, a/a b^+/-$, and $a/a b/b$. (As always, the use of the "–" after a wild-type allele signifies that the same phenotype results, whether the missing allele is a wild type or a mutant.) The testcross asked for is $a^+/a b^+/b \times a/a b/b$, and we can predict a 1:1:1:1 ratio of $a^+/a b^+/b:a^+/a b/b:a/a b^+/b:a/a b/b$. The first genotype will have the same phenotype as the 9/16 class of the F_2, but because of epistasis, the other

three genotypes will have the same phenotype as the 7/16 class of the F_2. In sum, the answer is a phenotypic ratio of 1:3 in the progeny of a testcross of the F_1.

b. We are asked to answer the same question for a 9:3:4 ratio in the F_2. Again, this question involves a modified dihybrid ratio, where two classes of the 9:3:3:1 have the same phenotype. Complete dominance for each of the two gene pairs occurs here also, so the F_1 individuals are $a^+/a b^+/b$. Perhaps both the $a^+/- b^+/b$ and $a/a b/b$ classes in the F_2 will have the same phenotype, while the $a^+/- b^+/-$ and $a/a b^+/-$ classes will have phenotypes distinct from each other and from the interaction class. The genotypic ratio of a testcross of the F_1 is the same as in part (a) of this question. Considering them in the same order as we did in part (a), the second and fourth classes would have the same phenotype, owing to epistasis. So there are only three possible phenotypic classes instead of the four found in the testcross of a dihybrid F_1, where there is complete dominance and no interaction. The phenotypic ratio here is 1:1:2.

c. This question is yet another example of epistasis. Since $15 + 1 = 16$, this number gives the outcome of an F_1 self of a dihybrid where there is complete dominance for each gene pair and interaction between the dominant alleles. In this case, the $a^+/- b^+/-, a^+/- b/b$, and $a/a b^+/-$ classes have one phenotype and include 15/16 of the F_2 progeny, and the $a/a b/b$ class has the other phenotype and 1/16 of the F_2. The genotypic results of a testcross of the F_1 are the same as in parts (a) and (b) of this question; that is, the F_2 progeny exhibit a 1:1:1:1 ratio of $a^+/a b^+/b:a^+/a b/b:a/a b^+/b:a/a b/b$. The first three classes have the same phenotype, which is the same as that of the 15/16 of the F_2s, and the last class has the other phenotype. The answer, then, is a 3:1 phenotypic ratio.

QUESTIONS AND PROBLEMS

4.1 In rabbits, C = agouti coat color, c^{ch} = chinchilla, c^h = Himalayan, and c = albino. The four alleles constitute a multiple allelic series. The agouti C is dominant to the three other alleles, c is recessive to all three other alleles, and chinchilla is dominant to Himalayan. Determine the phenotypes of progeny from the following crosses:

a. $C/C \times c/c$

b. $C/c^{ch} \times C/c$

c. $C/c \times C/c$

d. $C/c^h \times c^h/c$

e. $C/c^h \times c/c$

*4.2   If a given population of diploid organisms contains three, and only three, alleles of a particular gene (say w, $w1$, and $w2$), how many different diploid genotypes are possible in the populations? List all possible genotypes of diploids (consider *only* these three alleles).

4.3   The genetic basis of the ABO blood types seems most likely to be:
a.   multiple alleles.
b.   polyexpressive hemizygotes.
c.   allelically excluded alternates.
d.   three independently assorting genes.

4.4   In humans, the three alleles I^A, I^B, and i constitute a multiple allelic series that determines the ABO blood group system, as we described in this chapter. For the following problems, state whether the child mentioned can actually be produced from the marriage. Explain your answer.
a.   An O child from the marriage of two A individuals.
b.   An O child from the marriage of an A to a B.
c.   An AB child from the marriage of an A to an O.
d.   An O child from the marriage of an AB to an A.
e.   An A child from the marriage of an AB to a B.

4.5   A man is blood type O,M. A woman is blood type A,M and her child is type A,MN. The man cannot be the father of this child because:
a.   O men cannot have type A children.
b.   O men cannot have MN children.
c.   An O man and an A woman cannot have an A child.
d.   An M man and an M woman cannot have an MN child.

*4.6   A woman of blood group AB marries a man of blood group A whose father was group O. What is the probability that
a.   their two children will both be group A?
b.   one child will be group B and the other group O?
c.   the first child will be a son of group AB and the second child a son of group B?

4.7   If a mother and her child belong to blood group O, what blood group could the father *not* belong to?

*4.8   A man of what blood group could *not* be the father of a child of blood type AB?

4.9   In snapdragons, red flower color (C^R) is incompletely dominant to white (C^W); the C^R/C^W heterozygotes are pink. A red-flowered snapdragon is crossed with a white-flowered one. Determine the flower color of (a) the F_1; (b) the F_2; (c) the progeny of a cross of the F_1 to the red parent; (d) the progeny of a cross of the F_1 to the white parent.

*4.10   In Shorthorn cattle, the heterozygous condition of the alleles for red coat color (C^R) and white coat color

(C^W) is roan coat color. If two roan cattle are mated, what proportion of the progeny will resemble their parents in coat color?

4.11   What progeny will a roan Shorthorn have if bred to (a) a red; (b) a roan; (c) a white?

*4.12   In peaches, fuzzy skin (F) is completely dominant to smooth (nectarine) skin (f), and the heterozygous conditions of oval glands at the base of the leaves (G^O) and no glands (G^N) give round glands. A homozygous fuzzy, no-gland peach variety is bred to a smooth, oval-gland variety.
a.   What will be the appearance of the F_1?
b.   What will be the appearance of the F_2?
c.   What will be the appearance of the offspring of a cross of the F_1 back to the smooth, oval-glanded parent?

4.13   In guinea pigs, short hair (L) is dominant to long hair (l), and the heterozygous conditions of yellow coat (C^Y) and white coat (C^W) gives cream coat. A short-haired, cream guinea pig is bred to a long-haired, white guinea pig, and a long-haired, cream baby guinea pig is produced. When the baby grows up, it is bred back to the short-haired, cream parent. What phenotypic classes and in what proportions are expected among the offspring?

4.14   The shape of radishes may be long (S^L/S^L), oval (S^L/S^S), or round (S^S/S^S), and the color of radishes may be red (C^R/C^R), purple (C^R/C^W), or white (C^W/C^W). If a long, red radish plant is crossed with a round, white plant, what will be the appearance of the F_1 and the F_2?

4.15   In poultry, the dominant alleles for rose comb (R) and pea comb (P), if present together, give walnut comb. The recessive alleles of each gene, when present together in a homozygous state, give single comb. What will be the comb characters of the offspring of the following crosses?
a. $R/R\ P/p \times r/r\ P/p$
b. $r/r\ P/P \times R/r\ P/p$
c. $R/r\ p/p \times r/r\ P/p$

4.16   For the following crosses involving the comb character in poultry, determine the genotypes of the two parents:
a.   A walnut crossed with a single produces offspring that are 1/4 walnut, 1/4 rose, 1/4 pea, and 1/4 single.
b.   A rose crossed with a walnut produces offspring that are 3/8 walnut, 3/8 rose, 1/8 pea, and 1/8 single.
c.   A rose crossed with a pea produces five walnut and six rose offspring.
d.   A walnut crossed with a walnut produces one rose, two walnut, and one single offspring.

4.17   In poultry, feathered shanks (*F*) are dominant to clean (*f*), and white plumage of white leghorns (*I*) is dominant to black (*i*). Comb phenotypes and genotypes are given in Figure 4.4.

a.   A feathered-shanked, white, rose-combed bird crossed with a clean-shanked, white, walnut-combed bird produces these offspring: two feathered, white, rose; four clean, white, walnut; three feathered, black, pea; one clean, black, single; one feathered, white, single; two clean, white, rose. What are the genotypes of the parents?

b.   A feathered-shanked, white, walnut-combed bird crossed with a clean-shanked, white, pea-combed bird produces a single offspring that is clean-shanked, black, and single-combed. In additional offspring from this cross, what proportion may be expected to resemble each parent?

*4.18 F_2 plants segregate 9/16 colored : 7/16 colorless. If a colored plant from the F_2 is chosen at random and selfed, what is the probability that there will be *no* segregation of the two phenotypes among its progeny?

*4.19   In peanuts, a plant may be either "bunch" or "runner." Two different strains of peanut, V4 and G2, in which "bunch" occurred were crossed, with the following results:

V4 bunch × V4 bunch
↓
all bunch

G2 bunch × G2 bunch
↓
all bunch

The two true-breeding strains of bunch were crossed in the following way:

V4 bunch × G2 bunch
↓
F_1 runner

$F_1 × F_1$
↓
F_2 9 runner : 7 bunch

What is the genetic basis of the inheritance pattern of runner and bunch in the F_2?

*4.20   In rabbits, one enzyme (the product of a functional gene *A*) is needed to produce a substance required for hearing. Another enzyme (the product of a functional gene *B*) is needed to produce another substance required for hearing. The genes responsible for the two enzymes are not linked. Individuals homozygous for either one or both of the nonfunctional recessive alleles, *a* or *b*, are deaf.

a.   If a large number of matings were made between two double heterozygotes, what phenotypic ratio would be expected in the progeny?

b.   This phenotypic ratio is a result of what well-known phenomenon?

c.   What phenotypic ratio would be expected if rabbits homozygous recessive for trait A and heterozygous for trait B were mated to rabbits heterozygous for both traits?

4.21   In Doodlewags, the dominant allele *S* causes solid coat color; the recessive allele *s* results in white spots on a colored background. The black coat color allele *B* is dominant to the brown allele *b*, but these genes are expressed only in the genotype *a/a*. Individuals that are *A/−* are yellow regardless of *B* alleles. Six pups are produced in a mating between a solid yellow male and a solid brown female. Their phenotypes are: 2 solid black, 1 spotted yellow, 1 spotted black, and 2 solid brown.

a.   What are the genotypes of the male and female parents?

b.   What is the probability that the next pup will be spotted brown?

4.22   The allele *l* in *Drosophila* is recessive and sex-linked, and lethal when homozygous or hemizygous (the condition in the male). If a female of genotype *L/l* is crossed with a normal male, what is the probability that the first two surviving progeny will be males?

*4.23   A locus in mice is involved with pigment production; when parents heterozygous at this locus are mated, 3/4 of the progeny are colored and 1/4 are albino. Another phenotype concerns coat color; when two yellow mice are mated, 2/3 of the progeny are yellow and 1/3 are agouti. The albino mice cannot express whatever alleles they may have at the independently assorting agouti locus.

a.   When yellow mice are crossed with albino, they produce an F_1 consisting of 1/2 albino, 1/3 yellow, and 1/6 agouti. What are the probable genotypes of the parents?

b.   If yellow F_1 mice are crossed among themselves, what phenotypic ratio would you expect among the progeny? What proportion of the yellow progeny produced here would be expected to be true breeding?

4.24   In *Drosophila melanogaster,* a recessive autosomal allele, ebony (*e*), produces a black body color when homozygous, and an independently assorting autosomal allele, black (*b*), also produces a black body color when homozygous. Flies with genotypes *e/e b⁺/−*, *e⁺/− b/b*, and *e/e b/b* are phenotypically identical with respect to body color. Flies with genotype *e⁺/− b⁺/−*

have a grey body color. If true-breeding e/e b^+/b^+ ebony flies are crossed with true-breeding e^+/e^+ b/b black flies,

a. what will be the phenotype of the F_1 flies?

b. what phenotypes and what proportions would occur in the F_2 generation?

c. what phenotypic ratios would you expect to find in the progeny of these backcrosses: (1) $F_1 \times$ true-breeding ebony and (2) $F_1 \times$ true-breeding black?

*4.25 In four-o'clock plants, two genes, Y and R, affect flower color. Neither is completely dominant, and the two interact on each other to produce seven different flower colors:

Y/Y R/R = crimson Y/y R/R = magenta
Y/Y R/r = orange-red Y/y R/r = magenta-rose
Y/Y r/r = yellow Y/y r/r = pale yellow
y/y R/R, y/y R/r, and y/y r/r = white

a. In a cross of a crimson-flowered plant with a white one (y/y r/r), what will be the appearances of the F_1, the F_2, and the offspring of the F_1 backcrossed to the crimson parent?

b. What will be the flower colors in the offspring of a cross of orange-red $\times$ pale yellow?

c. What will be the flower colors in the offspring of a cross of a yellow with a y/y R/r white?

4.26 Two four-o'clock plants were crossed and gave the following offspring: 1/8 crimson, 1/8 orange-red, 1/4 magenta, 1/4 magenta-rose, and 1/4 white. Unfortunately, the person who made the crosses was color-blind and could not record the flower colors of the parents. From the results of the cross, deduce the genotypes and flower colors of the two parents.

*4.27 Genes A, B, and C are independently assorting and control production of a black pigment.

a. Assume that A, B, and C act in a pathway as follows:

$$\text{colorless} \xrightarrow{A} \xrightarrow{B} \xrightarrow{C} \text{black}$$

The alternative alleles that give abnormal functioning of these genes are designated a, b, and c, respectively. A black A/A B/B C/C is crossed with a

colorless a/a b/b c/c to give a black F_1. The F_1 is selfed. What proportion of the F_2 is colorless? (Assume that the products of each step except the last are colorless, so only colorless and black phenotypes are observed.)

b. Assume that C produces an inhibitor that prevents the formation of black by destroying the ability of B to carry out its function, as follows:

$$\text{colorless} \xrightarrow{A} \underset{\underset{C \text{ (inhibitor)}}{\uparrow}}{\xrightarrow{B}} \text{black}$$

A colorless A/A B/B C/C individual is crossed with a colorless a/a b/b c/c, giving a colorless F_1. The F_1 is selfed to give an F_2. What is the ratio of colorless to black in the F_2? (Only colorless and black phenotypes are observed, as in part [a].)

4.28 In *Drosophila*, a mutant strain has plum-colored eyes. A cross between a plum-eyed male and a plum-eyed female gives 2/3 plum-eyed and 1/3 red-eyed (wild-type) progeny flies. A second mutant strain of *Drosophila*, called stubble, has short bristles instead of the normal long bristles. A cross between a stubble female and a stubble male gives 2/3 stubble and 1/3 normal-bristled flies in the offspring. Assuming that the plum gene assorts independently from the stubble gene, what will be the phenotypes and their relative proportions in the progeny of a cross between two plum-eyed, stubble-bristled flies? (Both genes are autosomal.)

*4.29 In sheep, white fleece (W) is dominant over black (w), and horned (H) is dominant over hornless (h) in males but recessive in females. If a homozygous horned white ram is bred to a homozygous hornless black ewe, what will be the appearances of the F_1 and the F_2?

4.30 A horned black ram bred to a hornless white ewe produces the following offspring: Of the males, 1/4 are horned, white; 1/4 are horned, black; 1/4 are hornless, white; and 1/4 are hornless, black. Of the females, 1/2 are hornless and black, and 1/2 are hornless and white. What are the genotypes of the parents?

CHAPTER 5

GENETIC MAPPING IN EUKARYOTES

PRINCIPAL POINTS

~ Genetic recombinants result from physical exchanges between homologous chromosomes in meiosis. A chiasma is the site of crossing-over, and crossing-over is the reciprocal exchange of chromosome parts at corresponding positions along homologous chromosomes by symmetrical breakage and rejoining.

~ Crossing-over in eukaryotes occurs at the four-chromatid stage in prophase I of meiosis.

~ The map distance between two genes is based on the frequency of recombination between the two genes. The recombination frequency is an approximation of the frequency of crossovers between the two genes. As distance between genes increases, the incidence of multiple crossovers causes the recombination frequency to be an underestimate of the crossover frequency and hence of the map distance. Mapping functions may be used to correct for this problem and hence to give a more accurate estimate of map distance.

~ The occurrence of a chiasma between two chromatids may physically impede the occurrence of a second chiasma nearby, a phenomenon called chiasma interference.

~ A major goal of human genetics research is to construct a highly detailed map of the human genome. Two types of maps are being constructed: genetic linkage maps and physical maps. Genetic linkage maps show the relative locations of gene and/or DNA markers as determined by genetic recombination analysis. (*Marker* or *genetic marker* is another name for a mutation that gives a distinguishable phenotype. In other words, it is an allele that "marks" a chromosome or a gene.) Physical maps are maps of physically identifiable regions or markers along the DNA that are constructed without using genetic recombination analysis. The resolution of physical maps depends on the technique being used to construct the map. The markers on physical maps are used as landmarks for specific explorations of the genome and as starting points for sequencing efforts.

~ Tetrad analysis is a mapping technique that can be used to map the genes of certain haploid eukaryotic organisms in which the products of a single meiosis, the meiotic tetrad, are contained within a single structure. In these situations, map distance between genes is computed by analyzing the relative proportion of tetrad types rather than by analyzing individual progeny.

Genes on nonhomologous chromosomes assort independently during meiosis. In many instances, however, certain genes (and hence the phenotypes they control) are inherited together because they are located on the same chromosome. Genes on the same chromosome are said to be *syntenic*. Genes that do not appear to assort independently exhibit **linkage** and are called **linked genes**. Linked genes belong to a *linkage group*.

Genetic analysis is the dissection of the structure and function of the genetic material. In classic genetic analysis, progeny from crosses between parents with different genetic characters are analyzed to determine the frequency with which differing parental alleles are associated in new combinations. Progeny showing the parental combinations of alleles are called *parentals*, and progeny showing non-parental combinations of alleles are called *recombinants*. The process by which the recombinants are produced is called **genetic recombination**. When two genes do not assort independently, they are said to be linked. Through testcrosses we can determine which genes are linked to each other and can

then construct a *linkage map*, or *genetic map*, of each chromosome.

Classic genetic mapping has provided information that is useful in many aspects of genetic analysis. For example, knowing the locations of genes on chromosomes has been useful in recombinant DNA research and in experiments directed toward understanding the DNA sequences in and around genes. These days the focus of mapping studies is on constructing genetic maps of genomes using both gene markers and *DNA markers*. Definitionally, a *marker* or *genetic marker* is another name for a mutation that gives a distinguishable phenotype. In other words, it is an allele that "marks" a chromosome or a gene. Gene markers are alleles of the kind we have discussed to this point in the text. DNA markers are molecular markers, that is, DNA regions in the genome that differ sufficiently between individuals so that they can be detected by molecular analysis of DNA. The goal of these genome mapping studies is to generate high-resolution maps of the chromosomes that will be useful for investigating genes and their functions. The ultimate genetic maps will be of the base-pair sequence of organisms' genomes (see Human Genome Project, Chapter 14).

The goal of this chapter is to learn how genetic linkage affects Mendelian gene segregation patterns and how genes are mapped in eukaryotes.

DISCOVERY OF GENETIC LINKAGE

Our modern understanding of genetic linkage comes from the work of Thomas Hunt Morgan and his colleagues with linkage in *Drosophila melanogaster*, done around 1911.

Morgan's Linkage Experiments with *Drosophila*

By 1911 Morgan had identified a number of X-linked genes, including *w* (white eye) and *m* (miniature wing). Morgan crossed a female white miniature

($w\,m/w\,m$) fly with a wild-type male ($w^+\,m^+/Y$) (Figure 5.1). For the former genotype, the slash signifies the pair of homologous chromosomes and indicates that the genes on either side of the slash are linked. For the latter genotype, since the genes are X-linked, a slash indicates the X chromosome and the Y indicates a Y chromosome. We will also use another special genetic symbolism for genes on the same chromosome: $\dfrac{ab}{ab}$ signifies that genes a and b are on the same chromosome. With this system, X-linked genes in a female are indicated by appropriate allele symbols separated by one or two continuous lines to indicate the homologous chromosomes, for example,

$$\frac{w\,m}{w\,m} \quad \text{or} \quad \frac{w\,m}{w\,m}$$

and X-linked genes in a male are shown as, for example,

$$\underrightarrow{w\,m}$$

~ FIGURE 5.1

Morgan's experimental crosses of white-eye and miniature-wing variants of *Drosophila melanogaster*, showing evidence of linkage and recombination in the X chromosome.

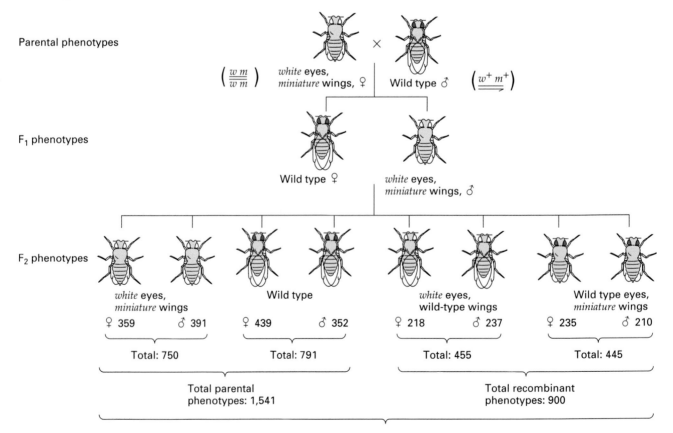

Parental phenotypes

$\left(\dfrac{w\,m}{w\,m}\right)$ *white* eyes, *miniature* wings, ♀   ×   Wild type ♂ $\left(\underrightarrow{w^+\,m^+}\right)$

F₁ phenotypes

Wild type ♀   *white* eyes, *miniature* wings, ♂

F₂ phenotypes

white eyes, *miniature* wings
♀ 359   ♂ 391
Total: 750

Wild type
♀ 439   ♂ 352
Total: 791

white eyes, wild-type wings
♀ 218   ♂ 237
Total: 455

Wild type eyes, *miniature* wings
♀ 235   ♂ 210
Total: 445

Total parental phenotypes: 1,541

Total recombinant phenotypes: 900

Total progeny: 1541 + 900 = 2,441

Percent recombinants: $^{900}/_{2,441} \times 100 = 36.9$

where the straight line indicates the X chromosome and the bent line indicates the Y chromosome. This genotype representation is the same as $w\ m//$ (where the bent slash is the Y chromosome) or $w\ m/Y$. (Note: If a discontinuous line is used between a series of allele pairs, the extent of each segment signifies a different chromosome.)

In the cross the F_1 males were white-eyed and had miniature wings (genotype $w\ m/Y$), while all females were heterozygous and wild type for both eye color and wing size (genotype $w^+\ m^+/w\ m$). The F_1 flies were interbred and 2,441 F_2 flies were analyzed. In crosses of X-linked genes set up as in Figure 5.1, the $F_1 \times F_1$ is equivalent to doing a testcross, since the F_1 males produce X-bearing gametes with recessive alleles of both genes and Y-bearing gametes that have no alleles for the genes being studied. In the F_2, the most frequent phenotypic classes in both sexes were the *grandparental phenotypes* of white eyes plus miniature wings or normal red eyes plus large wings. Conventionally, we refer to the original genotypes of the two chromosomes as **parental genotypes**, **parental classes**, or, more simply, **parentals**. The term is also used to describe phenotypes, so the original white miniature females and wild-type males in these particular crosses are defined as the parentals.

Morgan observed that 900 of the 2,441 F_2 flies, or 36.9 percent, had nonparental phenotypic combinations of white eyes plus normal wings and red eyes plus miniature wings. Nonparental combinations of linked genes are called **recombinants**. Fifty percent recombinant phenotypes was expected if independent assortment was the case; thus, the lower percentage observed is evidence for linkage of the two genes. To explain the recombinants, Morgan proposed that,

in meiosis, exchanges of genes had occurred between the two X chromosomes of the F_1 females.

Morgan's group analyzed a large number of other crosses of this type. *In each case the parental phenotypic classes were the most frequent, while the recombinant classes occurred much less frequently.* Approximately equal numbers of each of the two parental classes were obtained, and similar results were obtained for the recombinant classes. Morgan's general conclusion was that *during meiosis, alleles of some genes assort together because they lie near each other on the same chromosome.* To turn this around, the closer two genes are on the chromosome, the more likely they are to remain together during meiosis. This is because the recombinants are produced as a result of crossing-over between homologous chromosomes during meiosis, and the closer two genes are together the less likely there will be a recombination event between them.

The terminology relating to the physical exchange of homologous chromosome parts can be confusing. To clarify:

1. A chiasma (plural, *chiasmata*; see Figure 1.17) is the place on a homologous pair of chromosomes at which a physical exchange is occurring; it is the site of crossing-over.

2. Crossing-over is the actual process of reciprocal exchange of chromatid segments at corresponding positions along homologous chromosomes; the process involves symmetrical breakage of two chromatids and rejoining.

3. Crossing-over is also defined as the events leading to genetic recombination between linked genes in both prokaryotes and eukaryotes.

Figure 5.2 presents a very simplified diagram of the process of crossing-over. Crossing-over occurs at

---

~ FIGURE 5.2

Mechanism of crossing-over. A highly simplified diagram of a crossover between two nonsister chromatids during meiotic prophase, giving rise to recombinant (nonparental) combinations of linked genes.

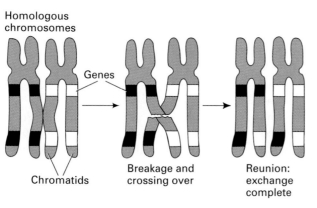

the four-chromatid stage in prophase I of meiosis. Each crossover involves two of the four chromatids and, along the length of a chromosome, all chromatids can be involved in crossing-over.

KEYNOTE

The production of genetic recombinants results from physical exchanges between homologous chromosomes during meiotic prophase I. A chiasma is the site of crossing-over. Crossing-over is the reciprocal exchange of chromatid parts at corresponding positions along homologous chromosomes by symmetrical breakage and rejoining. Crossing-over is also used to describe the events leading to genetic recombination between linked genes. Crossing-over in eukaryotes takes place at the four-chromatid stage in prophase I of meiosis.

CONSTRUCTING GENETIC MAPS

We have learned that the number of genetic recombinants produced is characteristic of the two linked genes involved. We now examine how genetic experiments can be used in **genetic mapping**, the process of determining the relative positions of genes on chromosomes.

Detecting Linkage Through Testcrosses

Before beginning experiments to construct a **genetic map** of the relative positions of genes on a chromosome (also called a **linkage map**), geneticists must show that the genes under consideration are linked. Unlinked genes assort independently. Therefore, a way to test for linkage is to analyze the results of crosses to see whether the data deviate significantly from those expected by independent assortment.

The best cross to use to test for linkage is the testcross, a cross of an individual with another individual homozygous recessive for all genes involved. We saw in Chapter 2 that a testcross between $a^+/a\ b^+/b$ and $a/a\ b/b$, where genes a and b are unlinked, gives a 1:1:1:1 ratio of the four possible phenotypic classes $a^+\ b^+ : a^+\ b : a\ b^+ : a\ b$. A significant deviation from this ratio in the direction of too many parental types and too few recombinant types would suggest that the two genes are linked. It is important to know how large a deviation must be in order to be considered "significant." The *chi-square test* can be used to make such a decision (see Chapter 2). Here we illustrate the use of the chi-square test for analyzing testcross data.

Consider data from a testcross involving fruit flies. In *Drosophila,* b is a recessive autosomal mutation that results in black body color, and *vg* is a recessive autosomal mutation that results in vestigial (short, crumpled) wings. Wild-type flies have grey bodies and long, uncrumpled (normal) wings. True-breeding black, normal ($b/b\ vg^+/vg^+$) flies were crossed with true-breeding grey, vestigial ($b^+/b^+\ vg/vg$) flies. F_1 grey, normal ($b^+/b\ vg^+/vg$) female flies were testcrossed to black, vestigial ($b/b\ vg/vg$) male flies. (The female is the heterozygote in this testcross because, in *Drosophila,* no crossing-over occurs between *any* homologous pair of chromosomes in males.) The testcross progeny data were:

283 grey, normal

1,294 grey, vestigial

1,418 black, normal

241 black, vestigial

Total   3,236 flies

We hypothesize that the two genes are unlinked (null hypothesis) and use the chi-square test to test the hypothesis, as shown in Table 5.1. We use a null hypothesis because the hypothesis must be testable; that is, we must be able to make exact predictions. An hypothesis that "two genes are linked" is not testable, because we cannot predict exactly what the progeny ratios would be.

If the two genes are unlinked, then a testcross should result in a 1:1:1:1 ratio of the four phenotypic classes. Column 1 lists the four phenotypes expected in the progeny of the cross, column 2 lists the observed (*o*) numbers for each phenotype, and column 3 lists the expected (*e*) number for each phenotypic class, given the total number of progeny (3,236) and the hypothesis being tested (1:1:1:1 in this case). Column 4 lists the deviation value (*d*) calculated by subtracting the expected number (*e*) from the observed number (*o*) for each class. The sum of the *d* values is always zero.

Column 5 lists the deviation squared (d^2), and column 6 lists the deviation squared divided by the expected number (d^2/e). The chi-square value, χ^2 (item 7 in the table), is given by the formula

$$\chi^2 = \Sigma \frac{d^2}{e}, \text{ where } d^2 = (o - e)^2$$

and where Σ means "sum of."

In Table 5.1, χ^2 is the sum of the four values in column 6. In our example, χ^2 is 1,489.99, rounded to 1,490. The last value in the table, item 8, is the degrees of freedom (df) for the set of data; there are $n - 1 = 3$ degrees of freedom in this case.

~ TABLE 5.1

Chi-Square Test Used with Testcross Data to Test the Hypothesis That Two Genes Are Unlinked

(1) PHENOTYPES	(2) OBSERVED NUMBER (o)	(3) EXPECTED NUMBER (e)	(4) d	(5) d^2	(6) d^2/e
grey, normal	283	809	−526	276,676	342.00
grey, vestigial	1,294	809	485	235,225	290.76
black, normal	1,418	809	609	370,881	458.44
black, vestigial	241	809	−568	322,624	398.79
Total	3,236	3,236			1,489.99

(7) $\chi^2 = 1,489.99$ (8) df 3

The χ^2 value and the degrees of freedom are used with a table of chi-square probabilities (see Table 2.5) to determine the probability (P) that the deviation of the observed values from the expected values is due to chance. For $\chi^2 = 1,490$ with 3 degrees of freedom, the P value is much lower than 0.001; in fact, it is not on the table. This is interpreted to mean that independent repetitions of this experiment would produce chance deviations from the expected as large as those observed in many fewer than 1 out of 1,000 trials. As a reminder, if the probability of obtaining the observed χ^2 values is greater than 5 in 100 ($P > 0.05$), the deviation is considered not statistically significant and could have occurred by chance alone. If $P \leq 0.05$, we consider the deviation from the expected values to be statistically significant and not due to chance alone; the hypothesis may well be invalid. If $P \leq 0.01$, the deviation is highly statistically significant; the data are not consistent with the hypothesis. Thus, in this case we would reject the independent assortment hypothesis for our data and think of an alternative hypothesis, for example, that the genes are linked.

THE CONCEPT OF A GENETIC MAP.   In an individual doubly heterozygous for the w and m alleles, for example, the alleles can be arranged in two ways:

$$\frac{w^+ \, m^+}{w \, m} \text{ or } \frac{w^+ \, m}{w \, m^+}$$

In the arrangement on the left, the two wild-type alleles are on one homolog and the two recessive mutant alleles are on the other homolog, an arrangement called **coupling** (or the *cis* configuration). Crossing-over between the two loci produces $w^+ \, m$ and $w \, m^+$

recombinants. In the arrangement on the right, each homolog carries the wild-type allele of one gene and the mutant allele of the other gene, an arrangement called **repulsion** (or the *trans* configuration). Crossing-over between the two genes produces $w^+ \, m^+$ and $w \, m$ recombinants.

The data obtained by Morgan from *Drosophila* crosses indicated that the frequency of crossing-over (and hence of recombinants) for linked genes is characteristic of the gene pairs involved: For the X-linked genes white (w) and miniature (m), the frequency of crossing over is 36.9 percent. Moreover, the frequency of recombinants for two linked genes is the same, regardless of whether the alleles of the two genes involved are in coupling or in repulsion. *While the actual phenotypes of the recombinant classes are different for the two arrangements, the percentage of recombinants among the total progeny will be the same in each case (within experimental error).*

In 1913 a student of Morgan's, Alfred Sturtevant, suggested that the percentage of recombinants could be used as a quantitative measure of the genetic distance between two genes on a genetic map. This distance is measured in **map units** (mu). A crossover frequency of 1 percent between two genes was defined as 1 map unit. That is, 1 map unit is the distance between genes for which 1 product out of 100 (1 percent) is recombinant. The map unit is sometimes called a **centimorgan** (cM) in honor of Morgan.

The genes on a chromosome, then, can be represented by a one-dimensional, genetic map that shows in linear order the genes belonging to the chromosome. Crossover and recombination values give the linear order of the genes on a chromosome and provide information about the genetic distance between

~ FIGURE 5.3

The relationship between recombination and map distance. The farther apart two genes are, the greater the number of possible sites for recombination. Thus, the probability of recombination occurring between genes *A* and *B* is much less than that between genes *B* and *C*. The percentage of recombinants can provide information about the relative genetic distance between two linked genes.

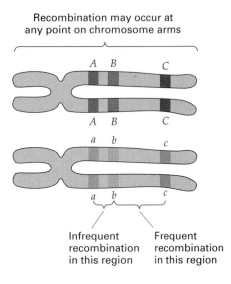

any two genes. The farther apart two genes are, the greater will be the *crossover frequency*. Thus, in Figure 5.3, the probability of recombination occurring between genes *A* and *B* is much less than between genes *B* and *C*, because *A* and *B* are closer together than are *B* and *C*. The first genetic map ever constructed was based on *recombination frequencies* from *Drosophila* crosses involving the sex-linked genes *w*, *m*, and *y*, where *w* gives white eyes, *m* gives miniature wings, and *y* gives yellow body. From these mapping experiments, the recombination frequencies for the *w* × *m*, *w* × *y*, and *m* × *y* crosses were established as 32.6, 1.3, and 33.9 percent, respectively. (Note that in this independent experiment the recombination frequency for *w* and *m* is a little lower than in the experiment discussed on pp. 100–101.) The percentages are quantitative measures of the distances between the genes involved.

The genetic map of these three genes is constructed to fit the recombination data. The recombination frequencies show that *w* and *y* are closely linked and that *m* is quite a distance away from the other two genes. Since the *w-m* genetic distance is less than the *y-m* distance (as shown by the smaller percentage of the recombinants in the *w* × *m* cross), the order of genes must be *ywm* (or *mwy*); thus, the three genes are ordered and spaced with 1.3 map units between *y* and *w*, and 32.6 map units between *w* and *m*:

Gene Mapping Using Two-Point Testcrosses

We have seen that the percentage of recombinants resulting from crossing-over is used to measure the genetic distance between two linked genes. By carrying out two-point testcrosses such as those shown in Figure 5.4, we can determine the relative numbers of parental and recombinant classes in the progeny. For autosomal recessives (as in Figure 5.4), a double heterozygote is crossed with a doubly homozygous recessive mutant strain. When the double heterozygous $a^+ b^+/a b$ F$_1$ progeny from a cross of $a^+ b^+/a^+ b^+$ with $a b/a b$ are testcrossed with $a b/a b$, four phenotypic classes are found among the F$_2$ progeny. Two of these classes have the parental phenotypes $a^+ b^+$ and $a b$. Since both classes result from chromosomes that have not crossed over between these genes, approximately equal numbers of these two types are expected.

The other two F$_2$ phenotypic classes have recombinant phenotypes $a^+ b$ and $a b^+$, which come from diploids in which a single crossover occurred between the chromosomes. We expect approximately equal numbers of these two recombinant classes. Because a single crossover event occurs more rarely than no crossing-over, an excess of parental phenotypes over recombinant phenotypes in the progeny of a testcross indicates linkage between the genes involved.

Two-point testcrosses for the purposes of mapping are set up in similar ways for genes showing other mechanisms of inheritance. For autosomal dominants, the double heterozygous individuals are testcrossed with an individual homozygous for the recessive alleles, which, in this case, are the wild-type alleles:

$$\frac{A \ B}{A^+ B^+} \times \frac{A^+ B^+}{A^+ B^+}$$

For X-linked recessives, a double heterozygous female is crossed with a hemizygous male carrying the recessive alleles:

$$\frac{a^+ b^+}{a \ b} \times \frac{a \ b}{\Longrightarrow}$$

And, for X-linked dominants, a doubly heterozygous female is crossed with a male carrying wild-type alleles on the X chromosome:

$$\frac{A \ B}{A^+ B^+} \times \frac{A^+ B^+}{}$$

For both X-linked cases, the females can be crossed with males of any genotype. If only male progeny are analyzed, the contribution of the male's X chromosome to the progeny is ignored.

In all cases a two-point testcross should yield a pair of parental types that occur with about equal frequencies, and a pair of recombinant types that also occur with about equal frequencies. As indicated previously, the actual phenotypes will depend, of course,

~ FIGURE 5.4

Testcross to show that two genes are linked. Genes a and b are recessive mutant alleles linked on the same autosome. A homozygous $a^+ b^+/a^+ b^+$ individual is crossed with a homozygous recessive $a b/a b$ individual, and the doubly heterozygous F_1 progeny ($a^+ b^+/a b$) are testcrossed with homozygous $a b/a b$ individuals.

Recessive linked autosomal mutant alleles

on the relative arrangement of the two allelic pairs in the homologous chromosomes; that is, whether they are in coupling (*cis*) or repulsion (*trans*). To get a count of the representatives of each progeny class (this will give the percentage of recombinant types), the following formula is used:

$$\frac{\text{number of recombinants}}{\text{total number of testcross progeny}} \times 100 = \frac{\text{percent}}{\text{recombinants}}$$

The value for the percentage of recombinants is usually directly converted into map units.

The two-point method of mapping is most accurate when the two genes examined are close together; when genes are far apart, there are inaccuracies, as we will see later. Large numbers of progeny must also be counted (scored) to ensure a high degree of accuracy.

Generating a Genetic Map

A genetic map is generated from estimating the number of times crossing-over occurred in a particular segment of the chromosome out of all meioses examined. In many cases the probability of a crossing-over event is not uniform along a chromosome, so we must be cautious about how far we extrapolate the genetic map (derived from data produced by genetic crosses) to the physical map of the chromosome (derived from determinations of the locations of genes along the chromosome itself; for example, from sequencing the DNA). Nonetheless, in practice it is common to assume that crossovers are randomly distributed along the chromosome.

The recombination frequencies observed between genes may also be used to predict the outcome of genetic crosses. For example, a recombination frequency of 20 percent between genes indicates that for a doubly heterozygous genotype (such as $a^+ b^+/a b$), 20 percent of the gametes produced, on average, will be recombinants ($a^+ b$ and $a b^+$ for the example, 10 percent of each expected). Further, if we assume for simplicity that crossing-over occurs randomly along a chromosome, more than one crossover may occur in a given region in a meiosis. The probabilities of these so-called **multiple crossovers** can be computed using the product rule: the probability of two independent events occurring simultaneously is equal to the product of the individual probabilities of two single events. That is, the probability of two crossovers (called a **double crossover**) occurring between the genes in our example is

$$0.2 \times 0.2 = 0.04$$

The probability of three crossovers (a triple crossover) is

$$0.2 \times 0.2 \times 0.2 = 0.008$$

and so on.

For any testcross, the percentage of recombinants in the progeny cannot exceed 50 percent. That is, if the genes are assorting independently, an equal number of recombinants and parentals are *expected* in the progeny, so the frequency of recombinants is 50 percent. If we get a recombinants value of 50 percent from a cross, then we state that the two genes are unlinked. Genes may be unlinked (that is, show 50 percent recombination) in two ways. First, the genes may be on different chromosomes, a case we discussed before. Second, *the genes may be far apart on the same chromosome.*

The second case can be explained by referring to Figure 5.5, which shows the effects of single crossovers and double crossovers on the production of parental and recombinant chromosomes. The loci are far apart on the same chromosome; as a result, in a given meiosis at least a single crossover (and usually more) always occurs between the two loci. Single crossovers between any pair of nonsister chromatids result in two parental and two recombinant chromosomes; that is, for two loci 50 percent of the products are recombinant (see Figure 5.5a).

Double crossovers can involve two, three, or all four of the chromatids (see Figure 5.5b). For double crossovers involving the same two nonsister chromatids (a *two-strand double crossover*), all four resulting chromosomes are parental for the two loci of interest. For *three-strand double crossovers* (double crossovers involving three of the four chromatids), two parental and two recombinant chromosomes result. For a *four-strand double crossover*, all four resulting chromosomes are recombinant. Considering all of the possible double crossover patterns together, 50 percent of the products are recombinant for the two loci. Similarly, for any multiple number of crossovers between loci that are far apart, examination of a large number of meioses will show that 50 percent of the resulting chromosomes are recombinant. This is the reason for the recombination frequency limit of 50 percent exhibited by unlinked genes on the same chromosome.

When genes are more closely linked together, no crossovers will occur between the two loci in some meioses, resulting in four parental chromosome products. In mapping linked genes, therefore, the map distance depends on the ratio of the meioses with no detectable crossovers to meioses with any number of crossovers between the loci.

The point is, if two genes show 50 percent recombination in a cross, they may not be on different chromosomes. More data would be needed to determine whether the genes are on the same chromosome or different chromosomes. One way to find out is to map

a number of other genes in the linkage group. For example, if a and m show 50 percent recombination, perhaps we will find that a shows 27 percent recombination with e, and e shows 36 percent recombination with m. This result would indicate that a and m are in the same linkage group approximately 63 mu apart, as shown here:

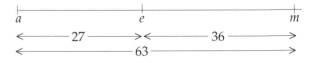

Double Crossovers

Consider a cross between a true-breeding yellow-bodied, white-eyed, miniature-winged female and a wild-type male *Drosophila*; that is, $y\,w\,m/y\,w\,m \times y^+\,w^+\,m^+/Y$. The F_1 flies are $y^+\,w^+\,m^+/y\,w\,m$ females (wild-type phenotype) and $y\,w\,m/Y$ males (yellow, white, miniature phenotypes). Table 5.2 shows the data for the F_2 progeny resulting from an $F_1 \times F_1$ cross. We have already shown that the gene order is y-w-m (see p. 104). The data show 29 recombinants for body color and eye color (y and w loci) and 719 recombinants for eye color and wing size (w and m loci), giving a total of 748. However, there are only 746 recombinants for body color and wing size. The discrepancy is due to the single normal-bodied, white-eyed, normal-winged fly. In the production of the egg from which this fly developed, a double crossover must have occurred, so that the body color and wing size genes (normal body color and normal-length wings) appeared together, as in one grandparent, although the intervening gene (for white eyes) was derived from the other grandparent. This example shows that large crossover frequencies give an inaccurate map. Unless intervening genes are available, there is no way of detecting double crossovers (or even-numbered multiple crossovers).

To see the significance of double crossovers in genetic-mapping experiments, consider a hypothetical case of two allelic pairs (a^+/a and b^+/b) in coupling and separated by quite a distance on the same chromosome. Figure 5.6a shows that a single crossover results in recombination of the two allelic pairs. Figure 5.6b, top, shows that a double crossover involving two of the four chromatids does not result in recombination of the allelic pairs, so only parental progeny result. Exactly the same progeny would result if there was no crossing-over in this region of the chromosome. However, the percentage of crossing-over between genes is a measure of the distance between them. Therefore, since the double crossover in Figure 5.6b, top, did not generate recombinants,

~ FIGURE 5.5

Demonstration that the recombination frequency between two genes located far apart on the same chromosome cannot exceed 50 percent. (a) Single crossovers produce one-half parental and one-half recombinant chromatids; (b) Double crossovers (two-strand, three-strand, and four-strand) collectively produce one-half parental and one-half recombinant chromatids.

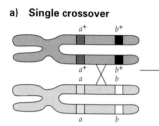

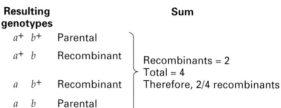
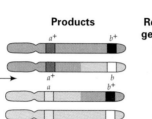
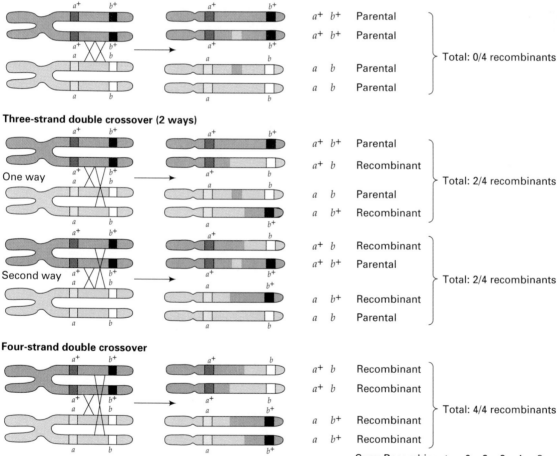

two crossover events will be uncounted, and the estimate of map distance between genes a and b will be low.

In sum, genetic map distance reflects only the *detectable* exchanges of genetic markers. If no multiple crossovers occurred between genes, there would be a

~ TABLE 5.2

Data for the Progeny of a Cross Between a $y^+ w^+ m^+/y\ w\ m$ Female (Wild Type Phenotype) and a $y\ w\ m$/Y Male (Yellow, White, Miniature) *Drosophila*

PHENOTYPES				CROSSOVER		
				BODY COLOR AND EYE COLOR	EYE COLOR AND WING SIZE	BODY COLOR AND WING SIZE
BODY COLOR	EYE COLOR	WING SIZE	NUMBER			
normal	red (normal)	long (normal)	758	—	—	—
yellow	white	miniature	700	—	—	—
normal	red	miniature	401	—	401	401
yellow	white	long	317	—	317	317
normal	white	miniature	16	16	—	16
yellow	red	long	12	12	—	12
normal	white	long	1	1	1	—
yellow	red	miniature	0	0	0	—
		Total	2,205	29	719	746
		Percentage	100	1.3	32.6	33.8

direct linear relationship between genetic map distance and the actual recombination frequency. In practice, we only see this relationship when genetic map distances are small; that is, when genes are closely linked. As the distance between genes increases, the chances of multiple crossovers between them increases, and there is no longer an exact linear relationship between map distance and recombination frequency. As a result, it is difficult to obtain an accurate measure of map distance when multiple crossovers are involved.

Fortunately, it has been possible to derive mathematical formulas, called **mapping functions**, to correct the observed recombination values for the incidence of multiple crossovers. Mapping functions all require some basic assumptions about the frequency of crossovers compared with distance between genes.

Three-Point Cross

How can we avoid this pitfall? Double crossovers occur quite rarely within distances of 10 mu or less. Consequently, one way to get accurate map distances is to study closely linked genes. Another efficient way to obtain such data is to use a **three-point testcross** involving three genes within a relatively short section of a chromosome.

The potential advantage of the three-point testcross can be seen in the theoretical case if we have a third allelic pair c^+/c between the a^+/a and b^+/b allelic pairs of Figure 5.6a, as diagrammed in Figure 5.6b, bottom. A double crossover between *a* and *b* might be detected by the recombination of the c^+/c allelic pair in relation to the other two allelic pairs. Three-point testcrosses are routinely used as an efficient way of mapping genes both for their order on the chromosome and for the distances between them, as we will see in the next section.

KEYNOTE

The map distance between two genes is based on the frequency of recombination between the two genes (an approximation of the frequency of crossovers between the two genes). By using genetic markers in crosses, geneticists are able to compute the recombination frequency between genes on chromosomes as the percentage of progeny showing the reciprocal recombinant phenotypes. The more closely the recombination frequency parallels the crossover frequency, the closer the genes are. But when the genetic distance increases, the incidence of multiple crossovers causes the recombination frequency to be an underestimate of the crossover frequency and hence of the map distance. Mapping functions may be used to correct for the effects of multiple crossovers and hence to give a more accurate estimate of map distance.

~ **FIGURE 5.6**

Progeny of single and double crossovers. (a) A single crossover between linked genes generates recombinant gametes. (b) *Top*: A double crossover between linked genes gives parental gametes. Thus, inaccurate map distances between genes result, since not all crossovers can be accounted for. *Bottom*: A possible solution to the double-crossover problem. The presence of a third allelic pair between the two allelic pairs in Figure 5.6a enables us to detect the double-crossover event. In a double crossover, the middle gene will change position relative to the outside genes.

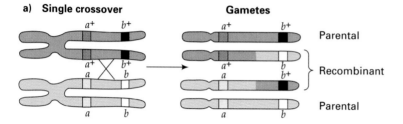

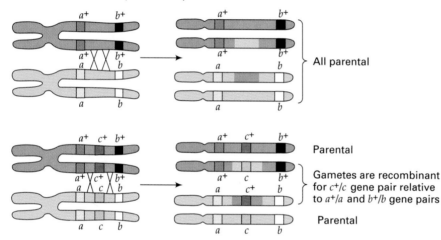

Mapping Chromosomes Using Three-Point Testcrosses

In diploid organisms the three-point testcross is a cross of a triple heterozygote with a triply homozygous recessive. If the mutant genes in the cross are all recessive, a typical three-point testcross might be:

$$\frac{a^+ \, b^+ \, c^+}{a \ \ b \ \ c} \times \frac{a \ b \ c}{a \ b \ c}$$

If any of the mutant genes are dominant to the wild type, the *triply homozygous recessive parent in the testcross will carry the wild-type allele of these genes.* For example, a testcross of a strain heterozygous for two recessive mutations (*a* and *c*) and one dominant mutation (*B*) might be

$$\frac{a^+ \, B^+ \, c^+}{a \ \ B \ \ c} \times \frac{a \ B^+ \, c}{a \ B^+ \, c}$$

In a testcross involving sex-linked genes, the female is the heterozygous strain (assuming that the female is the homogametic sex) and the male is hemizygous for the recessive alleles.

Three-point mapping is also done in haploid eukaryotic organisms, but in this case a testcross is not needed. Thus, in yeast we might cross two haploid strains to generate the triply heterozygous diploid cell. A cross of an *a b c* strain with an *a⁺ b⁺ c⁺* strain, for instance, would give an *a⁺ b⁺ c⁺/a b c* diploid cell. That cell goes through meiosis to generate haploid spores from which progeny yeast cultures grow. Since the cells are haploid, the phenotypes of the progeny are determined directly by the genotypes.

Suppose we have a hypothetical plant in which there are three linked genes, all of which control fruit phenotypes. A recessive allele *p* of the first gene determines purple fruit color versus yellow color of the wild type. A recessive allele *r* of the second gene results in a round fruit shape versus elongated fruit in the wild type. A recessive allele *j* of the third gene gives a juicy fruit versus the dry fruit of the wild-type. The task before us is to determine the order of the genes on the chromosome and the map distances between the genes. To do so, we make the appropriate testcross of a triple heterozygote ($p^+ r^+ j^+ / p r j$) with a triply homozygous recessive ($p r j / p r j$) and then count the different phenotypic classes in the progeny (Figure 5.7).

For each gene in the cross, two different phenotypes occur in the progeny; therefore, for the three genes a total of $(2)^3 = 8$ phenotypic classes will appear in the progeny, representing all possible combinations of phenotypes. In an actual experiment, not all the phenotypic classes may be generated. The absence of a phenotypic class is also important infor-

mation, and the experimenter should enter a 0 in the class for which no progeny are found.

ESTABLISHING THE ORDER OF GENES. The first step in mapping the three genes is to determine the order of the genes on the chromosome. One parent carries the recessive alleles for all three genes; the other is heterozygous for all three genes. Therefore, the phenotype of each of the progeny is determined by the alleles in the gamete from the triply heterozygous parent; the gamete from the other parent will carry only recessive alleles. We know from the genotypes of the original parents that all three genes are in coupling. Since the heterozygous parent in the test are parental progeny: Class 1 is produced by the fusion of

~ **FIGURE 5.7**

Three-point mapping, showing the testcross used and the resultant progeny.

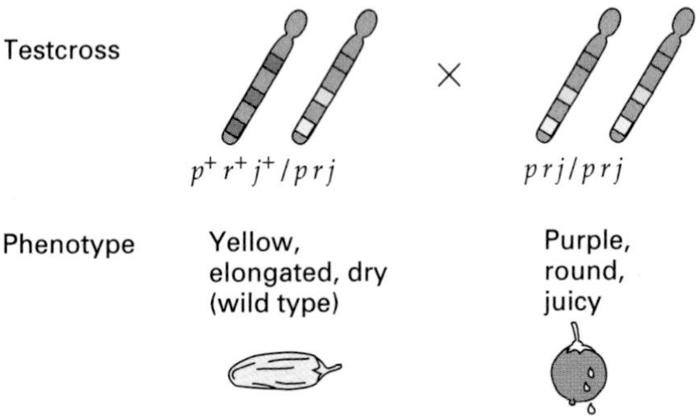

	Class	Phenotype		Number	Genotype of gamete from heterozygous parent
Parental phenotypes	1	Wild type (yellow, elongated, dry)	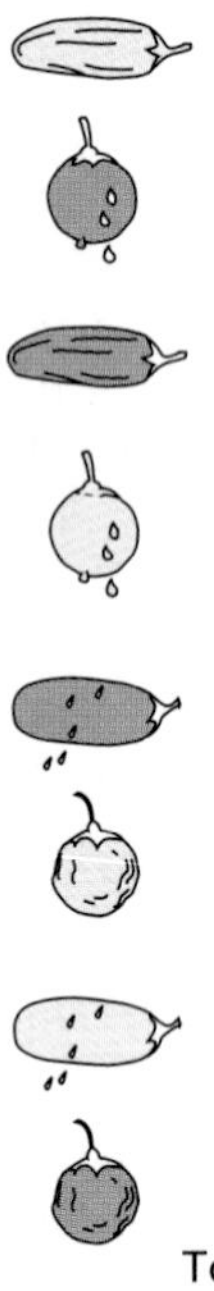	179	p^+ r^+ j^+
	2	purple, round, juicy		173	p r j
Recombinant phenotypes	3	purple, elongated, dry		52	p r^+ j^+
	4	yellow, round, juicy		46	p^+ r j
	5	purple, elongated, juicy		22	p r^+ j
	6	yellow, round, dry		22	p^+ r j^+
	7	yellow, elongated, juicy		4	p^+ r^+ j
	8	purple, round, dry		2	p r j^+

Total = 500

cross was $p^+ r^+ j^+/p r j$, classes 1 and 2 in Figure 5.7 a $p^+ r^+ j^+$ gamete with a $p r j$ gamete. Class 2 is produced by the fusion of a $p r j$ gamete from the heterozygous parent and a $p r j$ gamete. These classes are generated from meioses in which no crossing-over occurs in the region of the chromosome in which the three genes are located.

The progeny classes resulting from a double crossover involving the three loci can often be found by inspecting the numbers of each phenotypic class. Since the frequency of a double crossover in a region is expected to be lower than the frequency of a single crossover, *double-crossover gametes are the least frequent pair found*. Thus, to identify the double-crossover progeny, we can examine the progeny to find the *pair* of classes that have the lowest number of representatives. In Figure 5.7, classes 7 and 8 are such a pair. The genotypes of the gametes from the heterozygous parent that give rise to these phenotypes are $p^+ r^+ j$ and $p r j^+$.

Referring again to Figure 5.6b, bottom, we see that a double crossover changes the orientation of the gene in the center of the sequence (here, c^+/c) with respect to the two flanking allelic pairs. Therefore, genes p, r, and j must be arranged in such a way that the center gene switches to give classes 7 and 8. To determine the arrangement, we first check the relative organization of the genes in the parental heterozygote to be sure which genes are in coupling and which are in repulsion. In this example, the parental (noncrossover) gametes are $p^+ r^+ j^+$ and $p r j$, so all are in coupling. The double-crossover gametes are $p^+ r^+ j$ and $p r j^+$, so the only possible gene order compatible with the data is $p j r$, with the genotype of the heterozygous parent being $p^+ j^+ r^+/p j r$. Figure 5.8 illustrates the generation of the double-crossover gametes from that parent.

CALCULATING THE MAP DISTANCES BETWEEN GENES.
The cross data can be rewritten as shown in Figure 5.9 to reflect the newly determined gene order.

For convenience in the analysis, the region between genes p and j is called region I, and that between genes j and r is called region II.

Map distances can now be calculated. *The frequency of crossing-over (genetic recombination) is computed between two genes at a time.* For the *p-j* distance, all the crossovers that occurred in region I must be added together. Thus, we must consider the recombinant progeny resulting from a single crossover in that region (classes 3 and 4) *and* the double-crossover recombinant progeny. The double crossovers must be included since each double crossover includes a single crossover in region I and therefore involves recombination between genes p and j. From Figure 5.9 there are 98 recombinant progeny in classes 3 and 4, and 6 in classes 7 and 8, giving a total of 104 progeny that result from recombination in region I. There are 500 progeny in all, so the percentage of progeny generated by crossing over in region I is 20.8 percent, determined as follows (sco = single crossovers; dco = double crossovers):

$$\frac{\text{sco in region I } (p\text{-}j) + \text{dco}}{\text{total progeny}} \times 100\%$$

$$= \frac{(52 + 46) + (4 + 2)}{500} \times 100\%$$

$$= \frac{98 + 6}{500} \times 100\%$$

$$= \frac{104}{500} \times 100\%$$

$$= 20.8\%$$

In other words, the distance between genes p and j is 20.8 mu. This map distance, which is quite large, is chosen mainly for illustration. The map distance value obtained is in the range where double

~ FIGURE 5.8

Rearrangement of the three genes in Figure 5.7 to $p j r$. The evidence is that a double crossover involving the same two chromatids (as it is shown in this figure) generates the least frequent pair of recombinant phenotypes (in this case, class 7 with four progeny and class 8 with two progeny).

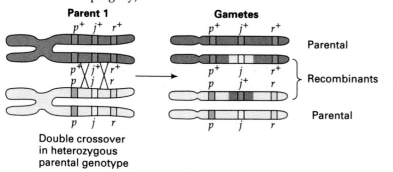

crossovers in the *p-j* region might occur at a significant frequency and go undetected. In an actual cross the value of 20.8 mu would probably underestimate the true distance.

The same method is used to obtain the map distance between genes *j* and *r*. That is, we calculate the frequency of crossovers in the cross that gave rise to progeny recombinant for genes *j* and *r* and directly relate that frequency to map distance. In this case, all the crossovers that occurred in region II (see Figure 5.9) must be added (classes 5, 6, 7, and 8). The percentage of crossovers is calculated in the following manner:

$$\frac{\text{sco in region II } (j\text{-}r) + \text{dco}}{\text{total progeny}} \times 100\%$$

$$= \frac{(22 + 22) + (4 + 2)}{500} \times 100\%$$

$$= \frac{44 + 6}{500} \times 100\%$$

$$= \frac{50}{500} \times 100\%$$

$$= 10.0\%$$

Thus, the map distance between genes *j* and *r* is 10.0 map units.

In summary, we have generated a genetic map of the three genes in the example (Figure 5.10). The example has illustrated that the three-point testcross is an effective way of establishing the order of genes and of calculating map distances.

To compute the map distance between the two outside genes, we simply add the two map distances. Thus, in the example the *p-r* distance is 20.8 + 10.0 = 30.8 mu. This map distance also can be computed directly from the data by combining the two formulas discussed previously:

$$\text{distance} = \frac{(\text{sco in region I}) + (\text{dco}) + (\text{sco in region II}) + (\text{dco})}{\text{total progeny}} \times 100\%$$

$$= \frac{(\text{sco in region I}) + (\text{sco in region II}) + (2 \times \text{dco})}{\text{total progeny}} \times 100\%$$

$$= \frac{(52 + 46) + (22 + 22) + 2(4 + 2)}{500} \times 100\%$$

$$= \frac{98 + 44 + 2(6)}{500} \times 100\%$$

$$= 30.8 \text{ map units}$$

~ FIGURE 5.9

Rewritten form of the testcross and testcross progeny in Figure 5.7, based on the actual gene order *p j r*.

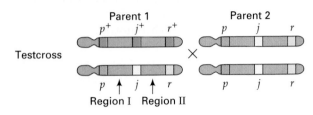

Testcross progeny

Class	Genotype of gamete from heterozygous parent			Number	Origin
1	p^+	j^+	r^+	179	Parentals, no crossover
2	p	j	r	173	
3	p	j^+	r^+	52	Recombinants, single crossover region I
4	p^+	j	r	46	
5	p	j	r^+	22	Recombinants, single crossover region II
6	p^+	j^+	r	22	
7	p^+	j	r^+	4	Recombinants, double crossover
8	p	j^+	r	2	

Total = 500

~ FIGURE 5.10

Genetic map of the *p-j-r* region of the chromosome computed from the recombination data in Figure 5.9.

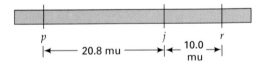

KEYNOTE

The map distance between genes is calculated from the results of testcrosses between strains carrying appropriate genetic markers. The most accurate testcross is the three-point testcross using a diploid organism in which a triple heterozygote is crossed with a homozygous recessive for all three genes. The most accurate map distances are obtained when the three genes are reasonably close to one another, with less than 10 map units between each locus. The unit of genetic distance is the map unit (mu), which is the distance between gene pairs for which 1 product out of 100 is recombinant; in other words, a recombinant frequency of 1 percent is 1 mu.

INTERFERENCE AND COINCIDENCE. The map distances determined by three-point mapping are useful in elaborating the overall organization of genes on a chromosome and in telling us a little about the recombination mechanisms themselves. For example, the map distance of 20.8 mu between genes *p* and *j* means that 20.8 percent of the gametes are predicted to result from crossing-over between the two gene loci. Similarly, the map distance of 10.0 mu between *j* and *r* indicates that 10.0 percent of the gametes are predicted from crossing-over between these two gene loci.

In the example of Figure 5.9, if crossing-over in region I is independent of crossing-over in region II, then the probability of a double crossover in the two regions is equal to the product of the probabilities of the two events occurring separately; that is,

$$\frac{\text{map distance, region I}}{100} \times \frac{\text{map distance, region II}}{100}$$

$$= 0.208 \times 0.100 = 0.0208$$

or 2.08 percent double crossovers are expected to occur. However, only $6/500 = 1.2$ percent double crossovers occurred in this cross (classes 7 and 8).

It is characteristic of mapping crosses that double-crossover progeny typically do not appear as often as the map distances between the genes lead us to expect. Thus, once a crossing-over event has occurred in one region of the meiotic bivalent (tetrad), the probability of another crossing-over event occurring nearby is reduced, most probably by physical interference caused by the breaking and rejoining of the chromatids. This phenomenon is called **chiasma interference** (also **chromosomal interference**).

The extent of interference is expressed as a **coefficient of coincidence**; that is

$$\frac{\text{coefficient of}}{\text{coincidence}} = \frac{\text{observed double crossover frequency}}{\text{expected double crossover frequency}}$$

and

$$\text{interference} = 1 - \text{coefficient of coincidence}$$

For the portion of the map in our example, the coefficient of coincidence is

$$0.012/0.0208 = 0.577$$

A coefficient of coincidence value of 1 means that in a given region, all double crossovers occurred that were expected on the basis of two independent events; there is no interference, so the interference value is zero. If the coefficient of coincidence is zero, none of the expected double crossovers occurred. Here there is total interference, with one crossover completely preventing a second crossover in the region under examination; the interference value is 1. These examples show that coincidence values and interference values are inversely related. In the example above, the coefficient of coincidence of 0.577 means that the interference value is 0.423. Only 57.7 percent of the expected double crossovers took place in the cross.

KEYNOTE

The occurrence of a chiasma between two chromatids may physically impede the occurrence of a second chiasma nearby, a phenomenon called chiasma interference. The extent of interference is expressed by the coefficient of coincidence, which is calculated by dividing the frequency of observed double crossovers by the frequency of expected double crossovers. The coefficient of coincidence typically ranges from zero to 1, and the extent of interference is measured as $1 -$ coefficient of coincidence.

MAPPING THE HUMAN GENOME

Classical genetic mapping experiments with numerous organisms have localized many genes to their chromosome locations. Gene localization is not an end point but a beginning point for further studies of the genes to determine their sequences and functions, and how their expression is regulated. Studies of the genes of organisms simpler than ourselves have been valuable in our studies of human genes, but an overriding desire has been to identify, map, and analyze functionally every gene in the human genome. The information we will get from such investigations of the human genome is of great importance; for example, it would enable us to understand more easily the bases of known genetic diseases and of any discovered in the future. Localizing all the estimated 50,000–100,000 genes in the human genome and obtaining the sequence of every base pair in the human genome is the goal of the Human Genome Project (HGP) (see Chapter 14). As part of the HGP, parallel studies are being done with selected model organisms, namely, *E. coli*, the yeast *Saccharomyces cerevisiae*, *Drosophila melanogaster*, the nematode *Caenorhabditis elegans*, and the mouse *Mus musculus*. Although much of the

HGP involves DNA sequencing (see Chapter 14), a lot of the groundwork involves mapping activities. In this section we provide a brief overview of some of the mapping activities involved in the HGP. These mapping procedures are not exclusive to the HGP.

Constructing Genetic Linkage Maps

A *genetic linkage map* is exactly the kind of map we have discussed to this point in the text, that is, a map showing the relative locations of specific genetic markers determined by the analysis of recombinants from genetic crosses. Markers may be genes or molecular markers, that is, DNA regions in the genome that differ sufficiently between individuals so that they can be detected by molecular analysis of DNA isolated from family members (see Chapter 14 for some general discussions of detecting DNA differences between genomes).

Human genetic linkage maps are constructed by analyzing the inheritance patterns of genetic markers and/or DNA markers in pedigrees for linkage. Due to the rarity of large, multigenerational pedigrees in which two genetic traits are segregating, it is hard to test for linkage and to calculate a map distance between markers. So in most cases a statistical test known as the **lod** (logarithm of **od**ds) **score method** is used to test for possible linkage between two loci. The lod score method is usually done by computer programs that use data pooled from a number of pedigrees. A full discussion of this method is beyond the scope of this text, so only a brief presentation is given here.

The lod score method calculates and compares the probability that the pedigree results would have been obtained if two markers showed a certain degree of linkage (calculated from the pedigree data) and the probability that the results would have been obtained even if the two markers were not linked. The results are expressed as the $\log_{10}$ of the ratio of the two probabilities; this is the lod score. By convention, a hypothesis of linkage between two genes cannot be rejected when the lod score is +3 or more because this means that the odds are 1,000:1 in favor of linkage between two genes or markers (the $\log_{10}$ of 1,000 is +3). Similarly, a hypothesis of linkage between two genes is rejected when the lod score reaches −2.

Once linkage is established between gene markers of any type, the map distance is computed from the recombination frequency giving the highest lod score. This is done by solving lod scores for a range of proposed map units. For the human genome, 1 mu corresponds, on average, to approximately 1 million

base pairs (1 megabase, or 1 Mb). One comprehensive genetic linkage map of the human genome that has been produced is based on 5,264 molecular markers and has an average distance between markers of 1.6 mu. A genetic linkage map of DNA markers is valuable for mapping the locations of disease and other genes, and for constructing the physical maps described in the next section.

Constructing Physical Maps

A **physical map** is a map of markers on genomic DNA that is constructed without genetic recombination analysis. To place a marker on a physical map, the researcher first localizes it to an individual chromosome and then to the smallest possible subregion of the chromosome that can be resolved with existing techniques. We discuss some examples of physical mapping briefly in this section.

SOMATIC CELL HYBRIDIZATION. One way to localize a marker to an individual human chromosome is to use the **somatic cell hybridization** technique. In this technique, cultured human cells are fused with cultured mouse cells to produce a *somatic cell hybrid*. Over several divisions the hybrid cells preferentially lose human chromosomes (the reason is not known) until eventually there is generated a *stable cell line* that contains all mouse chromosomes and a subset of human chromosomes. The number and particular set of human chromosomes vary with the cell line.

To map human genes by using somatic cell hybridization, we start with a human cell type that carries one or more genetic markers. We fuse the cells with mouse cells and then analyze various resulting stable hybrid cell lines for the presence or absence of the marker(s). We correlate these data for a number of somatic cell lines with the presence or absence of particular chromosomes. Given enough cell lines, we can show that a particular marker is present only when one particular chromosome is present, and absent when that chromosome is absent. Markers shown to be linked to the same chromosome through this experimental approach are called **syntenic** ("together thread").

Many genes have been localized to individual chromosomes in the human genome using somatic cell hybridization. For example, through a combination of molecular biology techniques and classical human genetic linkage analysis, a particular DNA marker called G8 was found to be closely linked to the Huntington disease (HD) gene. This tight linkage was a key to cloning the Huntington disease gene. To locate the chromosome on which G8 is found, hybrid human–mouse cells were made and allowed to produce stable lines. Those lines that contained the G8 DNA marker were analyzed to determine their human chromosome composition, with the following results:

HYBRID CELL LINE	HUMAN CHROMOSOMES PRESENT
WIL-5	4 17 18 21 X
WIL-6	4 5 6 7 8 10 11 14 17 19 20 21 X
NSL-15	2 4 5 7 8 12 13 14 15 17 18 19 21 22 X
ATR-13	1 2 3 4 5 6 7 8 10 12 13 14 15 16 17 18 19
XTR-22	2 4 5 6 8 10 11 18 19 20 21 22

If we look at the cell line with the fewest number of human chromosomes (WIL-5) and compare the chromosomes present with those in the other lines, we see that chromosome 4 is present in all five lines. We must be sure, though, that no other chromosome is also present in the five lines: chromosome 17 is not in XTR-22, chromosome 18 is not in WIL-6, chromosome 21 is not in ATR-13, and the X chromosome is not present in ATR-13 or XTR-22. Therefore, the G8 DNA marker—and therefore the Huntington disease gene—must be on chromosome 4.

Note that the somatic cell hybridization technique can only show on which chromosome genes are located. It cannot provide any information about the genetic distance between genes.

LOW-RESOLUTION PHYSICAL MAPPING: CHROMOSOMAL BANDING PATTERNS. At the subchromosomal level, *low-resolution physical mapping* has produced a map of human chromosomes, with each chromosome characterized by a banding pattern that can be seen under the microscope after using particular staining protocols. (See Chapter 10 for a discussion of chromosome banding techniques.) The average size of a band is about 10 Mb. A standard nomenclature has been established for the chromosomes based on the banding patterns so that scientists can talk about gene and marker locations with reference to specific regions and subregions. Remember that each chromosome has two arms separated by the centromere. The shorter arm is called p and the longer arm is called q. Numbered regions and numbered subregions are then assigned from the centromere outward; that is, region 1 is closest to the centromere. For example, the breast cancer susceptibility gene *BRCA1* is at location 17q21, meaning it is on the long arm of chromosome 17 in region 21.

MEDIUM- TO HIGH-RESOLUTION PHYSICAL MAPPING USING FISH. Individual eukaryotic chromosomes may be colored fluorescently at the location of specific gene or DNA sequences. In this method, human metaphase chromosomes on a microscope slide are treated to cause the two DNA strands to separate but stay in the same physical location. Specific DNA sequences are cloned and tagged with fluorescent chemicals. The tagged DNA sequences—called DNA probes—are added to the chromosomes where they will bind (hybridize) to the region from which they were cloned. This procedure is called *fluorescence in situ hybridization* and given the acronym FISH. (*In situ* hybridization is not a new technique. What is new is the use of fluorescent tags; the probes were traditionally radioactively labeled with tritium.) In this way the chromosome sites corresponding to the probe can be identified by the fluorescent emissions of the tag. By using chemicals that fluoresce at different wavelengths, we can use a number of different probes in the same experiment. We use computer imaging analysis of the sample examined under a fluorescence microscope to see where the probes have bound.

Figure 5.11 shows the results of FISH with six different probes. The probe colors are not the true colors from the fluorescence, but are pseudocolors generated by the computer. The complete chromosomes are seen and distinguished by staining them with a chemical that colors all the DNA blue (in this case). Each chromosome to which a probe has hybridized has two dots because the DNA in metaphase chromosomes is already duplicated in preparation for cell division. For example, the yellow dots identify an uncharacterized DNA sequence on chromosome 5, the green dots identify the Down syndrome region on chromosome 21, and the red dots identify the Duchenne muscular dystrophy gene on the X chromosome.

With metaphase chromosomes, FISH methods have a resolution of 2–5 Mb for DNA markers. (This means that two markers this far apart will be seen as two distinct dots.) Thus, it is possible to localize a marker to a subregion of a chromosome band using FISH. Much higher resolution physical mapping using FISH can be done with chromosomes that have been extended artificially; with this method, markers can be resolved in a range of 5–700 kb.

HIGH-RESOLUTION PHYSICAL MAPS OF SEQUENCE-TAGGED SITES. A sequence-tagged site (STS) is a short segment of DNA, about 60–1,000 bp long, that defines a unique position in the human genome. A particular STS is detected by using a procedure called the polymerase chain reaction (PCR), which is described in detail in Chapter 14. In brief, PCR permits the amplification of any segment of DNA that is flanked by a pair of DNA primers. (The primers are short DNA fragments needed for DNA synthesis to begin.) Thus, what makes each STS unique is the sequences of the pair of primers used in the PCR. The amplified DNA is detected by size separation of isolated DNA using gel electrophoresis (see Chapter 14).

Close to 30,000 STSs have been mapped, providing sufficient information to begin an orderly effort to sequence the human genome. Moreover, the STS map provides landmarks in the form of the STSs themselves (analogous to having a large number of individually identifiable pins on a printed map) that tell us exactly where we are in the genome, and so the map enables us to "move around" the genome with reference to those landmarks. Information about the STSs and the STS map itself has been made available via the World Wide Web to genome researchers worldwide, making it possible for any research group to explore regions of interest in the genome by starting from the nearest STSs.

Integrating Genetic Linkage Maps and Physical Maps

As the resolution of the physical maps increases, it will become easier to locate genes with respect to DNA markers and thereby integrate the genetic linkage maps with the physical maps. Newly discovered genes will then be more easily mapped to their locations on the chromosomes. Several disease genes have already been cloned by homing in on their chromosomal location through the use of DNA markers, a process called **positional cloning**. An example is the gene responsible for cystic fibrosis.

An example of the results of fluorescent *in situ* hybridization (FISH) in which fluorescently tagged DNA probes were hybridized to human metaphase chromosomes.

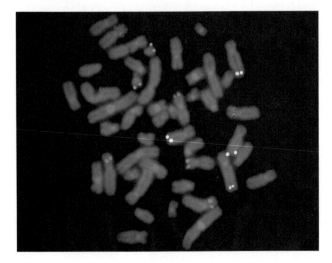

As a summary, Figure 5.12 shows a map of human chromosomes, indicating some of the details of the banding patterns and presenting a selection of the known locations of mutations causing disease.

KEYNOTE

Extensive efforts are being made to map the human genome. Two types of maps are being constructed: genetic linkage maps and physical maps. Physical maps have different resolutions, depending on the technique being used. A high-resolution physical map of thousands of sequence-tagged sites (STSs) that define unique positions in the genome has been constructed. The STSs are landmarks used as jumping-off points for further genome exploration and for sequencing efforts. Future integration of high-resolution physical maps with genetic linkage maps will make it easier to map newly discovered genes.

TETRAD ANALYSIS IN CERTAIN HAPLOID EUKARYOTES

Tetrad analysis is a special mapping technique that can be used to map the genes of those haploid eukaryotic organisms in which the products of a single meiosis, the meiotic tetrad, are contained within a single structure. The eukaryotic organisms in which this phenomenon occurs are either fungi or single-celled algae, all of which are haploid. The orange bread mold *Neurospora crassa* and the yeast *Saccharomyces cerevisiae* (both fungi) and *Chlamydomonas reinhardtii* (a single-celled alga) are frequently used in tetrad analysis. Tetrad analysis allows study of the details of events at meiosis that are not possible in any other system.

By analyzing the phenotypes of the meiotic tetrads, geneticists can directly infer the genotypes of each member of the tetrad. Haploid organisms exhibit no genetic dominance since there is only one copy of each gene; hence the genotype is expressed directly in the phenotype. We will outline this technique shortly.

Before we discuss the principles of tetrad analysis, let us learn a little about the life cycles of the organisms with which tetrad analysis can be done. For example, the life cycle of baker's yeast (also called the budding yeast), *Saccharomyces cerevisiae*, is

~ FIGURE 5.12

Genetic map of human chromosomes showing some of the details of the banding patterns. The map presents a selection of the known locations of mutations causing disease. Note: The chromosomes are not drawn to scale.

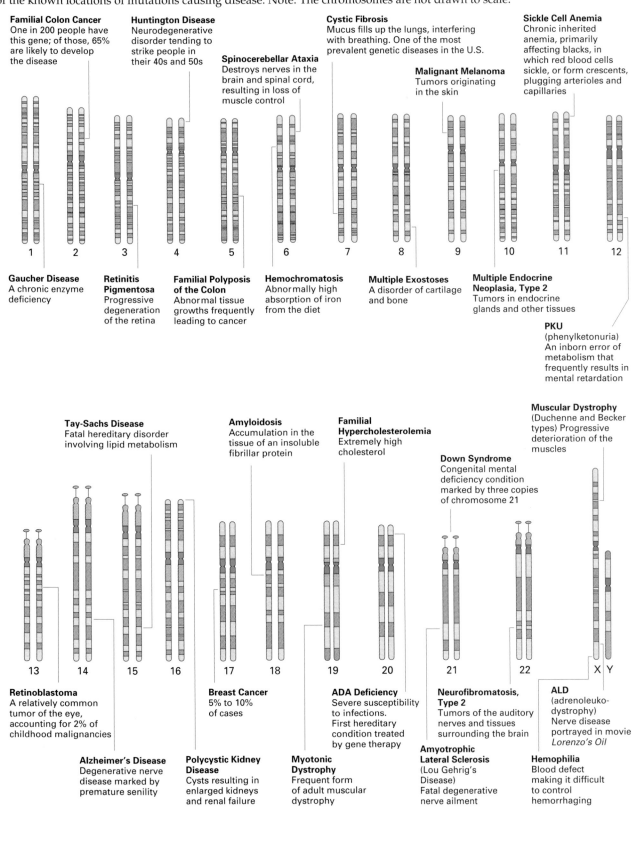

Familial Colon Cancer
One in 200 people have this gene; of those, 65% are likely to develop the disease

Huntington Disease
Neurodegenerative disorder tending to strike people in their 40s and 50s

Spinocerebellar Ataxia
Destroys nerves in the brain and spinal cord, resulting in loss of muscle control

Cystic Fibrosis
Mucus fills up the lungs, interfering with breathing. One of the most prevalent genetic diseases in the U.S.

Malignant Melanoma
Tumors originating in the skin

Sickle Cell Anemia
Chronic inherited anemia, primarily affecting blacks, in which red blood cells sickle, or form crescents, plugging arterioles and capillaries

Gaucher Disease
A chronic enzyme deficiency

Retinitis Pigmentosa
Progressive degeneration of the retina

Familial Polyposis of the Colon
Abnormal tissue growths frequently leading to cancer

Hemochromatosis
Abnormally high absorption of iron from the diet

Multiple Exostoses
A disorder of cartilage and bone

Multiple Endocrine Neoplasia, Type 2
Tumors in endocrine glands and other tissues

PKU
(phenylketonuria)
An inborn error of metabolism that frequently results in mental retardation

Tay-Sachs Disease
Fatal hereditary disorder involving lipid metabolism

Amyloidosis
Accumulation in the tissue of an insoluble fibrillar protein

Familial Hypercholesterolemia
Extremely high cholesterol

Down Syndrome
Congenital mental deficiency condition marked by three copies of chromosome 21

Muscular Dystrophy
(Duchenne and Becker types) Progressive deterioration of the muscles

Retinoblastoma
A relatively common tumor of the eye, accounting for 2% of childhood malignancies

Breast Cancer
5% to 10% of cases

ADA Deficiency
Severe susceptibility to infections. First hereditary condition treated by gene therapy

Neurofibromatosis, Type 2
Tumors of the auditory nerves and tissues surrounding the brain

ALD
(adrenoleuko-dystrophy)
Nerve disease portrayed in movie *Lorenzo's Oil*

Alzheimer's Disease
Degenerative nerve disease marked by premature senility

Polycystic Kidney Disease
Cysts resulting in enlarged kidneys and renal failure

Myotonic Dystrophy
Frequent form of adult muscular dystrophy

Amyotrophic Lateral Sclerosis
(Lou Gehrig's Disease)
Fatal degenerative nerve ailment

Hemophilia
Blood defect making it difficult to control hemorrhaging

~ FIGURE 5.13

Life cycle of the yeast *Saccharomyces cerevisiae.*

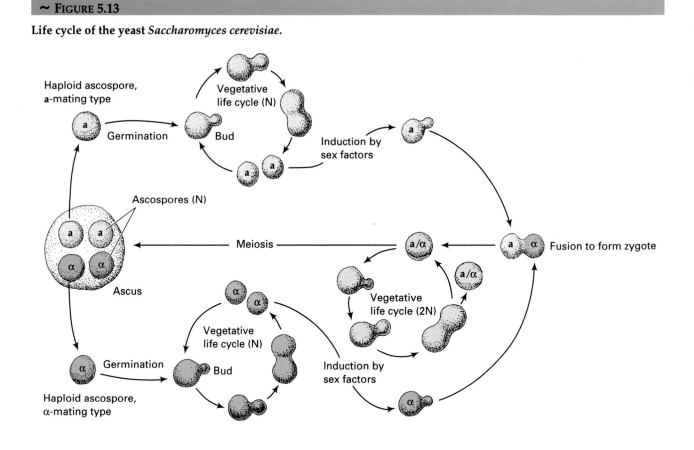

diagrammed in Figure 5.13. Two mating types occur in yeast, **a** and α. The haploid cells of this organism reproduce mitotically (the vegetative life cycle), with the new cell arising from the parental cell by budding. Fusion of haploid **a** and α cells produces a diploid cell that is stable and that also reproduces by budding. Diploid **a**/α cells *sporulate*; that is, they go through meiosis. The four haploid meiotic products of a diploid cell, the ascospores, are contained within a roughly spherical ascus. Two of these ascospores are of mating type **a**, and two are of mating type α. When the ascus is ripe, the ascospores are released, and they germinate to produce haploid cells that, on solid medium, grow and divide to produce a colony. In yeast the four ascospores are arranged randomly within the ascus. The green alga *Chlamydomonas reinhardtii* also produces four meiotic products that are arranged randomly in a sac.

The fungus *Neurospora crassa* has a somewhat similar life cycle, but in this organism the ascospores are arranged in a linear ascus. There are actually eight ascospores in this organism because each of the four meiotic products divides again by mitosis. The order of the four spore pairs within an ascus reflects exactly

the orientation of the four chromatids of each tetrad at the metaphase plate in meiosis I.

Using Tetrad Analysis to Map Two Linked Genes

Let us see how we can map the distance between two linked genes by analyzing meiotic tetrads.

By making an appropriate cross, a diploid zygote is constructed that is heterozygous for both genes, and after meiosis the resulting tetrads are analyzed. For yeast, a mechanical micromanipulator with a very fine needle is used to "dissect" each spore out of the ascus.

Consider the cross $a^+ b^+ \times a\,b$ in which a and b are linked genes. Figure 5.14 shows the three different tetrad types that result from no, one, or two crossovers. If no crossing-over (Figure 5.14a) occurs between the genes, then a **parental-ditype** (PD) tetrad results. The PD tetrads contain only two types of meiotic products, both of which are of the parental type (hence the name parental ditype). A single crossover (Figure 5.14b) produces a **tetratype** (T) tetrad. A T tetrad contains two parentals (one of each type) and two recombinants (one of each type).

For double crossovers we must take into account the chromatid strands involved. In a two-strand double crossover (Figure 5.14c), the two crossover events between the two genes involve the same two chromatids. This crossover results in a PD tetrad, since no recombinant progeny are produced. Three-strand double crossovers (Figure 5.14d) involve three of the four chromatids; there are two possible ways in which this can occur. In either case the result is a T tetrad. Lastly, in four-strand double crossovers

~ FIGURE 5.14

Origin of tetrad types for a cross $a\ b \times + + (a^+\ b^+)$ in which both genes are located on the same chromosome. (a) No crossover; (b) Single crossover; (c–e) Three types of double crossovers.

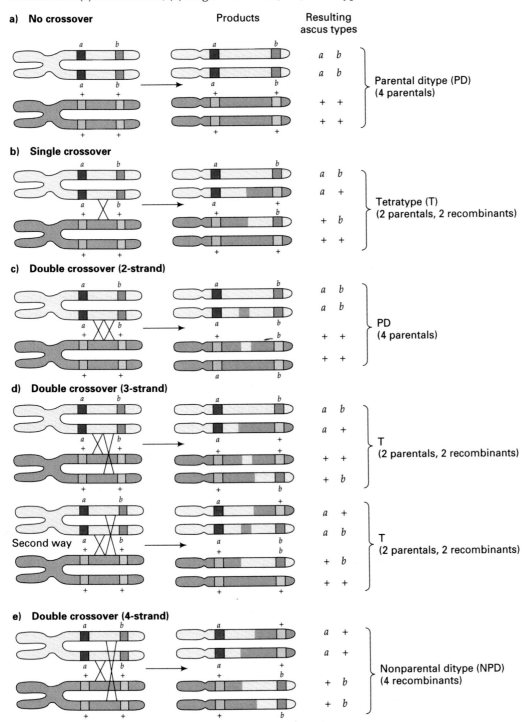

a) **No crossover**

Products   Resulting ascus types

a	b	
a	b	Parental ditype (PD)
$+$	$+$	(4 parentals)
$+$	$+$	

b) **Single crossover**

a	b	
a	$+$	Tetratype (T)
$+$	b	(2 parentals, 2 recombinants)
$+$	$+$	

c) **Double crossover (2-strand)**

a	b	
a	b	PD
$+$	$+$	(4 parentals)
$+$	$+$	

d) **Double crossover (3-strand)**

a	b	
a	$+$	T
$+$	$+$	(2 parentals, 2 recombinants)
$+$	b	

Second way

a	$+$	
a	b	T
$+$	b	(2 parentals, 2 recombinants)
$+$	$+$	

e) **Double crossover (4-strand)**

a	$+$	
a	$+$	Nonparental ditype (NPD)
$+$	b	(4 recombinants)
$+$	b	

(Figure 5.14e), each crossover event involves two distinct chromatids, so all four chromatids of the tetrad are involved. This crossover results in a **nonparental-ditype** (NPD) tetrad.

From the relative numbers of each type of meiotic tetrad, the distance between the two genes can be computed by using a modification of the basic mapping formula:

$$\frac{\text{number of recombinants}}{\text{total number of progeny}} \times 100$$

In tetrad analysis we analyze types of tetrads, rather than individual progeny. To convert the basic mapping formula into tetrad terms, the recombination frequency between genes *a* and *b* becomes

$$\frac{1/2\,\text{T} + \text{NPD}}{\text{total tetrads}} \times 100$$

In essence, we are looking at tetrads with recombinants and determining the proportion of spores in those tetrads that are recombinant. So, in the formula the 1/2 T and the NPD represent the recombinants from the cross; the other 1/2 T and the PD represent the nonrecombinants (the parentals). Thus, the formula does indeed compute the percentage of recombinants. For instance, if there are 200 tetrads with 140 PD, 48 T, and 12 NPD, the recombination frequency between the genes is

$$\frac{1/2(48) + 12}{200} \times 100 = 18\%$$

This conversion produces a formula for calculating the map distance between two linked genes by using the frequencies of the three possible types of tetrads rather than by analyzing individual progeny. If more than two genes are linked in a cross, the data may best be analyzed by considering two genes at a time and by classifying each tetrad into PD, NPD, and T for each pair.

KEYNOTE

In organisms in which all products of meiosis are contained within a single structure, the analysis of the relative proportion of tetrad types provides another way to compute the map distance between genes. The general formula when two linked genes are being mapped is

$$\frac{1/2\,\text{T} + \text{NPD}}{\text{total tetrads}} \times 100$$

SUMMARY

In this chapter we discussed linkage, crossing-over, and gene mapping in eukaryotes. The production of genetic recombinants results from physical exchanges between homologous chromosomes in meiosis. The site of crossing-over is called a chiasma, and the exchange of parts of chromatids is called crossing-over. Crossing-over is a reciprocal event that, in eukaryotes, occurs at the four-strand stage in prophase I of meiosis.

Genetic mapping is the process of locating the position of genes in relation to one another on the chromosome. To map genes in this way, it is first necessary to show that genes are linked (located on the same chromosome), which is indicated by the fact that they do not assort independently in crosses. The map distance between two linked genes is then calculated based on the frequency of recombination between the two genes. The recombination frequency is an approximation of the frequency of crossovers between the two genes. The most accurate map distances are determined for genes that are closely linked because as map distance increases, the incidence of multiple crossovers causes the recombination frequency to be an underestimate of the crossover frequency, and hence of the map distance.

A major research effort called the Human Genome Project is currently underway to map and sequence the human genome. Two types of maps are being constructed: genetic linkage maps and physical maps. Genetic linkage maps, constructed using genetic recombination analysis, show the relative positions of gene and/or DNA markers. Physical maps, constructed using nonrecombination analysis, show the locations of DNA markers. The ultimate physical map will be the complete sequence of the human genome. Physical maps currently being constructed have different resolutions depending on the method being used in their construction. In the future, high-resolution physical maps will be integrated with detailed genetic linkage maps, making it easier to map newly discovered genes.

Finally, we discussed a specialized mapping technique applicable to certain haploid organisms in which all four products of meiosis—the meiotic tetrad—are kept together in a single structure. With this technique, distances between linked genes are calculated by analyzing the relative proportions of tetrad types, which are reflective of recombination events, rather than by analyzing individual progeny.

ANALYTICAL APPROACHES FOR SOLVING GENETICS PROBLEMS

Q5.1 In corn, the gene for colored (C) seeds is completely dominant to the gene for colorless (c) seeds. Similarly, a single gene pair controls whether the endosperm (the part of the seed that contains the food stored for the embryo) is full or shrunken. Full (S) is dominant to shrunken (s). A true-breeding colored, full-seeded plant was crossed with a colorless, shrunken-seeded one. The F$_1$ colored, full plants were testcrossed to the doubly recessive type, that is, colorless and shrunken. The result was as follows:

colored, full	4,032
colored, shrunken	149
colorless, full	152
colorless, shrunken	4,035
Total	8,368

Is there evidence that the gene for color and the gene for endosperm shape are linked? If so, what is the map distance between the two loci?

A5.1 The best approach is to diagram the cross using gene symbols:

P: colored and full × colorless and shrunken

                CC SS   cc ss

                        ↓

F$_1$:   colored and full

                        Cc Ss

Testcross: colored and full × colorless and shrunken

                Cc Ss   cc ss

If the genes were unlinked, a 1:1:1:1 ratio of colored and full : colored and shrunken : colorless and full : colorless and shrunken would be the progeny of this testcross. By inspection, we can see that the actual progeny deviate a great deal from this ratio, showing a 27:1:1:27 ratio. If we did a chi-square test (using the actual numbers, not the percentages or ratios), we would see immediately that the hypothesis that the genes are unlinked is invalid, and we must consider the two genes to be linked in coupling. More specifically, the parental combinations (colored, full and colorless, shrunken) are more numerous than expected, while the recombinant types (colorless, full and colored, shrunken) are correspondingly less numerous than expected. This result comes directly from the inequality of the four gamete types produced by meiosis in the colored and full F$_1$ parent.

Given that the two genes are linked, the crosses may be diagrammed to reflect their linkage as follows:

P: $\dfrac{C\ S}{C\ S}$ × $\dfrac{c\ s}{c\ s}$

                        ↓

F$_1$: $\dfrac{C\ S}{c\ s}$

Testcross: $\dfrac{C\ S}{c\ s}$ × $\dfrac{c\ s}{c\ s}$

To calculate the map distance between the two genes, we need to compute the frequency of crossovers in that region of the chromosome during meiosis. We cannot do that directly, but we can compute the percentage of recombinant progeny that must have resulted from such crossovers:

Parental types:	colored, full	4,032
	colorless, shrunken	4,035
		8,067

Recombinant types:	colored, shrunken	149
	colorless, full	152
		301

This calculation gives about 3.6 percent recombinant types (301/8,368 × 100) and about 96.4 percent parental types (8,067/8,368 × 100). Since the recombination frequency can be used directly as an indication of map distance, especially when the distance is small, we can conclude that the distance between the two genes is 3.6 mu (3.6 cM).

We would get approximately the same result if the two genes were in repulsion rather than in coupling. That is, the crossovers are occurring between homologous chromosomes, regardless of whether or not there are genetic differences in the two homologs that we, as experimenters, use as markers in genetic crosses. This same cross in repulsion would be as follows:

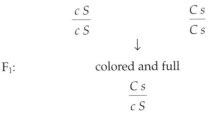

P: colorless and full × colored and shrunken

                $\dfrac{c\ S}{c\ S}$ $\dfrac{C\ s}{C\ s}$

                        ↓

F$_1$:   colored and full

                        $\dfrac{C\ s}{c\ S}$

Data from an actual testcross of the F_1 with colorless and shrunken (*cc ss*) gave 638 colored and full (recombinant) : 21,379 colored and shrunken (parental) : 21,906 colorless and full (parental) : 672 colorless and shrunken (recombinant) with a total of 44,595 progeny. Thus 2.94 percent were recombinants, for a map distance between the two genes of 2.94 mu, a figure reasonably close to the results of the cross made in coupling.

Q5.2 In the Chinese primrose, slate-colored flower (*s*) is recessive to blue flower (*S*); red stigma (*r*) is recessive to green stigma (*R*); and long style (*l*) is recessive to short style (*L*). All three genes involved are on the same chromosome. The F_1 of a cross between two true-breeding strains, when testcrossed, gave the following progeny:

PHENOTYPE	NUMBER OF PROGENY
slate flower, green stigma, short style	27
slate flower, red stigma, short style	85
blue flower, red stigma, short style	402
slate flower, red stigma, long style	977
slate flower, green stigma, long style	427
blue flower, green stigma, long style	95
blue flower, green stigma, short style	960
blue flower, red stigma, long style	27
Total	3,000

a. What were the genotypes of the parents in the cross of the two true-breeding strains?

b. Make a map of these genes, showing gene order and the distances between genes.

c. Derive the coefficient of coincidence for interference between these genes.

A5.2

a. With three gene pairs, eight phenotypic classes are expected, and eight are observed. The reciprocal pairs of classes with the most representatives are those resulting from no crossovers, and these pairs can tell us the genotypes of the original parents. The two classes are slate, red, long and blue, green, short. Thus, the F_1 triply heterozygous parent of this generation must have been *S R L/s r l*, so the true-breeding parents were *S R L/S R L* (blue, green, short) and *s r l/s r l* (slate, red, long).

b. The order of the genes can be determined by inspecting the reciprocal pairs of phenotypic classes that represent the results of double crossing-over. These classes have the least numerous representatives, so the double-crossover classes are slate, green, short (*s R L*) and blue, red, long (*S r l*). The gene pair that has changed its position relative to the other two

pairs of alleles is the central gene, *S/s* in this case. Therefore, the order of genes is *R S L* (or *L S R*). We can diagram the F_1 testcross as follows:

$$\frac{R\,S\,L}{r\,s\,l} \times \frac{r\,s\,l}{r\,s\,l}$$

A single crossover between the *R* and *S* genes gives the green, slate, long (*R s l*) and red, blue, short (*r S L*) classes, which have 427 and 402 members, respectively, for a total of 829. The double-crossover classes have already been defined, and they yield 54 progeny. The map distance between *R* and *S* is given by the crossover frequency in that region, which is the sum of the single crossovers and double crossovers divided by the total number of progeny, then multiplied by 100 percent. Thus

$$\frac{829 + 54}{3,000} \times 100\% = \frac{883}{3,000} \times 100\%$$

$$= 29.43\% \text{ or } 29.43 \text{ map units}$$

With similar logic the distance between *S* and *L* is given by the crossover frequency in that region, which is the sum of the single-crossover and double-crossover progeny classes divided by the total number of progeny. The single-crossover progeny classes are green, blue, long (*R S l*) and red, slate, short (*r s L*), which have 95 and 85 members, respectively, for a total of 180. The map distance is given by

$$\frac{180 + 54}{3,000} \times 100\% = \frac{234}{3,000} \times 100\%$$

$$= 7.8\% \text{ or } 7.8 \text{ map units}$$

The data we have derived give us the following map:

c. The coefficient of coincidence is given by

$$\frac{\text{frequency of observed double crossovers}}{\text{frequency of expected double crossovers}}$$

The frequency of observed double crossovers is $54/3,000 = 0.018$. The expected frequency of double crossovers is the product of the map distances between *r* and *s* and between *s* and *l*, that is, $0.294 \times 0.078 = 0.023$. The coefficient of coincidence, therefore, is $0.018/0.023 = 0.78$. In other words, 78 percent of the expected double crossovers did indeed take place; there was 22 percent interference.

QUESTIONS AND PROBLEMS

5.1 A cross $a^+a^+ b^+ b^+ \times aa\ bb$ results in an F_1 of phenotype $a^+ b^+$; the following numbers are obtained in the F_2 (phenotypes):

$a^+ b^+$	110
$a^+ b$	16
$a\ b^+$	19
$a\ b$	15
Total	160

Are genes at the a and b loci linked or independent? What F_2 numbers would otherwise be expected?

*5.2 In corn, a dihybrid for the recessives a and b is testcrossed. The distribution of the phenotypes was as follows:

$A\ B$	122
$A\ b$	118
$a\ B$	81
$a\ b$	79

Are the genes assorting independently? Test the hypothesis with a χ^2 test. Explain tentatively any deviation from expectation, and tell how you would test your explanation.

5.3 In *Drosophila*, the mutant black (b) has a black body, and the wild type has a grey body; the mutant vestigial (vg) has wings that are much shorter and crumpled compared to the long wings of the wild type. In the following cross, the true-breeding parents are given together with the counts of offspring of F_1 females × black and vestigial males:

P   black and normal × grey and vestigial
F_1 females × black and vestigial males

Progeny:	grey, normal	283
	grey, vestigial	1,294
	black, normal	1,418
	black, vestigial	241

From these data, calculate the map distance between the black and vestigial genes.

*5.4 Use the following two-point recombination data to map the genes concerned. Show the order and the length of the shortest intervals.

Gene Loci	% Recombination	Gene Loci	% Recombination
a,b	50	b,d	13
a,c	15	b,e	50
a,d	38	c,d	50
a,e	8	c,e	7
b,c	50	d,e	45

5.5 A corn plant known to be heterozygous at three loci is testcrossed. The progeny phenotypes and frequencies are as follows:

+	+	+	455
a	b	c	470
+	b	c	35
a	+	+	33
+	+	c	37
a	b	+	35
+	b	+	460
a	+	c	475
	Total		2,000

Give the gene arrangement, linkage relations, and map distances.

*5.6 Genes a and b are linked, with 10 percent recombination. What would be the phenotypes, and the probability of each, among progeny of the following cross?

$$\frac{a\ b^+}{a^+b} \times \frac{a\ b}{a\ b}$$

*5.7 Genes a and b are sex-linked and are located 7 mu apart on the X chromosome of *Drosophila*. A female of genotype $a^+ b / a\ b^+$ is mated with a wild-type ($a^+ b^+$).
a. What is the probability that one of her sons will be either $a^+ b^+$ or $a\ b^+$ in phenotype?
b. What is the probability that one of her daughters will be $a^+ b^+$ in phenotype?

5.8 In *Drosophila*, a and b are linked autosomal genes whose recombination frequency in females is 5 percent; c and d are X-linked genes, located 10 mu apart. A homozygous dominant female is mated to a recessive male, and the daughters are testcrossed. Which of the following would you expect to observe in the testcross progeny?
a. Different ratios in males and females.
b. Nearly equal frequency of $a^+ b\ c^+ d^+$, $a^+ b\ c\ d$, $a\ b^+ c^+ d^+$, and $a\ b^+ c\ d$ classes.
c. Independent segregation of some genes with respect to others involved in the cross.
d. Double-crossover classes less frequent than expected because of interference between the two marked regions.

*5.9   Genes *a* and *b* are on one chromosome, 20 mu apart; *c* and *d* are on another chromosome, 10 mu apart. Genes *e* and *f* are on yet another chromosome and are 30 mu apart. Cross a homozygous *A B C D E F* individual with an *a b c d e f* one, and cross the F_1 back to an *a b c d e f* individual. What are the chances of getting individuals of the following phenotypes in the progeny?

a. *A B C D E F*
b. *A B C d e f*
c. *A b c D E f*
d. *a B C d e f*
e. *a b c D e F*

*5.10   Genes *d* and *p* occupy loci 5 map units apart in the same autosomal linkage group. Gene *h* is a separate autosomal linkage group and therefore segregates independently of the other two. What types of offspring are expected, and what is the probability of each, when individuals of the following genotypes are testcrossed?

a. $\dfrac{D\,P}{d\,p}\,\dfrac{h}{h}$ **b.** $\dfrac{d\,P}{D\,p}\,\dfrac{H}{h}$

5.11   A hairy-winged (*h*) *Drosophila* female is mated with a yellow-bodied (*y*), white-eyed (*w*) male. The F_1 are all wild type. The F_1 progeny are then crossed, and the F_2 that emerge are as follows:

Females:	wild type	757
	hairy	243
Males:	wild type	390
	hairy	130
	yellow	4
	white	3
	hairy, yellow	1
	hairy, white	2
	yellow, white	360
	hairy, yellow, white	110

Give genotypes of the parents and the F_1, and note the linkage relations and distances where appropriate.

5.12   For each of the following tabulations of testcross progeny phenotypes and numbers, state which locus is in the middle, and reconstruct the genotype of the tested triple heterozygotes.

a.

A B C	191
a b c	180
A b c	5
a B C	5
A B c	21
a b C	31
A b C	104
a B c	109

b.

C D E	9
c d e	11
C d e	35
c D E	27
C D e	78
c d E	81
C d E	275
c D e	256

c.

F G H	110
f g h	114
F g h	37
f G H	33
F G h	202
f g h	185
F g H	4
f G h	0

5.13   The following numbers were obtained for testcross progeny in *Drosophila* (phenotypes):

+ m +		218
w + f		236
+ + f		168
w m +		178
+ m f		95
w + +		101
+ + +		3
w m f		1
	Total	1,000

Construct a genetic map.

*5.14   Three of the many recessive mutations in *Drosophila melanogaster* that affect body color, wing shape, or bristle morphology are black (*b*) body versus grey in the wild type; dumpy (*dp*), obliquely truncated wings versus long wings in the wild type; and hooked (*hk*) bristles at the tip versus not hooked in the wild type. From a cross of a dumpy female with a black, hooked male, all the F_1 were wild type for all three characters. The testcross of an F_1 female with a dumpy, black, hooked male gave the following results:

wild type		169
black		19
black, hooked		301
dumpy, hooked		21
hooked		8
hooked, dumpy, black		172
dumpy, black		6
dumpy		304
	Total	1,000

a. Construct a genetic map of the linkage group (or groups) these genes occupy. If applicable, show the order and give the map distances between the genes.

b. (1) Determine the coefficient of coincidence for the portion of the chromosome involved in the cross. (2) How much interference is there?

*5.15   Genes *a*, *b*, and *c* are recessive. Females heterozygous at these three loci are crossed to phenotypically wild-type males. The progeny are phenotypically as shown.

Daughters: all + + +

Sons:

+	+	+	23
a	b	c	26
+	+	c	45
a	b	+	54
+	b	c	427
a	+	+	424
a	+	c	1
+	b	+	0
	Total		1,000

a. What is known of the genotype of the females' parents with respect to these three loci? Give gene order and the arrangement in the homologs.

b. What is known of the genotype of the male parents?

c. Map the three genes.

5.16 Two normal-looking *Drosophila* are crossed and yield the following phenotypes among the progeny:

Females:	+	+	+	2,000
Males:	+	+	+	3
	a	b	c	1
	+	b	c	839
	a	+	+	825
	a	b	+	86
	+	+	c	90
	a	+	c	81
	+	b	+	75
		Total		4,000

Give parental genotypes, gene arrangement in the female parent, map distances, and the coefficient of coincidence.

5.17 The following questions make use of this genetic map:

10	5	20	10	20	5
a	k j		b	d	m f

Calculate:

a. the frequency of *j b* gametes from a *J B/j b* genotype.

b. the frequency of *A M* gametes from an *a M/A m* genotype.

c. the frequency of *J B D* gametes from a *j B d/J b D* genotype.

d. the frequency of *J B d* gametes from a *j B d/J b D* genotype.

e. the frequency of *j b d/j b d* genotypes in a *j B d/J b D* × *j B d/J b D* mating.

f. the frequency of *A k F* gametes from an *A K F/a k f* genotype.

*5.18 A female *Drosophila* carries the recessive mutations *a* and *b* in repulsion on the X chromosome (she is heterozy-

gous for both). She is also heterozygous for an X-linked recessive lethal allele, *l*. When she is mated to a true-breeding, normal male, she yields the following progeny:

Females:	1,000	+	+
Males:	405	a	+
	44	+	b
	48	+	+
	2	a	b

Draw a chromosome map of the three genes in the proper order and with map distances as nearly as you can calculate them.

5.19 A farmer who raises rabbits wants to break into the seasonal Easter market. He has stocks of two true-breeding lines. One is hollow and longeared but not chocolate, while the second is solid, short-eared, and chocolate. Hollow (*h*), long ears (*le*), and chocolate (*ch*) are all recessive, autosomal, and linked as in the following map:

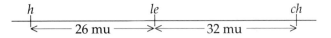

The farmer can generate a trihybrid by crossing his two lines, and at great expense he is able to obtain the services of a male homozygous recessive at all three loci to cross with his F₁ females.

The farmer has buyers for both solid and hollow bunnies; however, all must be chocolate and long-eared. Assuming that interference is zero, if he needs 25 percent of the progeny of the desired phenotypes to be profitable, should he continue with his breeding? Calculate the percent of the total progeny that will be the desired phenotypes.

5.20 In *Drosophila*, many different mutations have been isolated that affect a normally deep-red eye color caused by the deposition of brown and bright-red pigments. Two X-linked recessive mutations are *w* (white eyes, map position 1.5) and *cho* (chocolate-brown eyes, map position 13.0), with *w* epistatic to *cho*.

a. A white-eyed female is crossed to a chocolate-eyed male, and the normal, red-eyed F₁ females are crossed to either wild-type or white-eyed males. Determine the frequency of the progeny types produced in each cross.

b. The recessive mutation *st* causes scarlet (bright-red) eyes and maps to the third chromosome at position 44. Mutant flies with only *st* and *cho* alleles have white eyes, and *w* is epistatic to *st*. Suppose a true-breeding *w* male is crossed to a true-breeding *cho, st* female. Determine the frequency of the progeny types you would expect if the F₁ females are crossed to true-breeding scarlet-eyed males.

5.21 The accompanying table shows the only human chromosomes present in stable human–mouse cell hybrid lines.

		HUMAN CHROMOSONES			
		2	4	10	19
	A	−	+	+	−
Hybrid lines	B	+	−	+	+
	C	−	+	+	+
	D	+	+	−	−

The presence of four enzymes, I, II, III, and IV, was investigated: I was present @REV1a:in *A*, *B*, and *C* but absent in *D*; II was in *B* and *D* but absent in *A* and *C*; III was in *A*, *C*, and *D* but not in *B*; and IV was in *B* and *C* but not in *A* and *D*. On what chromosomes are the genes for the four enzymes?

5.22 In *Saccharomyces*, *Neurospora*, and *Chlamydomonas*, what meiotic events give rise to PD, NPD, and T tetrads?

5.23 Double exchanges between two loci can be of several types, called two-strand, three-strand, and four-strand doubles.

a. Four recombination gametes would be produced from a tetrad in which the first of two exchanges is depicted in the following figure. Draw in the second exchange.

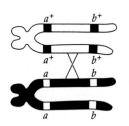

b. In the following figure, draw in the second exchange so that four nonrecombination gametes would result.

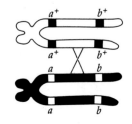

5.24 The following asci were obtained from the cross *leu* + × + *rib* in yeast. Draw the linkage map and determine the map distance.

110	45	6	39
leu +	*leu rib*	+ +	*leu* +
+ *rib*	*leu* +	*leu rib*	+ *rib*
leu +	+ +	*leu rib*	+ +
+ *rib*	+ *rib*	+ +	*leu rib*

*5.25 The genes *a*, *b*, and *c* are linked in *Neurospora crassa*. The following asci were obtained from the cross *a b* + × + + *c*

50	166	68	16
a b +	*a b* +	*a b* +	*a b* +
+ *b c*	*a b* +	*a* + *c*	*a b c*
a + +	+ + *c*	+ *b* +	+ + +
+ + *c*	+ + *c*	+ + *c*	+ + *c*

Determine the correct gene order. Calculate all gene-gene distances

CHAPTER *6*

GENETIC ANALYSIS IN BACTERIA AND BACTERIOPHAGES

127

PRINCIPAL POINTS

~ Conjugation is a process in which there is a unidirectional transfer of genetic information through direct cellular contact between a donor and a recipient bacterial cell. The donor state is conferred by the presence of a plasmid called an *F* factor. Conjugation results in the transfer of the *F* factor from donor to recipient.

~ The *F* factor can integrate into the bacterial chromosome. Strains in which this has occurred—*Hfr* strains—can conjugate with recipient strains, and transfer of the bacterial chromosome ensues. The sequence and distances between genes can be determined by the order and time of acquisition of the genes by the recipient from the donor during conjugation.

~ Transformation is the transfer of genetic material between organisms by small extracellular pieces of DNA. By genetic recombination, part of the transforming DNA molecule can exchange with a portion of the recipient's chromosomal DNA. Transformation can be used experimentally to determine gene order and map distances between genes.

~ Transduction is a process whereby bacteriophages (phages) mediate the transfer of bacterial DNA from one bacterium (the donor) to another (the recipient). Transduction can be used experimentally to map bacterial genes.

~ The same principles used to map eukaryotic genes are used to map phage genes. That is, a bacterial host is simultaneously infected with two strains of phages differing from each other in one or more gene loci. The percentages of recombinants are determined and the sequence and distances between genes are then postulated.

~ The same principles of recombinational mapping in eukaryotes can be applied to mapping the distance between mutational sites in different genes (intergenic mapping) and to mapping mutational sites within the same gene (intragenic mapping).

~ From fine-structure analysis of the *rII* region of bacteriophage T4, it was determined that the unit of mutation and of recombination is the DNA base pair.

~ The number of genes (units of function) that cause a particular mutant phenotype is determined by the complementation, or *cis-trans*, test. If two viral mutants, each carrying a mutation in a different gene, are combined in a single host cell, the mutations will make up for each other's defect (complement) and a wild-type phenotype will result. If two mutants, each carrying a mutation in the same gene, are combined, the mutations will not complement and the mutant phenotype will still be expressed.

*I*n Chapter 5 we considered the principles of genetic mapping in eukaryotic organisms. To map genes in bacteria and bacteriophages, geneticists use essentially the same classical experimental strategies. Crosses are made between strains that differ in genetic markers, and recombinants, the products of the exchange of genetic material, are detected and counted. Analysis of data obtained from such crosses is the same as for eukaryotes: the frequency with which exchanges occur between two sets of genes relates to the map distance between the two loci. (A locus is the specific place on a chromosome where a gene is located.) The major difference is the experimental techniques involved. The purpose of mapping bacterial and bacteriophage genes is similar to that for eukaryotic genes. For example, mapping is a primary step in localizing the genes. Information obtained will be helpful in cloning the genes in the future for a detailed analysis of their structures and functions.

Recently, emphasis in genetic research has shifted from localizing individual genes on chromosomes by making crosses to determining the sequence of bases in the DNA. With DNA base sequence analysis, scientists can identify genes directly so that the ultimate genetic map of a species can be constructed. When all the genes of an organism are identified, at least at the nucleotide level, the door is opened to investigating the function of each gene. In the case of pathogenic microorganisms, the genomic sequence information is an extremely valuable resource for efforts to identify and understand the genes responsible for pathogenesis. At this writing, complete genomic sequences have been determined for 11 eubacterial species and for 4 archaea. The eubacterial genomes sequenced include the 4.6 million-base-pair (4.6-megabase) genome of *E. coli*, the 1.44-megabase genome of the Lyme disease causative agent *Borrelia burgdorferi*, the 1.66-megabase genome of the stomach-ulcer-causing *Helicobacter pylori*, and the 1.14-megabase genome of the syphilis bacterium *Treponema pallidum*. The archaeon genomes sequenced include the 1.66-megabase genome of *Methanococcus jannaschii*. Incidentally, these and other studies have shown that not all bacteria have a single, circular chromosome as is typically described in textbooks. Rather, some species have more than one chromosome, and in some cases the chromosome is linear. *Helicobacter pylori*, for example, has one large circular chromosome and a small 58,000-base-pair (58-kb, where kb = 1,000 base pairs) chromosome.

Your goal for this chapter is to learn about the classic genetic studies of bacteria and bacteriophages. You will also learn about a series of classic genetic experiments in which researchers investigated the fine structure of the gene: that is, the detailed molecular organization of the gene as it relates to the mutational, recombinational, and functional events in which the gene is involved. A bacteriophage gene was the subject of these experiments.

GENETIC ANALYSIS OF BACTERIA

Genetic material can be transferred between bacteria by three main processes: conjugation, transformation, and transduction. It is possible to map bacterial genes by using any one of these methods. In each case, (1) transfer is unidirectional and (2) no true diploid zygote is formed (unlike in eukaryotes). However, not all methods can be used for all species of bacterium, and the size of the region that can be mapped varies according to method.

Among bacteria, *E. coli* is used extensively for genetic and molecular analysis. It is found commonly in the large intestines of most animals (including humans). This bacterium is a good subject for study since it can be grown on a simple, defined medium and can be handled with simple microbiological techniques.

E. coli is a cylindrical organism about 1–3 μm long and 0.5 μm in diameter (Figure 6.1); like most bacteria, it is small compared with eukaryotic cells.[1] Its protoplasm is full of ribosomes (particles on which proteins are made), and a single circular DNA chromosome is in a central region called the **nucleoid**. (Some bacteria have more than one chromosome or linear chromosomes.) As in all prokaryotes, there is no membrane between the nucleoid region and the rest of the cell.

Like other bacteria, *E. coli* can be grown both in a liquid culture medium and on the surface of growth medium solidified with agar. Genetic analysis of bacteria typically is done by spreading ("plating") cells on the surface of agar medium. Wherever a single bacterium lands on the agar surface it will grow and

[1]One of the largest bacteria is *Epulopiscium fishelsoni*, the surgeonfish symbiont. With a cell size that can reach 800 μm long and 60 μm wide, it is a million times larger than *E. coli* and larger than some eukaryotic cells.

~ **FIGURE 6.1**

Colorized scanning electron micrograph of *Escherichia coli* bacteria.

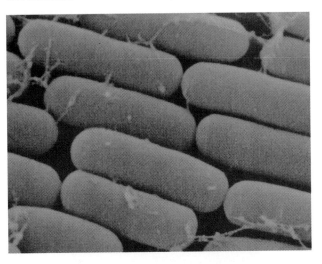

divide repeatedly, resulting in the formation of a visible cluster of genetically identical cells called a *colony* (Figure 6.2). Each colony consists of a clone of cells that are genetically identical to the parental cell that initiated the colony. The concentration of bacterial cells in a liquid culture (the *titer*) can be determined by spreading known volumes of the culture or of a known dilution of the culture on the agar surface, incubating the plates at a constant temperature, and then counting the number of resulting colonies. The number you obtain is converted to colony forming units (cfu) per milliliter (mL). For example, if 100 μl of a 1,000-fold dilution of a culture is spread on a plate and 165 colonies are produced, then this means there were 165 bacteria in 100 μl of the 1,000-fold dilution. Thus, in the original culture there were 165 (colonies) × 1,000 (dilution factor) × 10 (because 0.1 mL was plated) = 1,650,000 cfu/mL = 1.65×10^6 cfu/mL.

The composition of the culture medium used depends upon the experiment and the genotypes of the strains being used. Each bacterial species (or any other microorganism, such as yeast) has a characteristic *minimal medium* on which it will grow. A minimal medium contains only those nutrients absolutely required for the growth of wild-type cells. The minimal medium for wild-type *E. coli*, for example, consists of a sugar (a carbon source) and some salts and trace elements. From the minimal medium, the organism can synthesize all the other components it needs for growth and reproduction, including amino acids, vitamins, DNA, and RNA. By contrast, the *complete medium* for a microorganism supplies vitamins and amino acids and all kinds of substances that might be

~ **Figure 6.2**

Bacterial colonies growing on a nutrient medium in a Petri dish.

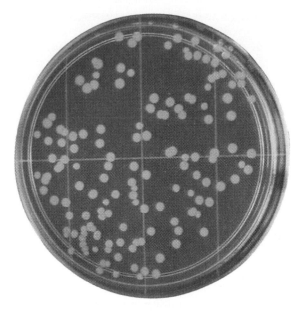

expected to be essential metabolites and whose biosynthesis might be interfered with by mutation.

Genetic analysis of bacteria (and other microorganisms) typically involves studying mutants defective in their abilities to make one or more molecules essential for growth, and perhaps also defective in genes affecting other metabolic processes. Strains that are unable to synthesize essential nutrients are called **auxotrophs** (also called *auxotrophic mutants*, *nutritional mutants*, or *biochemical mutants*). A strain that is wild type and thus can synthesize all essential nutrients is called a **prototroph**; prototrophs need no nutritional supplements in the growth medium. By definition, the wild type or prototroph grows on the minimal medium for that organism, whereas an auxotroph grows on complete medium or on minimal medium plus the appropriate nutritional supplement or supplements.

For example, consider the *E. coli* strain with the genotype *trp ade thi⁺*. Here the gene symbol indicates a particular nutritional requirement. If the genotype has a superscript +, then the gene is wild type, indicating that the bacteria are able to synthesize the nutritional substance. If no + superscript is present, the gene is mutant and the culture medium must be supplemented with this particular substance. Thus, the *trp ade thi⁺* strain can grow on the *E. coli* complete medium but not on the minimal medium. It will also grow on minimal medium supplemented with the amino acid tryptophan (*trp*) and the purine (needed for RNA and DNA precursor synthesis) adenine (*ade*), but without the vitamin thiamine (*thi*).

Some genes are not involved in biosynthetic pathways, but in utilization pathways. For example, there are a number of different genes for using various carbon sources such as lactose, arabinose, and maltose. In this case, the + next to the gene symbol means that the gene is wild type and, hence, that the bacterium can metabolize the substance. For example, a *lac⁺* strain can metabolize lactose, whereas a *lac* mutant strain cannot. So, by varying the nutrient composition of the medium it is possible to detect and quantify various genotypic classes in a genetic analysis.

In genetic experiments with microorganisms such as *E. coli*, crosses are made between strains differing in genotype (and, therefore, phenotype), and progeny are analyzed for parental and recombinant phenotypes. When auxotrophic mutations are involved, the determination of parental and progeny phenotypes (and, therefore, genotypes, since bacteria are haploid) involves testing colonies for their growth requirements. One convenient procedure for such testing is *replica plating* (see Figure 18.16). In the replica plating technique, samples from a liquid culture are plated onto a complete medium. On incubation, colonies grow wherever a cell has landed on the medium. The bacteria of the colonies are transferred onto a sterile velveteen cloth mounted on a replica plater by stamping onto it. Replicas of the original colony pattern on the cloth are then made by gently pressing new plates onto the velveteen. If the new plate contains minimal medium, only prototrophic colonies can grow. Then, by comparing the patterns on the original master plate with those on the minimal medium replica plate, researchers can readily identify auxotrophic colonies, since they will be on the master plate but not on the minimal medium plate. Using other plates containing minimal medium plus combinations of nutritional supplements appropriate for the strain or strains involved, the phenotypes/genotypes of all the auxotrophic colonies can be determined.

GENETIC MAPPING IN BACTERIA BY CONJUGATION

Discovery of Conjugation in *E. coli*

Conjugation is a process in which there is a unidirectional transfer of genetic information through direct cellular contact between a donor bacterial cell and a recipient bacterial cell. The contact is followed by formation of a physical bridge between the cells. Then a segment (rarely all) of the donor's chromosome may be transferred into the recipient and may undergo genetic recombination with a homologous chromosome segment of the recipient cell. Those recipients having incorporated a piece of donor DNA into their chromosomes are called **transconjugants**.

Conjugation was discovered in 1946 by Joshua Lederberg and Edward Tatum. They studied two *E. coli* strains that differed in their nutritional requirements. Strain A had the genotype *met bio thr⁺ leu⁺ thi⁺*, and strain B had the genotype *met⁺ bio⁺ thr leu thi*. Strain A can only grow on a medium supplemented with the amino acid methionine (*met*) and the vitamin biotin (*bio*) but does not need the amino acids threonine (*thr*) or leucine (*leu*) or the vitamin thiamine (*thi*). Strain B can only grow on a medium supplemented with threonine, leucine, and thiamine but does not require methionine or biotin.

Lederberg mixed *E. coli* strains A and B together and plated them onto minimal medium (Figure 6.3). The mixed culture, however, gave rise to some prototrophic colonies (*met⁺ bio⁺ thr⁺ leu⁺ thi⁺*). Since no colonies appeared on control plates, mutation was

ruled out as the cause of the prototrophic colonies. The mixing, then, is a genetic cross that produced recombinants.

In a separate experiment, Bernard Davis placed strains A and B in a liquid medium on either side of a U-tube apparatus (Figure 6.4) separated by a filter with pores too small to allow bacteria to move through. The medium was moved between compartments by alternating suction and pressure, and then the cells were plated on minimal medium to check for the appearance of prototrophic colonies. No prototrophic colonies appeared, meaning that cell-to-cell contact was required for the genetic exchange to occur. These experiments indicated that *E. coli* has the type of mating system called conjugation.

The Sex Factor *F*

In 1953 William Hayes showed that genetic exchange in *E. coli* occurs in only one direction, with one cell acting as a donor and the other cell acting as a recipient. Hayes proposed that the transfer of genetic material between the strains is mediated by a *sex factor* named *F* that the donor cell possesses (*F⁺*) and the recipient cell lacks (*F⁻*). The *F* factor found in *E. coli* is an example of a **plasmid**, a self-replicating, circular DNA found distinct from the main bacterial chromosome. About 1/40th of the size of the host chromosome, the *F* factor contains a region of DNA called the **origin** (or O), the point where DNA transfer to the recipient begins, as well as a number of genes, including those that specify hairlike host cell surface components called **F-pili** (singular, F-pilus), or sex-pili,

~**FIGURE 6.3**

Lederberg and Tatum experiment showing that sexual recombination occurs between cells of *E. coli*. After the cells from strain A and strain B have been mixed and the mixture plated, a few colonies grow on the minimal medium, indicating that they can now make the essential constituents. These colonies are recombinants produced by an exchange of genetic material between the strains.

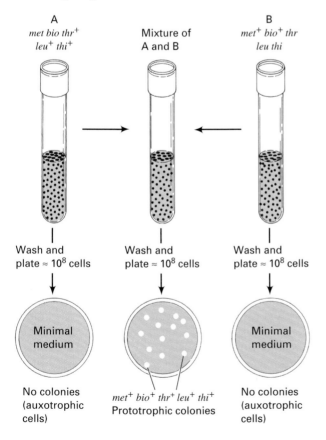

A
met bio thr⁺ leu⁺ thi⁺

Mixture of A and B

B
met⁺ bio⁺ thr leu thi

Wash and plate ≈ 10⁸ cells

Wash and plate ≈ 10⁸ cells

Wash and plate ≈ 10⁸ cells

Minimal medium

Minimal medium

No colonies (auxotrophic cells)

met⁺ bio⁺ thr⁺ leu⁺ thi⁺
Prototrophic colonies

No colonies (auxotrophic cells)

~**FIGURE 6.4**

Davis's U-tube experiment showing that physical contact between the two bacterial strains of the Lederberg and Tatum experiment was needed in order for genetic exchange to occur.

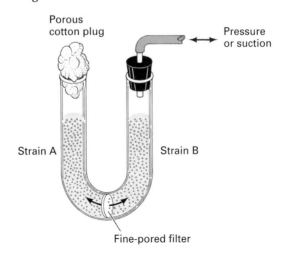

Porous cotton plug

Pressure or suction

Strain A

Strain B

Fine-pored filter

~ FIGURE 6.5

Electron micrograph of bacterial conjugation between an F^+ donor and F^- recipient E. coli bacterium. (The spherical structure on the upper F-pilus bridging the two bacteria is a donor-specific RNA phage MS-2.) Magnification 30,000×.

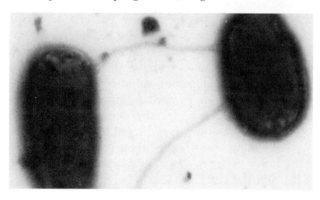

which allow the physical union of F^+ and F^- cells to take place.

When F^+ and F^- cells are mixed, they may conjugate ("mate"). Figure 6.5 is an electron micrograph of conjugation between an F^+ donor and an F^- recipient, and Figure 6.6a1 diagrams the event. No conjugation can occur between two cells of the same mating type (that is, two F^+ bacteria or two F^- bacteria). Genetic material is transferred from donor to recipient during conjugation beginning when one strand of the F factor is nicked at the origin, and DNA replication proceeds from that point (Figure 6.6a2). Beginning at the origin, a single strand of DNA is transferred to the F cell (Figure 6.6a3). Think of the process like unraveling a roll of paper towels. The origin is the first stretch of DNA unwound; as unwinding continues, replication maintains the remaining circular F factor in a double-stranded form. Once the F factor DNA enters the F^- recipient, the complementary strand is synthesized (Figure 6.6a4). When the complete F factor has been transferred, the F^- cell becomes an F^+ cell (Figure 6.6a5). In $F^+ \times F^-$ crosses, none of the bacterial chromosome is transferred; only the F factor is transferred.

KEYNOTE

Some E. coli bacteria possess a plasmid, called the F factor, that is required for mating. E. coli cells containing the F factor are designated F^+, and those without it are F^-. The F^+ cells (donors) can mate with F^- cells (recipients) in a process called conjugation, which leads to the one-way transfer of a copy of the F factor from donor to recipient during replication of the F factor. As a result, both donor and recipient are F^+. None of the bacterial chromosome is transferred during $F^+ \times F^-$ conjugation.

High-Frequency Recombination Strains of E. coli

To get recombinants for *chromosomal* genes by conjugation involves special derivatives of F^+ strains, called *Hfr*, or **high-frequency recombination**, strains. Discovered separately by William Hayes and Luca Cavalli-Sforza, *Hfr* strains originate by a rare crossover event in which the F factor integrates into the bacterial chromosome (Figure 6.6b1–2). Plasmids such as F that are also capable of integrating into the bacterial chromosome are called **episomes**. When the F factor is integrated, it no longer replicates independently but is replicated as part of the host chromosome.

Due to the F factor genes, *Hfr* cells can conjugate with F^- cells (Figure 6.6b3). When mating happens, events similar to those in the $F^+ \times F^-$ mating occur. The integrated F factor becomes nicked at the origin and replication begins (Figure 6.6b4). During replication, part of the F factor starting with the origin moves into the recipient cell, where the transferred strand is copied. In a short time the donor bacterial chromosome begins to be transferred into the recipient. If there are allelic differences between donor genes and recipient genes, recombinants can be isolated (Figure 6.6b5). The recombinants are produced by double crossovers between the linear donor DNA and the circular recipient chromosome. In a double crossover, a segment of donor DNA is exchanged for the homologous segment of recipient DNA.

In *Hfr* × F^- matings the F^- cell almost never acquires the *Hfr* phenotype. In order for the recipient cell to become *Hfr*, it must receive a complete copy of the F factor. However, only part of the F factor is transferred at the beginning of conjugation; the rest of the F factor is at the end of the donor chromosome. *All* of the donor chromosome would have to be transferred in order for a complete functional F factor to be found in the recipient, and that would require about 100 minutes at 37°C, the normal growth temperature for E. coli. This is an extremely rare event because all the while the bacteria are conjugating, they are "jiggling" around, so mating pairs typically break apart long before the second part of the F factor is transferred.

F' Factors

Hfr cells rarely will become F^+ cells by excision of the F factor from the chromosome. Occasionally, excision of the F factor is not precise and an F factor is produced with a small section of the host chromosome that was adjacent to the integrated F factor. Since the F factor integrates at one of many sites on the chromosome, many different host chromosome segments may be picked up in this way. Consider an

~ **FIGURE 6.6**

Transfer of genetic material during conjugation in *E. coli.* (a) Transfer of the *F* factor from donor to recipient cell during $F^+ \times F^-$ matings. (b) Production of *Hfr* strain by integration of *F* factor and transfer of bacterial genes from donor to recipient cell during $Hfr \times F^-$ matings.

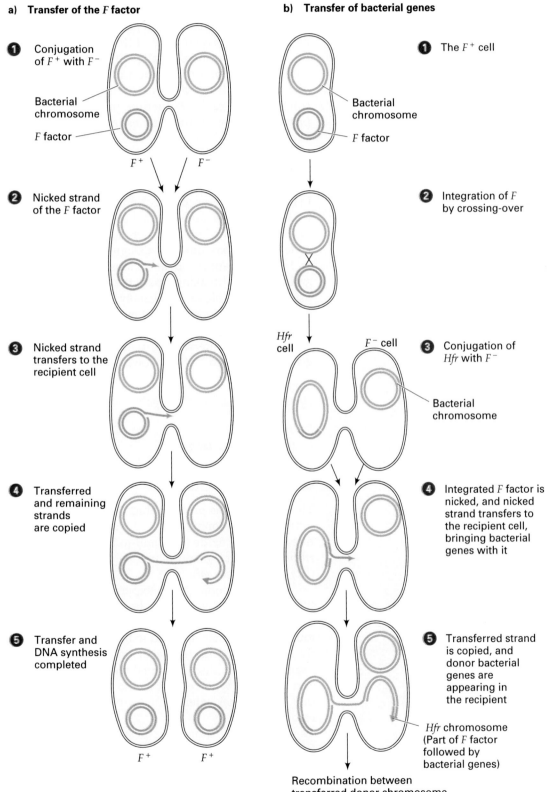

a) Transfer of the *F* factor

① Conjugation of F^+ with F^-

Bacterial chromosome

F factor

F^+ F^-

② Nicked strand of the *F* factor

③ Nicked strand transfers to the recipient cell

④ Transferred and remaining strands are copied

⑤ Transfer and DNA synthesis completed

F^+ F^+

b) Transfer of bacterial genes

① The F^+ cell

Bacterial chromosome

F factor

② Integration of *F* by crossing-over

Hfr cell F^- cell

③ Conjugation of *Hfr* with F^-

Bacterial chromosome

④ Integrated *F* factor is nicked, and nicked strand transfers to the recipient cell, bringing bacterial genes with it

⑤ Transferred strand is copied, and donor bacterial genes are appearing in the recipient

Hfr chromosome (Part of *F* factor followed by bacterial genes)

Recombination between transferred donor chromosome and recipient chromosome

E. coli strain in which the *F* factor has integrated next to the *lac*⁺ region, a set of genes required for the breakdown of lactose (Figure 6.7a). If the looping out is not precise, then the adjacent *lac*⁺ host chromosomal genes may be included in the loop (Figure 6.7b). Then, by a single crossover the looped-out DNA will be separated from the host chromosome (Figure 6.7c) to produce an *F* factor carrying the host's *lac*⁺ genes. *F* factors containing bacterial genes are called *F'* (*F* prime) factors, and they are named

for the genes they have picked up. An *F'* with the *lac* genes is called *F'* (*lac*).

Cells with *F'* factors can conjugate with *F*⁻ cells. As in *F*⁺ × *F*⁻ conjugation, a copy of the *F'* factor is transferred to the *F*⁻ cell, which then becomes *F'*. The recipient also receives a copy of the bacterial gene(s) on the *F* factor (*lac* in our example). Since the recipient has its own copy of that DNA, the resulting cell line will be partially diploid (*merodiploid*), having two copies of one or a few genes and only one copy of all the others. This particular type of conjugation is called **F-duction**, or *sexduction*, and it provides a way to study genes in a diploid state in *E. coli*.

Using Conjugation to Map Bacterial Genes

In the late 1950s François Jacob and Elie Wollman studied the transfer of chromosomal genes from *Hfr* strains to *F*⁻ cells that had allelic differences for a number of genes. Their experimental design involved making an *Hfr* × *F*⁻ mating and, at various times after conjugation began, breaking apart the conjugating pairs using a kitchen blender and analyzing the transconjugants for which donor genes they had received. This is called an *interrupted-mating* experiment.

The use of interrupted mating to map bacterial genes is illustrated by the following cross (Figure 6.8a):

Donor:
HfrH thr⁺ *leu*⁺ *azi*ʳ *ton*ʳ *lac*⁺ *gal*⁺

Recipient:
F⁻ *thr leu azi*ˢ *ton*ˢ *lac gal*

(The superscript ˢ means "sensitive" and the superscript ʳ means "resistant.")

The *HfrH* strain is prototrophic. The *F*⁻ strain is auxotrophic for threonine (*thr*) and leucine (*leu*), is sensitive to sodium azide (*azi*ˢ) and to infection by bacteriophage T1 (*ton*ˢ), and is unable to ferment lactose (*lac*) or galactose (*gal*).

In such a conjugation experiment the two cell types are mixed together in a liquid medium at 37°C. Samples are removed from the mating mixture at various times and then agitated to break the pairs apart. By plating on selective agar media, recombinant recipients (the transconjugants) are then searched for and analyzed with respect to the time at which the first donor genes entered the recipient and produced recombinants.

Figure 6.8b shows the results. The threonine (*thr*⁺) and leucine (*leu*⁺) genes are the first donor genes to be transferred to the *F*⁻, and we set their time of entry to 0 minutes. (The two genes are inseparable timewise in a conjugation experiment.) The next gene to be transferred is *azi*ʳ, and recombinants for this

~ **FIGURE 6.7**

Production of an F' factor. (a) Region of bacterial chromosome into which the *F* factor has integrated; (b) The *F* factor looping out incorrectly, so it includes a piece of bacterial chromosome, the *lac*⁺ genes; (c) Excision, in which a single crossover between the looped-out DNA segment and the rest of the bacterial chromosome results in an *F'* factor, called *F'* (*lac*).

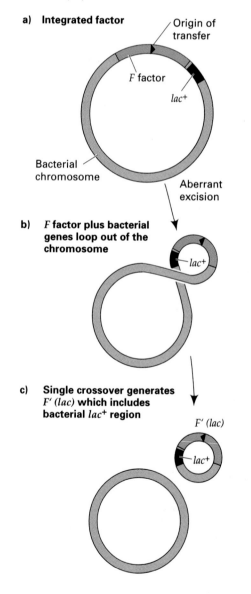

a) **Integrated factor**

Origin of transfer

F factor

lac⁺

Bacterial chromosome

Aberrant excision

b) ***F* factor plus bacterial genes loop out of the chromosome**

lac⁺

c) **Single crossover generates *F'* (*lac*) which includes bacterial *lac*⁺ region**

F' (*lac*)

lac⁺

~ FIGURE 6.8

Interrupted-mating experiment involving the cross *HfrH thr⁺ leu⁺ azi^r ton^r lac⁺ gal⁺ × F⁻ thr leu azi^s ton^s lac gal*. The progressive transfer of donor genes with time is illustrated. Recombinants are generated by an exchange of a donor fragment with the homologous recipient fragment resulting from a double crossover event. (a) At various times after mating commences, the conjugating pairs are broken apart and the transconjugant cells are plated on selective agar media to determine which genes have been transferred from the *Hfr* to the *F⁻*. (b) The graph shows the appearance of donor genetic markers in the *F⁻* cells as a function of time after the first genes that produced recombinants—*thr⁺* and *leu⁺*—have entered.

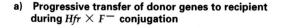

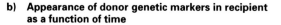

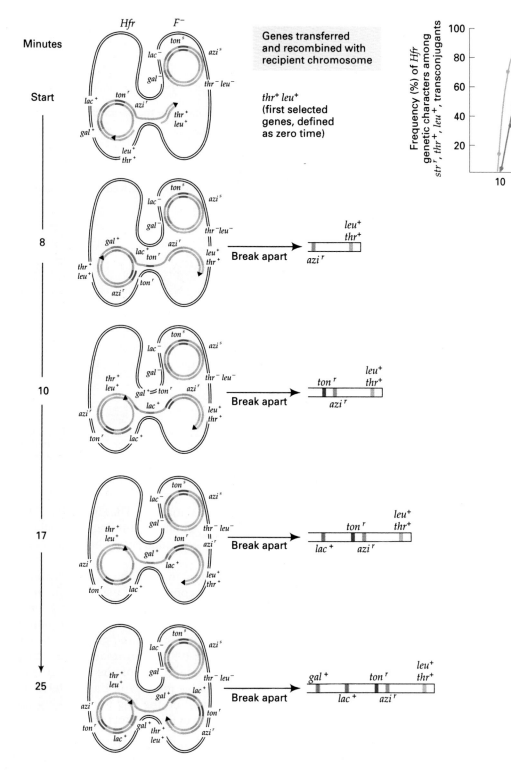

~ FIGURE 6.9

Genetic map of the genes in the experiment in Figure 6.8. The marker positions represent the time of entry of the genes into the recipient during the experiment, with the time of the first genes to enter—*thr*+ and *leu*+—set to 0 minutes. The *azi* marker, for example, entered about 8 minutes after *thr*+/*leu*+ entered, whereas the *gal* marker entered about 25 minutes after *thr*+/*leu*+.

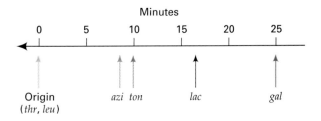

gene are seen at about 8 minutes after the *thr*+ and *leu*+ genes entered. Then *ton*r recombinants are seen at 10 minutes, followed by *lac*+ recombinants at about 17 minutes and *gal*+ recombinants at about 25 minutes. Note that the maximum frequency of recombinants becomes smaller the later the gene enters the recipient, because with time there is an increasing chance that mating pairs will break apart.

In this experiment, each gene from the *Hfr* bacterium appears in recombinants at a different but reproducible time after mating begins. Thus, from the time intervals for the experiment described, the genetic map in Figure 6.9 may be constructed. Again, the map units are in "minutes"; the entire *E. coli* chromosome requires about 100 minutes for transfer.

Circularity of the *E. coli* Map

Only one *F* factor is integrated in each *Hfr* strain. Different *Hfr* strains have the *F* factor integrated at different locations and in different orientations in the chromosome. Therefore, *Hfr* strains differ with respect to where the transfer of donor genes begins and the order of transfer of donor genes. Figure 6.10a shows the order of chromosomal gene transfer for four different *Hfr* strains, *H*, *1*, *2*, and *3*. The genetic distance in time units between a particular pair of genes is constant no matter which *Hfr* strain is used as donor.

From this sort of data, a genetic map of the chromosome is constructed by aligning the genes transferred by each *Hfr* as shown in Figure 6.10b. In view of the overlap of the genes, the simplest map that can be drawn from these data is a circular one, as shown in Figure 6.10c. The map, then, is a composite of the

~ FIGURE 6.10

Interrupted-mating experiments with a variety of *Hfr* strains, showing that the *E. coli* linkage map is circular. (a) Orders of gene transfer for the *Hfr* strains *H*, *1*, *2*, and *3*; (b) Alignment of gene transfer for the *Hfr* strains; (c) Circular *E. coli* chromosome map derived from the *Hfr* gene transfer data. The map is a composite showing various locations of integrated *F* factors. A given *Hfr* strain has only one integrated *F* factor.

a) Orders of gene transfer

Hfr strains:

H	origin–thr–pro–lac–pur–gal
1	origin–thr–thi–gly–his
2	origin–his–gly–thi–thr–pro–lac
3	origin–gly–his–gal–pur–lac–pro

b) Alignment of gene transfer for the *Hfr* strains

H	thr–pro–lac–pur–gal
1	his–gly–thi–thr
2	his–gly–thi–thr–pro–lac
3	pro–lac–pur–gal–his–gly

c) Circular *E. coli* chromosome map derived from *Hfr* gene transfer data

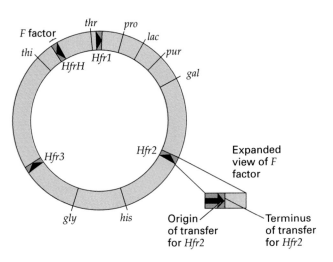

results of the individual matings. The circularity of the map was itself a truly significant finding, since all previous genetic maps—of eukaryotic chromosomes—were linear.

A complete genetic map of the *E. coli* chromosome was eventually constructed using conjugation experiments; it is 100 minutes long. As with genetic maps of other organisms, it provides information about the relative locations of *E. coli* genes on the circular chromosome. In 1997 the ultimate genetic map of *E. coli* was completed, that of the 4.6×10^6 (4.6 megabase) base-pair sequence of the bacterium's genome. This map will be the key to unlocking the function of every gene in this model bacterium.

The circular *F* factor can integrate into the circular bacterial chromosome by a single crossover event to produce *Hfr* (high-frequency recombination) strains. In *Hfr* × *F⁻* matings the chromosome is transferred in a one-way fashion from the *Hfr* cell to the *F⁻* cell, beginning at a specific site in the *F* factor. The farther a gene is from that site, the later it is transferred to the *F⁻*, and this is the basis for mapping genes by their times of entry into the *F⁻* cell. Conjugation and interrupted mating allow mapping of large chromosome segments.

GENETIC MAPPING IN BACTERIA BY TRANSFORMATION

Transformation is the unidirectional transfer of extracellular DNA into cells resulting in a phenotypic change in the recipient. Bacterial transformation is used to map the genes of certain bacterial species in which mapping by other methods (conjugation or transduction) is not possible. In mapping experiments using transformation, DNA from a donor bacterial strain is extracted, purified, and broken into small fragments. This DNA is then added to recipient bacteria with a different genotype. If the donor DNA is taken up by a recipient cell and recombines with the homologous parts of the recipient's chromosome, a recombinant chromosome is produced. Recipients whose phenotypes are changed by transformation are called **transformants.**

Bacterial species vary in their ability to take up DNA. To enhance the efficiency of transformation, cells typically are treated chemically or are exposed to a strong electric field in a process called *electroporation*, making the cell membrane more permeable to DNA. Cells prepared for taking up DNA by transformation are called *competent cells*.

Only a small proportion of the cells involved in transformation will actually take up DNA. Consider an example of transformation of *Bacillus subtilis* (Figure 6.11). (Other systems may differ in the details of the process.) The donor double-stranded DNA fragment is wild type (a^+) for a mutant allele *a* in the recipient cell (Figure 6.11a). During DNA uptake, one of the two DNA strands is degraded so that only one intact linear DNA strand is left inside the cell

Transformation in *Bacillus subtilis*. (a) Linear donor double-stranded bacterial DNA fragment carries the a^+ allele, and the recipient bacterium the *a* allele; (b) One donor DNA strand enters the recipient; (c) The single, linear DNA strand pairs with the homologous region of the recipient's chromosome, forming a triple-stranded structure; (d) A double crossover produces a recombinant a^+/a recipient chromosome and a linear *a* DNA fragment. The linear fragment is degraded and, by replication, one-half of the progeny are a^+ transformants and one-half are *a* nontransformants.

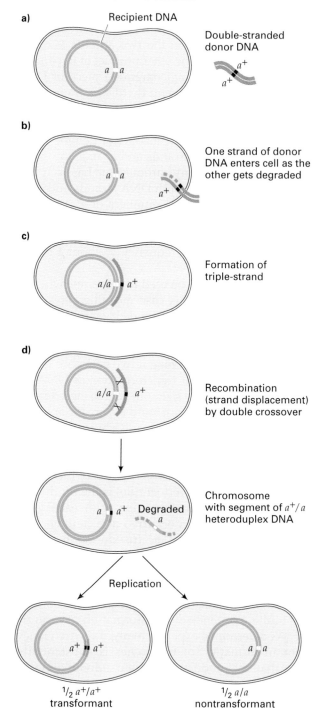

(Figure 6.11b). This single, linear strand pairs with the homologous DNA of the recipient cell's circular chromosome to form a triple-stranded region (Figure 6.11c). Recombination then occurs by a double crossover event involving the single-stranded DNA strand of the donor and the double-stranded DNA of the recipient (Figure 6.11d). The result is a recombinant recipient chromosome: in the region between the two crossovers, one DNA strand has the donor a^+ DNA segment, and the other strand has the recipient a DNA segment. In other words, in that region, *the two DNA strands are part donor, part recipient for the genetic information.* A region of DNA with different sequence information on the two strands is called **heteroduplex DNA**. (The other product of the double crossover event is a single-stranded piece of DNA carrying an a DNA segment; that DNA fragment is degraded.)

After replication of the recipient chromosome, one progeny chromosome has donor genetic information on both DNA strands and is an a^+ transformant. The other progeny chromosome has recipient genetic information on both DNA strands and is an a nontransformant. Equal numbers of a^+ transformants and a nontransformants are produced. Given highly competent recipient cells, transformation of most genes occurs at a frequency of about 1 cell in every 10^3 cells.

Transformation can be used to determine whether genes are linked (in this case meaning physically close to one another on the single bacterial chromosome), to determine the order of genes on the genetic map, and to determine map distance between genes. The principles of determining if two genes are linked are as follows. Efficient transformation of DNA involves fragments with a size sufficient to include only a few genes. If two genes, x^+ and y^+, are far apart on the donor chromosome, they will always be found on different DNA fragments. Thus, given an $x^+ y^+$ donor and an $x\ y$ recipient, the probability of simultaneous transformation (*cotransformation*) of the recipient to $x^+ y^+$ (from the product rule) is the product of the probability of transformation with each gene alone. If transformation occurred at a frequency of 1 in 10^3 cells per gene, $x^+ y^+$ transformants would be expected to appear at a frequency of 1 in 10^6 recipient cells ($10^{-3} \times 10^{-3}$). So, if two genes are close enough that they often are carried on the same DNA fragment, the cotransformation frequency would be close to the frequency of transformation of a single gene. As determined experimentally, if the frequency of cotransformation of two genes is substantially higher than the products of the two individual transformation frequencies, the two genes must be close together.

KEYNOTE

Transformation is the unidirectional transfer of extracellular DNA into cells resulting in a phenotypic change in the recipient. In bacterial transformation, DNA is extracted from a donor strain and added to recipient cells. A DNA fragment taken up by the recipient cell may associate with the homologous region of the recipient's chromosome. Part of the transforming DNA molecule can exchange with part of the recipient's chromosomal DNA. Frequent cotransformation of donor genes indicates close physical linkage of those genes. Transformation has been used to construct genetic maps for bacterial species for which conjugation or transduction analyses are not possible.

GENETIC MAPPING IN BACTERIA BY TRANSDUCTION

Transduction (literally "leading across") is a process by which bacteriophages (bacterial viruses: phages, for short) transfer genes from one bacterium (the donor) to another (the recipient); such phages are called **phage vectors**. Since the amount of DNA a phage can carry is limited, the amount of genetic material that can be transferred is usually less than 1 percent of that in the bacterial chromosome. Once the donor genetic material has been introduced into the recipient, it may undergo genetic recombination with a homologous region of the recipient chromosome. The recombinant recipients are called **transductants**.

Bacteriophages: An Introduction

Most bacterial strains can be infected by specific phages. For example, *E. coli* can be infected by DNA-containing phages such as T2, T4, T5, T6, T7, and λ (lambda).

A phage contains its genetic material (either DNA or RNA) in a single chromosome surrounded by a coat of protein molecules. Figure 6.12 shows two phages of great genetic significance, T4 (Figure 6.12a) and λ (Figure 6.12b). Phage T4 is one of a series of arbitrarily numbered phages with similar properties named T2, T4, and T6 (the T-even phages). It has a number of distinct protein structures: a head (which contains DNA), a core, a sheath, a base plate, and tail fibers. The last two structures enable the phage to attach itself to a bacterium.

~ **FIGURE 6.12**

Electron micrographs and diagrams of two bacteriophages. (a) T4 phage, which is representative of T-even phages (1 nm = 10^{-9} m); (b) λ phage.

a) T4 phage

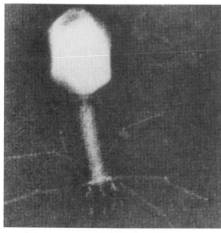

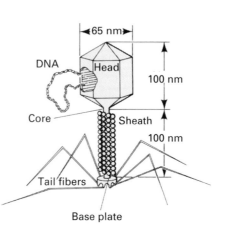

b) λ phage

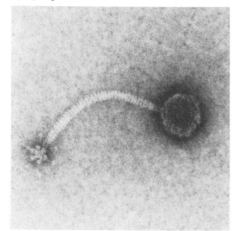

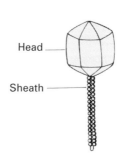

Figure 6.13 shows the T4 life cycle. First (1), the phage particle attaches to the surface of the bacterial cell, and the phage chromosome is injected into the bacterium by a springlike contraction of the sheath. Then (2–5), the phage takes over the bacterium, breaking down the bacterial chromosome and using the bacterial molecules to produce progeny phages. Finally (6), the progeny phages are released from the bacterium as the cell is broken open (lysed). The released progeny phages in solution is called a **phage lysate**. This type of phage life cycle is called the **lytic cycle,** and phages that always follow that lytic cycle when they infect bacteria are called **virulent phages**.

Phage λ has a structure similar to that of T4. When its DNA is injected into *E. coli*, there are two alternative paths that the phage can follow (Figure 6.14). One is a lytic cycle, exactly like that of the

T phages. The other is the **lysogenic pathway** (or *lysogenic cycle*). In the lysogenic pathway the λ chromosome does not replicate; instead, it inserts (integrates) itself physically into a specific region of the host cell's chromosome, much like *F* factor integration. In this integrated state the phage chromosome is called a **prophage**. Every time the host cell chromosome replicates, the integrated λ chromosome replicates as part of it. The bacterium that contains a phage in the prophage state is said to be **lysogenic** for that phage; the phenomenon of the insertion of a phage chromosome into a bacterial chromosome is called **lysogeny**. Phages that have a choice between lytic and lysogenic pathways are called *temperate phages*.

The prophage state is maintained by the action of a specific phage gene product (a repressor protein) that prevents expression of λ genes essential for the

~ FIGURE 6.13

Lytic life cycle of a virulent phage, such as T2 or T4.

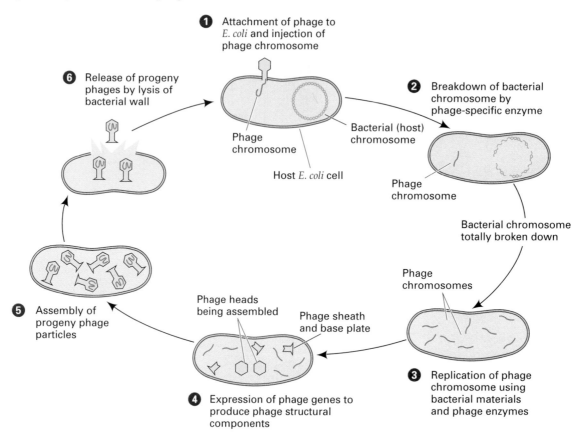

① Attachment of phage to *E. coli* and injection of phage chromosome

⑥ Release of progeny phages by lysis of bacterial wall

Phage chromosome

Bacterial (host) chromosome

Host *E. coli* cell

② Breakdown of bacterial chromosome by phage-specific enzyme

Phage chromosome

Bacterial chromosome totally broken down

Phage chromosomes

⑤ Assembly of progeny phage particles

Phage heads being assembled

Phage sheath and base plate

④ Expression of phage genes to produce phage structural components

③ Replication of phage chromosome using bacterial materials and phage enzymes

lytic cycle. When the repressor that maintains the prophage state is destroyed, for example by environmental factors such as ultraviolet light irradiation, then the lytic cycle is induced. When induction occurs, the integrated λ chromosome is excised from the bacterial chromosome and the lytic cycle begins, resulting in the production and release from the cell of progeny λ phages.

Transduction Mapping of Bacterial Chromosomes

Transduction is a useful mechanism for mapping bacterial genes. Two types of transduction occur: In **generalized transduction** any gene may be transferred between bacteria; in **specialized transduction** only specific genes are transferred. In specialized transduction, by a process similar to the production of an *F'* factor, an integrated phage genome excises imperfectly, picking up adjacent bacterial genes while leaving some phage genes behind. We discuss generalized transduction here.

Joshua and Esther Lederberg and Norton Zinder discovered transduction in 1952. These researchers tested whether conjugation occurred in the bacterial species *Salmonella typhimurium*. Their experiment was similar to the one showing that conjugation existed in *E. coli*. They mixed together two multiple auxotrophic strains and found that a low frequency of prototrophs was produced. But, unlike in the conjugation experiment, when they used the U-tube apparatus (which helped establish conjugation in *E. coli*), they still found prototrophs. This result indicated that recombinants were being produced by a mechanism that did not require cell-to-cell contact. The interpretation was that the agent responsible for the formation of recombinants was a *filterable agent*, since it could pass through a filter with pores small enough to block bacteria. In this particular case the filterable agent was identified as a temperate phage.

As an example, the mechanism for generalized transduction of *E. coli* by temperate phage P1 is shown in Figure 6.15. Normally, the P1 phage enters the lysogenic state when it infects *E. coli* (1), existing in the

~ FIGURE 6.14

Life cycle of a temperate phage, such as λ. When a temperate phage infects a cell, the phage may go through either the lytic or lysogenic cycle.

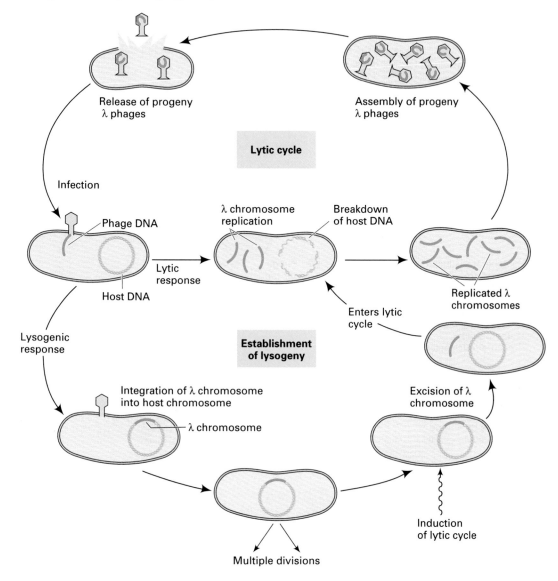

prophage state in these cells. If the lysogenic state is not maintained, the phage goes through the lytic cycle and produces progeny phages (2). During the lytic cycle, the bacterial DNA is degraded and, rarely, a piece of bacterial DNA is packaged into a phage head instead of phage DNA (3). These phages are called **transducing phages** since they are the vehicles by which genetic material is carried between bacteria. In the example shown in Figure 6.15, the transducing phages are those carrying the donor bacterial genes a^+, b^+, or c^+. The population of phages in the phage lysate

(4), consisting mostly of normal phages but with about 1 in 10^5 transducing phages present, can now be used to infect a new population of bacteria (5). The recipient bacteria are a in genotype. If a transducing phage carrying the a^+ gene infects the recipient, genetic exchange of the donor a^+ gene with the recipient a gene can occur by double crossing-over (6). The result is stable transduced bacteria called *transductants*; in this case they are a^+ transductants.

In *generalized transduction* the piece of bacterial DNA that the phage erroneously picks up is a *random*

~ FIGURE 6.15

Generalized transduction between strains of *E. coli*. (1) Wild-type donor cell of *E. coli* infected with the temperate bacteriophage P1; (2) The host cell DNA is broken up during the lytic cycle; (3) During assembly of progeny phages, some pieces of the bacterial chromosome are incorporated into some of the progeny phages to produce transducing phages; (4) Following cell lysis, a low frequency of transducing phages is found in the phage lysate; (5) The transducing phage infects an auxotrophic recipient bacterium; (6) A double-crossover event results in the exchange of the donor a^+ gene with the recipient mutant a gene; (7) The result is a stable a^+ transductant, with all descendants of that cell having the same genotype.

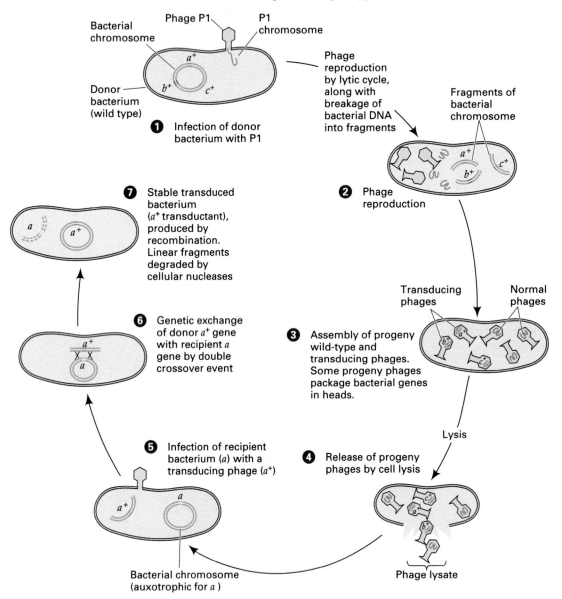

piece of the fragmented bacterial chromosome. Thus, any genes can be transduced; only an appropriate phage and bacterial strains carrying different genetic markers are needed. Just as in transformation, generalized transduction may be used to determine gene linkage and gene order and to map the distance between genes. The logic is identical for the two processes. That is, the size of the DNA fragment that can be contained within a phage head is relatively small, about 1.5 percent that of the *E. coli* chromosome. So, only if two genes are physically close on the chromosome will they be found in the same transducing phage. Consider a donor bacterium with genotype $a^+ b^+$ and a recipient bacterium with genotype $a\ b$. If the genes are close enough to be included in the transducing phage, then an $a^+ b^+$ transductant—called a *cotransductant*—can be produced through infection by a single transducing phage. If the genes are not closely linked, then an $a^+ b^+$ transductant can only be produced by the much rarer event of simultaneous infection of a recipient by two transducing phages, one carrying a^+ and the other carrying b^+. Therefore, if two genes are close enough that they often are packaged into the phage head on the same DNA fragment, the **cotransduction** frequency would be close to the frequency of transduction of a single gene; cotransduction of two or more genes is a good indication that the genes are closely linked.

KEYNOTE

Transduction is the process by which bacteriophages mediate the transfer of genetic information from one bacterium (the donor) to another (the recipient). The capacity of the phage particle is limited, so the amount of DNA transferred is usually less than 1.5 percent of that in the bacterial chromosome. In generalized transduction, any bacterial gene can be accidentally incorporated into the transducing phage during the phage life cycle and subsequently transferred to a recipient bacterium. Since only short segments of bacterial DNA can be packaged into phages, transduction can be used only to map small regions of the genome at a time.

MAPPING GENES OF BACTERIOPHAGES

The same principles used to map eukaryotic genes are used to map phage genes. Crosses are made between phage strains that differ in genetic markers, and the proportion of recombinants among the total progeny is determined. The basic procedure of mapping phage genes in two-, three-, or four-gene crosses involves the mixed infection of bacteria with phages of different genotypes.

Before doing a mapping experiment involving phages, a basic experimental design must be chosen to ensure that we can accomplish two goals. First, we must be able to count phage types. We do so by plating a mixture of phages and bacteria on a solid medium. The concentration of bacteria is chosen so that an entire "lawn" of bacteria grows. Phages are present in

~ FIGURE 6.16

Plaques of the *E. coli* bacteriophages (a) T4 and (b) λ.

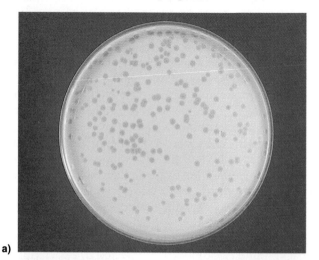

a)

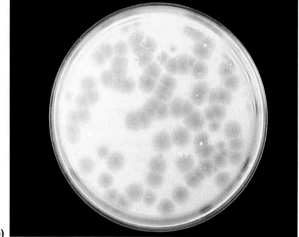

b)

much lower concentrations. Each phage infects a bacterium and goes through the lytic cycle. The released progeny phages infect neighboring bacteria, and the lytic cycle is repeated. The result is a clearing in the opaque lawn of bacteria. The clearing is called a **plaque**, and a single plaque derives from one of the original bacteriophages that was plated. Figure 6.16 shows plaques of the *E. coli* phages T4 and λ; T4 plaques are uniformly clear (completely transparent), while λ plaques are clear with a *turbid* (cloudy) center.

Next, we must be able to distinguish phage phenotypes easily. Several mutations affect the phage life cycle, giving rise to differences in the appearances of plaques on a bacterial lawn. For example, there are strains of T2 that differ in either plaque morphology (the size and shape of the edge of the *plaque*) and/or host range (which bacterial strain the phage can lyse).

Consider two phage strains. One has the genotype $h^+ r$, meaning that it is wild-type for the host range gene (h^+, able to lyse the B strain but not the $B/2$ strain of *E. coli*; that is, strain B is the *permissive host* and strain $B/2$ is the *nonpermissive host* for h^+ phages) and mutant for the plaque morphology gene (r, producing large plaques with distinct borders). The other phage strain has the genotype $h r^+$, meaning that it is mutant for the host range gene (able to lyse both the B and $B/2$ strains of *E. coli*) and wild-type for the plaque morphology gene (r^+, producing small plaques with fuzzy borders). When plated on a lawn containing both the B and $B/2$ strains, any phage carrying the mutant host range allele h (infects both B and $B/2$) produces clear plaques, while phages carrying the wild-type h^+ allele produce cloudy plaques. The latter characteristic arises because phages bearing the h^+ allele can only infect the B bacteria, leaving a background cloudiness of uninfected $B/2$ bacteria.

To map these two genes, we make a genetic cross. To do this we infect *E. coli* strain B with the two (parental) phages, $h^+ r$ and $h r^+$ (Figure 6.17a). Once the two genomes are within the bacterial cell, each will replicate (Figure 6.17b). If an $h^+ r$ and an $h r^+$ chromosome come together, a crossover can occur between the two gene loci to produce $h^+ r^+$ and $h r$ recombinant chromosomes (Figure 6.17c), which are assembled into progeny phages. When the bacterium lyses, the recombinant progeny are released into the medium, along with nonrecombinant (parental) phages (Figure 6.17d).

After the life cycle is completed, the progeny phages are plated onto a bacterial lawn containing a mixture of *E. coli* strains B and $B/2$. Four plaque phenotypes are found from the experiment in Figure 6.17: two parental types and two recombinant types. The parental type $h r^+$ gives a small clear plaque with a fuzzy border; the other parental $h^+ r$ gives a large cloudy plaque with a distinct border (Figure 6.18). The reciprocal recombinant types give recombined phenotypes: the $h^+ r^+$ plaques are cloudy and small with a fuzzy border, and the $h r$ plaques are clear and large with a distinct border (see Figure 6.18).

Once the progeny plaques are counted, the recombination frequency between h and r is given by

$$\frac{(h^+ r^+) + (h r) \text{ plaques}}{\text{total plaques}} \times 100$$

Making the same assumption used in mapping eukaryotic genes, namely, that crossovers occur randomly along the chromosome, the recombination frequency then represents the map distance (in map units) between the phage genes. A number of other T2 genes have been mapped in this way.

~ **FIGURE 6.17**

The principles of performing a genetic cross with bacteriophages. (a) Bacteria of *E. coli* strain B are coinfected with the two parental bacteriophages, $h^+ r$ and $h r^+$; (b) Replication of both parental chromosomes; (c) There is pairing of some chromosomes of each parental type, and crossing-over takes place between the two gene loci to produce $h^+ r^+$ and $h r$ recombinants; (d) Progeny phages are assembled and are released into the medium when the bacteria lyse; both parental and recombinant phages are found among the progeny.

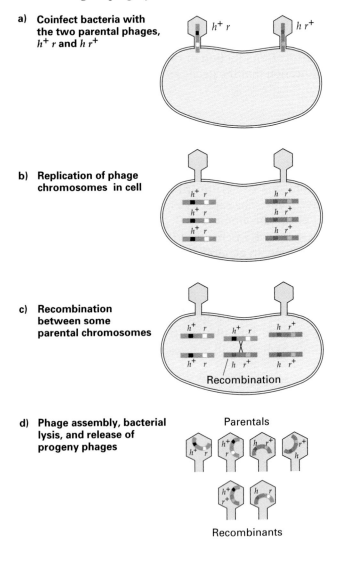

a) **Coinfect bacteria with the two parental phages, $h^+ r$ and $h r^+$**

b) **Replication of phage chromosomes in cell**

c) **Recombination between some parental chromosomes**

Recombination

d) **Phage assembly, bacterial lysis, and release of progeny phages**

Parentals

Recombinants

KEYNOTE

The same principles used to map eukaryotic genes are used to map phage genes. That is, genetic material is exchanged between strains differing in genetic markers and recombinants are detected and counted.

~ FIGURE 6.18

Plaques produced by progeny of a cross of T2 strains $h\,r^+ \times h^+\,r$. Four plaque phenotypes, representing both parental types and the two recombinants, may be discerned. The parental $h\,r^+$ phage produces a small clear plaque with a fuzzy border; the other parental $h^+\,r$ phage produces a large cloudy plaque with a distinct border. The recombinant $h^+\,r^+$ phage produces a small cloudy plaque with a fuzzy border, and the recombinant $h\,r$ phage produces a large clear plaque with a distinct border.

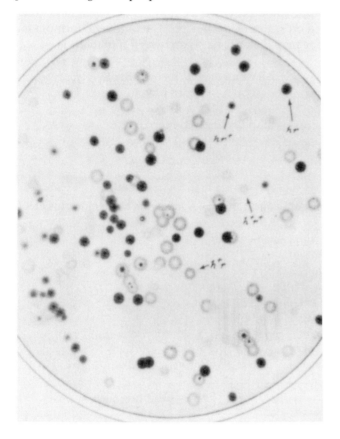

~ FIGURE 6.19

The r^+ and mutant rII plaques on a lawn of *E. coli* B. The r^+ plaque is turbid with a fuzzy edge; the rII plaque is larger, clear, and has a distinct boundary.

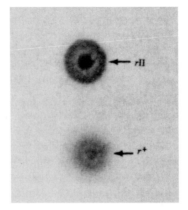

FINE-STRUCTURE ANALYSIS OF A BACTERIOPHAGE GENE

The recombinational mapping of the distance between genes, called *intergenic mapping* (inter = between), can be used to construct chromosome maps for both eukaryotic and prokaryotic organisms. The general principles of recombinational mapping can be applied to mapping the distance between mutational sites within the same gene, a process called *intragenic mapping* (intra = within).

Intragenic mapping is possible because each gene consists of many nucleotide pairs of DNA. Much of the impetus to analyze the fine details of gene struc-

ture came from the elegantly detailed work by Seymour Benzer in the 1950s and 1960s on the *rII* region of bacteriophage T4. His genetic experiments changed the concept of a "gene" as the unit of mutation and function. Benzer used phage T4 mainly because bacteriophages produce large numbers of progeny, which facilitates the potential determination of very low recombination frequencies. Benzer's initial experiments involved **fine-structure mapping**, the detailed genetic mapping of sites within a gene.

Benzer used strains of phage T4 carrying mutations of the *rII* region. *rII* mutants have both a distinct *plaque morphology* phenotype and distinct *host range properties*. Specifically, when cells of *E. coli* growing on solid medium are infected with wild-type (r^+) T4, small turbid plaques with fuzzy edges are produced, while plaques produced by *rII* mutants are large and clear (Figure 6.19). Regarding host range properties, wild-type T4 is able to grow in and lyse cells of either *E. coli* strain B or $K12(\lambda)$, while *rII* mutants can grow in B but not $K12(\lambda)$. That is, strain B is the permissive host for *rII* mutants, while strain $K12(\lambda)$ is the nonpermissive host.

Recombination Analysis of *rII* Mutants

Benzer realized that the growth defect of *rII* mutants on *E. coli* $K12(\lambda)$ could serve as a powerful selective tool for detecting the presence of a very small proportion of r^+ phages within a large population of *rII* mutants. Initially, Benzer set out to construct a fine-structure genetic map of the *rII* region. He crossed 60 independently isolated *rII* mutants in all possible combinations using *E. coli* B as the permissive host, and then collected the progeny phages once the cells had lysed (Figure 6.20). For each cross he plated a sample of the phage progeny on *E. coli* B, the permissive host, in order to count the total number of progeny phage per milliliter. He plated another sample on *E. coli* K12(λ), the nonpermissive host, to find the frequency of r^+ recombinants resulting from genetic recombination between the pairs of *rII* mutants used in each cross. In this way Benzer was able to calculate the percentage of very rare r^+ recombinants between closely linked genetic sites.

For each cross, such as *rII*1 and *rII*2 (Figure 6.21), four genetic classes of progeny are possible: the two parental *rII* types (*rII*1 and *rII*2) and the two recombinant types. One of the recombinant types is the wild type (r^+), and the other is a double mutant carrying both *rII* mutations (*rII*1, 2), which has an *r* phenotype and is phenotypically indistinguishable from either of

the parental *rII* mutants. The map distance between two mutations is given by the percentage of recombinant progeny among all the progeny produced by crosses of the two mutants. For the crosses of *rII* mutants, a single crossover event between the two mutant genes produces the r^+ and double *rII* mutant recombinants (see Figure 6.21). Therefore, the frequencies of the two recombinant classes of phages are expected to be the same. Consequently, the total number of recombinants from each *rII* × *rII* cross is approximately twice the number of r^+ plaques counted on plates of strain K12(λ). The general formula for the map distance between two *rII* mutations is

$$\frac{2 \times \text{number of } r^+ \text{ recombinants}}{\text{total number of progeny}} \times 100\%$$

$$= \text{map distance (in map units)}$$

~ FIGURE 6.20

Benzer's general procedure for determining the number of r^+ recombinants from a cross involving two *rII* mutants of T4.

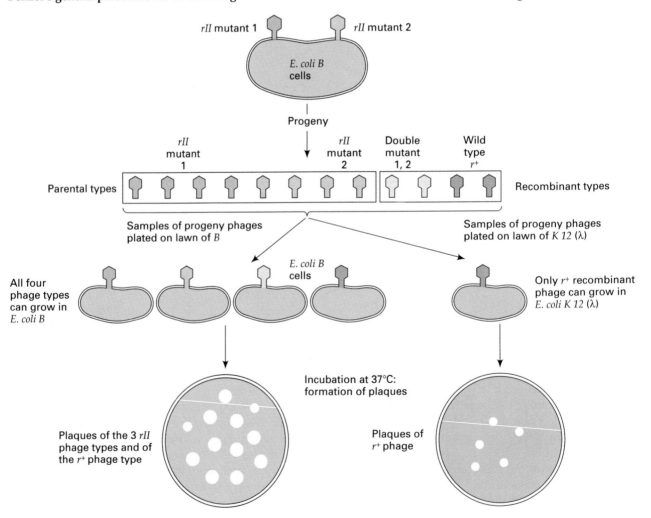

~ **FIGURE 6.21**

Production of parental and recombinant progeny from a cross of two *rII* mutants with mutations in different sites within the *rII* region. Progeny phages of parental genotype are produced if no crossing-over occurs, whereas progeny phages with recombinant genotypes are produced if a single crossover occurs between the sites of the two mutations.

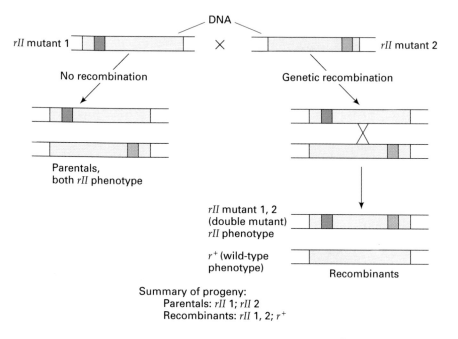

Summary of progeny:
Parentals: *rII* 1; *rII* 2
Recombinants: *rII* 1, 2; *r*⁺

An important control was set up for each cross. Each *rII* parent alone was used to infect the permissive *E. coli* B host and the progeny tested on plates of B and of *K12*(λ). Just as a mutation can occur to produce an *rII* mutant from the *r*⁺, so a mutation can occur for an *rII* mutant to change back (revert) to the *r*⁺. Thus, it is extremely important to calculate the reversion frequencies for the two *rII* mutations in a cross and subtract the combined value from the computed recombination frequency. Fortunately for enabling the experimental results to be analyzed to give accurate recombination frequencies, the reversion frequency for an *rII* mutation is at least an order of magnitude lower than the smallest recombination frequency that was found.

From the recombination data obtained from all possible pairwise crosses of the 60 *rII* mutants, Benzer constructed a linear genetic map. Some pairs produced no *r*⁺ recombinants when they were crossed. This result was interpreted to mean that those pairs carried mutations at exactly the same site. That is, the same nucleotide pair in the DNA had been changed; there was no possibility of recombination between the mutations. Mutations that change the same nucleotide pair within a gene are termed *homoallelic*. However, most pairs of *rII* mutants did produce *r*⁺ recombinants when crossed, indicating that they car-

ried different altered nucleotide pairs in the DNA. Mutations that change different nucleotide pairs within a gene are termed *heteroallelic*.

Benzer followed up these initial experiments with a more detailed mapping of more than 3,000 *rII* mutants. He showed from this that the *rII* region is subdivisible into more than 300 mutable sites that were separable by recombination (Figure 6.22). The distribution of mutants is not random; certain sites (called *hot spots*) are represented by a large number of independently isolated mutants.

The finished map showed that the lowest frequency with which *r*⁺ recombinants were formed in any pairwise crosses of *rII* mutants involving heteroallelic mutations was 0.01 percent. The minimum map distance of 0.01 percent can be used to make a rough calculation of the molecular distance—the distance in nucleotide pairs—between mutant markers. The genetic map of phage T4 is about 1,500 map units. If two *rII* mutants produce 0.01 percent *r*⁺ recombinants, this means that the mutations are separated by 0.02 map units, or by about $0.02/1,500 = 1.3 \times 10^{-5}$ of the total T4 genome. The total T4 genome contains about 2×10^5 nucleotide pairs (base pairs), so the smallest recombination distance that was observed was $(1.3 \times 10^{-5}) \times (2 \times 10^5)$, or about 3 base pairs. That means Benzer's data had shown that

~ FIGURE 6.22

Fine-structure map of the *rII* region derived from Benzer's experiments. The number of independently isolated mutations that mapped to a given site is indicated by the number of blocks at the site. Hot spots are represented by a large number of blocks.

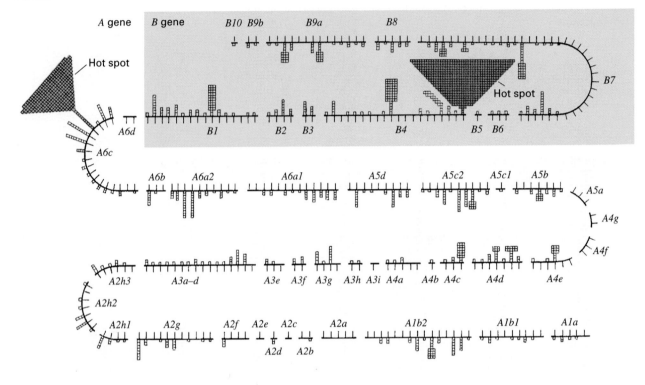

genetic recombination can occur within distances on the order of 3 base pairs. Later experiments by others showed conclusively that recombination can occur between mutations that affect adjacent base pairs in the DNA. That is, genetic experiments have shown that the base pair is both the *unit of mutation* and the *unit of recombination*. These definitions replaced the classical definitions that the gene was the unit of mutation and the unit of recombination; that is, that the gene was indivisible by the processes of mutation and recombination.

KEYNOTE

The general principles of recombinational mapping can be applied to mapping the distance between mutational sites in different genes (intergenic mapping) and to mapping mutational sites within the same gene (intragenic mapping). As a result of fine-structure analysis of the *rII* region of bacteriophage T4 as well as other experiments, it was determined that the unit of mutation and of recombination is the base pair in DNA.

Defining Genes by Complementation (*Cis-Trans*) Tests

From the classical point of view, the gene is a *unit of function*; that is, each gene specifies one function. Benzer designed genetic experiments to determine whether this classical view was true of the *rII* region. To find out whether two different *rII* mutants belonged to the same gene (unit of function), Benzer used the *cis-trans* or **complementation test**. As you read the following discussion, it will help you to know that the complementation tests showed that the *rII* region actually consists of two genes (units of function), *rIIA* and *rIIB*. A mutation anywhere in either gene will produce the *rII* plaque morphology phenotype and host range property. In other words, *rIIA* and *rIIB* each specify a product needed for growth in *E. coli K12(λ)*.

The complementation test is used to establish how many units of function (genes) are defined by a given set of mutations that express the same mutant phenotypes. In Benzer's work with the *rII* mutants, the nonpermissive strain *K12(λ)* was infected with a pair of *rII* mutant phages to see whether the two mutants, each unable by itself to grow in strain *K12(λ)*, are able to

~ FIGURE 6.23

Complementation tests for determining the units of function in the *rII* region of phage T4; the nonpermissive host *E. coli K12*(λ) is infected with two different *rII* mutants. (a) Complementation occurs; (b) Complementation does not occur.

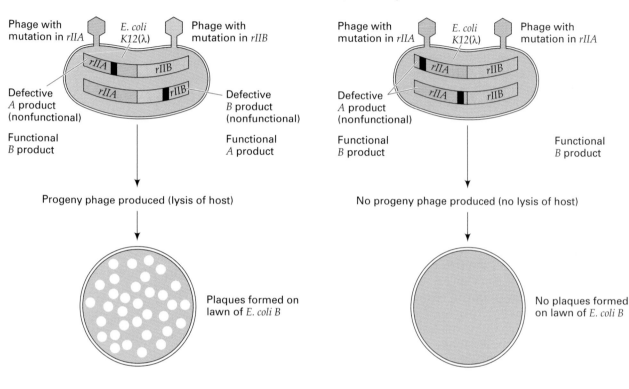

"work together" to produce progeny phages. If the phages do produce progeny, the two mutants are said to complement each other, meaning that the two mutations must be in different genes (units of function) that encode different products. That is, those two products work together to allow progeny to be produced. If no progeny phages are produced, the mutants are not complementary, indicating that the mutations are in the same functional unit. Here, both mutants produce the same defective product, so the phage life cycle cannot proceed and no progeny phages result. (Note that genetic recombination is not necessary for complementation to occur; if genetic recombination does take place, a few plaques may occur on the lawn, but if complementation occurs, the entire lawn of bacteria will be lysed.)

These two situations are diagrammed in Figure 6.23. In the first case the bacterium is infected with two phage genomes, one with a mutation in the *rIIA* gene and the other with a mutation in the *rIIB* gene (Figure 6.23a). The *rIIA* mutant makes a nonfunctional *A* product and a functional *B* product, while the *rIIB* mutant makes a functional *A* product and a

nonfunctional *B* product. Complementation occurs because the *rIIA* mutant still makes a functional *B* product and the *rIIB* mutant makes a functional *A* product. In the second case the bacterium is infected with two phage genomes, each with a different mutation in the same gene, *rIIA* (Figure 6.23b). Here, no complementation occurs because, while both produce a functional *rIIB* product, neither of the mutants makes functional *A* product, so the *A* function cannot take place.

On the basis of the results of such complementation tests, Benzer found that each *rII* mutant falls into one of two units of function, *rIIA* and *rIIB* (also called complementation groups, which directly correspond to genes). That is, all *rIIA* mutants complement all *rIIB* mutants, but *rIIA* mutants fail to complement other *rIIA* mutants, and *rIIB* mutants fail to complement other *rIIB* mutants. The dividing line between the *rIIA* and *rIIB* units of function is indicated in the fine-structure map in Figure 6.22.

For the complementation test examples shown in Figure 6.23, each phage that coinfects the nonpermissive *E. coli* strain *K12*(λ) carries one *rII* mutation, a configuration of mutations called the *trans* configuration. In this configuration, the two mutations are carried by different phages. As a control, it is usual to coinfect *E. coli K12*(λ) with an *r*⁺ (wild-type) phage and an *rII* mutant phage carrying both mutations to see whether the expected wild-type function results. When both mutations under investigation are carried on the same chromosome, the configuration is called the *cis* configuration of mutations. (Because of the *cis* and *trans* configurations of mutations used, the complementation test is also called the *cis-trans* test.) In the *cis* test, the *r*⁺ is expected to be dominant over the two mutations carried by the *rII* mutant phage, so progeny phages will be produced. Therefore, failure to produce progeny would not prove that the mutations are in different functional genes.

Benzer referred to the genetic unit of function revealed by the *cis-trans* test as the *cistron*. A cistron may be defined as the smallest segment of DNA that encodes a piece of RNA. At the present time, *gene* is commonly used and *cistron* is being used less. Genetically, the *rIIA* cistron is about 6 map units and 800 base pairs long, and the *rIIB* cistron is about 4 map units and 500 base pairs long.

The principles for performing a complementation test are the same in other organisms; only the practical details of performing the test are organism-specific. For example, in yeast one could select two haploid cells that are of different mating types (**a** and α) and that carry different mutations conferring the same mutant phenotype. Mating these two types would produce a diploid, which would then be analyzed for complementation of the two mutations. In animal cells, two cells, each exhibiting the same mutant phenotype, can be fused together and analyzed; a wild-type phenotype indicates that complementation has occurred. Again, in neither of these cases is recombination necessary for complementation to occur.

KEYNOTE

The complementation, or *cis-trans*, test is used to determine how many units of function (genes) define a given set of mutations expressing the same mutant phenotypes. If two mutants, each carrying a mutation in a different gene, are combined, the mutations will complement and a wild-type function will result. If two mutants, each carrying a mutation in the same gene, are combined, the mutations will not complement and the mutant phenotype will be exhibited.

SUMMARY

In this chapter we have seen how genes are mapped in prokaryotes such as bacteria and in bacteriophages. The same experimental strategy is used for all gene mapping; that is, genetic material is exchanged between strains differing in genetic markers, and recombinants are detected and counted. In bacteria, the mechanism of gene transfer may be transformation, conjugation, or transduction. In each process, there is a donor strain and a recipient strain.

Conjugation is a plasmid-mediated process in which there is a unidirectional transfer of genetic information through direct cellular contact between a donor and a recipient. Transformation is the unidirectional transfer of extracellular DNA into cells resulting in a phenotypic change in the recipient. Transformation is the transfer of genetic material as small extracellular pieces of DNA between organisms. Transduction is a process whereby bacteriophages mediate the transfer of bacterial DNA from the donor bacterium to the recipient. Bacteriophage DNA may be mapped by infecting bacteria simultaneously with two phage strains and analyzing the resulting phage progeny for parental and recombinant phenotypes. Formally, this method of mapping is the same as that used for mapping genes in haploid eukaryotes.

Insights into the relationships between mapping and gene structure were obtained from a fine-structure analysis of the bacteriophage T4 *rII* region. The mutational sites within a gene were mapped through intragenic mapping. The resulting map indicated that the unit of mutation and the unit of recombination are the same—that is, the base pair in DNA. These definitions replaced the classical definition of a gene as a unit of mutation, recombination, and function.

The number of units of function (genes) is determined by complementation tests. Given a set of mutations expressing the same mutant phenotype, two mutants are combined and the phenotype is determined. If the phenotype is wild type, the two mutations have complemented and must be in different units of function. If the phenotype is mutant, the two mutations have not complemented and must be in the same unit of function.

ANALYTICAL APPROACHES FOR SOLVING GENETICS PROBLEMS

Q6.1 In *E. coli,* the following *Hfr* strains donate the genes shown in the order given:

Hfr STRAIN	ORDER OF GENE TRANSFER
1	*G E B D N A*
2	*P Y L G E B*
3	*X T J F P Y*
4	*B E G L Y P*

All the *Hfr* strains were derived from the same *F*⁺ strain. What is the order of genes in the original *F*⁺ chromosome?

A6.1 This question is an exercise in piecing together various segments of the circumference of a circle. The best approach is to draw a circle and label it with the genes transferred from one *Hfr* and then see which of the other *Hfr*s transfers an overlapping set. For example, *Hfr* 1 transfers *E*, then *B*, then *D*, and so on; and *Hfr* 4 transfers *B* then *E*, and so forth. Now we can juxtapose the two sets of genes transferred by the two *Hfr*s and deduce that the polarities of transfer are opposite:

Hfr 1	*G E B D N A*
Hfr 2	*P Y L G E B*

Extending this reasoning to the other *Hfr*s, we can draw an unambiguous map (see the following figure), with the arrowheads indicating the order of transfer.

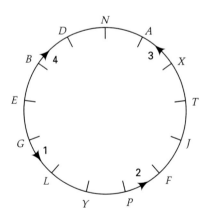

The same logic would be used if the question gave the relative time units of entry of each of the genes. In that case we would expect that the time "distance" between any two genes would be approximately the same regardless of the order of transfer or how far the genes were from the origin.

QUESTIONS AND PROBLEMS

*6.1 In $F^+ \times F^-$ crosses the F^- recipient is converted to a donor with very high frequency. However, it is rare for a recipient to become a donor in $Hfr \times F^-$ crosses. Explain why.

*6.2 With the technique of interrupted mating, four *Hfr* strains were tested for the sequence in which they transmitted a number of different genes to an F^- strain. Each *Hfr* strain was found to transmit its genes in a unique sequence, as shown in the accompanying table (only the first six genes transmitted were scored for each strain).

ORDER OF TRANSMISSION	*Hfr* STRAIN			
	1	2	3	4
First	*O*	*R*	*E*	*O*
	F	*H*	*M*	*G*
	B	*M*	*H*	*X*
	A	*E*	*R*	*C*
	E	*A*	*C*	*R*
Last	*M*	*B*	*X*	*H*

What is the gene sequence in the original strain from which these *Hfr* strains derive? Indicate on your diagram the origin and polarity of each of the four *Hfr*s.

6.3 At time zero an *Hfr* strain (strain 1) was mixed with an F^- strain, and at various times after mixing, samples were removed and agitated to separate conjugating cells. The cross may be written as:

$$Hfr\ 1:\quad a^+\ b^+\ c^+\ d^+\ e^+\ f^+\ g^+\ h^+$$
$$F^-:\quad a\ \ b\ \ c\ \ d\ \ e\ \ f\ \ g\ \ h$$

(No order is implied in listing the markers.)

The samples were then plated onto selective media to measure the frequency of h^+ recombinants that had received certain genes from the *Hfr* cell. The graph of the number of recombinants against time is shown in the accompanying figure.

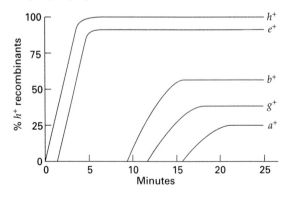

a. Indicate whether each of the following statements is true or false.

i. All F^- cells that received a^+ from the *Hfr* in the chromosome transfer process must also have received b^+.

ii. The order of gene transfer from *Hfr* to F^- was a^+ (first), then g^+, then b^+, then e^+, then h^+.

iii. Most e^+ recombinants are likely to be *Hfr* cells.

iv. None of the b^+ recombinants plated at 15 minutes are also a^+.

b. Draw a linear map of the *Hfr* chromosome indicating the

i. point of nicking (the origin) and direction of DNA transfer;

ii. order of the genes a^+, b^+, e^+, g^+, and h^+;

iii. shortest distance between consecutive genes on the chromosomes.

*6.4 If an *E. coli* auxotroph *A* could grow only on a medium containing thymine, and an auxotroph *B* could grow only on a medium containing leucine, how would you test whether DNA from *A* could transform *B*?

6.5 Distinguish between generalized and specialized transduction.

*6.6 Wild-type phage T4 grows on both *E. coli* B and *E. coli* K12(λ), producing turbid plaques. The *rII* mutants of T4 grow on *E. coli* B, producing clear plaques, but do not grow on *E. coli* K12(λ). This host range property permits the detection of a very low number of r^+ phages among a large number of *rII* phages. With this sensitive system it is possible to determine the genetic distance between two mutations within the same gene, in this case the *rII* locus. Suppose *E. coli* B is mixedly infected with *rIIx* and *rIIy*, two separate mutants in the *rII* locus. Suitable dilutions of progeny phages are plated on *E. coli* B and *E. coli* K12(λ). A 0.1-mL sample of a thousandfold dilution plated on *E. coli* B produced 672 plaques. A 0.2-mL sample of undiluted phage plated on *E. coli* K12(λ) produced 470 turbid plaques. What is the genetic distance between the two *rII* mutations?

6.7 Construct a map from the following two-factor phage cross data (show map distance):

CROSS	% RECOMBINATION
$r1 \times r2$	0.10
$r1 \times r3$	0.05
$r1 \times r4$	0.19
$r2 \times r3$	0.15
$r2 \times r4$	0.10
$r3 \times r4$	0.23

*6.8 The following two-factor crosses were made to analyze the genetic linkage between four genes in phage λ: *c*, *mi*, *s*, and *co*.

PARENTS	PROGENY
$c + \times + mi$	1,213 c +, 1,205 + mi, 84 + +, 75 c mi
$c + \times + s$	566 c +, 808 + s, 19 + +, 20 c s
$co + \times + mi$	5,162 co +, 6,510 + mi, 311 + +, 341 co mi
$mi + \times + s$	502 mi +, 647 + s, 65 + +, 56 mi s

Construct a genetic map of the four genes.

6.9 Three gene loci in T4 that affect plaque morphology in easily distinguishable ways are *r* (rapid lysis), *m* (minute), and *tu* (turbid). A culture of *E. coli* is mixedly infected with two types of phage: *r m tu* and $r^+ m^+ tu^+$. Progeny phage are collected and the following genotype classes are found:

CLASS	NUMBER
$r^+\ m^+\ tu^+$	3,729
$r^+\ m^+\ tu$	965
$r^+\ m\ tu^+$	520
$r\ m^+\ tu^+$	172
$r^+\ m\ tu$	162
$r\ m^+\ tu$	474
$r\ m\ tu^+$	853
$r\ m\ tu$	3,467
	10,342

Construct a map of the three genes. What is the coefficient of coincidence, and what does the value suggest?

*6.10 The *rII* mutants of bacteriophage T4 grow in *E. coli* B but not in *E. coli* K12(λ). The *E. coli* strain B is doubly infected with two *rII* mutants. A 6×10^7 dilution of the lysate is plated on *E. coli* B. A 2×10^5 dilution is plated on *E. coli* K12(λ). Twelve plaques appeared on strain K12(λ), and 16 on strain B. Calculate the amount of recombination between these two mutants.

6.11 Wild-type (r^+) strains of T4 produce turbid plaques, whereas *rII* mutant strains produce larger, clearer plaques. Five *rII* mutations (*a–e*) in the *A* cistron of the *rII* region of T4 give the following percentages of wild-type recombinants in two-point crosses:

CROSS	% OF WILD-TYPE RECOMBINANTS	CROSS	% OF WILD-TYPE RECOMBINANTS
$a \times b$	0.2	$e \times d$	0.7
$a \times c$	0.9	$e \times c$	1.2
$a \times d$	0.4	$e \times b$	0.5
$b \times c$	0.7	$b \times d$	0.2
$e \times a$	0.3	$d \times c$	0.5

What is the order of the mutational sites, and what are the map distances between the sites?

6.12 Some adenine-requiring mutants of yeast are pink because of the intracellular accumulation of a red pigment. Diploid strains were made by mating haploid mutant strains. The diploids exhibited the following phenotypes:

CROSS	DIPLOID PHENOTYPES
1 × 2	pink, adenine-requiring
1 × 3	white, prototrophic
1 × 4	white, prototrophic
3 × 4	pink, adenine-requiring

How many genes are defined by the four different mutants? Explain.

*6.13   In *Drosophila*, mutants *A, B, C, D, E, F,* and *G* all have the same phenotype: the absence of red pigment in the eyes. In pairwise combinations in complementation tests the following results were produced, where "+" = complementation and "−" = no complementation.

	A	B	C	D	E	F	G
G	+	−	+	+	+	+	−
F	−	+	+	−	+	−	
E	+	+	−	+	−		
D	−	+	+	−			
C	+	+	−				
B	+	−					
A	−						

a. How many genes are present?

b. Which mutants have defects in the same gene?

6.14   A large number of mutations in *Drosophila* alter the normal deep-red eye color. As we discussed in Chapter 4, even alleles at one gene (*white, w*) can display a variety of phenotypes.

You are given a wild-type, deep-red strain and six independently isolated, true-breeding mutant strains that have varying shades of brown eyes, with the assurance that each mutant strain has only a single mutation. How would you proceed to determine:

a. whether the mutation in each strain is dominant or recessive?

b. how many different genes are affected in the six mutant strains?

c. which mutants, if any, are allelic?

d. whether any of these mutants are alleles of genes already known to affect eye color?

CHAPTER 7

CHROMOSOMAL MUTATIONS

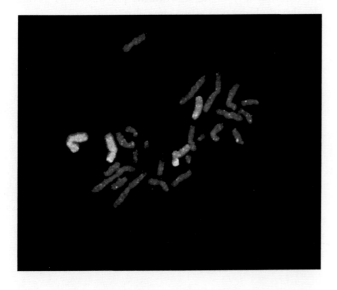

PRINCIPAL POINTS

~ Chromosomal mutations are variations from the normal condition in either chromosome number or chromosome structure. Chromosomal mutations can occur spontaneously, or they can be induced by chemicals or radiation.

~ Deletion is the loss of a DNA segment, duplication is the addition of one or more extra copies of a DNA segment, inversion is a reversal of orientation of a DNA segment in a chromosome, and translocation is the movement of a DNA segment to another chromosomal location in the genome.

~ Variations in the chromosome number of a cell or an organism include aneuploidy, monoploidy, and polyploidy. In aneuploidy there are one, two, or more whole chromosomes greater or fewer than the diploid number.

In monoploidy, each body cell of the organism has only one set of chromosomes, and in polyploidy there are more than two sets of chromosomes present.

~ A change in chromosome number or chromosome structure may have serious, even lethal, consequences to the organism. In eukaryotes, abnormal phenotypes typically result from abnormal chromosome segregation during meiosis, from gene disruptions where chromosomes break, or from altered gene expression levels when the number of copies of a gene or genes (gene dosage) is altered or when the rearrangement separates a gene from its regulatory sequence. Some human tumors, for example, have chromosomal mutations associated with them, either a change in the number of chromosomes or a change in chromosome structure.

Chromosomal mutations—changes in normal chromosome structure or chromosome number—affect both prokaryotes and eukaryotes as well as viruses. The association of genetic defects with changes in chromosome structure or chromosome number indicates that not all genetic defects result from simple mutations of single genes. The study of normal and mutated chromosomes and their behavior is called *cytogenetics*. Your goal in this chapter is to learn about the various types of chromosomal mutations in eukaryotes, and about some of the human disease syndromes that result from chromosomal mutations.

TYPES OF CHROMOSOMAL MUTATIONS

Chromosomal mutations (or **chromosomal aberrations**) are *variations from the normal (wild-type) condition in either chromosome structure or chromosome number*. In bacteria, archaea, and eukarya, chromosomal mutations arise spontaneously or can be induced experimentally by certain chemicals or radiation. Chromosomal mutations are detected by genetic analysis, that is, by observing changes in the linkage arrangements of genes. In eukaryotes, chromosomal mutations can also often be detected under the microscope during mitosis and meiosis. In this chapter we limit our discussion to chromosomal changes in eukaryotes.

We often have the impression that reproduction in humans usually occurs without significant problems affecting chromosome structure or number. After all, the vast majority of babies appear normal, as does the majority of the adult population. However, chromosomal mutations are more common than we once thought, and they contribute very significantly to spontaneously aborted pregnancies and stillbirths. For example, major chromosomal mutations are present in about half of spontaneous abortions, and a visible chromosomal mutation is present in about 6 out of 1,000 live births. Other studies have shown that about 11 percent of men with serious fertility problems and about 6 percent of people institutionalized with mental deficiencies have chromosomal mutations. Indeed, chromosomal mutations are significant causes of developmental disorders.

KEYNOTE

Chromosomal mutations are variations from the normal (wild-type) condition in either chromosome structure or chromosome number. Chromosomal mutations can occur spontaneously, or they can be induced by exposure to chemicals or radiation.

VARIATIONS IN CHROMOSOME STRUCTURE

There are four common types of chromosomal mutations involving changes in chromosome structure: *deletions* and *duplications* (both of which involve a change in the amount of DNA on a chromosome), *inversions* (which involve a change in the arrangement of a chromosomal segment), and *translocations* (which involve a change in the location of a chromosomal segment).

All four classes of chromosomal structure mutations begin by one or more breaks in the chromosome. If a break occurs within a gene, then the function of that gene may be lost. Wherever the break occurs, the breakage process leaves broken ends without the specialized sequences found at the ends of chromosomes (the telomeres) that prevent their degradation and "stickiness." As a result, the broken end of a chromosome is "sticky" and may adhere to other broken chromosome ends. This stickiness can help us understand the formation of the types of chromosomal structure mutations we will discuss.

We have learned a lot about changes in chromosome structure from the study of **polytene chromosomes**, a special type of chromosome found in certain insects such as *Drosophila*. Polytene chromosomes consist of chromatid bundles resulting from repeated cycles of chromosome duplication without nuclear or cell division. Polytene chromosomes may be a thousand times the size of corresponding chromosomes at meiosis or in the nuclei of ordinary somatic cells, and are easily detectable under the microscope. In each polytene chromosome, the homologous chromosomes are tightly paired; therefore, the observed number of polytene chromosomes per cell is reduced to half the diploid number of chromosomes. The polytene chromosomes are joined together at their centromeres by a proteinaceous structure called the chromocenter.

As a result of the intimate pairing of the multiple copies of chromatids, characteristic banding patterns are easily seen, enabling cytogeneticists to identify any segment of a polytene chromosome. In *Drosophila melanogaster*, for example, more than 5,000 bands and interbands can be counted in the four polytene chromosomes. Each band contains an average of 30,000 base pairs (30 kb) of DNA, enough to encode several average-sized proteins. DNA cloning and sequencing studies have shown that many bands contain up to seven genes. Genes are also found in the interbands. Polytene chromosomes will appear at times in this chapter, as it is relatively easy to see the different types of chromosomal mutations in *Drosophila* salivary gland polytene chromosomes.

Deletion

A **deletion** is a chromosomal mutation in which part of a chromosome is missing (Figure 7.1). A deletion starts where breaks occur in chromosomes. These breaks may be induced by agents such as heat, radiation (especially ionizing radiation; see Chapter 18), viruses, chemicals, transposable elements (see Chapter 19), or by errors in recombination. Because a segment of chromosome is missing, deletion mutations cannot revert to the wild-type state.

The consequences of a deletion depend on the genes or parts of genes that have been removed. In diploid organisms an individual heterozygous for a deletion may be normal. However, if the homolog contains recessive genes with deleterious effects, the consequences can be severe. If the deletion involves the loss of the centromere of a chromosome, for example, the result is an acentric chromosome, which is usually lost during meiosis. This leads to the deletion of an entire chromosome from the genome, which may have very serious or lethal consequences, depending on the organism. For example, there are no known living humans who have one whole chromosome of a homologous pair of autosomes deleted from the genome. (Recall from Chapter 3 that XO human females are viable.)

In organisms in which karyotype analysis (analysis of the chromosome complement; see Chapter 3) is practical, deletions can be detected by that procedure if the losses are large enough. In that case, a mismatched pair of homologous chromosomes is seen, with one shorter than the other. In heterozygous individuals, deletions result in unpaired loops when the two homologous chromosomes pair at meiosis. Figure 7.2 diagrams such a loop in polytene chromosomes, which are paired during interphase in salivary gland cells.

~ **FIGURE 7.1**

A deletion of a chromosome segment (here, *D*).

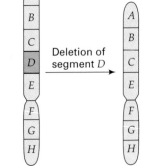

~ FIGURE 7.2

Cytological effects at meiosis of heterozygosity for a deletion. Paired *Drosophila* salivary gland polytene chromosomes in a strain with a heterozygous deletion, showing the unpaired region; numbers refer to bands, which are known to include genes.

Paired polytene chromosomes

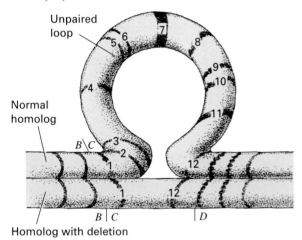

A number of human disorders are caused by deletions of chromosome segments. In many cases the abnormalities are found in heterozygous individuals; homozygotes for deletions usually die if the deletion is large. This tells us that, in humans at least, the number of copies of genes is important for normal development and function. One human disorder caused by a heterozygous deletion is *cri-du-chat syndrome*, which results from an observable deletion of part of the short arm of chromosome 5, one of the larger human chromosomes (Figure 7.3). About 1 infant in 50,000 live births has cri-du-chat syndrome. Children with cri-du-chat syndrome are severely mentally retarded, have a number of physical abnormalities, and cry with a sound like the meow of a cat (hence the name, which is French for "cry of the cat").

Another example is *Prader-Willi syndrome*, which results from heterozygosity for a deletion of part of the long arm of chromosome 15. Many individuals with the syndrome go undiagnosed, so its frequency of occurrence is not known accurately, although it is estimated to affect between 1 in 10,000 and 1 in 25,000 people, predominantly males. Infants with this syndrome are weak because their sucking reflex is poor, making feeding difficult. As a result, growth is poor. By age 5 to 6, for reasons not yet understood, Prader-Willi children become compulsive eaters, and this produces obesity and related health problems. If untreated, afflicted individuals may feed themselves to death. Other phenotypes associated with the syndrome include poor sexual development in males, behavioral problems, and mental retardation.

Duplication

A **duplication** is a chromosomal mutation that results in the doubling of a segment of a chromosome

~ FIGURE 7.3

Cri-du-chat syndrome results from the deletion of part of one of the copies of human chromosome 5. (a) Karyotype of individual with cri-du-chat syndrome; (b) A child with cri-du-chat syndrome.

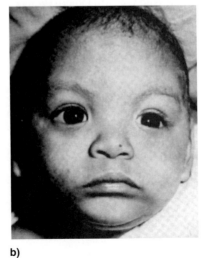

a)

b)

~ FIGURE 7.4

Duplication, with a chromosome segment (here, *BC*) repeated.

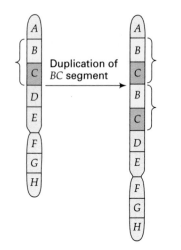

(Figure 7.4). The size of the duplicated segment may vary considerably, and duplicated segments may occur at different locations in the genome or in a *tandem* configuration (that is, adjacent to each other). Heterozygous duplications result in unpaired loops similar to those described for chromosome deletions and, hence, may be detected cytologically.

Duplications of particular genetic regions can have unique phenotypic effects, as in the *Bar* mutant on the X chromosome of *Drosophila melanogaster*. In females homozygous for the *Bar* mutation (not to be confused with the Barr body), the number of facets of the compound eye is far fewer than for the normal eye (shown in Figure 7.5a), giving the eye a bar-shaped (narrow) rather than an oval appearance (shown in Figure 7.5b). Females heterozygous for *Bar* have more facets and hence a somewhat larger bar-shaped eye (sometimes referred to as "wide-*Bar*" eye) than do females homozygous for *Bar*. Males hemizygous for *Bar* have very small eyes like those of homozygous *Bar* females. The *Bar* trait is the result of a duplication of a small segment (16A) of the X chromosome (see Figure 7.5b).

Duplications have played an important role in the evolution of multiple genes with related functions (a *multigene family*). For example, hemoglobin molecules contain two copies each of two different subunits, the α-globin polypeptide and the β-globin polypeptide. At different developmental stages from the embryo to the adult, a human individual has different hemoglobin molecules assembled from different types of α-globin and β-globin polypeptides. The genes for each of the α-globin type of polypeptides are clustered together on one chromosome, while the genes for each of the β-globin type of polypeptides are clustered together on another chromosome. (See

~ FIGURE 7.5

Chromosome constitutions of *Drosophila* strains, showing the relationship between duplications of region 16A of the X chromosome and the production of reduced-eye-size phenotypes. (a) Wild type; (b) Homozygous *Bar* mutant.

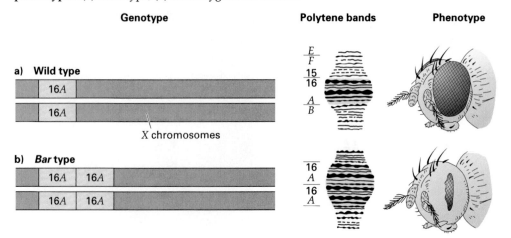

Chapter 16 for further discussion.) The sequences of the α-globin genes are all very similar, as are the sequences of the β-globin genes. It is thought that each assembly of genes evolved from a different ancestral gene by duplication and subsequent sequence divergence.

Inversion

An **inversion** is a chromosomal mutation that results when a segment of a chromosome is excised and then reintegrated in an orientation 180 degrees from the original orientation (Figure 7.6). There are two types of inversions: a **pericentric inversion** includes the centromere (Figure 7.6a), and a **paracentric inversion** does not include the centromere (Figure 7.6b).

In general, genetic material is not lost when an inversion takes place, although there can be phenotypic consequences when the breakpoints (inversion ends) occur within genes or within regions that control gene expression. Homozygous inversions can be seen because of the non-wild-type linkage relationships that result for the genes within the inverted segment and the genes that flank the inverted segment. For example, if the order of genes on the normal chromosome is *ABCDEFGH* and the *BCD* segment is inverted (shown now in bold), the gene order will now be *ADCBEFGH*, with *D* now more closely linked to *A* than to *E*, and *B* now more closely linked to *E* than to *A* (see Figure 7.6b).

The meiotic consequences of a chromosome inversion depend on whether the inversion occurs in a homozygote or a heterozygote. If the inversion is homozygous (say, *ADCBEFGH/ADCBEFGH*, where the *BCD* segment is the inverted segment), then meiosis will be normal and there are no problems related to gene duplications or deletions. However, crossing-over within inversion heterozygotes (say, *ABCDEFGH/ADCBEFGH*, where one chromosome has an inverted *BCD* segment) has serious genetic consequences. Moreover, the recombinant chromosomes are different for crossing-over in paracentric as opposed to pericentric inversion heterozygotes. In paracentric inversion heterozygotes, the homologous chromosomes attempt to pair so that the best possible base-pairing occurs. Because of the inverted segment on one homolog, pairing of homologous chromosomes requires the formation of loops containing the inverted segments, called inversion loops (Figure 7.7a; also see Figure 7.8). Inversion heterozygotes, then, may be identified by looking for those loops.

Paracentric inversion heterozygotes can be identified in genetic experiments, since viable recombinants are reduced significantly or suppressed. That is, the frequency of crossing-over is not diminished in inversion heterozygotes in comparison with normal cells, but gametes or zygotes derived from recombined chromatids are inviable. The inviability results from an *unbalanced* set of genes in the gamete—one or more genes are missing and/or one or more genes are present in two copies instead of one (see Figure 7.7b). If no crossovers occur in the inversion loop of an inversion heterozygote, then all resulting gametes receive a complete set of genes (two gametes with a normal gene order, *ABCDEFGH*, and two gametes with the inverted segment, *ADCBEFGH*), and they

~ FIGURE 7.6

Inversions. (a) Pericentric inversion; (b) Paracentric inversion.

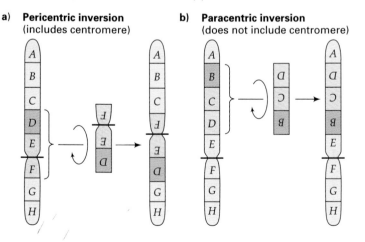

a) **Pericentric inversion**
     (includes centromere)

b) **Paracentric inversion**
     (does not include centromere)

~ FIGURE 7.7

Consequences of a paracentric inversion. (a) Photomicrograph of an inversion loop in polytene chromosomes of a strain of *Drosophila melanogaster* that is heterozygous for a paracentric inversion; (b) Meiotic products resulting from a single crossover within a heterozygous, paracentric inversion loop. Crossing-over occurs at the four-strand stage involving two nonsister homologous chromatids.

a)

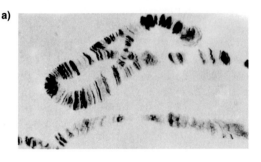

b)   Products of meiotic crossover

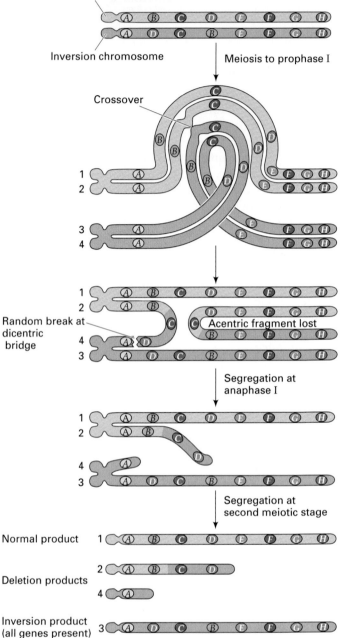

are all viable. Figure 7.7b shows the effects of a single crossover in the inversion loop of an individual heterozygous for a paracentric inversion. During the first meiotic anaphase, the two centromeres migrate to opposite poles of the cell. Because of the crossover between genes *B* and *C* in the inversion loop, one recombinant chromatid becomes stretched across the cell as the two centromeres begin to migrate in anaphase, forming a **dicentric bridge**, that is, a chromosome with two centromeres (a **dicentric chromosome**). With continued migration, the dicentric bridge breaks because of tension. The other recombinant product of the crossover event is a chromosome without a centromere (an acentric fragment). This acentric fragment is unable to continue through meiosis and is usually lost (it is not found in the gametes).

~ FIGURE 7.8

Meiotic products resulting from a single crossover within a heterozygous, pericentric inversion loop. Crossing-over occurs at the four-strand stage involving two nonsister homologous chromatids.

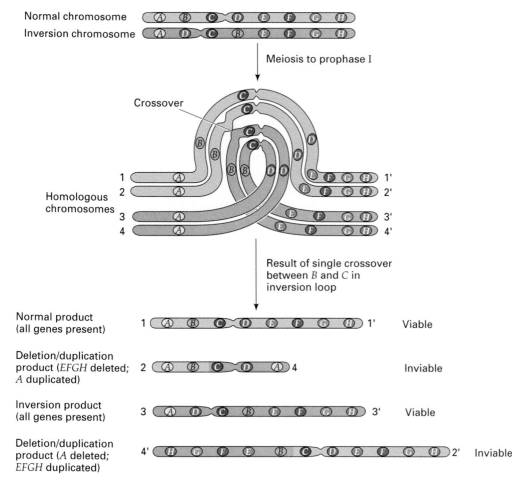

In the second meiotic division, each daughter cell receives a copy of each chromosome. Two of the gametes have complete sets of genes and are viable: the gamete with the normal order of genes (*ABCDEFGH*) and the gamete with the inverted segment of genes (*ADCBEFGH*). The other two gametes are inviable because they are unbalanced: many genes are deleted. Thus, *the only gametes that can give rise to viable progeny are those containing the chromosomes that did not involve crossing-over.* However, in many cases in female animals, the dicentric chromosomes or acentric fragments arising as a result of inversion are shunted to the polar bodies, so the reduction in fertility may not be so great.

The consequences of a single crossover in the inversion loop of an individual heterozygous for a pericentric inversion are shown in Figure 7.8. The crossover event and the ensuing meiotic divisions result in two viable gametes with the nonrecombinant

chromosomes *ABCDEFGH* (normal) and *ADCBEFGH* (inversion), and in two recombinant gametes that are inviable, each as a result of the deletion of some genes and the duplication of other genes.

Some crossover events within an inversion loop do not affect gamete viability. For example, a double crossover close together and involving the same two chromatids (a two-strand double crossover; see Chapter 5) produces four viable gametes. Also, recent studies with mammals show that inverted segments may remain unpaired. Since crossing-over cannot occur between unpaired segments, the generation of inviable gametes is avoided.

Translocation

A **translocation** is a chromosomal mutation in which there is a change in position of chromosome segments and the genes they contain. No gain or loss of genetic material is involved in a translocation. A translocation may involve the movement of a segment within a chromosome, a movement of a segment to another chromosome, or a reciprocal exchange of segments on two chromosomes. The last is a *reciprocal translocation* (Figure 7.9).

Translocations typically affect the products of meiosis. In many cases, some of the gametes produced are unbalanced in that they have duplications and/or deletions, and in many cases they are inviable. In other cases, such as familial Down syndrome, gametes are viable (see later in the chapter).

In strains *homozygous* for a reciprocal translocation, meiosis takes place normally, since all chromosome pairs can pair properly and crossing-over does not produce any abnormal chromatids. In strains *heterozygous* for a reciprocal translocation, however, all homologous chromosome parts pair as best they can. Because one set of normal chromosomes (N) and one set of translocated chromosomes (T) are involved, the result is a crosslike configuration in meiotic prophase I (Figure 7.10). These crosslike figures consist of four associated chromosomes, each chromosome being partially homologous to two other chromosomes in the group.

Segregation at anaphase I may occur in three different ways. (We are ignoring crossing-over in this discussion.) In one way, termed *alternate segregation*, alternate centromeres segregate to the same pole (Figure 7.10, left: N_1 and N_2 to one pole, T_1 and T_2 to the other pole). This produces two gametes, each of which is viable because it contains a complete set of genes, no more, no less. One of these gametes has two normal chromosomes and the other has two translocated chromosomes. In the second way, termed *adjacent 1 segregation*, adjacent *nonhomologous* centromeres migrate to the same pole (Figure 7.10, middle: N_1 and T_2 to one pole, N_2 and T_1 to the other pole). Both

gametes produced contain gene duplications and deletions and are often inviable. Adjacent 1 segregation occurs about as frequently as alternate segregation. In the third way, termed *adjacent 2 segregation*, different pairs of adjacent *homologous* centromeres migrate to the same pole (Figure 7.10, right: N_1 and T_1 to one pole, N_2 and T_2 to the other pole). Both products have gene duplications and deletions and are always inviable. Adjacent 2 segregation seldom occurs. In sum, of the six theoretically possible gametes, the two from alternate segregation are functional, the two from adjacent 1 segregation are usually inviable (because of gene duplications and deficiencies), and the two from adjacent 2 seldom occur and are inviable if they do. And, since alternate segregation and adjacent 1 segregation occur with about equal frequency, the term *semisterility* is applied to this condition. (This term is also used for inversion heterozygotes.)

In practice, animal gametes that have large duplicated or deleted chromosome segments may function, but the zygotes formed by such gametes typically die. The gametes may function normally and viable offspring may result if the duplicated and deleted chromosome segments are small. In plants, pollen grains with duplicated or deleted chromosome segments typically do not develop completely and hence are nonfunctional.

Some human tumors are associated with consistent chromosome translocations. Examples are chronic myelogenous leukemia (reciprocal translocation involving chromosomes 9 and 22) and Burkitt's lymphoma (reciprocal translocation involving chromosomes 8 and 14). The former is described here.

Chronic myelogenous leukemia (CML) is an invariably fatal cancer involving uncontrolled replication of myeloblasts (stem cells of white blood cells). Ninety percent of chronic myelogenous leukemia patients have a chromosomal mutation in the leukemic cells called the *Philadelphia chromosome* (Ph^1), so named because the discovery was made in the city of Philadelphia. The Philadelphia chromosome results from a reciprocal translocation involving the movement of part of the long arm of chromosome 22 (the second smallest human chromosome) to chromosome 9, and the movement of a small part from the tip of the long arm of chromosome 9 to chromosome 22 (Figure 7.11). This reciprocal translocation event apparently activates genes (called *oncogenes*; see Chapter 17) that cause the transition from a differentiated cell to a tumor cell with an uncontrolled pattern of growth. Specifically, the c-*abl* ("c-able"—cellular *Abel*son) oncogene, normally located on chromosome 9, is translocated to chromosome 22 in CML patients (see Figure 7.11). The

~ FIGURE 7.9

Reciprocal translocation.

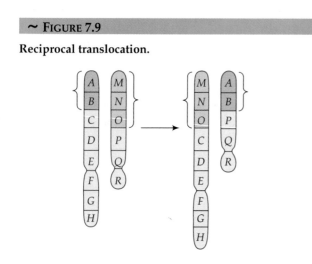

~ FIGURE 7.10

Meiosis in a translocation heterozygote in which no crossover occurs.

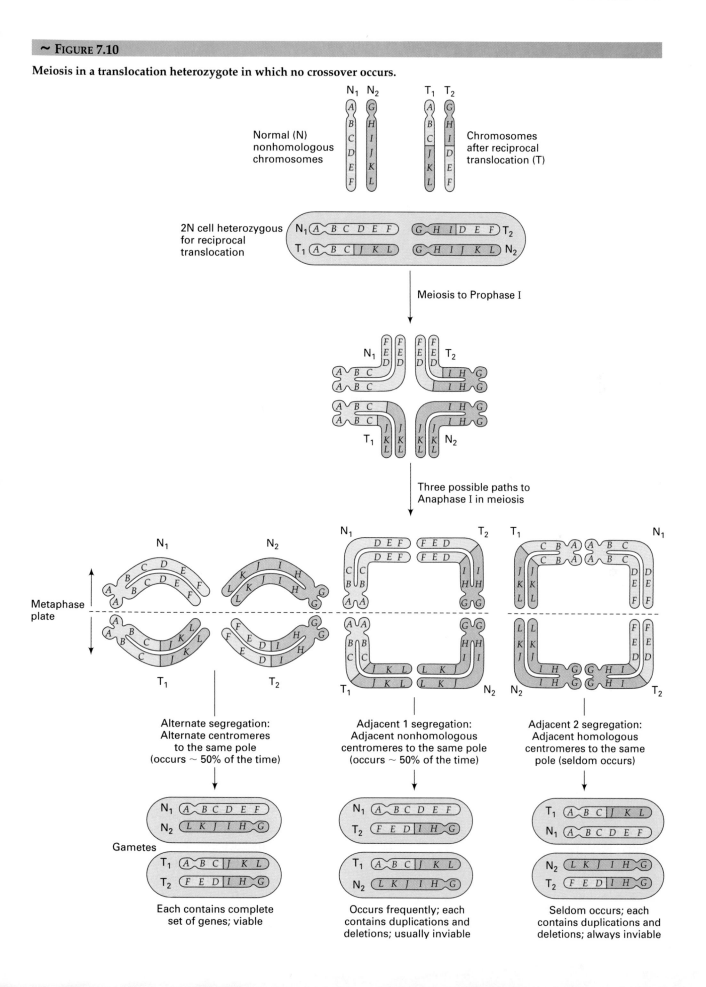

Alternate segregation:
Alternate centromeres
to the same pole
(occurs ~ 50% of the time)

Adjacent 1 segregation:
Adjacent nonhomologous
centromeres to the same pole
(occurs ~ 50% of the time)

Adjacent 2 segregation:
Adjacent homologous
centromeres to the same
pole (seldom occurs)

Each contains complete
set of genes; viable

Occurs frequently; each
contains duplications and
deletions; usually inviable

Seldom occurs; each
contains duplications and
deletions; always inviable

~ FIGURE 7.11

Origin of the Philadelphia chromosome in chronic myelogenous leukemia by a reciprocal translocation involving chromosomes 9 and 22. The arrows show the sites of the breakage points.

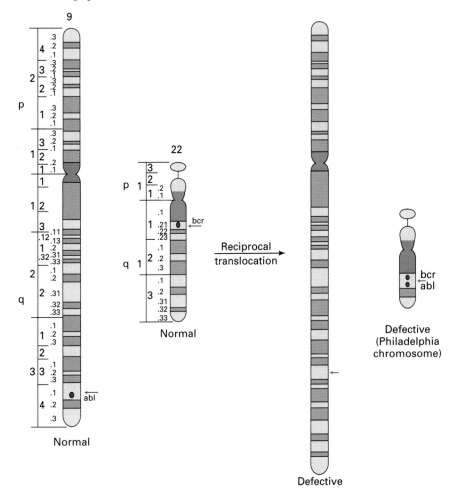

translocation event positions *c-abl* within the *bcr* (*breakpoint cluster region*) gene. This hybrid oncogene arrangement somehow causes a new gene product, producing leukemia.

KEYNOTE

Chromosomal mutations may involve parts of individual chromosomes rather than whole chromosomes or sets of chromosomes. The four major types of structural alterations are deletion (loss of a DNA segment), duplication (the addition of one or more extra copies of a DNA segment), inversion (reversal of orientation of a DNA segment in a chromosome), and translocation (movement of a DNA segment to another chromosomal location in the genome).

Fragile Sites and Fragile X Syndrome

When human cells are grown in culture, some of the chromosomes develop narrowings or unstained areas (gaps) called *fragile sites*. Over 40 fragile sites have been identified since the first one was discovered in 1965. One human condition related to a fragile site is *fragile X syndrome,* in which the X chromosome is prone to breakage at a site in the long arm of the X chromosome at position Xq27.3, as shown in Figure 7.12. After Down syndrome, fragile X syndrome is the leading genetic cause of mental retardation in the United States, with an incidence of about 1 in 1,250 males and 1 in 2,500 females (heterozygotes). As with all recessive X-linked traits, males predominantly exhibit this type of mental retardation.

The fragile X chromosome is inherited as a typical Mendelian gene. Male offspring of carrier females

~ FIGURE 7.12

Fragile site on the X chromosome. (a) Scanning electron micrograph and (b) Diagram of a human X chromosome showing the location of the fragile site responsible for fragile X syndrome. (From Gerald Stine, *The New Human Genetics.* Copyright © 1989. Reproduced by permission of The McGraw-Hill Companies.)

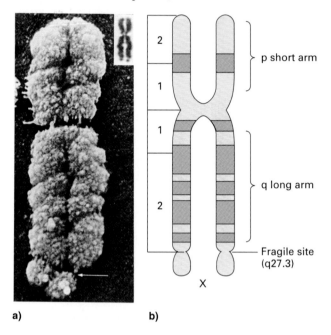

a)   b)

have a 50 percent chance of receiving a fragile X chromosome. However, only 80 percent of those males receiving a fragile X chromosome are mentally retarded; the rest are normal. These phenotypically normal males are called *normal transmitting males* and carry a "premutation" because they can pass on the fragile X chromosome to their daughters. The sons of those daughters frequently show symptoms of mental retardation. Female offspring of carrier (heterozygous) females also have a 50 percent chance of inheriting a fragile X chromosome. Up to 33 percent of the carrier females show mild mental retardation.

Modern molecular techniques brought to bear on this disease have resulted in an understanding of the disease at the DNA level. There is a repeated three-base-pair sequence, CGG, in a gene called *FMR-1* (*frag*ile X *m*ental *r*etardation-1) located at the fragile X site. Normal individuals have an average of 29 CGG repeats (the range is from 6 to 54) in the coding region of the *FMR-1* gene. Phenotypically normal transmitting males and their daughters, as well as some carrier females, have a significantly larger number of CGG repeats, ranging from 55 to 200 copies. Since these individuals do not show symptoms of fragile X syndrome, the increased number of repeats they have is called *pre*-

mutations. A premutation could be considered a silent mutation. Males and females with fragile X syndrome have even larger numbers of the CGG repeats, ranging from 200 to 1,300 copies; these are considered to be the full mutations. In other words, there is a mechanism that results in the tandem duplication (amplification) of the triplet repeat, CGG, in the *FMR-1* gene, and below a certain threshold number of copies (about 200 or fewer) there are no clinical symptoms, while above that threshold number of copies (greater than 200) clinical symptoms are seen. Interestingly, amplification of the CGG repeats does not occur in males, only in females. Therefore, a phenotypically normal transmitting male (who has the premutation) transmits his X chromosome to his daughter. During a slipped mispairing process during DNA replication in his daughter, perhaps, the triplets may amplify, and she can transmit the amplified X to her son. Thus, affected males inherit the mutation from their grandfather. As yet, the function of the *FMR-1* gene, in which the triplet repeat amplification occurs, is unknown, so we do not yet understand how such amplification produces mental retardation. A protein is made by the gene but has unidentified function.

Triplet repeat amplification has also been shown to cause other human diseases such as myotonic dystrophy, spinobulbar muscular atrophy (also called Kennedy disease), and Huntington disease (see Chapter 4). Each of these cases differs from fragile X syndrome in that the amplification can occur in both sexes at each generation. For each, there is a threshold number of triplet repeat copies above which symptoms of the disease are produced.

VARIATIONS IN CHROMOSOME NUMBER

When an organism or cell has one complete set of chromosomes, or an exact multiple of complete sets, that organism or cell is said to be **euploid**. Thus, eukaryotic organisms that are normally diploid (such as humans and fruit flies) and eukaryotic organisms that are normally haploid (such as yeast) are all euploids. Chromosome mutations that result in variations in the number of chromosome sets occur in nature, and the resulting organism or cells are also euploid. Chromosome mutations resulting in variations in the number of individual chromosomes are examples of **aneuploidy**. An aneuploid organism or cell has a chromosome number that is not an exact multiple of the haploid set of chromosomes. Both euploid and aneuploid variations affecting whole chromosomes are discussed in this section.

Changes in One or a Few Chromosomes

GENERATION OF ANEUPLOIDY.   Changes in chromosome number can occur in both diploid and haploid organisms. Nondisjunction (see Chapter 3) of one or more chromosomes during meiosis I or meiosis II is typically responsible for generating gametes with abnormal numbers of chromosomes. Refer to Figure 3.5) and consider just one particular chromosome. Nondisjunction at meiosis I produces four abnormal gametes; that is, two gametes with a chromosome duplicated and two gametes with the corresponding chromosome missing. Nondisjunction at meiosis I can produce gametes with two sets of homologs of a particular chromosome (and thus possible heterozygotes) and some other gametes with that chromosome missing. Fusion of the former gamete type with a normal gamete will produce a zygote with three copies of the particular chromosome instead of the normal two and, unless nondisjunction has involved other chromosomes, two copies of all other chromosomes. Similarly, fusion of the latter gamete type with a normal gamete will produce a zygote with only one copy of the particular chromosome instead of the normal two, and two copies of all other chromosomes. Nondisjunction in meiosis II (see Figure 3.5c) also results in two normal gametes and two abnormal gametes, that is, a single gamete with two sister chromosomes and one gamete with that same chromosome missing. Fusion of these abnormal gametes with normal gametes will give the zygote types just discussed. Unlike the case with nondisjunction at meiosis I, some normal gametes are produced by nondisjunction at meiosis II; specifically, two of the four gametes are normal.

TYPES OF ANEUPLOIDY.   In aneuploidy, one or more chromosomes are lost from or added to the normal set of chromosomes (Figure 7.13). Aneuploidy can occur, for example, from the loss of individual chromosomes in meiosis or (very rarely) in mitosis by nondisjunction. In most cases, autosomal aneuploidy is lethal in animals, so in mammals it is detected mainly in aborted fetuses. Aneuploidy "is tolerated" more by plants, especially in species that are considered polyploidy.

In diploid organisms, there are four main types of aneuploidy (see Figure 7.13):

1. **Nullisomy** (a nullisomic cell) involves a loss of one homologous chromosome pair; that is, the cell is 2N − 2. (This could arise, for example, if nondisjunction occurs for the same chromosome in meiosis in both parents, producing gametes with no copies of that chromosome and one copy of all other chromosomes in the set.)

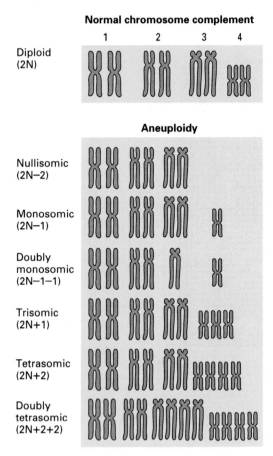

Normal (theoretical) set of metaphase chromosomes in a diploid (2N) organism (top) and examples of aneuploidy (bottom).

2. **Monosomy** (a monosomic cell) involves a loss of a single chromosome; that is, the cell is 2N − 1. (This can arise, for example, if nondisjunction in meiosis in a parent produces a gamete with no copies of a particular chromosome and one copy of all other chromosomes in the set.)

3. **Trisomy** (a trisomic cell) involves a single extra chromosome; that is, the cell has three copies of a particular chromosome and two copies of other chromosomes. A trisomic cell is 2N + 1. (This can arise, for example, if nondisjunction in meiosis in a parent produces a gamete with two cop-ies of a particular chromosome and one copy of all other chromosomes in the set.)

4. **Tetrasomy** (a tetrasomic cell) involves an extra chromosome pair, resulting in the presence of four copies of one particular chromosome and two copies of other chromosomes. A tetrasomic cell is 2N + 2. (This could arise, for example, if

nondisjunction occurs for the same chromosome in meiosis in both parents producing gametes with two copies of that chromosome and one copy of all other chromosomes in the set.)

Aneuploidy may involve the loss or the addition of more than one specific chromosome or chromosome pair. For example, a *double monosomic* has two separate chromosomes present in only one copy each; that is, it is $2N - 1 - 1$. A *double tetrasomic* has two chromosomes present in four copies each; that is, it is $2N + 2 + 2$. In both of these cases, meiotic nondisjunction involved two different chromosomes in one parent's gamete production.

Most forms of aneuploidy have serious consequences in meiosis. Monosomics, for example, will produce two kinds of haploid gametes, N and N − 1. Alternatively, the odd, unpaired chromosome in the 2N − 1 cell may be lost during meiotic anaphase and not be included in either daughter nucleus, thereby producing two N − 1 gametes. There are more segregation possibilities for trisomics in meiosis. Consider

a trisomic of genotype $+/+/a$ in an organism that can tolerate trisomy, and assume no crossing-over between the *a* locus and its centromere. As shown in Figure 7.14, random segregation of the three types of chromosomes produces four genotypic classes of gametes, as follows: $2(+a):2(+):1(++):1(a)$. In a cross of a $+/+/a$ trisomic to an a/a individual, the predicted phenotypic ratio among the progeny is 5 wild type : 1 mutant (a). This ratio is seen in many actual crosses of this kind.

In the following sections we examine some examples of aneuploidy as they are found in the human population. Table 7.1 summarizes various aneuploid abnormalities for autosomes and for sex chromosomes in the human population. Examples of aneuploidy of the X and Y chromosomes are discussed in Chapter 3. Recall that, in mammals, aneuploidy of the sex chromosomes is more often found in adults than aneuploidy of the autosomes because of a dosage compensation mechanism (lyonization) by which excess X chromosomes are inactivated.

In humans, autosomal monosomy is only found rarely. Presumably, monosomic embryos do not develop significantly and are lost early in pregnancy. In contrast, autosomal trisomy accounts for about one-half of chromosomal abnormalities producing fetal deaths. In fact, only a few autosomal trisomies are seen in live births. Most of these (trisomy-8, -13, and -18) result in early death. Only in trisomy-21 (Down syndrome) does survival to adulthood occur.

~ **FIGURE 7.14**

Meiotic segregation possibilities in a trisomic individual. Shown is segregation in an individual of genotype $+/+/a$, when two chromosomes migrate to one pole and one goes to the other pole, and assuming no crossing-over between the *a* locus and its centromere.

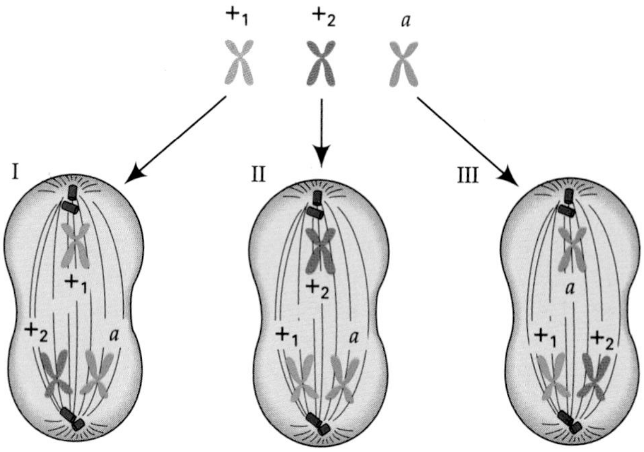

Gametes produced
after 2nd meiotic division

	haploid	disomic
I	$+_1$	$+_2/a$
II	$+_2$	$+_1/a$
III	a	$+_1/+_2$

In sum: $2 +/a : 2 + : 1 + /+ : 1 a$

~ **TABLE 7.1**

Aneuploid Abnormalities in the Human Population

CHROMOSOMES	SYNDROME	FREQUENCY AT BIRTH
Autosomes		
Trisomic 21	Down	14.3/10,000
Trisomic 13	Patau	2/10,000
Trisomic 18	Edwards	2.5/10,000
Sex Chromosomes, Females		
XO, monosomic	Turner	4/10,000 females
XXX, trisomic	Viable; most are fertile	14.3/10,000 females
XXXX, tetrasomic		
XXXXX, pentasomic		
Sex Chromosomes, Males		
XYY, trisomic	Normal	25/10,000 males
XXY, trisomic		
XXYY, tetrasomic	Klinefelter	40/10,000
XXXY, tetrasomic		

~ FIGURE 7.15

Trisomy-21 (Down syndrome). (a) Karyotype; (b) Individual.

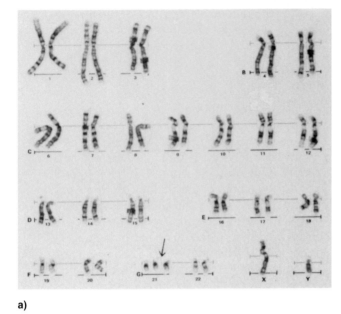

a)

b)

Trisomy-21. Trisomy-21 occurs when there are three copies of chromosome 21 (Figure 7.15a). Trisomy-21 occurs with a frequency of about 3,510 per 1 million conceptions and about 1,430 per 1 million live births. Individuals with trisomy-21 have Down syndrome (Figure 7.15b) and have such abnormalities as low IQ, epicanthal folds over the eyes, short and broad hands, and below-average height.

The relationship between the age of the mother and the probability of her having a trisomy-21 individual is indicated in Table 7.2. (There is no correlation with age of the father.) During the development of a female fetus before birth, the primary oocytes in the ovary undergo meiosis but stop at prophase I. In a fer-

tile female, each month at ovulation the nucleus of a secondary oocyte (see Chapter 3) begins the second meiotic division but progresses only to metaphase, when division again stops. If a sperm penetrates the secondary oocyte, the second meiotic division is completed. The probability of nondisjunction increases with the length of time the primary oocyte is in the ovary. It is important, then, that older mothers-to-be consider testing to determine whether the fetus has a normal complement of chromosomes.

Down syndrome individuals can also result from a different sort of chromosomal mutation called *centric fusion* or **Robertsonian translocation**, which produces three copies of the long arm of chromosome 21. This form of Down syndrome is called *familial Down syndrome*. A Robertsonian translocation is a type of nonreciprocal translocation in which two nonhomologous acrocentric chromosomes (chromosomes with centromeres near their ends) break at their centromeres and then the long arms become attached to a single centromere (Figure 7.16). The short arms also join to form the reciprocal product, which typically contains nonessential genes and is usually lost within a few cell divisions. In humans, when a Robertsonian translocation joins the long arm of chromosome 21 with the long arm of chromosome 14 (or 15), the heterozygous carrier is phenotypically normal, since there are two copies of all major chromosome arms and hence two copies of all essential genes involved.

~ TABLE 7.2

Relationship Between Age of Mother and Risk of a Trisomy-21

AGE OF MOTHER	RISK OF TRISOMY-21 IN CHILD
16–26	7.7/10,000
27–34	4/10,000
35–39	29/10,000
40–44	100/10,000
45–47	333/10,000
All mothers combined	14.3/10,000

~ **FIGURE 7.16**

Robertsonian translocation. Production of a Robertsonian translocation (centric fusion) by breakage of two acrocentric chromosomes at their centromeres (indicated by arrows) and fusion of the two large chromosome arms and of the two small chromosome arms.

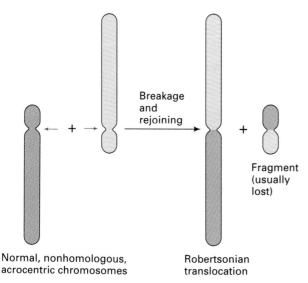

~ **FIGURE 7.17**

The three segregation patterns of a heterozygous Robertsonian translocation involving the human chromosomes 14 and 21. Fusion of the resulting gametes with gametes from a normal parent produces zygotes with various combinations of normal and translocated chromosomes.

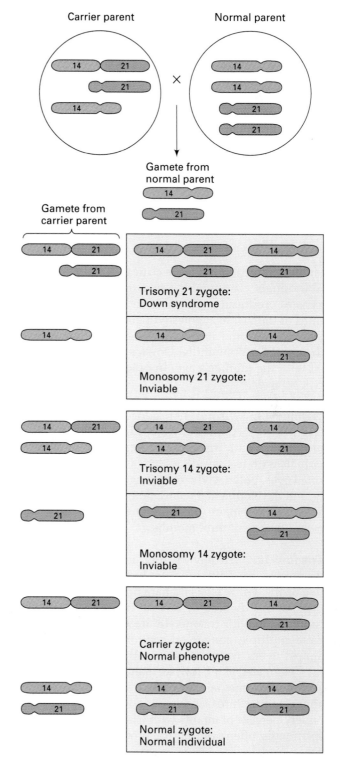

There is a high risk of Down syndrome among the offspring of pairings between heterozygous carriers and normal individuals (Figure 7.17). The normal parent produces gametes with one copy each of chromosomes 14 and 21. The heterozygous carrier parent produces three reciprocal pairs of gametes, each pair produced as a result of different segregation of the three chromosomes involved. As Figure 7.17 shows, the zygotes produced by pairing these gametes with gametes of normal chromosomal constitution theoretically are as follows: one-sixth have normal chromosomes 14 and 21; one-sixth are heterozygous carriers like the parent and are phenotypically normal; one-sixth are inviable because of monosomy for chromosome 14; one-sixth are inviable because of monosomy for chromosome 21; one-sixth are inviable because of trisomy for chromosome 14; and one-sixth are trisomy-21 and therefore produce a Down syndrome individual. (These latter individuals actually have the normal diploid number of 46 chromosomes but, because of the Robertsonian translocation, they have three copies of the long arm of chromosome 21, which is sufficient to give the Down syndrome symptoms. Similarly, the trisomy-14 zygotes shown in Figure 7.17 have 46 chromosomes, but they have three copies of the long arm of chromosome 14. Apparently the dosage of the genes involved on this

~ FIGURE 7.18

Trisomy-13 (Patau syndrome). (a) Karyotype; (b) Individual.

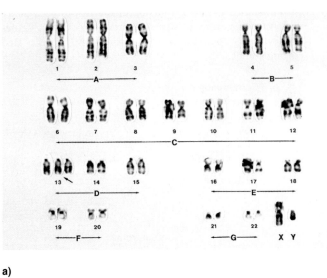

a)

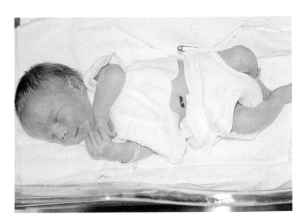

b)

larger chromosome is more critical, since trisomy-14 individuals are inviable.) In sum, one-half of the zygotes produced are inviable and theoretically one-third of the viable zygotes will give rise to a familial Down syndrome individual—a much higher risk than that for nonfamilial Down syndrome associated with the mother's age. In practice, the observed risk is lower.

Trisomy-13. Trisomy-13 produces Patau syndrome (Figure 7.18). About 2 in 10,000 live births produce individuals with trisomy-13. Characteristics of individuals with trisomy-13 include cleft lip and palate, small eyes, polydactyly (extra fingers and toes), mental and developmental retardation, and cardiac anomalies, among many other abnormalities. Most die before the age of three months.

Trisomy-18. Trisomy-18 produces Edwards syndrome (Figure 7.19). It occurs in about 2.5 in 10,000 live births. For reasons that are not known, about 80 percent of Edwards syndrome infants are female. Individuals with trisomy-18 are small at birth and have multiple congenital malformations affecting almost every organ system in the body. Clenched fists, elongated skull, low-set malformed ears, mental and developmental retardation, and many other abnormalities are associated with the syndrome. Ninety percent of infants with trisomy-18 die within six months, often from cardiac problems.

Changes in Complete Sets of Chromosomes

Monoploidy and **polyploidy** involve variations from the normal state in the number of complete sets of chromosomes. Since the number of complete sets of chromosomes is involved in each case, monoploids and polyploids are both euploids. Monoploidy and polyploidy are lethal for most animal species but are tolerated more readily by plants. Both have played significant roles in plant speciation.

Changes in complete sets of chromosomes can result, for example, when the first or second meiotic division is abortive (lack of cytokinesis), or when meiotic nondisjunction occurs for all chromosomes. If such nondisjunction occurs at meiosis I, one-half of the gametes will have no chromosome sets, and the other half will have two chromosome sets (refer to Figure 3.5b). If such nondisjunction occurs at meiosis II, one-half of the gametes will have the normal one set of chromosomes, one-quarter will have two sets of chromosomes, and one-quarter will have no chromosome sets (see Figure 3.5b). Fusion of a gamete with two chromosome sets with a normal gamete will produce a polyploid zygote, in this case one with three sets of chromosomes, which is a *triploid* (3N). Similarly, fusion of two gametes, each with two chromosome sets, will produce a *tetraploid* (4N) zygote. Polyploidy of somatic cells can also occur following mitotic nondisjunc-

Trisomy-18 (Edwards syndrome). (a) Karyotype; (b) Individual.

a)

b)

tion of complete chromosome sets. Monoploid (haploid) individuals, by contrast, typically develop from unfertilized eggs.

MONOPLOIDY.   A monoploid individual has only one set of chromosomes instead of the usual two sets (Figure 7.20a). Monoploidy is sometimes called haploidy, although the term *haploidy* is typically used to describe the chromosome complement of gametes. Some fungi and males of haploid/diploid species (ants, bees, wasps) are haploid, for example.

Monoploidy is seen only rarely in normally adult diploid organisms. Because of the presence of recessive lethal mutations (which are usually counteracted by dominant alleles in heterozygous individuals) in the chromosomes of many diploid eukaryotic organisms, many monoploids probably do not survive. Certain species produce monoploid organisms as a normal part of their life cycle. Some male wasps, ants, and bees, for example, are monoploid because they develop from unfertilized eggs.

Monoploids are used in plant-breeding experiments. Normal haploid cells produced by meiosis in plant anthers can be isolated and induced to grow to produce monoploid cultures for study. The single chromosome set of these monoploids then can be doubled using the chemical colchicine (which inhibits the formation of the mitotic spindle, thereby resulting in nondisjunction for all chromosomes) to produce completely homozygous diploid breeding lines.

Cells of a monoploid individual are very useful for producing mutants because there is only one dose of each of the genes. Thus, mutants can be isolated directly.

POLYPLOIDY.   Polyploidy is the chromosomal constitution of a cell or organism having three or more sets of chromosomes (Figure 7.20b). Polyploids may arise spontaneously or be induced experimentally (for instance, by using colchicine—see above). They often occur as a result of a breakdown of the spindle apparatus in one or more meiotic divisions or in mitotic divisions. Virtually all plants and animals probably have some polyploid tissues. For example, the endosperm of plants is triploid, the liver of mammals and perhaps other vertebrates is polyploid, and the giant abdominal neuron of the sea hare *Aplysia* has about 75,000 copies of the genome. Plants that are completely polyploid include wheat, which is hexaploid. Some animal species such as the North American sucker (a freshwater fish), salmon, and some salamanders are polyploid.

There are two general classes of polyploids: those that have an *even* number of chromosome sets and those that have an *odd* number of sets. Polyploids with an even number of chromosome sets have a better chance of being at least partially fertile, since there is

~ FIGURE 7.20

Variations in number of complete chromosome sets. (a) Monoploidy (only one set of chromosomes instead of two); (b) Polyploidy (three or more sets of chromosomes).

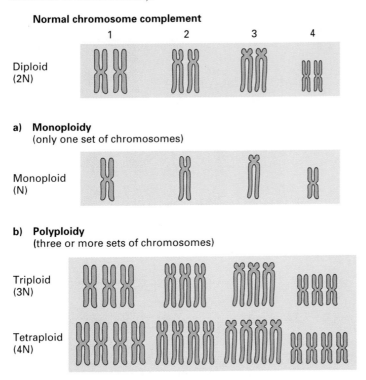

the potential for homologs to be segregated equally during meiosis. Polyploids with an odd number of chromosome sets always have an unpaired chromosome for each chromosome type, so the probability of producing a balanced gamete is extremely low; such organisms are usually sterile or have increased zygote abortion.

In triploids, the nucleus of a cell has three sets of chromosomes. As a result, triploids are very unstable in meiosis because, as in trisomics, two of the three homologous chromosomes go to one pole, and the other goes to the other pole. The segregation of each chromosome from its homologs in the triploid is at random, so the probability of producing balanced gametes with either a haploid or a diploid set of chromosomes is small; many of the gametes will be unbalanced with one copy of one chromosome, two copies of another, and so on. In general, the probability of a triploid producing a haploid gamete is $(1/2)^n$, where n is the number of chromosomes.

In humans, the most common type of polyploidy is triploidy. Triploidy is always lethal. Triploidy is seen in 15 to 20 percent of spontaneous abortions and in about 1 in 10,000 live births, but most affected infants die within one month. Triploid infants have many abnormalities, including a characteristically enlarged head. Tetraploidy in humans is also always lethal, usually before birth. It is seen in about 5 percent of spontaneous abortions. Very rarely a tetraploid human will be born, but such an individual does not survive long.

Plants are more "tolerant" of polyploidy for two reasons. First, sex determination is less sensitive to polyploidy in plants than in animals. Second, many plants undergo self-fertilization, so if a plant is produced with an even polyploid number of chromosome sets (for example, 4N) it can still produce fertile gametes and reproduce.

Two types of polyploidy are encountered in plants. In **autopolyploidy** all the sets of chromosomes originate in the same species. The condition probably results from a defect in meiosis that leads to diploid or triploid gametes. If a diploid gamete fuses with a normal haploid gamete, the zygote and the organism that develops from it will have three sets of chromosomes: it will be triploid. The cultivated banana is an example of a triploid autopolyploid plant. Because it has an odd number of chromosome sets, the gametes have a variable number of chromosomes and few fertile seeds are set, thereby making most bananas seedless and highly palatable. Because of the triploid state, cultivated bananas are propagated vegetatively (by cuttings). In general, the development of "seedless" fruits such as seedless grapes and seedless watermelons relies on "odd-number" polyploidy. Triploidy has also been found in grasses, garden flowers, crop plants, and forest trees.

In **allopolyploidy** the sets of chromosomes involved come from different, though usually related, species. This situation can arise if two different species interbreed to produce an organism with one haploid set of each parent's chromosomes and then both chromosome sets double. For example, fusion of haploid gametes of two diploid plants that can cross may produce an $N_1 + N_2$ hybrid plant that will have a haploid set of chromosomes from plant 1 and a haploid set from plant 2. However, because of the differences between the two chromosome sets, pairing of chromosomes does not occur at meiosis and no viable gametes are produced. As a result, the hybrid plants are sterile. Rarely, through a division error, the two sets of chromosomes double, producing tissues of $2N_1 + 2N_2$ genotype. (That is, the cells in the tissue have a diploid set of chromosomes from plant 1 and a diploid set from plant 2.) Each diploid set can function normally in meiosis, so that gametes produced from the $2N_1 + 2N_2$ plant are $N_1 + N_2$. Fusion of two gametes like this can produce fully fertile, allotetraploid, $2N_1 + 2N_2$ plants.

A classic example of allopolyploidy resulted from crosses made between cabbages (*Brassica oleracea*) and radishes (*Raphanus sativus*) by Karpechenko in 1928. Both parents have a chromosome number of 18, and the F_1 hybrids also have 18 chromosomes, nine from each parent. These hybrids are morphologically intermediate between cabbages and radishes. The F_1 plants are mostly sterile because of the failure of chromosomes to pair at meiosis. However, a few seeds are produced through meiotic errors, and some of those seeds are fertile. The somatic cells of the plants produced from the seeds have 36 chromosomes; that is, full diploid sets of chromosomes from both the cabbage and the radish. These plants are completely fertile and belong to a breeding species named *Raphanobrassica*, a fusion of the two parental genus names.

Lastly, all commercial grains, most crops, and many common commercial flowers are polyploid. In fact, polyploidy is the rule rather than the exception in agriculture. For example, the cultivated bread wheat, *Triticum aestivum*, is an allohexaploid with 42 chromosomes. This plant species is descended from three distinct species, each with a diploid set of 14 chromosomes. Meiosis is normal because only homologous chromosomes pair, so the plant is fertile.

KEYNOTE

Variations in the chromosome number of a cell or an organism give rise to aneuploidy, monoploidy, or polyploidy. In aneuploidy a cell or organism has one, two, or a few whole chromosomes more or less than the basic number of the species under study. In monoploidy an organism that is usually diploid has only one set of chromosomes. And in polyploidy an organism has three or more complete sets of chromosomes. Any or all of these abnormal conditions may have serious consequences to the organism.

SUMMARY

In this chapter we have considered several kinds of chromosomal mutations. Chromosomal mutations are variations from the normal (wild-type) condition in either chromosome structure or chromosome number. They may occur spontaneously, or their frequency can be increased by exposure to radiation or chemical mutagens. There are four major types of chromosomal structural mutations: (1) deletion, in which a DNA segment is lost; (2) duplication, in which there are one or more extra copies of a DNA segment; (3) inversion, in which there is a reversal of orientation of a DNA segment in a chromosome; and (4) translocation, in which a DNA segment has moved to a new location in the genome.

The consequences of these kinds of structural mutations depend on the specific mutation involved. Firstly, each kind of mutation involves one or more breaks in a chromosome. If a break occurs within a gene, then a gene mutation has been produced. In deletions, genes may be lost and multiple mutant phenotypes may result. In some cases deletions and duplications result in lethality or severe defects as a consequence of a deviation from the normal gene dosage. Secondly, chromosomal mutations in the heterozygous condition can result in production of some gametes that are inviable because of duplications and/or deficiencies. Commonly this is seen following meiotic crossovers that produce inversions and translocations in heterozygotes.

Variations in chromosome number involve departure from the normal diploid (or haploid) state of the organism. For diploids, the three classes of such mutations are: (1) aneuploidy, in which one to a few whole chromosomes are lost from or added to the normal chromosome set; (2) monoploidy, in which only one set of chromosomes is present in a usually diploid organism; and (3) polyploidy, in which a cell or organism has three or more sets of chromosomes. The consequences of these chromosomal mutations depend on the organism. In general, plants are more "tolerant" than animals of variations in the number of chromosome sets; for example, wheat is hexaploid. While some animal species are naturally polyploid, in the majority of instances monoploidy and polyploidy are lethal, probably because gene expression problems occur when abnormal numbers of gene copies are present. Even in viable individuals, viable gametes may not result because of segregation problems during meiosis.

ANALYTICAL APPROACHES FOR SOLVING GENETICS PROBLEMS

Q7.1 Diagram the meiotic pairing behavior of the four chromatids in an inversion heterozygote $a\ b\ c\ d\ e\ f\ g/$ $a'\ b'\ f'\ e'\ d'\ c'\ g'$. Assume that the centromere is to the left of gene a. Next, diagram the early anaphase configuration if a crossover occurred between genes d and e.

A7.1 This question requires a knowledge of meiosis and the ability to draw and manipulate an appropriate inversion loop. Part (a) of Figure 7.A to the right shows the diagram for the meiotic pairing.

Note that the lower pair of chromatids (a', b', etc.) must loop over in order for all the genes to align; this looping is characteristic of the pairing behavior expected for an inversion heterozygote.

Once the first diagram has been constructed, answering the second part of the question is straightforward. We diagram the crossover, then trace each chromatid from the centromere end to the other end. It is convenient to distinguish maternal and paternal genes, perhaps by a' versus a, and so on, as we did in part (a) of the figure. The result of the crossover between d and e is shown in part (b).

In anaphase I of meiosis, the two centromeres, each with two chromatids attached, migrate toward the opposite poles of the cell. At anaphase the noncrossover chromatids (top and bottom chromatids in the figure) segregate to the poles normally. As a result of the single crossover between the other two chromatids, however, unusual chromatid configurations are produced, and these configurations are found by tracing the chromatids from left to right. If we begin by tracing the second chromatid from the top, we get ($_\circ a\ b\ c\ d\ e'\ f'\ b'\ a'_\circ$), a dicentric chromatid (where $\circ$ is a centromere); in other words, we have a single chromatid attached to two centromeres. This chromatid also has duplications and deletions for some of the genes. Thus, during anaphase this so-called dicentric chromosome becomes stretched between the two poles of the cells as the centromeres separate, and the chromosome will eventually break at

FIGURE 7.A

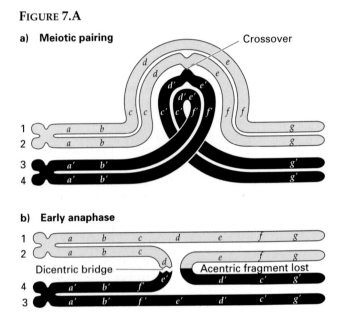

a) Meiotic pairing

b) Early anaphase

a random location. The other product of the single crossover event is an acentric fragment (without a centromere) that can be traced starting from the right with the second chromatid from the top. This chromatid ($g\ f\ e\ d'\ c'\ g'$) contains neither a complete set of genes nor a centromere—it is an acentric fragment that will be lost as meiosis continues.

Thus, the consequence of a crossover event within the inversion in an inversion heterozygote is the production of gametes with duplicated or deleted genes. These gametes often will be inviable. Viable gametes are produced, however, from the noncrossover chromatids: One of these chromatids (1 in part [b] of the figure) has the normal gene sequence, and the other (3 in part [b] of the figure) has the inverted gene sequence.

Q7.2 Eyeless is a recessive gene (*ey*) on chromosome 4 of *Drosophila melanogaster*. Flies homozygous for *ey* have tiny eyes or no eyes at all. A male fly trisomic for chromosome 4 with the genotype $+/+/ey$ is crossed with a normal diploid, eyeless female of genotype ey/ey. What expected genotypic and phenotypic ratios would result from random assortment of the chromosomes to the gametes?

A7.2 To answer this question, we must apply our understanding of meiosis to the unusual situation of a trisomic cell. Regarding the ey/ey female, only one gamete class can be produced, namely, eggs of genotype ey. Gamete production with respect to the trisomy for chromosome 4 occurs by a random segregation pattern in which two chromosomes migrate to one pole and the other chromosome migrates to the other pole during meiosis I. (This pattern is similar to the meiotic segregation pattern shown in secondary nondisjunction of XXY cells; see Chapter 3.) Three types of segregation are possible in the formation of gametes in the trisomy, as shown in part (a) of Figure 7.B. The union of these sperm at random with eggs of genotype ey occurs as shown in part (b).

The resulting genotypic and phenotypic ratios are illustrated in part (c).

FIGURE 7.B

a) Segregation

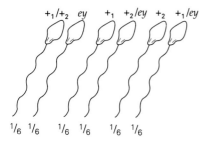

$+_1/+_2$ ey $+_1$ $+_2/ey$ $+_2$ $+_1/ey$

$^1/_6$ $^1/_6$ $^1/_6$ $^1/_6$ $^1/_6$ $^1/_6$

b) Union

		Eggs ey	Phenotype
	$+/+$	$+/+/ey$	$+$
	ey	ey/ey	ey
Sperm	$+$	$+/ey$	$+$
	$+/ey$	$+/ey/ey$	$+$
	$+$	$+/ey$	$+$
	$+/ey$	$+/ey/ey$	$+$

c) Summary of genotypes and phenotypes

Ratios:

Genotypes	Phenotype
$^1/_6$ $+/+/ey$	$^5/_6$ wild type
$^1/_3$ $+/ey/ey$	$^1/_6$ eyeless
$^1/_3$ $+/ey$	
$^1/_6$ ey/ey	

QUESTIONS AND PROBLEMS

*7.1 A normal chromosome has the following gene sequence:

$$A B C D \, _\circ \, E F G H$$

Determine the chromosomal mutation illustrated by each of the following chromosomes:

a. $A B C F E \, _\circ \, D \, G \, H$

b. $A D \, _\circ \, E F B C G H$

c. $A B C D \, _\circ \, E F E F G H$

d. $A B C D \, _\circ \, E F F E G H$

e. $A B D \, _\circ \, E F G H$

*7.2 Distinguish between pericentric and paracentric inversions.

7.3 What would be the effect on protein structure if a small inversion were to occur within the amino-acid-coding region of a gene?

*7.4 Inversions are known to affect crossing-over. The following homologs with the indicated gene order are given (the filled and open circles are homologous centromeres):

$$\bullet A B C D E$$

$$_\circ A D C B E$$

a. Diagram the alignment of these chromosomes during meiosis.
b. Diagram the results of a single crossover between homologous genes B and C in the inversion.
c. Considering the position of the centromere, what is this sort of inversion called?

7.5 Single crossovers within the inversion loop of inversion heterozygotes give rise to chromatids with duplications and deletions. What happens if, within the inversion loop, there is a two-strand double crossover in such an inversion heterozygote when the centromere is outside the inversion loop?

*7.6 A particular species of plant that had been subjected to radiation for a long time in order to produce chromosome mutations was then inbred for many generations until it was homozygous for all of these mutations. It was then crossed to the original unirradiated plant, and the meiotic process of the F_1 hybrids was examined. It was noticed that the following structures occurred, at low frequency, in anaphase I of the hybrid: a cell with a dicentric chromosome (bridge) and a fragment.

a. What kind of chromosome mutation occurred in the irradiated plant? In your answer, indicate where the centromeres are.

b. Explain in words and with a clear diagram where crossover(s) occurred, and how the bridge chromosome of the cell arose.

*7.7   Mr. and Mrs. Lambert have not yet been able to produce a viable child. They have had two miscarriages and one severely defective child who died soon after birth. Studies of banded chromosomes of father, mother, and child showed all chromosomes were normal except for pair number 6. The number 6 chromosomes of mother, father, and child are shown in the following figure:

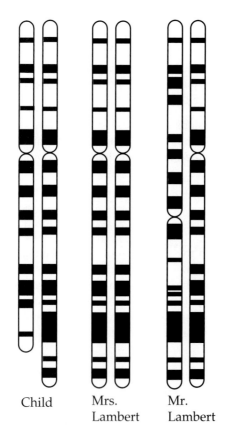

Child   Mrs.   Mr.
           Lambert   Lambert

a. Does either parent have an abnormal chromosome? If so, what is the abnormality?

b. How did the chromosomes of the child arise? Be specific as to what events in the parents gave rise to these chromosomes.

c. Why is the child not phenotypically normal?

d. What can be predicted about future conceptions by this couple?

7.8   Mr. and Mrs. Simpson have been trying for years to have a child but have been unable to conceive. They consulted a physician, and tests revealed that Mr. Simpson had a markedly reduced sperm count. His chromosomes were studied, and a testicular biopsy was done as well. His chromosomes proved to be normal, except for pair 12. The following figure shows Mrs. Simpson's normal pair of number 12 chromosomes and Mr. Simpson's number 12 chromosomes.

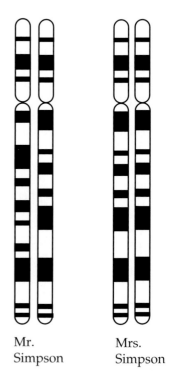

Mr.   Mrs.
Simpson   Simpson

a. What is the nature of the abnormality in Mr. Simpson's chromosomes of pair number 12?

b. What abnormal feature would you expect to see in the testicular biopsy (cells in various stages of meiosis can be seen)?

c. Why is Mr. Simpson's sperm count low?

d. What can be done about Mr. Simpson's low sperm count?

*7.9   Chromosome I in maize has the gene sequence *ABCDEF*, whereas chromosome II has the sequence *MNOPQR*. A reciprocal translocation resulted in *ABCPQR* and *MNODEF*. Diagram the expected pachytene (see Chapter 1, p. 17) configuration of the F$_1$ of a cross of homozygotes of these two arrangements (see p. 17).

7.10   Diagram the pairing behavior at prophase of meiosis I of a translocation heterozygote that has normal chromosomes of gene order *abcdefg* and *tuvwxyz* and has the translocated chromosomes *abcdvwxyz* and *tuefg*. Assume that the centromere is at the left end of all chromosomes.

*7.11 Mr. and Mrs. Denton have been trying for several years to have a child. They have experienced a series of miscarriages, and last year they had a child with multiple congenital defects. The child died within days of birth. The birth of this child prompted the Dentons' physician to order a chromosome study of parents and child. The results of the study are shown in the figure below. Chromosome banding was done, and all chromosomes were normal in these individuals except some copies of number 6 and number 12. The number 6 and number 12 chromosomes of mother, father, and child are shown in the figure (the number 6 chromosomes are the larger pair).

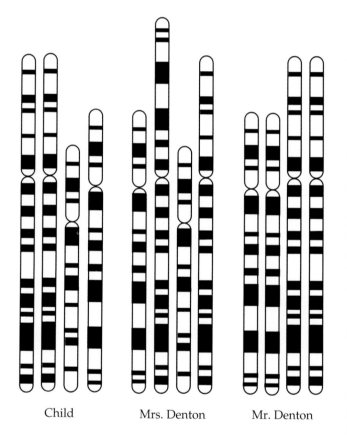

Child   Mrs. Denton   Mr. Denton

a. Does either parent have an abnormal karyotype? If so, which parent has it, and what is the nature of the abnormality?

b. How did the child's karyotype arise (what pairing and segregation events took place in the parents)?

c. Why is the child phenotypically defective?

d. What can this couple expect to occur in subsequent conceptions?

e. What medical help, if any, can be offered to them?

7.12 Irradiation of *Drosophila* sperm produces translocations between the X chromosome and autosomes, between the Y chromosome and autosomes, and between different autosomes. Translocations between the X and Y chromosomes are not produced. Explain the absence of X-Y translocations.

7.13 Define the terms *aneuploidy, monoploidy,* and *polyploidy.*

7.14 If a normal diploid cell is 2N, what is the chromosome content of the following?

a. a nullisomic   **e.** a double trisomic
b. a monosomic   **f.** a tetraploid
c. a double monosomic   **g.** a hexaploid
d. a tetrasomic

*7.15 In humans, how many chromosomes would be typical of nuclei of cells that are

a. monosomic   **d.** triploid
b. trisomic   **e.** tetrasomic
c. monoploid

*7.16 An individual with 47 chromosomes, including an additional chromosome 15, is said to be

a. triplet   **c.** triploid
b. trisomic   **d.** tricycle

*7.17 A color-blind man marries a homozygous normal woman, and after four joyful years of marriage they have two children. Unfortunately, both children have Turner syndrome, although one has normal vision and one is color-blind. The type of color blindness involved is a sex-linked recessive trait.

a. For the color-blind child with Turner syndrome, did nondisjunction occur in the mother or the father? Explain your answer.

b. For the Turner child with normal vision, in which parent did nondisjunction occur? Explain your answer.

7.18 Assume that *x* is a new mutant gene in corn. A female *x/x* plant is crossed with a triplo-10 individual (trisomic for chromosome 10) carrying only dominant alleles at the *x* locus. Trisomic progeny are recovered and crossed back to the *x/x* female plant.

a. What ratio of dominant to recessive phenotypes is expected if the *x* locus is *not* on chromosome 10?

b. What ratio of dominant to recessive phenotypes is expected if the *x* locus *is* on chromosome 10?

7.19 Why are polyploids with even multiples of the chromosome set generally more fertile than polyploids with odd multiples of the chromosome set?

7.20 Select the correct answer from the key below for the following statement: One plant species (N = 11) and another (N = 19) produced an allotetraploid.

I. The chromosome number of this allotetraploid is 30.

II. The number of nuclear linkage groups of this allotetraploid is 30.

KEY:

A. Statement I is true and Statement II is true.
B. Statement I is true, but Statement II is false.
C. Statement I is false, but Statement II is true.
D. Statement I is false and Statement II is false.

*7.21 How many chromosomes would be found in somatic cells of an allotetraploid derived from two plants, one with N = 7 and the other with N = 10?

7.22 Plant species A has a haploid complement of four chromosomes. A related species B has five. In a geographical region where A and B are both present, C plants are found that have some characters of both species and somatic cells with 18 chromosomes. What is the chromosome constitution of the C plants likely to be? With what plants would they have to be crossed in order to produce fertile seed?

CHAPTER *8*

GENE CONTROL OF PROTEINS

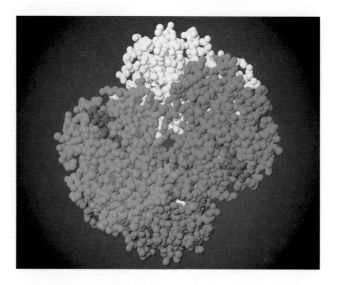

PRINCIPAL POINTS

~ There is a specific relationship between genes and enzymes, historically embodied in the "one gene–one enzyme" hypothesis stating that each gene controls the synthesis or activity of a single enzyme. Since we now know that enzymes may consist of more than one polypeptide and that genes code for individual polypeptide chains, a more modern name for this hypothesis is "one gene–one polypeptide."

~ Many human genetic diseases are caused by deficiencies in enzyme activities. Although some of these diseases are inherited as dominant traits, most are inherited as recessive traits.

~ From the study of alterations in proteins other than enzymes, convincing evidence was obtained that genes control the structures of all proteins.

~ Genetic counseling is the analysis of the risk that prospective parents may produce a child with a genetic defect, and the presentation to appropriate family members of the available options for avoiding or minimizing those possible risks. Early detection of a genetic disease is done by carrier detection and by fetal analysis.

---

*I*n this chapter we examine gene function. We present some of the classical evidence that genes code for enzymes and for other proteins. In particular, we examine the involvement of certain sets of genes in directing and controlling a particular biochemical pathway, the series of enzyme-catalyzed steps required to break down or synthesize a particular chemical compound. Instead of thinking about the gene in isolation, we will see that the gene must often work in cooperation with other genes in order for cells to function properly.

GENE CONTROL OF ENZYME STRUCTURE

Garrod's Hypothesis of Inborn Errors of Metabolism

In 1902, Archibald Garrod, an English physician, provided the first evidence of a specific relationship between genes and enzymes. Garrod studied *alkaptonuria*, a human disease characterized by urine that turns black upon exposure to the air and by a tendency to develop arthritis later in life. Because of the urine phenotype, affected individuals are easily detected soon after birth.

Garrod and geneticist William Bateson concluded that alkaptonuria is a genetically controlled trait because (1) several members of the same families frequently had alkaptonuria, and (2) the disease was much more common among children of marriages

involving first cousins than among children of marriages between unrelated partners. This finding was significant because first cousins have many alleles in common, and therefore the chances are greater for recessive alleles to be homozygous in children of first-cousin marriages.

Garrod found that people with alkaptonuria excrete homogentisic acid (HA) in their urine, whereas normal people do not. Moreover, he showed that HA in urine turns black in air. Garrod reasoned that alkaptonuria is a genetic disease caused by the absence of a particular enzyme necessary for the metabolism of HA. Figure 8.1 shows part of the phenylalanine-tyrosine metabolic pathway including the HA-to-maleylacetoacetic acid step, which is blocked in people with alkaptonuria. In Garrod's terms, this disease is an example of an *inborn error of metabolism*. The mutation responsible for alkaptonuria is recessive, so only people homozygous for the mutant gene express the defect. Later analysis has pinpointed the location of this gene on chromosome 3.

The One Gene–One Enzyme Hypothesis

George Beadle and Edward Tatum in 1942 heralded the beginnings of biochemical genetics, a branch of genetics that combines genetics and biochemistry to explain the nature of metabolic pathways. Results of their studies involving the haploid fungus *Neurospora crassa* (orange bread mold) showed a direct relationship between genes and enzymes and led to the *one gene–one enzyme hypothesis*, a landmark in the history

~ **FIGURE 8.1**

Phenylalanine-tyrosine metabolic pathways. Individuals with alkaptonuria cannot metabolize homogentisic acid (HA) to maleylacetoacetic acid, and HA accumulates. Individuals with PKU cannot metabolize phenylalanine to tyrosine, and phenylpyruvic acid accumulates. Individuals with albinism cannot synthesize much melanin from tyrosine.

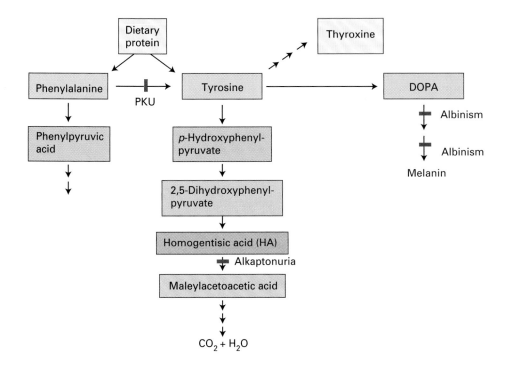

of genetics. Beadle and Tatum received the Nobel Prize for this work in 1958.

ISOLATION OF NUTRITIONAL MUTANTS OF *NEUROSPORA.*

Wild-type *Neurospora* can grow on a simple minimal medium containing only inorganic salts (including a source of nitrogen), an organic carbon source (such as glucose or sucrose), and the vitamin biotin. Beadle and Tatum reasoned that *Neurospora* synthesized the other materials it needed for growth (amino acids, nucleotides, vitamins, nucleic acids, proteins, etc.) from the various chemicals present in the minimal medium. They saw that nutritional mutants (auxotrophs) of *Neurospora* were not able to grow on minimal medium, but would grow if that medium was supplemented with the appropriate nutrient.

Figure 8.2 shows how Beadle and Tatum isolated and characterized nutritional mutants. They treated asexual spores (conidia) with X rays to increase the frequency of genetic mutants and then crossed them with a wild-type (prototrophic) strain to be sure that any nutritional mutant they isolated would have a genetic basis.

The progeny spores were inoculated into a complete medium that contained all necessary amino acids, purines, pyrimidines, and vitamins, in addition to the sucrose, salts, and biotin found in minimal medium. Thus, any strain that could not make one or more necessary compounds from the basic ingredients in minimal medium could still grow by using the compounds supplied in the growth medium. Each culture grown on the complete medium was then tested for growth on minimal medium. Those strains that did not grow were the auxotrophic mutants. These mutants were, in turn, individually tested for their ability to grow on minimal medium plus amino acids, and minimal medium plus vitamins. Theoretically, an amino acid auxotroph—a mutant strain that has lost the ability to synthesize a particular amino acid—would grow on minimal medium plus amino acids but not on minimal medium plus vitamins or on minimal medium alone. Similarly, vitamin auxotrophs would only grow on minimal medium plus vitamins.

Beadle and Tatum next did a second round of screening to determine which specific substance each auxotrophic strain needed for growth. Suppose an amino acid auxotroph was identified. To determine which of the 20 amino acids was required, the strain was inoculated into 20 tubes, each containing minimal medium plus one of the 20 amino acids. In the

~ FIGURE 8.2

Method devised by Beadle and Tatum to isolate auxotrophic mutations in *Neurospora*. Here, the mutant strain isolated is a tryptophan auxotroph.

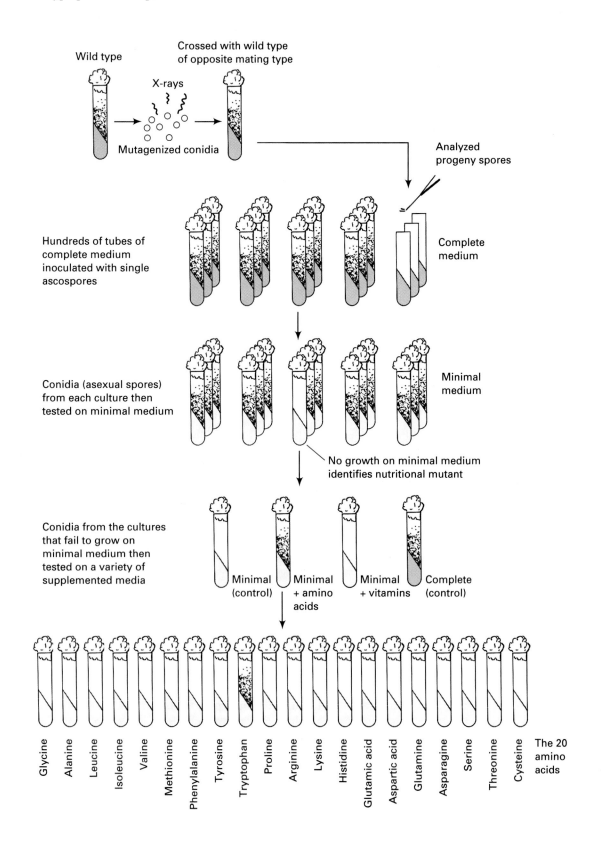

example shown in Figure 8.2, a tryptophan auxotroph was identified because it grew only in the tube containing minimal medium plus tryptophan. Crosses between each mutant and wild type segregated in the 1:1 ratio expected for a single gene defect.

GENETIC DISSECTION OF A BIOCHEMICAL PATHWAY.

Once Beadle and Tatum had isolated and identified nutritional mutants, they investigated the biochemical pathways affected by the mutations. They assumed that wild-type *Neurospora* converted the constituents of minimal medium into amino acids and other required compounds by a series of reactions that were organized into pathways. In this way the synthesis of cellular components occurred by a series of small steps, each catalyzed by an enzyme. An example of Beadle and Tatum's analysis that led to an understanding of the relationship between genes and enzymes is the genetic dissection of the pathway for the biosynthesis of the amino acid arginine in *Neurospora crassa*.

Starting with a set of arginine auxotrophs, genetic analysis showed Beadle and Tatum that four genes are involved; a mutation in any one of them gives rise to auxotrophy for arginine, that is, none of the auxotrophs will grow on minimal medium, and all will grow on minimal medium plus arginine. These four genes in a wild-type cell are designated $argE^+$, $argF^+$, $argG^+$, and $argH^+$. Next, Beadle and Tatum determined the growth pattern of the four mutant strains on media supplemented with chemicals thought to be intermediates involved in the arginine biosynthetic pathway—ornithine, citrulline, and argininosuccinate—with the results shown in Table 8.1.

The sequence of steps in a pathway can be deduced from the pattern of growth supplementation. The principles are as follows: The farther along in a pathway a mutant strain is blocked, the fewer intermediate compounds permit the strain to grow. If a mutant strain is blocked at early steps, a larger

number of intermediates enable the strain to grow. Thus, in these analyses, not only is the pathway deduced, but the steps controlled by each gene are determined. In addition, a genetic block in a pathway may lead to an accumulation of the intermediate compound used in the step that is blocked.

Considering the results in Table 8.1 with this in mind, the *argH* mutant strain grows when supplemented with arginine but not when supplemented with any of the intermediates. This means that the *argH* gene must control the last step in the pathway, which leads to the formation of arginine. The *argG* mutant strain grows on media supplemented with arginine or argininosuccinate, so argininosuccinate must be immediately before arginine in the pathway, and the *argG* gene must control the synthesis of argininosuccinate from another chemical. The *argF* mutant strain grows on media supplemented with arginine, argininosuccinate, or citrulline, so citrulline must precede argininosuccinate in the pathway and the *argF* gene must control the synthesis of citrulline from another compound. The *argE* strain grows on media supplemented with either arginine, argininosuccinate, citrulline, or ornithine, so ornithine must precede citrulline in the pathway and the *argE* gene must control the synthesis of ornithine from another compound. The whole pathway involved here is shown in Figure 8.3. Gene $argE^+$ encodes the enzyme for the conversion of acetylornithine to ornithine, so mutants for this gene can grow on minimal medium plus either ornithine, citrulline, argininosuccinate, or arginine. Gene $argF^+$ codes for the enzyme that converts ornithine to citrulline, so an *argF* mutant strain can grow on minimal medium plus either citrulline, argininosuccinate, or arginine, and so on.

From the results of experiments such as these, Beadle and Tatum proposed that a specific gene encodes each enzyme. This proposed relationship between an organism's genes and the enzymes that

~ TABLE 8.1					
Growth Responses of Arginine Auxotrophs					
	GROWTH RESPONSE ON MINIMAL MEDIUM AND . . .				
MUTANT STRAINS	NOTHING	ORNITHINE	CITRULLINE	ARGININO-SUCCINATE	ARGININE
Wild-type	+	+	+	+	+
argE	–	+	+	+	+
argF	–	–	+	+	+
argG	–	–	–	+	+
argH	–	–	–	–	+

~ FIGURE 8.3

Arginine biosynthetic pathway, showing the four genes in *Neurospora crassa* that code for the enzymes that catalyze each reaction. (The genes are not on the same chromosome.)

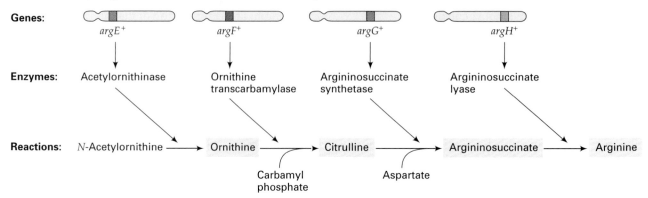

catalyze the steps in a biochemical pathway is called the **one gene–one enzyme hypothesis**. Gene mutations that result in the loss of enzyme activity lead to the accumulation of precursors in the pathway (and to possible side reactions), as well as to the absence of the end product of the pathway. With the approach described, then, a biochemical pathway can be dissected genetically—that is, we can determine the sequence of steps in the pathway and relate each step to a specific gene or genes.

However, more than one gene may control each step in a pathway. An enzyme[1] may have two or more different polypeptide chains, each of which is coded for by a specific gene. In this case, more than one gene specifies that enzyme and thus that step in the pathway. Therefore, the one gene–one enzyme hypothesis has been changed to the **one gene–one polypeptide hypothesis**. We should also be aware that some biochemical pathways are branched and, hence, their analysis is more complicated.

KEYNOTE

A number of classical studies indicated the specific relationship between genes and enzymes, eventually embodied in the one gene–one enzyme hypothesis, which states that each gene controls the synthesis or activity of a single enzyme. Since we now know that protein enzymes may consist of more than one polypeptide and that genes code for individual polypeptide chains, a more correct description for this hypothesis is the one gene–one polypeptide hypothesis.

[1] We will learn later in the book that some enzymes are not proteins, but RNA.

GENETICALLY BASED ENZYME DEFICIENCIES IN HUMANS

Many human genetic diseases are caused by a single gene mutation that alters the function of an enzyme (Table 8.2). In general, an enzyme deficiency caused by a mutation may have either simple or **pleiotropic** (wide-reaching) consequences. Studies of these diseases have offered further evidence that many genes code for enzymes. Some genetic diseases are discussed in the following sections.

Phenylketonuria

An individual with *phenylketonuria* (PKU) is called a *phenylketonuric*. PKU occurs in about 1 in 12,000 Caucasian births; it is most commonly caused by a recessive mutation of a gene on chromosome 1 (an autosome), and individuals must therefore be homozygous for the mutation in order to exhibit the condition. The mutation is in the gene for phenylalanine hydroxylase. The absence of that enzyme activity prevents the conversion of the amino acid phenylalanine to the amino acid tyrosine (see Figure 8.1). Phenylalanine is one of the *essential amino acids*; that is, it is an amino acid that must be included in the diet since humans are unable to synthesize it. Phenylalanine is required to make our own proteins, but excess amounts are harmful and are converted to tyrosine for further metabolism. Children born with PKU accumulate the phenylalanine they ingest, and that phenylalanine is converted to phenylpyruvic acid, which drastically affects the cells of the central nervous system and produces serious symptoms: severe mental retardation, a slow growth rate, and early death. (PKU children are unaffected prior to birth or at birth because

~ TABLE 8.2

Selected Human Genetic Disorders with Demonstrated Enzyme Deficiencies

GENETIC DEFECT	ENZYME DEFICIENCY
Acatalasia I and II[a]	Catalase
Acid phosphatase deficiency	Acid phosphatase
Alkaptonuria	Homogentisic acid oxidase
Ataxia, intermittent[a]	Pyruvate decarboxylase
Cystic fibrosis	Cystic fibrosis transmembrane conductance regulator (CFTR)
Cataract	Galactokinase
Citrullinemia[a]	Argininosuccinate synthetase
Disaccharide intolerance	Invertase
Fructose intolerance	Fructose-1-phosphate aldolase
Fructosuria	Liver fructokinase
Galactosemia[a]	Galactose-1-phosphate uridyl transferase
Gaucher disease[a]	Glucocerebroside oxidase
G6PD deficiency (favism)[a]	Glucose-6-phosphate dehydrogenase
Glycogen storage disease I	Glucose-6-phosphatase
Glycogen storage disease II[a]	α-1,4-Glucosidase
Glycogen storage disease III[a]	6-Phosphofructokinase
Glycogen storage disease IV[a]	Amylo-1:4, 1:6-transglucosidase
Gout, primary	Hypoxanthine-guanine phosphoribosyltransferase
Hemolytic anemia[a]	Glutathione peroxidase or glutathione reductase or glutathione synthetase or hexokinase or pyruvate kinase
Hypervalinemia[a]	Valine transaminase
Hypoglycemia and acidosis	Fructose-1,6-diphosphatase
Immunodeficiency	Uridine monophosphate kinase
Intestinal lactase deficiency (adult)	Lactase
Ketoacidosis[a]	Succinyl CoA:3-ketoacid CoA-transferase
Kidney tubular acidosis with deafness	Carbonic anhydrase B
Leigh's necrotizing encephalomelopathy[a]	Pyruvate carboxylase
Lesch-Nyhan syndrome[a]	Hypoxanthine phosphoribosyltransferase
Lysine intolerance	Lysine: NAD-oxidoreductase
Male pseudohermaphroditism	Testicular 17,20-desmolase
Maple sugar urine disease[a]	Keto acid decarboxylase
Muscular dystrophy	Dystrophin absent or defective; serum acetylcholinesterase or acetylcholine transferase or creatine phosphokinase elevated
Niemann-Pick disease[a]	Sphingomyelin hydrolase
Orotic aciduria[a]	Orotidylic decarboxylase
Phenylketonuria[a]	Phenylalanine hydroxylase
Porphyria, acute[a]	Uroporphyrinogen III synthetase
Porphyria, congenital erythropoietic[a]	Uroporphyrinogen III cosynthetase
Pulmonary emphysema	α-1-Antitrypsin
Pyridoxine-dependent infantile convulsions	Glutamic acid decarboxylase
Pyridoxine-responsive anemia	λ-Aminolevulinic synthetase
Ricketts, vitamin D-dependent	25-Hydroxycholecalciferol 1-hydroxylase
Tay-Sachs disease[a]	Hexosaminidase A
Thyroid hormone synthesis, defect in	Iodide peroxidase or deiodinase
Tyrosinemia	p-Hydroxyphenylpyruvate oxidase
Xeroderma pigmentosum[a]	DNA-specific endonuclease (repair enzyme)

[a]Prenatal diagnosis possible.

excess phenylalanine that accumulates is metabolized by maternal enzymes.)

PKU has pleiotropic effects. A phenylketonuric cannot make tyrosine, an amino acid required for protein synthesis, production of the hormones thyroxine and adrenaline, and production of the skin pigment melanin. This aspect of the phenotype is not very serious because tyrosine can be obtained from food. Yet food does not normally contain a lot of tyrosine. As a result, people with PKU make relatively little melanin and, hence, tend to have very fair skin and blue eyes (even if they have brown-eye genes). In addition, PKU individuals have relatively low adrenaline levels.

The symptoms of PKU depend on the amount of phenylalanine that accumulates, so the disease can be managed by controlling the dietary intake of phenylalanine. A mixture of individual amino acids with a very controlled amount of phenylalanine is used as a protein substitute in the PKU diet. However, since the diet contains some naturally occurring protein, it does contain some phenylalanine. The diet is also expensive, costing more than $5,000 per year, and needs to be continued for life. Additionally, female phenylketonurics are usually advised to maintain the restricted diet strictly until after completion of reproduction because of the potential harm high maternal phenylalanine levels may cause the developing fetus. Given the very serious consequences of allowing PKU to go untreated, all U.S. states require that newborns be screened for PKU. The screen—the Guthrie test—is conducted by placing a drop of blood on a filter-paper disc and placing the disc on solid culture medium containing the bacterium *Bacillus subtilis* and the chemical β-2-thienylalanine. The β-2-thienylalanine inhibits the growth of the bacterium. If phenylalanine is present, the inhibition is prevented; hence, continued growth of the bacterium is evidence for the presence of high levels of phenylalanine in the blood and indicates the need for further tests to show that the individual has PKU.

Some foods and drinks containing the artificial sweetener NutraSweet carry a warning that people with PKU should not use them. NutraSweet is aspartame, a chemical consisting of the amino acid aspartic acid attached to the amino acid phenylalanine. This combination signals to your taste receptors that the substance is sweet, yet it is not sugar and does not have the calories of sugar. Once ingested, aspartame is broken down to aspartic acid and phenylalanine, so given enough aspartame there can be very serious effects for a phenylketonuric.

Albinism

Albinism (see Figure 2.17) is caused by an autosomal recessive mutation. About 1 in 33,000 Caucasians and

1 in 28,000 African-Americans in the United States have albinism. The mutation affects a gene for an enzyme used in the pathway from tyrosine to the brown pigment melanin (see Figure 8.1). Melanin absorbs light in the UV range and serves to protect the skin against harmful UV radiation from the sun. Persons with albinism produce no melanin, so they have white skin, white hair, and red eyes (because of a lack of pigment in the iris) and are very light-sensitive. No apparent problems result from the accumulation of precursors in the pathway prior to the block.

There are at least two kinds of albinism, since at least two biochemical steps can be blocked to prevent melanin formation. Thus, two parents with albinism that has resulted from different enzyme deficiencies for two different steps of the pathway can produce normal children as a result of complementation of the two nonallelic mutations (see Chapter 6).

Lesch-Nyhan Syndrome

Lesch-Nyhan syndrome is an ultimately fatal human trait caused by a recessive mutation in an X chromosome gene located at Xq26–27. The gene spans 44 kb (44,000 base pairs) and encodes a polypeptide of 218 amino acids. An estimated 1 in 10,000 males exhibit Lesch-Nyhan syndrome. Because this disease is often lethal before reproductive age, virtually no Lesch-Nyhan females (who would be homozygous mutants) are seen.[2] Carrier females (heterozygotes) may, however, have symptoms if the normal Lesch-Nyhan allele has undergone random X inactivation (lyonization; see Chapter 3).

Lesch-Nyhan syndrome is the result of a deficiency in the enzyme hypoxanthine-guanine phosphoribosyl transferase (HGPRT), an enzyme essential to the utilization of chemicals called purines. When the biochemical pathway is highly impaired, as in this case, excess purines accumulate and are converted to uric acid. At birth, Lesch-Nyhan individuals are healthy and develop normally for several months. The uric acid excreted in the urine leads to the deposition of orange uric acid crystals in the diapers, an indication of disease. From three to eight months, delays in motor development occur that lead to weak muscles. Later, the muscle tone changes radically, producing uncontrollable movements and involuntary spasms, seriously affecting feeding activities.

After two or three years, as a result of severe neurological complications, Lesch-Nyhan children begin to show extremely bizarre activity, such as compul-

---

[2] The statement refers to normal XX females. XO females could have Lesch-Nyhan syndrome because only one copy of the gene is present.

sive biting of fingers, lips, and the inside of the mouth. This self-mutilation is difficult to control and painful. Typically, behavior toward others becomes aggressive. There is no clear explanation for how an HGPRT deficiency gives rise to self-mutilating activity, and there are no available therapies for these neurological complications. In intelligence tests, Lesch-Nyhan patients score in the severely retarded region, although the difficulties they show in communicating with others may be a major contributing factor to the low scores. Most die before they reach their 20s, usually from infection, kidney failure, or uremia (uric acid in the blood).

Tay-Sachs Disease

Lysosomes are membrane-bound organelles in the cell containing 40 or more different digestive enzymes that catalyze the breakdown of nucleic acids, proteins, polysaccharides, and lipids. A number of human diseases are caused by mutations in genes that code for lysosomal enzymes. Such diseases, collectively called *lysosomal-storage diseases*, are generally caused by recessive mutations.

The best-known genetic disease of this type is *Tay-Sachs disease*, which is caused by homozygosity for a rare recessive mutation of a gene on chromosome 15. While Tay-Sachs disease is rare in the population as a whole, it has a relatively high incidence in Ashkenazi Jews of Central European origin, with about 1 in 3,600 children having the disease.

The gene in question (*hexA*) codes for the enzyme N-acetylhexosaminidase A, which converts a brain chemical (a particular ganglioside) to a different form. (A ganglioside is one of a group of complex glycolipids found mainly in nerve membranes.) In Tay-Sachs infants the enzyme is nonfunctional, so the unprocessed ganglioside accumulates in the brain cells and causes a number of different clinical symptoms. Typically, the symptom first recognized is an unusually enhanced reaction to sharp sounds. A cherry-colored spot on the retina surrounded by a white halo also facilitates early diagnosis of the disease. About a year after birth, there is rapid neurological degeneration as the unprocessed ganglioside accumulates and the brain begins to lose control over normal function and activities. This degeneration involves generalized paralysis, blindness, a progressive loss of hearing, and serious feeding problems. By two years of age the infants are essentially immobile, and death occurs at about three to four years of age, often from respiratory infections. There is no known cure for Tay-Sachs disease, but since carriers can be detected, the incidence of affected individuals can be controlled.

GENE CONTROL OF PROTEIN STRUCTURE

While most enzymes are proteins, not all proteins are enzymes. To understand completely how genes function, we next look at the experimental evidence that genes are responsible for the structure of nonenzymatic proteins such as hemoglobin.

Sickle-Cell Anemia

Sickle-cell anemia (SCA) is a genetic disease affecting hemoglobin, the oxygen-transporting protein in red blood cells. Red blood cells from individuals with the disease lose their characteristic disc shape in conditions of low oxygen tension and assume the shape of a sickle (Figure 8.4). The sickled red blood cells are more fragile than normal red blood cells and break easily, hence the "anemia." Sickled cells are also not as flexible as normal cells, and therefore tend to clog the capillaries rather than squeeze through them. As a result, blood circulation is impaired and tissues become deprived of oxygen. Although oxygen deprivation occurs particularly at the extremities, the heart, lungs, brain, kidneys, gastrointestinal tract, muscles, and joints can also become oxygen deprived and be damaged. An individual with SCA may therefore suffer

~ **FIGURE 8.4**

Colorized scanning electron micrographs of (left) normal and (right) sickled red blood cells.

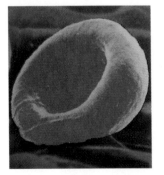

from a pleiotropy of health problems, including heart failure, pneumonia, paralysis, kidney failure, abdominal pain, and rheumatism. Some individuals have a milder form of the disease called *sickle-cell trait*.

In 1949, E. A. Beet and J. V. Neel independently hypothesized that sickling was caused by a single mutant allele that was homozygous in sickle-cell anemia and heterozygous in sickle-cell trait. In the same year, Linus Pauling and coworkers showed that the hemoglobins of normal, sickle-cell anemia, and sickle-cell trait individuals differ when they are subjected to electrophoresis, a technique used to separate molecules based on their electrical charges. They found that hemoglobin from normal people (called Hb-A) migrated more slowly than the hemoglobin from people with sickle-cell anemia (called Hb-S) (Figure 8.5). Hemoglobin from sickle-cell trait individuals had a 1:1 mixture of Hb-A and Hb-S, indicating that heterozygous individuals make both types of hemoglobin.

Hemoglobin, the molecule affected in sickle-cell anemia, consists of four polypeptide chains—two α polypeptides and two β polypeptides—each of which is associated with a heme group (involved in the binding of oxygen) (Figure 8.6). In 1956, V. M. Ingram analyzed some of the amino acid sequences of Hb-A and Hb-S and found that the molecular defect in the Hb-S hemoglobin is a change from the acidic amino acid glutamic acid (negative electric charge) at the sixth position from the N-terminal end of the β polypeptide to the neutral amino acid valine (no electrical charge) (Figure 8.7). We have learned that this particular substitution of amino acids causes the β polypeptide to fold up in a different way. In turn, this leads to sickling of the red blood cells in individuals with sickle-cell anemia, and mild sickling of the red blood cells in individuals with sickle-cell trait.

Let us outline the genetics and the products of the genes involved. The β-polypeptide sickle-cell mutant allele is β^S, and it is codominant with the wild-type allele β^A. Homozygous $\beta^A\beta^A$ individuals make normal Hb-A with two normal α chains and two normal β chains encoded by the wild-type α-globin gene and the wild-type β-globin gene (β^A). Homozygous $\beta^S\beta^S$ individuals make Hb-S, the defective hemoglobin, with two normal α chains specified by wild-type α-globin genes and two abnormal β chains specified by the mutant β-globin gene β^S. These individuals have SCA. Heterozygous $\beta^A\beta^S$ individuals make both Hb-A and Hb-S and have sickle-cell trait. Under normal conditions they usually show few symptoms of the disease. However, after a sharp drop in oxygen tension (as in an unpressurized aircraft climbing into the atmosphere), sickling of red blood cells may occur, giving rise to symptoms similar to those found in people with severe anemia.

~ FIGURE 8.5

Electrophoresis of hemoglobin variants. Hemoglobin found (left) in normal $\beta^A\beta^A$ individuals, (center) in $\beta^A\beta^S$ individuals with sickle-cell trait, and (right) in $\beta^S\beta^S$ individuals who exhibit sickle-cell anemia. The two hemoglobins migrate to different positions in an electric field and hence must differ in electric charge.

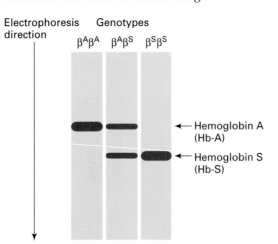

~ FIGURE 8.6

The hemoglobin molecule. The diagram shows the two α polypeptides and two β polypeptides, each of which is associated with a heme group.

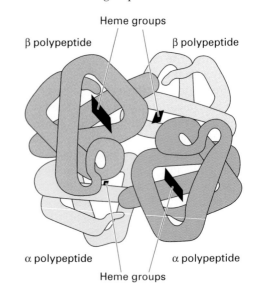

~ FIGURE 8.7

The first seven N-terminal amino acids in normal and sickled hemoglobin β polypeptides. There is a single amino acid change from glutamic acid to valine at the sixth position in the sickled hemoglobin polypeptide.

Normal
β polypeptide, Hb-A

	1	2	3	4	5	6	7
H_3N^+ —	Val —	His —	Leu —	Thr —	Pro —	Glu —	Glu ···

Changes to ↓

Sickle-cell
β polypeptide, Hb-S H_3N^+ — Val — His — Leu — Thr — Pro — Val — Glu ···

Other Hemoglobin Mutants

Over 200 hemoglobin mutants have been detected in general screening programs in which hemoglobin is isolated from red blood cells and analyzed by electrophoresis. Some mutations affect the α chain and others the β chain, and there is a wide variety in the types of amino acid substitutions that occur.

The identified hemoglobin mutants have various effects, depending on the amino acid substitution involved and its position in the polypeptide chains. Most have effects that are not as drastic as the SCA mutant. For example, in the Hb-C hemoglobin molecule, the same β-polypeptide glutamic acid that is altered in SCA is changed to a lysine. Unlike the Hb-S change, this change is not as serious a defect because both amino acids are hydrophilic ("water loving"), so the conformation of the hemoglobin molecule is not as drastically altered. Therefore, people homozygous for the $β^C$ mutation experience only a mild form of anemia.

Biochemical Genetics of the Human ABO Blood Groups

The mechanism of inheritance of the human ABO blood groups was mentioned in Chapter 4. To summarize, the blood group depends on the presence of chemical substances called antigens on the red blood cell surface. When antigens are injected into a host organism, they may be recognized as foreign and removed from the circulation by the host's antibodies.

The I^A, I^B, and i alleles of the ABO locus specify the ABO blood groups. People of blood group A (genotypes I^A/I^A or I^A/i) have the A antigen on their red blood cells, people of blood group B (genotypes I^B/I^B or I^B/i) have the B antigen on their red blood cells, people of blood group AB (genotype I^A/I^B) have both the A and B antigens on their red blood cells, and people of blood group O (genotype i/i) have neither the A nor the B antigens on their red blood cells.

The ABO locus encodes *glycosyltransferases*, which are enzymes that add sugar groups to a preexisting polysaccharide. The polysaccharides here are those that have combined with lipids to form glycolipids. These glycolipids are the blood group antigens found on the surface of red blood cells. Most individuals produce a glycolipid called the H antigen. The I^A allele produces a glycosyltransferase enzyme that adds the sugar α-N-acetylgalactosamine to the H antigen to produce the A antigen. The I^B allele produces a different glycosyltransferase, which adds galactose to the H antigen to produce the B antigen. In both cases, some H antigen remains unconverted. Figure 8.8 shows the genetic control of the glycolipids involved in the ABO blood group system.

For the I^A/I^B heterozygote, both enzymes are produced, so some H antigen is converted to the A antigen and some to the B antigen. The red blood cell has both antigens on the surface, so the individual is of blood group AB. It is the presence of both surface antigens that provides the molecular basis of the codominance of the I^A and I^B alleles (see Chapter 4).

~ FIGURE 8.8

Human ABO blood type antigens. Conversion of the H antigen to the A antigen by the I^A allele product and to the B antigen by the I^B product.

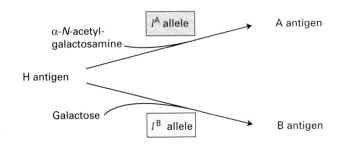

People who are homozygous for the *i* allele produce no enzymes to convert the H antigen glycolipid. As a consequence, their red blood cells only carry the H antigen. The H antigen does not elicit an antibody response in people of other blood groups because its polysaccharide component is the basic component of the A and the B antigens as well, so it is not detected as a foreign substance. People who are heterozygous for the *i* allele have the blood type of the other allele. For example, in I^B/i individuals the I^B allele results in the conversion of some of the H antigen to the B antigen, determining the blood group of the individual.

The H antigen is produced by the action of the dominant *H* allele at a locus distinct from the ABO locus. Individuals homozygous for the recessive mutant allele, *h*, do not make the H antigen, and therefore, regardless of the presence of I^A or I^B alleles at the ABO locus, no A or B antigens can be produced. These very rare *h/h* individuals are like blood group O individuals in the sense that they lack A and B antigens; they are said to have the Bombay blood type. However, individuals with the Bombay blood group produce anti-O antibodies (antibodies against the H antigen), while individuals with blood group O do not. As a consequence of the Bombay blood group phenomenon, it is theoretically possible for two blood group O individuals to produce a child who has blood group A or B. For example, if one parent is *h/h I^B/–* (*h/h I^B/I^B* or *h/h I^B/i*) and the other is *H/H i/i*, the child could be *H/h I^B/i*, and, therefore, be of blood group B.

Cystic Fibrosis

Cystic fibrosis (CF) is a human disease that causes pancreatic, pulmonary, and digestive dysfunction in children and young adults. Typical of the disease is an abnormally high viscosity of secreted mucus. In some male patients, the vas deferens does not form properly, resulting in sterility. CF is managed by pounding the chest and back of a patient to help shake mucus free in different parts of the lungs, and by giving antibiotics to treat any infections that develop. CF is a lethal disease; life expectancy is about 40 years.

CF is caused by homozygosity for an autosomal recessive mutation located on the long arm of chromosome 7. CF is the most common lethal autosomal recessive disease among Caucasians, with about 1 in 2,000 newborns having the disease. About 1 in 23 Caucasians is estimated to be a heterozygous carrier. In the African-American population, about 1 in 17,000 newborns have CF, and in Asians the CF frequency is 1 in 90,000 newborns.

The defective gene product in CF patients was not identified by biochemical analysis, as was the case for PKU and many other diseases, but by a combination of genetic and modern molecular biology techniques. Researchers studied pedigrees of families in which the disease appeared and were able to correlate the segregation of particular DNA fragments (generated by cutting nuclear DNA with a restriction enzyme) with the inheritance of the CF mutation. (This method is called restriction fragment length polymorphism [RFLP] mapping, and is described in more detail in Chapter 14.) One fragment that showed linkage with the CF mutation enabled investigators to locate the CF gene to chromosome 7. Subsequent molecular experiments homed in on the CF gene, and it was cloned from both a normal individual and from a CF patient. (The cloning of the CF gene is described in more detail in Chapter 14.)

Analysis of the DNA sequences of the cloned genes showed that in patients with a serious form of CF, the most common mutation—ΔF508—is the deletion of three consecutive base pairs in the ATP-binding, nucleotide-binding fold (NBF) region toward the left end of the gene. Since each amino acid in a protein is specified by three base pairs in the DNA, one amino acid is missing in the CF protein of the patients. But what does the CF protein do? From the DNA sequence of the gene, researchers deduced the amino acid sequence of the protein and then made some predictions about the type and three-dimensional structure of that protein. Their analysis indicated that the 1,480-amino acid CF protein is associated with membranes. The proposed structure for the CF protein—called cystic fibrosis transmembrane conductance regulator (CFTR)—is shown in Figure 8.9. By comparing the amino acid sequence of the CF protein with the amino acid sequences of other proteins in a computer database, CFTR protein was found to be homologous to a large family of proteins involved in active transport of materials across cell membranes. This protein has now been shown to be a chloride channel in certain cell membranes. In people with CF, the mutated CF gene results in an abnormal CFTR protein, and as a result ion transport across membranes is impaired. Because of this, the symptoms of CF occur, starting with abnormal mucus secretion and accumulation.

KEYNOTE

From the study of alterations in proteins other than enzymes, such as those in hemoglobin responsible for sickle-cell anemia, convincing evidence was obtained that genes control the structures of all proteins.

~ FIGURE 8.9

Proposed structure for cystic fibrosis transmembrane conductance regulator (CFTR). The protein has two hydrophobic segments that span the plasma membrane, and after each segment is a nucleotide-binding fold (NBF) region that binds ATP. The site of the amino acid deletion resulting from the three-nucleotide-pair deletion in the CF gene most commonly seen in patients with severe cystic fibrosis is in the first (toward the amino end) NBF; this is the ΔF508 mutation. The central portion of the molecule contains sites that can be phosphorylated by the enzymes protein kinase A and protein kinase C.

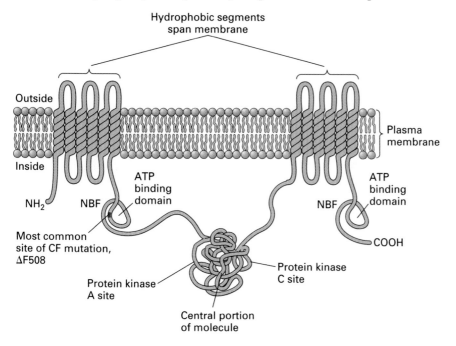

GENETIC COUNSELING

We have learned that many human genetic diseases are caused by enzyme or protein defects; those defects ultimately result from mutations at the DNA level. Many other genetic diseases arise from chromosome defects. We can now test for many enzyme or protein deficiencies and for many of the DNA changes associated with genetic diseases and thereby determine whether or not an individual has a genetic disease or is a carrier for that disease. It is also possible to determine whether individuals have any chromosomal defects. **Genetic counseling** is the analysis of the probability that individuals have a genetic defect, or of the risk that prospective parents may produce a child with a genetic defect. In the latter case, genetic counseling involves presenting the available options for avoiding or minimizing those possible risks. If a serious genetic defect is identified in a fetus, one obvious option is abortion. Thus, genetic counseling provides people with an understanding of the genetic problems that are or may be in their families or prospective families.

Genetic counseling employs a wide range of information on human heredity. In many instances the risk of having a child with a genetic condition may be stated in terms of rather precise probabilities; in others, where the role of heredity is not completely clear, the risk is estimated only generally. It is the responsibility of genetic counselors to supply their clients with clear, unemotional, and nonprescriptive statements based on the family history and on their knowledge of all relevant scientific information and of the probable risks of giving birth to a child with a genetic defect. In sum, genetic counseling is a lot more than the simple presentation of risk facts and figures to patients; it is the prevention of disease, the relief of pain, and the maintenance of health, all of which are goals of the medical profession in general.

Genetic counseling generally starts with pedigree analysis of both families to determine the likelihood that a particular allele is present in either family. (Pedigree analysis is described in Chapters 2 and 3.) If it is, prospective parents need to be informed of the probability that they will produce a child with that allele. Early detection of a genetic condition occurs at one or both of two levels. One is detecting heterozy-gotes (carriers) of recessive mutations, and the other is determining whether or not the developing fetus shows the condition. Assays for enzyme activities or protein amounts are limited to genetic diseases in which the biochemical condition is expressed in the parents and/or the developing fetus. Tests that measure actual changes in the DNA do not depend on expression of the gene in the parents or the fetus.

Although we can identify carriers of many mutant alleles and can determine if fetuses have a genetic condition, in most cases there is no way to change the phenotype. Carrier detection and fetal analysis serve mainly to inform parents of the risks and probabilities of having a child with the mutation.

Carrier Detection

Carrier detection is the detection of individuals who are heterozygous for a recessive gene mutation. The heterozygous carrier of a mutant gene usually is normal in phenotype. If homozygosity for the mutation results in serious deleterious effects, there is great value in determining whether two people who are contemplating having a child are both carriers, because in that situation one-quarter of the children would be born with the trait. Carrier detection can be used in those cases in which a gene product (protein or enzyme) can be assayed. In those cases, the heterozygote (carrier) is expected to have approximately half of the enzyme activity or protein amount as homozygous normal individuals. In Chapter 14 we see how carriers can be detected by DNA tests.

Fetal Analysis

A second important aspect of genetic counseling is finding out whether a fetus is normal. **Amniocentesis** is one way in which this can be done (Figure 8.10). As a fetus develops in the amniotic sac, amniotic fluid surrounds it, serving as a cushion against shock. In amniocentesis a syringe needle is carefully inserted through the mother's uterine wall and into the amniotic sac, and a sample of amniotic fluid is taken. The fluid contains cells that have sloughed off the fetus's skin; these cells can be cultured in the laboratory and then examined for protein or enzyme alterations or deficiencies, DNA changes, and chromosomal abnormalities. Amniocentesis is possible at any stage of pregnancy, but the small quantity of amniotic fluid available and the increased risk to the fetus makes it impractical to perform the procedure before the twelfth week of pregnancy. Because amniocentesis is

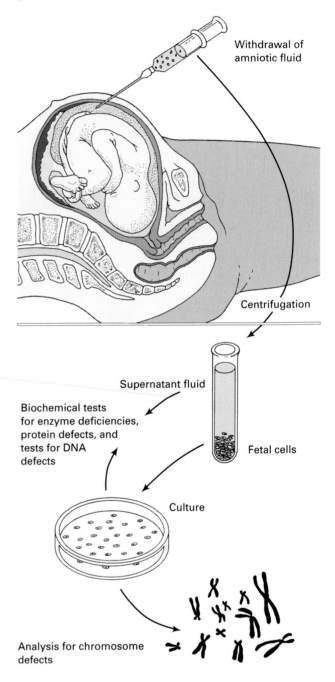

~ FIGURE 8.10

Amniocentesis, a procedure used for prenatal diagnosis of genetic defects.

Withdrawal of amniotic fluid

Centrifugation

Supernatant fluid

Biochemical tests for enzyme deficiencies, protein defects, and tests for DNA defects

Fetal cells

Culture

Analysis for chromosome defects

complicated and costly, it is primarily used in high-risk cases, such as when a mother is over 35.

Another method for fetal analysis is **chorionic villus sampling** (Figure 8.11). The procedure can be done between the eighth and twelfth weeks of pregnancy,

~ FIGURE 8.11

Chorionic villus sampling, a procedure used for early prenatal diagnosis of genetic defects.

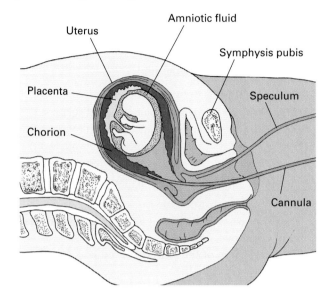

Uterus

Amniotic fluid

Symphysis pubis

Placenta

Speculum

Chorion

Cannula

which is earlier than for amniocentesis. The chorion is a membrane layer surrounding the fetus consisting entirely of embryonic tissue. A chorionic villus tissue sample may be taken from the developing placenta through the abdomen (as in amniocentesis) or, preferably, via the vagina using biopsy forceps or a flexible catheter, aided by ultrasound. Once the tissue sample is obtained, the analysis is similar to that used in amniocentesis. Advantages of the technique are that the time of testing permits the parents to learn if the fetus has a genetic defect earlier in the pregnancy than with amniocentesis, and that it is not necessary to culture cells to obtain enough to do the biochemical assays. Fetal death and inaccurate diagnoses due to the presence of maternal cells are more common in chorionic villus sampling than in amniocentesis, however.

KEYNOTE

Genetic counseling is the analysis of the risk that prospective parents may produce a child with a genetic defect, and the presentation to appropriate family members of the available options for avoiding or minimizing those possible risks. Early detection of a genetic disease is done by carrier detection and by fetal analysis.

SUMMARY

In this chapter we discussed gene function—the evidence that genes code for enzymes and for nonenzymatic proteins. A number of the experiments described in this context were done prior to the unequivocal proof that DNA is the genetic material and before the elucidation of DNA structure.

As early as 1902 there was evidence for a specific relationship between genes and enzymes. This evidence was obtained by Archibald Garrod in his investigations of "inborn errors of metabolism," human genetic diseases that resulted in enzyme deficiencies. The specific relationship between genes and enzymes is historically embodied in the one gene–one enzyme hypothesis, which states that each gene controls the synthesis or activity of a single enzyme. Since some protein enzymes and nonenzymatic proteins (such as hemoglobin) consist of more than one polypeptide, a more modern maxim is one gene–one polypeptide.

In this chapter we also discussed a number of examples of genetically based enzyme and protein deficiencies that give rise to genetic diseases in humans. Many are the result of homozygosity for recessive mutant genes. The severity of the disease depends on the effects of the loss of function of the particular enzyme involved. Thus, albinism is a relatively mild genetic disease, while Tay-Sachs disease and cystic fibrosis are lethal. To make our understanding of gene function more complete, we examined experimental evidence that genes control the structure of nonenzymatic proteins.

Given the knowledge we currently have about a large number of genetically based enzyme and protein deficiencies in humans, it is possible to make some predictions about the existence of certain genetic diseases in individuals or in families. Genetic counseling is the analysis of the probability that individuals have a genetic defect, or of the risk that prospective parents may produce a child with a genetic defect, and the presentation to the individuals involved of any available options for avoiding or minimizing those possible risks. Techniques available to genetic counselors include amniocentesis and chorionic villus sampling.

ANALYTICAL APPROACHES FOR SOLVING GENETICS PROBLEMS

Q8.1 k^+, l^+, and m^+ are independently assorting genes that control the production of a red pigment. These three genes act in a biochemical pathway as follows:

$$\text{colorless 1} \xrightarrow{k^+} \text{colorless 2} \xrightarrow{l^+} \text{orange} \xrightarrow{m^+} \text{red}$$

The mutant alleles that produce abnormal functioning of these genes are k, l, and m; each is recessive to its wild-type counterpart. A red individual homozygous for all three wild-type alleles is crossed with a colorless individual that is homozygous for all three recessive mutant alleles. The F_1 is red. The F_1 is then selfed to produce the F_2 generation.

a. What proportion of the F_2 are colorless?
b. What proportion of the F_2 are orange?
c. What proportion of the F_2 are red?

A8.1a. There are two ways to answer this question. One is to determine all of the genotypes that can specify the colorless phenotypes, and the other is to use subtractive logic, in which the proportions of orange and red are first calculated and then subtracted from 1 to give the proportion of colorless progeny in the F_2. We will first consider the second method.

To produce an orange phenotype, three things are needed: (1) at least one wild-type k^+ allele must be present so that the colorless 1 to colorless 2 step can occur; (2) at least one wild-type l^+ allele must be present so that the colorless 2 to orange step can occur; and (3) the individual must be m/m so that the orange to red step cannot proceed. Thus, an orange phenotype results from the genotype $k^+/- l^+/- m/m$. From an $F_1 \times F_1$ cross, the probability of getting an individual with that genotype is $3/4 \times 3/4 \times 1/4 = 9/64$.

To produce a red phenotype three things are needed: (1) at least one wild-type k^+ allele must be present so that the colorless 1 to colorless 2 step can occur; (2) at least one wild-type l^+ allele must be present so that the colorless 2 to orange step can occur; and (3) at least one m^+ allele must be present so that the orange to red step can proceed. Thus, a red phenotype results from the genotype $k^+/- l^+/- m^+/-$. From an $F_1 \times F_1$ self, the probability of getting an individual with that genotype is $3/4 \times 3/4 \times 3/4 = 27/64$.

By default, all other F_2 individuals are colorless. The proportion of F_2 individuals that are colorless is, therefore, $1 - 9/64 - 27/64 = 1 - 36/64 = 28/64$.

Using the first approach to calculate the proportion of F_2 colorless, we must determine all genotypes that

will give a colorless phenotype and then add up the probabilities of each occurring. The genotypes and their probabilities are as follows:

GENOTYPES	PROBABILITIES	
$k/k\ l/l\ m/m$	1/64	(cannot convert colorless 1)
$k/k\ l/l\ m^+/-$	3/64	(cannot convert colorless 1)
$k/k\ l^+/-\ m/m$	3/64	(cannot convert colorless 1)
$k/k\ l^+/-\ m^+/-$	9/64	(cannot convert colorless 1)
$k^+/-\ l/l\ m/m$	3/64	(cannot convert colorless 2)
$k^+/-\ l/l\ m^+/-$	9/64	(cannot convert colorless 2)
Total	28/64	

b. The proportion of the F_2 that are orange was calculated in (a), above, i.e., 9/64.

c. The proportion of the F_2 that are red was calculated in (a), above, i.e., 27/64.

Q8.2 A number of auxotrophic mutant strains were isolated from wild-type, haploid yeast. These strains responded to the addition of certain nutritional supplements to minimal culture medium either by growth (+) or no growth (0). The following table gives the growth patterns for single gene mutant strains:

MUTANT STRAINS	SUPPLEMENTS ADDED TO MINIMAL CULTURE MEDIUM				
	B	A	R	T	S
1	+	0	+	0	0
2	+	+	+	+	0
3	+	0	+	+	0
4	0	0	+	0	0

Diagram a biochemical pathway that is consistent with the data, indicating where in the pathway each mutant strain is blocked.

A8.2 The data to be analyzed are very similar to those discussed in the text for Beadle and Tatum's analysis of *Neurospora* auxotrophic mutants, from which they proposed the one gene–one enzyme hypothesis. Recall that the later in the pathway a mutant is blocked, the fewer nutritional supplements need to be added to allow growth. In the data given, we must assume that the nutritional supplements are not necessarily listed in the order in which they appear in the pathway.

Analysis of the data indicates that all four strains will grow if given R, and that none will grow if given S. From this we can conclude that R is likely to be the end product of the pathway (all mutants should grow if given the end product) and that S is likely to be the first compound in the pathway (none of the mutants should grow given the first compound in the pathway). Thus, the pathway as deduced so far is:

$$S \longrightarrow [B,A,T] \longrightarrow R$$

where the order of B, A, and T is as yet undetermined.

Now let us consider each of the mutant strains and see how their growth phenotypes can help define the biochemical pathway:

Strain 1 will grow only if given B or R. Therefore, the defective enzyme in strain 1 must act somewhere prior to the formation of B and R, and after the substances A, T, and S. Since we have deduced that R is the end product of the pathway, we can propose that B is the immediate precursor to R, and that strain 1 cannot make B. The pathway so far is:

$$S \longrightarrow [A,T] \xrightarrow{1} B \longrightarrow R$$

Strain 2 will grow on all compounds except S, the first compound in the pathway. Thus, the defective enzyme in strain 2 must act to convert S to the next compound in the pathway, which is either A or T. We do not know yet whether A or T follows S in the pathway, but the growth data at least allow us to conclude where strain 2 is blocked in the pathway, that is:

$$S \xrightarrow{2} [A,T] \xrightarrow{1} B \longrightarrow R$$

Strain 3 will grow on B, R, and T, but not on A or S. We know that R is the end product and S is the first compound in the pathway. This mutant strain allows us to determine the order of A and T in the pathway. That is, since strain 3 grows on T but not on A, T must be later in the pathway than A, and the defective enzyme in 3 must be blocked in the yeast's ability to convert A to T. The pathway now is:

$$S \xrightarrow{2} A \xrightarrow{3} T \xrightarrow{1} B \longrightarrow R$$

Strain 4 will grow only if given the deduced end product R. Therefore, the defective enzyme produced by the mutant gene in strain 4 must act before the formation of R, and after the formation of A, T, and B from the first compound S. The mutation in 4 must be blocked in the last step of the biochemical pathway in the conversion of B to R. The final deduced pathway, and the positions of the mutant blocks, are:

$$S \xrightarrow{2} A \xrightarrow{3} T \xrightarrow{1} B \xrightarrow{4} R$$

QUESTIONS AND PROBLEMS

8.1 Phenylketonuria (PKU) is an inheritable metabolic disease of humans; its symptoms include mental deficiency. This phenotypic effect is due to:
a. accumulation of phenylketones in the blood
b. accumulation of sugar in the blood
c. deficiency of phenylketones in the blood
d. deficiency of phenylketones in the diet

*8.2 If a person were homozygous for both PKU (phenylketonuria) and AKU (alkaptonuria), would you expect him or her to exhibit the symptoms of PKU, or AKU, or both? Refer to the following pathway.

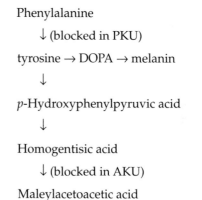

Phenylalanine

↓ (blocked in PKU)

tyrosine → DOPA → melanin

↓

p-Hydroxyphenylpyruvic acid

↓

Homogentisic acid

↓ (blocked in AKU)

Maleylacetoacetic acid

8.3 Refer to the pathway shown in Question 8.2. What effect, if any, would you expect PKU (phenylketonuria) and AKU (alkaptonuria) to have on pigment formation?

*8.4 a^+, b^+, c^+, and d^+ are independently assorting Mendelian genes controlling the production of a black pigment. The alternate alleles that give abnormal functioning of these genes are a, b, c, and d. A black individual of genotype $a^+/a^+ \ b^+/b^+ \ c^+/c^+ \ d^+/d^+$ is crossed with a colorless individual of genotype $a/a \ b/b \ c/c \ d/d$ to produce a black F_1. $F_1 \times F_1$ crosses are then done. Assume that a^+, b^+, c^+, and d^+ act in a pathway as follows:

$$\text{colorless} \xrightarrow{a^+} \text{colorless} \xrightarrow{b^+} \text{colorless} \xrightarrow{c^+} \text{brown} \xrightarrow{d^+} \text{black}$$

a. What proportion of the F_2 are colorless?
b. What proportion of the F_2 are brown?

8.5 Using the genetic information given in Problem 8.4, now assume that a^+, b^+, and c^+ act in a pathway as follows:

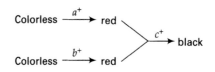

Black can be produced only if both red pigments are present; that is, c^+ converts the two red pigments together into a black pigment.

a. What proportion of the F_2 are colorless?
b. What proportion of the F_2 are red?
c. What proportion of the F_2 are black?

*8.6 Three genes on different chromosomes are responsible for three enzymes that catalyze the same reaction in corn:

$$\text{colorless compound} \xrightarrow{a^+, b^+, c^+} \text{red compound}$$

The normal functioning of any one of these genes is sufficient to convert the colorless compound to the red compound. The abnormal functioning of these genes is designated by a, b, and c, respectively.

a. A red $a^+/a^+ b^+/b^+ c^+/c^+$ is crossed with a colorless $a/a\ b/b\ c/c$ to give a red F_1, $a^+/a\ b^+/b\ c^+/c$. The F_1 is selfed. What proportion of the F_2 are colorless?
b. It turns out that another step is involved in the pathway. It is controlled by gene d^+, which assorts independently of a^+, b^+, and c^+:

$$\underset{\substack{\text{colorless} \\ \text{compound 1}}}{} \xrightarrow{d^+} \underset{\substack{\text{colorless} \\ \text{compound 2}}}{} \xrightarrow{a^+, b^+, c^+} \underset{\substack{\text{red} \\ \text{compound}}}{}$$

The inability to convert colorless 1 to colorless 2 is designated d. A red $a^+/a^+ b^+/b^+ c^+/c^+ d^+/d^+$ is crossed with a colorless $a/a\ b/b\ c/c\ d/d$. The F_1 are all red. The red F_1s are now selfed. What proportion of the F_2 are colorless?

8.7 In *Drosophila*, the recessive allele bw causes a brown eye, and the (unlinked) recessive allele st causes a scarlet eye. Flies homozygous for both recessives have white eyes. The genotypes and corresponding phenotypes, then, are as follows:

$bw^+/-$	$st^+/-$	red eye
bw/bw	$st^+/-$	brown
$bw^+/-$	st/st	scarlet
bw/bw	st/st	white

Outline a hypothetical biochemical pathway that would produce this type of gene interaction. Demonstrate why each genotype shows its specific phenotype.

*8.8 In J. R. R. Tolkien's *The Lord of the Rings*, the Black Riders of Mordor ride steeds with eyes of fire. As a geneticist, you are very interested in the inheritance of the fire-red eye color. You discover that the eyes contain two types of pigments, brown and red, that are usually bound to core granules in the eye. In wild-type steeds, precursors are converted by these granules to the above pigments, but in steeds homozygous for the recessive X-linked gene w (white eye), the granules remain unconverted and a white eye results. The metabolic pathways for the synthesis of the two pigments are shown in Figure 8.A. Each step of the pathway is controlled by a gene: a mutation v results in vermilion eyes; cn results in cinnabar eyes; st results in scarlet eyes; bw results in brown eyes; and se results in black eyes. All these mutations are recessive to their wild-type alleles and all are unlinked. For the following genotypes, show the proportions of steed eye phenotypes that would be obtained in the F_1 of the given matings.

a. $w/w\ bw^+/bw^+ st/st \times w^+/Y\ bw/bw\ st^+/st^+$
b. $w^+/w^+ se/se\ bw/bw \times w/Y\ se^+/se^+ bw^+/bw^+$
c. $w^+/w^+ v^+/v^+ bw/bw \times w/Y\ v/v\ bw/bw$
d. $w^+/w^+ bw^+/bw\ st^+/st \times w/Y\ bw/bw\ st/st$

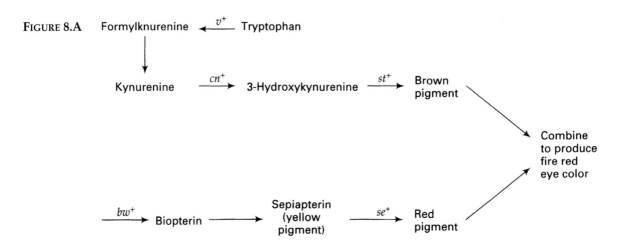

FIGURE 8.A

*8.9   Upon infection of *E. coli* with bacteriophage T4, a series of biochemical pathways result in the formation of mature progeny phages. The phages are released following lysis of the bacterial host cells. Suppose that the following pathway exists:

$$A \xrightarrow{\text{enzyme}} B \xrightarrow{\text{enzyme}} \text{mature phage}$$

Also suppose that we have two temperature-sensitive mutants that involve the two enzymes catalyzing these sequential steps. One of the mutations is cold-sensitive (*cs*) in that no mature phages are produced at 17°C. The other is heat-sensitive (*hs*) in that no mature phages are produced at 42°C. Normal progeny phages are produced when phages carrying either of the mutations infect bacteria at 30°C. However, let us assume that we do not know the sequence of the two mutations. Two models are therefore possible:

$$(1) \ A \xrightarrow{hs} B \xrightarrow{cs} \text{phage}$$
$$(2) \ A \xrightarrow{cs} B \xrightarrow{hs} \text{phage}$$

Outline how you would experimentally determine which model is the correct model without artificially lysing phage-infected bacteria.

8.10   Four mutant strains of *E. coli* (*a, b, c,* and *d*) all require substance X in order to grow. Four plates were prepared, as shown in Figure 8.B. In each case the medium was minimal, with just a trace amount of substance X, to allow a small amount of growth of the mutant cells. On plate a, cells of mutant strain *a* were spread over the agar and grew to form a thin lawn. On plate b the lawn is composed of mutant *b* cells, and so on. On each plate, cells of the four mutant types were inoculated over the lawn, as indicated by the circles. Dark circles indicate luxuriant growth. This experiment tests whether the bacterial strain spread on the plate can "feed" the four strains inoculated on the plate, allowing them to grow. What do these results show about the relationship of the four mutants to the metabolic pathway leading to substance X?

*8.11   The following growth responses (where + = growth and 0 = no growth) of mutants *1–4* were seen on the related biosynthetic intermediates A, B, C, D, and E. Assume that all intermediates are able to enter the cell, that each mutant carries only one mutation, and that all mutants affect steps after B in the pathway.

	GROWTH ON				
MUTANT	A	B	C	D	E
1	+	0	0	0	0
2	0	0	0	+	0
3	0	0	+	0	0
4	0	0	0	+	+

Which of the schemes in the figure best fits the data with regard to the biosynthetic pathway?

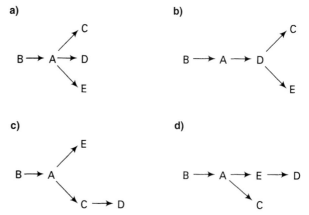

FIGURE 8.B

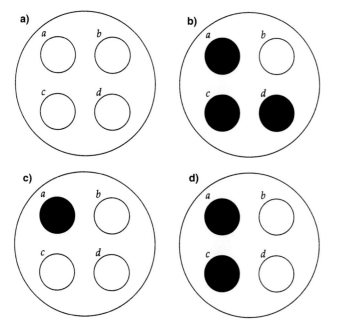

*8.12   Four strains of the haploid fungus *Neurospora*, all of which require arginine but have unknown genetic constitution, have the following nutrition and accumulation characteristics:

	GROWTH ON				
STRAIN	MINIMAL MEDIUM	ORNITHINE	CITRULLINE	ARGININE	ACCUMULATES
1	−	−	−	+	+ Ornithine
2	−	−	−	−	+ Citrulline
3	−	−	−	−	+ Citrulline
4	−	−	−	−	+ Ornithine

The pairwise complementation tests of the four strains gave the following results (+ = growth on minimal medium and 0 = no growth on minimal medium):

	4	3	2	1
1	0	+	+	0
2	0	0	0	
3	0	0		
4	0			

Crosses among mutants yielded prototrophs in the following percentages:

1 × 2:	25 percent
1 × 3:	25 percent
1 × 4:	none detected among 1 million ascospores
2 × 3:	0.002 percent
2 × 4:	0.001 percent
3 × 4:	none detected among 1 million ascospores

Analyze the data and answer the following questions.

a. How many distinct mutational sites are represented among these four strains?

b. In this collection of strains, how many types of polypeptide chains (normally found in the wild type) are affected by mutations?

c. Write the genotypes of the four strains, using a consistent and informative set of symbols.

d. Determine the map distances between all pairs of linked mutations.

e. Determine the percentage of prototrophs that would be expected among ascospores of the following types: (1) strain 1 × wild type; (2) strain 2 × wild type; (3) strain 3 × wild type; (4) strain 4 × wild type

8.13   A breeder of Irish setters has a particularly valuable show dog that he knows is descended from the famous bitch Rheona Didona, who carried a recessive gene for atrophy of the retina. Before he puts the dog to stud, he must ensure that it is not a carrier for this allele. How should he proceed?

*8.14   Suppose you were on a jury to decide the following case:

The Jones family claims that Baby Jane, given to them at the hospital, belongs not to them but to the Smith family, and that the Smiths' baby Joan really belongs to the Jones family. It is alleged that the two babies were accidentally exchanged soon after birth. The Smiths deny that such an exchange has been made. Blood group determinations show the following results:

Mrs. Jones, AB
Mr. Jones, O
Mrs. Smith, A
Mr. Smith, O
Baby Jane, A
Baby Joan, O

Which baby belongs to which family?

*8.15   Glutathione (GSH) is important for a number of biological functions, including prevention of oxidative damage in red blood cells, synthesis of deoxyribonucleotides from ribonucleotides, transport of some amino acids into cells, and maintenance of protein conformation. Mutations that have lowered levels of glutathione synthetase (GSS), a key enzyme in the synthesis of glutathione, result in one of two clinically distinguishable disorders. The severe form is characterized by massive urinary excretion of 5-oxoproline (a chemical derived from a synthetic precursor to glutathione), metabolic acidosis (an inability to regulate physiological pH appropriately), anemia, and central nervous system damage. The mild form is characterized solely by anemia. The characterization of GSS activity and the GSS protein in two affected patients, each with normal parents, is shown below.

PATIENT	DISEASE FORM	GSS ACTIVITY IN FIBROBLASTS (PERCENT OF NORMAL)	EFFECT OF MUTATION ON GSS PROTEIN
1	severe	9%	Arginine at position 267 replaced by tryptophan
2	mild	50%	Aspartate at position 219 replaced by glycine

a. What pattern of inheritance do you expect these disorders to exhibit?

b. Explain the relationship of the form of the disease to the level of GSS activity.

c. How can two different amino acid substitutions lead to dramatically different phenotypes?

d. Why is 5-oxoproline only produced in significant amounts in the severe form of the disorder?

e. Is there evidence that the mutations causing the severe and mild forms of the disease are allelic (in the same gene)?

f. How might you design a test to aid in prenatal diagnosis of this disease?

8.16   Some methods used to gather fetal material for prenatal diagnosis are invasive and therefore pose a small but very real risk to the fetus.

a. What specific risks and problems are associated with chorionic villus sampling and amniocentesis?

b. How are these risks balanced with the benefits of each procedure?

c. Fetal cells are reportedly present in maternal circulation after about eight weeks of pregnancy. However, the number of cells is very low, perhaps no more than one in several million maternal cells. To date, it has not been possible to isolate fetal cells from maternal blood in sufficient quantities for routine genetic analysis. If the problems associated with isolating fetal cells from maternal blood were overcome, and sufficiently sensitive methods were developed to perform genetic tests on a small number of cells, what would be the benefits of performing genetic tests on these fetal cells?

8.17 In evaluating my teacher, my sincere opinion is that:

a. He/she is a swell person whom I would be glad to have as a brother-in-law/sister-in-law.

b. He/she is an excellent example of how tough it is when you do not have either genetics or environment going for you.

c. He/she may have okay DNA to start with, but somehow all the important genes got turned off.

d. He/she ought to be preserved in tissue culture for the benefit of other generations.

CHAPTER 9

DNA: THE GENETIC MATERIAL

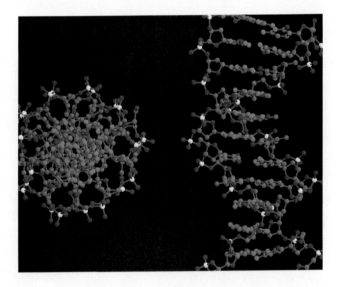

PRINCIPAL POINTS

~ Genetic material must contain all the information for the cell structure and function of an organism. Genetic material must also replicate accurately so that progeny cells have the same genetic information as the parental cell. In addition, genetic material must be capable of variation, a basis for evolutionary change. A series of experiments proved that the genetic material of organisms consists of one of two types of nucleic acids, DNA or RNA. Of the two, DNA is the genetic material of all living organisms and of some viruses; only certain viruses have RNA as their genetic material.

~ DNA and RNA are macromolecules composed of smaller building blocks called nucleotides. Each nucleotide consists of a 5-carbon sugar (deoxyribose in DNA, ribose in RNA) to which are attached one of four nitrogenous bases and a phosphate group.

~ In DNA the four nitrogenous bases (purines and pyrimidines) are adenine, guanine, cytosine, and thymine, while in RNA the four bases are adenine, guanine, cytosine, and uracil.

~ According to Watson and Crick's model, the DNA molecule consists of two polynucleotide chains (polymers of nucleotides) joined by hydrogen bonds between pairs of bases (A and T, G and C) in a double helix. The diameter of the helix is 2 nm, and there are 10 base pairs in each complete turn (3.4 nm).

~ Three major types of double-helical DNA have been identified by X-ray diffraction analysis, the right-handed A- and B-DNAs, and the left-handed Z-DNA. The common form found in cells is B-DNA. A-DNA probably does not exist in cells. Z-DNA may exist in cells in stretches of DNA that are particularly rich in guanine and cytosine. The functional significance of Z-DNA is unknown.

---

*I*n the previous chapters we learned much about how traits are inherited, how recombination brings about variations in allele organization along chromosomes, and how genes are mapped. In all our discussions, however, we have taken as given the existence and the function of the genetic material. In the next several chapters we explore the molecular structure and function of genetic material—both **DNA (deoxyribonucleic acid)** and **RNA (ribonucleic acid)**—and examine the molecular mechanisms by which genetic information is transmitted from generation to generation. You will see exactly what the genetic message is, and you will learn how DNA directs the manufacture of proteins, including enzymes.

Your goals in this chapter are to learn how DNA was shown to be the genetic material and to learn about the structure of DNA.

THE GENETIC MATERIAL

Before the molecular nature of the genetic material was understood, geneticists stated that it must have three principal characteristics:

1. It must contain, in a stable form, *the information* for an organism's cell structure, function, development, and reproduction.

2. It must *replicate accurately* so that progeny cells have the same genetic information as the parental cell.

3. It must be capable of *variation*. Without variation (such as through mutation and recombination), organisms would be incapable of change and adaptation, and evolution could not occur.

From experiments in the mid- to late 1800s and the early 1900s, many scientists thought that the genetic material was protein. Chemical analysis over the first 40 years of the twentieth century revealed that the nucleus contained deoxyribonucleic acid (DNA). This DNA was believed to be a molecular framework for a theoretical class of special proteins that carried genetic information. However, evidence gathered from experiments in the middle of the twentieth century showed unequivocally that the genetic material of living organisms and many viruses consists not of protein but of double-stranded DNA; in other viruses the genetic material may be single-stranded DNA, double-stranded RNA, or single-stranded RNA. The discovery that

DNA was the genetic material led to an explosion of knowledge about molecular aspects of biology.

Griffith's Transformation Experiment

One of the first studies (published in 1928) that ultimately led to the identification of DNA as the genetic material involved the bacterium *Streptococcus pneumoniae* (also called pneumococcus) (Figure 9.1). Two strains of pneumococcus were used. The smooth (*S*) strain is infectious (virulent) and kills the infected animal. Although this was not known in 1928, each bacterial cell of the *S* type is surrounded by a polysaccharide coat or capsule that gives the strain its infectious properties and results in the smooth, shiny appearance of *S* colonies. The rough (*R*) strain is noninfectious (avirulent) because *R* cells lack the polysaccharide coat; colonies of this strain are nonshiny.

The *S* strains can be classified further into *IIS* and *IIIS*, which have differences in the polysaccharide coat. The differences between *II* and *III* are genetically determined, as are the differences between *S* and *R*. Occasionally, *S* cells mutate to the *R* type and, when they do, genetically, they will be *II* or *III* depending on what type the *S* cell was, even though *R* cells do not make a polysaccharide coat.

~ FIGURE 9.1

Transmission electron micrograph of the bacterium *Streptococcus pneumoniae*. (Many of the cells are in the process of dividing.)

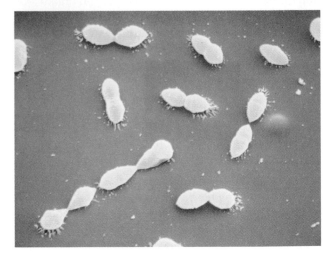

In 1928, Frederick Griffith injected mice with *R* bacteria derived by mutation from *IIS* bacteria (Figure 9.2). The *R* bacteria did not affect the mice, and after a while the bacteria disappeared from the animals'

~ FIGURE 9.2

Griffith's transformation experiment. Mice injected with type *IIIS* pneumococcus died, whereas mice injected with either type *IIR* or heat-killed type *IIIS* bacteria survived. When injected with a mixture of living type *IIR* and heat-killed type *IIIS* bacteria, however, the mice died.

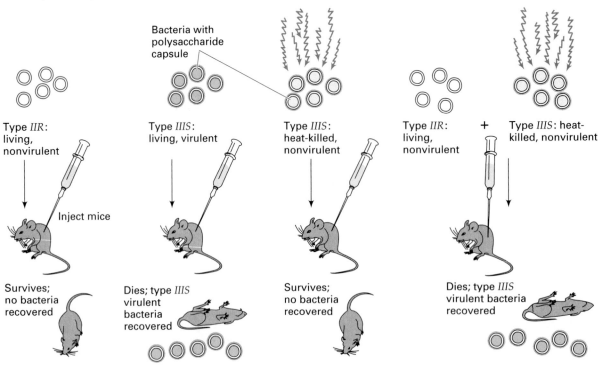

bloodstreams. Griffith also injected mice with living *IIIS* bacteria. Those animals died, and living *IIIS* bacteria could be isolated from their blood. If the *IIIS* bacteria were heat-killed before injecting them, however, the mice survived. These two experiments showed that the bacteria had to be alive and had to possess the polysaccharide coat in order for them to be infectious and kill the mice.

Lastly, Griffith injected mice with a mixture of living *R* bacteria (derived from *IIS*) and heat-killed *IIIS* bacteria. In this case the mice died, and living *S* bacteria were present in the blood. These bacteria were all of type *IIIS* and therefore could not have arisen by mutation of the *R* bacteria, since mutation would have produced *IIS* colonies. Griffith concluded that some *R* bacteria had somehow been transformed into smooth, infectious *IIIS* cells by interaction with the dead *IIIS* cells. Griffith believed that the unknown agent responsible for the change in genetic material was a protein. He referred to the agent as the *transforming principle*. (See Chapter 6 for a discussion of bacterial transformation.)

From a modern molecular perspective, the genetic material of the virulent bacteria was released from the heat-killed cells and entered the living avirulent cells, where recombination occurred to generate a virulent transformant.

The Nature of the Transforming Principle

In 1944, Oswald T. Avery, Colin M. MacLeod, and Maclyn McCarty studied the transformation of bacteria from *R* to *S* type in the test tube. They broke open *IIIS* cells and separated the cell extract into the various cellular macromolecular components such as lipids, polysaccharides, proteins, and the nucleic acids RNA and DNA. They tested each component to see whether it contained the transforming principle by checking whether it could transform living *R* bacteria derived from *IIS*. The nucleic acids (at this point not separated into DNA and RNA) were the only components from *IIIS* cells that could transform the *R* cells into *IIIS*.

Avery and his colleagues then used specific nucleases—enzymes that degrade nucleic acids—to determine whether DNA or RNA was the transforming principle (Figure 9.3). When they treated the nucleic acids with **ribonuclease** (**RNase**), which degrades RNA but not DNA, the transforming activity was still present. However, when they used **deoxyribonuclease** (**DNase**), which degrades only

---

~ **FIGURE 9.3**

Experiment that showed DNA, and not RNA, was the transforming principle. When the nucleic acid mixture of DNA and RNA was treated with ribonuclease (RNase), *S* transformants still resulted. However, when the DNA and RNA mixture was treated with deoxyribonuclease (DNase), no *S* transformants resulted. (Note: *R* colonies resulting from untransformed cells were present on both plates; they have been omitted from the drawings.)

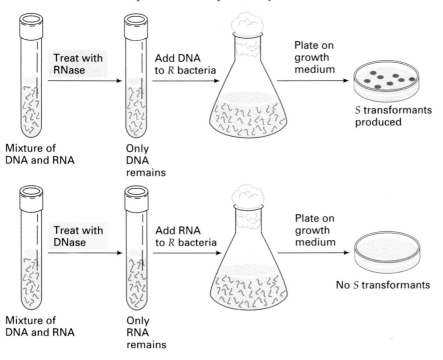

DNA, no transformation resulted. These results strongly suggested that DNA was the genetic material. Although Avery's work was important, it was criticized at the time because the nucleic acids isolated from the bacteria were contaminated by proteins.

The Hershey and Chase Bacteriophage Experiments

In 1953, Alfred D. Hershey and Martha Chase published a paper showing DNA to be genetic material. They were studying the replication of bacteriophage T2 to see which phage components were needed to complete the life cycle. (A bacteriophage is a virus that infects bacteria. See Figure 6.12 for an EM of T4, a virtual "twin" of T2, and Figure 6.13 for the life cycle of T2.) At the time it was known that T2 infected *Escherichia coli* by first attaching itself with its tail fibers, then injecting into the bacterium genetic material that somehow controlled the production of progeny T2 particles. Ultimately, the bacterium lysed, releasing 100 to 200 phages that could infect other bacteria. However, the nature of T2's genetic material was not known.

Hershey and Chase knew that T2 phages consisted of only DNA and protein, so one of them must be the genetic material. They grew cells of *E. coli* in media containing either a radioactive isotope of phosphorus (^{32}P) or a radioactive isotope of sulfur (^{35}S) (Figure 9.4a), because DNA contains phosphorus but no sulfur, and protein contains sulfur but no phosphorus. They infected the bacteria with T2 and collected the progeny phages. At this point, Hershey and Chase had two batches of T2; one had the proteins radioactively labeled with ^{35}S, and the other had the DNA labeled with ^{32}P.

They then infected *E. coli* with the two types of radioactively labeled T2 (Figure 9.4b). When the infecting phage was ^{32}P-labeled, most of the radioactivity was found within the bacteria soon after infection. Very little was found in protein parts of the phage (the phage ghosts) released from the cell surface after agitating the cells in a kitchen blender. After lysis, some of the ^{32}P was found in the progeny phages. In contrast, after infecting *E. coli* with ^{35}S-labeled T2, virtually none of the radioactivity appeared within the cell, and none was found in the progeny phage particles. Most of the radioactivity could be found in the phage ghosts released after treating the cultures in the kitchen blender.

Since genes serve as the blueprint for making the progeny virus particles, it was also presumed that the blueprint must get into the bacterial cell in order for new phage particles to be built. Therefore, since it *was DNA and not protein that entered the cell*, as evidenced by the presence of ^{32}P and the absence of ^{35}S, Hershey and Chase reasoned that DNA *must be the material responsible for the function and reproduction of phage T2*. The protein, they hypothesized, provided a structural framework to contain the DNA and the specialized structures required to inject the DNA into the bacterial cell.

Most of the organisms and viruses discussed in this book (such as humans, *Drosophila*, yeast, *E. coli*, and phage T2) have DNA as their genetic material. However, some bacterial viruses (for example, Qβ), some animal viruses (for example, poliovirus), and some plant viruses (for example, tobacco mosaic virus) have RNA as their genetic material. No known prokaryotic or eukaryotic organism has RNA as its genetic material.

KEYNOTE

Genetic material must contain all the information for the cell structure and function of an organism and must replicate accurately so that progeny cells have the same genetic information as the parental cell. In addition, genetic material must be capable of variation, one of the bases for evolutionary change. A series of experiments proved that the genetic material of organisms consists of one of two types of nucleic acids, DNA or RNA. Of the two, DNA is more common; only certain viruses have RNA as their genetic material.

THE CHEMICAL COMPOSITION OF DNA AND RNA

Both DNA and RNA are **macromolecules**, meaning that they have a molecular weight of at least a few thousand daltons (1 dalton is equivalent to a twelfth of the mass of the carbon 12 atom, or 1.67×10^{-24} g). Both DNA and RNA are polymeric molecules made up of four different monomeric units called **nucleotides**. Each nucleotide consists of three distinct parts: (1) a **pentose** (5-carbon) **sugar**, (2) a **nitrogenous** (nitrogen-containing) **base**, and (3) a **phosphate group**. Because they can be isolated from nuclei, and because they are acidic, these macromolecules are called *nucleic acids*. Macromolecules also include proteins, which are made of amino acid monomers, and

~ FIGURE 9.4

Hershey-Chase experiment. (a) The production of T2 phages either with (1) ^{32}P-labeled DNA or with (2) ^{35}S-labeled protein; (b) The experimental evidence showing that DNA is the genetic material in T2: (1) The ^{32}P is found within the bacteria and appears in progeny phages, while (2) the ^{35}S is not found within the bacteria and is released with the phage ghosts.

a) Preparation of radioactively labeled T2 bacteriophage

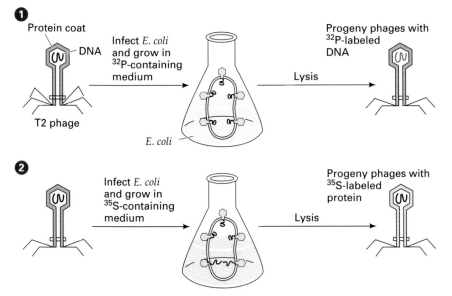

b) Experiment that showed DNA to be the genetic material of T2

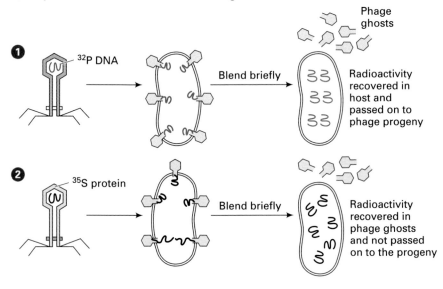

polysaccharides, which are made of monosaccharide (simple sugar) monomers.

For DNA the pentose sugar is **deoxyribose,** and for RNA it is **ribose** (Figure 9.5). The two sugars differ by the chemical groups attached to the 2' carbon: a hydrogen atom (H) in deoxyribose and a hydroxyl

group (OH) in ribose. (The carbon atoms in the pentose sugar are numbered 1' to 5' to distinguish them from the numbered carbon and nitrogen atoms in the rings of the bases.)

There are two classes of bases, the **purines** and the **pyrimidines**. In DNA the purines are **adenine**

~ FIGURE 9.5

Structures of deoxyribose and ribose, the pentose sugars of DNA and RNA, respectively. The difference between the two sugars is highlighted.

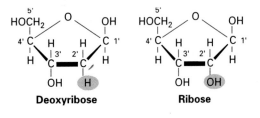

sugar in both DNA and RNA. Examples of a DNA nucleotide (a **deoxyribonucleotide**) and an RNA nucleotide (a **ribonucleotide**) are shown in Figure 9.7a. For future discussions, remember that a *nucleotide*, the basic building block of the DNA and RNA molecules, consists of the sugar, a base, and a phosphate group. The sugar plus base only is called a *nucleoside*, so a nucleotide is also a **nucleoside phosphate**. A complete listing of the names for the bases, nucleosides, and nucleotides is in Table 9.1.

Because DNA contains the sugar deoxyribose and the pyrimidine thymine, while RNA contains the sugar ribose and the pyrimidine uracil, these two nucleic acids have different chemical and biological properties. For example, the cellular enzymes that catalyze nucleic acid synthesis (polymerases) and nucleic acid degradation (nucleases) are usually DNA-specific or RNA-specific. The differences between the two molecules permit them to be separated and purified relatively easily for study in the laboratory.

(A) and **guanine** (G), and the pyrimidines are **thymine** (T) and **cytosine** (C). RNA also contains adenine, guanine, and cytosine, but the pyrimidine **uracil** (U) replaces thymine. The chemical structures of the five bases are shown in Figure 9.6. (The carbons and nitrogens of the purine rings are numbered 1 to 9, and those of the pyrimidines are numbered 1 to 6.)

In DNA and RNA, bases are always attached to the 1' carbon of the pentose sugar by a covalent bond. The purine bases are bonded at the 9 nitrogen, while the pyrimidines bond at the 1 nitrogen. The phosphate group (PO_4^{2-}) is attached to the 5' carbon of the

To form polynucleotides of either DNA or RNA, nucleotides are linked together by a covalent bond between the phosphate group (which is attached to the 5' carbon of the sugar ring) of one nucleotide and the 3' carbon of the sugar of another nucleotide. These 5'-3' phosphate linkages are called **phosphodiester bonds**. A short polynucleotide chain is diagrammed

~ FIGURE 9.6

Structures of the nitrogenous bases in DNA and RNA. The parent compounds are purine (top left), and pyrimidine (bottom left). Differences between the bases are highlighted.

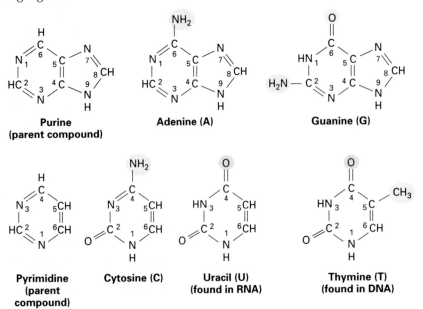

~ **FIGURE 9.7**

Chemical structures of DNA and RNA. (a) Basic structures of DNA and RNA nucleosides (sugar plus base) and nucleotides (sugar plus base plus phosphate group), the basic building blocks of DNA and RNA molecules. Here the phosphate groups are orange, the sugars are red, and the bases are brown. (b) A segment of a polynucleotide chain, in this case a single strand of DNA. The deoxyribose sugars are linked by phosphodiester bonds (shaded) between the 3' carbon of one sugar and the 5' carbon of the next sugar.

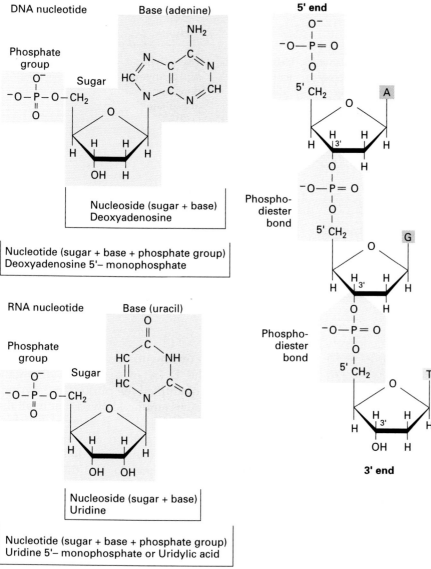

a) DNA and RNA nucleotides

b) DNA polynucleotide chain

in Figure 9.7b. The phosphodiester bonds are relatively strong, and as a consequence the repeated sugar-phosphate-sugar-phosphate backbone of DNA and RNA is a stable structure.

To understand how a polynucleotide chain is synthesized (which we will study in another chapter), we must be aware of one more feature of the chain: The two ends of the chain are not the same; that is, the chain has a 5' carbon (with a phosphate group on it) at one end and a 3' carbon (with a hydroxyl group on it) at the other end, as shown in Figure 9.7b. This asymmetry is referred to as *polarity* of the chain.

~ TABLE 9.1

Names of the Base, Nucleoside, and Nucleotide Components Found in DNA and RNA

		BASE: PURINES (PU)		BASE: PYRIMIDINES (PY)		
		ADENINE (A)	GUANINE (G)	CYTOSINE (C)	THYMINE (T) (DEOXYRIBOSE ONLY)	URACIL (U) (RIBOSE ONLY)
DNA	Nucleoside: deoxyribose + base	Deoxy-adenosine (dA)	Deoxy-guanosine (dG)	Deoxycytidine (dC)	Thymidine (dT)	
	Nucleotide: deoxyribose + base + phosphate group	Deoxyadenylic acid or deoxy-adenosine monophos-phate (dAMP)	Deoxyguanylic acid or deoxy-guanosine monophos-phate (dGMP)	Deoxycytidylic acid or deoxy-cytidine monophos-phate (dCMP)	Thymidylic acid or thymi-dine mono-phosphate (TMP)	
RNA	Nucleoside: ribose + base	Adenosine (A)	Guanosine (G)	Cytidine (C)		Uridine (U)
	Nucleotide: ribose + base + phosphate group	Adenylic acid or adenosine monophos-phate (AMP)	Guanylic acid or guanosine monophos-phate (GMP)	Cytidylic acid or cytidine monophos-phate (CMP)		Uridylic acid or uridine monophos-phate (UMP)

KEYNOTE

DNA and RNA occur in nature as macromolecules composed of smaller building blocks called nucleotides. Each nucleotide consists of a 5-carbon sugar (deoxyribose in DNA, ribose in RNA) to which is attached a phosphate group and one of four nitrogenous bases—adenine, guanine, cytosine, and thymine (in DNA) or adenine, guanine, cytosine, and uracil (in RNA).

The DNA Double Helix

In 1953, James D. Watson and Francis H. C. Crick (Figure 9.8) published a paper in which they proposed a model for the physical and chemical structure of the DNA molecule. In generating their model, Watson and Crick used three main pieces of data:

1. The DNA molecule is composed of bases, sugars, and phosphate groups linked together as a **polynucleotide** (deoxyribonucleotide) chain.

2. By chemical treatment, Erwin Chargaff had hydrolyzed the DNA of a number of organisms and had quantified the purines and pyrimidines released. His studies showed that in all the (double-stranded) DNAs, 50 percent of the bases were purines and 50 percent were pyrimidines. More important, the amount of adenine (A) was equal to that of thymine (T), and the amount of guanine (G) was equal to that of cytosine (C). These equivalencies have become known as Chargaff's rules. In comparisons of double-stranded DNAs from different organisms, the A/T ratio is 1, the G/C ratio is 1, but the $(A + T)/(G + C)$ ratio (typically presented as %GC) varies. Because purines = pyrimidines, the $(A + G)/(C + T)$ ratio = 1.

3. Rosalind Franklin, working with Maurice H. F. Wilkins (Figure 9.9a), studied isolated fibers of DNA by using the X-ray diffraction technique, a procedure in which a beam of parallel X rays is aimed at molecules. The beam is diffracted ("broken up") by the atoms in a pattern that is characteristic of the atomic weight and the spatial arrangement of the molecules. The diffracted

James Watson (left) and Francis Crick (right) in 1993 at a fortieth anniversary celebration of their discovery of the structure of DNA, and in 1953 with the model of DNA structure.

X-ray diffraction analysis of DNA. (a) Rosalind Franklin and Maurice H. F. Wilkins (photographed in 1962, the year he received the Nobel Prize shared with Watson and Crick); (b) The X-ray diffraction pattern of DNA that Watson and Crick used in developing their double-helix model. The dark areas that form an X shape in the center of the photograph indicate the helical nature of DNA. The dark crescents at the top and bottom of the photograph indicate the 0.34-nm distance between the base pairs.

a)

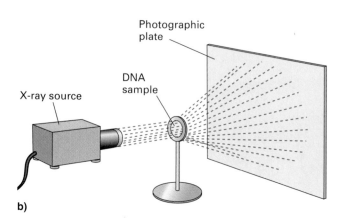

b)

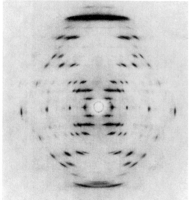

X-ray diffraction pattern

X rays are recorded on a photographic plate (Figure 9.9b). By analyzing the photograph, Franklin obtained information about the molecule's atomic structure. In particular, she concluded that DNA is a helical structure with two distinctive regularities of 0.34 nm and 3.4 nm along the axis of the molecule [1 nanometer (nm) = 10^{-9} meter = 10 Ångstrom units (Å); 1 Å = 10^{-10} meter].

Watson and Crick used these data to build three-dimensional models for the structure of DNA. The model they devised, which fit all the known data on the composition of the DNA molecule, is the now-famous double-helix model for DNA. Unquestionably, the determination of the structure of DNA was a momentous occasion in biology, leading directly to the transformation in our understanding of all aspects of the life sciences. Figure 9.10a shows a three-

dimensional model of the DNA molecule, and Figure 9.10b is a diagram of the DNA molecule, showing the ar-rangement of the sugar-phosphate backbone and base pairs in a stylized way.

Watson and Crick's double-helical model of DNA has the following main features:

1. The DNA molecule consists of two polynucleotide chains wound around each other in a right-handed double helix; that is, viewed on end (from either end), the two strands wind around each other in a clockwise (right-handed) fashion.

2. The external diameter of the helix is 2 nm.

3. The two chains are *antiparallel* (show *opposite polarity*); that is, the two strands are oriented in opposite directions, with one strand oriented in the 5' to 3' way, while the other strand is oriented 3'

~ FIGURE 9.10

Molecular structure of DNA. (a) Three-dimensional molecular model of DNA as prepared by Watson and Crick; (b) stylized representation of the DNA double helix.

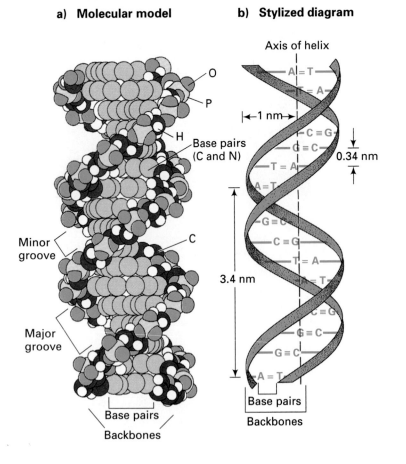

a) Molecular model

b) Stylized diagram

to 5'. To put it in simpler terms, if the 5' end is the "head" of the chain and the 3' end is the "tail" of the chain, antiparallel means that the head of one chain is against the tail of the other chain and vice versa.

4. The sugar-phosphate backbones are on the outsides of the double helix, with the bases oriented toward the central axis (see Figure 9.10). The bases of both chains are flat structures oriented perpendicularly to the long axis of the DNA; that is, the bases are stacked like pennies on top of one another following the "twist" of the helix.

5. The bases of the opposite strands are bonded together by hydrogen bonds, which are relatively weak chemical bonds. The specific pairings observed are A with T (two hydrogen bonds; Figure 9.11a) and G with C (three hydrogen bonds; Figure 9.11b). The hydrogen bonds make it relatively easy to separate the two strands of the DNA, for example, by heating. Breaking the A-T base pair by heating is easier than breaking the G-C base pair because A-T has two hydrogen bonds and G-C has three hydrogen bonds. The A-T and G-C base pairs are the only ones that can fit the physical dimensions of the helical model, and they are totally in accord with Chargaff's rules. The specific A-T and G-C pairs are called **complementary base pairs**, so the nucleotide sequence in one strand dictates the nucleotide sequence of the other. For instance, if one chain has the sequence 5'-TAT TCCGA-3', then the opposite, antiparallel chain must bear the sequence 3'-ATAAGGCT-5'.

6. The base pairs are 0.34 nm apart in the DNA helix. A complete (360°) turn of the helix takes 3.4 nm; therefore, there are 10 base pairs per turn.

7. Because of the way the bases bond with each other, the two sugar-phosphate backbones of the double helix are not equally spaced along the helical axis. This results in grooves of unequal size between the backbones called the *major* (wider) *groove* and the *minor* (narrower) *groove* (see Figure 9.10a). Both of these grooves are large enough to allow protein molecules to make contact with the bases. The phenomenon of proteins "reading" specific base-pair sequences is common to many molecular processes; we will encounter many examples in other chapters.

~ **FIGURE 9.11**

Structures of the complementary base pairs found in DNA. In both cases a purine pairs with a pyrimidine: (a) The adenine-thymine bases, which pair through two hydrogen bonds; (b) The guanine-cytosine bases, which pair through three hydrogen bonds.

a) **Adenine-thymine base (Double hydrogen bond)**

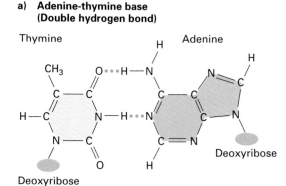

b) **Guanine-cytosine base (Triple hydrogen bond)**

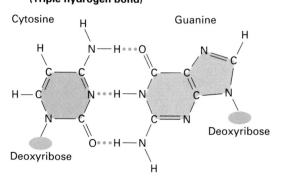

KEYNOTE

According to Watson and Crick's model, the DNA molecule consists of two polynucleotide chains joined by hydrogen bonds between pairs of bases (A and T, G and C) in a double helix. The diameter of the helix is 2 nm, and there are 10 base pairs in each complete turn (3.4 nm). The double-helix model was proposed by Watson and Crick from chemical and physical analyses of DNA.

Different DNA Structures

In recent years, methods have been developed to synthesize short DNA molecules (called **oligomers**, *oligo* meaning "few") of defined sequences. The pure DNA oligomers can be crystallized and analyzed by X-ray diffraction. These more modern approaches have revealed that DNA can exist in several different forms, notably the A-, B-, and Z-DNA forms (Figure 9.12).

~ **FIGURE 9.12**

Space-filling models of different forms of DNA. (a) A-DNA. (b) B-DNA. (c) Z-DNA.

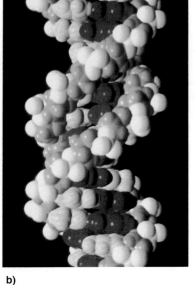

a)   b)   c)

Watson and Crick's double-helical model was made of B-DNA. Like B-DNA, A-DNA is a right-handed helix, but Z-DNA is a left-handed helix with a zigzag sugar-phosphate backbone. (The latter property gave this DNA form its "Z" designation.) Some of the key structural features of each DNA type are listed in Table 9.2.

A-DNA AND B-DNA.   A-DNA and B-DNA are right-handed double helices with 10.9 and 10.0 base pairs per 360° turn of the helix, respectively. The A-DNA double helix is short and wide with a narrow, very deep major groove and a wide, shallow minor groove. (Think of these descriptions in terms of canyons—narrow versus wide describes the distance from rim to rim, and shallow vs. deep describes the distance from the rim down to the bottom of the canyon.) The B-DNA double helix is thinner and longer for the same number of base pairs than A-DNA, with a wide major groove and a narrow minor groove; both grooves are of similar depths.

Z-DNA.   Z-DNA is a left-handed helix with 12.0 base pairs per complete helical turn. The Z-DNA helix is thin and elongated with a deep, minor

~ **TABLE 9.2**

Properties of A-DNA, B-DNA, and Z-DNA

PROPERTY	A-DNA	B-DNA	Z-DNA
Helix direction	Right-handed	Right-handed	Left-handed
Base pairs per helix turn	10.9	10.0	12.0
Overall morphology	Short and wide	Longer and thinner	Elongated and thin
Major groove	Extremely narrow and very deep	Wide and of intermediate depth	Flattened out on helix surface
Minor groove	Very wide and shallow	Narrow and of intermediate depth	Extremely narrow and very deep
Helix axis location	Major groove	Through base pairs	Minor groove
Helix diameter	2.2 nm	2.0 nm	1.8 nm

groove. The major groove is very near the surface of the helix, so it is not really distinct.

DNA IN THE CELL.   In solution, DNA usually is found in the B form and hence B-DNA is the form mostly found in cells. A-DNA is found only when the DNA is dehydrated, so it is unlikely that any lengthy sections of A-DNA exist within cells. Whether Z-DNA exists in living cells has long been a topic of debate among scientists. Some evidence for the existence of Z-DNA in cells includes the presence of Z-DNA–binding proteins in the nuclei of *Drosophila*, human, wheat, and bacterial cells. These proteins may stabilize the DNA in the Z-DNA form. The function of Z-DNA is not clearly understood, although roles in DNA replication, genetic recombination, and transcription regulation have been proposed.

KEYNOTE

X-ray diffraction analysis of crystals of DNA oligomers has produced detailed molecular models of three DNA types: A-DNA, B-DNA, and Z-DNA. A-DNA and B-DNA are both right-handed double helices, while Z-DNA is a left-handed double helix. The number of base pairs per helical turn are 10.9, 10.0, and 12.0 for A-, B-, and Z-DNA, respectively. A-DNA is a short and broad molecule, B-DNA is a longer and thinner molecule, and Z-DNA is elongated and thin. The three DNA forms differ in the dimensions of the major and minor grooves and location of the helix axis. B-DNA is the form found predominantly in cells. Z-DNA has also been found in the chromosomes of some organisms, but its role in cellular function, if any, is not clear. Since A-DNA forms only under relatively dehydrated conditions, this DNA type does not exist in a stable form in cells.

SUMMARY

In this chapter we have learned that DNA or RNA can be the genetic material. All prokaryotic and eukaryotic organisms, and most viruses, have DNA as their genetic material, while some viruses have RNA as their genetic material. The form of the genetic material varies among organisms and their viruses. In living organisms, the DNA is always double-stranded, while in viruses the genetic material may be double- or single-stranded DNA or RNA, depending on the virus.

Chemical analysis has revealed that DNA and RNA are macromolecules composed of building blocks called nucleotides. Each nucleotide consists of a 5-carbon sugar (deoxyribose in DNA, ribose in RNA) to which is attached one of four nitrogenous bases and a phosphate group. In DNA the four nitrogenous bases are adenine, guanine, cytosine, and thymine, while in RNA the four bases are adenine, guanine, cytosine, and uracil. Thymine and uracil differ only in a single methyl group, which is present in thymine and absent in uracil. From chemical and physical analysis it was determined that the DNA molecule consists of two polynucleotide chains joined by hydrogen bonds between pairs of bases (A and T, G and C) in a double helix. (The double helix model was first proposed by Watson and Crick.) The diameter of the helix is 2 nm, and there are 10 base pairs in each complete turn of the helix (3.4 nm).

A number of different types of double-helical DNA have been identified by X-ray diffraction analysis. The three major types are the right-handed A- and B-DNAs and the left-handed Z-DNA. The common form found in cells is B-DNA (the form analyzed by Watson and Crick). A-DNA probably does not exist in cells. Z-DNA may exist in cells but its functional role is unknown.

ANALYTICAL APPROACHES FOR SOLVING GENETICS PROBLEMS

Q9.1   The linear chromosome of phage T2 is 52 μm long. The chromosome consists of double-stranded DNA, with 0.34 nm between each base pair. The average weight of a base pair is 660 daltons. What is the molecular weight of the T2 molecule?

A9.1   This question involves the careful conversion of different units of measurement. The first step is to put the

lengths in the same units: 52 μm is 52 millionths of a meter, or $52,000 \times 10^{-9}$ m, or 52,000 nm. One base occupies 0.34 nm in the double helix, so the number of base pairs in this chromosome is 52,000 divided by 0.34, or 152,941 base pairs. Each base pair, on the average, weighs 660 daltons, and therefore the molecular weight of the chromosome is $152,941 \times 660 = 1.01 \times 10^8$ daltons, or 101 million daltons.

The human genome contains 3×10^9 bp of DNA, for a total length of about 1 meter distributed among 23 chromosomes. The average length of the double helix in a human chromosome is 3.8 cm, which is 3.8 hundredths of a meter or 38 million nm—substantially longer than the T2 chromosome! There are over 111.7 million base pairs in the average human chromosome.

Q9.2 The accompanying table lists the relative percentages of bases of nucleic acids isolated from different species. For each one, what type of nucleic acid is involved? Is it double- or single-stranded? Explain your answer.

SPECIES	ADENINE	GUANINE	THYMINE	CYTOSINE	URACIL
(i)	21	29	21	29	0
(ii)	29	21	29	21	0
(iii)	21	21	29	29	0
(iv)	21	29	0	29	21
(v)	21	29	0	21	29

A9.2 This question focuses on the base-pairing rules and the difference between DNA and RNA. In analyzing the data, we should determine first whether the nucleic acid is RNA or DNA, and then whether it is double- or single-stranded. If the nucleic acid has thymine, it is DNA; if it has uracil, it is RNA. Thus species (i), (ii), and (iii) must have DNA as their genetic material, and species (iv) and (v) must have RNA as their genetic material.

Next, we must analyze the data for strandedness. Double-stranded DNA must have equal percentages of A and T and of G and C. Similarly, double-stranded RNA must have equal percentages of A and U and of G and C. Hence species (i) and (ii) have double-stranded DNA, while species (iii) must have single-stranded DNA since the base-pairing rules are violated, with A = G and T = C but A ≠ T and G ≠ C. As for the RNA-containing species, (iv) contains double-stranded RNA since A = U and G = C, and (v) must contain single-stranded RNA.

QUESTIONS AND PROBLEMS

9.1 Griffith's experiment injecting a mixture of dead and live bacteria into mice demonstrated (choose the correct answer):
a. that DNA is double-stranded.
b. that mRNA of eukaryotes differs from mRNA of prokaryotes.
c. bacterial transformation.
d. that bacteria can recover from heat treatment if live helper cells are present.

9.2 In the 1920s while working with *Streptococcus pneumoniae*, the agent that causes pneumonia, Griffith injected mice with different types of bacteria. For each of the following bacteria type(s) injected, indicate whether the mice lived or died:
a. type *IIR*
b. type *IIIS*
c. heat-killed *IIIS*
d. type *IIR* + heat-killed *IIIS*

9.3 Several years after Griffith described the transforming principle, Avery, MacLeod, and McCarty investigated the same phenomenon.
a. Describe their experiments.
b. What did their experiments demonstrate beyond Griffith's?
c. How were enzymes used as a control in their experiments?

*9.4 The Hershey-Chase experiment with T2 phages demonstrated that (choose the correct answer):
a. the coat material of the phage controls the kind of DNA replicated in the host cell.
b. the nucleic acid of the phage contains the genetic information.
c. the prophage state is necessary for generalized transduction.
d. conjugation is dependent upon some of the cells being *Hfr*.
e. a metaphase chromosome is composed of two chromatids, each containing a single DNA molecule.

*9.5 Hershey and Chase showed that when phages were labeled with ^{32}P and ^{35}S, the ^{35}S remained outside the cell and could be removed without affecting the course of infection, whereas the ^{32}P entered the cell and could be recovered in progeny phages. What distribution of isotope would you expect to see if parental phages were labeled with isotopes of
a. C?
b. N?
c. H?
Explain your answer.

*9.6 In DNA and RNA, which carbon atoms of the sugar molecule are connected by a phosphodiester bond?

9.7 Which base is unique to DNA, and which base is unique to RNA?

9.8 How do nucleosides and nucleotides differ?

9.9 What chemical group is found at the 5' end of a DNA chain? At the 3' end of a DNA chain?

9.10 What evidence do we have that in the helical form of the DNA molecule the base pairs are composed of one purine and one pyrimidine?

*9.11 What evidence is there to substantiate the statement: "There are only two base pair combinations in DNA, A-T and C-G"?

9.12 How many different kinds of nucleotides are there in DNA molecules?

*9.13 What is the base sequence of the DNA strand that would be complementary to the following single-stranded DNA molecules?

a. 5' A G T T A C C T G A T C G T A 3'
b. 5' T T C T C A A G A A T T C C A 3'

*9.14 Is an adenine-thymine or a guanine-cytosine base pair harder to break apart? Explain your answer.

9.15 The double-helix model of DNA, as suggested by Watson and Crick, was based on data gathered on DNA by other researchers. The facts fell into the following two general categories; give two examples of each:
a. chemical composition
b. physical structure

*9.16 For double-stranded DNA, which of the following base ratios always equals 1?
a. $(A + T)/(G + C)$
b. $(A + G)/(C + T)$
c. C/G
d. $(G + T)/(A + C)$
e. A/G

9.17 Explain whether or not the $(A + T)/(G + C)$ ratio in double-stranded DNA is expected to be the same as the $(A + C)/(G + T)$ ratio.

*9.18 The percent cytosine in a double-stranded DNA is 17. What is the percent adenine in that DNA?

9.19 A double-stranded DNA molecule was found to contain 32 percent thymine. What percent of this same molecule would be made up of cytosine?

9.20 A sample of double-stranded DNA has 62 percent GC. What is the percentage of A in the DNA?

9.21 Analysis of DNA from a bacterial virus indicates that it contains 33 percent A, 26 percent T, 18 percent G, and 23 percent C. Interpret these data.

9.22 The genetic material of bacteriophage ΦX174 is single-stranded DNA. What base equalities or inequalities might we expect for single-stranded DNA?

9.23 Through X-ray diffraction analysis of crystallized DNA oligomers, different forms of DNA have been identified. These forms include A-DNA, B-DNA, and Z-DNA, and each has unique molecular attributes.
a. How do the major and minor grooves of B-DNA and Z-DNA differ?
b. How do the helical properties of these forms of DNA differ?
c. What functions might Z-DNA have?

9.24 If a virus particle contains double-stranded DNA with 200,000 base pairs, how many complete 360° turns occur in this molecule?

*9.25 A double-stranded DNA molecule is 100,000 base pairs (100 kilobases) long.
a. How many nucleotides does it contain?
b. How many complete turns are there in the molecule?
c. How long is the DNA molecule?

9.26 Organisms have vastly different amounts of genetic material. *E. coli* has about 4.6 million base pairs of DNA in one circular chromosome, the haploid budding yeast (*S. cerevisiae*) has 12,057,500 base pairs of DNA in 17 chromosomes, and the gametes of humans have about 2.75 billion base pairs of DNA in 23 chromosomes.
a. If all of the DNA were B-DNA, what would be the average length of a chromosome in these cells?
b. On average, how many complete turns would be in each chromosome?
c. Would your answers to (a) and (b) be significantly different if the DNA were composed of, say, 20% Z-DNA and 80% B-DNA?
d. What implications do your answers to these questions have for the packaging of DNA in cells?

CHAPTER *10*

DNA: ORGANIZATION IN CHROMOSOMES

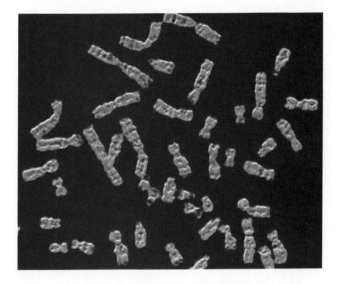

PRINCIPAL POINTS

~ In bacteria, archaea, and viruses in which the genetic material is DNA, chromosomes consist of DNA with some associated proteins.

~ Bacteria and archaea have one or a very few chromosomes. In *E. coli* the chromosome is a circular, double-stranded DNA molecule that is compacted by supercoiling of the DNA helix to produce about 100 looped domains.

~ The complete set of metaphase chromosomes in a eukaryotic cell is called its karyotype. The karyotype is species-specific.

~ The nuclear chromosomes of eukaryotes are complexes of DNA and histone and nonhistone chromosomal proteins. Each chromosome consists of one linear, unbroken, double-stranded DNA molecule running throughout its length; the DNA is variously coiled and folded. The histones are constant from cell to cell within an organism, while the nonhistones vary significantly among cell types.

~ The large amount of DNA present in the eukaryotic chromosome is condensed by its association with his-

tones in nucleosomes, and by higher levels of folding of the nucleosomes into chromatin fibers. Each chromosome contains a large number of looped domains of 30-nm chromatin fibers attached to a protein scaffold. The more condensed a part of a chromosome is, the less likely it is that the genes in that region will be active.

~ The centromere region of each eukaryotic chromosome is responsible for the accurate segregation of the replicated chromosome to the daughter cells during mitosis and meiosis. The DNA sequences of centromeres are species-specific.

~ The ends of chromosomes, the telomeres, often are associated with the nuclear envelope. Telomeres consist of simple, relatively short, tandemly repeated sequences that are species-specific.

~ Prokaryotic genomes consist mostly of unique-sequence DNA, with only a few repeated sequences and genes. Eukaryotes have both unique and repetitive sequences in the genome. The spectrum of complexity of repetitive DNA sequences among eukaryotes is extensive.

The genetic material of the cell is organized into physical structures called chromosomes. Your goal in this chapter is to learn about chromosome structure. The chromosomes of eukaryotes are much more complex than the chromosomes of bacteria and archaea. Eukaryotic chromosomes consist of a highly ordered complex of DNA and proteins, with special regions—centromeres and telomeres—that are of particular importance for chromosome function. Since gene expression is related to how the DNA is organized in chromosomes (discussed in Chapter 16), an understanding of chromosome structure will help us unravel the details of how gene expression is regulated.

CHROMOSOMES OF BACTERIA, ARCHAEA, AND VIRUSES

In the following section we discuss the chromosomes of bacteria, archaea, and viruses.

Chromosomes of Bacteria and Archaea

Many bacteria, such as *E. coli*, have a single, double-stranded DNA chromosome. Other bacteria have two or a few chromosomes that may be circular or linear. For example, *Borrelia burgdorferi*, the causative agent of Lyme disease, has a 0.91-Mb linear chromosome and at least 17 small linear and circular chromosomes with a combined size of 0.53 Mb, and *Agrobacterium tumefaciens*, the causative agent of crown gall disease in some plants, has a 3.0-Mb circular chromosome and a 2.1-Mb linear chromosome.

Chromosome organization also varies among the archaea, although no linear chromosomes have yet been found. For example, *Methanococcus jannaschii* has 1.66-Mb, 58-kb, and 16-kb circular chromosomes, and *Archaeoglobus fulgidus* has a single 2.2-Mb circular chromosome.

If an *E coli* cell is lysed (broken open) gently, its DNA is released in a highly folded state (Figure 10.1). The double-stranded DNA is present as a single, 4.6-Mb circular chromosome, approximately 1,100 μm long. The chromosome is in a central region of the cell

~ FIGURE 10.1

Chromosome released from a lysed *E. coli* cell.

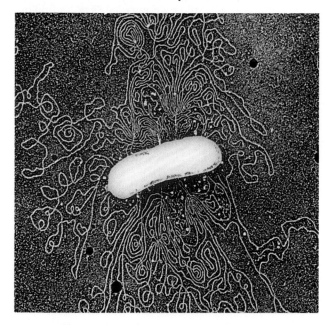

~ FIGURE 10.2

Electron micrographs of a circular DNA molecule showing relaxed and supercoiled states. (a) Relaxed (nonsupercoiled) DNA; (b) Supercoiled DNA. Both molecules are shown at the same magnification.

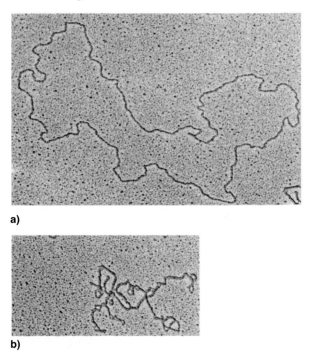

a)

b)

called the **nucleoid**. As in all bacteria and archaea, there is no membrane between the nucleoid region and the rest of the cell.

The length of DNA in the *E. coli* chromosome is approximately 1,000 times the length of the *E. coli* cell. The DNA fits into the nucleoid area because it is *supercoiled*; that is, the double helix has been twisted in space about its own axis. Figure 10.2 pictures supercoiled circular DNA to show how much more compact a supercoiled molecule is compared to a non-supercoiled (relaxed) molecule. There are two types of supercoiling, *negative supercoiling* and *positive supercoiling*. To visualize supercoiling, think of the DNA double helix as a spiral staircase that turns in a clockwise direction. If you untwist the spiral staircase by one complete turn, you have *the same number of stairs to climb, but you have one less 360° turn to make*; this is a negative supercoil. If, instead, you twist the spiral staircase by one more complete turn, you have *the same number of stairs to climb, but now there is one more 360° turn to make*; this is a positive supercoil. Either type of supercoiling causes the DNA to become more compact. The amount and type of DNA supercoiling is brought about by enzymes called **topoisomerases**. Topoisomerases are found in all organisms.

The *E. coli* chromosome is organized into loops. These loops are formed by the association of certain DNA-binding proteins with the chromosome. Two of the proteins—HU and H—resemble two of the histone structural proteins found associated with eukaryotic DNA. Figure 10.3 shows a model for the

E. coli chromosome with the DNA organized into about 100 independent *domains*. Each domain consists of a loop of about 40,000 base pairs (40 kb) of supercoiled DNA. The ends of each domain are held, presumably by proteins (by unknown means), such that the supercoiled DNA state in one domain is not affected by events that influence supercoiling of DNA in the other domains.

Viral Chromosomes

Viruses infect all types of living organisms. In this section we focus on the viruses that infect bacteria, the bacteriophages.

T2, T4, and T6, the T-even bacteriophages, have similar structures (see Figure 6.12a). The characteristics of these phages, along with features of other viruses, are summarized in Table 10.1. The T-even phages contain a single, linear, double-stranded DNA chromosome surrounded by a protein coat. These phages are virulent, meaning that infection by them almost always results in the lysis of the infected bacterial cell.

The virulent phage ΦX174 ("fi-X-one-seventy-four") is a DNA phage that infects *E. coli*. The ΦX174 phage is an icosahedron consisting of protein subunits surrounding the genetic material (Figure 10.4). The

~ FIGURE 10.3

Model for the structure of a bacterial chromosome. The chromosome is organized into looped domains, the bases of which are anchored in an unknown way.

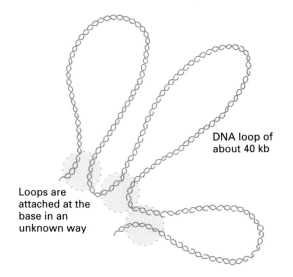

DNA loop of about 40 kb

Loops are attached at the base in an unknown way

DNA of ΦXI74 has a base ratio of 25A:33T:24G:18C, a composition that does not fit the A = T and G = C complementary base-pairing rules. This is because the 5,386-nucleotide ΦX174 chromosome is single-stranded. In addition, if the chromosome is treated with an exonuclease (an enzyme that digests DNA from free ends), the DNA is not affected. The reason is that the ΦXI74 chromosome is circular.

The last bacteriophage we will consider is λ (lambda). Bacteriophage λ is a temperate phage, meaning that when it infects *E. coli*, it has a choice of entering a pathway that results in the lysis of the cell or of following a pathway to a quiescent state that does not result in phage reproduction (see Chapter 6). The phage λ chromosome is linear, double-stranded DNA. The two ends of the DNA molecule have 12-nucleotide single-stranded segments that are complementary. Regardless of which pathway λ follows when it infects a cell, the first step after infection is the conversion of the linear molecule into a circular molecule. This circularization is brought about by pairing of the naturally complementary ends ("sticky ends").

KEYNOTE

In prokaryotes in which the genetic material is DNA, the chromosome consists of DNA with a number of associated proteins. Many bacteria such as *E. coli* have a single, circular, double-stranded DNA chromosome that is compacted by supercoiling. Other bacteria have two or a few chromosomes, which may be circular or linear. In the *E. coli* chromosome, about 100 independent looped domains of supercoiled DNA have been identified. In bacteriophages the genetic material may be double-stranded or single-stranded DNA or RNA. Phage chromosomes can be linear or circular.

~ TABLE 10.1

Characteristics of the Chromosomes of Selected Viruses

VIRUS	HOST	STRUCTURE AND TYPE OF GENETIC MATERIAL	CHROMOSOME DESCRIPTION	LENGTH OF CHROMOSOME (μm)	%GC
T-even phages	*E. coli*	Double-stranded DNA	Linear; circularly permuted and terminally redundant	60	35
λ	*E. coli*	Double-stranded DNA	Linear; single-stranded "sticky ends"	16	49
ΦX174	*E. coli*	Single-stranded DNA	Circular	1.8	A25 G24 T33 C18
Qβ	*E. coli*	Single-stranded RNA	Linear	1.4	A22 G24 T29 C25
Murine leukemia virus (Moloney)	Mouse	Single-stranded RNA	Linear	2.8	A25 G25 U23 C27
Tobacco mosaic virus (TMV)	Tobacco	Single-stranded RNA	Linear	2.2	A30 G25 U26 C19

~ FIGURE 10.4

Bacteriophage ΦX174. (a) Electron micrograph of ΦX174 phage particles; (b) Model of the ΦX174 phage particle.

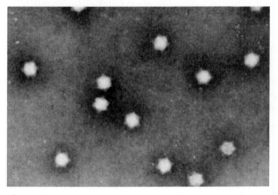

a)

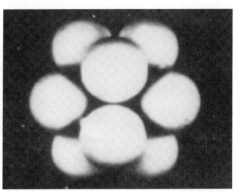

b)

EUKARYOTIC CHROMOSOMES

Eukaryotes have genetic material in the nucleus as well as in mitochondria and (if a plant) in chloroplasts. In the nucleus, the genomic DNA is distributed among a number of linear chromosomes. In humans, the diploid number of chromosomes is 46, with one haploid (N) set of 23 chromosomes deriving from the egg and another haploid set of 23 chromosomes deriving from the sperm. In this section we learn about the structural organization of eukaryotic chromosomes.

~ FIGURE 10.5

Human male metaphase chromosomes arranged as a karyotype. Note that groups of chromosomes with similar morphologies are arranged under letter designations (A through G). This arrangement is based on the size of the condensed chromosomes. In the male, all chromosomes except the X and Y sex chromosomes are present in pairs.

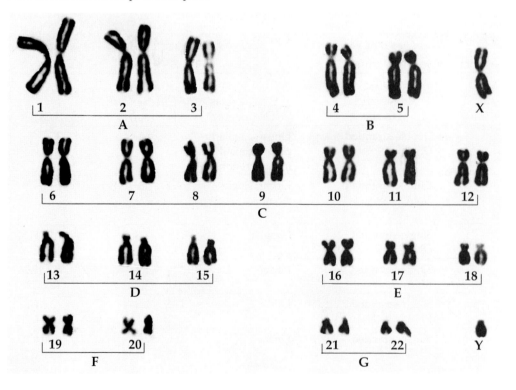

The Karyotype

A complete set of all the metaphase chromosomes in a cell is called its **karyotype** ("carry-o-type"; literally, "nucleus type"). Metaphase chromosomes are used because they are the most compact form of eukaryotic chromosomes and therefore are easy to see under the microscope after staining. The karyotype is species-specific, so a wide range of number, size, and shape of metaphase chromosomes is seen among eukaryotic organisms. Even closely related organisms may have quite different karyotypes. Table 10.2 lists the chromosome numbers of selected eukaryotes. The diploid number of chromosomes in humans is 46, in cats it is 38, and in garden peas it is 14. In the fruit fly, *Drosophila melanogaster*, it is 8. Examples of chromosome numbers for haploid organisms are 16 in the green alga *Chlamydomonas reinhardtii* and 17 in baker's yeast, *Saccharomyces cerevisiae*.

Figure 10.5 shows the karyotype for the cell of a normal human male. It is customary, particularly with human chromosomes, to arrange chromosomes in order according to size and position of the centromere. This karyotype shows 46 chromosomes: two pairs of each of the 22 autosomes and one of each of the X and Y sex chromosomes. In a human karyotype, the chromosomes are numbered for easy identification. Conventionally, the largest pair of homologous chromosomes is designated 1, the next largest 2, and so on. Although chromosome 21 is actually smaller than chromosome 22, it is called 21 for historical reasons. As shown in Figure 10.5, the human X chromosome is a large metacentric chromosome, and the Y chromosome is a much smaller chromosome.

~ TABLE 10.2

Chromosome Numbers in Eukaryotic Organisms

SCIENTIFIC NAME	COMMON NAME	NUMBER OF CHROMOSOMES[a]
DIPLOID ORGANISMS		
Animals		
Homo sapiens	Human	46
Pan troglodytes	Chimpanzee	48
Equus caballus	Horse	64
Canis familiaris	Dog	78
Felis domesticus	Cat	38
Mus musculus	House mouse	40
Xenopus laevis	Toad	36
Caenorhabditis elegans	Nematode	11 ♂/12 hermaphrodite
Drosophila melanogaster	Fruit fly	8
Plants		
Pisum sativum	Garden pea	14
Solanum tuberosum	Potato	48
Nicotiana tabacum	Tobacco	48
Triticum aestivum	Bread wheat	42
Zea mays	Maize	20
HAPLOID ORGANISMS		
Chlamydomonas reinhardtii	Unicellular alga	16
Neurospora crassa	Orange bread mold	7
Aspergillus nidulans	Green bread mold	8
Saccharomyces cerevisiae	Brewer's yeast	17

[a]For the diploid organisms, the diploid number of chromosomes is given.

An analysis of karyotypes permits geneticists to identify certain chromosome mutations that correlate with congenital abnormalities or dysfunctions. For example, on rare occasions chromosome nondisjunction (Chapter 3) or chromosomal mutations (Chapter 7) may occur. As you can see from Figure 10.5, it is hard to distinguish the different chromosomes unambiguously when they are stained evenly. Fortunately, a number of procedures stain certain regions or *bands* of the chromosomes more intensely than other regions. Banding patterns are specific for each chromosome, enabling us to distinguish each chromosome in the karyotype clearly. One of these staining techniques is called **G banding**. In this procedure, chromosomes are treated with mild heat or proteolytic enzymes (enzymes that digest proteins) to digest the chromosomal proteins partially, and then stained with Giemsa stain to produce dark bands called G bands (Figure 10.6). In humans, approx-

imately 300 G bands can be distinguished in metaphase chromosomes, while approximately 2,000 G bands can be distinguished in chromosomes from prophase. Conventionally, drawings of human chromosomes show the G banding pattern.

KEYNOTE

In eukaryotes the complete set of metaphase chromosomes in a cell is called its karyotype. The karyotype is species-specific.

The Molecular Structure of the Eukaryotic Chromosome

The total amount of DNA in the haploid genome is characteristic of each living species, prokaryote or eukaryote, and is known as its **C value**. Figure 10.7 (page 224) presents a summary of the range of C values found in a variety of animals. The "tree" on the left of the figure represents the evolutionary relationships between the animals. Within any given taxon, there is a general trend toward increasing amounts of DNA as organisms are further removed from the original ancestor. The C value data show that the amount of DNA found among organisms varies considerably. You can see that there may or may not be significant variation in DNA amount among related organisms. For example, mammals, birds, and reptiles each show little variation, while amphibians, insects, and plants each vary over a wide range, often tenfold or so. There is not a direct relationship between the C value and the structural or organizational complexity of the organism. At least one reason for this is variation in the amount of repetitive-sequence DNA in the genome (see pp. 000-000).

C value data indicate that a human cell, for example, has more than a thousand times as much DNA as does *E. coli*. Without the compacting of the 5.5×10^9 base pairs of DNA in the diploid nucleus, the DNA of the chromosomes of a single human cell would be over 6 feet long (about 200 cm) if the molecules were placed end to end!

The compacting of eukaryotic DNA is brought about by the association of the DNA with specific proteins. In fact, each eukaryotic chromosome consists of one linear, double-stranded DNA molecule running throughout its length and contains about twice as much protein by weight as DNA. **Chromatin**

~ FIGURE 10.6

G banding in a karyotype of human male metaphase chromosomes.

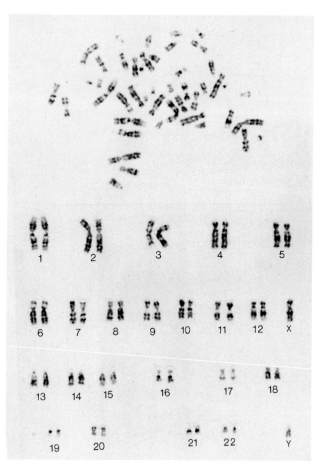

~ FIGURE 10.7

The amount of DNA per haploid chromosome set in a variety of animals. Variation within certain phyla is quite wide (as in insects and amphibians). The "tree" on the left of the figure indicates the evolutionary relationship of the animals.

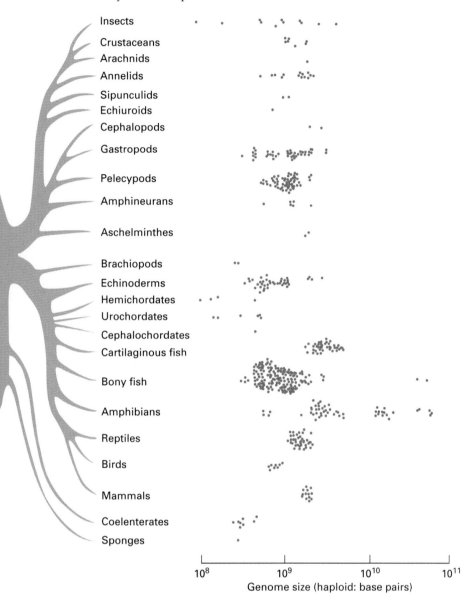

is the term used for the complex of DNA, chromosomal proteins, and RNA in the chromosome. The fundamental structure of chromatin is essentially identical in all eukaryotes.

There are two types of chromatin: **euchromatin** and **heterochromatin**. Euchromatin is chromatin that stains lightly. It is uncoiled during interphase but becomes condensed during mitosis. Most of the genome consists of euchromatin. Hetero-chromatin is chromatin that stains darkly. This occurs because the chromatin region involved is more highly condensed than is the case with euchromatin. Functionally, euchromatin is genetically active (that is, it contains genes that are being expressed), whereas heterochromatin is genetically inactive, either because it contains no genes or because the genes it does contain cannot be expressed.

Heterochromatin is found in all eukaryotic species near centromeres, at *telomeres* (the ends of the chromosomes), and elsewhere in a species-specific manner. There are two classes of heterochromatin. **Constitutive heterochromatin** is always genetically inactive. Chromatin at centromeres and telomeres is constitutive heterochromatin. **Facultative heterochromatin** is chromatin that can become condensed to the heterochromatin state. It may contain genes that are inactivated when the chromatin becomes condensed. Barr bodies (inactivated X chromosomes in female mammals—see Chapter 3) are examples of facultative heterochromatin.

Two major types of proteins are associated with DNA in chromatin: **histones** and **nonhistones**. Histones and nonhistones play an important role in determining the physical structure of the chromosome. The DNA is wrapped around a core of histone molecules, and the nonhistones are somehow associated with that complex. Some of the nonhistones have a structural role in the chromosomes. If the histones are removed from the chromosome, for example, the DNA unravels and is displaced from the complex, but a skeleton of nonhistone proteins in the shape of the chromosome remains.

THE HISTONES.   The histones are the most abundant proteins associated with chromosomes. They are relatively small basic proteins; that is, the histones have a net positive charge, thus facilitating their binding to the negatively charged DNA.

Five main types of histones are associated with eukaryotic DNA: H1, H2A, H2B, H3, and H4. Comparison of amino acid sequences of H2A, H2B, H3, and H4 histones in various organisms has indicated, in some cases, that the sequences are highly conserved (very similar) among distantly related species. For example, there are only two amino acid differences in the H4 histones of cows and peas, and there is only one amino acid difference between sea urchin and calf thymus histone H3. Their evolutionary sequence conservation is a strong indicator that histones perform the same basic role in organizing the DNA in the chromosomes of all eukaryotes.

NONHISTONE CHROMOSOMAL PROTEINS.   Nonhistones are all the proteins associated with DNA apart from the histones. There are many different types of nonhistone chromosomal proteins. Some nonhistones play a structural role, and others are only transiently associated with chromatin, such as the enzymes and proteins involved with replication, transcription, and the regulation of gene expression.

The nonhistones are very different from the histones. Nonhistones are usually acidic proteins—that is, proteins with a net negative charge—and are likely to bind to positively charged histones in the chromatin. In contrast to the histones, the nonhistone proteins differ markedly in number and type from cell type to cell type within an organism, at different times in the same cell type, and from organism to organism.

KEYNOTE

The chromosomes of eukaryotes are complexes of DNA, histone proteins, and nonhistone chromosomal proteins. Each chromosome consists of one linear, unbroken, double-stranded DNA molecule running throughout the length of the chromosome. The five main types of histones—H1, H2A, H2B, H3, and H4—are constant from cell to cell within an organism. Nonhistones, of which there are a large number, vary significantly among cell types, both within and among organisms as well as at different times in the same cell type.

NUCLEOSOMES.   Several levels of packing enable chromosomes several millimeters or even centimeters long to fit into a nucleus that is a few micrometers in diameter. The simplest level of packing involves the winding of DNA around histones in a structure called a **nucleosome**, and the most highly packed state involves higher-order coiling and looping of the DNA-histone complexes, as seen in metaphase chromosomes. In general, the expression of a gene is facilitated by a change in the chromosome from a high level of coiling to a lower level of coiling. That is, the more condensed a part of a chromosome is, the less likely it is that the genes in that region will be active.

In one recent model for the nucleosome, there are two molecules of each of the four core histones, H2A, H2B, H3, and H4 (called the *histone octamer*), a single molecule of the linker histone H1, and about 180 bp of DNA. In the test tube the core histones self-assemble into a histone octamer that is a flat-ended, cylindrical particle about 11 nm in diameter and 5.7 nm thick. The DNA winds around the outside of the nucleosome core about one and three-quarters times, and this condenses the DNA by a factor of about seven. The exact path of the DNA around the histone octamer, the position of the H1 linker histone in the nucleosome, and the position of the DNA that con-

~ FIGURE 10.8

A possible nucleosome structure.

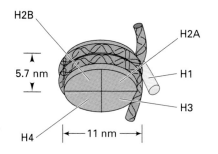

~ FIGURE 10.10

Electron micrograph of a 30-nm chromatin fiber.

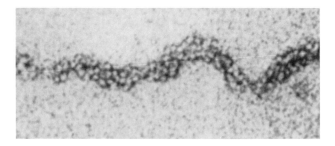

nects adjacent nucleosomes—called *linker DNA*—remain to be elucidated. A diagram of a possible nucleosome structure is shown in Figure 10.8.

With an electron microscope we see the DNA-protein complex as 10-nm chromatin fibers (10-nm **nucleofilaments**), that is, fibers with a diameter of about 10 nm. In its most unraveled state, the chromatin fiber of a nucleofilament has the appearance of "beads on a string," where the beads are the nucleosomes (Figure 10.9) and the thinner "thread" connecting the "beads" is naked linker DNA.

HIGHER-ORDER STRUCTURES IN CHROMATIN.

In the cell, chromatin is more highly condensed than the "beads on a string" nucleofilament. The details of these higher levels of folding and of the changes in chromatin folding that take place as a cell goes from interphase to mitosis (or meiosis) and back to interphase are incompletely understood. There is general agreement that the next level of packing above the nucleosome is the *30-nm chromatin fiber* (Figure 10.10). One model for the 30-nm fiber is shown in Figure 10.11. We know that histone H1 plays an important role in the formation of the 30-nm fiber, since chromatin from which histone H1 has been removed can form 10-nm fibers, but not 30-nm fibers.

Coiling of the 10-nm nucleofilament to produce the 30-nm fiber condenses the DNA another sixfold. However, this does not provide sufficient packing of the chromosomes to explain the degree to which chromosomes are condensed in the cell nucleus. That is, an average human chromosome would extend approximately 1 mm as a 30-nm fiber, and, as such, would be 200 times longer than the diameter of the nucleus (about 5 μm). However, the next levels of packing beyond the 30-nm chromatin fiber are not clearly understood. An interphase chromosome may have a diameter of 300 nm, while a metaphase chromosome may have a diameter of 700 nm.

A major step in modeling the structure of metaphase chromosomes came when researchers showed that metaphase chromosomes from which the histones have been removed still retain a residual folded structure with *looped domains* of 30-nm fiber DNA extending from a condensed protein lattice (Figure 10.12). The central structure, called the *chromosome scaffold*, actually has the shape of the metaphase chromosome (Figure 10.13). The amount of DNA in each loop ranges from tens to hundreds of kilobase pairs, so an average human chromosome has approximately 2,000 looped domains. The looped domains extend at an angle from the main chromosome

~ FIGURE 10.9

Electron micrograph of unraveled chromatin showing the nucleosomes in a "beads on a string" morphology.

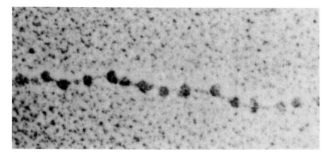

~ FIGURE 10.11

A model for the packaging of nucleosomes into the 30-nm chromatin fiber.

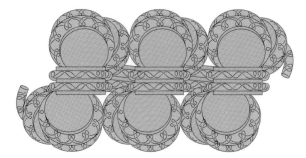

axis and are anchored to a filamentous structural framework of protein inside the nuclear envelope called the *nuclear matrix*.

The DNA sequences attached to the proteins in the matrix are called *MARs* (*m*atrix *a*ttachment *r*egions). Most MARs border transcriptionally active genes and actively replicating regions. The organization of chromatin into loops appears to be important, then, both for condensing the chromatin fiber in the chromosome and for the regulation of gene expression.

Figure 10.14 shows the different orders of DNA packing that could give rise to the highly condensed metaphase chromosome. Interphase chromosomes would be packed less tightly than this.

KEYNOTE

The large amount of DNA present in the eukaryotic chromosome is compacted by its association with histones in nucleosomes and by higher levels of folding of the nucleosomes into chromatin fibers. Each chromosome contains a large number of looped domains of 30-nm chromatin fibers attached to a protein scaffold. The functional state of the chromosome is related to the extent of coiling: the more condensed a part of a chromosome is, the less likely it is that the genes in that region will be active.

~ **FIGURE 10.12**

Electron micrograph of a metaphase chromosome depleted of histones. The chromosome maintains its general shape by a nonhistone protein scaffolding from which loops of DNA protrude.

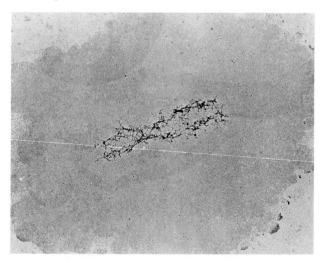

~ **FIGURE 10.13**

Schematic model for the organization of 30-nm chromatin fiber into looped domains that are anchored to a nonhistone protein chromosome scaffold.

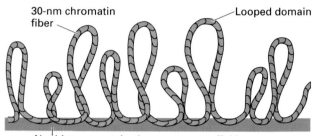

~ **FIGURE 10.14**

Schematic drawing of the many different orders of chromatin packing that are thought to give rise to the highly condensed metaphase chromosome.

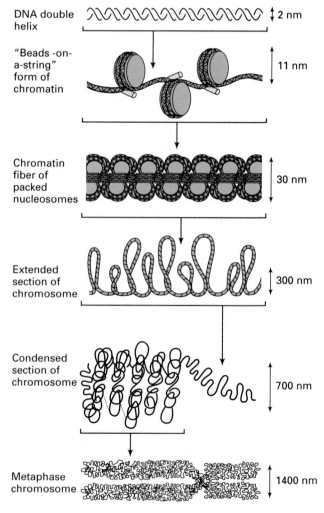

Centromeres and Telomeres

All eukaryotic chromosomes have two areas of special function: the centromere and the telomere, the end of the chromosome. We learned earlier (Chapter 1) that the behavior of chromosomes in mitosis and meiosis depends on the kinetochores that form on the centromeres. The telomeres also have a special role, in this case in the replication of the DNA in the chromosome.

CENTROMERES. Centromeres are the DNA sequences found near the attachment point of mitotic or meiotic spindle fibers. Cytologically, the centromere is seen as a constricted region at one point along the chromosome. The region contains the kinetochore to which the kinetochore microtubules attach during cell division. In most eukaryotes, several microtubules attach to each centromere, whereas in the yeast *Saccharomyces cerevisiae*, the organism in which centromeres have been best characterized, only one microtubule attaches to each centromere.

The DNA sequences (called *CEN* sequences, after the *cen*tromere) of the yeast centromeres have been determined. Though each centromere has the same function, the *CEN* regions are very similar, but not identical, to one another in nucleotide sequence and organization. Centromere sequences have been determined for a number of other organisms; they are different both from those of yeast and from each other. Thus, while centromeres carry out the same function in all eukaryotes, there is no common sequence that is responsible for that function.

TELOMERES. A telomere is required for replication and stability of that chromosome. Telomeres are characteristically heterochromatic. In most organisms that have been examined, the telomeres are positioned just inside the nuclear envelope and are often found associated with each other as well as with the nuclear envelope.

All telomeres in a given species share a common sequence. Most telomeric sequences may be divided into two types:

1. **Simple telomeric sequences** are at the extreme ends of the chromosomal DNA molecules. These sequences are species-specific and consist of simple, tandemly repeated DNA sequences. In the ciliate *Tetrahymena*, for example, reading the sequence toward the end of one DNA strand, the repeated sequence is 5'-TTGGGG-3' (Figure 10.15), and in humans the repeated sequence is 5'-TTAGGG-3'. Note in Figure 10.15 that the

~ **FIGURE 10.15**

Telomere sequence at the ends of *Tetrahymena* chromosomes.

DNA is not double stranded all the way out to the end. We will see in the next chapter the role these telomere sequences have in the replication of eukaryotic chromosomes.

2. **Telomere-associated sequences** are regions near but not at the ends of chromosomes. These sequences often contain repeated but still complex DNA sequences extending many thousands of base pairs in from the chromosome end.

While the telomeres of most eukaryotes contain short, simple, repeated sequences, the telomeres of *Drosophila* are quite different. *Drosophila* telomeres consist of *transposons*, DNA sequences that can move to other locations in the genome.

KEYNOTE

The centromere region of each eukaryotic chromosome is responsible for the accurate segregation of the replicated chromosome to the daughter cells during both mitosis and meiosis. The DNA sequences of yeast centromeres (*CEN* sequences) are very similar, but not identical.

The ends of chromosomes, the telomeres, often are associated with the nuclear envelope. Telomeres in a species share a common sequence. Characteristically, sequences at or very close to the extreme ends of the chromosomal DNA consist of simple, relatively short, tandemly repeated sequences. Repeated, often complex, DNA sequences, called telomere-associated sequences, are found farther in from the chromosome ends.

UNIQUE-SEQUENCE AND REPETITIVE-SEQUENCE DNA IN EUKARYOTIC CHROMOSOMES

Now that we know about the basic structure of DNA and its organization in chromosomes, next we can think about the distribution of certain sequences in the genomes of prokaryotes and eukaryotes. From molecular analyses, geneticists have found that some sequences are present only once in the genome, while other sequences are repeated. To date we have relatively sketchy information about the distribution of the various classes of sequences in the genome. As the complete DNA sequences of more and more eukaryotic genomes are determined (that of yeast was reported in 1996), we will develop a precise understanding of the molecular organization of genomes.

For convenience, three different types of DNA sequences have been named: **unique sequences** (or single-copy sequences, present in one to a few copies per genome), **moderately repetitive sequences** (present in a few to as many as 10^3 to 10^5 times in the genome), and **highly repetitive sequences** (present as many as 10^5 to 10^7 times), although there are really no discrete boundaries between them. In prokaryotes, with the exception of the ribosomal RNA genes, transfer RNA genes, and a few other sequences, all of the genome is present as unique-sequence DNA. Eukaryotic genomes, on the other hand, consist of both unique-sequence and repetitive-sequence DNA, with the latter typically being quite complex in number of types, number of copies, and distribution. Using the three sequence types just defined, the human genome can be described as consisting of about 64 percent unique-sequence DNA, 25 percent moderately repetitive DNA, and 10 percent highly repetitive DNA. By contrast, the green frog has about 22 percent unique-sequence DNA, 67 percent moderately repetitive DNA, and 9 percent highly repetitive DNA. However, it is more instructive to discuss the classes of DNA with respect to their arrangement in the chromosome and to their functions, rather than with respect to their relative abundance in the genome. This we will do in the following sections.

Unique-Sequence DNA

Unique sequences (sometimes called single-copy sequences) are defined as sequences present as single copies in the genome. (Thus, there are two copies per diploid cell.) Actually, in current usage the term usually applies to sequences that have one to a few copies per genome. Most of the genes that we know about—those that code for proteins in the cell—are in the unique-sequence class of DNA. But not all unique-sequence material contains protein-coding sequences.

Repetitive-Sequence DNA

In this section we divide the subject into two and discuss tandemly repeated DNA sequences and dispersed repeated DNA sequences.

TANDEMLY REPEATED DNA.　*Tandemly repeated DNA* refers to reiterated sequences that are tandemly arranged in the genome. Tandemly repeated DNA is quite common in eukaryotic genomes. The genes for ribosomal RNA (rRNA; see Chapter 12) and transfer RNA (tRNA; see Chapter 12), for example, are tandemly repeated in one or more clusters in most eukaryotes. For instance, there are about 160–200 copies of the major class of rRNA genes in humans, 260 copies in the sea urchin, and 3,900 copies in the garden pea. The histone genes are repeated in eukaryotes, and except for birds and mammals, where they are dispersed in the genome, the repeated copies tend to be linked. As for copy number, yeast has two copies of each of the histone genes, *Drosophila* has 100 copies, and humans have 5 to 20 copies.

Examples are also known of tandemly repeated genes that are related, but not identical. In vertebrates, hemoglobin consists of two copies each of an α-globin and a β-globin polypeptide. These polypeptides are encoded by multiple genes—a *multigene family*—with each gene encoding a related but different polypeptide. Thus, there is an α-globin family of genes and a β-globin family of genes. In humans, the three genes of the α-globin family are located in a 25-kb region of DNA on chromosome 16, and the five genes of the β-globin family are located in a 65-kb region of DNA on chromosome 11.

Much of the tandemly repeated DNA is not associated with genes. The greatest amount is associated with centromeres and telomeres. At each centromere hundreds to thousands of copies of simple, relatively short, tandemly repeated sequences (highly repetitive sequences) are found. Indeed, a significant proportion of the eukaryotic genome may consist of the highly repeated sequences found at centromeres: 8 percent in the mouse, about 50 percent in the kangaroo rat, and about 5 percent in humans.

DISPERSED REPEATED SEQUENCES.　*Dispersed repeated sequences* refers to reiterated sequences that are dispersed in the genome rather than being tandemly arranged. Examples of both gene sequences and nongene sequences are found in this class. For example, members of some multigene families may be dispersed through the genome. As previously mentioned, the repeated histone genes of mammals and birds are dispersed in the genome. Typically, dispersed repeated genes are present in relatively few copies, often fewer than 50. Transposons—DNA elements that can move to other locations in the genome—are examples of another type of dispersed repeated sequence. Also, many eukaryotic DNAs contain dispersed repeated sequences that are most likely not genes; these sequences may be reiterated as many as 1 million times. We focus our attention on this latter class in the following discussion.

The distribution of dispersed repeated sequences has been studied in a variety of organisms. The picture that has emerged is one of families of repeated sequences interspersed through the genome with unique-sequence DNA. Often small numbers of families have very high copy numbers and make up most of the dispersed repeated sequences in the genome. Two general interspersion patterns are encountered. In one, the families have sequences 100 to 500 bp long—these sequences are called **SINEs** for **short *in*terspersed repeated sequences**. In the other, the families have sequences about 5,000 bp or more long—these sequences are called **LINEs** for **long *in*terspersed repeated sequences**. All eukaryotic organisms have SINEs and LINEs, although the relative proportions vary considerably. *Drosophila* and birds, for example, have mostly LINEs, while humans and frogs have mostly SINEs.

The SINEs pattern of repeated sequences is found in a diverse array of eukaryotic species, including mammals, amphibians, and sea urchins. Each species with SINEs has its own characteristic array of SINE families. A well-studied SINE family is the Alu family of certain primates. This family is named for the cleavage site for the restriction enzyme *Alu*I ("Al-you-one") typically found in the repeated sequence. In humans, the Alu family is the most abundant SINE family in the genome, consisting of 200–300-bp sequences repeated as many as 9×10^5 times, and making up about 9 percent of the total haploid DNA. One Alu repeat is located about every 5,000 bp in the genome!

Mammalian genomes have many copies of a particular LINE family of repeated sequence, the LINE-1 family, in addition to many SINE families. Other LINE families may be present also, but they are much less abundant than LINE-1. LINE-1 family members are up to 7,000 bp long, although a wide range of sizes is seen. Some of the longer LINE-1 family sequences are transposons.

While we have a lot of information about the sequence organization and distribution of SINEs and LINEs, we have very little knowledge about the function(s) of those sequences. One hypothesis is that most of these sequences have no function at all. Another hypothesis is that some of the repeated sequences, or their transcripts, or both, are somehow involved in mechanisms for regulating gene expression.

KEYNOTE

Prokaryotic genomes consist mostly of unique-sequence DNA, with only a few sequences and genes repeated; eukaryotes have both unique and repetitive sequences in the genome. In eukaryotes, the spectrum of complexity of the repetitive DNA sequences is extensive. The highly repetitive sequences tend to be localized to heterochromatic regions, for example, around centromeres, whereas the unique and moderately repetitive sequences tend to be interspersed.

SUMMARY

In this chapter we have learned about aspects of the molecular organization of prokaryotic, viral, and eukaryotic chromosomes. Such structural information is important to know in order to understand gene regulation mechanisms more completely.

The chromosomes of prokaryotic organisms consist of circular, double-stranded DNA molecules that are complexed with a number of proteins. The chromosome is condensed within the cell by supercoiling of the DNA helix and formation of looped domains of supercoiled DNA. Examples of chromosome organization in bacteriophages were discussed, including the linear double-stranded DNA chromosomes in T-even phages, sticky ends in λ double-stranded DNA and their involvement in the phage life cycle, and circular, single-stranded DNA in ΦX174.

A distinguishing feature of eukaryotic chromosomes is that DNA is distributed among a number of chromosomes. The set of metaphase chromosomes in a cell is called its karyotype. We saw in an earlier chapter how karyotype analysis is useful for diagnosing certain human chromosome aberrations. This analysis is aided by our ability to stain human chromosomes so that each chromosome is distinguishable by its banding pattern.

The DNA in the genome specifies an organism's structure, function, and reproduction. The amount of DNA in a genome (made up of a prokaryotic chromosome or of the haploid set of chromosomes in eukaryotes) is called the C value. There is no direct relationship between the C value and the structural or organizational complexity of an organism.

The organization of DNA in eukaryotic chromosomes was described in this chapter. The chromosomes of eukaryotes are complexes of DNA with histone and nonhistone chromosomal proteins. Each chromosome consists of one linear, double-stranded

DNA molecule that runs through its length. As a class, the histones are constant from cell to cell within an organism and have been evolutionarily conserved. Nonhistone chromosomal proteins, on the other hand, vary significantly among cell types and among organisms. The DNA in a eukaryotic chromosome is highly condensed by its association with histones to form nucleosomes and by the several higher levels of folding of nucleosomes into chromatin fibers. The most highly condensed structure is seen in metaphase chromosomes, while the least condensed structure is seen in interphase chromosomes. The factors controlling the transitions between different levels of chromosome folding are not known. Like prokaryotic chromosomes, eukaryotic chromosomes are organized into a large number of looped domains. These loops are attached to a protein scaffold.

Two special eukaryotic chromosome structures are centromeres and telomeres. Significant progress has been made in defining the sequences of centromeres and telomeres; for example, the DNA sequences of centromeres are relatively complex and are species-specific, and the DNA sequences of telomeres are simple, relatively short, tandemly repeated sequences that are also species-specific.

Molecular analysis has provided information about the distribution of DNA sequences in genomes. Such analysis has revealed that prokaryotic and viral genomes consist mostly of unique-sequence DNA, with only a few repeated sequences. In contrast, the genomes of eukaryotes contain both unique-sequence DNA and repeated-sequence DNA, and there is a wide spectrum of complexity of the repeated DNA sequences. Many genes are found in unique-sequence DNA, but not all unique-sequence DNA contains genes.

Repeated DNA sequences may be tandemly arranged or interspersed with unique-sequence DNA in the eukaryotic genome. While some tandemly repeated DNA is composed of gene families, the greatest amounts of tandemly repeated DNA are not associated with genes but with centromeres and telomeres. At each centromere, for example, hundreds to thousands of copies of simple, relatively short, tandemly repeated sequences may be found. Dispersed repeated sequences include both gene sequences and nongene sequences. Dispersed repeated gene sequences typically are present in relatively low copy numbers (fewer than 50), while dispersed nongene sequences may be present in hundreds of thousands of copies.

Eukaryotic species have characteristic families of these repeated sequences interspersed through the genome with unique-sequence DNA. The two general interspersion families are SINEs (short *inter*spersed repeats), in which the repeated sequences are 100–500 bp long, and LINEs (*long inter*spersed repeats), in which the repeated sequences are about 5,000 bp or more long. SINEs and LINEs are found in all eukaryotes, although the relative proportions vary considerably. While much knowledge has been accumulated about SINEs and LINEs in a number of eukaryotic genomes, little is known about the functions of these sequences.

ANALYTICAL APPROACHES FOR SOLVING GENETICS PROBLEMS

Q10.1 An organism has a haploid genome of 10^{10} nucleotide pairs, of which 70 percent is unique-sequence DNA with a copy number of one, 20 percent is moderately repetitive DNA with an average copy number of 1,000, and 10 percent is highly repetitive DNA with an average copy number of 10^6. Assuming that an average DNA sequence is 10^3 nucleotide pairs, how many different sequences are in each of the three DNA classes?

A10.1

a. *Unique-sequence DNA:* the total DNA in this class is 70 percent of 10^{10} nucleotide pairs $= 0.7 \times 10^{10} = 7 \times 10^9$. Only one of each sequence exists, so given an average DNA sequence length of 10^3 bp, the number of different sequences $= (7 \times 10^9)/10^3 = 7 \times 10^6$.

b. *Moderately repetitive DNA:* DNA in this class is 20 percent of 10^{10} bp $= 0.2 \times 10^{10} = 2 \times 10^9$. Given an average DNA sequence length of 10^3 bp, there are $(2 \times 10^9)/10^3 = 2 \times 10^6$ sequences in this class. Since the average copy number is 1,000, the number of *different* sequences $= (2 \times 10^6)/10^3 = 2 \times 10^3 = 2,000$.

c. *Highly repetitive DNA:* DNA in this class is 10 percent of 10^{10} bp $= 0.1 \times 10^{10} = 1 \times 10^9$. Given an average DNA sequence length of 10^3 bp, there are $(1 \times 10^9)/10^3 = 1 \times 10^6$ sequences in this class. Since the average copy number is 10^6, the number of different sequences $= (1 \times 10^6)/10^6 = 1$.

QUESTIONS AND PROBLEMS

10.1   What are topoisomerases?

*10.2   In typical human fibroblasts in culture, the G_1 period of the cell cycle lasts about 10 h, S lasts about 9 h, G_2 takes 4 h, and M takes 1 h. Imagine you were to do an experiment in which you added radioactive (^{3}H) thymidine to the medium and left it there for 5 min, and then washed it out and put in ordinary medium.
a.   What percentage of cells would you expect to become labeled by incorporating the ^{3}H-thymidine into their DNA?
b.   How long would you have to wait after removing the ^{3}H medium before you would see labeled metaphase chromosomes?
c.   Would one or both chromatids be labeled?
d.   How long would you have to wait if you wanted to see metaphase chromosomes containing ^{3}H in the regions of the chromosomes that replicated at the beginning of the S period?

*10.3   Assume you did the experiment in Question 10.2 but left the radioactive medium on the cells for 16 h instead of 5 min. How would your answers to the above questions change?

*10.4   Karyotype analysis performed on cells cultured from an amniotic fluid sample reveals that the cells contain 47 chromosomes. The stained chromosomes are classified into groups, and the arrangement shows 6 chromosomes in A, 4 in B, 16 in C, 6 in D, 6 in E, 4 in F, and 5 in the G group. Based on this information:
a.   What could be the genotype of the fetus? If more than one possibility exists, give all.
b.   How would you proceed to distinguish among the possibilities?

10.5   Eukaryotic chromosomes contain (choose the best answer):
a.   protein.
b.   DNA and protein.
c.   DNA, RNA, histone, and nonhistone protein.
d.   DNA, RNA, and histone.
e.   DNA and histone.

*10.6   In a particular eukaryotic chromosome (choose the best answer):
a.   Heterochromatin and euchromatin are regions where genes make functional gene products (where they are active).
b.   Heterochromatin is active but euchromatin is inactive.
c.   Heterochromatin is inactive but euchromatin is active.
d.   Both heterochromatin and euchromatin regions are inactive.

10.7   List four major features that distinguish eukaryotic chromosomes from prokaryotic chromosomes.

*10.8   From the following list, identify
a.   three features that both eukaryotic and bacterial chromosomes have in common.
b.   four features that eukaryotic chromosomes have but that are not found in bacterial chromosomes.
c.   one feature that bacterial chromosomes have but that is not found in eukaryotic chromosomes.

Features

A. centromeres	F. nonhistone protein scaffolds
B. hexose sugars	G. DNA
C. amino acids	H. nucleosomes
D. supercoiling	I. circular chromosome
E. telomeres	J. looping

10.9   A nucleosome core particle is composed of
a.   two molecules each of H2A, H2B, H3, and H4.
b.   one molecule each of H1, H2A, H2B, H3, and H4.
c.   two molecules each of H1, H2, H3, and H4.

10.10   Discuss the structure and role of nucleosomes.

*10.11   Answer the following questions after setting up the following "rope trick." Start with a belt (representing a DNA molecule; imagine the phosphodiester backbones lying along the top and bottom edges of the belt) and a soda can. Holding the belt buckle at the bottom of the can, wrap the belt flat against the side of the can, counterclockwise three times around the can. Now remove the "core" soda can and, holding the ends of the belt, pull the ends of the belt taut. After some reflection, answer the following questions.
a.   Did you make a left- or a right-handed helix?
b.   How many helical turns were present in the coiled belt before it was pulled taut?
c.   How many helical turns were present in the coiled belt after it was pulled taut?
d.   Why does the belt appear more twisted when pulled taut?
e.   About what percentage of the length of the belt was decreased by this packaging?
f.   Is the DNA of a linear chromosome that is coiled around histones supercoiled?
g.   Why are topoisomerases necessary to package linear chromosomes?

10.12   Arrange the following figures in increasing order of eukaryotic chromosome condensation. Designate the simplest chromosome organization as number 1 and the most complex chromosome organization as number 6.

A. Chromatin fiber of packed nucleosomes

B. Metaphase chromosome

C. DNA double helix

D. Extended section of chromosome

E. "Beads-on-a-string" form of chromatin

F. Condensed section of chromosome

*10.13   What are telomeres?

10.14   Would you expect to find most protein-coding genes in unique-sequence DNA, in moderately repetitive DNA, or in highly repetitive DNA?

10.15   Would you expect to find ribosomal RNA genes in unique-sequence DNA, in moderately repetitive DNA, or in highly repetitive DNA?

*10.16   Both histone and nonhistone proteins are essential for DNA packaging in eukaryotic cells. However, these classes of proteins are fundamentally dissimilar in a number of ways. Describe how these classes of proteins differ in terms of

a.   their protein characteristics.

b.   their presence and abundance in cells.

c.   their interactions with DNA.

d.   their role in DNA packaging.

*10.17   Many different types of moderately repetitive DNA sequences exist. Some of these have been associated with specific functions; for others, either the function is not known or they simply have no function.

a.   List three specific types of moderately repetitive DNA sequences and give their characteristics and function(s), if known.

b.   What functional significance might there be to the fact that some genes are repeated?

CHAPTER *11*

DNA REPLICATION

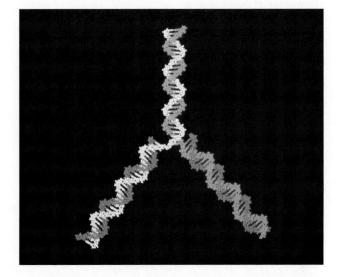

PRINCIPAL POINTS

~ DNA replication in prokaryotes and eukaryotes occurs by a semiconservative mechanism in which the two strands of a DNA double helix are separated and a new complementary strand of DNA is synthesized on each of the two parental template strands. This mechanism ensures faithful copying of genetic information at each cell division.

~ The enzymes called DNA polymerases catalyze the synthesis of DNA. All DNA polymerases make new strands in the 5'-to-3' direction using deoxyribonucleoside 5'-triphosphate (dNTP) precursors.

~ DNA polymerases cannot initiate the synthesis of a new DNA strand. All new DNA strands need a short primer of RNA, the synthesis of which is catalyzed by the enzyme DNA primase.

~ DNA replication in *E. coli* requires two DNA polymerases and several other enzymes and proteins. In both prokaryotes and eukaryotes, synthesis of DNA on one template strand is continuous, while it is discontinuous on the other template strand, a process called semidiscontinuous replication.

~ In eukaryotes, DNA replication occurs in the S phase of the cell cycle and is biochemically and molecularly very similar to replication in prokaryotes. To enable long chromosomes to replicate efficiently, DNA replication is initiated at many sites (origins) along the chromosomes and proceeds bidirectionally (in both directions).

~ Special enzymes—telomerases—replicate the ends of chromosomes in eukaryotes. A telomerase is a complex of proteins and RNA. The RNA acts as a template for synthesis of the complementary telomere repeat of the chromosome. In mammals, telomerase activity is limited to immortal cells (such as tumor cells). The absence of telomerase activity in nontumor cells results in progressive shortening of chromosome ends as the cell divides, thereby limiting the number of cell divisions before the cell dies.

~ In both prokaryotes and eukaryotes, genetic recombination involves the breakage and rejoining of homologous DNA double helices.

~ Numerous models have been proposed to describe the molecular events involved with genetic recombination. The Holliday model for genetic recombination involves a precise alignment of homologous DNA sequences of two parental double helices, followed by a series of enzyme-catalyzed reactions, which can lead to the generation of DNA molecules that are recombinant for loci flanking the recombination site.

*I*n this chapter your goal is to learn about the mechanism of DNA replication and chromosome duplication in prokaryotes and eukaryotes and about some of the enzymes and other proteins required for replication. Some of these enzymes are also involved in the repair of damage to DNA, a topic we discuss in Chapter 18. Also, you will learn some of the basic molecular details of DNA recombination.

SEMICONSERVATIVE DNA REPLICATION

When Watson and Crick proposed their double-helix model for DNA in 1953, they realized that replication of the DNA would be straightforward if their model was correct. That is, by unwinding the DNA molecule and separating the two strands, each strand would be a template for the synthesis of a new, complementary strand of DNA that would remain bound to the parental strand. This model for DNA replication is known as the **semiconservative model** since each progeny molecule retains one of the parental strands (Figure 11.1a).

At the time, two other models for DNA replication were proposed, the **conservative model** (Figure 11.1b) and the **dispersive model** (Figure 11.1c). In the conservative model, the two parental strands of DNA remain together or pair again after replication and as a whole serve as a template for the synthesis of new progeny DNA double helices. Thus, one of the two progeny DNA molecules is actually the parental double-stranded DNA molecule, and the other consists of totally new material. In the dispersive model, the parental double helix is cleaved into double-stranded DNA segments that act as templates for the synthesis of new double-stranded DNA

~ FIGURE 11.1

Three models for the replication of DNA. (a) Semiconservative model (the correct model); (b) Conservative model; (c) Dispersive model. The parental strands are shown in taupe, and the newly synthesized strands are shown in red.

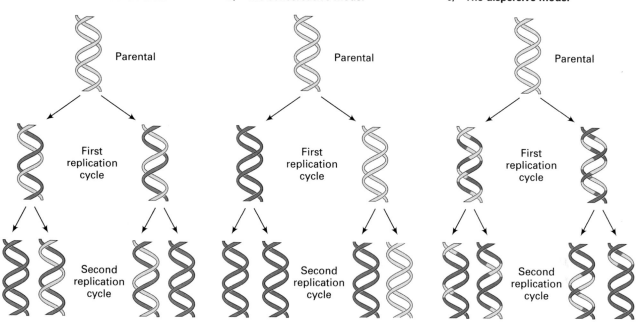

a) **The semiconservative model**
b) **The conservative model**
c) **The dispersive model**

segments. Somehow, the segments reassemble into complete DNA double helices, with parental and progeny DNA segments interspersed. Thus, while the two progeny DNAs are identical with respect to base pair sequence, double-stranded parental DNA has actually become dispersed throughout both progeny molecules. Logically, it is hard to imagine how the DNA sequences of chromosomes could be kept the same with such a mechanism. We include this model only for historical completeness.

The Meselson-Stahl Experiment

In 1958, Matthew Meselson and Frank Stahl obtained experimental evidence that the semiconservative replication model was correct. Meselson and Stahl grew *E. coli* in a medium in which the only nitrogen source was $^{15}NH_4Cl$ (ammonium chloride) (Figure 11.2). In this compound the normal isotope of nitrogen, ^{14}N, is replaced with ^{15}N, the heavy isotope. (Note: Density is weight/volume, so ^{15}N, with one extra neutron in its nucleus, is 1/14 denser than ^{14}N.) As a result, all the bacteria's nitrogen-containing compounds, including its DNA, contained ^{15}N instead of ^{14}N. ^{15}N DNA can be separated from ^{14}N

DNA by using equilibrium density gradient centrifugation (described in Box 11.1). Briefly, in this technique a solution of cesium chloride (CsCl) is centrifuged at high speed, causing the cesium chloride to form a density gradient. If DNA is present in the solution, it forms a band at a position where its buoyant density is the same as that of the surrounding cesium chloride.

Next, the ^{15}N-labeled bacteria were transferred into a medium containing nitrogen in the normal ^{14}N form, and the bacteria were allowed to reproduce for several generations. During this time, samples of *E. coli* were taken and the DNA was extracted and analyzed in CsCl density gradients (see Figure 11.2). After one replication cycle (one generation) in ^{14}N medium, all the DNA had a density that was exactly intermediate between that of totally ^{15}N DNA and totally ^{14}N DNA. After two replication cycles, half the DNA was of the intermediate density and half was of the density of DNA containing entirely ^{14}N. These observations, presented in Figure 11.2, and those for subsequent replication cycles were exactly what the semiconservative model predicted.

If the conservative model for DNA replication was correct, after one replication cycle there would be

~ FIGURE 11.2

The Meselson-Stahl experiment. The demonstration of semiconservative replication in *E. coli*. Cells were grown in ^{15}N-containing medium for several replication cycles, and then transferred to ^{14}N-containing medium. At various times over several replication cycles, samples were taken; the DNA was extracted and analyzed by CsCl equilibrium density gradient centrifugation. Shown in the figure are a schematic interpretation of the DNA composition after various replication cycles, photographs of the DNA bands, and densitometric scans of the bands.

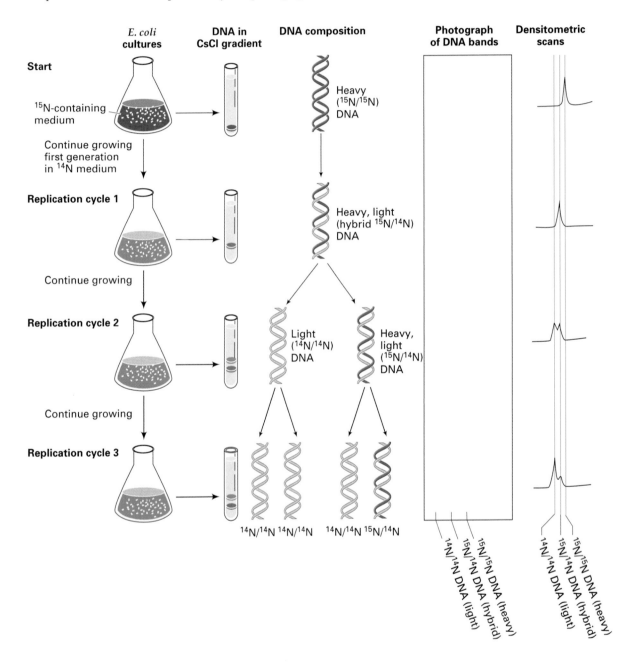

two bands of DNA, one in the heavy-density position of the gradient containing parental DNA molecules with both strands labeled with ^{15}N, and the other in the light-density position containing progeny DNA molecules with both strands labeled with ^{14}N (see Figure 11.1b). The heavy parental DNA band would be seen at each subsequent replication cycle, and in the amount found at the start of the experiment. All

BOX 11.1

EQUILIBRIUM DENSITY GRADIENT CENTRIFUGATION

In equilibrium density gradient centrifugation, a concentrated solution of cesium chloride (CsCl) is centrifuged at high speed to produce a linear concentration gradient of the CsCl. The actual densities of CsCl at the extremes of the gradient are related to the CsCl concentration that is centrifuged.

For example, to examine DNA of density 1.70 g/cm³ (a typical density for DNA), one makes a gradient that spans that density, for example from 1.60 to 1.80 g/cm³. If DNA is mixed with the CsCl and the mixture is centrifuged, the DNA will come to equilibrium at the point in the gradient where its buoyant density equals the density of the surrounding CsCl (see the figure below). The DNA is said to have banded in the gradient. If DNAs are present that have different densities, as is the case with ¹⁵N-DNA and ¹⁴N-DNA, then they will band (come to equilibrium) in different positions. The DNA is detected in the gradient by its UV absorption.

~ BOX FIGURE 11.1

Schematic diagram for separating DNAs of different buoyant densities by equilibrium centrifugation in a cesium chloride density gradient. The separation of ¹⁴N-DNA and ¹⁵N-DNA is illustrated.

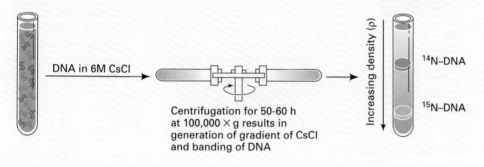

new DNA molecules would have both strands labeled with ¹⁴N. Hence, the amount of DNA in the light-density position would increase with each replication cycle. In the conservative model of DNA replication, then, the most significant prediction was that *at no time would any DNA of intermediate density be found*. The fact that intermediate-density DNA *was* found ruled out the conservative model.

If the dispersive model for DNA replication was correct, all DNA present after one replication cycle in ¹⁴N medium would have been of intermediate density (see Figure 11.1c), and this was seen in the Meselson-Stahl experiment. After a second replication cycle in ¹⁴N medium, the dispersive model predicted that DNA segments from the first replication cycle would be dispersed throughout the progeny DNA double helices produced. Thus, the ¹⁵N-¹⁵N DNA segments dispersed among new ¹⁴N-¹⁴N DNA after one replication cycle would then be distributed among twice as many DNA molecules after two

replication cycles. As a result, the DNA molecules would be found in *one* band located halfway between the intermediate-density and light-density band positions. With subsequent replication cycles, there would continue to be one band, and it would become lighter in density with each replication cycle. The results of the Meselson-Stahl experiment did not match this prediction, and therefore the dispersive model was ruled out.

Semiconservative DNA Replication in Eukaryotes

Semiconservative replication of DNA in eukaryotic chromosomes can be visualized using a special staining procedure. The experimental system involves Chinese hamster ovary (CHO) cells growing in tissue culture. At the beginning of the experiment, the chemical 5-bromodeoxyuridine (BUdR) is added. BUdR is a *base analog*, with a structure very similar to a normal

~ FIGURE 11.3

Visualization of semiconservative DNA replication in eukaryotes. Shown are harlequin chromosomes in Chinese hamster ovary cells that have been allowed to go through two rounds of DNA replication in the presence of the base analog 5-bromodeoxyuridine, followed by staining. Arrows indicate the sites where crossing-over has occurred.

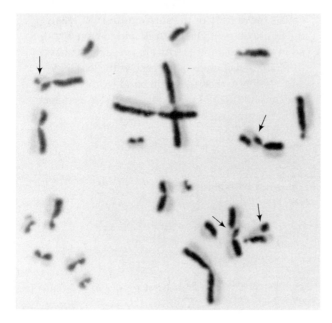

base in DNA, in this case thymine. During replication, wherever a T is called for in the new strand, BUdR can be incorporated instead. After two cycles of replication in the presence of BUdR, mitotic chromosomes are stained with a fluorescent dye and Giemsa stain, with the result shown in Figure 11.3. Sister chromatids are visible, one darkly stained and one lightly stained. The dark-light staining pattern brings to mind the costumes of harlequins, so the chromosomes are called **harlequin chromosomes**. A chromatid with a DNA double helix consisting of two BUdR-labeled strands stains less intensely than a chromatid with a DNA double helix consisting of one BUdR-labeled strand and one T-labeled strand. So, if semiconservative replication occurs, after one replication cycle two progeny DNAs will be produced, each with one T-labeled strand and one BUdR-labeled strand. Then, after a second replication cycle in the presence of BUdR, we will get one sister chromatid consisting of a DNA with one T-labeled and one BUdR-labeled strand (darkly staining), and the other sister chromatid consisting of a DNA with two BUdR-labeled strands (lightly staining). This is the staining pattern that was observed, showing that semiconservative DNA replication also is the correct model in eukaryotes.

KEYNOTE

DNA replication in prokaryotes and eukaryotes occurs by a semiconservative mechanism in which the strands of a DNA double helix are separated and a new complementary strand of DNA is synthesized on each of the two parental template strands. Semiconservative replication results in two double-stranded DNA molecules, each of which has one strand from the parent molecule and one newly synthesized strand. This mechanism ensures the faithful copying of the genetic information at each cell division.

ENZYMES INVOLVED IN DNA SYNTHESIS

In 1955, Arthur Kornberg and his colleagues set out to find the enzymes necessary for DNA replication so that they could dissect the reactions involved in detail. Their work focused on bacteria because it was assumed that bacterial replication machinery would be less complex than that of eukaryotes.

DNA Polymerase I

Kornberg's approach was to identify all the ingredients required for the synthesis of *E. coli* DNA in vitro. The first successful synthesis of DNA was accomplished in a reaction mixture containing DNA fragments, a mixture of four deoxyribonucleoside 5'-triphosphate precursors (dATP, dGTP, dTTP, and dCTP, collectively abbreviated dNTP for *deoxyribonucleoside triphosphate*), and a lysate prepared from *E. coli* cells. To measure the minute quantities of DNA expected to be synthesized in the reaction, Kornberg used radioactively labeled dNTPs.

Kornberg analyzed the lysate and isolated an enzyme that was capable of DNA synthesis. This enzyme was originally called the *Kornberg enzyme*, but it is now most commonly called **DNA polymerase I**. (By definition, enzymes that catalyze the synthesis of DNA are called **DNA polymerases**.)

With DNA polymerase I isolated, more detailed information could be obtained about DNA synthesis in vitro. Researchers found that four components were needed. If any one of the following four components was omitted, DNA synthesis would not occur:

1. All four dNTPs (If any one dNTP is missing, no synthesis occurs.) These are the precursors for the

nucleotide (phosphate-sugar-base) building blocks of DNA described in Chapter 9 (pp. 206–208);

2. A fragment of DNA to act as a template;

3. DNA polymerase I;

4. Magnesium ions (Mg^{2+}) (required for optimal DNA polymerase activity).

Subsequent experiments showed that the fragment of DNA acted as a template for synthesis of the new DNA; that is, the new DNA made in vitro was a faithful base-pair–for–base-pair copy of the original DNA.

Roles of DNA Polymerases

All DNA polymerases catalyze the polymerization of nucleotide precursors (dNTPs) into a DNA chain (Figure 11.4a). The same reaction in shorthand notation is shown in Figure 11.4b. The reaction has two main features:

1. At the growing end of the DNA chain, DNA polymerase catalyzes the formation of a phosphodiester bond between the 3'-OH group of the deoxyribose on the last nucleotide and the 5'-phosphate of the dNTP precursor. The energy for the formation of the phosphodiester bond comes from the release of two of three phosphates from the dNTP. The important concept here is that *the lengthening DNA chain acts as a primer in the reaction*, where a primer is a preexisting polynucleotide chain in DNA replication to which a new nucleotide can be added at the free 3'-OH.

2. At each step in lengthening the new DNA chain, DNA polymerase finds the correct precursor dNTP that can form a complementary base pair with the nucleotide on the template strand of DNA. Nucleotides can be added as rapidly as 800/second. This does not occur with 100 percent accuracy, but the error frequency is extremely low.

E. coli contains not just one but three DNA polymerases, the properties of which are summarized in Table 11.1. Both DNA polymerases I and III replicate DNA in the 5'→3' direction in the cell, but the role of DNA polymerase II is unknown. All three *E. coli* DNA polymerases have 3'→5' exonuclease activity—they can remove nucleotides from the 3' end of a DNA chain. This means that the DNA polymerases check the accuracy of the most recently assembled base pair. If an error has been made (which occurs at a frequency of about 10^{-6} for both DNA polymerase I and DNA polymerase III), the 3'→5' exonuclease activity can excise the erroneous nucleotide on the new strand. This process resembles that of the backspace delete key on a computer keyboard. The DNA polymerase then resumes the forward direction and inserts the correct character. Thus, in DNA replication, 3'→5' exonuclease activity is a **proofreading** mechanism that helps keep the frequency of DNA replication errors very low. With proofreading, the frequency of replication errors by DNA polymerase I or III is reduced to lower than 10^{-9}.

DNA polymerase I has 5'→3' exonuclease activity and can remove nucleotides from the 5' end of a DNA strand or of an RNA primer strand. This activity, also important in DNA replication and repair, will be examined later in this chapter.

KEYNOTE

The enzymes that catalyze the synthesis of DNA are called DNA polymerases. Three DNA polymerases, I, II, and III, have been identified in *E. coli*: I and III are known to be involved in DNA replication. The three enzymes differ in a number of properties, including number of molecules per cell and proofreading ability.

~ TABLE 11.1

Comparison of the Structural and Functional Characteristics of the *E. coli* DNA Polymerases I, II, and III

DNA POLYMERASE	PROPERTIES			
	POLYMERIZATION: 5'→3'	EXONUCLEASE: 3'→5'	EXONUCLEASE: 5'→3'	MOLECULES PER CELL (APPROXIMATELY)
I	Yes	Yes	Yes	400
II	Yes	Yes	No	?
III	Yes	Yes	No	10–20

~ FIGURE 11.4

DNA chain elongation catalyzed by DNA polymerase. (a) Mechanism at molecular level; (b) The same mechanism, using a shorthand method to represent DNA.

a)   Mechanism of DNA elongation

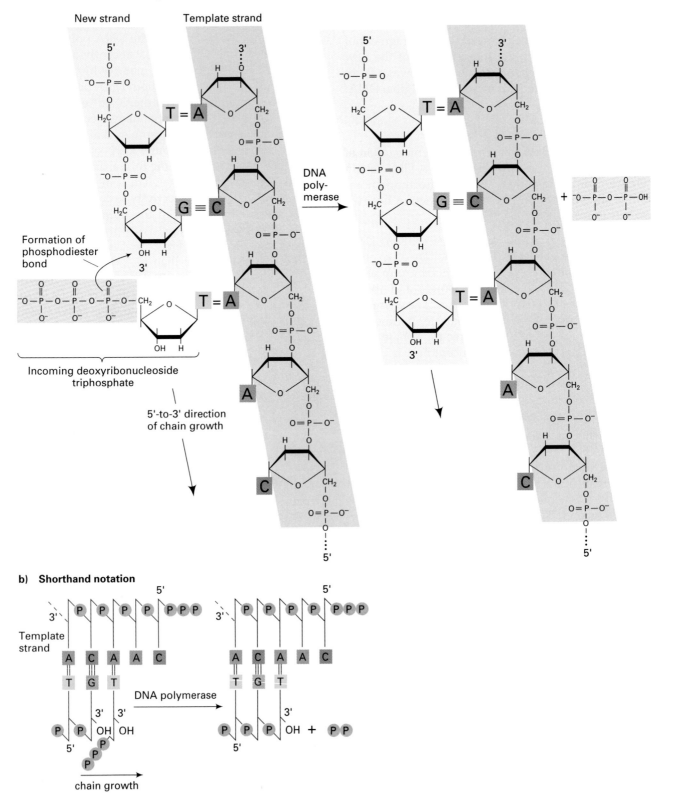

b)   Shorthand notation

MOLECULAR MODEL OF DNA REPLICATION

Table 11.2 presents the functions of some of the *E. coli* replication genes and key DNA sequences involved in the replication process. In this section we discuss a molecular model of DNA replication involving these genes and sequences.

Initiation of Replication

The initiation of DNA replication in *E. coli* occurs at a specific **origin of replication** sequence called *oriC*. At that site the double helix is denatured into single strands, exposing the bases for the synthesis of new strands. This produces a **replication bubble**—a region in the chromosome where the DNA is single stranded and from which replication proceeds bidirectionally (in both directions). The single strands are the **template strands** upon which new DNA synthesis occurs.

Figure 11.5 shows a model for the formation of a replication bubble at the origin of replication in *E. coli* and the initiation of the new DNA strand. First, *gyrase* (a form of topoisomerase; see Chapter 10, p. 218) relaxes the supercoiled DNA. Then, initiator proteins wrap the DNA around them at the origin of replication to form a large complex (Figure 11.5, part 1). **DNA helicase** then binds to the initiator proteins (Figure 11.5, part 2) and is loaded onto the DNA (Figure 11.5, part 3). The DNA helicase untwists (denatures) the DNA, and then **primase** (**DNA primase**, a form of RNA polymerase) binds to the helicase, forming a complex called the **primosome** (Figure 11.5, part 4).

Primase is important in DNA replication because no known DNA polymerases can initiate the synthesis of a DNA strand; they can only add nucleotides to a preexisting strand. Primase synthesizes a short **RNA primer** (about 11 nucleotides) to which new nucleotides can be added by DNA polymerase, and so the first new DNA chain is made (Figure 11.5, part 5). The RNA primer is removed later and replaced with DNA; we will return to discuss this event further. Next, another set of replication proteins bind to produce two replication complexes and two replication forks that move away from the replication origin in opposite directions, enabling replication to occur bidirectionally (Figure 11.6).

We must be clear about the difference between *template* and *primer* with respect to DNA replication. A template strand is the one on which the new strand is synthesized according to complementary base-pairing rules. A primer is a short segment of nucleotides bound to the template strand. The

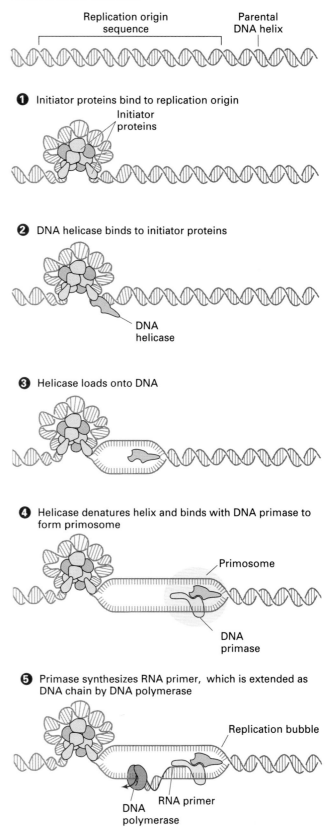

~ FIGURE 11.5

Schematic model for the formation of a replication bubble at a replication origin in *E. coli* and the initiation of the new DNA strand.

Replication origin sequence

Parental DNA helix

❶ Initiator proteins bind to replication origin

Initiator proteins

❷ DNA helicase binds to initiator proteins

DNA helicase

❸ Helicase loads onto DNA

❹ Helicase denatures helix and binds with DNA primase to form primosome

Primosome

DNA primase

❺ Primase synthesizes RNA primer, which is extended as DNA chain by DNA polymerase

Replication bubble

DNA polymerase

RNA primer

~ **TABLE 11.2**

Functions of Some of the Genes and DNA Sequences Involved in DNA Replication in *E. coli*

GENE/SEQUENCE	ENZYME OR FUNCTION IN REPLICATION
polA	DNA polymerase I
polB	DNA polymerase II
dnaE, dnaQ, dnaX, dnaN, holA→E	DNA polymerase III subunits
dnaA	Initiator protein; binds to *oriC*
dnaB	Helicase—unwinds DNA at replication fork
dnaC	Binds to helicase and delivers it to DNA
dnaG	Primase—makes RNA primer for initiation of DNA replication
gyrA, gyrB	Gyrase subunits—gyrase, a form of topoisomerase, prevents DNA ahead of replication fork from getting overtwisted
lig	DNA ligase—joins segments of DNA during replication
ssb	Single-stranded binding (SSB) proteins—bind to single-stranded DNA during replication
oriC	Origin of chromosomal replication—where replication starts in *E. coli*
ter	Terminus of chromosomal replication—where replication ends in *E. coli*

primer acts as a substrate for DNA polymerase, which extends the primer as a new DNA strand, the sequence of which is complementary to the template strand.

~ **FIGURE 11.6**

Schematic diagram of the formation at a replication origin sequence of two replication forks that move in opposite directions. The upper diagram is a DNA-only version of Figure 11.5, part 5.

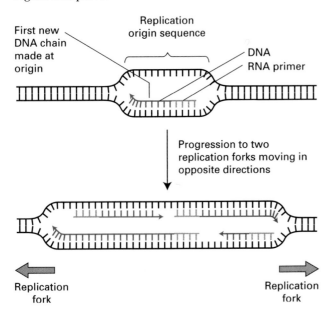

First new DNA chain made at origin

Replication origin sequence

DNA

RNA primer

Progression to two replication forks moving in opposite directions

Replication fork

Replication fork

KEYNOTE

No DNA polymerase can initiate the synthesis of a new DNA chain. Instead, the initiation of DNA synthesis first involves the denaturation of double-stranded DNA at an origin of replication, catalyzed by DNA helicase. Next, DNA primase binds to the helicase and the denatured DNA, forming a primosome that synthesizes a short RNA primer. The RNA primer is extended by DNA polymerase as new DNA is made. The RNA primer is later removed.

Semidiscontinuous DNA Replication

When DNA unwinds to expose the two single-stranded template strands for DNA replication, a Y-shaped structure called a **replication fork** is formed. A replication fork moves in one direction. When DNA unwinds in the middle of a DNA molecule, as in a circular chromosome, there are two replication forks head to head—think of this as two Ys joined together at their tops. In many cases each replication fork is active, so that DNA replication proceeds bidirectionally. We will consider what happens at one replication fork.

Model for the events occurring around a single replication fork of the _E. coli_ chromosome. (a) Initiation; (b) Further untwisting and elongation of the new DNA strands; (c) Further untwisting and continued DNA synthesis; (d) Removal of the primer by DNA polymerase I; (e) Joining of adjacent DNA fragments by the action of DNA ligase. Green = RNA; red = new DNA.

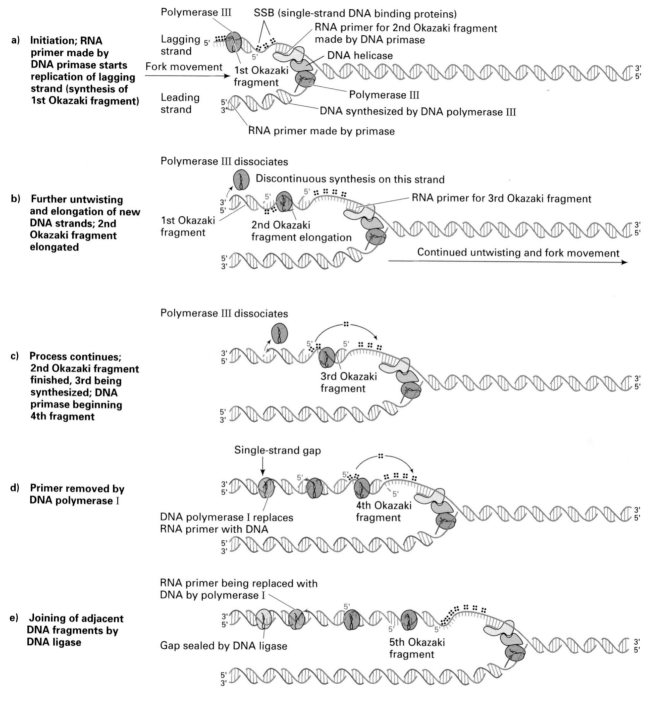

a) **Initiation; RNA primer made by DNA primase starts replication of lagging strand (synthesis of 1st Okazaki fragment)**

b) **Further untwisting and elongation of new DNA strands; 2nd Okazaki fragment elongated**

c) **Process continues; 2nd Okazaki fragment finished, 3rd being synthesized; DNA primase beginning 4th fragment**

d) **Primer removed by DNA polymerase I**

e) **Joining of adjacent DNA fragments by DNA ligase**

As we just discussed (see Figure 11.5), helicase untwists the DNA to produce single-stranded template strands. **Single-strand DNA-binding (SSB) proteins** bind to the single-stranded DNA, stabilizing it (Figure 11.7) and preventing reannealing. Primase synthesizes an RNA primer on each template strand.

These RNA primers are lengthened by DNA polymerase III, which synthesizes new DNA complementary to the template strands (Figure 11.7a). Recall that DNA polymerases can only make DNA in the 5'-to-3' direction; however, the two DNA strands are of opposite polarity. To maintain the 5'-to-3' polarity of DNA synthesis on each template, and one overall direction of replication fork movement, DNA is made in opposite directions on the two template strands (see Figure 11.7a). The new strand being made 5'-to-3' in the *same* direction as the movement of the replication fork is called the **leading strand**, and the new strand being made in the *opposite* direction as the movement of the replication fork is called the **lagging strand**. The leading strand, then, requires a single RNA primer for its synthesis, while the lagging strand requires a series of primers.

As the replication fork moves, helicase untwists more DNA (Figure 11.7b). Gyrase (a form of topoisomerase) relaxes the tension produced in the DNA ahead of the replication fork. This tension could be considerable given that the replication fork rotates at about 3,000 rpm. On the leading-strand template (bottom strand in Figure 11.7), the leading strand is synthesized continuously toward the replication fork. Since DNA synthesis can only proceed in the 5'-to-3' direction, however, lagging-strand synthesis has gone as far as it can. For DNA replication to continue on the lagging-strand template (top strand in Figure 11.7), a new initiation of DNA synthesis must occur. An RNA primer is made by the primase still bound to the helicase at the replication fork (see Figure 11.7b). DNA polymerase III then adds DNA to the RNA primer to make another DNA fragment. Because the leading strand is being made continuously, while the lagging strand can only be made in pieces, or discontinuously, DNA replication as a whole occurs in a **semidiscontinuous** manner. The fragments of lagging strand made in this process are called **Okazaki fragments** after their discoverers, Reiji and Tuneko Okazaki and colleagues.

In Figure 11.7c the process repeats itself: helicase untwists the DNA, continuous DNA synthesis occurs on the leading-strand template, and discontinuous DNA synthesis occurs on the lagging-strand template. Eventually, the unconnected Okazaki fragments on the lagging-strand template are joined into a continuous DNA strand. This requires the activities of DNA polymerase I and **DNA ligase**. Consider two adjacent Okazaki fragments. The 3' end of the newer DNA fragment is adjacent to, but not joined to, the primer at the 5' end of the previously made fragment. DNA polymerase III leaves the DNA, and DNA polymerase I continues the 5'-to-3' synthesis of the DNA fragment, simultaneously removing the primer sec-

tion of the older fragment by its 5'→3' exonuclease activity (Figure 11.7d). When DNA polymerase I has completed replacement of RNA primer nucleotides with DNA nucleotides, a single-stranded gap is left between the two fragments. The two fragments are joined by DNA ligase to produce a longer DNA strand (Figure 11.7e). The whole sequence of events is repeated until all the DNA is replicated.

KEYNOTE

Replication of DNA in *E. coli* requires two DNA polymerases and several other enzymes and proteins. The DNA helix is untwisted by DNA helicase to provide single-stranded templates for the synthesis of new DNA. Since new DNA is made in the 5'-to-3' direction, chain growth is continuous on one strand and discontinuous (that is, in segments that are later joined) on the other strand.

DNA REPLICATION IN EUKARYOTES

The biochemistry and molecular biology of DNA replication are very similar in prokaryotes and eukaryotes. However, an added complication in eukaryotes is that DNA is distributed among many chromosomes rather than just one. In this section we summarize some of the important aspects of DNA replication in eukaryotes.

DNA Replication and the Cell Cycle

In every cell cycle, all chromosomes must be faithfully duplicated and a copy of each distributed to each of the two progeny cells. This means that both the DNA and the histones must be doubled with each cell cycle. Recall from Chapter 1 that the cell cycle in most somatic cells of higher eukaryotes is divided into four stages: gap 1 (G_1), synthesis (S), gap 2 (G_2), and mitosis (M) (see Figure 1.10). DNA replication and chromosome duplication occur during the S phase, and the progeny chromosomes segregate into daughter cells during the M phase.

Progression through the cell cycle is tightly controlled by the activities of many genes in an elaborate system of checks and balances (Figure 11.8). As a cell proceeds through G_1, it is gearing up for DNA replication and chromosome duplication in the S phase. A major checkpoint in G_1 called *START* in yeasts and

~ FIGURE 11.8

Some of the molecular events that control progression through the cell cycle in yeasts.

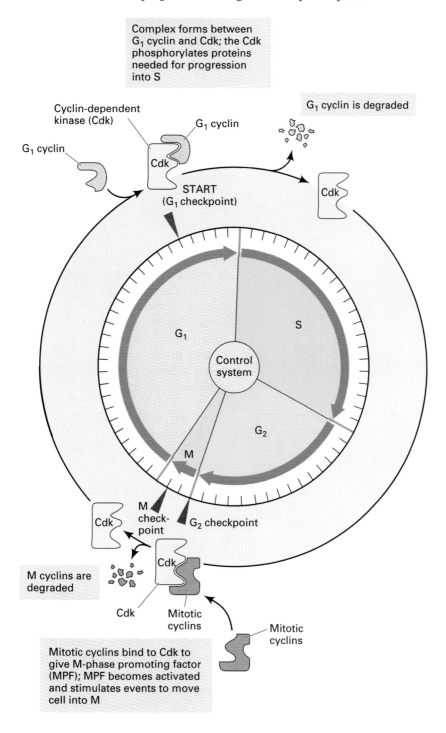

G_1 *checkpoint* in mammalian cells determines whether the cell is able to or should continue into S. Unless a cell grows to a large enough size and the environment is favorable, it will stay in G_1. Another major check-

point, called the G_2 *checkpoint* both in yeasts and mammalian cells, occurs at the junction between G_2 and M. Unless all the DNA has replicated, the cell is big enough, and the environment is favorable, the cell is unable to enter the mitotic phase of the cell cycle. A third checkpoint occurs during M: the chromosomes must be attached properly to the mitotic spindle to trigger the separation of chromatids and the completion of mitosis.

Key components in the regulatory events that occur at checkpoints are proteins known as *cyclins* (named because their concentration increases and decreases in a regular pattern through the cell cycle) and enzymes known as *cyclin-dependent kinases* (Cdks). In yeasts, a single Cdk functions at both the G_1 and the G_2 checkpoints, while in mammalian cells at least two Cdks are involved at each checkpoint.

At START, one or more *G_1 cyclins* bind to the Cdk kinase and activate it. The Cdk then phosphorylates (adds a phosphate to) key proteins that are needed for progression into S. Once the cyclin has activated the Cdk, the cyclin level decreases as a result of increased proteolysis (breakdown by enzymes). A similar process occurs at the G_2 checkpoint, when one or more *mitotic cyclins* bind to the Cdk to form the *M-phase promoting factor (MPF)*. Then, when other enzymes phosphorylate and dephosphorylate it, MPF is activated and stimulates the cell to move into M. During mitosis, just after metaphase, the mitotic cyclin is degraded, which leads to MPF inactivation and allows the cell to complete mitosis.

The regulatory events controlling the cell cycle are complex. Many more steps and pathways are involved than have been described. We return to this subject when we discuss the genetics of cancer in Chapter 17, for some genes that normally control steps in the cell cycle are often mutated in cancer.

Eukaryotic Replication Enzymes

As we saw earlier, many of the enzymes and proteins involved in prokaryote DNA replication have been identified. Less is known about the enzymes and proteins involved in eukaryotic DNA replication. It is clear, however, that the steps described for DNA synthesis in prokaryotes also occur for DNA synthesis in eukaryotes, namely, denaturation of the DNA double helix and the semiconservative, semidiscontinuous replication of DNA. New DNA strands are initiated by RNA primers, which DNA polymerases extend as DNA chains.

Five different DNA polymerases have been identified in mammalian cells, designated α (alpha), β (beta), δ (delta), γ (gamma), and ∈ (epsilon). The α, β, δ, and ∈ polymerases are located in the nucleus, while the γ polymerase is found in the mitochondrion. DNA polymerases α and δ are the two enzymes responsible for nuclear DNA replication. DNA poly-merase β serves a DNA repair function, and DNA polymerase ∈ appears to serve a similar function. DNA polymerase γ replicates mitochondrial DNA. Both nuclear polymerases α and δ and the mitochondrial polymerase γ use RNA primers for the initiation of DNA synthesis. Only δ, ∈, and γ have been shown to have proofreading (3'-to-5' exonuclease) activity.

Replicons

Each eukaryotic chromosome consists of one linear DNA double helix. If there was only one origin of replication per chromosome, the replication of each chromosome would take many, many hours. For example, there are about 3×10^9 base pairs of DNA in the haploid human genome (23 chromosomes), and the average chromosome is roughly 10^8 base pairs long. With a replication rate of 2 kilobases (2,000 bases) per minute, it would take approximately 830 hours to replicate one chromosome.

Actual measurements show that the chromosomes in eukaryotes replicate much faster than would be the case with only one origin of replication per chromosome. The diploid set of chromosomes in *Drosophila* embryos, for example, replicates in 3 minutes. This is six times faster than the replication of the *E. coli* chromosome, even though there is about 100 times more DNA in *Drosophila* than there is in *E. coli*.

Eukaryotic chromosomes duplicate rapidly because DNA replication initiates at many origins of replication throughout the genome. At each origin of replication, the DNA denatures (as in *E. coli*), and the replication proceeds bidirectionally. Eventually, each replication fork runs into an adjacent replication fork, initiated at an adjacent origin of replication. In eukaryotes the stretch of DNA from the origin of replication to the two termini of replication (where adjacent replication forks fuse) on each side of the origin is called a **replicon** or a **replication unit** (Figure 11.9). Table 11.3 presents the number of replicons, their average size, and the rate of replication fork movement for six organisms. Note that the replicon size is much smaller, and the rate of fork movement is much slower, in eukaryotic organisms than in bacteria.

DNA replication does not occur simultaneously in all the replicons in an organism's genome. Instead, there is a cell-specific timing of initiation of replication at the various origins (Figure 11.10). The figure shows one segment of one chromosome in which there are three replicons that always begin replicating at distinct times. When the replication forks fuse at the margins of adjacent replicons, the chromosome has replicated into two sister chromatids.

~ FIGURE 11.9

Replicating DNA of *Drosophila melanogaster*. (a) Electron micrograph showing replication units (replicons); (b) An interpretation of the electron micrograph shown in (a).

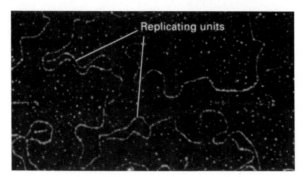

a)

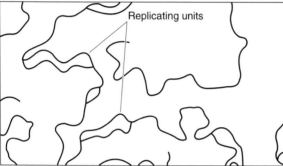

b)

Origins of Replication

In the yeast *Saccharomyces cerevisiae*, specific chromosomal sequences have been identified that appear to act as origins of replication during chromosome duplication. These sequences are called **autonomously replicating sequences**, or **ARSs**. In mammals and other complex eukaryotes, origins of replication sequences are not as well defined.

KEYNOTE

In eukaryotes, DNA replication occurs in the S phase of the cell cycle and is similar to the replication process in prokaryotic cells. Synthesis of DNA is initiated by RNA primers, occurs in the 5'-to-3' direction, is catalyzed by DNA polymerases, requires a large number of other enzymes and proteins, and is a semiconservative and semidiscontinuous process. Replication of DNA is initiated at a large number of sites throughout the chromosomes and proceeds bidirectionally.

~ FIGURE 11.10

Temporal ordering of DNA replication initiation events in replication units of eukaryotic chromosomes.

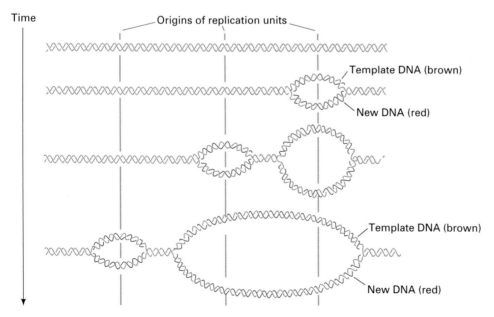

Time

Origins of replication units

Template DNA (brown)

New DNA (red)

Template DNA (brown)

New DNA (red)

~ TABLE 11.3

Comparison of Bacterial and Eukaryote Replicons

ORGANISM	NO. OF REPLICONS	AVERAGE LENGTH	FORK MOVEMENT
Bacterium (*E. coli*)	1	4,200 kb	50,000 bp/min
Yeast (*Saccharomyces cerevisiae*)	500	40 kb	3,600 bp/min
Fruit fly (*Drosophila melanogaster*)	3,500	40 kb	2,600 bp/min
Toad (*Xenopus laevis*)	15,000	2,000 kb	500 bp/min
Mouse (*Mus musculus*)	25,000	150 kb	2,200 bp/min
Plant (bean: *Vicia faba*)	35,000	300 kb	n.a.[a]

[a]n.a.—value not available.

Replicating the Ends of Chromosomes

Because DNA polymerases can only synthesize new DNA by extending a primer, there are special problems in replicating the ends of eukaryotic chromosomes (Figure 11.11). A parental chromosome (Figure 11.11a) is replicated, resulting in two new DNA molecules each of which has an RNA primer at the 5' end of the newly synthesized strand in the telomere region (Figure 11.11b). The RNA primers are removed, leaving a gap at the 5' end of the new strand. The gaps left by removal of the RNA primers remain because DNA polymerase cannot initiate new DNA synthesis. If nothing is done about these gaps, the chromosomes would get shorter and shorter with each replication cycle.

There is a special mechanism for replicating the ends of chromosomes. Recall that most eukaryotic chromosomes have tandemly repeated, species-specific, simple sequences at their telomeres (see Chapter 10). The work of Elizabeth Blackburn and Carol W. Greider has shown that an enzyme called *telomerase* maintains chromosome lengths by adding telomere repeats to the chromosome ends.

Figure 11.12 shows the mechanism deduced for the protozoan *Tetrahymena* in a simplified way. The repeated sequence in *Tetrahymena* is 5'-TTGGGG-3' reading toward the end of the DNA on the top strand in Figure 11.12. Telomerase acts at the stage shown in Figure 11.11c; that is, where a chromosome end has been produced with a gap at the end of the chromosome at the 5' end of the new DNA (Figure 11.12a). Telomerase is an enzyme made up of both protein and RNA. The RNA component includes a base sequence that is complementary to the telomere repeat unit of the organism in which it is found. Because of this, the telomerase binds specifically to the overhanging telomere repeat at the end of the chromosome (Figure 11.12b). Next, the telomerase catalyzes the synthesis of three nucleotides of new

~ FIGURE 11.11

The problem of replicating completely a linear chromosome in eukaryotes. (a) Schematic diagram of a parent double-stranded DNA molecule representing the full length of a chromosome; (b) After semiconservative replication, new DNA segments hydrogen-bonded to the template strands have RNA primers at their 5' ends; (c) The RNA primers are removed, DNA polymerase fills the resulting gaps, and DNA ligase joins the adjacent fragments. However, at the two telomeres there are still gaps at the 5' ends of the new DNA resulting from RNA primer removal because no new DNA synthesis could fill them in.

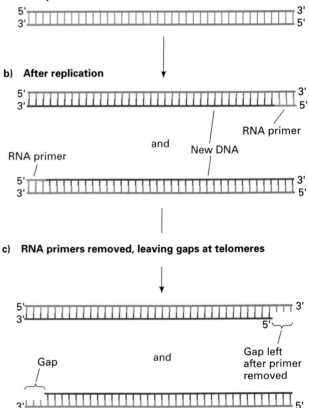

~ FIGURE 11.12

Synthesis of telomeric DNA by telomerase. The example is for *Tetrahymena* telomeres. The process is described in the text. (a) Starting point: chromosome end with 5' gap left after primer removal; (b) Binding of telomerase to the overhanging telomere repeat at the end of the chromosome; (c) Synthesis of three-nucleotide DNA segment at chromosome end using the RNA template of telomerase; (d) The telomerase moves so that the RNA template can bind to the newly synthesized TTG in a different way; (e) Telomerase catalyzes the synthesis of a new telomere repeat using the RNA template. The process is repeated to add more telomere repeats. (f) After telomerase has left, new DNA is made on the template starting with an RNA primer. After the primer is removed (g), the result is a longer chromosome than at the start, with a new 5' gap.

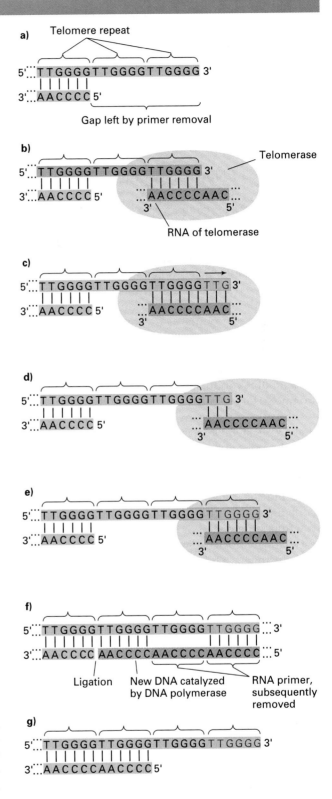

DNA—TTG—using the telomerase RNA as a template (Figure 11.12c). The telomerase then slides toward the end of the chromosome so that its AAC at the 3' end of the RNA template now pairs with the newly synthesized TTG on the DNA (Figure 11.12d). Telomerase then makes the rest of the TTGGGG telomere repeat (Figure 11.12e). The process is repeated to add more telomere repeats. In this way, the chromosome is lengthened by the addition of a number of telomere repeats. Then, by primer synthesis and DNA synthesis catalyzed by DNA polymerase in the conventional way, the former gap is filled in, and the new chromosomal DNA is lengthened (Figure 11.12f). After removal of the RNA primer, a new 5' gap is left (Figure 11.12g), but any net shortening of the chromosome has been averted.

Telomere length, while not identical from chromosome end to chromosome end, is regulated to an average length for the organism and cell type. In wild-type yeast, for example, the telomere occupies an average of about 300 bp. Genes that control telomere length have been identified. For example, deletion mutations of the *TLC1* gene (that encodes the telomerase RNA) or mutations of the *EST1* (*Ever Shorter Telomeres*) gene cause telomeres to shorten continuously until the cells die. This phenotype provides evidence that telomerase activity is required for long-term viability of yeast cells. Interestingly, telomerase activity in mammals is limited to immortal cells (such as tumor cells). The absence of telomerase activity in other cells results in progressive shortening of chromosome ends during successive divisions because of the failure to replicate those ends, and in a limited number of cell divisions before the cell will die.

Assembly of New DNA into Nucleosomes

Eukaryotic DNA is complexed with histones in nucleosome structures, which are the basic units of chromosomes (see Chapter 10). Therefore, when the DNA is replicated, the histone complement must be doubled so that all nucleosomes are duplicated. This involves two processes: the synthesis of new histone proteins and the assembly of new nucleosomes.

Most histone synthesis is coordinated with DNA replication. The transcription of the genes for the five histones is initiated near the end of the G_1 phase, just prior to S. Translation of the histone mRNAs occurs throughout S, producing the histones to be assembled into nucleosomes as the chromosomes are duplicated.

Electron microscopy studies have shown that newly replicated DNA is assembled into nucleosomes virtually immediately. Nonetheless, for replication to proceed, nucleosomes must disassemble during the short time when a replication fork passes. Measurements indicate that there is a nucleosome-free zone of about 200 to 300 base pairs around replication forks. Data from different kinds of experiments strongly suggest that new nucleosomes are made from an all-new set of histones, while the old nucleosomes are conserved. Completed nucleosomes go randomly to the two DNA strands after the replication fork.

DNA RECOMBINATION

In the mid-1960s, Robin Holliday proposed a model for reciprocal recombination. Since then, the *Holliday model* has been refined by other geneticists, notably Matthew Meselson and Charles Radding, and T. Orr-Weaver and Jack Szostak. To give a flavor for the recombination process at the molecular level, we present the Holliday model here.

The Holliday model is diagrammed in Figure 11.13 for genetically distinguishable homologous chromosomes, one with alleles a^+ and b^+ at opposite ends, and the other with alleles a and b. Genetic recombination occurs between the two DNAs. The first stage of the recombination process is *recognition and alignment*

(Figure 11.13, part 1), in which the two DNAs become aligned precisely. In the second stage, one strand of each double helix breaks; each broken strand then "invades" the opposite double helix and base-pairs with the complementary nucleotides of the invaded helix (Figure 11.13, part 2). Enzymes are responsible for each of these steps. DNA polymerase and DNA ligase seal the gaps that are left, producing what is called a *Holliday intermediate*, with an internal branch point (Figure 11.13, part 3). The hybrid DNA molecules at this stage are called *heteroduplexes*; that is, the two strands of the double-stranded DNA molecules do not have completely complementary sequences. The two DNA double helices in the Holliday intermediate can rotate, causing the branch point to move to the right or to the left. Figure 11.13, part 4, shows a branch migration event that has occurred to the right. The four-armed structure for the DNA strands is produced simply by pulling the four chromosome ends apart. Branch migration generates complementary regions of hybrid DNA in both double helices (diagrammed as stretches of DNA helices with two different colors in Figure 11.13, parts 5 through 8).

The *cleavage and ligation* phase of recombination is best visualized if the Holliday intermediate is redrawn so that no DNA strand passes over or under another DNA strand. Thus, if the four-armed Holliday intermediate following Figure 11.13, part 4 is taken as a starting point and the lower two arms are rotated 180° relative to the upper arms, the structure shown in Figure 11.13, part 5 is produced.

Next, enzymes cut the Holliday intermediate at two points in the single-stranded DNA region of the branch point (Figure 11.13, part 5). The cuts can be in either the horizontal or vertical planes; both kinds of cuts occur with equal probability. Endonuclease cleavage in the horizontal plane (Figure 11.13, part 6, left) produces the two double helices shown in Figure 11.13, part 7, left. In each helix is a single-stranded gap. DNA ligase seals the gaps to produce the double helices shown in Figure 11.13, part 8, left. Since each of the resulting helices contains a segment of single-stranded DNA from the other helix, flanked by nonrecombinant DNA, these double helices are called *patched duplexes*.

If the endonuclease cleavage in Figure 11.13, part 5 is in the vertical plane (Figure 11.13, part 6, right), the gapped double helices of Figure 11.13, part 7, right are produced. In this case, there are segments of hybrid DNA in each duplex, but they are formed by what looks like a splicing together of two helices. The result is the double helices in Figure 11.13, part 8, right, which are called *spliced duplexes*.

In the example diagrammed in Figure 11.13, the parental duplexes contained different genetic markers

~ FIGURE 11.13

Holliday model for reciprocal genetic recombination. Shown are two homologous DNA double helices that participate in the recombination process. Inset: Electron micrograph of a Holliday intermediate with some single-stranded DNA in the branch point region.

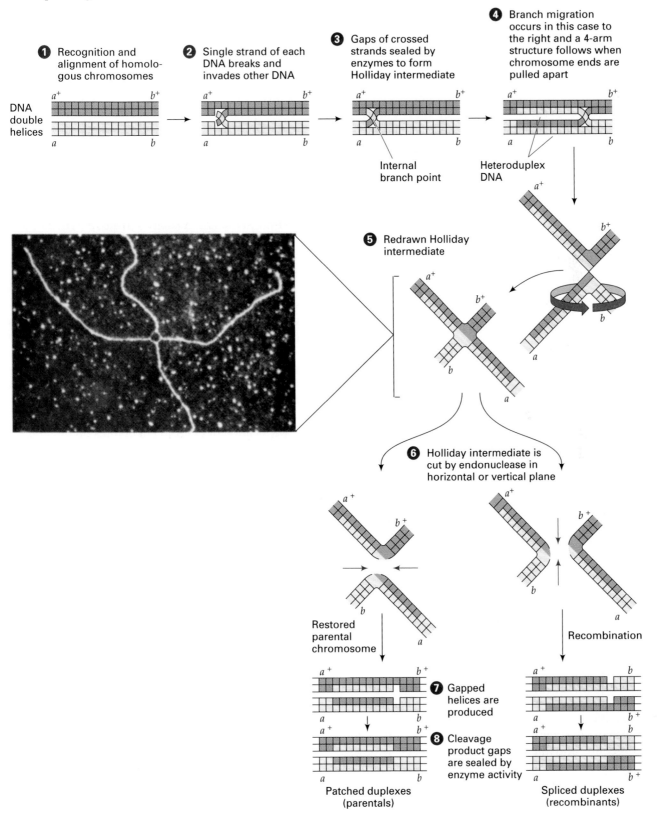

① Recognition and alignment of homologous chromosomes

② Single strand of each DNA breaks and invades other DNA

③ Gaps of crossed strands sealed by enzymes to form Holliday intermediate

④ Branch migration occurs in this case to the right and a 4-arm structure follows when chromosome ends are pulled apart

DNA double helices

a^+ b^+ a^+ b^+ a^+ b^+ a^+ b^+

a b a b a b a b

Internal branch point

Heteroduplex DNA

a^+
b^+
b
a

⑤ Redrawn Holliday intermediate

a^+
b^+
b
a

⑥ Holliday intermediate is cut by endonuclease in horizontal or vertical plane

a^+
b^+
b
a

a^+
b^+
b
a

Restored parental chromosome

Recombination

⑦ Gapped helices are produced

a^+ b^+ a^+ b
a b a b^+

⑧ Cleavage product gaps are sealed by enzyme activity

a^+ b^+ a^+ b
a b a b^+

Patched duplexes (parentals)

Spliced duplexes (recombinants)

at the ends of the molecules. One parent was $a^+ b^+$ and the other was $a\,b$, as would be the case for a doubly heterozygous parent. However, the reciprocal recombination events shown in Figure 11.13 result in different products. In the patched duplexes (Figure 11.13, part 8, left) the markers are $a^+ b^+$ and $a\,b$, which is the parental configuration. However, in the spliced duplexes (Figure 11.13, part 8, right) the markers are recombinant, $a^+ b$ and $a\,b^+$. Since the enzymes that cut in the branch region during the final cleavage and ligation phase (Figure 11.13, part 6, left and right) cut randomly with respect to the plane of the cut (that is, horizontal or vertical), the Holliday model predicts that a physical exchange between two gene loci on homologous chromosomes should result in the genetic exchange of the outside chromosome markers about half of the time.

While the basic features of the Holliday model are generally accepted, a number of other models attempt to explain recombination either at a more detailed level or in special systems. Discussion of these other models is beyond the scope of this text.

KEYNOTE

The Holliday model for genetic recombination involves a precise alignment of homologous DNA sequences of two parental double helices, endonucleolytic cleavage, invasion of the other helix by each broken end, ligation to produce the Holliday intermediate, branch migration, and finally, cleavage and ligation to resolve the Holliday intermediate into the recombinant double helices. Depending on the orientation of the cleavage events, the resulting double helices will be patched duplexes or spliced duplexes. If the recombination events occur between heterozygous loci, the patched duplexes produced are parental, whereas the spliced duplexes are recombinant for the two loci.

SUMMARY

In this chapter we discussed DNA replication and DNA recombination. Many aspects of DNA replication are similar in prokaryotes and eukaryotes, for example, a semiconservative and semidiscontinuous mechanism, synthesis of new DNA in the 5'-to-3' direction, and the use of RNA primers to initiate DNA chains. The enzymes that catalyze the synthesis of DNA are the DNA polymerases. In *E. coli* there are three DNA polymerases, two of which are known to

be involved in DNA replication along with several other enzymes and proteins. The DNA polymerases have 3'-to-5' exonuclease activity, which permits proofreading to take place during DNA synthesis if an incorrect nucleotide is inserted opposite the template strand. In eukaryotes, two DNA polymerases are involved in nuclear DNA replication. Neither has associated proofreading activity, that function presumably being the property of a separate protein.

In prokaryotes, DNA replication begins at specific chromosomal sites. Such sites are known also for yeast. A prokaryotic chromosome has one initiation site for DNA replication, while a eukaryotic chromosome has several initiation sites dividing the chromosome into replication units, or replicons. It is not clear whether the initiation sites used are the same sequences from cell generation to cell generation. The existence of replicons means that DNA replication of the entire set of chromosomes in a eukaryotic organism can proceed relatively quickly, in some cases faster than with the single *E. coli* chromosome, despite the presence of orders of magnitude more DNA.

Since eukaryotic chromosomes are linear, there is a special problem of maintaining the lengths of chromosomes because removal of RNA primers results in a shorter new DNA strand. This problem is overcome by special enzymes called telomerases that maintain the length of chromosomes. Telomerases are a combination of proteins and RNA. The RNA component acts as a template to guide the synthesis of new telomere repeat units at the chromosome ends.

Replication of DNA in eukaryotes occurs in the S phase of the cell division cycle. Eukaryotic chromosomes are complexes of DNA with histones and nonhistone chromosome proteins, so not only must DNA be replicated but the chromosome structure must also be duplicated. In particular, the nucleosome organization of chromosomes must be duplicated as the replication forks migrate. It is known that nucleosomes are "mature" soon after a replication fork passes, and evidence suggests that nucleosomes are conserved in chromosome duplication, with nucleosomes after the replication fork consisting of all new histones or all old histones.

A number of models have been proposed to describe the molecular events involved in the breakage and rejoining of DNA during crossing-over. One model, developed by Holliday, is described in this chapter. All molecular models for genetic recombination involve new DNA synthesis in small regions participating in crossing-over. In Chapter 18 we see that DNA synthesis may also be involved in the repair of certain genetic damage. Thus, three major cellular processes involve DNA synthesis: DNA replication, genetic recombination, and DNA repair.

ANALYTICAL APPROACHES FOR SOLVING GENETICS PROBLEMS

Q11.1 What would be the effect on chromosome replication in *E. coli* strains carrying deletions of the following genes?
a. *dnaE*
b. *polA*
c. *dnaG*
d. *lig*
e. *ssb*
f. *oriC*

A11.1 When genes are deleted, the function encoded by those genes is lost. All of the genes listed in the question are involved in DNA replication in *E. coli,* and their functions are briefly described in Table 11.2 and discussed further in the text.
a. *dnaE* encodes a subunit of DNA polymerase III, the principal DNA polymerase in *E. coli* that is responsible for elongation of DNA chains. A deletion of the *dnaE* gene would undoubtedly lead to a nonfunctional DNA polymerase III. In the absence of DNA polymerase III activity, DNA strands could not be synthesized from RNA primers; hence, synthesis of new DNA strands could not occur, and there would be no chromosome replication.
b. *polA* encodes DNA polymerase I, which is used in DNA synthesis to extend DNA chains made by DNA polymerase III while simultaneously excising the RNA primer by 5'-to-3' exonuclease activity. As discussed in the text, in mutant strains lacking the originally studied DNA polymerase, DNA polymerase I, chromosome replication still occurred. Thus, chromosome replication would occur normally in an *E. coli* strain carrying a deletion of *polA*.
c. *dnaG* encodes DNA primase, the enzyme that synthesizes the RNA primer on the DNA template. Without the synthesis of the short RNA primer, DNA polymerase III cannot initiate DNA synthesis and, therefore, chromosome replication will not take place.
d. *lig* encodes DNA ligase, the enzyme that catalyzes the ligation of Okazaki fragments. In a strain carrying a deletion of *lig*, DNA synthesis would occur, but stable progeny chromosomes would not result because the Okazaki fragments could not be ligated together, so the lagging strand synthesized discontinuously on the lagging-strand template would be in fragments.
e. *ssb* encodes the single-strand binding proteins that bind to and stabilize the single-stranded DNA regions produced as the DNA is unwound at the replication fork. In the absence of single-strand binding proteins, impeded or absent DNA replication would result because the replication bubble could not be kept open.
f. *oriC* is the origin of replication region in *E. coli*, that is, the location at which chromosome replication initiates. Without the origin, the initiator protein cannot bind, no replication bubble can form, and therefore chromosome replication cannot take place.

QUESTIONS AND PROBLEMS

11.1 Compare and contrast the conservative and semiconservative models of DNA replication.

11.2 Describe the Meselson-Stahl experiment, and explain how it showed that DNA replication is semiconservative.

*11.3 In the Meselson-Stahl experiment, ^{15}N-labeled cells were shifted to ^{14}N medium, at what we can designate as generation 0.
a. For the semiconservative model of replication, what proportion of ^{15}N-^{15}N, ^{15}N-^{14}N, and ^{14}N-^{14}N would you expect to find after 1, 2, 3, 4, 6, and 8 replication cycles?
b. Answer part a in terms of the conservative model of DNA replication.

11.4 Suppose *E. coli* cells are grown on an ^{15}N medium for many generations. Then they are quickly shifted to an ^{14}N medium, and DNA is extracted from the samples taken after one, two, and three replication cycles. The extracted DNA is subjected to equilibrium density gradient centrifugation in CsCl. In figures (a) and (b), using the reference positions of pure ^{15}N and pure ^{14}N DNA as guides, indicate where the bands of DNA would be if replication were semiconservative or conservative.

a) Semiconservative model

b) Conservative model

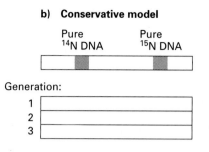

11.5 A spaceship lands on Earth, and with it a sample of extraterrestrial bacteria. You are assigned the task of determining the mechanism of DNA replication in this organism.

You grow the bacteria in unlabeled medium for several generations, then grow it in presence of ^{15}N for exactly one generation. You extract the DNA and subject it to CsCl centrifugation. The banding pattern you find is as follows:

It appears to you that this is evidence that DNA replicates in the semiconservative manner, but you are wrong. Why? What other experiment could you perform (using the same sample and technique of CsCl centrifugation) that would further distinguish between semiconservative and dispersive modes of replication?

*11.6 The elegant Meselson-Stahl experiment was among the first experiments to contribute to what is now a highly detailed understanding of DNA replication. Reconsider this experiment in light of current molecular models by answering the following questions.
a. Does the fact that DNA replication is semiconservative mean that it must be semidiscontinuous?
b. Does the fact that DNA replication is semidiscontinuous ensure that it is also semiconservative?
c. Do any properties of known DNA polymerases ensure that DNA is synthesized semiconservatively?

*11.7 List the components necessary to make DNA in vitro by using the enzyme system isolated by Kornberg.

*11.8 Kornberg isolated DNA polymerase I from *E. coli*. DNA polymerase I has an essential function in DNA replication. Which of the following is that function?
a. filling gaps left by the removal of RNA primer
b. filling in gaps where introns are removed
c. the formation of stem loops in tRNA
d. recognition of rho factor for the initiation of transcription
e. production of poly(A) tails on eukaryotic mRNAs

11.9 Assume you have a DNA molecule with the base sequence T-A-T-C-A going from the 5' to the 3' end of one of the polynucleotide chains. The building blocks of the DNA are drawn as in the following figure. Use this shorthand system to diagram the completed double-stranded DNA molecule, as proposed by Watson and Crick.

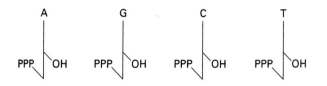

*11.10 Base analogs are compounds that resemble the natural bases found in DNA and RNA but are not normally found in those macromolecules. Base analogs can replace their normal counterparts in DNA during in vitro DNA synthesis. Four base analogs were studied for their effects on in vitro DNA synthesis using *E. coli* DNA polymerase. The results were as follows, with the amounts of DNA synthesized expressed as percentages of the DNA synthesized from normal bases only.

ANALOG	NORMAL BASES SUBSTITUTED BY THE ANALOG			
	A	T	C	G
A	0	0	0	25
B	0	54	0	0
C	0	0	100	0
D	0	97	0	0

Which bases are analogs of adenine? of thymine? of cytosine? of guanine?

11.11 Describe the semidiscontinuous model of DNA replication. What is the evidence showing that DNA synthesis is discontinuous on at least one template strand?

*11.12 Distinguish between a primer strand and a template strand.

11.13 In *E. coli*, the replication fork moves forward at 500 nucleotide pairs per second. How fast is the DNA ahead of the replication fork rotating?

*11.14 A diploid organism has 4.5×10^8 base pairs in its DNA. This DNA is replicated in 3 minutes. Assuming all replication forks move at a rate of 10^4 base pairs per minute, how many replicons (replication units) are present in this organism's genome?

11.15 The following events, steps, or reactions occur during *E. coli* DNA replication. For each entry in Column A, select the appropriate entry in Column B.

Each entry in A may have more than one answer, and each entry in B can be used more than once.

Column A	Column B
____ a. Unwinds the double helix	A Polymerase I
____ b. Prevents reassociation of complementary bases	B Polymerase III
	C Helicase
____ c. Is an RNA polymerase	D Primase
____ d. Is a DNA polymerase	E Ligase
____ e. Is the "repair" enzyme	F SSB protein
____ f. Is the major elongation enzyme	G Gyrase
	H None of these
____ g. A 5'-to-3' polymerase	
____ h. A 3'-to-5' polymerase	
____ i. Has 5'-to-3' exonuclease function	
____ j. Has 3'-to-5' exonuclease function	
____ k. Bonds free 3'-OH end of a polynucleotide to a free 5' monophosphate end of polynucleotide	
____ l. Bonds 3'-OH end of a polynucleotide to a free 5' nucleotide triphosphate	
____ m. Separates daughter molecules and causes supercoiling	

11.16 Compare and contrast the three *E. coli* DNA polymerases with respect to their enzymatic activities.

*11.17 In *E. coli*, distinguish among the activities of primase; single-strand binding protein; helicase; DNA ligase; DNA polymerase I; and DNA polymerase III in DNA replication.

11.18 What properties would you expect an *E. coli* cell to have if it had a temperature-sensitive mutation in the gene for DNA ligase?

*11.19 Chromosome replication in *E. coli* commences from a constant point, called the origin of replication. It is known that DNA replication is bidirectional. Devise a biochemical experiment to prove that the *E. coli* chromosome replicates bidirectionally. (*Hint*: Assume that the amount of gene product is directly proportional to the number of genes.)

11.20 What property of DNA replication was indicated by the presence of Okazaki fragments?

*11.21 A space probe returns from Jupiter and brings with it a new microorganism for study. It has double-stranded DNA as its genetic material. However, studies of replication of the alien DNA reveal that, while the process is semiconservative, DNA synthesis is continuous on both the leading-strand and the lagging-strand templates. What conclusion(s) can you draw from that result?

11.22 Draw a eukaryotic chromosome as it would appear at each of the following cell cycle stages. Show both DNA strands, and use different line styles for old and newly synthesized DNA.
a. G_1
b. anaphase of mitosis
c. G_2
d. anaphase of meiosis I
e. anaphase of meiosis II

*11.23 Autoradiography is a technique that allows radioactive areas of chromosomes to be observed under the microscope. The slide is covered with a photographic emulsion, which is exposed by radioactive decay. In regions of exposure the emulsion forms silver grains on being developed. The tiny silver grains can be seen on top of the (much larger) chromosomes. Devise a method for finding out which regions in the human karyotype replicate during the last 30 min of the S period. (Assume a cell cycle in which the cell spends 10 h in G_1, 9 h in S, 4 h in G_2, and 1 h in M.)

11.24 A mutant *Tetrahymena* has an altered repeated sequence in its telomeric DNA. What change in the telomerase enzyme would have this phenotype?

*11.25 What evidence is there that telomere length is under genetic control? Why might such control be important?

CHAPTER *12*

GENE EXPRESSION: TRANSCRIPTION

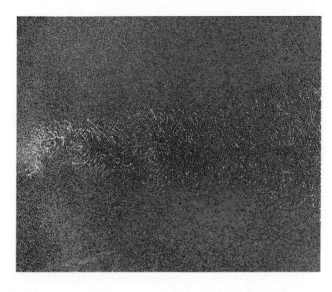

PRINCIPAL POINTS

~ Transcription, the process of copying DNA base-pair sequences into RNA base sequences, is similar in prokaryotes and eukaryotes. The DNA unwinds in a short region next to a gene, and an RNA polymerase catalyzes the synthesis of an RNA molecule. Only one strand of the double-stranded DNA is transcribed into an RNA molecule.

~ In *E. coli*, the initiation of transcription of protein-coding (structural) genes requires a complex of RNA polymerase and the sigma factor protein binding to the promoter. Once transcription has begun, the sigma factor dissociates and RNA synthesis is completed by the RNA polymerase core enzyme. Termination of transcription is signaled by specific sequences in the DNA.

~ In *E. coli*, a single RNA polymerase synthesizes mRNA, tRNA, and rRNA. Eukaryotes have three distinct nuclear RNA polymerases, each of which transcribes different gene types: RNA polymerase I transcribes the genes for the 28S, 18S, and 5.8S ribosomal RNAs; RNA polymerase II transcribes mRNA genes and some snRNA genes; and RNA polymerase III transcribes genes for the 5S rRNAs, the tRNAs, and the other snRNAs.

~ The transcripts of protein-coding genes are linear precursor mRNAs (pre-mRNAs). Prokaryotic mRNAs are modified little once they are transcribed, while eukaryotic mRNAs are modified by the addition of a 5' cap and a 3' poly(A) tail. Most eukaryotic pre-mRNAs contain sequences, called introns (for *inter-vening* sequences), that do not code for amino acids. The introns are removed as the primary transcript is processed to produce the mature, functional mRNA molecule. Segments of RNA that remain in the mRNA are exons (for *expressed* sequences).

~ Introns are removed from pre-mRNAs in a series of well-defined steps. Intron removal begins with the cleavage of the pre-mRNA at the 5' splice junction. The free 5' end of the intron loops back and bonds to a nucleotide in another region of the intron called the branch-point consensus sequence. Cleavage at the 3' splice junction releases the intron, which is shaped like a lariat. The exons that flanked the intron are then spliced together. The removal of introns from eukaryotic pre-mRNA occurs in the nucleus in complexes called spliceosomes. The spliceosome consists of several small nuclear ribonucleoprotein particles (snRNPs) bound specifically to each intron.

~ Ribosomes are the cellular organelles on which protein synthesis takes place. In both prokaryotes and eukaryotes, ribosomes consist of two unequally sized subunits. Each subunit consists of a complex between one or more rRNA molecules and many ribosomal proteins. Eukaryotic ribosomes are larger and more complex than prokaryotic ribosomes.

~ In eukaryotes, three of the four rRNAs are encoded by tandem arrays of transcription units that can number in the thousands. Each transcription unit produces a single pre-rRNA molecule, with the three rRNAs separated by spacer sequences. The individual rRNAs are generated by processing the pre-rRNA to remove the spacers. The fourth rRNA is encoded by separate genes.

~ In the precursor rRNAs of some organisms there are introns, the RNA sequences of which fold into a secondary structure that excises itself, a process called self-splicing. This process does not involve protein enzymes.

~ tRNA molecules bring amino acids to the ribosomes, where the amino acids are polymerized into a polypeptide chain. All tRNAs are about the same length, contain a number of modified bases, and have similar three-dimensional shapes.

~ For 5S rRNA genes and tRNA genes, the promoter for RNA polymerase III is located within the transcribed region of the gene. The internal promoter is called the internal control region. Transcription factors and RNA polymerase III bind to the internal control region for transcription of the genes.

*T*he structure, function, development, and reproduction of an organism depend on the properties of the proteins present in each cell and tissue. A protein consists of one or more chains of amino acids. Each chain of amino acids is called a polypeptide, and the sequence of amino acids in a polypeptide chain is coded for by a gene. Two major steps occur in the process of protein synthesis:

transcription and translation. **Transcription** is the synthesis of a single-stranded RNA copy of a segment of DNA. **Translation** (protein synthesis) is the conversion of the messenger RNA base sequence information into the amino acid sequence of a polypeptide. Your goal in this chapter is to learn about the process of transcription.

GENE EXPRESSION: AN OVERVIEW

In 1956, three years after Watson and Crick proposed their double-helix model of DNA, Crick gave the name *Central Dogma* to the two-step process of DNA → RNA → protein (transcription followed by translation). Transcription is the synthesis of an RNA copy of a segment of DNA. The RNA is synthesized by RNA polymerase.

Not all genes encode proteins, so not all gene transcripts are the kind of RNA that is translated. In fact, there are four different types of RNA molecules, each encoded by its own type of gene:

1. **mRNA—messenger RNA**: Encodes the amino acid sequence of a polypeptide. mRNAs are the transcripts of *protein-coding genes*, also called **structural genes**.
2. **tRNA—transfer RNA**: Brings amino acids to ribosomes during translation.
3. **rRNA—ribosomal RNA**: With ribosomal proteins, makes up the ribosomes, the organ-

elles that translate the mRNA to produce a polypeptide.
4. **snRNA—small nuclear RNA**: With proteins, forms complexes that are used in eukaryotic RNA processing.

In the remainder of the chapter we will learn about the transcription process and more about these four types of RNA.

THE TRANSCRIPTION PROCESS

In this section we discuss how an RNA chain is synthesized.

RNA Synthesis

Transcription is also referred to as *gene expression*. Associated with each gene are sequences called **gene regulatory elements**, which are involved in the regulation of transcription.

In both prokaryotes and eukaryotes, the enzyme **RNA polymerase** catalyzes the process of transcription (Figure 12.1). The DNA double helix unwinds for a short region next to the gene before transcription can begin. Only one of the two DNA strands is transcribed into an RNA.

In transcription, RNA is synthesized in the 5'-to-3' direction. The 3'-to-5' DNA strand that is read to make the RNA strand is called the *template strand*. The 5'-to-3' DNA strand complementary to the template strand that has the *same* polarity as the resulting RNA strand is called the *nontemplate strand*.

~ **FIGURE 12.1**

Transcription process. The DNA double helix is denatured by RNA polymerase in prokaryotes, or by other proteins in eukaryotes. RNA polymerase then catalyzes the synthesis of a single-stranded RNA chain, beginning at the "start of transcription" point. The RNA chain is made in the 5'-to-3' direction, using only one strand of the DNA as a template to determine the base sequence.

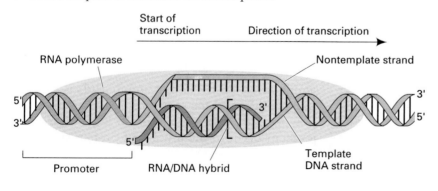

~ **FIGURE 12.2**

Chemical reaction involved in the RNA polymerase-catalyzed synthesis of RNA on a DNA template strand.

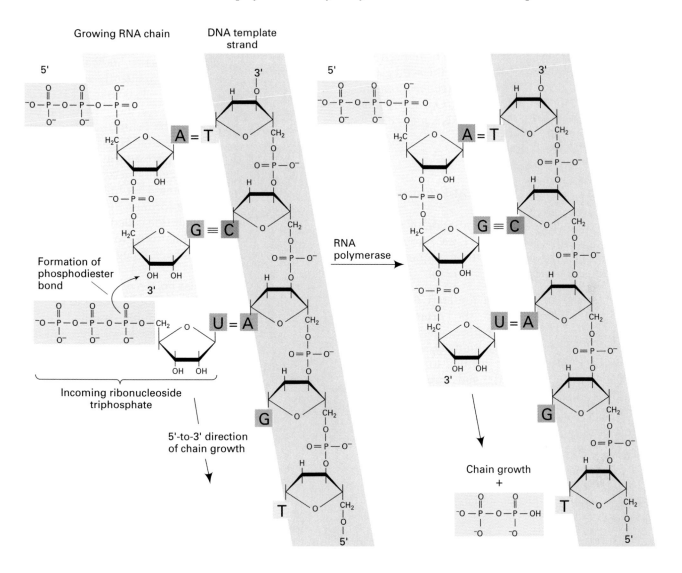

The RNA precursors for transcription are the ribonucleoside triphosphates ATP, GTP, CTP, and UTP, collectively called NTPs (*nucleoside triphosphates*). RNA synthesis occurs by polymerization reactions that are very similar to the polymerization reactions involved in DNA synthesis (Figure 12.2; DNA polymerization is shown in Figure 11.4). RNA polymerase selects the next nucleotide to be added to the chain by its ability to pair with the exposed base on the DNA template strand. Unlike DNA polymerases, RNA polymerases can initiate new polynucleotide chains (no primer is needed), but they have no proofreading abilities.

Recall that RNA chains contain nucleotides with the base uracil instead of thymine, and that uracil pairs with adenine. Therefore, where an A nucleotide occurs on the DNA template chain, a U nucleotide is placed in the RNA chain instead of a T.

As an example, if the template DNA strand reads

3'-ATACTGGAC-5',

then the RNA chain will be synthesized in the 5'-to-3' direction and will have the sequence

5'-UAUGACCUG-3'.

KEYNOTE

Transcription, the process of copying DNA base-pair sequences into RNA base sequences, is similar in prokaryotes and eukaryotes. The DNA unwinds in a short region next to the sequence to be transcribed, and an RNA polymerase catalyzes the synthesis of an RNA molecule in the 5'-to-3' direction along the 3'-to-5' template strand of the DNA. Only one strand of the DNA is transcribed into an RNA molecule.

Initiation of Transcription at Promoters

In this section we discuss the initiation of transcription in prokaryotes, focusing on *E. coli*. A prokaryotic gene may be divided into three sequences with respect to its transcription (Figure 12.3):

1. A sequence upstream of the start of the RNA coding sequence called the **promoter**, with which the RNA polymerase interacts to begin transcription. Conventionally, upstream sequences are numbered relative to +1, the first nucleotide transcribed, and preceded by a minus (−) sign.

2. The RNA-coding sequence; that is, the DNA sequence transcribed by RNA polymerase into the RNA transcript.

3. A **terminator sequence** (or more simply, **terminator**), which is downstream of the end of the RNA-coding sequence, and specifies where transcription will stop.

From comparisons of sequences upstream of coding sequences and from studies of the effects of specific base-pair changes, two DNA sequences in most promoters of *E. coli* genes have been shown to be critical for specifying the initiation of transcription. These sequences generally are found at −35 and −10; that is, centered at 35 and 10 base pairs upstream from +1, the base pair at which transcription starts. The **consensus sequence** (the sequence indicating which nucleotides are found most frequently at each position) for the −35 region (the −35 box) is 5'-TTGACA-3'. The consensus sequence for the −10 region (the **−10 box**, formerly called the **Pribnow box** after the researcher who first discovered it) is 5'-TATAAT-3'.

For transcription to begin, a form of RNA polymerase called the *holoenzyme* (or *complete enzyme*) must bind to the promoter. This holoenzyme consists of the **core enzyme** form of RNA polymerase bound to another polypeptide called the *sigma factor* (σ). The core enzyme has four polypeptides. The sigma factor is essential for recognition of a promoter sequence.

The RNA polymerase holoenzyme binds to a promoter in two steps. First, it binds loosely to the −35 box while the DNA is still double stranded (Figure 12.4a). Next, the RNA polymerase binds more tightly to the DNA as the DNA untwists for about 17 base pairs centered around the −10 box (Figure 12.4b). Once the RNA polymerase is bound at the −10 box, it is oriented properly to begin transcription at the correct nucleotide (see Figure 12.4b).

Promoters differ slightly in their actual sequence, so the binding efficiency of RNA polymerase varies. As a result, the rate at which transcription is initiated varies from gene to gene, which explains in part why different genes have different levels of expression.

Several different sigma factors in *E. coli* play important roles in regulating gene expression. Each type of sigma factor binds to the core RNA

~ **FIGURE 12.3**

Promoter, RNA-coding sequence, and terminator regions of a gene. The promoter is termed "upstream" of the coding sequence, and the terminator is termed "downstream" of the coding sequence. The coding sequence begins at nucleotide +1.

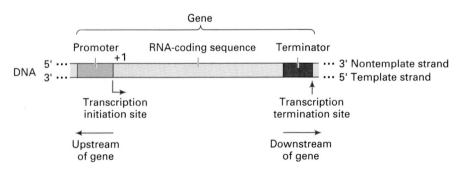

~ FIGURE 12.4

Action of _E. coli_ RNA polymerase in the initiation and elongation stages of transcription. (a) In initiation, the RNA polymerase holoenzyme first binds loosely to the promoter at the −35 region. (b) As initiation continues, RNA polymerase binds more tightly to the promoter at the −10 region, accompanied by a local untwisting of about 17 bp around the −10 region. At this point, the RNA polymerase is correctly oriented to begin transcription at +1. (c) After 8 to 9 nucleotides have been polymerized, the sigma factor dissociates from the core enzyme. (d) As the RNA polymerase elongates the new RNA chain, the enzyme untwists the DNA ahead of it; as the double helix reforms behind the enzyme, the RNA is displaced away from the DNA.

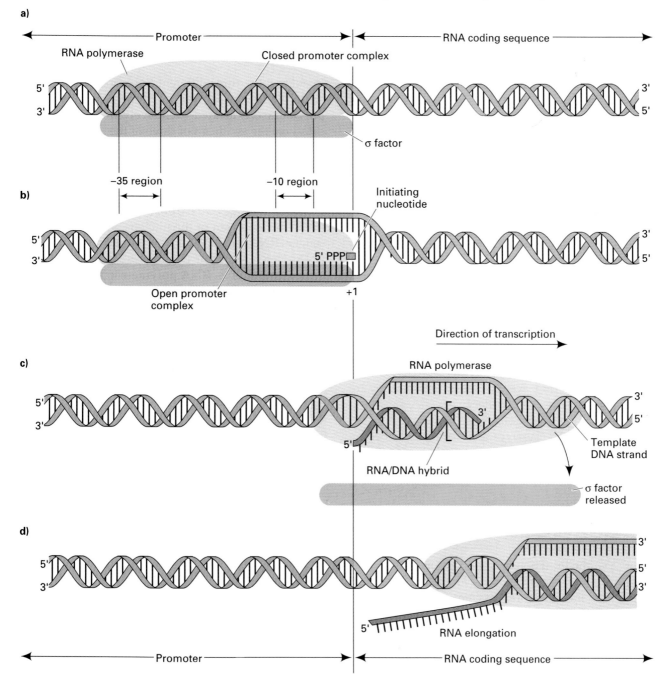

polymerase and permits the holoenzyme to recognize different promoters. Most promoters have the recognition sequences we have just discussed, and these are recognized by a sigma factor called σ^{70}.

Under conditions of high heat (heat shock) and other forms of stress, a different sigma factor called σ^{32} increases in amount. This directs some RNA polymerase molecules to bind to the promoters of genes that encode proteins required to cope with the stress. Such promoters have recognition sequences specific for the σ^{32} factor. Other sigma factors control expression of yet other types of genes under various conditions.

Elongation and Termination of an RNA Chain

RNA synthesis takes place in a region of the DNA that has separated to form a transcription bubble. Once 8 or 9 RNA nucleotides have been linked together, the sigma factor dissociates from the RNA polymerase core enzyme (Figure 12.4c) and is used again in other transcription initiation reactions. The core enzyme completes the transcription of the gene.

As the core RNA polymerase moves along, it untwists the DNA double helix ahead of it (Figure 12.4d). Within the untwisted region, some bases of RNA are base-paired to the DNA in a temporary RNA-DNA hybrid, while the rest of the RNA is displaced away from the DNA as the DNA double helix reforms behind the enzyme. Transcription proceeds at a rate averaging 30 to 50 nucleotides per second.

Termination of prokaryotic gene transcription is signaled by *terminators*. One important protein involved in the termination of transcription of some *E. coli* genes is *rho* (ρ). The terminators of such genes are called *rho-dependent terminators*. At many other terminators the core RNA polymerase itself carries out the termination events. Those terminators are called *rho-independent terminators*.

Since only one type of RNA polymerase is found in prokaryotes, all classes of genes are transcribed by it, namely protein-coding genes, tRNA genes, and rRNA genes.

KEYNOTE

In *E. coli*, the initiation of transcription of protein-coding genes requires the holoenzyme form of RNA polymerase (core enzyme + sigma factor) binding to the promoter. Once transcription has begun, the sigma factor dissociates, and RNA synthesis is completed by the RNA polymerase core enzyme. Termination of transcription is signaled by specific sequences in the DNA. Two types of termination sequences are found, and a particular gene will have one or the other. One type of terminator is recognized by RNA polymerase in association with the *rho* factor, and the other type is recognized by RNA polymerase alone.

TRANSCRIPTION IN EUKARYOTES

Transcription is more complicated in eukaryotes than in prokaryotes because eukaryotes possess three different classes of RNA polymerases, and because of the way the processing of transcripts to their functional forms occurs.

Eukaryotic RNA Polymerases

In eukaryotes, three different RNA polymerases transcribe the genes for the four types of RNAs. **RNA polymerase I**, located exclusively in the nucleolus, catalyzes the synthesis of three of the RNAs found in ribosomes: the 28S, 18S, and 5.8S ribosomal RNA (rRNA) molecules. The S values derive from the rate at which the rRNA molecules sediment during centrifugation; the S values give a very rough indication of molecular sizes. **RNA polymerase II**, found only in the nucleoplasm of the nucleus, synthesizes messenger RNAs (mRNAs) and some small nuclear RNAs (snRNAs), some of which are involved in RNA proc-essing events. **RNA polymerase III**, found only in the nucleoplasm, synthesizes: (1) the transfer RNAs (tRNAs), which bring amino acids to the ribosome; (2) 5S rRNA, a small rRNA molecule found in each ribosome; and (3) the snRNAs not made by RNA polymerase II.

KEYNOTE

In *E. coli*, a single RNA polymerase synthesizes mRNA, tRNA, and rRNA. Eukaryotes have three distinct nuclear RNA polymerases, each of which transcribes different gene types: RNA polymerase I transcribes the genes for the 28S, 18S, and 5.8S ribosomal RNAs; RNA polymerase II transcribes mRNA genes and some snRNA genes; and RNA polymerase III transcribes genes for the 5S rRNAs, the tRNAs, and the remaining snRNAs.

Transcription of Protein-Coding Genes by RNA Polymerase II

In eukaryotes, RNA polymerase II transcribes protein-coding genes. The product of transcription is a **precursor-mRNA (pre-mRNA)** molecule, a transcript that must be modified and/or processed to produce the mature, functional mRNA molecule. In this section we discuss the sequences and molecular events involved in transcribing a protein-coding gene.

Promoters of protein-coding genes are analyzed in two principal ways. One way is to examine the effect of mutations that delete or alter base sequences upstream from the start point of transcription, and to see if those mutants affect transcription. Mutations that significantly affect transcription define important promoter elements. The second way is to compare the DNA sequences upstream of a number of protein-coding genes to see if there are any regions with similar sequences. The results of these experiments show that the promoters of protein-coding genes contain *basal promoter elements* and *promoter proximal elements*.

The best characterized basal promoter elements are the **TATA box** or **TATA element** (also called the **Goldberg-Hogness box** after its discoverers) located at about position −25, and a pyrimidine-rich sequence near the transcription start site called the *initiator element* or simply *Inr*. The TATA box has the seven-nucleotide consensus sequence TATAAAA. It is found in many genes, so it has only a general activity in transcription initiation. AT-rich DNA is easy to denature to single strands, so the TATA box probably facilitates strand separation for the initiation of transcription.

Promoter proximal elements are further upstream from the TATA box, between about 50 and 200 nucleotides from the start of transcription. Examples of these elements are the **CAAT** ("cat") **box**, named for its consensus sequence and located at about −75, and the **GC box**, consensus sequence GGGCGG, located at about −90. Both the CAAT box and GC box work in either orientation and, like the TATA box, they have only general activities in transcription initiation.

Promoters contain various combinations of basal promoter elements and promoter proximal elements. In other words, a number of different elements can contribute to promoter function, and no one element is essential for transcription to take place.

Accurate initiation of transcription of protein-coding genes involves the assembly of RNA polymerase II and a number of proteins called **basal transcription factors** (TFs) on the basal promoter elements. All three eukaryotic RNA polymerases require basal transcription factors for transcription initiation. The basal transcription factors are numbered for the RNA polymerase with which they work and lettered to reflect their order of discovery. For example, TFIID is the fourth basal transcription factor (D) discovered that works with RNA polymerase II.

For protein-coding genes, the basal transcription factors and RNA polymerase II bind to promoter elements in a particular order (Figure 12.5a). First, TFIID binds to the TATA box to form the *initial committed complex*. The multisubunit TFIID has one subunit called the *TATA-binding protein* (TBP) that actually

recognizes the TATA box sequence, and a number of other proteins called *TBP-associated factors* (TAFs). The TFIID-TATA box complex acts as a binding site for TFIIB, which then recruits RNA polymerase II and TFIIF to produce the *minimal transcription initiation complex*. (RNA polymerase II, like all eukaryotic RNA polymerases, cannot directly recognize and bind to promoter elements.) Next, TFIIE and TFIIH bind to produce the *complete transcription initiation complex*, also called the *preinitiation complex* or PIC because it is ready to begin transcription.

The initiation complex is sufficient for only a low level of transcription. For a high level of transcription to occur, other transcription factors called **activators** control which promoters are transcribed actively. Activators bind to regulatory elements called **enhancers**, sequences required for maximal transcription. Figure 12.5b shows a simplified model in which an activator binds to an enhancer and, through an interaction with another protein called an adapter, forms a bridge to the preinitiation complex. As a result of this bridging the DNA between the basal promoter elements and the enhancer becomes looped. The resulting interaction of activator proteins and the preinitiation complex stimulates transcription.

Enhancers are found singly or in multiple copies. They function in either orientation and at a large distance from the gene, either upstream, downstream, or in the gene itself. In most cases the enhancers are upstream of the gene. Similar elements that have essentially the same properties as enhancer elements, except that they repress rather than activate gene transcription, are called **silencer elements**. Silencers are much less common than enhancers; they function when transcription factors called *repressors* bind to them. Thus, the level of transcription resulting from activities at the basal promoter elements is modulated by the effects of other transcription factors binding to enhancer and silencer elements. Because the activators and repressors are cell- and tissue-specific, the interactions of these transcription factors at enhancers and silencers are responsible for cell- and tissue-specific gene expression.

KEYNOTE

Protein-coding genes in eukaryotes are transcribed by RNA polymerase II. The promoter for these genes consists of different combinations of promoter elements. Low levels of transcription occur when basal transcription factors bind to promoter elements. The rate of transcription is then controlled by the binding of other transcription factors to enhancers and silencers. Only when activators bind to enhancers can maximal transcription occur.

~ FIGURE 12.5

Events that may occur during the initiation of transcription catalyzed by RNA polymerase II. (a) Assembly of the preinitiation complex of RNA polymerase II and transcription factors on the basal promoter elements. (b) Simplified model for the stimulation of transcription by the binding of an activator transcription factor at an enhancer. In actuality, many more proteins are involved than are shown.

a) Assembly of preinitiation complex

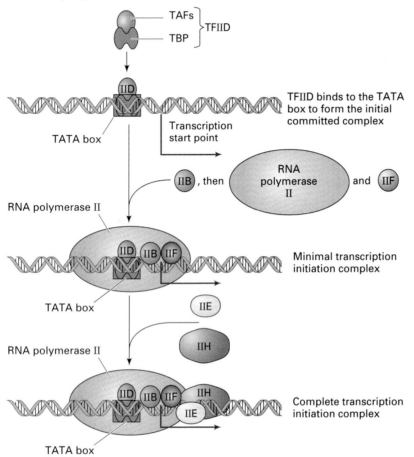

TAFs ⎫
TBP ⎬ TFIID

TFIID binds to the TATA box to form the initial committed complex

TATA box

Transcription start point

IIB , then   RNA polymerase II   and IIF

RNA polymerase II

Minimal transcription initiation complex

TATA box

IIE

IIH

RNA polymerase II

Complete transcription initiation complex

TATA box

b) Stimulation of transcription by activator binding to an enhancer

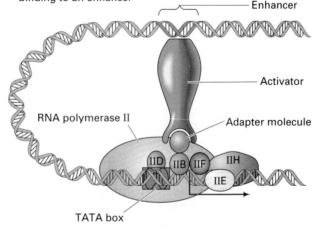

Enhancer

Activator

RNA polymerase II

Adapter molecule

TATA box

Eukaryotic mRNAs

Figure 12.6 shows the general structure of the mature, biologically active mRNA as it exists in both prokaryotic and eukaryotic cells. The mRNA molecule has three main parts. At the 5' end is a **leader sequence**, or 5' untranslated region (5' UTR), which varies in length between mRNAs of different genes. Following the 5' leader sequence is the actual **coding sequence** of the mRNA, the sequence that determines the amino acid sequence of a protein during translation. The coding sequence varies in length, depending on the length of the protein for which it codes. Following the amino acid–coding sequence is a **trailer sequence**, or 3' untranslated region (3' UTR). The trailer sequence also varies in length from mRNA to mRNA.

~ FIGURE 12.6

~ FIGURE 12.6

General structure of mRNA found in both prokaryotic and eukaryotic cells.

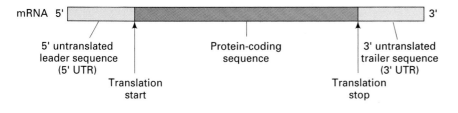

The production of mRNA is different in prokaryotes and eukaryotes. In prokaryotes (Figure 12.7a), the RNA transcript functions directly as the mRNA molecule. Thus, the base pairs of a prokaryotic gene are colinear with the bases of the translated mRNA. In eukaryotes (Figure 12.7b), the RNA transcript (the pre-mRNA) must be modified in the nucleus by a series of events known as *RNA processing* to produce the mature mRNA. In addition, since prokaryotes lack a nucleus, an mRNA begins to be translated on ribosomes before it has been completely transcribed; this process is called *coupled transcription and translation*. In eukaryotes, the mRNA must migrate from the nucleus to the cytoplasm (where the ribosomes are located) before it can be translated. Thus, a eukaryotic mRNA is always transcribed completely and processed before it is translated.

Another fundamental difference between prokaryotic and eukaryotic mRNAs is that prokaryotic mRNAs often are *polycistronic*, meaning they contain the amino acid–coding information from more than one gene, while eukaryotic mRNAs are always *monocistronic*, meaning they contain the amino acid–coding information from just one gene. This concept is not shown in Figure 12.7.

~ FIGURE 12.7

Processes for synthesis of functional mRNA in prokaryotes and eukaryotes. (a) In prokaryotes, the mRNA synthesized by RNA polymerase does not have to be processed before it can be translated by ribosomes. Also, since there is no nuclear membrane, translation of the mRNA can begin while transcription continues, resulting in a coupling of transcription and translation. (b) In eukaryotes, the primary RNA transcript is a precursor-mRNA (pre-mRNA) molecule, which is processed in the nucleus by the addition of a 5' cap and a 3' poly(A) tail and the removal of introns. Only when that mRNA is transported to the cytoplasm can translation occur.

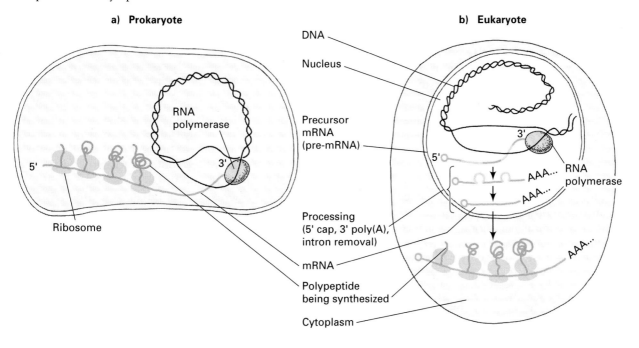

PRODUCTION OF MATURE mRNA IN EUKARY-OTES. Unlike prokaryotic mRNAs, eukaryotic mRNAs are modified at both the 5' and 3' ends. In addition, most protein-coding genes have insertions of non-amino acid–coding sequences called **introns** between the other sequences present in mRNA—the **exons**. The term *intron* is derived from *int*ervening sequence, a sequence that is *not* translated into an amino acid sequence, and the term *exon* is derived from *ex*pressed sequence. Exons include the 5' and 3' UTRs as well as the amino acid–coding portions. The introns are removed in the processing of pre-mRNA to the mature mRNA molecule.

5' and 3' Modifications. Once RNA polymerase II has made about 20 to 30 nucleotides of pre-mRNA, a *capping enzyme* adds a modified nucleotide to the 5' end in a process called **5' capping**. Modifications to several bases near the 5' end of the RNA are also made. The 5' cap consists of a modified GTP (guanosine triphosphate), and it remains throughout processing and is present in the mature mRNA. The cap is essential for the ribosome to bind to the 5' end of the mRNA, an initial step of translation.

Most eukaryotic pre-mRNAs become modified at their 3' ends by the addition of a sequence of about 50 to 250 adenine nucleotides called a **poly(A) tail**. There is no DNA template for the poly(A) tails, and the poly(A) tail remains while pre-mRNA is processed to mature mRNA.

Addition of the poly(A) tail marks the 3' end of the mRNA. That is, in many eukaryotes there are no termination sequences in the DNA to signal the end of transcription. Instead, mRNA transcription continues, in some cases for hundreds or thousands of nucleotides, past a site called the **poly(A) site**. At the poly(A) site, the pre-mRNA is cleaved to generate a 3' OH end to which the poly(A) tail is added by the enzyme **poly(A) polymerase** (PAP), which adds A nucleotides one by one using ATP as the substrate. The poly(A) addition consensus sequence is AAUAAA and is located 10 to 30 nucleotides upstream of the poly(A) site.

Introns. Pre-mRNAs often contain a number of introns. Introns must be excised from each pre-mRNA to generate a mature mRNA that can be translated into a complete polypeptide. The mature mRNA, then, contains in a contiguous form the exons that in the gene were separated by introns.

At the time introns were discovered, it was known that the nucleus contains a large population of RNA molecules of various sizes. These RNA molecules are called **heterogeneous nuclear RNA**, or **hnRNA**, and it was correctly assumed that hnRNAs

included pre-mRNA molecules. One of the early experiments showing the existence of introns was done by Philip Leder's group in 1978. Leder's group studied the β-globin gene in cultured mouse cells. The β-globin gene encodes the 146-amino-acid β-globin polypeptide that is part of a hemoglobin protein molecule. A 1.5-kb RNA molecule was isolated from nuclear hnRNA that was the β-globin pre-mRNA. Like the 0.7-kb mature mRNA, the pre-mRNA has a 5' cap and a 3' poly(A) tail. Leder's group demonstrated that the 1.5-kb pre-mRNA is colinear with the gene that encoded it, whereas the 0.7-kb β-globin mRNA is not. They interpreted their results to mean that the β-globin gene has an intron of about 800 nucleotide pairs. Transcription of the gene results in a 1.5-kb pre-mRNA containing both exon and intron sequences. This RNA is found only in the nucleus. The intron sequence is excised by processing events, and the flanking exon sequences are spliced together to produce a mature mRNA. Subsequent research showed that the β-globin gene contains two introns; the second, smaller intron was not detected in the early research.

The scientific community was shocked by the discovery of introns. At the time it was accepted that the gene sequence was completely colinear with the amino acid sequence of the encoded protein. Thus, finding that genes could be "in pieces" was one of those highly significant discoveries that changed our thinking about genes. In the years following the discovery of introns, we have learned that most eukaryotic protein-coding genes contain introns. Interestingly, some bacteriophage genes also contain introns.

KEYNOTE

The transcripts of protein-coding genes are messenger RNAs in prokaryotes or their precursors in eukaryotes. These molecules vary widely in length according to the size of the polypeptides they specify and whether they contain introns. Prokaryotic mRNAs are not modified once they are transcribed, whereas most eukaryotic mRNAs are modified by the addition of a cap at the 5' end and a poly(A) tail at the 3' end. Many eukaryotic pre-mRNAs contain non-amino acid–coding sequences called introns, which must be excised from the mRNA transcript to make a mature, functional mRNA molecule. The sequences in mRNA separated by introns are called exons.

~ FIGURE 12.8

General sequence of steps in the formation of eukaryotic mRNA. Not all steps are necessary for all mRNAs.

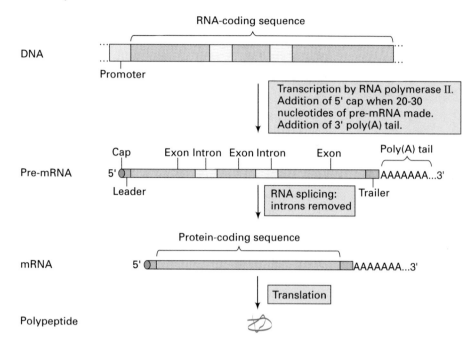

Processing of Pre-mRNA to Mature mRNA. A model for mRNA production from genes with introns is diagrammed in Figure 12.8. The steps are the transcription of the gene by RNA polymerase II, the addition of the 5'-cap and the poly(A) tail to produce the pre-mRNA molecule, and finally the processing of the pre-mRNA in the nucleus to remove the introns and splice the exons together to produce the mature mRNA.

Introns in pre-mRNAs are removed and exons joined in a process called **RNA splicing.** How does the system know where an intron starts and ends? Introns typically begin with 5'-GU and end with AG-3'. More than just those nucleotides are needed to specify a junction between an intron and an exon; the 5' splice junction probably involves at least seven nucleotides, and the 3' splice junction involves at least ten nucleotides of intron sequence.

Splicing of pre-mRNA occurs in the nucleus. The splicing events occur in complexes called **spliceosomes,** which consist of the pre-mRNA bound to **small nuclear ribonucleoprotein particles (snRNPs;** sometimes called *snurps*). These are small nuclear RNAs (snRNAs) associated with proteins. There are six principal snRNAs (named U1–U6), and each is associated with several proteins to form the snRNPs. A simplified model of splicing for two exons sepa-

rated by an intron is shown in Figure 12.9. The steps are as follows:

1. U1 snRNP, a snRNP containing U1 snRNA, binds to the 5' splice junction of the intron. This binding is primarily the result of base-pairing of the U1 snRNA to the 5' splice junction.

2. U2 snRNP binds to a sequence upstream of the 3' splice junction called the **branch-point sequence.**

3. A U4/U6 snRNP and a U5 snRNP interact, and the combination binds to the U1 and U2 snRNPs, causing the intron to loop, thereby bringing the two ends of the intron closer together.

4. U4 snRNP dissociates from the complex, and this results in the formation of the active spliceosome.

5. The snRNPs in the spliceosome cleave the intron from exon 1 at the 5' splice junction, and the now-free 5' end of the intron is bonded to a particular nucleotide (often an A nucleotide) in the branch-point sequence. The intron is now folded into a closed loop called an *RNA lariat structure.*

6. Next, the intron is excised (still in lariat shape) by cleavage at the 3' splice junction, and then exons 1 and 2 are ligated together. The snRNPs are released at this time. The process is repeated for each intron.

~ FIGURE 12.9

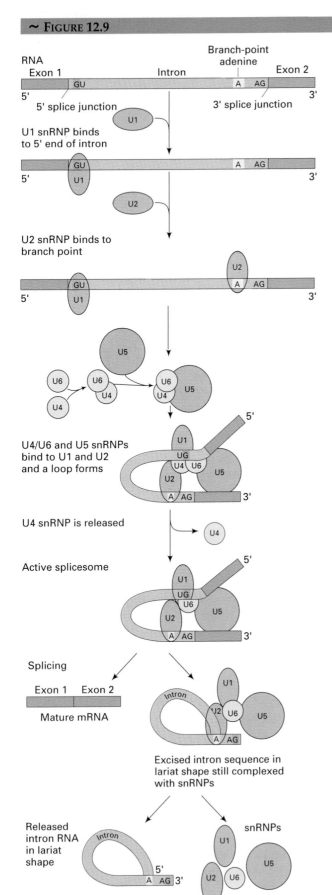

Model for intron removal by the spliceosome.

KEYNOTE

Introns are removed from pre-mRNAs in a series of well-defined steps. Introns begin with a 5'-GU and end with an AG-3'. Intron removal begins with the cleavage of the pre-mRNA at the 5' splice junction. The free 5' end of the intron loops back and binds to a particular nucleotide in the branch-point consensus sequence, which is located upstream of the 3' splice junction. Cleavage at the 3' splice junction releases the intron, now shaped like a lariat. The exons that flanked the intron are spliced together once the intron is excised. The removal of introns from eukaryotic pre-mRNA occurs in the nucleus in complexes called spliceosomes. The spliceosome consists of several small nuclear ribonucleoprotein particles (snRNPs) bound specifically to each intron.

Transcription of Other Genes

In this section we discuss the transcription of nonprotein-coding genes.

RIBOSOMAL RNA AND RIBOSOMES.   Protein synthesis takes place on ribosomes. Each cell contains thousands of ribosomes. Ribosomes bind to mRNA and facilitate the binding of the tRNA to the mRNA so that a polypeptide chain can be synthesized.

Ribosome Structure.   In both prokaryotes and eukaryotes, the ribosomes consist of two unequally sized subunits—the large and small ribosomal subunits—each of which consists of a complex between RNA molecules and proteins. Each subunit contains one or more ribosomal RNA (rRNA) molecules and a large number of **ribosomal proteins**.

We can use the *E. coli* ribosome as a model of a bacterial ribosome. It has a size of 70S (Figure 12.10), with distinctly shaped subunits of 50S (large subunit) and 30S (small subunit). The 50S subunit has 34 different proteins, plus a 23S rRNA (2,904 nucleotides) and a 5S rRNA (120 nucleotides). The small 30S ribosomal subunit has 20 different proteins and a 16S rRNA (1,542 nucleotides). Transcription of the ribosomal protein genes occurs by the mechanism we

~ **FIGURE 12.10**

Two views of a model of the complete (70S) ribosome of *E. coli.* The small (30S) ribosomal subunit is yellow, and the large (50S) ribosomal subunit is red.

~ **FIGURE 12.11**

Composition of whole ribosomes and of ribosomal subunits in mammalian cells.

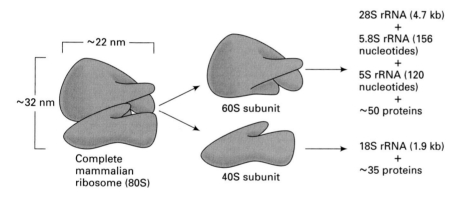

have already discussed for protein-coding genes of bacteria.

Eukaryotic ribosomes are larger and more complex than their prokaryotic counterparts, and they vary in size and composition among eukaryotic organisms. We will use the mammalian ribosome as a model for discussion. Mammalian ribosomes have a size of 80S and consist of a large 60S subunit and a small 40S subunit (Figure 12.11). The 40S subunit contains 18S rRNA (~1,900 nucleotides) and about 35 ribosomal proteins, and the 60S subunit contains 28S (~4,700 nucleotides), 5.8S (156 nucleotides), and 5S rRNAs (120 nucleotides) and about 50 ribosomal proteins. Transcription of the genes that code for ribosomal proteins occurs by the mechanism already discussed for protein-coding genes of eukaryotes.

KEYNOTE

Ribosomes, the organelles within the cell on which protein synthesis takes place, consist of two unequally sized subunits in both prokaryotes and eukaryotes. Each subunit contains one or more ribosomal RNA molecules and many ribosomal proteins.

Transcription of rRNA Genes. In prokaryotes and eukaryotes, the regions of DNA that contain the genes for rRNA are called **ribosomal DNA (rDNA)** or *rRNA transcription units*. Figure 12.12a shows the general organization of an rRNA transcription unit,

~ FIGURE 12.12

rRNA genes and rRNA production in *E. coli*. (a) General organization of an *E. coli* rRNA transcription unit (an *rrn* region); (b) The scheme for the synthesis and processing of a precursor rRNA (p30S) to the mature 16S, 23S, and 5S rRNAs of *E. coli*. (The tRNAs have been omitted from this depiction of the processing scheme.)

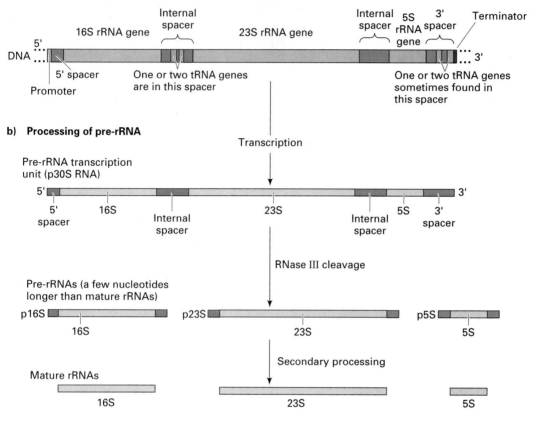

a) *E. coli* rRNA transcription unit

b) Processing of pre-rRNA

named an *rrn* region. There are seven *rrn* regions scattered in the *E. coli* chromosome. Each *rrn* contains one copy each of the 16S, 23S, and 5S rRNA coding sequences arranged in the order 16S-23S-5S. There are one or two tRNA genes in the spacer region between the 16S and 23S rRNA sequences, and another one or two tRNA genes in the 3' spacer region between the end of the 5S rRNA sequence and the 3' end of the transcription unit.

The rRNA transcription units are transcribed by RNA polymerase to produce a single **precursor rRNA (pre-rRNA)** molecule, the 30S pre-rRNA (p30S), that contains all three rRNAs separated by spacers (Figure 12.12b). RNase III cleaves p30S to produce the p16S, p23S, and p5S precursors. Other processing enzymes release the tRNAs from the spacers. As the rRNA genes are being transcribed, the pre-rRNA transcript rapidly becomes associated with ribosomal proteins. Cleavage of the transcript takes

place within a complex formed between the rRNA transcript (as it is being transcribed) and ribosomal proteins. In this way, the functional ribosomal subunits are assembled.

Most eukaryotes have many copies of the genes for each of the four rRNA species 18S, 5.8S, 28S, and 5S. The genes for 18S, 5.8S, and 28S rRNAs are found adjacent to one another in the order 18S-5.8S-28S, with each set of three genes typically repeated 100 to 1,000 times (depending on the organism) to form tandem arrays called **rDNA repeat units** (Figure 12.13a). Human cells, for example, have 1,250 gene sets. The rDNA repeat units are organized into one or more clusters in the genome and, as a result of active transcription, a nucleolus forms around each cluster. Typically the multiple nucleoli so formed fuse to form one nucleolus. Within the nucleolus, the 18S, 5.8S, and 28S rRNAs are synthesized, and they associate with 5S rRNA transcribed from multiple 5S rRNA genes

~ Figure 12.13

rRNA genes and rRNA production in eukaryotes. (a) Generalized diagram of a eukaryotic, ribosomal DNA repeat unit. The coding sequences for 18S, 5.8S, and 28S rRNAs are indicated in light brown. NTS = nontranscribed spacer; ETS = external transcribed spacer; ITS = internal transcribed spacer. (b) The transcription and processing of 45S pre-rRNA to produce the mature 18S, 5.8S, and 28S rRNAs.

a) rDNA repeat unit

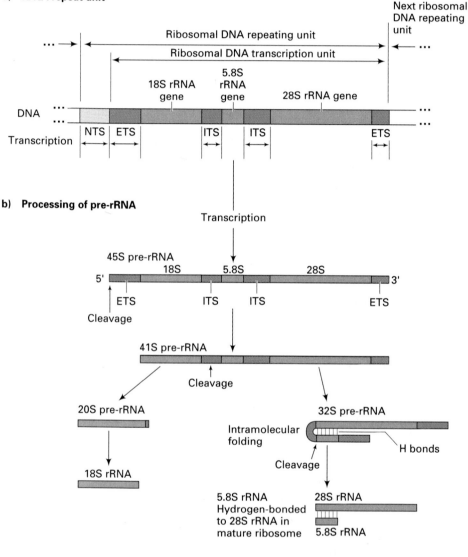

b) Processing of pre-rRNA

located elsewhere in the genome, and with ribosomal proteins, to produce the ribosomal subunits.

Each eukaryotic rDNA repeat unit is transcribed by RNA polymerase I to produce a pre-rRNA molecule (45S pre-rRNA in human cells) (Figure 12.13b). The pre-rRNA contains the 18S, 5.8S, and 28S rRNA sequences and sequences between, and flanking, those three sequences—the **spacer sequences.** The external transcribed spacers—ETSs—are located upstream of the 18S sequence and downstream of the

28S sequence. The internal transcribed spacers—ITSs—are located on each side of the 5.8S sequence; that is, between the 18S sequence and the 5.8S sequence, and between the 5.8S sequence and the 28S rRNA sequence. Between each rDNA repeat unit is the **nontranscribed spacer (NTS) sequence,** which is not transcribed (see Figure 12.13a).

The promoter for RNA polymerase I is upstream of the rRNA genes. Like other eukaryotic RNA polymerases, RNA polymerase I does not bind

~ FIGURE 12.14

Transcription factors involved in the initiation of human rDNA transcription by RNA polymerase I. (For explanation, see text.)

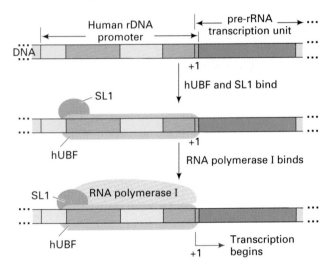

directly to the promoter; instead, specific transcription factors bind and form a complex to which RNA polymerase I binds to begin transcription. For transcription of the human rDNA repeat unit (Figure 12.14), two transcription factors bind to the promoter: the human upstream binding factor (hUBF) binds to two key promoter elements (dark green in the figure), and SL1 binds to the complex. SL1 is needed for promoter recognition and transcription initiation by RNA polymerase I. Termination of transcription of the pre-rRNA involves specific termination sites located downstream of the transcribed sequences of the rDNA.

KEYNOTE

RNA polymerase I transcribes the 18S, 5.8S, and 28S rRNA sequences into a single precursor molecule. A tandem array of 18S-5.8S-28S rRNA transcription units is found in most eukaryotic organisms. Each transcription unit is separated from the next by a nontranscribed spacer (NTS) sequence. The promoter and terminator for each transcription unit are located within the NTS. As for the other eukaryotic RNA polymerases, specific transcription factors are required for the initiation of transcription by RNA polymerase I.

To produce the 18S, 5.8S, and 28S rRNAs, the transcribed pre-rRNA is cleaved at specific sites to remove ITS and ETS sequences. As an example, Figure 12.13b shows how pre-rRNA is processed in human cells. First, the 5' ETS sequence is removed, producing a precursor molecule containing all three rRNA sequences. The next cleavage produces the 20S precursor to 18S rRNA, and the 32S precursor is processed to 28S and 5.8S rRNAs. Removal of the ITS sequence generates the 18S rRNA from the 20S precursor.

All the pre-rRNA-processing events take place in complexes formed between the pre-rRNA, 5S rRNA, and the ribosomal proteins. The 5S rRNA is produced by transcription of the 5S rRNA genes by RNA polymerase III (described later), and the ribosomal proteins are produced by transcription of the ribosomal protein genes by RNA polymerase II and the subsequent translation of the mRNAs. As pre-rRNA processing proceeds, the complexes undergo shape changes resulting in the formation of the 60S and 40S ribosomal subunits. The ribosomal subunits are transported to the cytoplasm, where they function in protein synthesis.

KEYNOTE

Three of the four rRNAs in eukaryotic ribosomes, the 18S, 5.8S, and 28S rRNAs, are transcribed from rDNA onto a single pre-rRNA molecule. In addition to the rRNA coding sequences, the pre-rRNA contains spacer sequences located at the ends of the molecule (external transcribed spacers) and internally between the rRNA sequences (internal transcribed spacers). 5S rRNA, which is transcribed elsewhere in the nucleus, and ribosomal proteins associate with the pre-rRNA in the nucleolus. The pre-rRNA is processed within the complex and shape changes occur, resulting in the formation of the 60S and 40S ribosomal subunits.

Self-Splicing of Introns in *Tetrahymena* Pre-rRNA. In some species of the protozoan *Tetrahymena,* the genes for the 28S rRNA are all interrupted by a 413-bp intron. The intron sequence is removed during processing of the pre-rRNA to produce the mature rRNAs. The excision of the intron—called a group 1 intron—unexpectedly was shown to occur by a *protein-independent reaction* in which the RNA intron folds into a secondary structure that promotes its own excision. This process is called **self-splicing** and was

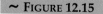

~ **FIGURE 12.15**

Self-splicing reaction for the group I intron in *Tetrahymena* pre-rRNA.

Tetrahymena **pre-rRNA**

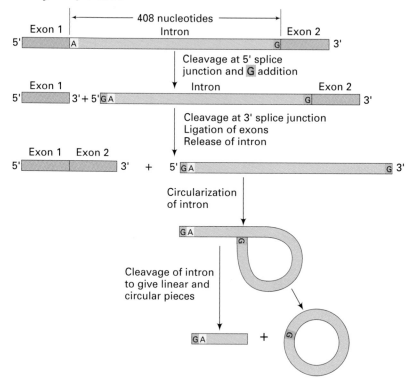

discovered in 1982 by Tom Cech and his research group. In 1989, Cech received the Nobel Prize for his research. Figure 12.15 diagrams the self-splicing reaction for this group I intron in *Tetrahymena* pre-rRNA. The steps are as follows:

1. The pre-rRNA is cleaved at the 5' splice junction as guanosine is added to the 5' end of the intron.

2. The intron is cleaved at the 3' splice junction.

3. The two exons are spliced together.

4. The excised intron circularizes to produce a lariat molecule, which is cleaved to produce a circular RNA and a short linear piece of RNA.

In the processing of *Tetrahymena* pre-rRNA, then, two separate events occur: (1) the pre-rRNA is cleaved to remove spacer sequences from the mature rRNAs, and (2) the intron in the 28S rRNA sequence is removed and the two 28S rRNA parts spliced together. The two events are not identical. *Removal of the spacer sequences releases rRNAs that remain separate. Intron removal, by contrast, results in the splicing together of the RNA sequences that flanked the intron.*

The self-splicing activity of the intron RNA sequence cannot be considered an enzyme activity.

That is, while the RNA carries out the reaction, it is not regenerated in its original form at the end of the reaction, as is the case with protein enzymes. Modified forms of the *Tetrahymena* intron RNA and of other self-cleaving RNAs that function catalytically have been produced in the lab. These **RNA enzymes** are called **ribozymes**; they can be used experimentally to cleave RNA molecules at specific sequences.

The discovery that RNA can act like a protein was an important landmark in biology. The discovery has revolutionized theories about the origin of life. Previous theories proposed that proteins were required for replication of the first nucleic acid molecules. The new theories propose that the first nucleic acid was self-replicating through ribozyme-like activity carried in the molecule.

KEYNOTE

In some precursor rRNAs there are introns, the RNA sequences of which fold into a secondary structure that excises itself. This is called self-splicing. The self-splicing reaction does not involve any proteins.

TRANSCRIPTION OF GENES BY RNA POLYMERASE III. RNA polymerase III transcribes eukaryotic 5S rRNA genes, tRNA genes, and some snRNA genes. 5S rRNA is a 120-nucleotide RNA found in the large subunit of the eukaryotic ribosome. Transfer RNAs (tRNAs) are 75- to 90-nucleotide RNAs that function in both prokaryotes and eukaryotes to bring amino acids to the ribosome-mRNA complex, where they are polymerized into protein chains in the translation (protein synthesis) process.

Prokaryotic 5S rRNA genes are part of the rDNA, while eukaryotic 5S rRNA genes are found in multiple copies in the genome, usually separate from the rDNA repeat units. Prokaryotic tRNA genes are found in one or at most a few copies in the genome, while eukaryotic tRNA genes are repeated many times in the genome. In *Xenopus*, for example, there are about 200 copies of each tRNA gene. Each type of tRNA molecule has a different nucleotide sequence, although all tRNAs have the sequence CCA at their 3' ends. This CCA sequence is added to the tRNA posttranscriptionally. The differences in nucleotide sequences explain the ability of a particular tRNA molecule to bind a particular amino acid. All tRNA molecules are also extensively modified chemically by enzyme reactions after transcription.

The nucleotide sequences of all tRNAs can be arranged into what is called a *cloverleaf* (Figure 12.16). The cloverleaf results from complementary base-pairing between different sections of the molecule, which results in four base-paired "stems" separated by four loops, I, II, III, and IV. Loop II contains the three-nucleotide **anticodon** sequence, which pairs with a codon (three-nucleotide sequence) in mRNA by complementary base-pairing during translation. This codon-anticodon pairing is crucial for adding the amino acid specified by the mRNA to the growing polypeptide chain. Figure 12.17 shows the tertiary structure for phenylalanine tRNA from yeast. All other tRNAs that have been examined show similar three-dimensional structures. The cloverleaf is folded into a compact shape, like an upside-down "L." In this L-shaped structure the 3' end of the tRNA—the end to which the amino acid attaches—is at the opposite end of the "L" from the anticodon loop.

KEYNOTE

Molecules of tRNA (transfer RNA) bring amino acids to the ribosomes, where the amino acids are polymerized into a protein chain. All tRNA molecules are between 75 and 90 nucleotides long, contain a number of modified bases, have similar three-dimensional shapes, and contain a CCA sequence at their 3' ends.

~ FIGURE 12.16

Cloverleaf structure of yeast alanine tRNA. Py = pyrimidine. Modified bases are: I = inosine; T = ribothymidine; Ψ = pseudouridine; D = dihydrouridine; GMe = methylguanosine; GMe₂ = dimethylguanosine; IMe = methylinosine.

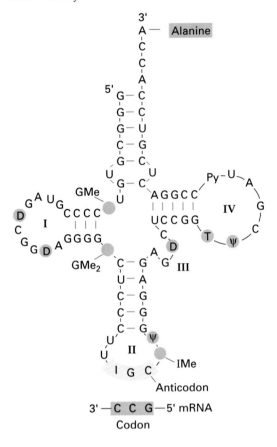

How does RNA polymerase III transcribe the genes under its control? The 5S rRNA genes were the first eukaryotic genes for which the promoter structure and transcription factor requirements were determined. Researchers found that the promoter for RNA polymerase III is *within the transcribed region.* This internal promoter is called the **internal control region (ICR)**. The tRNA genes also have internal promoters, while snRNA genes transcribed by RNA polymerase III typically have promoters upstream of the genes.

Like transcription initiation by RNA polymerases I and II, transcription initiation of 5S rRNA and tRNA genes involves the binding of transcription factors to the promoter, in this case TFIIIs to the ICR, to facilitate the binding of RNA polymerase III. The enzyme then initiates transcription at the beginning of the transcribed region of the gene. Transcription of a

~ **FIGURE 12.17**

tRNA structures. (a) Schematic of the three-dimensional structure of yeast phenylalanine tRNA as determined by X-ray diffraction of tRNA crystals. Note the characteristic L-shaped structure. (b) Photograph of a space-filling molecular model of yeast phenylalanine tRNA. The CCA end of the molecule is at the upper right, and the anticodon loop is at the bottom.

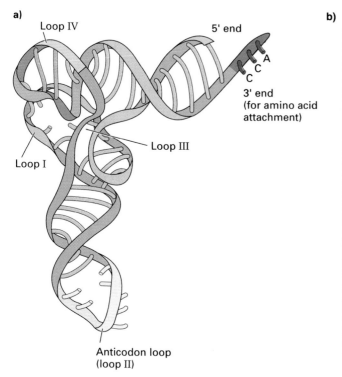

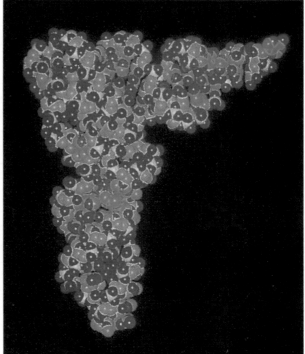

5S rRNA gene directly produces the mature 5S rRNA; there are no sequences that must be removed. Transcription of tRNA genes produces a **pre-tRNA molecule**, which has extra sequences at each end that are subsequently removed to produce the mature tRNA. In yeast, some tRNAs contain an intron in the anticodon loop; a tRNA-specific splicing mechanism is used to cut out the intron and to splice the adjacent tRNA segments together.

KEYNOTE

RNA polymerase III transcribes 5S rRNA genes, tRNA genes, and some snRNA genes. For 5S rRNA genes and tRNA genes, the promoter for RNA polymerase III is located within the gene itself. The internal promoter is called the internal control region. Transcription factors and RNA polymerase III bind to the internal control region for transcription of the genes.

SUMMARY

Transcription

When a gene is expressed, the DNA base-pair sequence is transcribed into the base sequence of an RNA molecule. Transcription of four classes of genes produces messenger RNA (mRNA), transfer RNA (tRNA), ribosomal RNA (rRNA), and small nuclear RNA (snRNA). snRNA is found only in eukaryotes, while the other three classes are found in both prokaryotes and eukaryotes. Only mRNA is translated to produce a protein molecule.

When a gene is transcribed by RNA polymerase, only one of the two DNA strands is copied. The direction of RNA synthesis is 5' to 3'. In addition to coding sequences, genes contain other sequences important for the regulation of transcription, including promoter sequences and terminator sequences. Promoter sequences specify where transcription of the gene is to begin, and terminator sequences specify where transcription is to stop.

Only one type of RNA polymerase is found in bacteria, hence all classes of RNA are synthesized by the same enzyme. Consequently, the promoters for all three classes of genes are very similar. The promoter is recognized by a complex between the RNA polymerase core enzyme and a protein factor called sigma. Once transcription is initiated correctly, the sigma factor dissociates from the enzyme and is reused in other transcription initiation events. Termination is signaled by one of two possible termination sequences.

In eukaryotes there are three different RNA polymerases located in the nucleus. RNA polymerase I, in the nucleolus, transcribes the 18S, 5.8S, and 28S rRNA sequences. These rRNAs are part of ribosomes. RNA polymerase II, in the nucleoplasm, transcribes protein-coding genes (into mRNA precursors) and some snRNA genes. RNA polymerase III, in the nucleoplasm, transcribes tRNA genes, 5S rRNA genes, and the other snRNA genes. The promoters for the three types of RNA polymerase differ but, in each case, none of the three eukaryotic RNA polymerases directly binds to the promoter. Instead, the promoter sequences for the genes they transcribe are first recognized by transcription factors. The transcription factors bind to the DNA and facilitate binding of the polymerase to the transcription factor–DNA complex for correct initiation of transcription. For 5S rRNA genes and tRNA genes transcribed by RNA polymerase III, the promoter is located within the gene itself.

While transcription is very similar in prokaryotes and eukaryotes, the molecular components of the process differ considerably. Even within eukaryotes, three different promoters have evolved along with three distinct RNA polymerases. Much remains to be learned about the associated transcription factors and regulatory proteins before we have a complete understanding of transcription and its regulation.

RNA Molecules and RNA Processing

We discussed the structure, synthesis, and function of mRNA, tRNA, and rRNA. Each mRNA encodes the amino acid sequence of a polypeptide chain. The nucleotide sequence of the mRNA is translated into the amino acid sequence of the polypeptide chain. Ribosomes consist of two unequal-sized subunits, each of which contains both rRNA and protein molecules. The amino acids that are assembled into proteins are brought to the ribosome bound to tRNA molecules.

mRNAs have three main parts: a 5' untranslated leader sequence, the amino acid coding sequence, and the 3' untranslated trailer sequence. In prokaryotes the gene transcript functions directly as the mRNA molecule, while in eukaryotes the RNA transcript must be modified in the nucleus to produce mature mRNA. Modifications include the addition of a 5' cap and a 3' poly(A) tail and the removal of any introns. Intron removal is known as RNA splicing and involves specific interactions with snRNPs in structures called spliceosomes. Only when all processing events have been completed is the mRNA functional; at that point, it can be translated in the cytoplasm once it leaves the nucleus.

Ribosomal RNAs are important structural components of ribosomes. The smaller subunit contains one rRNA molecule, 16S in prokaryotes and 18S in eukaryotes. The prokaryotic larger subunit contains 23S and 5S rRNAs, while the eukaryotic larger subunit contains 28S, 5.8S, and 5S rRNAs.

In prokaryotes, the genes for the 16S, 23S, and 5S rRNAs are transcribed into a single pre-rRNA molecule. Some tRNA genes are found in the spacer regions of each transcription unit. The pre-rRNA molecules are processed to remove the noncoding sequences found at the ends of the molecules and spacers between the rRNA sequences. At the same time the tRNAs are released. All of the processing events occur while the pre-rRNA is associating with the ribosomal proteins so that, when processing is complete, the functional 50S and 30S subunits have been assembled.

In eukaryotes, the 18S, 5.8S, and 28S rRNA genes constitute a transcription unit, and many such transcription units are organized in a tandem array. The 5S rRNA genes are also present in many copies but are usually located elsewhere in the genome. The 18S, 5.8S, and 28S sequences are transcribed into pre-rRNA molecules that, in addition to the rRNA sequences, contain spacer sequences at the ends and between the rRNA sequences. Transcription of tandem arrays of these rRNA genes results in the formation of a nucleolus around each cluster. Assembly of the 60S and 40S ribosomal subunits occurs in the nucleolus. That is, the spacers are removed by specific processing events that take place while the rRNA sequences are associated with ribosomal proteins and with 5S rRNA, which is transcribed elsewhere in the nucleus. The completed ribosomal subunits exit the nucleus and participate in protein synthesis in the cytoplasm.

In the pre-rRNA of *Tetrahymena*, the 28S rRNA sequence is interrupted by an intron. The intron is removed during processing of the pre-rRNA in the nucleolus. The excision of this intron occurs by a protein-independent reaction in which the RNA sequence of the intron folds into a secondary structure that promotes its own excision. This process is called self-splicing.

All tRNAs bring amino acids to the ribosomes. Thus, all tRNAs are very similar in length (75 to 90 nucleotides) and in secondary and tertiary structure. Introns are present in some tRNA genes of yeast. Removal of these introns involves a different mechanism from the other RNA splicing mechanisms described.

ANALYTICAL APPROACHES FOR SOLVING GENETICS PROBLEMS

Q12.1 If two RNA molecules have complementary base sequences, they can hybridize to form a double-stranded helical structure just as DNA can. Imagine that, in a particular region of the genome of a certain bacterium, one DNA strand is transcribed to give rise to the mRNA for protein A, while the other DNA strand is transcribed to give rise to the mRNA for protein B.
a. Would there be any problem in expressing these genes?
b. What would you see in protein B if a mutation occurred that affected the structure of protein A?

A12.1
a. mRNA A and mRNA B would have complementary sequences, so they might hybridize with each other and not be available for translation.
b. Every mutation in gene A would also be a mutation in gene B, so protein B might also be abnormal.

Q12.2 Compare and contrast the following two events in terms of what their consequences would be. Event (1): an incorrect nucleotide is inserted into the new DNA strand during replication, and not corrected by the proofreading or repair systems before the next replication. Event (2): an incorrect nucleotide is inserted into an mRNA during transcription.

A12.2 Event (1) would result in a mutation, assuming it occurred within a gene. The mistake would be inherited by future generations and would affect the structure of all mRNA molecules transcribed from the region; therefore, all molecules of the corresponding protein could be affected.

Event (2) would produce a single aberrant mRNA. This could produce a few aberrant protein molecules. Additional normal protein molecules would exist because other, normal mRNAs would have been transcribed. The abnormal mRNA would soon be degraded. The mRNA mistake would not be hereditary.

QUESTIONS AND PROBLEMS

*12.1 Describe the differences between DNA and RNA.

12.2 Compare and contrast DNA polymerases and RNA polymerases.

12.3 All base pairs in the genome are replicated during the DNA synthesis phase of the cell cycle, but only *some* of the base pairs are transcribed into RNA. How is it determined *which* base pairs of the genome are transcribed into RNA?

*12.4 Discuss the similarities and differences between the *E. coli* RNA polymerase and eukaryotic RNA polymerases.

*12.5 Which classes of RNA do each of the three eukaryotic RNA polymerases synthesize? What are the functions of the different RNA types in the cell?

12.6 What is the −10 (Pribnow) box? the Goldberg-Hogness box (TATA element)?

*12.7 What is an enhancer?

12.8 Compare and contrast the structures of prokaryotic and eukaryotic mRNAs.

*12.9 Compare the structures of mRNA, rRNA, and tRNA.

12.10 Many eukaryotic mRNAs, but not prokaryotic mRNAs, contain introns. Describe how these sequences are removed during the production of mature mRNA.

12.11 Distinguish among leader sequence, trailer sequence, coding sequence, intron, spacer sequence, nontranscribed spacer sequence, external transcribed spacer sequence, and internal transcribed sequence. Give examples of actual molecules in your answer.

*12.12 Discuss the posttranscriptional modifications and processing events that take place on the primary transcripts of eukaryotic rRNA and protein-coding genes.

12.13 Describe the organization of the ribosomal DNA repeating unit of a higher eukaryotic cell.

*12.14   Which of the following kinds of mutations would be likely to be recessive lethals in humans? Explain your reasoning.
a.   deletion of the U1 genes
b.   deletion within intron 2 of β-globin
c.   deletion of four bases at the end of intron 2 and three bases at the beginning of exon 3 in β-globin

12.15   Figure 12.A shows the transcribed region of a typical eukaryotic protein-coding gene. What is the size (in bases) of the fully processed, mature mRNA? Assume in your calculations a poly(A) tail of 200 As.

*12.16   Most human obesity does not follow Mendelian inheritance patterns, as body fat content is determined by a number of interacting genes and environmental variables. Insights into how specific genes function to regulate body fat content have come from studies of mutant, obese mice. In one mutant strain, *tubby* (*tub*), obesity is inherited as a recessive trait. Comparison of the DNA sequence of the *tub*⁺ and *tub* alleles has revealed a single base-pair change: within the transcribed region, a 5' GC base pair has been mutated to a TA base pair. The mutation causes an alteration of the initial 5' base of the first intron. Therefore, in the *tub/tub* mutant, a longer transcript is found. Propose a molecularly based explanation for how a single base change causes a nonfunctional gene product to be produced, why a longer transcript is found in *tub/tub* mutants, and why the *tub* mutant is recessive.

*12.17   Which of the following could occur in a single mutational event in a human? Explain.
a.   Deletion of 10 copies of the 5S ribosomal RNA genes only
b.   Deletion of 10 copies of the 18S rRNA genes only

c.   Simultaneous deletion of 10 copies of the 18S, 5.8S, and 28S rRNA genes only
d.   Simultaneous deletion of 10 copies each of the 18S, 5.8S, 28S, and 5S rRNA genes

12.18   During DNA replication in a mammalian cell a mistake occurs: 10 wrong nucleotides are inserted into a 28S rRNA gene, and this mistake is not corrected. What will likely be the effect on the cell?

*12.19   Give the correct answers, noting that each blank may have more than one correct answer, and that each answer (1 through 4) could be used more than once.

Answers:
1.   Eukaryotic mRNAs
2.   Prokaryotic mRNAs
3.   Transfer RNAs
4.   Ribosomal RNAs

a.   _____   Have a cloverleaf structure

b.   _____   Are synthesized by RNA polymerases

c.   _____   Display an anticodon each

d.   _____   Are the template of genetic information during protein synthesis

e.   _____   Contain exons and introns

f.   _____   Are of four types in eukaryotes and only three types in *E. coli*

g.   _____   Are capped on their 5' end and polyadenylated on their 3' end

FIGURE 12.A

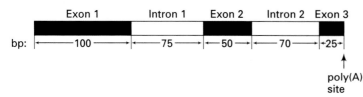

poly(A) site

CHAPTER *13*

GENE EXPRESSION: TRANSLATION

PRINCIPAL POINTS

~ A protein consists of one or more subunits called polypeptides, which are themselves composed of smaller building blocks called amino acids. The amino acids are linked together in the polypeptide by peptide bonds.

~ The primary amino acid sequence of a protein determines its secondary, tertiary, and quaternary structures, and hence its functional state.

~ The genetic code is a triplet code in which each three-nucleotide codon in an mRNA specifies one amino acid. Some amino acids are represented by more than one codon. The code is almost universal, and it is read without gaps in successive, nonoverlapping codons.

~ Translation of the mRNA into a polypeptide chain occurs on ribosomes. Amino acids are brought to the ribosome on tRNA molecules. The correct amino acid sequence is achieved by specific binding of each amino acid to its specific tRNA, and by specific binding between the codon of the mRNA and the complementary anticodon of the tRNA.

~ In prokaryotes and eukaryotes, AUG (methionine) is the initiator codon for the start of translation.

Elongation of the protein chain involves peptide bond formation between the amino acid on the tRNA in the A site of the ribosome and the growing polypeptide on the tRNA in the adjacent P site. Once the peptide bond has formed, the ribosome translocates one codon along the mRNA in preparation for the next tRNA. The incoming tRNA with its amino acid binds to the next codon occupying the A site.

~ Translation continues until a chain-terminating codon (UAG, UAA, or UGA) is reached in the mRNA. These codons are read by release factor proteins, and then the polypeptide is released from the ribosome and the other components of the protein synthesis machinery dissociate.

~ In eukaryotes, proteins are found free in the cytoplasm and in various cell compartments such as the nucleus, mitochondria, chloroplasts, and secretory vesicles. Mechanisms exist to sort proteins to their appropriate cell compartments. For example, proteins to be secreted have N-terminal signal sequences that facilitate their entry into the endoplasmic reticulum for later sorting in the Golgi apparatus and beyond.

The information for the proteins found in a cell is encoded in the structural genes of the cell's genome. Expression of a protein-coding gene occurs by transcription of the gene to produce an mRNA (discussed in Chapter 12), followed by **translation** of the mRNA—the conversion of the mRNA base sequence information into the amino acid sequence of a polypeptide. The nucleotide information that specifies the amino acid sequence of a polypeptide is called the **genetic code**. In this chapter your goal is to learn how the nucleotide sequence of mRNA is translated into the amino acid sequence of a polypeptide.

PROTEINS

Chemical Structure of Proteins

A **protein** is a high-molecular-weight compound of complex shape and composition. Each cell type has a characteristic set of proteins that give it its functional properties. A protein consists of one or more subunits called **polypeptides**, which are themselves composed of smaller building blocks, the **amino acids**, linked together to form long linear chains. The sequence of amino acids gives the polypeptide its three-dimensional shape and its properties in the cell.

With the exception of proline, the amino acids have a common structure, which is shown in Figure 13.1. The structure consists of a central carbon atom (α-carbon, α meaning alpha) to which is bonded an amino group (NH_2), a carboxyl group (COOH), and a hydrogen atom. The figure shows an amino acid at the pH commonly found within cells, where the NH_2 and COOH groups of free amino acids are in a charged state, $-NH_3^+$ and $-COO^-$, respectively. Also bound to the α-carbon is the *R group*. The R group varies from one amino acid to another and gives each amino acid its distinctive properties. Since different polypeptides have different sequences and propor-

~ **FIGURE 13.1**

General structural formula for an amino acid.

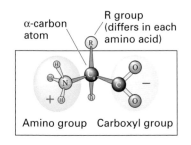

Structures common to all amino acids

tions of amino acids, the organization of the R groups gives a polypeptide its structural and functional properties.

There are 20 amino acids used to make proteins; their names, three-letter and one-letter abbreviations, and chemical structures are shown in Figure 13.2. The 20 amino acids are divided into subgroups based on whether the R group is acidic, basic, neutral and polar, or neutral and nonpolar.

Amino acids of a polypeptide are joined by a **peptide bond**, a covalent bond formed between the carboxyl group of one amino acid and the amino group of an adjacent amino acid (Figure 13.3). Every polypeptide has a free amino group at one end (called the N terminus, or the N-terminal end) and a free carboxyl group at the other end (called the C terminus, or the C-terminal end). The N-terminal end is defined as the beginning of a polypeptide chain.

Molecular Structure of Proteins

Proteins may have four levels of structural organization, as shown in Figure 13.4.

1. The *primary structure* of a polypeptide chain is the amino acid sequence (Figure 13.4a).

2. The *secondary structure* of a protein refers to the regular folding and twisting of a single polypeptide chain into a variety of shapes. A polypeptide's secondary structure is the result of weak bonds, such as electrostatic or hydrogen bonds, between NH and CO groups of amino acids that are near each other on the chain. One type of secondary structure found in regions of many polypeptides is the α-*helix* (Figure 13.4b), a structure discovered by Linus Pauling and Robert Corey in 1951.

Another type of secondary structure is the β-pleated sheet (not illustrated). The β-pleated sheet involves a polypeptide chain or chains folded in a zigzag way, with parallel regions or chains linked by hydrogen bonds. Many proteins contain a mixture of α-helical and β-pleated sheet regions.

3. A protein's *tertiary structure* (Figure 13.4c) is the three-dimensional structure of a single polypeptide chain. Tertiary folding is a direct property of the amino acid sequence of the chain. Figure 13.4c shows the tertiary structure of the β polypeptide of hemoglobin (see below).

4. The *quaternary structure* is the complex of polypeptide chains in a multi-subunit protein, hence quaternary structure is found only in proteins having more than one polypeptide chain (Figure 13.4d). Shown in Figure 13.4d is the quaternary structure of a multimeric ("many subunit") protein, the oxygen-carrying protein hemoglobin, which consists of four polypeptide chains (two 141-amino-acid α polypeptides and two 146-amino-acid β polypeptides), each of which is associated with a heme group that is involved in the binding of oxygen. In the quaternary structure of hemoglobin, each α chain is in contact with each β chain, but there is little interaction between the two α chains or between the two β chains.

KEYNOTE

A protein consists of one or more subunits called polypeptides, which are themselves composed of smaller building blocks, the amino acids, linked together by peptide bonds to form long chains. The primary amino acid sequence of a protein determines its secondary, tertiary, and quaternary structures and hence its function.

THE NATURE OF THE GENETIC CODE

How do nucleotides in the mRNA molecule specify the amino acid sequence in proteins? Genetic experiments done in the 1960s by Francis Crick, Leslie Barnett, Sidney Brenner, and R. Watts-Tobin showed that the genetic code is a triplet code.

~ Figure 13.2

Structures of the 20 naturally occurring amino acids organized according to chemical type. Below each amino acid name are its three-letter and one-letter abbreviations.

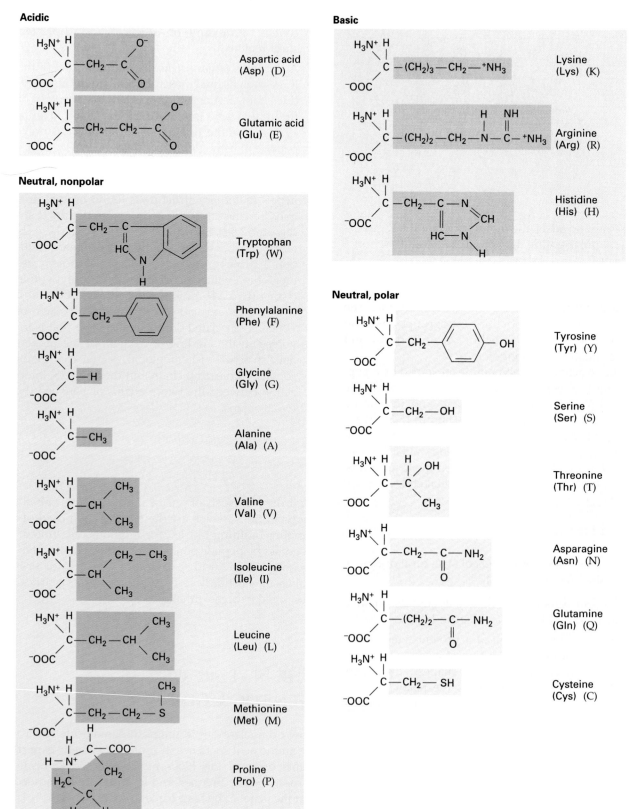

~ Figure 13.3

Mechanism for peptide bond formation between the carboxyl group of one amino acid and the amino group of another amino acid.

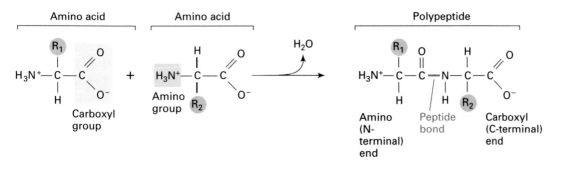

~ Figure 13.4

Four levels of protein structure. (a) Primary, the sequence of amino acids in a polypeptide chain. (b) Secondary, the folding and twisting of a single polypeptide chain into a variety of shapes. Shown is one type of secondary structure: the α-helix. Both structures are stabilized by hydrogen bonds. (c) Tertiary, the specific three-dimensional folding of the polypeptide chain. Shown here is the β polypeptide chain of hemoglobin, a heme-containing polypeptide that carries oxygen in the blood. (d) Quaternary, the aggregate of polypeptide chains. Shown here is hemoglobin; it consists of two α chains, two β chains, and four heme groups.

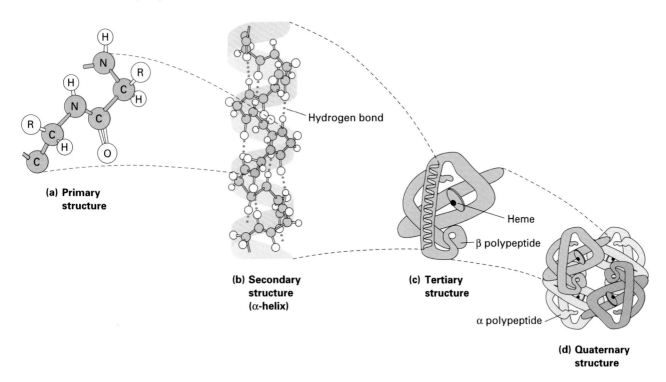

Deciphering the Genetic Code

The exact relationship of the 64 codons to the 20 amino acids was determined by experiments done mostly in the laboratories of Marshall Nirenberg and Ghobind Khorana. Essential to these experiments was the use of **cell-free, protein-synthesizing systems** with components isolated and purified from *E. coli*. These systems contain ribosomes, tRNAs with amino acids attached, and all the necessary protein factors for polypeptide synthesis. Protein synthesis in these systems is inefficient, so radioactively labeled amino acids are used to measure the incorporation of amino acids into new proteins.

In one approach to establish which codons specify which amino acids, synthetic mRNAs containing one, two, or three different types of bases were made and added to cell-free, protein-synthesizing systems. The polypeptides made in these systems were then analyzed. When the synthetic mRNA contained only one type of base, the results were unambiguous. Synthetic poly(U) mRNA, for example, directed the synthesis of a polypeptide consisting of a chain of phenylalanines. Since the genetic code is a triplet code, this result indicated that UUU is a codon for phenylalanine. Similarly, a synthetic poly(A) mRNA directed the synthesis of a lysine chain, and poly(C) directed the synthesis of a proline chain, indicating that AAA is a codon for lysine and CCC is a codon for proline.

Synthetic mRNAs made by the random incorporation of two different bases (called *random copolymers*) were also analyzed in the cell-free, protein-synthesizing systems. When mixed copolymers are made, the bases are incorporated into the synthetic molecule in a random way. Thus, poly(AC) molecules can contain the eight different codons CCC, CCA, CAC, ACC, CAA, ACA, AAC, and AAA. In the cell-free, protein-synthesizing system, poly(AC) synthetic mRNAs caused the incorporation of asparagine, glutamine, histidine, and threonine into polypeptides, in addition to the lysine expected from AAA codons and the proline expected from CCC codons. The proportions of asparagine, glutamine, histidine, and threonine incorporated into the polypeptides produced depended on the A/C ratio used to make the mRNA, and these observations were used to deduce information about the codons that specify the amino acids. For example, since an AC random copolymer containing much more A than C resulted in the incorporation of many more asparagines than histidines, researchers concluded that asparagine is coded by two As and one C and histidine by two Cs and one A. With experiments of this kind, the base composition (*but not the base sequence*) of the codons for a number of amino acids was determined.

Another experimental approach also used copolymers, but these copolymers had been synthesized so that they had a known sequence, not a random one. For example, a repeating copolymer of U and C gives a synthetic mRNA of UCUCUCUCUC. When this copolymer is tested in a cell-free, protein-synthesizing system, the resulting polypeptide had a repeating amino acid pattern of leucine-serine-leucine-serine. Therefore, UCU and CUC specify leucine and serine, although which coded for which cannot be determined from the result.

Yet another approach used a *ribosome-binding assay*, developed in 1964 by Nirenberg and Philip Leder. This assay depends on the fact that, in the absence of protein synthesis, specific tRNA molecules will bind to ribosome–mRNA complexes. For example, when synthetic mRNA poly(U) is mixed with ribosomes, it forms a poly(U)-ribosome complex, and only a phenylalanine tRNA (the tRNA with an AAA anticodon that will bring phenylalanine to an mRNA) will bind to the UUU codon. This trinucleotide-binding property made it possible to determine the specific relationships between many codons and the amino acids for which they code. Note that in this particular approach, the *specific nucleotide sequence of the codon is determined*. Using the ribosome-binding assay, many ambiguities that had arisen from other approaches were resolved. For example, UCU was found to be a codon for serine, and CUC was found to be a codon for leucine. All in all, about 50 codons were clearly identified using this approach.

In sum, no single approach produced an unambiguous set of codon assignments, but information obtained through all the approaches enabled all 64 codons to be assigned (Figure 13.5). Each codon is written *as it appears in mRNA* and reads in a 5'-to-3' direction.

Characteristics of the Genetic Code

The characteristics of the genetic code are as follows:

1. *The code is a triplet code.* Each mRNA codon that specifies an amino acid in a polypeptide chain consists of three nucleotides.

2. *The code is comma-free; that is, it is continuous.* The mRNA is read continuously, three nucleotides at a time, without skipping any nucleotides of the message.

3. *The code is nonoverlapping.* The mRNA is read in successive groups of three nucleotides.

4. *The code is almost universal.* Almost all organisms share the same genetic language. Hence we can isolate an mRNA from one organism, translate it using the machinery from another organism, and produce the protein as if it had been translated in the original organism. The code, however, is not completely universal. For example, the mitochondria of some organisms, such as mammals, have minor changes in the code, as does the nuclear genome of the protozoan *Tetrahymena*.

5. *The code is degenerate.* With two exceptions (AUG, which codes for methionine, and UGG, which codes for tryptophan), more than one

~ Figure 13.5

The genetic code. Of the 64 codons, 61 specify one of the 20 amino acids. One of those 61 codons, AUG, is used in the initiation of protein synthesis. The other three codons are chain-terminating codons and do not specify any amino acid.

= Chain termination codon (stop)

= Initiation codon

codon occurs for each amino acid. This multiple coding is called the **degeneracy** of the code. There are particular patterns in this degeneracy (see Figure 13.5). When the first two nucleotides in a codon are identical and the third letter is U or C, the codon always codes for the same amino acid. For example, UUU and UUC specify phenylalanine, and CAU and CAC specify histidine. Also, when the first two nucleotides in a codon are identical and the third letter is A or G, the same amino acid is often specified. For example, UUA and UUG specify leucine, and AAA and AAG specify lysine. In a few cases, when the first two nucleotides in a codon are identical and the base in the third position is U, C, A, or G, the same amino acid is often specified. For example, CUU, CUC, CUA, and CUG all code for leucine.

6. *The code has start and stop signals.* Specific start and stop signals for protein synthesis are contained in the code. In both eukaryotes and prokaryotes,

AUG (which codes for methionine) is almost always the start codon for protein synthesis.

Only 61 of the 64 codons specify amino acids; these codons are called **sense codons** (see Figure 13.5). The other three codons—UAG (amber), UAA (ochre), and UGA (opal)—do not specify an amino acid, and no tRNAs in normal cells carry the appropriate anticodons. (The three-nucleotide anticodon pairs with the codon in the mRNA by complementary base-pairing during translation.) These three codons are the **stop codons**, also called **nonsense codons**, or **chain-terminating codons**. They are used singly or in tandem groups (UAG UAA, for example) to specify the end of translation of a polypeptide chain. Thus, when we read a particular mRNA sequence, we look for a stop codon located at a multiple of three nucleotides—*in the same reading frame*—from the AUG start codon to determine where the amino acid–coding sequence for the polypeptide ends.

7. *Wobble occurs in the anticodon.* Since 61 sense codons specify amino acids in mRNA, a total of 61 tRNA molecules could have the appropriate anticodons. According to the **wobble hypothesis** proposed by Francis Crick, the complete set of 61 sense codons can be read by fewer than 61 distinct tRNAs because of pairing properties of the bases in the anticodon (Table 13.1). Specifically, the base at the 5' end of the anticodon complementary to the base at the 3' end of the codon, or the third letter, is not as constrained three-dimensionally as the other two bases. This feature allows for less exact base-pairing, so that the base at the 5' end of the anticodon can pair with more than one type of base at the 3' end of the codon; it can wobble. As Table 13.1 shows, no single tRNA molecule can recognize four

~ TABLE 13.1

Wobble in the Genetic Code

NUCLEOTIDE AT 5' END OF ANTICODON		NUCLEOTIDE AT 3' END OF CODON
G	can pair with	U or C
C	can pair with	G
A	can pair with	U
U	can pair with	A or G
I (inosine)	can pair with	A, U, or C

~ Figure 13.6

Example of base-pairing wobble. Two different leucine codons (CUC, CUU) can be read by the same leucine tRNA molecule, contrary to regular base-pairing rules.

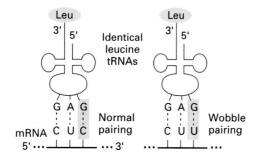

different codons. But if the tRNA molecule contains the modified purine inosine at the 5' end of the anticodon, then that tRNA can recognize three different codons. Figure 13.6 gives an example of how a single leucine tRNA can read two different leucine codons by base-pairing wobble.

KEYNOTE

The genetic code is a triplet code in which each codon in an mRNA specifies one amino acid. Since 64 code words are possible and only 20 amino acids exist, some amino acids are specified by more than one codon. The genetic code in mRNA is read without gaps in successive, nonoverlapping codons. The code is nearly universal: the same codons specify the same amino acids with very few exceptions. Protein synthesis is usually initiated by the codon AUG (methionine) and is terminated by three codons singly or in combination: UAG, UAA, UGA. These codons do not code for any amino acid.

TRANSLATION: THE PROCESS OF PROTEIN SYNTHESIS

Protein synthesis takes place on ribosomes, where the genetic message encoded in mRNA is translated. The mRNA is translated in the 5'-to-3' direction, and the polypeptide is made in the N-terminal to C-terminal direction. Amino acids are brought to the ribosome

bound to tRNA molecules. The correct amino acid sequence is achieved as a result of (1) the binding of each amino acid to its own specific tRNA, and (2) the binding between the codon of the mRNA and the complementary anticodon in the tRNA.

Charging tRNA

The correct amino acid is attached to the tRNA by a type of enzyme called an **aminoacyl-tRNA synthetase**. The process is called aminoacylation, or **charging,** and produces an **aminoacyl-tRNA (or charged tRNA)**. Aminoacylation uses energy from ATP hydrolysis. Since there are 20 different amino acids, there are at least 20 different aminoacyl-tRNA synthetases. The recognition site for an aminoacyl-tRNA synthetase is enzyme-specific; in some cases it is the anticodon region of the tRNA, but in other cases sequences elsewhere in the tRNA are used.

Figure 13.7 shows charging of a tRNA molecule to produce valine-tRNA (Val-tRNA). First, the amino acid and ATP bind to the specific aminoacyl-tRNA synthetase enzyme. The enzyme then catalyzes a reaction in which the ATP loses two phosphates and is coupled to the amino acid as AMP to form aminoacyl-AMP. Next, the tRNA molecule binds to the enzyme, which transfers the amino acid from the aminoacyl-AMP to the tRNA and then displaces the AMP. The aminoacyl-tRNA molecule produced is then released from the enzyme. Chemically, the amino acid attaches at the 3' end of the tRNA by a covalent linkage between the carboxyl group of the amino acid and the 3'-OH or 2'-OH group of the ribose of the adenine nucleotide found at the end of every tRNA.

KEYNOTE

Protein synthesis occurs on ribosomes, where the genetic message encoded in mRNA is translated. Amino acids are brought to the ribosome on charged tRNA molecules; tRNAs are charged by aminoacyl-tRNA synthetases.

The three basic stages of protein synthesis—*initiation, elongation,* and *termination*—are similar in prokaryotes and eukaryotes. In the following sections we discuss each of these stages in turn, concentrating on the processes in *E. coli*. In the discussions we note where significant differences in translation occur between prokaryotes and eukaryotes.

~ Figure 13.7

Charging of a tRNA molecule by aminoacyl-tRNA synthetase to produce an aminoacyl-tRNA (charged tRNA).

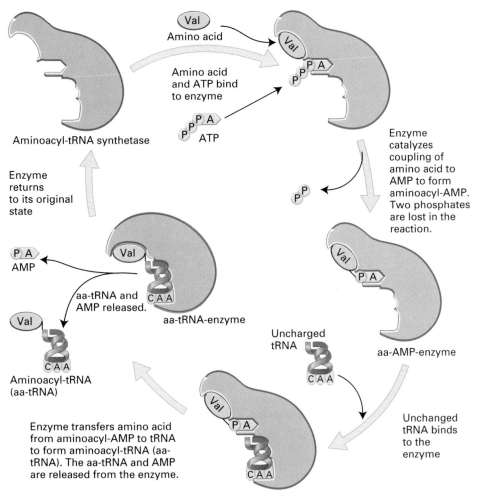

Initiation of Translation

The initiation of translation is shown in Figure 13.8. Initiation involves an mRNA molecule, a ribosome, a specific initiator tRNA, three different protein **initiation factors**, GTP (guanosine triphosphate), and magnesium ions.

In prokaryotes, the first step in translation is the binding of the 30S small ribosomal subunit to the region of the mRNA with the AUG initiation codon. The ribosomal subunit comes to the mRNA bound to three initiation factors, IF1, IF2, and IF3, along with a molecule of GTP and magnesium ions. The AUG initiation codon is not sufficient to "tell" the ribosomal subunit where to bind to the mRNA. There is a sequence upstream (to the 5' side in the leader of the mRNA) of the AUG initiation codon that is complementary to a short sequence at the 3' end of the 16S rRNA found in the 30S ribosomal subunit. The mRNA sequence—AGGAGG—is called the **Shine-Dalgarno sequence** after its discoverers. The model is that the formation of complementary base pairs between the mRNA and 16S rRNA allows the ribosome to locate the start codon in the mRNA for the initiation of translation.

The next step is the binding of the initiator tRNA to the AUG start codon to which the 30S subunit is bound. In both prokaryotes and eukaryotes, the AUG initiator codon specifies methionine. As a result, newly made proteins in both types of organisms begin with methionine. In many cases, the methionine is subsequently removed. In prokaryotes, the initiator methionine is a modified form of methionine,

~ **Figure 13.8**

Initiation of protein synthesis in prokaryotes. A 30S ribosomal subunit, complexed with initiation factors and GTP, binds to mRNA and fMet-tRNA to form a 30S initiation complex. Next, the 50S ribosomal subunit binds, forming a 70S initiation complex. During this event, the initiation factors are released and GTP is hydrolyzed.

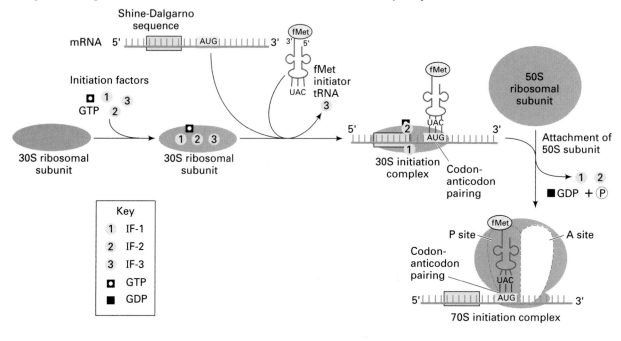

called **formylmethionine** (abbreviated fMet), in which a formyl group has been added to the methionine's amino group. The initiator tRNA brings the fMet to the 30S–mRNA complex. When the fMet-tRNA binds to that complex, IF3 is released, and at this point the *30S initiation complex* has been formed. Next, the 50S ribosomal subunit binds, leading to GTP hydrolysis and release of IF1 and IF2. The final complex is called the *70S initiation complex*. The 70S ribosome has two binding sites for aminoacyl-tRNA, the peptidyl (P) and aminoacyl (A) sites; the fMet-tRNA is bound to the mRNA in the P site, and the A site is vacant.

Translation initiation is very similar in eukaryotes. The main differences are: (1) the initiator methionine is unmodified, although a special initiator tRNA brings it to the ribosome, and (2) Shine-Dalgarno sequences are not found in eukaryotic mRNAs. Instead, the eukaryotic ribosome uses another way to find the AUG initiation codon. First, a eukaryotic initiator factor eIF4A, a multimer of several proteins including the *cap binding protein* (CBP), binds to the cap at the 5' end of the mRNA (see Chapter 12). Then, a complex of the 40S ribosomal subunit with the initiator Met-tRNA, several eIF proteins, and GTP binds, along with other eIFs, and

moves along the mRNA, scanning for the initiator AUG codon. The AUG codon is embedded in a short sequence—called the Kozak sequence—that indicates it is the initiator codon. This is called the *scanning model* for initiation. The AUG codon is almost always the first AUG codon from the 5' end of the mRNA. Once it finds this AUG, the 40S subunit binds to it, and then the 60S ribosomal subunit binds, displacing the eIFs, producing the *80S initiation complex*. Protein synthesis then begins at the AUG codon. Like its prokaryotic counterpart, the eukaryotic 80S ribosome has a P site and an A site, and the initiator Met-tRNA is bound to the mRNA in the P site.

Elongation of the Polypeptide Chain

Figure 13.9 depicts the elongation events as they take place in prokaryotes. This phase has three steps:

1. the binding of aminoacyl-tRNA (charged tRNA) to the ribosome

2. the formation of a peptide bond

3. the movement (translocation) of the ribosome along the mRNA, one codon at a time

BINDING OF AMINOACYL-tRNA.   At the start of elongation, the anticodon of fMet-tRNA is hydrogen-bonded to the AUG initiation codon in the peptidyl (P) site of the ribosome (Figure 13.9, part 1). The next codon in the mRNA is in the aminoacyl (A) site. In Figure 13.9 this codon (UCC) specifies serine (Ser).

Next, the appropriate aminoacyl-tRNA (here, Ser-tRNA) binds to the codon in the A site (Figure

~ Figure 13.9

Elongation stage of translation in prokaryotes.

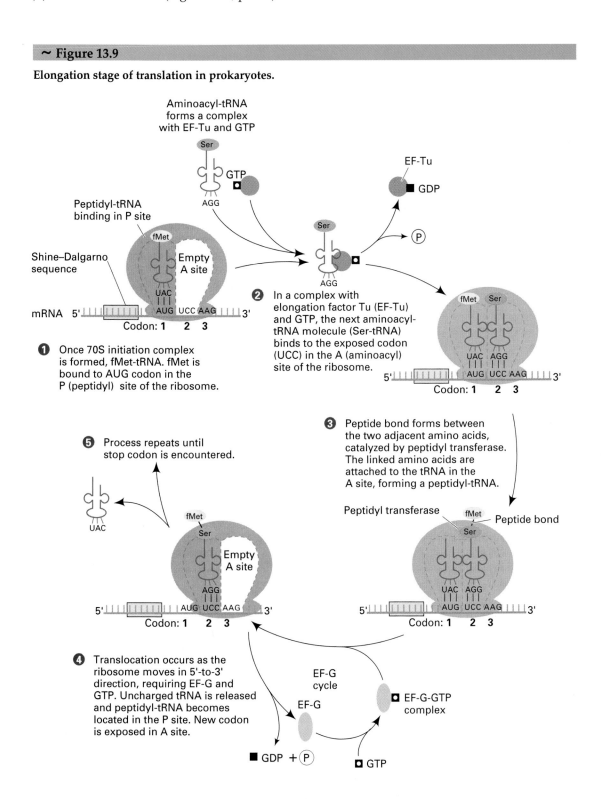

13.9, part 2). This aminoacyl-tRNA is brought to the ribosome bound to the protein elongation factor EF-Tu and a molecule of GTP. When the aminoacyl-tRNA binds to the codon in the A site, the hydrolysis of GTP releases EF-Tu–GDP. The EF-Tu is separated from the GDP and is recycled in other elongation events.

PEPTIDE BOND FORMATION. The ribosome maintains the two aminoacyl-tRNAs in the P and A sites in the correct positions, so that a peptide bond can form between the two amino acids (Figure 13.9, part 3). Two steps are involved in the formation of this peptide bond. First, the bond between the amino acid and the tRNA in the P site is cleaved. In this case the breakage is between the fMet and its tRNA. Second, the peptide bond is formed between the now-freed fMet and the Ser attached to the tRNA in the A site. This reaction is catalyzed by **peptidyl transferase.** For many years this enzyme activity was thought to be a result of the interaction of a few ribosomal proteins of the 50S ribosomal subunit. However, newer data indicate that the 23S rRNA molecule of the large ribosomal subunit is the enzyme. This exciting finding has expanded our understanding of the role particular RNA molecules can have in catalytic activities previously thought to be the function only of proteins.

Once the peptide bond has formed (see Figure 13.9, part 3), a tRNA without an attached amino acid (an uncharged tRNA) is left in the P site. The tRNA in the A site, now called peptidyl-tRNA, has the first two amino acids of the polypeptide chain attached to it, in this case fMet-Ser.

TRANSLOCATION. In the last step in the elongation cycle, **translocation** (Figure 13.9, part 4), the ribosome moves one codon along the mRNA toward the 3' end. In prokaryotes, translocation requires the activity of another protein elongation factor, EF-G. An EF-G–GTP complex binds to the ribosome, and translocation occurs along with displacement of the uncharged tRNA from the P site, hydrolysis of GTP to GDP, and release of EF-G and GDP. The EF-G is recycled in other translocation events.

After translocation is completed, the A site is vacant. An aminoacyl-tRNA with the correct anticodon binds to the newly exposed codon in the A site, repeating the process already described. The whole process is repeated until translation terminates at a stop codon (Figure 13.9, part 5).

In eukaryotes, the elongation and translocation steps are similar to those in prokaryotes, although there are differences in the number and properties of

elongation factors and in the exact sequences of events.

In both prokaryotes and eukaryotes, once the ribosome moves away from the initiation site on the mRNA, another initiation event occurs. Thus, many ribosomes may simultaneously be translating each mRNA. The complex between an mRNA molecule and all the ribosomes that are translating it simultaneously is called a **polyribosome** or **polysome** (Figure 13.10). Simultaneous translation enables a large amount of protein to be produced from each mRNA molecule.

KEYNOTE

The AUG (methionine) initiator codon signals the start of translation in prokaryotes and eukaryotes. Elongation proceeds when a peptide bond forms between the amino acid attached to the tRNA in the A site of the ribosome and the growing polypeptide attached to the tRNA in the P site. This reaction is catalyzed by peptidyl transferase, which is an activity of the 23S rRNA molecule of the large ribosomal subunit. Translocation occurs when the now uncharged tRNA in the P site is released from the ribosome and the ribosome moves one codon down the mRNA.

Termination of Translation

Elongation continues until the polypeptide coded for in the mRNA is completed. The termination of translation is signaled by one of three stop codons, UAG, UAA, and UGA, which are the same in prokaryotes and eukaryotes (Figure 13.11, part 1). The stop codons do not code for any amino acid, and so no tRNAs in the cell have anticodons for them. The ribosome recognizes a stop codon with the help of proteins called **termination factors,** or **release factors** (RF), which read the codons (Figure 13.11, part 2) and then initiate a series of specific termination events. Different release factors are found in prokaryotes and eukaryotes.

The specific termination events triggered by the release factors are: (1) release of the polypeptide from the tRNA in the P site of the ribosome in a reaction catalyzed by peptidyl transferase (Figure 13.11, part 3), (2) release of the tRNA from the ribosome (Figure

~ Figure 13.10

Electron micrograph and diagram of a polysome, a number of ribosomes each translating the same mRNA sequentially.

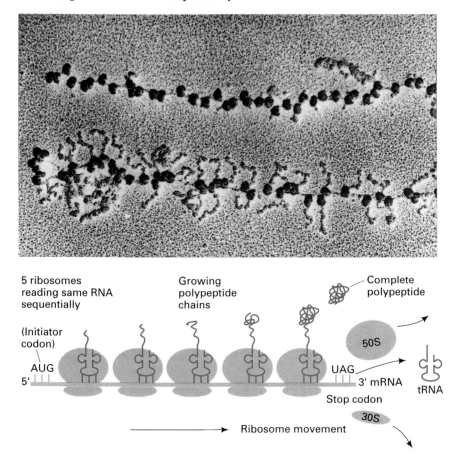

~ Figure 13.11

Termination of translation. The ribosome recognizes a chain termination codon (UAG) with the aid of release factors. The release factor reads the stop codon, and this initiates a series of specific termination events leading to the release of the completed polypeptide.

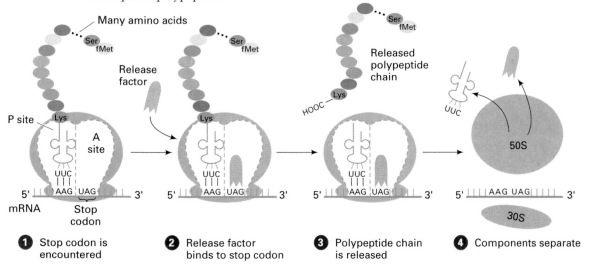

13.11, part 4), and then (3) dissociation of the two ribosomal subunits and the RF from the mRNA (Figure 13.11, part 4).

KEYNOTE

Protein synthesis continues until a stop codon is located in the A site of the ribosome. These codons are read by one or more release factors. Then the polypeptide is cleaved from the tRNA and the tRNA is released from the ribosome, and the ribosomal subunits disengage from the mRNA.

PROTEIN SORTING IN THE CELL

In prokaryotes and eukaryotes, some proteins may be secreted, and in eukaryotes some other proteins need to be placed in different cell compartments such as the nucleus, mitochondrion, chloroplast, and lysosome. The sorting of proteins to their appropriate compartments is under genetic control in that specific "signal" or "leader" sequences on the proteins direct them to the correct organelles. Similarly, in prokaryotes, certain proteins become localized in the membrane and others are secreted.

Let us briefly describe protein distribution by the endoplasmic reticulum (ER) and Golgi apparatus in eukaryotes. Proteins synthesized on the rough ER are extruded into the space between the two membranes of the ER (the cisternal space), modified by the attachment of sugar groups to produce glycoproteins, and then transferred to the Golgi apparatus in vesicles. The Golgi apparatus is the main distribution center where most of the sorting occurs. Proteins targeted for the plasma membrane are carried from the Golgi apparatus to the plasma membrane in transport vesicles. Proteins destined to be secreted are packaged into secretory storage vesicles. The secretory vesicles migrate to the cell surface, where they fuse with the plasma membrane and release their packaged proteins to the outside of the cell (Figure 13.12). Lysosomal proteins are collected into a specific class of Golgi export vesicles, which eventually become functional lysosomes.

In 1975, Gunther Blobel, B. Dobberstein, and his colleagues found that secreted proteins initially contain extra amino acids at the amino terminal end. This is also the case for other proteins sorted by the Golgi. Blobel's work led to the **signal hypothesis**, which states that proteins sorted by the Golgi bind to the ER by a hydrophobic, amino terminal extension (the **signal sequence**) to the membrane that is subsequently removed and degraded (Figure 13.13).

In more detail, when the signal sequence of a protein destined for the ER is translated and exposed on the ribosome surface, a cytoplasmic **signal recognition particle** (SRP, a complex of a small RNA molecule with six proteins) recognizes the signal sequence, binds to it, and blocks further translation of the mRNA. Translation stops until the growing polypeptide-SRP-ribosome-mRNA complex binds to the ER. The SRP binds to a membrane protein of the ER called the **docking protein,** and this causes the firm binding of the ribosome to the ER and the insertion of the signal sequence into the membrane. As the ribosome binds, the SRP is released and translation resumes. The growing polypeptide extends through the ER membrane into the cisternal space of the ER as a loop from the signal sequence. The signal

~ Figure 13.12

Movement of secretory proteins through the cell membrane system. These proteins move from their site of synthesis on the rough endoplasmic reticulum (RER), through the Golgi apparatus, into vesicles, and then into the extracellular space. The arrow shows the general path of the proteins to be secreted.

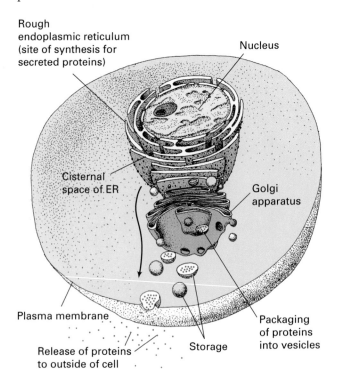

Rough endoplasmic reticulum (site of synthesis for secreted proteins)

Nucleus

Cisternal space of ER

Golgi apparatus

Plasma membrane

Release of proteins to outside of cell

Storage

Packaging of proteins into vesicles

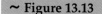

~ **Figure 13.13**

Model for the translocation of proteins into the endoplasmic reticulum in eukaryotes.

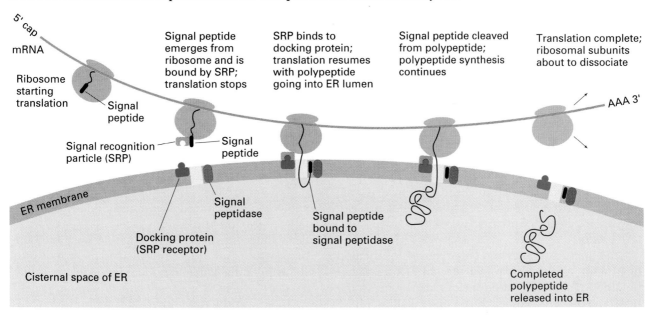

sequence is removed from the polypeptide by the **signal peptidase**, an enzyme embedded in the ER membrane. The transport of the protein molecule into the ER simultaneously with its synthesis is **cotranslational transport**.

Proteins sorted to other organelles such as the nucleus, mitochondrion, or chloroplast use a different mechanism. These proteins are translated completely and then, by virtue of specific amino acid sequences they have, they are transported posttranslationally into the organelle where they are destined to function.

KEYNOTE

Eukaryotic proteins that enter the endoplasmic reticulum are distinguished from proteins that remain free in the cytoplasm by the presence of a specific signal sequence at their N-terminal ends. Proteins destined for the ER are translocated into the cisternal space of the ER while they are being synthesized on the ribosomes (cotranslationally). Once in the ER, the signal sequence is removed by signal peptidase. The proteins are then sorted to their final destinations by the Golgi apparatus.

SUMMARY

In this chapter we discussed features of the genetic code and the translation of an mRNA to produce a protein. We learned that a protein consists of one or more subunits called polypeptides, which are themselves composed of smaller building blocks called amino acids. The amino acids are linked together in the polypeptide by peptide bonds. The amino acid sequence of a protein determines its secondary, tertiary, and quaternary structures, and hence its functional state.

RNA consists of four different subunits (nucleotides), while protein consists of 20 different subunits (amino acids). The genetic code has the following features: it is a triplet code (meaning that a sequence of three nucleotides specifies one amino acid), it is continuous, it is nonoverlapping, it is almost universal, and it is degenerate. Of the 64 codons, 61 specify amino acids, while the other three are stop codons. One codon, AUG (for methionine), is almost always used to specify the first amino acid in a protein chain.

Protein synthesis occurs on ribosomes, where the genetic message encoded in mRNA is translated. Amino acids are brought to the ribosome on charged tRNA molecules. Each tRNA has an anticodon that binds specifically to a codon in the mRNA. As a result, the correct amino acid sequence is achieved by (1) the specific binding of each amino acid to its own specific tRNA and (2) the specific binding between the codon of the mRNA and the complementary anticodon in the tRNA.

In both prokaryotes and eukaryotes, an AUG codon is the initiator codon for the start of protein synthesis. In prokaryotes, the initiation of protein synthesis requires a sequence upstream of the AUG codon, to which the small ribosomal subunit binds. This sequence is the Shine-Dalgarno sequence, and it binds specifically to the 3' end of the 16S rRNA of the small ribosomal subunit, thereby associating the small subunit with the mRNA. No functionally equivalent sequence occurs in eukaryotic mRNAs; instead, the ribosomes load onto the mRNA at its 5' end and scan toward the 3' end, initiating translation at the first AUG codon.

In both prokaryotes and eukaryotes, the initiation of protein synthesis requires protein factors called initiation factors. Initiation factors are bound to the ribosome-mRNA complex during the initiation phase and dissociate once the polypeptide chain has been initiated. During the elongation phase, the polypeptide chain is lengthened one amino acid at a time. This occurs simultaneously with the movement of the ribosome toward the 3' end of the mRNA one codon at a time. Protein factors called elongation factors play important catalytic roles in this part of translation.

The signal for polypeptide chain growth to stop is the presence of a stop codon (UAG, UAA, or UGA) in the mRNA. No naturally occurring tRNA has an anticodon that can read a stop codon. Instead, specific protein factors called release factors read the stop codon and initiate the events characteristic of protein synthesis termination; namely, the release of the completed polypeptide from the ribosome, the release of the tRNA from the ribosome, and the dissociation of the two ribosomal subunits from the mRNA.

In eukaryotes, certain proteins are secreted, while others become localized in different cell compartments. The sorting of proteins to their appropriate compartments is brought about by the presence of specific sequences in the proteins. Proteins destined for the ER, for example, have a signal sequence at their amino ends. A special process exists to recognize the signal sequence, block translation until the growing polypeptide-ribosome-mRNA complex binds to the ER, and then cotranslationally insert the polypeptide into the cisternal space of the ER, where the signal sequence is removed. Once inside the ER, the protein is modified and then transported to the Golgi apparatus where it undergoes further modifications before being sorted to its final location.

In sum, protein synthesis is a complex process involving interaction between three major classes of RNA (mRNA, tRNA, and rRNA) and a large number of accessory protein factors that act catalytically in the process. By repeated translation of an mRNA molecule by a string of ribosomes (producing a polysome), a large number of identical protein molecules can be produced. Thus, from a single gene, large quantities of a protein can be produced by two steps: (1) the production of multiple mRNAs from the gene, and (2) the production of many protein molecules by repeated translation of each mRNA.

ANALYTICAL APPROACHES FOR SOLVING GENETICS PROBLEMS

Q13.1
a. How many of the 64 codon permutations can be made from the three nucleotides A, U, and G?
b. How many of the 64 codon permutations can be made from the four nucleotides A, U, G, and C, with one or more Cs in each codon?

A13.1
a. This question involves probability. There are four bases, so the probability of a cytosine at the first position in a codon is 1/4. Conversely, the probability of a base other than cytosine in the first position is $(1 - 1/4) = 3/4$. These same probabilities apply to the other two positions in the codon. Therefore the probability of a codon without a cytosine is $(3/4)^3 = 27/64$.

b. This question involves the relative frequency of codons that have one or more cytosines. We have already calculated the probability of a codon not having a cytosine, so all the remaining codons have one or more cytosines. The answer to this question, therefore, is $(1 - 27/64) = 37/64$.

Q13.2 Random copolymers were used in some of the experiments directed toward deciphering the genetic code. For a ribonucleotide mixture of 2U:1C, give the expected codons and their frequencies, and give the expected proportions of the amino acids that would be found in a polypeptide directed by the copolymer in a cell-free, protein-synthesizing system.

A13.2 The probability of a U at any position in a codon is 2/3, and the probability of a C at any position in a codon is 1/3. Thus, the codons, their relative frequencies, and the amino acids for which they code are:

$$UUU = (2/3)(2/3)(2/3) = 8/27 = 0.296 = 29.6\% \quad Phe$$
$$UUC = (2/3)(2/3)(1/3) = 4/27 = 0.148 = 14.8\% \quad Phe$$
$$UCC = (2/3)(1/3)(1/3) = 2/27 = 0.0743 = 7.43\% \quad Ser$$
$$UCU = (2/3)(1/3)(2/3) = 4/27 = 0.148 = 14.8\% \quad Ser$$
$$CUU = (1/3)(2/3)(2/3) = 4/27 = 0.148 = 14.8\% \quad Leu$$
$$CUC = (1/3)(2/3)(1/3) = 2/27 = 0.0743 = 7.43\% \quad Leu$$
$$CCU = (1/3)(1/3)(2/3) = 2/27 = 0.0743 = 7.43\% \quad Pro$$
$$CCC = (1/3)(1/3)(1/3) = 1/27 = 0.037 = 3.7\% \quad Pro$$

In sum, 44.4% Phe, 22.23% Ser, 22.23% Leu, and 11.13% Pro. (Does not quite add up to 100% because of rounding off.)

QUESTIONS AND PROBLEMS

*13.1 The form of genetic information used directly in protein synthesis is (choose the correct answer):
a. DNA
b. mRNA
c. rRNA
d. Ribosomes

*13.2 Proteins are (choose the correct answer):
a. Branched chains of nucleotides
b. Linear, folded chains of nucleotides
c. Linear, folded chains of amino acids
d. Invariably enzymes

13.3 The process in which ribosomes engage is (choose the correct answer):
a. Replication
b. Transcription
c. Translation
d. Disjunction
e. Cell division

13.4 What are the characteristics of the genetic code?

*13.5 Base-pairing wobble occurs in the interaction between the anticodon of the tRNAs and the codons. On a theoretical level, determine the minimum number of tRNAs needed to read the 61 sense codons.

13.6 Antibiotics have been very useful in elucidating the steps of protein synthesis. If you have an artificial messenger RNA with the sequence AUGUUUUUUUUUUUUU . . . , it will produce the following polypeptide in a cell-free, protein-synthesizing system: fMet-Phe-Phe-Phe . . . In your search for new antibiotics you find one called putyermycin, which blocks protein synthesis. When you try it with your artificial mRNA in a cell-free system, the product is fMet-Phe. What step in protein synthesis does putyermycin affect? Why?

13.7 Describe the reactions involved in the amino-acylation (charging) of a tRNA molecule.

13.8 Compare and contrast protein synthesis initiation in prokaryotes and eukaryotes.

*13.9 Random copolymers were used in some of the experiments that revealed the characteristics of the genetic code. For each of the following ribonucleotide mixtures, give the expected codons and their frequencies, and give the expected proportions of the amino acids that would be found in a polypeptide directed by the copolymer in a cell-free, protein-synthesizing system:
a. 4 A : 6 C
b. 1 A : 3 U : 1 C

*13.10 What would the minimum WORD (CODON) SIZE be if the number of different bases in the mRNA were, instead of four:
a. two
b. three
c. five

*13.11 Suppose that at stage A in the evolution of the genetic code only the first two nucleotides in the coding triplets led to unique differences and that any nucleotide could occupy the third position. Then, suppose there was a stage B in which differences in meaning arose depending on whether a purine (A or G) or pyrimidine (C or U) was present at the third position. Without reference to the number of amino acids or multiplicity of tRNA molecules, how many triplets of different meaning can be constructed out of the code at stage A? at stage B?

*13.12 A gene encodes a polypeptide 30 amino acids long containing an alternating sequence of phenylalanine and tyrosine. What are the sequences of nucleotides corresponding to this sequence in the following:
a. The DNA strand that is read to produce the mRNA, assuming Phe = UUU and Tyr = UAU in mRNA
b. The DNA strand that is not read
c. tRNAs

*13.13 Two populations of RNAs are made by the random combination of nucleotides. In population A the RNAs contain only A and G nucleotides (3 A : 1 G), while in population B the RNAs contain only A and U nucleotides (3 A : 1 U). In what ways *other than amino acid content* will the proteins produced by translating the population A RNAs differ from those produced by translating the population B RNAs?

13.14 In *E. coli,* a particular tRNA normally has the anticodon 5'-GGG-3', but because of a mutation in the tRNA gene, the tRNA has the anticodon 5'-GGA-3'.
a. What codon would the normal tRNA recognize?
b. What codon would the mutant tRNA recognize?

*13.15 A protein found in *E. coli* normally has the N-terminal amino acid sequence Met-Val-Ser-Ser-Pro-Met-Gly-Ala-Ala-Met-Ser... A mutation alters the anticodon of a tRNA from 5'-GAU-3' to 5'-CAU-3'. What would be the N-terminal amino acid sequence of this protein in the mutant cell? Explain your reasoning.

13.16 The gene encoding an *E. coli* tRNA containing the anticodon 5'-GUA-3' mutates so that the anticodon now is 5'-UUA-3'. What will be the effect of this mutation? Explain your reasoning.

*13.17 The normal sequence of the coding region of an mRNA is shown below, along with six mutant versions of the same mRNA. Indicate what protein would be formed in each case. [(. . .) = a multiple of three unspecified bases.]

Normal: AUGUUCUCUAAUUAC(. . .)AUGGGGUGGGUGUAG
Mutant *a*   AUGUUCUCUAAUUAG(. . .)AUGGGGUGGGUGUAG
Mutant *b*   AGGUUCUCUAAUUAC(. . .)AUGGGGUGGGUGUAG
Mutant *c*   AUGUUCUCGAAUUAC(. . .)AUGGGGUGGGUGUAG
Mutant *d*   AUGUUCUCUAAAUAC(. . .)AUGGGGUGGGUGUAG
Mutant *e*   AUGUUCUCUAAUUC(. . .)AUGGGGUGGGUGUAG
Mutant *f*   AUGUUCUCUAAUUAC(. . .)AUGGGGUGGGUGUGG

*13.18 The normal sequence of a particular protein is given below, along with several mutant versions of it. For each mutant, explain what mutation occurred in the coding sequence of the gene. [(. . .) = a multiple of three unspecified bases.]

Normal: Met-Gly-Glu-Thr-Lys-Val-Val- (. . .) -Pro
Mutant 1: Met-Gly
Mutant 2: Met-Gly-Glu-Asp
Mutant 3: Met-Gly-Arg-Leu-Lys
Mutant 4: Met-Arg-Glu-Thr-Lys-Val-Val- (. . .) -Pro

13.19 The N-terminus of a protein has the sequence Met-His-Arg-Arg-Lys-Val-His-Gly-Gly. A molecular biologist wishes to synthesize a DNA chain that could encode this portion of the protein. How many possible DNA sequences could encode this polypeptide?

13.20 In the recessive condition in humans known as sickle cell anemia, the β-globin polypeptide of hemoglobin is found to be abnormal. The only difference between it and the normal β-globin is that the sixth amino acid from the N-terminal end is valine, whereas the normal β-globin has glutamic acid at this position. Explain how this occurred.

*13.21 Cystic fibrosis is an autosomal recessive disease in which the cystic fibrosis transmembrane conductance regulator (CFTR) protein is abnormal. The transcribed portion of the cystic fibrosis gene spans about 250,000 base pairs of DNA. The CFTR protein, with 1,480 amino acids, is translated from an mRNA of about 6,500 bases. The most common mutation in this gene results in a protein that is missing a phenylalanine at position 508 (ΔF508).
a. Why is the RNA coding sequence of this gene so much larger than the mRNA from which the CFTR protein is translated?
b. About what percent of the mRNA together makes up 5' untranslated leader and 3' untranslated trailer sequences?
c. At the DNA level, what alteration would you expect to find in the ΔF508 mutation?
d. What consequences might you expect if the DNA alteration that you describe in (c) were to occur at random in the protein-coding region of the cystic fibrosis gene?

13.22 In the last several years, a set of diseases have been shown to be caused by triplet repeat amplification (for review, see Chapter 7, p. 165). Amplification of the following triplet repeats has been associated with disease phenotypes: 5'-CAG-3', 5'-CGG-3', 5'-GAA-3', and 5'-CTG-3'. CGG expansions have been found on the 5' side (upstream) of the coding sequence of a gene, CAG expansions have been found within coding regions, GAA expansions have been found in intronic regions, and CTG and CGG expansions have been found to the 3' side (downstream) of the gene's coding sequence. Speculate how each of these triplet repeat expansions might result in abnormal gene function.

CHAPTER *14*

CLONING AND MANIPULATION OF DNA

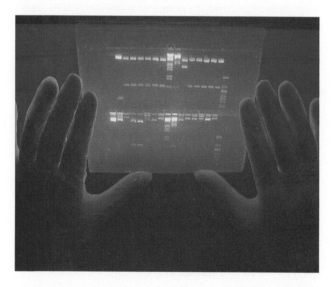

PRINCIPAL POINTS

~ Cloning in the molecular biology sense is making many copies of a segment of DNA such as a gene. Cloning makes possible the generation of large amounts of pure DNA for manipulation. DNA is cloned by splicing a DNA fragment into a cloning vector (a DNA molecule capable of replication in a host organism) to make a recombinant DNA molecule, and then introducing that molecule into a host cell in which it will replicate. Essen-tial to cloning are restriction enzymes. These enzymes recognize specific nucleotide-pair sequences in DNA (restriction sites) and cleave at a specific point within the sequence.

~ Different kinds of cloning vectors have been developed. Cloning vectors typically replicate within one or more host organisms, have one or more restriction sites into which foreign DNA can be inserted, and have one or more selectable markers to use in detecting cells that contain the vectors. Two important classes of cloning vectors are plasmids and bacteriophages. Shuttle vectors have the above properties but also are able to replicate in more than one type of host. Yeast artificial chromosomes (YACs) and bacterial artical chromosomes (BACs) enable DNA fragments several hundred kilobase pairs long to be cloned in yeast.

~ Genomic libraries are collections of clones that contain one copy of every DNA sequence in an organism's genome. Genomic libraries are useful for isolating specific genes and for studying the organization of the genome.

~ Individual chromosomes can be isolated and chromosome-specific libraries made from them. If a gene has been localized to a specific chromosome by genetic means, a chromosome-specific library makes it easier to isolate a clone of the gene.

~ DNA copies, called complementary DNAs (cDNAs), can be made from mRNA molecules isolated from eukaryotic cells. The cDNAs can be cloned.

~ Specific sequences in genomic libraries and cDNA libraries can be identified using a number of approaches, including the use of specific DNA or cDNA probes, specific antibodies, or complementation of mutations.

~ Genes and cloned DNA sequences can be analyzed to determine the arrangement and specific locations of restriction sites, a process called restriction mapping. Gene transcripts can be analyzed to determine tissue specificity and the level of gene expression using recombinant DNA procedures.

~ Methods have been developed for determining the se-quence of a cloned piece of DNA. A commonly used method, the dideoxy procedure, uses enzymatic synthesis of a new DNA chain on a template DNA strand. With this procedure, synthesis of new strands is stopped by the in-corporation of a dideoxy analog of the normal deoxyribonucleotide. Using four different dideoxy analogs, the new strands stop at all possible nucleotide positions, thereby allowing the complete DNA sequence to be determined.

~ The polymerase chain reaction (PCR) uses synthetic oligonucleotide primers to amplify a specific segment of DNA many thousandfold in an automated procedure. A major benefit of PCR is that small amounts of DNA are needed. Thus, DNA from a single cell can be amplified using PCR. Increasing applications both in research and in the commercial arena are being found for PCR, including the generation of specific DNA segments for cloning or for sequencing and the amplification of DNA to detect specific genetic defects.

~ There are many applications for recombinant DNA technology and related procedures. For example, with appropriate probes, it is possible to detect specific genetic diseases and to develop clinical, veterinary, and agricultural products. An international effort is underway to determine the complete sequence of the human genome. The knowledge obtained will expand greatly our understanding of human genetics. With continued advances in genetic engineering of plants, we can anticipate many improvements such as increased yields and disease resistance.

*T*he field of molecular genetics changed radically in the 1970s when procedures were developed that enabled researchers to construct **recombinant DNA molecules** and to clone (make many copies of) those molecules. Cloning generates large amounts of pure DNA, such as genes, which can then be manipulated in various ways, including mapping, sequencing, mutating, and transforming cells. For example, suppose we want to study the gene for a particular human protein in order to determine its DNA sequence and how its expression is regulated. Each human cell contains only two copies of that gene, making it a daunting task to isolate enough copies of the gene for analysis. By contrast, an essentially unlimited number of copies of the gene can be produced by cloning. Your goal in this chapter is to learn about DNA cloning and some ways that cloned DNA can be manipulated. Applications of this technology in medicine, agriculture, and commerce are also discussed.

DNA CLONING

In brief, DNA is cloned by the following steps:

1. Isolate DNA from an organism.

2. Cut the DNA into pieces with *restriction enzymes* and splice each piece individually into a **cloning vector** to make a recombinant DNA molecule. A cloning vector is an artificially constructed DNA molecule capable of replication in a host organism, such as a bacterium.

3. Introduce (transform) the recombinant DNA molecule into a host such as *E. coli*, yeast, an animal cell, or a plant cell. Replication of the recombinant DNA molecule (**molecular cloning**) occurs in the host cell, producing many identical copies called *clones*. As the host organism reproduces, the recombinant DNA molecules are passed on to all the progeny, giving rise to a cell population all carrying the cloned sequence.

In this section we describe how DNA can be cloned.

Restriction Enzymes

A **restriction enzyme** (or **restriction endonuclease**) recognizes a specific base-pair sequence in DNA called a *restriction site* and cleaves the DNA within that sequence. Restriction enzymes are used to produce a pool of DNA fragments to be cloned.

Most restriction enzymes are found naturally in bacteria, although one restriction enzyme has been found in the green alga *Chlorella*. In bacteria, restriction enzymes protect the organism against viruses by cutting up—restricting—invading viral DNA. The bacterium's own DNA is protected from its restriction enzyme(s) by methylation of these sequences.

Over 400 different restriction enzymes have been isolated. They are named for the organism from which they are isolated. Conventionally, a three-letter system is used, italicized or underlined, followed by roman numerals. Additional letters are sometimes added to signify a particular bacterial strain from which the enzymes were obtained. For example, *Eco*RI is from *E. coli* strain RY13, and *Hind*III is from *Haemophilus influenzae* strain Rd. The names are pronounced in ways that follow no set pattern. For example, *Bam*HI is "bam-H-one," *Bgl*II is "bagel-two," *Eco*RI is "echo-R-one," *Hind*III is "hin-D-three," *Hha*I is "ha-ha-one," and *Hpa*II is "hepa-two."

Many restriction sites have an axis of symmetry through the midpoint. Figure 14.1 shows this symmetry for the *Eco*RI restriction site—the base sequence from 5' to 3' on one DNA strand is the same as the base sequence from 5' to 3' on the complementary DNA strand. Thus, the sequences are said to have *twofold rotational symmetry* or to be *palindromes*. A number of restriction sites are shown in Table 14.1. The most commonly used restriction enzymes recognize four base pairs (for example, *Hha*I) or six nucleotide pairs (for example, *Bam*HI, *Eco*RI). Some enzymes recognize eight-nucleotide-pair sequences (for example, *Not*I ["not-one"]). Based on probability principles, the frequency of a short nucleotide-pair sequence in the genome will be greater than the frequency of a long nucleotide-pair sequence, so an enzyme that recognizes a four-nucleotide-pair sequence will cut a DNA sequence more frequently than one that recognizes a six-nucleotide-pair sequence, and that enzyme will cut more frequently than one that recognizes an eight-nucleotide-pair sequence.

As Table 14.1 indicates, some enzymes, such as *Sma*I ("sma-one"), cut both strands of DNA between the same two base pairs to produce *blunt ends* (Figure 14.2a), while others, such as *Bam*HI, make staggered cuts in the symmetrical nucleotide pair sequence to produce *sticky* or *staggered ends*. The staggered ends may either have an overhanging 5' end, as in the case

~ FIGURE 14.1

Restriction site in DNA, showing symmetry of the sequence around the center point. The sequence is a palindrome, reading the same from left to right (5'-to-3') on the top strand (GAATTC, here) as it does from right to left (5'-to-3') on the bottom strand. Shown here is the restriction site for *Eco*RI.

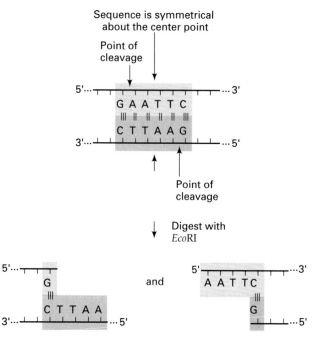

~ TABLE 14.1

Characteristics of Some Restriction Endonucleases

	ENZYME NAME	PRONUNCIATION	ORGANISM IN WHICH ENZYME NATURALLY OCCURS	RECOGNITION SEQUENCE AND POSITION OF CUT[a]
Enzymes with 6-bp Recognition Sequences	BamHI	"bam-H-one"	Bacillus amyloliquefaciens H	5' G↓G A T C C 3' 3' C C T A G↑G 5'
	EcoRI	"echo-R-one"	E. coli RY13	G↓A A T T C C T T A A↑G
	HindIII	"hin-D-three"	Haemophilus influenzae R_d	A↓A G C T T T T C G A↑A
	PstI	"P-S-T-one"	Providencia stuartii	C T G C A↓G G↑A C G T C
	SmaI	"sma-one"	Serratia marcescens	C C C↓G G G G G G↑C C C
Enzymes with 4-bp Recognition Sequences	HhaI	"ha-ha-one"	Haemophilus hemolyticus	G C G↓C C↑G C G
	HpaII	"hepa-two"	Haemophilus parainfluenzae	C↓C G G G G C↑C
	Sau3A	"sow-three-A"	Staphylococcus aureus 3A	↓G A T C C T A G↑
Enzymes with 8-bp Recognition Sequences	NotI	"not-one"	Nocardia otitidis-caviarum	G C↓G G C C G C C G C C G G↑C G
Enzymes with Recognition Sequences That Are Not Symmetrical	BstXI	"b-s-t-x-one"	Bacillus stearothermophilus	C C A N N N N N↓N T G G G G T N N N N N↑N A C C

[a]In this column, the two strands of DNA are shown with the sites of cleavage indicated by arrows. Since there is an axis of twofold rotational symmetry in each recognition sequence, the DNA molecules resulting from the cleavage are symmetrical.   Key: N = any base

of cleavage with BamHI (Figure 14.2b) or EcoRI, or an overhanging 3' end, as in the case of cleavage with PstI ("P-S-T-one") (Figure 14.2c).

Restriction enzymes that produce staggered ends are of particular value in cloning DNA because every DNA fragment generated by cutting a piece of DNA with the same restriction enzyme has the same base sequence at the two staggered ends. That is, if the ends of two pieces of DNA produced by the action of the same restriction enzyme (such as EcoRI)—for example, a cloning vector and a chromosomal DNA fragment—

come together in solution, base-pairing occurs; the two single-stranded DNA ends are said to *anneal* (Figure 14.3). Using DNA ligase, the two DNAs can be covalently linked (ligated) to produce a whole DNA molecule with reconstituted restriction sites. Even DNA fragments with blunt ends can be ligated together by DNA ligase at high concentrations of the enzyme. The ligation of two DNA fragments is the principle behind the formation of recombinant DNA molecules.

Since each restriction enzyme cuts DNA at an enzyme-specific sequence, the number of cuts the

~ FIGURE 14.2

Examples of how restriction enzymes cleave DNA. (a) *Sma*I results in blunt ends; (b) *Bam*HI results in overhanging ("sticky") 5' ends; (c) *Pst*I results in overhanging ("sticky") 3' ends.

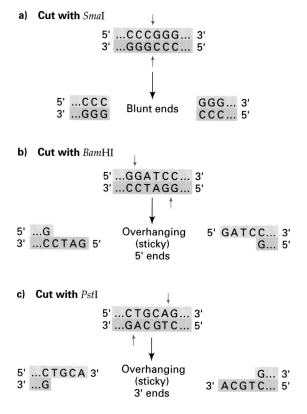

a) **Cut with** *Sma*I

b) **Cut with** *Bam*HI

c) **Cut with** *Pst*I

enzyme makes in a particular DNA molecule depends on the number of times that particular restriction site occurs. When we cut a number of copies of the same genome with a particular restriction enzyme, the DNA will be cleaved at the enzyme's specific restriction sites, which are distributed throughout the genome. Although this will produce millions of fragments of different sizes, *all* of the identical chromosomal DNAs in the multiple genome copies will be cut at identical recognition sequences.

KEYNOTE

DNA is cloned by splicing a DNA fragment into a cloning vector to make a recombinant DNA molecule, and then introducing that molecule into a host cell in which it will replicate. Essential to cloning are restriction enzymes, or restriction endonucleases. A restriction enzyme recognizes a specific nucleotide-pair sequence in DNA (restriction site) and cleaves at a specific point within the sequence. Cleavage of the DNA with a restriction enzyme can be either staggered (producing DNA fragments with single-stranded, "sticky" ends) or blunt.

Cloning Vectors and the Cloning of DNA

Plasmids and bacteriophages are two types of vectors commonly used for cloning DNA. The two types differ in the way they must be manipulated to clone

~ FIGURE 14.3

Cleavage of DNA by the restriction enzyme *Eco*RI. *Eco*RI makes staggered, symmetrical cuts in DNA, leaving "sticky" ends. A DNA fragment with a sticky end produced by *Eco*RI digestion can bind by complementary base-pairing (anneal) to any other DNA fragment with a sticky end produced by *Eco*RI cleavage. The gaps may then be sealed by DNA ligase.

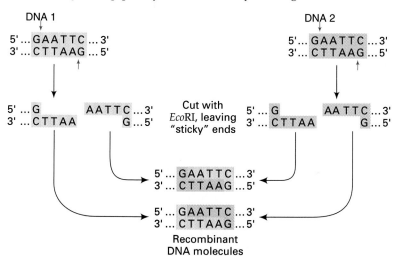

DNA and in the maximum size of DNA that can be cloned into each.

PLASMID CLONING VECTORS. Bacterial plasmids are circular extrachromosomal elements that replicate autonomously within cells (see Chapter 6). Plasmid cloning vectors are derivatives of these natural plasmids "engineered" to have features useful for cloning DNA. We focus here on features of *E. coli* plasmid cloning vectors. DNA fragments of up to a few kilobase pairs are efficiently cloned in plasmid vectors, although they may accept fragments of 5 to 10 kb. Plasmids carrying larger DNA fragments are often unstable and tend to lose most of the inserted DNA.

An *E. coli* plasmid cloning vector must have three features:

1. An *ori* (origin of DNA replication) sequence, needed for the plasmid to replicate in *E. coli*.

2. A *selectable marker*, which enables *E. coli* cells that carry the plasmid to be easily distinguished from cells that lack the plasmid. The usual selectable marker is a gene for resistance to an antibiotic, such as the *amp*^R gene for ampicillin resistance, or the *tet*^R gene for tetracycline resistance. When plasmids carrying antibiotic-resistance genes are added to a population of plasmid-free and therefore antibiotic-sensitive *E. coli*, those cells that take up the plasmid can be selected for by culturing the cells on a medium containing the appropriate antibiotic; only cells with the plasmid will grow on the medium.

3. *Unique restriction enzyme cleavage sites*—sites present just once in the vector—for the insertion of the DNA fragments to be cloned. Cloning most commonly involves cutting the plasmid at one of the unique sites with the appropriate restriction enzyme and splicing into that site a piece of DNA that has been cut with the same enzyme.

As an example, Figure 14.4 diagrams the plasmid vector pUC19 ("puck-19"). This 2,686-bp vector has the following features that make it useful for cloning DNA:

1. It has a high copy number, approaching 100 copies per cell.

2. It has the *amp*^R selectable marker.

3. It has a number of unique restriction sites clustered in one region, called a **polylinker** or **multiple cloning site**. By cutting pUC19 with a restriction enzyme that has a site in the polylinker and ligating with DNA fragments produced by cutting high-molecular-weight DNA with the same restriction enzyme, recombinant DNA molecules are produced (Figure 14.5).

~ **FIGURE 14.4**

The plasmid cloning vector pUC19. This plasmid has an origin of replication (*ori*), an *amp*^R selectable marker, and a polylinker located within part of the β-galactosidase gene, *lacZ*^+.

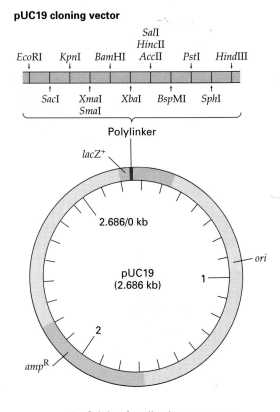

ori = Origin of replication sequence
amp^R = Ampicillin resistance gene
lacZ^+ = β-galactosidase gene

4. The polylinker is inserted into part of the *E. coli* β-galactosidase (*lacZ*^+) gene (see Figure 14.4). The insertion was engineered so that functional β-galactosidase is produced when pUC19 is introduced into a *lacZ*^− mutant *E. coli* cell that makes nonfunctional β-galactosidase. Then, when a piece of DNA is cloned into the polylinker, the β-galactosidase reading frame is disrupted, so functional β-galactosidase cannot be produced in *E. coli* (see Figure 14.5). The chemical X-gal is included in the medium on which the cells are plated so that, if β-galactosidase is made by a colony, the colony will turn blue, whereas if no β-galactosidase is made, the colony will be white. This simple color test is used to select colonies of *E. coli* containing pUC19 with inserted DNA because they are white, in contrast to colonies containing pUC19 with no inserted DNA, which are blue.

~ FIGURE 14.5

Insertion of a piece of DNA into the plasmid cloning vector pUC19 to produce a recombinant DNA molecule.
pUC19 contains several unique restriction enzyme sites localized in a polylinker that are convenient for constructing recombinant DNA molecules. Insertion of a DNA fragment into the polylinker disrupts part of the β-galactosidase (*lacZ*⁺) gene, leading to nonfunctional β-galactosidase in *E. coli*. The color selection test described in the text can be used to select for vectors with or without inserts.

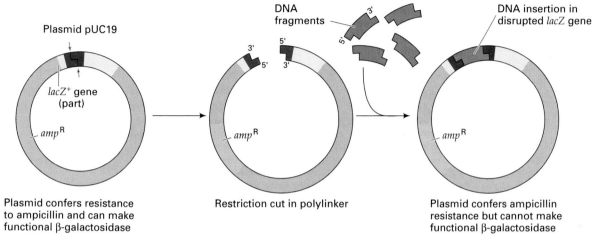

Plasmid confers resistance to ampicillin and can make functional β-galactosidase

Restriction cut in polylinker

Plasmid confers ampicillin resistance but cannot make functional β-galactosidase

PHAGE CLONING VECTORS. Commonly used phage cloning vectors are derivatives of bacteriophage λ (lambda; see Chapter 6) that have been engineered so that the lytic cycle is possible, but lysogeny is not possible. The λ cloning vectors possess restriction sites useful for cloning DNA fragments into them. One such vector, the lambda replacement vector (Figure 14.6), has a chromosome in which there is a "left" arm and a "right" arm that collectively contain all the essential genes for the lytic cycle. Between the two arms is a segment of DNA that is "disposable" because it does not contain any genes needed for phage propagation. The junctions between the disposable central segment and the two arms each have an *Eco*RI restriction site in this case.

Cloning a DNA fragment using a λ replacement vector is achieved as follows: First the λ vector is cut, in this case, with the enzyme *Eco*RI, to separate the two arms from the replaceable segment. Next, DNA fragments to be cloned are generated by cutting high-molecular-weight DNA from an organism with *Eco*RI. The foreign DNA fragments are then mixed with the λ DNA fragments and the pieces ligated together by DNA ligase.

The ligated DNA is mixed in vitro with the various protein parts of the lambda phage, resulting in insertion of the DNA into phage heads and assembly of complete particles. This process is called *packaging*. The phage head can accommodate DNA fragments in the size range 37–52 kb, so fragments outside of that range cannot be packaged. The assembled particles are used to infect a culture of *E. coli*. Only those phages in which the foreign DNA is inserted between the left arm and the right arm will replicate, because only then are all the genes necessary for phage reproduction present. Thus, each DNA fragment is cloned by the repeated rounds of infection and lysis that each original functional phage goes through in the culture. Eventually, the culture becomes transparent as all the bacteria have been lysed, and a population of progeny λ phages, with a concentration of 10^{10} to 10^{11} phages/mL, is produced, with many, many representatives of each of the original recombinant DNA molecules.

SHUTTLE VECTORS. The cloning vectors we have discussed thus far are used to clone DNA within *E. coli* cells. Other vectors have been developed to introduce recombinant DNA molecules into a variety of prokaryotic and eukaryotic organisms. There are vectors that can be used to transform mammalian cells in culture, as well as vectors to transform other animal cells, plant cells, and yeast cells, among others. Often these are **shuttle vectors,** that is, cloning vectors that can be introduced into two or more host organisms. For example, there are shuttle vectors that can be transformed into and replicate in *E. coli* (selected for using antibiotic resistance) and also be transformed into yeast (selected for by a nutritional marker, such as the *URA3* gene conferring uracil-independent growth on a *ura3* mutant yeast cell). Different "flavors" of yeast–*E. coli* shuttle vectors have been developed, some of which replicate and some of which do not. The latter vectors integrate into the host cell's chromosome and are replicated when that chromosome replicates.

~ FIGURE 14.6

Scheme for using phage λ DNA as a cloning vector. Foreign DNA fragments of approximately 15 kb in length can be cloned in λ vectors.

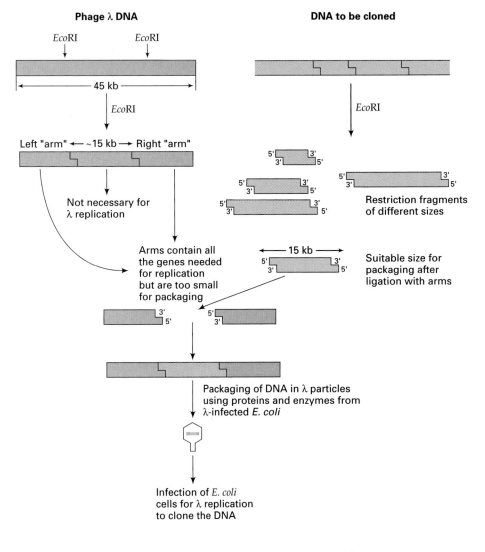

~ FIGURE 14.7

Example of a yeast artificial chromosome (YAC) cloning vector. It contains a yeast telomere (*TEL*) at each end, a yeast centromere sequence (*CEN*), a yeast selectable marker for each arm (here, *TRP1* and *URA3*), a sequence that allows autonomous replication in yeast (*ARS*), and restriction sites for cloning.

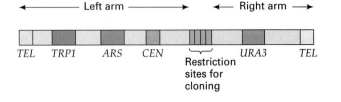

YEAST ARTIFICIAL CHROMOSOMES (YACs).

Yeast artificial chromosomes (YACs; "yaks") are cloning vectors that enable artificial chromosomes to be made and cloned in yeast cells. YACs are linear vectors that have the following features (Figure 14.7):

1. A yeast telomere (*TEL*) at each end (recall that all eukaryotic chromosomes have a telomere at each end);

2. A yeast centromere sequence (*CEN*);

3. A selectable marker on each arm for detecting the plasmid in yeast (for example, *TRP1* and *URA3* for tryptophan and uracil independence in *trp1* and *ura3* mutant strains, respectively);

4. An origin of replication sequence—*ARS* (autonomously replicating sequence)—that allows the vector to replicate in a yeast cell;

5. Restriction sites unique to the YAC that can be used for inserting foreign DNA.

YAC vectors can accommodate DNA fragments that are several hundred kilobase pairs long, much longer than the fragments that can be cloned in the other vectors we have discussed. Thus, YACs are used to clone very large DNA fragments (up to about 1,000 kb), for example, in creating physical maps of large genomes such as the human genome. YAC clones are made by ligating very-high-molecular-weight DNA pieces to YAC arms generated by cutting with a restriction enzyme in the cloning site. Clones are introduced into yeast by transformation; by selecting for both *TRP1* and *URA3*, it can be ensured that the clone has both the left and right arms of the YAC vector.

BACTERIAL ARTIFICIAL CHROMOSOMES (BACs).

Bacterial artificial chromosomes (BACs; "backs") are useful for cloning large DNA fragments up to about 300 kb in *E. coli*. BACs are modified versions of the *E. coli* F factor that is responsible for conjugation in this bacterium (see Chapter 6, pp. 131–132). The modifications include unique restriction sites for cloning and a selectable marker. While YACs can accommodate larger DNA inserts, BACs have the advantage that they can be handled like regular bacterial plasmids.

KEYNOTE

Different kinds of vectors have been developed to construct and clone recombinant DNA molecules. Many vectors replicate within their host organism, have one or more restriction enzyme cleavage sites at which foreign DNA fragments can be inserted, and have one or more selectable markers. Two important types of cloning vectors are used to clone recombinant DNA in *E. coli*: plasmids, which are introduced by transformation into the *E. coli* host, where they replicate; and bacteriophage λ, which are packaged into λ phage particles, which, in turn, inject the DNA into the *E. coli* host. In λ clones, the injected λ DNA replicates and progeny phages are produced.

Shuttle vectors can also be used to introduce cloned recombinant DNA into more than one host. Yeast artificial chromosomes and bacterial artificial chromosomes enable DNA fragments several hundred kilobase pairs long to be cloned in yeast and *E. coli*, respectively.

RECOMBINANT DNA LIBRARIES

Typically, researchers want to study a particular gene or DNA fragment. When genomic DNA is isolated from an organism, cut with a restriction enzyme, and the population of DNA fragments cloned in a vector, we have what is called a **genomic library**, a collection of clones containing at least one copy of every DNA sequence in the genome.

Genomic libraries have been made for many organisms, including humans (see the discussion of the Human Genome Project, p. 324). Related to genomic libraries are *chromosome libraries*, which are collections of clones of fragments of individual chromosomes, and **complementary DNA (cDNA)** libraries, which are collections of clones of DNA copies of mRNAs isolated from cells. The following sections describe these different types of libraries.

Genomic Libraries

A genomic library is a collection of clones that, when successfully made, contains at least one copy of every DNA sequence in the genome. The genomic library can be used to isolate and study a particular clone, such as that for a gene of interest. We will see how this can be done later. In this section we focus on the construction of genomic libraries of eukaryotic DNA.

Genomic libraries are made using the cloning procedures already described. A restriction enzyme is used to cut up the genomic DNA, and a vector is chosen so that the entire genome is represented in a manageable number of clones. The number of clones needed to include all sequences in the genome depends on the size of the genome being cloned and the average size of the DNA fragments inserted into the vector. The probability of having at least one copy of any DNA sequence in the genomic library can be calculated from the formula

$$N = \frac{\ln(1 - P)}{\ln(1 - f)}$$

where N is the necessary number of recombinant DNA molecules, P is the probability desired, f is the fractional proportion of the genome in a single recombinant DNA molecule (that is, f is the average size, in kilobase pairs, of the fragments used to make the library divided by the size of the genome, in kilobase pairs), and ln is the natural logarithm. For example, for a 99 percent chance that a particular yeast DNA fragment is represented in a library of 15-kb fragments in a lambda vector-based genomic library, where the yeast genome size is about 12,000 kb, 3,682 recombinant DNA molecules would be needed. For the approximately 3,000,000-kb human genome, over 920,000 clones would be needed, hence the use of YAC or BAC vectors for making libraries of large genomes. Whatever the genome or vector, to have confidence that all genomic sequences are represented, it is the practice to make a library with many times more than the calculated minimum number of clones.

KEYNOTE

A genomic library is a collection of clones that contains at least one copy of every DNA sequence in an organism's genome. Like regular book libraries, genomic libraries are great resources of information—in this case, the information is about the genome. They are used for isolating specific genes and for studying the organization of the genome, among many other things.

Chromosome Libraries

It is very time consuming to screen for a sequence of interest in a genomic library made from an organism with a large genome. One approach for reducing the screening time is to make libraries of the individual chromosomes in the genome. In humans, this gives 24 different libraries, one each for the 22 autosomes, the X, and the Y. Then, if a gene has been localized to a chromosome by genetic means, researchers can focus their attention on the library of that chromosome when they search for its DNA sequence.

Individual chromosomes can be separated if their morphologies and sizes are distinct enough, as is the case for human chromosomes. In one current procedure, *flow cytometry*, chromosomes from cells in mitosis are stained with a fluorescent dye and passed through a laser beam connected to a light detector. This system sorts the chromosomes based on differences in dye binding and the resulting light scattering. Once the chromosomes have been sorted and collected, a library of each chromosome type can be made.

KEYNOTE

In certain organisms, individual chromosomes can be isolated and chromosome-specific libraries made from them. If a gene has been localized to a specific chromosome by genetic means, the existence of chromosome-specific libraries makes it easier to isolate a clone of the gene.

cDNA Libraries

DNA copies, called *complementary DNA (cDNA)*, can be made from all mRNA molecules present in a population of eukaryotic cells at a particular time. These cDNA molecules can then be cloned to produce a **cDNA library**. Since a cDNA library reflects the gene activity of the cell type at the time the mRNAs are isolated, the construction and analysis of cDNA libraries are useful for comparing gene activities in different cell types of the same organism.

Figure 14.8 shows how a cDNA molecule can be made from eukaryotic mRNA molecules. Key to this synthesis is the presence of the poly(A) tail at the 3' ends of the mRNAs. The first step in cDNA synthesis is annealing a short oligo(dT) primer to the poly(A) tail. The primer is extended by **reverse transcriptase** (RNA-dependent DNA polymerase) to make a DNA copy of the mRNA strand. The result is a DNA-mRNA double-stranded molecule. Next, RNase H ("R-N-aze H": a type of ribonuclease), DNA polymerase I, and DNA ligase are used to synthesize the second DNA strand. RNase H degrades the RNA strand in the hybrid DNA-mRNA, DNA polymerase I makes new DNA fragments using the partially degraded RNA fragments as primers, and finally DNA ligase ligates the new DNA fragments together to make a complete chain. The result is a double-stranded cDNA molecule that is a faithful DNA copy of the starting mRNA.

Figure 14.9 illustrates the cloning of cDNA using a **restriction site linker**, or **linker**, which is a relatively short, double-stranded piece of DNA (oligodeoxyribonucleotide) about 8 to 12 nucleotide pairs long that includes a restriction site, in this case *Bam*HI. Both the cDNA molecules and the linkers have blunt ends, and they can be ligated together at high concentrations of T4 DNA ligase. Sticky ends are produced in the cDNA molecule by cleaving the cDNA (with linkers now at each end) with *Bam*HI. The resulting DNA is inserted into a cloning vector that has also been cleaved with *Bam*HI, and the recombinant DNA molecule produced is transformed into an *E. coli* host cell for cloning.

The clones in a cDNA library represent the *mature mRNAs* found in the cell. In eukaryotes, mature mRNAs are processed molecules, so the cDNA clones obtained are *not* equivalent to gene clones because promoters, terminators, and intron sequences are typically present in gene clones but *not* in cDNA clones. For any mRNA, cDNA clones can be useful for subsequently isolating the gene that codes for that mRNA. The gene clone can provide more information than can the cDNA clone, for example, on the presence and arrangement of introns and on the regulatory sequences associated with the gene.

~ FIGURE 14.8

~ FIGURE 14.8

The synthesis of double-stranded complementary DNA (cDNA) from a polyadenylated mRNA, using reverse transcriptase, RNase H, DNA polymerase I, and DNA ligase.

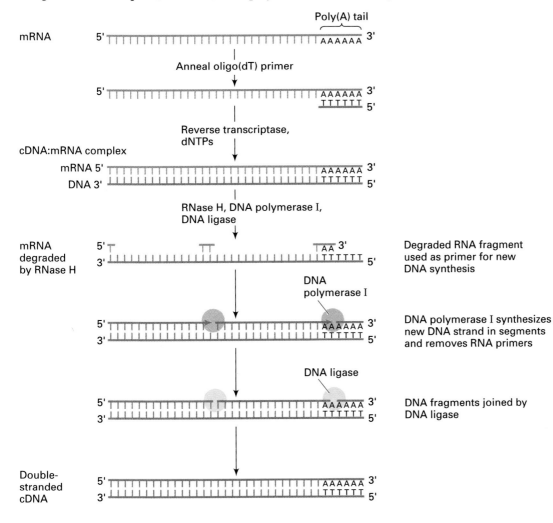

FINDING A SPECIFIC CLONE IN A LIBRARY

Unlike libraries of books, clone libraries have no catalog, so they must be searched through—screened—to find a clone of interest. Fortunately, a number of screening procedures have been developed, and we discuss some of them in this section.

Screening a cDNA Library

We can screen a cDNA library in a number of ways to identify a cDNA clone we are interested in studying. One way is to select for a cDNA clone that encodes a

KEYNOTE

Given a population of mRNAs purified from a cell, it is possible to make DNA copies of those mRNA molecules. First, the enzyme reverse transcriptase makes a single-stranded DNA copy of the mRNA; then RNase H, DNA polymerase I, and DNA ligase are used to make a double-stranded DNA copy called complementary DNA (cDNA). This cDNA can be spliced into cloning vectors using restriction site linkers.

~ **FIGURE 14.9**

The cloning of cDNA using *Bam*HI linkers.

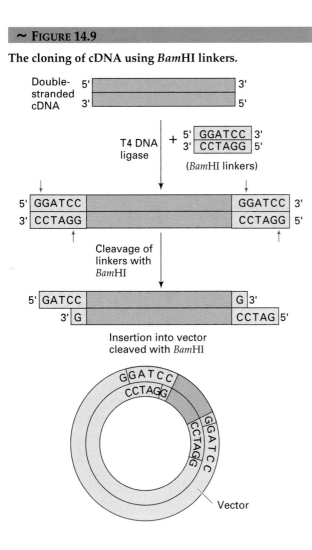

specific protein (Figure 14.10). This approach requires using antibodies that can bind to the protein. It also requires that the cDNAs be cloned in a special kind of vector called an *expression vector* (Figure 14.10, part 1), in which the cDNA is inserted between a promoter and a transcription termination signal. In the host, an mRNA is transcribed corresponding to the cDNA, and the mRNA is translated to produce the encoded protein.

For screening, first *E. coli* is transformed with cDNA clones made in an expression vector (Figure 14.10, part 2), and then the cells are plated so that each bacterium gives rise to a colony (Figure 14.10, part 3). These clones are preserved, for example, by picking each colony off the plate and placing it into the medium in a well of a microtiter dish (Figure 14.10, part 4). Replicas of the set of clones are placed (printed) onto a membrane filter that has been placed on a Petri plate of selective medium appropriate for the recombinant molecules—for example, ampicillin for plasmids carrying the ampicillin-resistance gene

(Figure 14.10, part 5). Colonies grow on the filter in the same pattern as the clones in the microtiter dish. The filter is peeled from the dish and the cells are lysed *in situ* (Figure 14.10, part 6). The proteins within the cell, including those expressed from the cDNA, become stuck to the filter. The filter is then incubated with an antibody to the protein of interest (Figure 14.10, part 7).

If the antibody is radioactively labeled, any clones that expressed the protein of interest can be identified by placing the dried filter against X-ray film, leaving it in the dark for a period of time (from 1 hour to overnight) to produce an *autoradiogram* (Figure 14.10, part 8). The process is called *autoradiography*. When the film is developed, dark spots are seen wherever the radioactive probe is bound to the filter in the antibody reaction. (The dark spots result from the decay of the radioactive atoms, which changes silver grains in the film.) Once a cDNA clone for a protein of interest has been identified, it can be used, for example, to analyze the genome of the same or other organisms for homologous sequences, to isolate the nuclear gene for the mRNA from a genomic library, or to quantify mRNA production.

Screening a Genomic Library

Given the existence of a probe, such as a cloned cDNA, it is possible to identify in a genomic library the cloned gene that codes for the mRNA molecule from which the cDNA was made, and then to isolate it for characterization. Here we discuss the screening of a genomic library made in a plasmid vector.

FINDING A SPECIFIC CLONE IN A PLASMID GENOMIC LIBRARY. The process is similar to that just described for screening a cDNA library. First, *E. coli* cells are transformed with the genomic library (Figure 14.11, part 1), and the cells are plated on selective medium, where colonies are produced (Figure 14.11, part 2). Then the colonies are replica plated onto another plate of selective medium, this one with a membrane filter on its surface (Figure 14.11, part 3). (Replica plating involves pressing a pad of sterile velveteen onto the original, master plate to pick up some of each colony in the pattern they grew on that plate, and then pressing the velveteen gently onto the new plate, thereby "inoculating" it with cells from each colony in its original pattern.) Colonies grow on the membrane filter, which is then lifted off the plate and processed to lyse the bacterial cells, denature the DNA to single strands, and then bind that DNA firmly to the filter (Figure 14.11, part 4).

~ FIGURE 14.10

Screening for specific cDNA plasmids in a cDNA library by using an antibody probe.

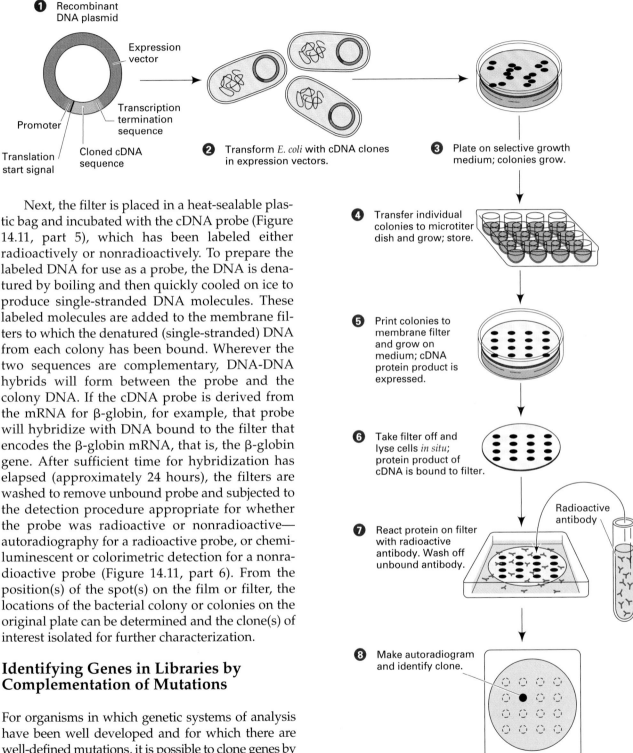

① Recombinant DNA plasmid

Expression vector

Promoter

Transcription termination sequence

Translation start signal

Cloned cDNA sequence

② Transform *E. coli* with cDNA clones in expression vectors.

③ Plate on selective growth medium; colonies grow.

④ Transfer individual colonies to microtiter dish and grow; store.

⑤ Print colonies to membrane filter and grow on medium; cDNA protein product is expressed.

⑥ Take filter off and lyse cells *in situ*; protein product of cDNA is bound to filter.

⑦ React protein on filter with radioactive antibody. Wash off unbound antibody.

Radioactive antibody

⑧ Make autoradiogram and identify clone.

Next, the filter is placed in a heat-sealable plastic bag and incubated with the cDNA probe (Figure 14.11, part 5), which has been labeled either radioactively or nonradioactively. To prepare the labeled DNA for use as a probe, the DNA is denatured by boiling and then quickly cooled on ice to produce single-stranded DNA molecules. These labeled molecules are added to the membrane filters to which the denatured (single-stranded) DNA from each colony has been bound. Wherever the two sequences are complementary, DNA-DNA hybrids will form between the probe and the colony DNA. If the cDNA probe is derived from the mRNA for β-globin, for example, that probe will hybridize with DNA bound to the filter that encodes the β-globin mRNA, that is, the β-globin gene. After sufficient time for hybridization has elapsed (approximately 24 hours), the filters are washed to remove unbound probe and subjected to the detection procedure appropriate for whether the probe was radioactive or nonradioactive—autoradiography for a radioactive probe, or chemiluminescent or colorimetric detection for a nonradioactive probe (Figure 14.11, part 6). From the position(s) of the spot(s) on the film or filter, the locations of the bacterial colony or colonies on the original plate can be determined and the clone(s) of interest isolated for further characterization.

Identifying Genes in Libraries by Complementation of Mutations

For organisms in which genetic systems of analysis have been well developed and for which there are well-defined mutations, it is possible to clone genes by complementation of those mutations. (Complementation is discussed in Chapter 6.) This can be done, for example, with the yeast *Saccharomyces cerevisiae*.

~ FIGURE 14.11

Using DNA probes to screen plasmid genomic libraries for specific DNA sequences.

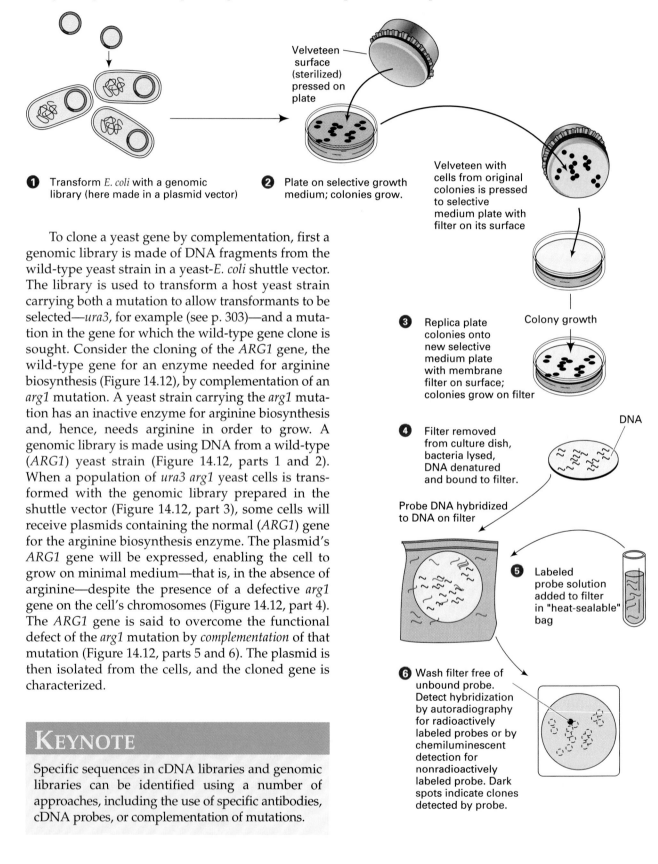

❶ Transform *E. coli* with a genomic library (here made in a plasmid vector)

❷ Plate on selective growth medium; colonies grow.

Velveteen surface (sterilized) pressed on plate

Velveteen with cells from original colonies is pressed to selective medium plate with filter on its surface

❸ Replica plate colonies onto new selective medium plate with membrane filter on surface; colonies grow on filter

Colony growth

❹ Filter removed from culture dish, bacteria lysed, DNA denatured and bound to filter.

DNA

Probe DNA hybridized to DNA on filter

❺ Labeled probe solution added to filter in "heat-sealable" bag

❻ Wash filter free of unbound probe. Detect hybridization by autoradiography for radioactively labeled probes or by chemiluminescent detection for nonradioactively labeled probe. Dark spots indicate clones detected by probe.

To clone a yeast gene by complementation, first a genomic library is made of DNA fragments from the wild-type yeast strain in a yeast-*E. coli* shuttle vector. The library is used to transform a host yeast strain carrying both a mutation to allow transformants to be selected—*ura3*, for example (see p. 303)—and a mutation in the gene for which the wild-type gene clone is sought. Consider the cloning of the *ARG1* gene, the wild-type gene for an enzyme needed for arginine biosynthesis (Figure 14.12), by complementation of an *arg1* mutation. A yeast strain carrying the *arg1* mutation has an inactive enzyme for arginine biosynthesis and, hence, needs arginine in order to grow. A genomic library is made using DNA from a wild-type (*ARG1*) yeast strain (Figure 14.12, parts 1 and 2). When a population of *ura3 arg1* yeast cells is transformed with the genomic library prepared in the shuttle vector (Figure 14.12, part 3), some cells will receive plasmids containing the normal (*ARG1*) gene for the arginine biosynthesis enzyme. The plasmid's *ARG1* gene will be expressed, enabling the cell to grow on minimal medium—that is, in the absence of arginine—despite the presence of a defective *arg1* gene on the cell's chromosomes (Figure 14.12, part 4). The *ARG1* gene is said to overcome the functional defect of the *arg1* mutation by *complementation* of that mutation (Figure 14.12, parts 5 and 6). The plasmid is then isolated from the cells, and the cloned gene is characterized.

KEYNOTE

Specific sequences in cDNA libraries and genomic libraries can be identified using a number of approaches, including the use of specific antibodies, cDNA probes, or complementation of mutations.

~ FIGURE 14.12

Example of cloning a gene by complementation of mutations: the cloning of the yeast *ARG1* gene.

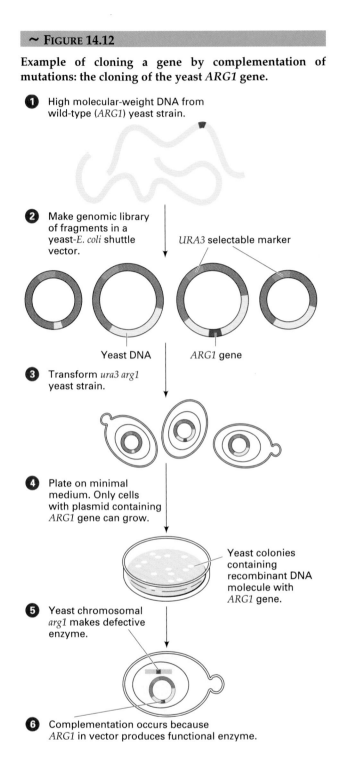

1 High molecular-weight DNA from wild-type (*ARG1*) yeast strain.

2 Make genomic library of fragments in a yeast-*E. coli* shuttle vector.

URA3 selectable marker

Yeast DNA   *ARG1* gene

3 Transform *ura3 arg1* yeast strain.

4 Plate on minimal medium. Only cells with plasmid containing *ARG1* gene can grow.

Yeast colonies containing recombinant DNA molecule with *ARG1* gene.

5 Yeast chromosomal *arg1* makes defective enzyme.

6 Complementation occurs because *ARG1* in vector produces functional enzyme.

ANALYSIS OF GENES AND GENE TRANSCRIPTS

Cloned DNA sequences are resources for experiments designed to answer many kinds of biological questions. The following experimental techniques are described in this section:

1. *The cloned DNA may be mapped with respect to the number and arrangement of restriction sites.* The resulting map, analogous to the arrangement of genes on a linkage map, is called a **restriction map**. This is commonly done for both cloned genes and cloned cDNAs. The restriction maps produced can be useful for making clones of subsections of the gene or the cDNA, or for comparing the gene to the cDNA.

2. *A cloned cDNA or a cloned gene may be used to analyze transcription of the corresponding gene in the cell.* For example, we can study the size of the initial transcript of the gene, the processing steps it goes through (if any) to produce a mature mRNA, the amount of the gene transcript, the time of expression in the cell cycle, and the time and amount of expression in different tissues and/or during development.

3. *The complete sequence of the cloned DNA may be determined.* In the case of a cloned gene, the sequence information can be useful in studying how the expression of the gene is regulated. Comparison of DNA sequences to other DNA sequences in a computer database can be used to determine the extent of similarity between related genes. If the cloned DNA sequence is not a known gene, a computer search may provide insights into what kind of gene the sequence might be. Further, the DNA sequence of a protein-coding gene can be "translated" by computer to provide information about the properties of the protein for which it codes. Such information can be helpful for an investigator who wishes to isolate and study an unknown protein product of a gene for which a clone is available.

Restriction Enzyme Analysis of Cloned DNA Sequences

Because cloned DNA sequences represent a homogeneous population of DNA molecules, restriction enzymes cleave cloned DNA into a relatively small number of discretely sized DNA fragments. These DNA fragments can be visualized using agarose gel electrophoresis and ethidium bromide staining (described in the next paragraph). With such gels, restriction maps can be constructed without the need for hybridization with a labeled probe and subsequent detection.

Let us assume that we have cloned a 5.0-kb piece of DNA and wish to construct a restriction map of it (Figure 14.13, part 1). One sample of the DNA is digested with *Eco*RI, a second sample is digested with *Bam*HI, and a third sample is digested with both

~ FIGURE 14.13

Construction of a restriction map for *Eco*RI and *Bam*HI in a DNA fragment.

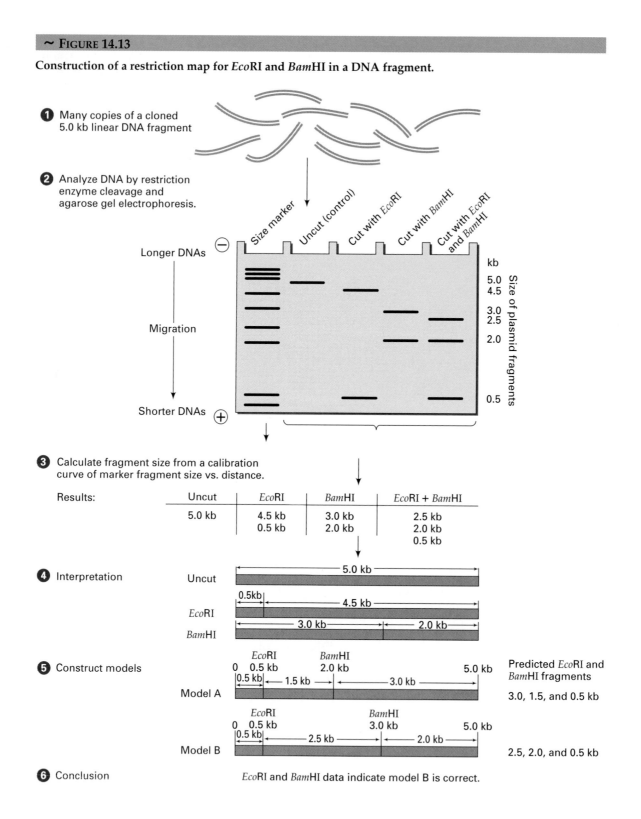

❶ Many copies of a cloned 5.0 kb linear DNA fragment

❷ Analyze DNA by restriction enzyme cleavage and agarose gel electrophoresis.

Size marker
Uncut (control)
Cut with *Eco*RI
Cut with *Bam*HI
Cut with *Eco*RI and *Bam*HI

⊖ Longer DNAs

Migration

Shorter DNAs ⊕

kb
5.0
4.5
3.0
2.5
2.0

0.5

Size of plasmid fragments

❸ Calculate fragment size from a calibration curve of marker fragment size vs. distance.

Results:

Uncut	*Eco*RI	*Bam*HI	*Eco*RI + *Bam*HI
5.0 kb	4.5 kb	3.0 kb	2.5 kb
	0.5 kb	2.0 kb	2.0 kb
			0.5 kb

❹ Interpretation

Uncut — 5.0 kb

*Eco*RI — 0.5kb | 4.5 kb

*Bam*HI — 3.0 kb | 2.0 kb

❺ Construct models

Model A

*Eco*RI
0   0.5 kb
0.5 kb | 1.5 kb
*Bam*HI
2.0 kb
3.0 kb
5.0 kb

Predicted *Eco*RI and *Bam*HI fragments

3.0, 1.5, and 0.5 kb

Model B

*Eco*RI
0   0.5 kb
0.5 kb | 2.5 kb
*Bam*HI
3.0 kb
2.0 kb
5.0 kb

2.5, 2.0, and 0.5 kb

❻ Conclusion

*Eco*RI and *Bam*HI data indicate model B is correct.

*Eco*RI and *Bam*HI. The DNA restriction fragments of each reaction are separated according to their molecular size by agarose gel electrophoresis; controls are a sample of the same DNA uncut with any enzyme, and DNA fragments of known size—"DNA fragment size markers," or simply "size markers"—so that the sizes of the unknown DNA fragments can be computed (Figure 14.13, part 2). The gel is a rectangular, horizontal slab of agarose (a firm, gelatinous material), which has a matrix of pores through which DNA passes in an electric field. Each gel is made by boiling a buffered agarose solution and allowing it to cool in a mold. A toothed comb is used to form discrete wells in the gels so that different samples can be analyzed simultaneously.

Since DNA is negatively charged due to its phosphates, the DNA migrates toward the positive pole in an electric field. Migration is in a straight line from the well (called a lane). Because small DNA fragments can move more readily through the pores in the gel, small DNA fragments move through the gel more rapidly than large fragments.

After electrophoresis, the DNA is stained with ethidium bromide. The DNA complexed with ethidium bromide fluoresces under ultraviolet light, and the gel is photographed. (The chapter opener photograph shows such a gel, with DNA bands illuminated with ultraviolet light.) From the photograph the distance each DNA band migrated can be measured. The molecular size of each DNA band in the size markers is known, so a calibration curve can be drawn of DNA size versus migration distance. The migration distances for the DNA bands from the uncut and cut DNA are then used with the calibration curve to determine the molecular sizes of the DNA fragments in the bands (Figure 14.13, part 3).

The results are analyzed as follows (Figure 14.13, parts 4–6):

1. When the 5.0-kb DNA is cut with *Eco*RI, 4.5-kb and 0.5-kb DNA fragments are obtained, indicating that there is one restriction site for *Eco*RI in the DNA located 0.5 kb from one end of the molecule.

2. Using similar logic, there is one restriction site for *Bam*HI located 2.0 kb from one end of the molecule.

3. At this point, we know there is one restriction site for each enzyme, but we do not know the relationship between the two. We can, however, make two models (see Figure 14.13, part 5). In model A, the *Eco*RI site is 0.5 kb from one end, and the *Bam*HI site is 2.0 kb from that same end. In model B, the *Eco*RI site is 0.5 kb from one end, and the *Bam*HI site is 3.0 kb from that end (that is, 2.0 kb from the other end). Model A predicts that cutting with *Eco*RI and *Bam*HI will produce three fragments of 0.5, 1.5, and 3.0 kb (going from left to right along the DNA), while model B predicts that cutting with both enzymes will produce three fragments of 0.5, 2.5, and 2.0 kb. The actual data show three fragments with sizes 2.5, 2.0, and 0.5 kb, validating model B.

In real situations, restriction mapping involves data that are much more complicated—for example, involving more restriction enzymes and a number of sites for each enzyme.

Restriction Enzyme Analysis of Genes in the Genome

As part of the analysis of genes, it can be helpful to determine the arrangement and specific locations of restriction sites. This information is useful, for example, *for comparing homologous genes in different species, for analyzing intron organization, or for planning experiments to clone parts of a gene, such as its promoter or controlling sequences, into a vector.* The arrangement of restriction sites in a gene can be analyzed without actually cloning the gene by using a cDNA probe or by using as a probe the same gene cloned from a closely related organism. The process of analysis is as follows:

1. Samples of genomic DNA are cut with different restriction enzymes (Figure 14.14, parts 1 and 2), each of which will produce DNA fragments of different lengths depending on the locations of the restriction sites.

2. The DNA restriction fragments are separated by size using agarose gel electrophoresis (Figure 14.14, part 3).

   After electrophoresis, the DNA is stained with ethidium bromide so that it can be seen under ultraviolet light. When genomic DNA is digested with a restriction enzyme, the result is a continuous smear of fluorescence down the length of the gel lane due to the fact that the enzyme produces fragments of all sizes.

3. The DNA fragments are transferred to a membrane filter (Figure 14.14, part 4). The transfer to the membrane filter is done by the **Southern blot technique** (named after its inventor, Edward Southern). In brief, the gel is treated to denature the double-stranded DNA into single strands. The gel is then placed on a piece of blotting paper that spans a glass plate. The ends of the paper are in a container of buffer and act as wicks. A piece of membrane filter is laid down so that it covers the gel. Sheets of blotting paper (or paper towels) and a weight are stacked on top of the membrane filter. The buffer solution in the bottom tray is wicked up by the blotting paper, passing through the gel and the membrane filter. During this process, the DNA fragments are picked up by the buffer and transferred from the gel to the membrane filter, to which they bind. The fragments on the filter are arranged in exactly the same way as they were in the gel (Figure 14.14, part 5).

~ FIGURE 14.14

Southern blot procedure for analyzing cellular DNA for the presence of sequences complementary to a labeled probe, such as a cDNA molecule made from an isolated mRNA molecule. The hybrids, shown as three bands in this theoretical example, are visualized by autoradiography or chemiluminescence.

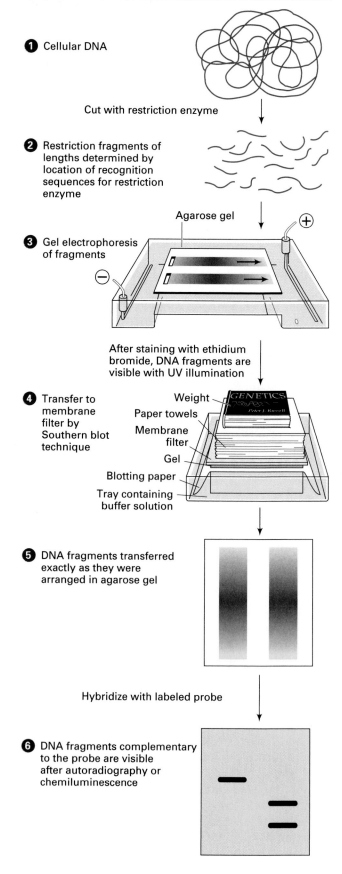

① Cellular DNA

Cut with restriction enzyme

② Restriction fragments of lengths determined by location of recognition sequences for restriction enzyme

Agarose gel

③ Gel electrophoresis of fragments

After staining with ethidium bromide, DNA fragments are visible with UV illumination

④ Transfer to membrane filter by Southern blot technique

Weight
Paper towels
Membrane filter
Gel
Blotting paper
Tray containing buffer solution

⑤ DNA fragments transferred exactly as they were arranged in agarose gel

Hybridize with labeled probe

⑥ DNA fragments complementary to the probe are visible after autoradiography or chemiluminescence

4. A labeled probe is added to the membrane filter; it will hybridize to any complementary DNA fragment (Figure 14.14, part 6). Detection of the probe is carried out in a way appropriate for whether the probe is radioactive or nonradioactive to determine the position(s) of the hybrids (Figure 14.14, part 6). If DNA size markers are separated in a different lane in the agarose gel electrophoresis process, the sizes of the genomic restriction fragments that hybridized with the probe can be calculated. From the fragment sizes obtained, a restriction map can be generated to show the relative positions of the restriction sites. Suppose, for example, that using only *Bam*HI produces a DNA fragment of 3 kb that hybridizes with the radioactive probe. If a combination of *Bam*HI and *Pst*I is then used and produces two DNA fragments, one of 1 kb and the other of 2 kb, we would deduce that the 3-kb *Bam*HI fragment contains a *Pst*I restriction site 1 kb from one end and 2 kb from the other end. Further analysis with other enzymes, individually and combined, enables the researcher to construct a map of all the enzyme sites relative to all other sites.

Analysis of Gene Transcripts

A blotting technique related to the Southern blot technique—called **northern blot analysis**—has been developed to analyze RNA rather than DNA. In this case, the name is not derived from a person but to indicate that the technique is related to the Southern blot technique. In northern blot analysis, RNA extracted from a cell is separated by size using gel electrophoresis, and the RNA molecules are transferred and bound to a filter in a procedure that is essentially identical to Southern blotting. After hybridization with a labeled probe and use of the appropriate detection system, bands show the locations of RNA fragments that were complementary to the probe. Given appropriate RNA size markers, the sizes of the RNA fragments identified with the probe can be determined.

Northern blot analysis is useful for revealing the size or sizes of the mRNA encoded by a gene. In some cases, a number of different mRNA species encoded by the same gene have been identified in this way, suggesting the use of different promoter sites and/or different terminator sites, or that alternative mRNA processing can occur. Northern blot analysis can also be used to investigate whether or not an mRNA is present in a cell type or tissue and how much of it is present. This type of experiment is useful for determining levels of gene activity, for instance during development, in different cell types of an organism, or in cells before and after they are subjected to various physiological stimuli.

KEYNOTE

Genes and cloned DNA sequences are often analyzed to determine the arrangement and specific locations of restriction sites. The analytical process involves cleaving the DNA with restriction enzymes, separating the resulting DNA fragments by agarose gel electrophoresis, and then staining the DNA fragments with ethidium bromide so that they may be visualized with ultraviolet light. The DNA fragments produced by cleavage of cloned DNA sequences can be seen as discrete bands, enabling restriction maps to be constructed based on the calculated molecular lengths of the DNA in the bands. DNA fragments produced by cleavage of genomic DNA show a wide range of sizes, resulting in a continuous smear of DNA fragments in the gel. In this case, specific gene fragments can only be visualized by transferring the DNA fragments to a membrane filter in a procedure called the Southern blot technique, hybridizing a specific labeled probe to the DNA fragments, and detecting the hybrids by autoradiography or some other label detection method. With these data, a restriction map can be made.

DNA Sequencing

Cloned DNA fragments may be analyzed to determine the nucleotide-pair sequence of the DNA. This is the most detailed information one can obtain about a DNA fragment. The information is useful, for example, for identifying gene sequences and regulatory sequences within the fragment and for comparing the sequences of homologous genes from different organisms.

DIDEOXY SEQUENCING. The most commonly used method of DNA sequencing, called **dideoxy sequencing**, involves extension of a short primer by DNA polymerase (Figure 14.15). Four separate reactions are set up. Each reaction contains the single-stranded DNA to be sequenced, the primer annealed

~ **FIGURE 14.15**

Dideoxy DNA sequencing of a theoretical DNA fragment.

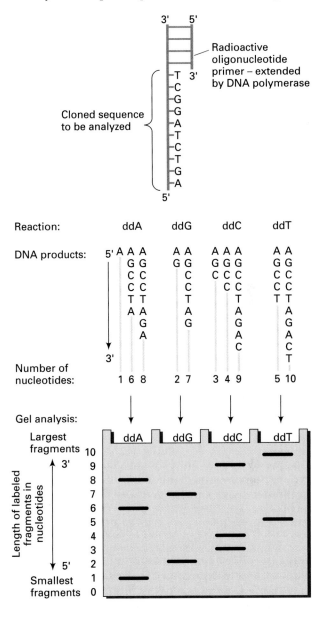

Sequence deduced from banding pattern of autoradiogram made from gel:

5' A-G-C-C-T-A-G-A-C-T 3'

to that DNA, DNA polymerase, the four normal deoxynucleotide precursors, and a small amount of a modified nucleotide precursor called a **dideoxynucleotide** (Figure 14.16). A dideoxynucleotide differs from a normal deoxynucleotide in that it has a 3'-H rather than a 3'-OH on the deoxyribose sugar.

The four reactions differ by which dideoxynucleotide is present, meaning whether it has A, T, G, or C as the base. When the primer is extended, occasionally DNA polymerase inserts a dideoxynucleotide

~ FIGURE 14.16

A dideoxynucleotide DNA precursor.

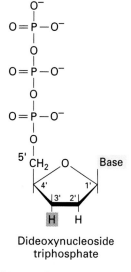

Dideoxynucleoside
triphosphate

(Normal DNA precursor
has OH at 3' position)

~ FIGURE 14.17

Autoradiogram of a dideoxy sequencing gel. The letters over the lanes (A, C, G, and T) correspond to the particular dideoxy nucleotide used in the sequencing reaction analyzed in the lane.

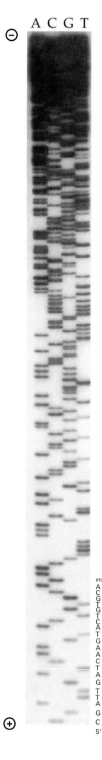

instead of the normal deoxynucleotide. Once that happens, no further DNA synthesis can occur because the *absence of a 3'-OH prevents the formation of a phosphodiester bond with an incoming DNA precursor.*

For example, if an A is specified by the DNA template strand, a dideoxy A nucleotide (ddA) could be incorporated rather than the normal A nucleotide in the reaction mixture, and elongation of the chain would stop. In a population of molecules in the same DNA synthesis reaction, new DNA chains will stop at *all* possible positions where the nucleotide is required because of the incorporation of the dideoxynucleotide. In the ddA reaction, the many different chains produced all end with ddA; all chains end with ddG in the ddG reaction, and so on (see Figure 14.15).

The DNA chains in each reaction mixture are separated by polyacrylamide gel electrophoresis, and the locations of the DNA bands are revealed by autoradiography (in the case of radioactive DNA sequencing experiments). The DNA sequence of the newly synthesized strand is determined from the autoradiogram by reading from the bottom (shortest DNA fragment) to the top (longest DNA fragment) to give the sequence in 5'-to-3' orientation. In the example, the band that moved the farthest ended with ddA, the band that moved the second farthest ended with ddG, and so on. The complete sequence determined is 5'-AGCCTAGACT-3'; this is *complementary* to the sequence on the template strand (see Figure 14.15). An example of a dideoxy sequencing gel result is shown in Figure 14.17.

~ FIGURE 14.18

Results of automated DNA sequence analysis using fluorescent dyes. The procedure is described in the text. The automated sequencer generates the curves shown in the figure from the fluorescing bands on the gel. The colors are generated by the machine and indicate the four bases: A is green, G is black, C is blue, and T is red. Where bands cannot be distinguished clearly, an N is listed.

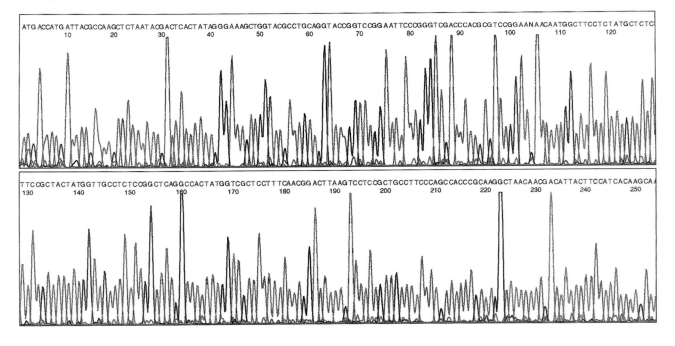

KEYNOTE

Methods have been developed for determining the sequence of a cloned piece of DNA. A commonly used method, the dideoxy procedure, employs enzymatic synthesis of a new DNA chain on a cloned template DNA strand. With this procedure, synthesis of new strands is stopped by the incorporation of a dideoxy analog of the normal deoxyribonucleotide. Using four different dideoxy analogs, the new strands stop at all possible nucleotide positions, thereby allowing the complete DNA sequence to be determined.

Recently, automated procedures have been developed that enable DNA sequencing to proceed much more rapidly than with the manual dideoxy method. The automated procedures involve one reaction containing four differently colored dideoxynucleotides, one for each of the four bases. The DNA fragments synthesized are separated by electrophoresis in a single gel lane, and that gel lane is then scanned by a laser device that excites the fluorescent labels and determines which particular fluorescent label is present at each position. The output is a series of colored peaks corresponding to each nucleotide position in the sequence (Figure 14.18); the output is converted to a sequence of bases by computer. Such procedures are of great utility as research teams proceed to determine the complete sequences of various genomes, including that of humans.

ANALYSIS OF DNA SEQUENCES. Sequences determined by any sequencing method are entered into computer databases. Computer programs have been written to analyze DNA sequences for restriction site locations, for example, and to compare a variety of sequences, homologous regions, transcription regulatory sequences, and so on. Programs can search DNA sequences for possible protein-coding regions by looking for an initiator codon in a frame with a stop codon (called an open reading frame, or ORF). Other programs can be used to translate a cloned DNA sequence theoretically into an amino acid sequence and to make predictions about the structure and function of the protein. Increasingly, much of this computer analysis of sequences can be done using the Internet.

POLYMERASE CHAIN REACTION (PCR)

The generation of large numbers of identical copies of DNA by the construction and cloning of recombinant DNA molecules was made possible in the 1970s. Recombinant DNA techniques revolutionized molecular genetics by making it possible to analyze genes and their functions in new ways. However, cloning DNA is time consuming. In the mid-1980s the **polymerase chain reaction (PCR)** was developed, and this has resulted in yet a new revolution in the way genes may be analyzed. PCR is a method for producing an extremely large number of copies of a *specific* DNA sequence from a DNA mixture *without* having to clone it, a process called *amplification*. Kary Mullis, who developed PCR, shared the Nobel Prize in chemistry in 1993 with M. Smith.

The starting point for PCR is the double-stranded DNA containing the sequence to be amplified and a pair of oligonucleotide ("few nucleotides") primers that flank that DNA (Figure 14.19). The primers are usually 20 or more nucleotides long and are made synthetically, so some information must be available about the sequence of interest. In brief, the PCR procedure is as follows:

1. Denature the double-stranded DNA to single strands by heating at 94–95°C (Figure 14.19, part 1). Cool, and anneal the primers (A and B in the figure) at 37–65°C, depending on how well the base sequences of the primers match the base sequence of the DNA. Since the primers anneal to the opposite strands of the template DNA, their 3' ends face each other.

2. Extend the primers with DNA polymerase at 70–75°C (Figure 14.19, part 2). For this, a special heat-resistant DNA polymerase such as *Taq* ("tack") *DNA polymerase* is used. This enzyme is the DNA polymerase of a thermophilic bacterium, *Thermus aquaticus.*

3. Repeat the heating cycle to denature the DNA to single strands, and cool to anneal the primers again (Figure 14.19, part 3). (The further amplification of the original strands is omitted in the remainder of the figure.)

4. Repeat the primer extension with *Taq* DNA polymerase (Figure 14.19, part 4). In each of the two double-stranded molecules produced in the fig-

ure, one strand is of unit length; that is, it is the length of DNA between the 5' end of primer A and the 5' end of primer B—the length of the target DNA. The other strand in both molecules is longer than unit length.

5. Repeat the denaturation of DNA and annealing of new primers (Figure 14.19, part 5). (For simplification, the further amplification of the longer-than-unit-length strands continues with linear increase only and is omitted in the rest of the figure.)

6. Repeat the primer extension with *Taq* DNA polymerase (Figure 14.19, part 6). This produces unit-length, double-stranded DNA. Note that it took three cycles to produce the two molecules of target-length DNA. Repeated denaturation, annealing, and extension cycles result in the geometric increase in the amount of the unit-length DNA.

Using PCR, the amount of new DNA generated increases geometrically. Starting with one molecule of DNA, one cycle of PCR produces two molecules, two cycles produce four molecules, and three cycles produce eight molecules, two of which are the target DNA. A further 10 cycles produce 1,024 copies (2^{10}) of the target DNA, and in 20 cycles there will be 1,048,576 copies (2^{20}) of the target DNA! The procedure is rapid, each cycle taking only a few minutes using a *thermal cycler*, a machine that automatically cycles the reaction through the temperature changes in a programmed way.

There are many applications for PCR, including amplifying DNA for cloning, amplifying DNA from genomic DNA preparations for sequencing without cloning, mapping DNA segments, disease diagnosis, sex determination of embryos, forensics, and studies of molecular evolution. In disease diagnosis, for example, PCR can be used to detect bacterial pathogens or viral pathogens such as HIV (human immunodeficiency virus, the causative agent of AIDS) and hepatitis B virus. PCR can also be used in genetic disease diagnosis, which is discussed in the next section.

In forensics, PCR can be used, for example, to amplify trace amounts of DNA in samples such as hair, blood, or semen collected from a crime scene. The amplified DNA can be analyzed and compared with DNA from a victim and a suspect, and the results can be used to implicate or exonerate suspects in the crime. This analysis of DNA—called *DNA typing*—is discussed in more detail on pp. 325–327. Another interesting application of PCR is amplifying "ancient" DNA for analysis, using samples of tissues preserved hundreds or thousands of years ago, such as 440,000-year-old mammoths. PCR makes it possible to amplify selected DNA sequences and then analyze the sequences of those DNA molecules for comparison

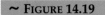

~ FIGURE 14.19

The polymerase chain reaction (PCR) for selective amplification of DNA sequences.

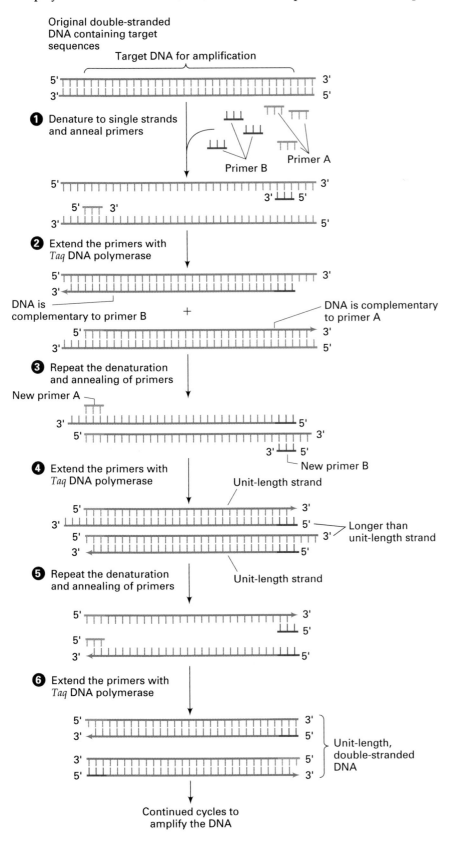

with contemporary DNA samples. These analyses enable us to make evolutionary comparisons between ancient forebears and present-day descendants.

KEYNOTE

The polymerase chain reaction (PCR) uses specific primers to amplify a specific segment of DNA many thousandfold in an automated procedure. PCR is finding increasing applications both in research and in the commercial arena, including generating specific DNA segments for cloning, for sequencing, and for amplifying DNA to detect the presence of specific genetic defects.

APPLICATIONS OF RECOMBINANT DNA AND PCR TECHNIQUES

Recombinant DNA and PCR techniques have many applications, including the diagnosis of human and animal genetic diseases, the synthesis of commercially important products such as human insulin and human growth hormone, in vitro modifications of genes, and genetic engineering of plants. Some important applications are briefly outlined in this section.

Analysis of Biological Processes

Recombinant DNA and PCR techniques have fundamental and widespread applications in basic research exploring biological functions. Questions in most areas of biology are now being addressed with these techniques. In genetics, researchers are investigating such things as the functional organization of genes and the regulation of gene expression. In developmental genetics, key regulatory genes and target genes responsible for developmental events are being discovered and analyzed, and gene changes associated with aging and cancer are being investigated. In evolutionary biology, DNA sequence analysis is adding new information about the evolutionary relationships between organisms.

Diagnosis of Human Genetic Diseases by DNA Analysis

For an increasing number of genetic diseases, including Huntington disease, hemophilia, cystic fibrosis, Tay-Sachs disease, and sickle-cell anemia, we can screen individuals for the actual DNA mutation, rather than for the resulting biochemical change. Such

information is of obvious use in *genetic counseling* (see Chapter 8).

Recombinant DNA and/or PCR approaches for detecting genetic diseases require cellular DNA as the starting point. Such DNA can be isolated from fetal cells obtained by amniocentesis or chorionic villus sampling (see Chapter 8) and from blood samples of children and adults. The DNA can be digested with a restriction enzyme, and the restriction fragments analyzed as described earlier, that is, separated by agarose gel electrophoresis, transferred to a membrane filter by Southern blotting, and hybridized with a specific labeled DNA probe. Alternatively, specific regions of the isolated DNA can be amplified by PCR and then, with or without restriction enzyme digestion, analyzed by gel electrophoresis. In this case, no blotting or probing is required.

These analytical procedures are most useful when the genetic mutation that causes a disease is associated with the loss or addition of a restriction site, either within the gene or in a flanking region. The different patterns of restriction sites result in **restriction fragment length polymorphisms**, or **RFLPs** ("rifflips"), which are restriction fragments of different lengths. (*Polymorphism* means the existence of many different forms.) A RFLP can be used as a genetic marker in the same way as a "conventional" genetic marker. In this case we assay the DNA—that is, determine the genotype—directly in the form of a restriction map. Moreover, because we are looking directly at DNA, both parental types are seen in heterozygotes, so carriers can easily be identified.

An example of the use of RFLP analysis in human disease diagnosis concerns *sickle-cell anemia* (discussed in more detail in Chapter 8). Sickle-cell anemia results from a single base-pair change in the gene for the hemoglobin's β-globin polypeptide, so that an abnormal form of hemoglobin, Hb-S, is made instead of the normal Hb-A form. The mutation changes the sixth codon from GAG to GTG, resulting in the substitution of a valine for glutamic acid, which in turn produces abnormal associations of hemoglobin molecules, sickling of the red blood cells, tissue damage, and often death.

Using a cDNA probe for human β-globin, a RFLP has been shown for the restriction enzyme *Dde*I ("D-D-E-one"). In the normal β-globin gene there are three *Dde*I sites, one upstream of the start of the gene and the other two within the coding sequence, and in the sickle-cell mutant β-globin gene the *Dde*I site in the beginning of the coding sequence is not present, leaving only two *Dde*I sites (Figure 14.20a). When DNA from normal individuals is cut with *Dde*I, the fragments separated by gel electrophoresis and transferred to a membrane filter by the Southern blot technique, and then probed with the 5' end of a cloned β-globin gene, two fragments of 175 bp and 201 bp

~ FIGURE 14.20

Detection of sickle-cell gene by the *Dde*I restriction fragment length polymorphism. (a) DNA segments showing the *Dde*I restriction sites; (b) Results of analysis of DNA cut with *Dde*I, subjected to gel electrophoresis, blotted, and probed with a β-globin probe.

a) *Dde* I restriction sites

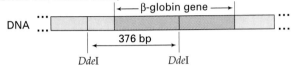

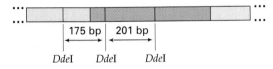

b) *Dde* I fragments detected on a Southern blot by probing with beginning of β-globin gene

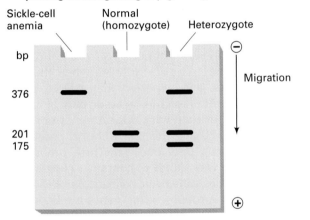

are seen (Figure 14.20b). DNA from individuals with sickle-cell anemia analyzed in the same way gives one fragment of 376 bp because of the loss of the *Dde*I site. Heterozygotes are detected by the presence of three bands of 376 bp, 201 bp, and 175 bp.

Other examples of human genetic diseases for which recombinant DNA technology can or will soon provide early detection include PKU, four types of thalassemia (hemoglobin diseases resulting in anemia), α-antitrypsin deficiency (a deficiency of a plasma protein), hemophilia A, hemophilia B, Huntington disease, cystic fibrosis, and Duchenne muscular dystrophy (a progressive disease resulting in muscle atrophy and muscle dysfunction).

The beauty of a DNA-based analysis procedure is that it directly tests for a DNA genotype (a RFLP), so detection does not depend on the expression of the gene (phenotype). To illustrate the present power of the approach, it is now possible to isolate DNA from one cell of an eight-cell-stage blastula, diagnose whether or not a mutant gene for a genetic disease is present using PCR and probing, and then reimplant the blastula if it is normal.

Isolation of Human Genes

With a defined gene product isolated by biochemical procedures, it is relatively easy to clone the gene, for example by using antibodies against the gene product to screen a cDNA library made in an expression vector (see p. 308 and Figure 14.10). However, for most human genetic diseases, the gene product that is altered is unknown, so cloning the gene involved is difficult. Fortunately, a number of approaches are available to solve this problem. One approach is to use linkage analysis to identify a RFLP marker that is genetically linked to the disease phenotype, and then to home in on the gene starting from the marker location on the chromosome. The isolation of a gene associated with a genetic disease on the basis of its approximate chromosomal position is called **positional cloning**. Some of the techniques used in positional cloning are illustrated in the story of how the cystic fibrosis (CF) gene was cloned.

CLONING THE CF GENE. Cystic fibrosis is a disease caused by an autosomal recessive mutation. Disease symptoms and genetic properties of the disease are described in Chapter 8. The CF gene was the first human disease gene to be cloned solely by positional cloning. The effort took four years and the involvement of many researchers in many laboratories.

Identifying RFLP Markers Linked to the CF Gene. Many individuals in CF pedigrees were screened with a large number of RFLPs to determine if any RFLPs were linked genetically to the CF gene. This was done by tracking the inheritance of the CF gene (in both homozygotes and heterozygotes) in the families and simultaneously analyzing their DNA by Southern blot analysis and hybridizing with probes to identify any RFLP marker (detected as characteristic DNA fragment sizes) that showed genetic linkage to the CF locus. One RFLP showed weak linkage to the CF locus.

Identifying the Chromosome on Which the CF Gene Is Located. The RFLP marker was used to identify the chromosome on which the CF gene is located. This was done by *in situ hybridization*, a technique in which chromosomes are spread on a microscope slide and hybridized with a labeled probe. Using this technique, it was shown that chromosome 7 was the location of the CF gene.

Identifying the Chromosome Region Where the CF Gene Is Located.   The next step was to use other RFLPs on chromosome 7 to find those most closely linked to the CF gene. Two closely linked flanking markers (one marker on each side of the CF gene) called *met* and D7S8 were found. The *met* marker is a known gene, and the D7S8 marker is a randomly cloned DNA fragment (also called an *anonymous probe*). The two flanking markers were known to be located at region 7q31-q32 (7 = chromosome 7; q = the long arm; 31-32 = subregions 31 and 32), so this localized the CF gene to that stretch of chromosome 7.

Cloning the CF Gene Between the Flanking Markers.   Flanking markers are used as starting points for cloning the DNA in between that contains the gene of interest, in this case the CF gene. However, this is not always as simple as it seems. In this particular case, the two flanking markers are about 1.5 map units (cM) apart. In the human genome, 1 map unit on average is equivalent to approximately 1 million base pairs (1 megabase) of DNA. Fortunately, in subsequent research two other cloned DNA probes detected RFLPs linked to CF and reduced the targeted span of DNA to about 500 kb.

An approach often used to find a gene between flanking markers is **chromosome walking**, a process used to identify adjacent clones in a genomic library (Figure 14.21). A chromosome walk is done as follows: an initial cloned DNA fragment—for example, one of the flanking markers—is used to begin the walk. A labeled end piece of the initial clone (right end in the figure) is used to screen a genomic library in, for example, a lambda vector, for clones that hybridize with it. The labeled probe should find all clones that overlap the original clone. These clones can be analyzed by restriction mapping to determine the extent of overlap. Then, a new labeled probe can be made from the right end of a clone with minimal overlap, and the library is screened again. By repeating this process over and over, we can walk along the chromosome clone by clone.

A procedure related to chromosome walking is also used to move along chromosomes. Called *chromosome jumping*, it is a technique to cross large amounts of DNA. So, while in chromosome walking each "step" is an overlapping DNA clone, in chromosome jumping each "jump" is from one chromosome location to another without "touching down" on the intervening DNA.

For cloning the CF gene, seven chromosome jumps were made toward the locus, and chromosome walks were made from each jump site to identify overlapping clones. In the end, a large number of clones were isolated that spanned over 500 kb of DNA of the CF region.

Identifying the CF Gene in the Cloned DNA.   The clones spanning the region between the RFLP markers were known to include the CF gene itself, but how does one identify the particular gene of interest in the set of clones? This is a general problem gene cloners face.

One approach to home in on genes in the clones is to use cloned DNA as probes to see if they can hybridize with sequences in other species. The reasoning is that genes are conserved in sequence among related species, whereas nongene sequences are likely not to be conserved. The procedure is to isolate DNA from other organisms (such as mouse, hamster, or chicken), digest it with restriction enzymes, separate the fragments by agarose gel electrophoresis, transfer the fragments to a membrane filter by Southern blotting, and hybridize them with a labeled probe. Because the blot contains DNA from a variety of organisms, it is often called a *zoo blot*. In the CF project, five probes from the CF region cross-hybridized with DNA sequences from other organisms, identifying them as possible candidates for the CF gene. Two of the probes were ruled out based on linkage analysis, and a third was ruled out because it proved to be a pseudogene, that is, a sequence resembling a functional gene but lacking appropriate expression signals.

Another obvious property of protein-coding genes is that they produce mRNAs when they are expressed. Thus, it is expected that a DNA probe made from a protein-coding gene or part of such a gene should hybridize with mRNAs on a northern blot (see p. 314). Using this test, a fourth probe was ruled out, leaving only one. This fifth probe was sequenced and was found to contain a cluster of C and G nucleotides called *CpG islands*. Since the promoters of many protein-coding genes are known to contain CpG islands, this discovery was an encouraging sign that the CF gene was nearby.

A probe made from this fifth sequence was used to screen a cDNA library made from mRNA molecules isolated from cultured normal sweat gland cells. This tissue type was used as the source of mRNA because the symptoms of CF were believed to result from a defect in sodium and chloride transport, and such transport is an important function of sweat gland cells. The probe identified a single, positive clone on northern blots, and the subsequent study of that clone and overlapping clones established a candidate cDNA clone of about 6,500 bp indicating, of course, an mRNA of the same size. The cDNA clone was used to analyze the genomic clones in more detail, with the result that the candidate CF gene was shown to span approximately 250 kb of DNA and involve 24 exons. Confirmation that the CF gene had been cloned came from comparing the DNA sequences of the candidate CF gene in a normal indi-

~ **FIGURE 14.21**

Chromosome walking. (From *Biochemistry* by Donald Voet and Judith G. Voet. Copyright © 1990 Donald Voet and Judith G. Voet. Reprinted by permission of John Wiley & Sons, Inc.)

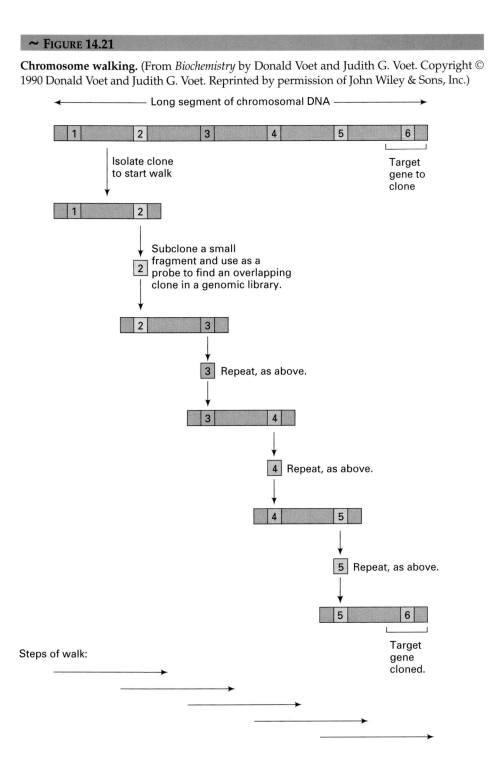

vidual and in a CF patient. The expectation was that mutational changes would be obvious if the correct gene had been identified. This proved to be the case— a 3 base-pair deletion was detected in the CF patient.

The Gene Defects in Cystic Fibrosis.   With the CF gene identified, researchers then investigated more fully the nature of the mutations responsible for CF. Sixty-eight percent of patients had the 3-bp deletion mentioned above, which results in the loss of the amino acid phenylalanine in the protein encoded by the gene. The remaining CF patients show over 60 different mutations, so the prospect of developing a simple DNA-based diagnostic test for CF is daunting.

KEYNOTE

The isolation of human genes, particularly those associated with disease, is possible using an array of molecular techniques. Where the gene product is not known, the starting point for cloning is knowledge of the genetic linkage between the disease locus and a DNA marker or markers. The isolation of a gene associated with a genetic disease on the basis of its approximate chromosomal position is called positional cloning.

The Human Genome Project

The **Human Genome Project** (HGP) is an extensive, collaborative effort to map all of the estimated 50,000 to 100,000 human genes and to obtain the sequence of the complete approximately 3 billion (3×10^9) nucleotide pairs of the genome. In the United States, the Human Genome Project is being coordinated by the National Center of Human Genome Research (a part of the National Institutes of Health) and the Department of Energy. Officially, the Human Genome Project began on October 1, 1990, with a goal for completion of 15 years. Associated with the Human Genome Project are parallel efforts to obtain gene maps and complete sequences of the genomes of a number of other model organisms, including *E. coli*, yeast, *Drosophila melanogaster*, the plant *Arabidopsis thaliana*, the nematode *Caenorhabditis elegans*, and the mouse.

At least 17 countries outside the United States have established human genome research programs. Coordinating the international collaborative effort on the Human Genome Project is the Human Genome Organisation (HUGO). Numerous other genome projects are also being done independently of the official HGP, including the dog genome project, the maize genome project, and the rice genome project. In August 1998, the company Perkin-Elmer announced its intention to determine the entire human genome sequence within three years, that is, more quickly and more cheaply than the federally funded effort.

At this writing (June 1999), 14 eubacterial genomes, 4 archaeon genomes, and 1 eukaryotic genome (yeast) have been sequenced completely. Soon the sequencing of the *Caenorhabditis elegans* genome will be complete, providing the first complete genome of a metazoan. At least 59 bacterial genomes, 9 archaeon genomes, and 14 eukaryotic genomes are in the process of being sequenced. The latter include those from the mouse, the fruit fly, the plant *Arabidopsis thaliana,* and the dog. Many of the eubacterial genomes sequenced or being sequenced are associated with diseases in humans, such as *Treponema pallidum*, the cause of syphilis, and *Helicobacter pylori*, the cause of stomach ulcers, and the hope is that the sequence information will lead to a better understanding of their pathogenesis.

The complete yeast genome sequence was obtained in early 1996. The genome is 12,057 kb (excluding repetitive DNA) distributed among 16 chromosomes; it contains an estimated 6,000 genes. Late in 1996, the complete 1,660-kb genome sequence of the microbe *Methanococcus jannaschii* was reported. This archaeon was isolated from a deep-sea vent in the Pacific Ocean where the temperature is close to the boiling point of water and where the pressure is 245 times greater than at sea level. Fifty-six percent of the 1,738 genes the organism contains are entirely new to science. In early 1997, the complete 4,600-kb genome of *E. coli* was reported. The sequence indicates the presence of 4,288 genes, although at this point 38 percent have no known function.

While obtaining a complete sequence of the human genome is the major goal of the HGP, the path to that goal involves constructing detailed genetic maps and detailed physical maps, and then coalescing the two to provide a framework for the sequencing effort (see Chapter 5). To date, over 7,000 human genes have been mapped to particular chromosomes, and over 8,000 markers have been placed on the physical map (about one-quarter of the way toward mapping 30,000 markers). Since the focus of the HGP to date has mostly been on developing efficient and automated procedures for DNA sequencing, only about 4 percent of the genome has been sequenced.

How will the HGP benefit us? Most importantly, the complete sequence of the human genome will be a key resource for many, many years of research to understand the structure, organization, and function of DNA in chromosomes. Of particular interest will be the identification of genes involved in genetic diseases, since the study of those genes is necessary to developing therapies and cures. Having complete sequences of other organisms' genomes will enable comparative studies to be done, thereby leading to better understanding of biological functions. Numerous other benefits will result through the application of the new technologies developed as part of the HGP. Optimistically, the information obtained from the HGP and other genome projects will revolutionize the future of biological research, both basic and applied.

~ FIGURE 14.22

The concept involved in using VNTRs (variable number of tandem repeats) as DNA markers.

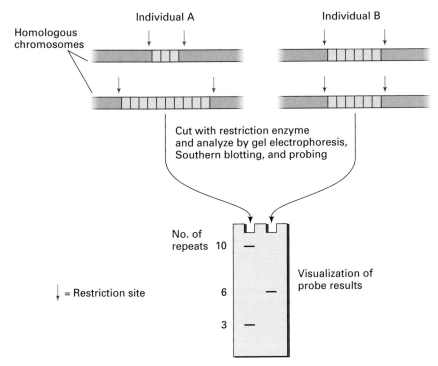

DNA Typing

Everyone is familiar with the use of fingerprints in forensic science. The principle is that no two individuals have the same fingerprints, so fingerprints left at the scene of a crime are important evidence in a criminal investigation. Similarly, no two human individuals (except identical twins) have exactly the same genome, base pair for base pair, and this has led to the development of DNA techniques for use in forensic science, in paternity and maternity testing, and elsewhere.

For **DNA typing** (also called **DNA fingerprinting**, or *DNA profiling*)—the use of DNA analysis to identify an individual—scientists may use *highly polymorphic markers* scattered throughout the genome. Each marker consists of a restriction fragment within which are short, identical segments of DNA tandemly arranged head to tail. Differences between individuals result from a great variation in the number of tandem repeats, called *variable number of tandem repeats*, or VNTRs (also called *minisatellite sequences*). Depending on the VNTR, the repeating unit is 15–100 base pairs long.

The use of VNTRs as markers is illustrated in Figure 14.22. Individual A is heterozygous for two alleles at a VNTR locus flanked by restriction sites. One allele has three copies of the repeat sequence,

and the other allele has 10 copies. Individual B is homozygous for a six-repeat allele. When genomic DNA is digested with the restriction enzyme, restriction fragments of different sizes are produced because of the VNTRs. The results of the digests can be visualized by using a probe for the particular repeat sequence at the marker locus. (A probe that is specific for the VNTR sequences at one locus in the genome is called a *monomorphic probe*. Probes that detect VNTR sequences at a number of loci in the genome are known as *polymorphic probes*.) In this way, two DNA bands would be detected for individual A, and one DNA band would be detected for individual B. (Note that if sequence information is available for the polymorphic region, PCR can be used to analyze the size alleles, and then no probing would be needed.)

Let us now consider an example of using DNA typing in a paternity case. In this fictional scenario, a mother of a new baby has accused a particular man of being the father of her child, and the man denies it. The court will decide the case based on evidence from DNA typing. The DNA typing proceeds as follows (Figure 14.23): DNA samples are obtained from each individual involved in the case (Figure 14.23, part 1). In a paternity case, the usual source of DNA is from a blood sample. The DNA is cut with the appropriate restriction enzyme, the resulting fragments are

~ FIGURE 14.23

Procedure for DNA typing (DNA fingerprinting) as used for a paternity case.

① DNA is obtained from the mother, the baby, and the alleged father. In separate analyses, the DNA is cut into fragments with a restriction enzyme.

② Gel electrophoresis of DNAs from each sample and of standards.

③ Southern blot prepared from the gel.

④ Filter from the blot is incubated with a radioactive DNA probe. DNA probe binds to specific DNA sequences on the filter.

⑤ Excess probe is washed away, leaving hybridized radioactive probe on filter.

⑥ Autoradiogram is prepared. The banding pattern for each sample is a DNA fingerprint.

separated by electrophoresis (Figure 14.23, part 2), transferred to a membrane filter by Southern blotting (Figure 14.23, part 3), and probed with a labeled VNTR probe (Figure 14.23, parts 4 and 5). The DNA banding pattern after autoradiography or chemiluminescence detection is then analyzed to compare the samples (Figure 14.23, part 6).

The data may be interpreted as follows: two DNA fragments are detected for the mother, so she is heterozygous for one particular pair of alleles at the VNTR locus under study. Likewise, two DNA fragments are detected for the baby, so the baby is also heterozygous. One of the fragments for the baby matches the larger of the fragments for the mother, while the other fragment for the baby is much larger, indicating many more repeats in that allele. The baby receives one allele from its mother and one from its father. The important question is whether the paternal allele of the baby matches an allele from the alleged father. Inspection of that lane in the autoradiogram leads us to conclude that the answer is "yes," for there are two fragments, one in common with the larger fragment of the baby, and the other distinct from other fragments already discussed.

The data indicate that the accused shares an allele with the baby, but they do not prove that he contributed that allele to the genome of the baby, though he obviously could have. If the man had no alleles in common with the baby, then the DNA typing data would have proved that he is not the father. To establish positive identity through DNA typing is more difficult. It boils down to calculating the relative odds that the allele came from the accused or from another person. This calculation depends on knowing the frequencies of VNTR alleles identified by the probe in the ethnic population from which the accused comes. Most of the legal arguments stem from this matter, since good estimates of VNTR allele frequencies are known for only a limited array of ethnic groups, so that calculations of probability of paternity give numbers of questionable accuracy in many cases. To minimize possible inaccuracy, investigators use a number of different probes (often five or more) so that the combined probabilities calculated for the set of VNTRs can be high enough to convince the court that the accused is guilty, even allowing for problems with knowing true VNTR allele frequencies for the population in question.

It is these combined probabilities that you hear or read about in the media with respect to DNA typing in court cases. In our paternity case, we would likely be more persuaded that the accused was the father of the child if the data for each of five different monomorphic probes indicated that he contributed a particular allele to the child.

DNA typing is more and more commonly used in forensic analysis. In murder, rape, and other violent cases, DNA is taken from crime scenes and compared with that of victims and suspects. For example, for a murder, DNA can be isolated from blood or hair at the scene, and for a rape DNA can be taken from a semen sample. If only minuscule amounts of DNA can be collected, PCR is used to amplify the DNA for typing experiments.

DNA typing is not yet a generally accepted method for proving guilt in U.S. courts. The scientific basis for the method is not in question; rather, DNA evidence is most commonly rejected for reasons such as possible errors in evidence collection or processing, or weak population statistics. There are, however, many cases in which DNA typing has excluded an accused individual because of mismatches, and such conclusions are now widely accepted by U.S. courts. DNA typing is also used extensively outside of human applications—for example, in population genetics studies, in animal poaching cases, in proving pedigree status in certain breeds of horses, and in conservation biology studies of endangered species.

Gene Therapy

Is it possible to treat genetic diseases? These diseases result from the phenotypes caused by the mutant gene in somatic cells. Theoretically, two types of gene therapy are possible: (1) somatic cell therapy, in which somatic cells are modified genetically to correct a genetic defect; and (2) germ-line cell therapy, in which the germ-line cells are modified to correct a genetic defect. Somatic cell therapy results in a treatment for the genetic disease in the individual, but not a cure, since progeny could still inherit the mutant gene. Germ-line cell therapy, however, would cure the disease since the mutant gene(s) would be replaced by the normal gene. While both somatic cell therapy and germ-line cell therapy have been demonstrated in nonhuman organisms, only somatic cell therapy has been attempted in humans because of ethical issues raised by germ-line cell therapy.

Gene therapy involving somatic cells proceeds as follows: (1) a sample of the individual's mutant cells is taken; (2) normal, wild-type copies of the mutant gene are introduced into the cells, and the cells are reintroduced into the individual. There, it is hoped, the cells will produce a normal gene product and the symptoms of the genetic disease will be treated successfully.

The source of the mutant cells will vary with the genetic disease. For example, blood disorders, such as thalassemia or sickle-cell anemia, require modification of blood-line cells isolated from the bone marrow. For genetic diseases affecting circulating proteins, a promising approach is the gene therapy of skin fibroblasts, constituents of the dermis (the lower layer of the skin). Modified fibroblasts can easily be implanted back into the dermis, where blood vessels invade the tissue, allowing gene products to be distributed.

A cell that has had a gene introduced into it by artificial means is called a *transgenic* cell, and the gene involved is called a *transgene*. The introduction of normal genes into a mutant cell poses several problems. First, procedures to introduce DNA into cells (transformation) typically are inefficient, so a large population of cells is needed to attempt gene therapy. Present procedures use special virus-related vectors to introduce the transgene. Second, in cells that take up the cloned gene, the fate of the "foreign" DNA cannot be predicted. In some cases the mutant gene will be replaced by the normal gene, while in others the normal gene will integrate into the genome elsewhere. In the first case, the gene therapy will be successful, provided that the gene is expressed. In the second case, successful treatment of the disease will only result if (1) the introduced gene is expressed; and (2) the resident mutant gene is recessive, so that it does not interfere with the normal gene.

Somatic gene therapy has been repeatedly demonstrated in experimental animals such as mouse, rat, and rabbit. In recent years, a few gene therapy trials have been done with humans. For example, in 1990, a four-year-old girl suffering from severe combined immunodeficiency (SCID), which results from a deficiency in adenosine deaminase (ADA), an enzyme required for normal function of the immune system, was treated by somatic gene therapy. T cells (cells involved in the immune system) were isolated from the girl and grown in the laboratory, and the normal ADA gene was introduced using a viral vector. The "engineered" cells were then reintroduced into the patient. Since T cells have a finite life in the body, continued infusions of engineered cells have been necessary. The introduced ADA gene is expressed, probably throughout the life of the T cell. As a result, the patient's immune system is functioning more normally, and she now gets no more than the average number of infections, compared with many more than the average number before the therapy. The gene therapy treatment has enabled her to live a more normal life.

With time, many other genetic diseases will be treatable with somatic gene therapy. Human genetic diseases that could be early candidates for gene therapy include sickle-cell anemia, thalassemias, phenylketonuria, Lesch-Nyhan syndrome, cancer, and cystic fibrosis. For example, after successful experiments with rats, human clinical trials are in process for transferring the normal CF gene to cystic fibrosis patients. As "tricks" are learned for targeting genes to replace their mutant counterparts and regulating the expression of the introduced genes, increasing success in treating genetic diseases is expected. However, many scientific, ethical, and legal questions will have to be addressed before the routine implementation of gene therapy.

Commercial Products

The development of cloning and other DNA manipulation techniques has spawned the formation of many biotechnology companies, which focus on using those techniques for making a wide array of commercial products. Some examples are:

1. Tissue plasminogen activator (TPA)—used to prevent or dissolve blood clots, therefore preventing strokes, heart attacks, or pulmonary embolisms.

2. Human growth hormone—to treat pituitary dwarfism.

3. Human blood clotting factor VIII—to treat hemophiliacs.

4. Human insulin ("humulin")—to treat insulin-dependent diabetes.

5. DNase—to treat cystic fibrosis.

6. Recombinant vaccines—to treat human and animal viral diseases (such as hepatitis B in humans).

7. Genetically engineered bacteria and other microorganisms for improved production of, for example, industrial enzymes (such as amylases to break down starch to glucose), citric acid (flavoring), and ethanol.

8. Genetically engineered bacteria that can accelerate the degradation of oil pollutants or of certain chemicals in toxic wastes (such as dioxin).

Genetic Engineering of Plants

For many centuries the traditional genetic engineering of plants involved selective breeding experiments in which plants with desirable traits were used to produce offspring with those traits. As a result, humans have produced hardy varieties of plants (for example, corn, wheat, and oats) as well as those with increased yields, all using standard plant breeding techniques. Now, vectors developed by recombinant DNA technology are available for transforming cells of crop plants; this has made possible the genetic engineering of plants for agricultural use.

With the use of these more modern techniques, an increasing number of transgenic plants will be developed. Of particular value will be crop plants with greater yield, insect pest resistance, and herbicide tolerance (so fields can be sprayed to kill weeds without harm to the crop). Already, rice has been genetically engineered to produce strains resistant to the damaging rice stripe virus, and wheat has been engineered to be resistant to a particular herbicide.

Let us briefly consider approaches to generating transgenic plants that are tolerant to the broad-spectrum herbicide Roundup™. This herbicide contains the active ingredient glyphosate, which kills plants by inhibiting EPSPS, a chloroplast enzyme required for the biosynthesis of essential aromatic amino acids. Roundup™ is used widely because it is active in relatively low doses and is degraded rapidly in the environment by microbes in the soil. Approaches for making transgenic, Roundup™-tolerant plants include: (1) introducing a mutated bacterial form of EPSPS that is resistant to the herbicide, so that the aromatic amino acids can still be synthesized even when the chloroplast enzyme is inhibited; and (2) introducing genes that encode enzymes for converting the herbicide to an inactive form. Monsanto brought Roundup™ Ready soybeans to market in 1996, although their use has been controversial because of opposition by groups questioning the safety of genetically engineered plants for human consumption.

With more sophisticated approaches it will be possible to make transgenic plants that control the expression of genes in different tissues. Examples would be controlling the rate at which cut flowers die or the time at which fruit ripens. Already approved for market is the "Flavr Savr" tomato (see Figure 1.2), genetically engineered by Calgene Inc. in collaboration with the Campbell Soup Company. Commercially produced tomatoes are picked while unripe so they can be shipped without bruising. Prior to shipping, they are exposed to ethylene gas, which initiates the ripening process so that they arrive in the ripened state at the store. Such prematurely picked, artificially ripened tomatoes do not have the flavor of tomatoes picked when they are ripe. Calgene scientists devised a way to block the tomato from making the normal amount of polygalacturonase (PG), a fruit-softening enzyme. They introduced into the plant a copy of the PG gene that was backward in its orientation with respect to the promoter. When this gene is transcribed, the mRNA is complementary to the mRNA produced by the normal gene: it is called an *antisense* mRNA.[1] In the

[1]The use of antisense mRNA to prevent or inhibit the translation of a natural mRNA is called antisense technology. This technology is being tested in a number of systems as a means of controlling genetic diseases.

cell, the antisense mRNA binds to the normal, "sense" mRNA, preventing much of it from being translated. As a result, much less PG enzyme is produced, and the tomato can remain longer on the vine without getting too soft for handling. Once picked, the Flavr Savr tomato is also less susceptible to bruising in shipping or to overripening in the store. The Flavr Savr tomato is advertised as tasting better than store-ripened tomatoes and more like home-grown tomatoes.

KEYNOTE

Recombinant DNA technology is finding ever-increasing applications throughout the world. In the basic research laboratory, processes across all areas of biology are being studied. With appropriate probes, a number of genetic diseases can now be diagnosed using recombinant DNA methods, genes can be isolated even if only linkage information is known, and DNA fingerprinting is being used in forensic science. In addition, many products in the clinical, veterinary, and agricultural areas can be synthesized in commercial quantities using recombinant DNA procedures. Genetic engineering of plants is also readily possible using recombinant DNA technology. It is expected that many types of improved crops will result from applications of this new technology.

ETHICS AND GENETICS

Many of the experiments and experimental procedures discussed in this chapter have raised ethical issues. For instance, when recombinant DNA cloning techniques were first developed, many scientists were concerned that the introduction of foreign genes into bacteria could produce a dangerous, pathogenic strain. Several scientists organized a meeting at Asilomar in California to discuss the possible hazards of cloning. Out of this meeting came an agreement to halt such experiments until safety issues were studied. Eighteen months later it was determined that the procedures were generally safe, and recombinant DNA research resumed under agreed-upon guidelines.

The Human Genome Project is also raising ethical issues. With the entire human genome sequence in hand, we will be able to identify and isolate all human genes that cause diseases. This will lead to the development of tests for many gene defects that we cannot test for now, including those that may lead to the development of a disease later in life, such as cancer. However, these tests will become available before we have developed a cure for those diseases, as is already the case for most testable genetic diseases. A number of ethical questions arise from this scenario, such as: Should a patient be told if a test for an incurable genetic disease is positive? (This issue applies now for Huntington disease.) Should employers be able to ask for a genetic test if the individual does not wish to know? Should health insurance companies and/or employers have access to genetic testing data and, if so, how can the patient protect his/her insurability and employability? Should states be able to collect genetic data on their populace? The last two questions raise fundamental privacy issues.

Fortunately, these issues are not being ignored. The federal agencies funding the HGP are devoting 3 to 5 percent of their annual budgets to study the ethical, legal, and social issues (ELSI) related to the availability of genetic information. This amounts to the world's largest bioethics program. Four areas are currently being emphasized by the ELSI program: (1) privacy of genetic information, (2) safe and effective introduction of genetic information in the clinical setting, (3) fairness in the use of genetic information, and (4) professional and public education. Appropriate laws and regulations are expected to be developed as a result of the activities of the ELSI program and of continuing dialogues among scientists, physicians, lawmakers, and members of the public.

SUMMARY

In this chapter we have discussed some of the procedures involved in recombinant DNA technology and the manipulation of DNA. Collectively, these procedures are also referred to as genetic engineering. We have seen how it is possible to cut DNA at specific sites using restriction enzymes, how DNA can be cloned into specially constructed vectors, and how the cloned DNA can be analyzed in various ways. Through the construction of genomic libraries and cDNA libraries, and the application of screening procedures to those libraries, a large number of genes from a wide variety of organisms have been cloned and identified.

Restriction mapping analysis has provided detailed molecular maps of genes and chromosomes analogous to the genetic maps constructed on the basis of recombination analysis. DNA sequencing methods have given us an enormous amount of information about the DNA organization of genes, both their coding sequences and regulatory sequences. The amount of DNA sequence information available is growing at an extremely rapid rate, and computer databases of such sequences are available for researchers to analyze. For example, when a new gene is sequenced, we can determine if it has any sequences in common with genes already in the database.

In the mid-1980s, a new technique called polymerase chain reaction (PCR) was developed. Given some sequence information about a DNA fragment, synthetic oligonucleotide primers can be made and used to amplify large amounts of the DNA fragment from the genome in a repeated cycle of DNA denaturation (strand separation), annealing of the primers, and extension of the primers with a special DNA polymerase. Numerous applications have been rapidly found for PCR, including cloning rare pieces of DNA, preparing DNA for sequencing without cloning, and genetic disease diagnosis.

Recombinant DNA technology and PCR are being widely applied in both basic research and commerce. Basic biology has been revolutionized by these new molecular techniques. The Human Genome Project has the mandate to generate a complete map of all the genes in the human genome, and to obtain the complete sequence of the human genome. The knowledge obtained from this ambitious, long-range project will contribute markedly to our understanding of human genetics.

Recombinant DNA and PCR techniques are also being used to develop new pharmaceuticals (including drugs, other therapeutics, and vaccines); to develop new tools for diagnosing infectious and genetic diseases; for human gene therapy; in forensic analysis (for instance, analyzing DNA from a crime scene to match with a suspect); and in agriculture (as in improving disease resistance and yields of livestock and crops). While products generated by genetic engineering have been slower coming to the market than originally expected, there is an enthusiasm for the genesis and commercialization of a wide array of useful products in the future

ANALYTICAL APPROACHES FOR SOLVING GENETICS PROBLEMS

Although this is a rather descriptive area, it is often necessary to interpret data derived from restriction enzyme analysis of DNA fragments in order to generate a restriction map, that is, a map of the locations of restriction enzymes. The logic used for this type of analysis is very similar to that used in generating a genetic map of loci from two-point mapping crosses.

Q14.1 A piece of DNA 900 bp long is cloned and then cut out of the vector for analysis. Digestion of this linear piece of DNA with three different restriction enzymes singly and in all possible pairs gave the following restriction fragment size data:

ENZYME(S)	RESTRICTION FRAGMENT SIZES
EcoRI	200 bp, 700 bp
HindIII	300 bp, 600 bp
BamHI	50 bp, 350 bp, 500 bp
EcoRI + HindIII	100 bp, 200 bp, 600 bp
EcoRI + BamHI	50 bp, 150 bp, 200 bp, 500 bp
HindIII + BamHI	50 bp, 100 bp, 250 bp, 500 bp

Construct a restriction map from these data.

A14.1 The approach to this kind of problem is to consider a pair of enzymes and to analyze the data from the single and double digestions. First, consider the EcoRI and HindIII data. Cutting with EcoRI produces two fragments, one of 200 bp and the other of 700 bp, while cutting with HindIII also produces two fragments, one of 300 bp and the other of 600 bp. Thus, we know that both restriction sites are asymmetrically located along the linear DNA fragment, with the EcoRI site 200 bp from an end and the HindIII site 300 bp from an end. When we consider the EcoRI + HindIII data, we can determine the positions of these two restriction sites relative to one another. If, for example, the EcoRI site is 200 bp from the fragment end, and the HindIII site is 300 bp from that same end, then we would predict that cutting with both enzymes would produce three fragments of sizes 200 bp (end to EcoRI site), 100 bp (EcoRI site to HindIII site), and 600 bp (HindIII site to other end). On the other hand, if the EcoRI site is 200 bp from one fragment end and the HindIII site is 300 bp from the other fragment end, then cutting with both enzymes would produce three fragments of sizes 200 bp (end to EcoRI site), 400 bp (EcoRI site to HindIII site), and 300 bp (HindIII site to end). The actual data support the first model.

Now we pick another pair of enzymes, HindIII and BamHI. (We could have picked EcoRI and BamHI.) Cutting with HindIII produces fragments of 300 bp and 600 bp as we have seen, and cutting with BamHI produces three fragments of sizes 50 bp, 350 bp, and 500 bp, indicating that there are two BamHI sites in the DNA fragment. Again the double digestion products are useful in locating the sites. Double digestion with HindIII and BamHI produces four fragments of 50 bp, 100 bp, 250 bp, and 500 bp. The simplest interpretation of the data is that the 300-bp HindIII fragment is cut into the 50-bp and 250-bp fragments by BamHI, and that the

600-bp *Hind*III fragment is cut into the 100-bp and 500-bp fragments by *Bam*HI. Thus, the restriction map shown in the accompanying figure can be drawn:

The *Bam*HI + *Eco*RI data are compatible with this model.

Q14.2 The recessive allele *bw*, when homozygous, results in brown eyes in *Drosophila*, in contrast to the wild-type bright red eye color. A restriction fragment length polymorphism (RFLP) for a particular DNA region results in either two restriction fragments (type I) or one restriction fragment (type II) when *Drosophila* DNA is cut with restriction enzyme C and the fragments are separated by DNA electrophoresis, blotted to a membrane filter, and probed with a particular DNA probe.

A true-breeding brown-eyed fly with type I DNA was crossed with a true-breeding, wild-type fly with type II DNA. The F_1 flies had wild-type eye color and exhibited both type I and type II DNA patterns. The F_1 flies were crossed with true-breeding brown-eyed flies with type I DNA, and the progeny were scored for eye color and RFLP type. The results were as follows:

CLASS	PHENOTYPES	NUMBER
1	Red eyes, type I and II DNA	184
2	Red eyes, type I DNA	21
3	Brown eyes, type I DNA	168
4	Brown eyes, type I and II DNA	27
	Total progeny	400

Analyze these data.

A14.2 The eye color mutation is a familiar genetic marker. The restriction fragment length polymorphisms (RFLPs) are also genetic markers and can be analyzed just like any gene marker. If we symbolize the type I DNA as I and the type II DNA as II, the F_1 cross is:

$$\frac{bw^+ \text{ II}}{bw \text{ I}} \times \frac{bw \text{ I}}{bw^+ \text{ II}}$$

This is a testcross with the exception that DNA markers do not exhibit dominance or recessiveness. The cross is drawn as if the markers were linked; of course, we have yet to show this. If the eye color and RFLP markers are unlinked, the result would be equal numbers of the four progeny classes. However, the data show a great excess of two classes: (a) red eyes, type I and II DNA; and (b) brown eyes, type I DNA. Their origin was the pairing of F_1 bw^+ II and bw I gametes with bw I gametes to give bw^+ II/bw I and bw I/bw I progeny

genotypes, respectively. Similarly, the progeny (a) red eyes, type I DNA and (b) brown eyes, type I and II DNA derive from pairing bw^+ I and bw II gametes with bw I gametes to give bw^+ I/bw I and bw II/bw I progeny, respectively. These latter two classes occur with about equal frequency, that is, a frequency much lower than those for the other two classes. The simplest explanation is that the brown eye color gene and the RFLP marker are linked, so the F_1 cross was a mapping cross, much like those we analyzed in Chapter 5. Classes 1 and 3 are the parentals, and classes 2 and 4 are the recombinants. The map distribution between *bw* and the RFLP is, therefore:

$$\frac{21 + 27}{\text{total}} \times 100\%$$

$$= \frac{48}{400} \times 100\%$$

$$= 12 \text{ map units}$$

QUESTIONS AND PROBLEMS

14.1 What features of plasmid cloning vectors make them useful for constructing and cloning recombinant DNA molecules?

14.2 Considerable effort has been spent on developing cloning vectors that replicate in organisms other than *E. coli*.
a. Describe several different reasons one might want to clone DNA in an organism other than *E. coli*.
b. What is a shuttle vector and why is it used?
c. Describe the salient features of a vector that could be used for cloning DNA in yeast.

*14.3 Genomic libraries are important resources for isolating genes of interest and for studying the functional organization of chromosomes. List the steps you would use to make a genomic library of yeast in a lambda vector.

14.4 The human genome contains about 3×10^9 bp of DNA. How many 40-kb pieces would you have to clone into a library if you wanted to be 90 percent certain of including a particular sequence?

14.5 What is a cDNA library, and from what cellular genetic material is it derived? How is a cDNA library used in cloning particular genes?

*14.6 Suppose you wanted to produce human insulin (a peptide hormone) by cloning. Assume that this could be done by inserting the human insulin gene into a bacterial host, where, given the appropriate conditions, the human gene would be transcribed and then translated into human insulin. Which would be better to use as your source of the gene, human genomic insulin DNA or a cDNA copy of this gene? Explain your choice.

14.7 You are given a genomic library of yeast prepared in a bacterial plasmid vector. You are also given a cloned cDNA for human actin, a protein that is conserved in protein sequences among eukaryotes. Outline how you would use these resources to attempt to identify the yeast actin gene.

14.8 A researcher digests genomic DNA with the restriction enzyme *Eco*RI, separates it by size on an agarose gel, and transfers the DNA fragments in the gel to a membrane filter using the Southern blot procedure. What result would she expect to see if the source of the DNA and the probe for the blot is as described below?

a. The genomic DNA is from a normal human. The probe is a 2.0-kb DNA fragment obtained by excision with the enzyme *Eco*RI from a plasmid containing single-copy genomic DNA.

b. The genomic DNA is from a normal human. The probe is a 5.0-kb DNA fragment that is a copy of a LINE sequence that has an internal *Eco*RI site.

c. The genomic DNA is from a normal human. The probe is a 5.0-kb DNA fragment that is a copy of a LINE sequence that lacks an internal *Eco*RI site.

d. The genomic DNA is from a human heterozygous for a translocation between chromosomes 14 and 21. The probe is a 3.0-kb DNA fragment that is obtained by excision with the enzyme *Eco*RI from a plasmid containing single-copy genomic DNA from a normal chromosome 14. The translocation breakpoint on chromosome 14 lies within the 3.0-kb genomic DNA fragment.

e. The genomic DNA is from a normal female. The probe is a 5.0-kb DNA fragment containing part of the *test is determining factor* gene (see Chapter 3, p. 64–65).

14.9 Restriction endonucleases are used to construct restriction maps of linear or circular pieces of DNA. The DNA is usually produced in large amounts by recombinant DNA techniques. The generation of restriction maps is similar to the process of putting the pieces of a jigsaw puzzle together. Suppose we have a circular piece of double-stranded DNA that is 5,000 base pairs long. If this DNA is digested completely with restriction enzyme I, four DNA fragments are generated: fragment *a* is 2,000 base pairs long; *b* is 1,400 base pairs long; *c* is 900 base pairs long; and *d* is 700 base pairs long. If, instead, the DNA is incubated with the enzyme for a short time, the result is incomplete digestion of the DNA: not every restriction enzyme site in every DNA molecule will be cut by the enzyme, and all possible combinations of adjacent fragments can be produced. From an incomplete digestion experiment of this type, fragments of DNA were produced from the circular piece of DNA that contained the following combinations of the above fragments: *a-d-b*, *d-a-c*, *c-b-d*, *a-c*, *d-a*, *d-b*, and *b-c*. Lastly, after digesting the original circular DNA to completion with restriction enzyme I, the DNA fragments were treated with restriction enzyme II under conditions conducive to complete digestion. The resulting fragments were 1,400, 1,200, 900, 800, 400, and 300 bp. Analyze all the data to locate the restriction enzyme sites as accurately as possible.

*14.10 A piece of DNA 5,000 bp long is digested with restriction enzymes A and B, singly and together. The DNA fragments produced were separated by DNA electrophoresis and their sizes were calculated, with the following results:

DIGESTION WITH		
A	B	A + B
2,100 bp	2,500 bp	1,900 bp
1,400 bp	1,300 bp	1,000 bp
1,000 bp	1,200 bp	800 bp
500 bp		600 bp
		500 bp
		200 bp

Each A fragment was extracted from the gel and digested with enzyme B, and each B fragment was extracted from the gel and digested with enzyme A. The sizes of the resulting DNA fragments were determined by gel electrophoresis, with the following results:

A FRAGMENT	FRAGMENTS PRODUCED BY DIGESTION WITH B	B FRAGMENT	FRAGMENTS PRODUCED BY DIGESTION WITH A
2,100 bp →	1,900, 200 bp	2,500 bp →	1,900, 600 bp
1,400 bp →	800, 600 bp	1,300 bp →	800, 500 bp
1,000 bp →	1,000 bp	1,200 bp →	1,000, 200 bp
500 bp →	500 bp		

Construct a restriction map of the 5,000-bp DNA fragment.

*14.11 Draw the banding pattern you would expect to see on a DNA-sequencing gel if you annealed the primer 5'-C-T-A-G-G-3' to the following single-stranded DNA fragment and carried out a dideoxy sequencing experiment. Assume the dNTP precursors were all labeled.

3'-G-A-T-C-C-A-A-G-T-C-T-A-C-G-T-A-T-A-G-G-C-C-5'.

14.12 Imagine that you have been able to clone the structural gene for an enzyme in a catecholamine biosynthetic pathway from the adrenal gland of rats.

How could you use this cloned DNA as a probe to determine whether this same gene functions in the rat brain?

14.13 Imagine that you find a RFLP in the rat genomic region homologous to your cloned catecholamine synthetic gene from question 14.12, and that in a population of rats displaying this polymorphism there is also a behavioral variation. You find that some of the rats are normally calm and placid, but others are hyperactive, nervous, and easily startled. Your hypothesis is that the behavioral difference seen is caused by variations in your gene. How could you use your cloned sequence to test this hypothesis?

*14.14 The maps of the sites for restriction enzyme R in the wild type and the mutated cystic fibrosis genes are shown schematically in the following figure:

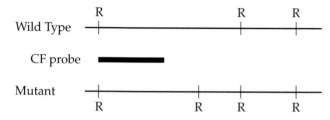

Samples of DNA obtained from a fetus (F) and her parents (M and P) were analyzed by gel electrophoresis followed by the Southern blot technique and hybridization with the radioactively labeled probe designated "CF probe" in the previous figure. The autoradiographic results are shown in the following figure:

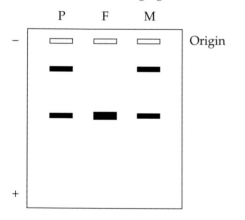

Given that cystic fibrosis is a recessive mutation, will the fetus be affected? Explain.

14.15 The polymerase chain reaction (PCR) is widely applied in molecular genetics. It has often provided a faster, more sensitive means of answering a question than more traditional methods such as Southern blotting or DNA cloning.

a. Describe the polymerase chain reaction, including the components and products involved.

b. Describe a specific application of PCR in the cloning or analysis of a prokaryotic or eukaryotic gene. How might the use of PCR in this application replace, augment, speed, improve the sensitivity of, or simplify an alternative method of analysis?

14.16 One application of DNA fingerprinting technology has been to identify stolen children and return them to their parents. Bobby Larson was taken from a supermarket parking lot in New Jersey in 1978, when he was 4 years old. In 1990, a 16-year-old boy called Ronald Scott was found in California, living with a couple named Susan and James Scott, who claimed to be his parents. Authorities suspected that Susan and James might be the kidnappers, and that Ronald Scott might be Bobby Larson. DNA samples were obtained from Mr. and Mrs. Larson, and from Ronald, Susan, and James Scott. Then DNA fingerprinting was done, using a polymorphic probe for a particular VNTR family, with the results shown in Figure 14.A. From the information in the figure, what can you say about the parentage of Ronald Scott? Explain.

FIGURE 14.A

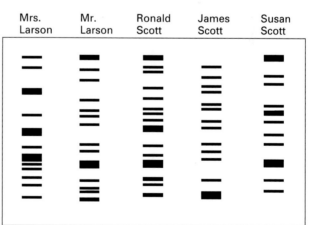

CHAPTER *15*

REGULATION OF GENE EXPRESSION IN BACTERIA AND BACTERIOPHAGES

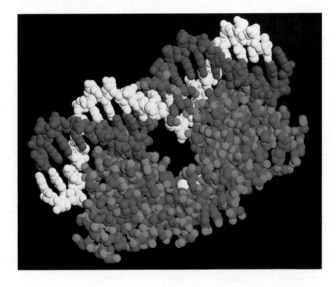

PRINCIPAL POINTS

~ In the lactose system of *E. coli*, the addition of lactose to cells brings about a rapid synthesis of three enzymes. In the absence of lactose the synthesis of the three enzymes is turned off. The genes for the enzymes are contiguous on the *E. coli* chromosome and are adjacent to a controlling site (an operator) and a single promoter. The genes, the operator, and the promoter constitute an operon. Transcription of the genes results in a single polycistronic mRNA. A regulatory gene is associated with an operon. To turn on gene expression in the lactose system, a lactose metabolite binds with a repressor protein (the product of the regulatory gene), inactivating it and preventing it from binding to the operator. As a result, RNA polymerase can bind to the promoter and transcribe the three genes as a single polycistronic mRNA. Operons are commonly involved in the regulation of gene expression in a large number of prokaryotic and bacteriophage systems.

~ Expression of a number of bacterial amino acid synthesis operons is controlled by a repressor-operator system and through attenuation at a second controlling site called an attenuator. The repressor-operator system

functions essentially like that for the *lac* operon, except that the addition of amino acid to the cell activates the repressor, thereby turning the operon off. An attenuator is located between the operator region and the first structural gene; it is a transcription termination site that allows only a fraction of RNA polymerases to transcribe the rest of the operon. Attenuation requires a coupling between transcription and translation, and formation of particular RNA secondary structures that signal whether or not transcription can continue.

~ Bacteriophages such as lambda are especially adapted for undergoing reproduction within a bacterial host. Many genes related to the production of progeny phages, or to the establishment or reversal of lysogeny in temperate phages, are organized into operons. These operons, like bacterial operons, are controlled through the interaction of regulatory proteins with operators that are adjacent to clusters of structural genes. Phage lambda has been an excellent model for studying the genetic switch that controls the choice between lytic and lysogenic pathways in a lysogenic phage.

*B*acteria are free-living organisms that grow by increasing in mass and then divide by binary fission. Growth and division are under the control of genes, the expression of which must be regulated appropriately. Your goal in this chapter is to learn about some of the mechanisms by which gene expression is regulated in bacteria and bacteriophages. Significantly, genes that encode proteins that work together in the cell typically are organized into particular regulated clusters called *operons*. The genes are transcribed together onto a single mRNA molecule called **polycistronic mRNA**. Regulation of synthesis of this mRNA depends on interactions between a regulatory protein and a regulatory site called an operator that is next to the gene array. Of course, much remains to be done to understand completely the regulation of gene expression in bacteria. The 4.6-megabase (4.6×10^6 base pair) genome of *E. coli*, for example, has 4,288 protein-coding genes according to the genomic sequence, and the functions of over 35 percent of those genes are not yet known.

THE *lac* OPERON OF *E. COLI*

When gene expression is "turned on" in a bacterium by adding a substance (such as lactose) to the medium, the genes involved are said to be *inducible*. The regulatory substance that brings about this gene induction is called an **inducer**, and the phenomenon of producing a gene product in response to an inducer is called **induction**. The inducer is an example of a class of small molecules, called **effectors** or **effector molecules**, that help control expression of many regulated genes. Transcription of an inducible gene occurs in response to a regulatory event occurring at a specific DNA sequence—a **controlling site**—adjacent to or near the protein-coding sequence (Figure 15.1). The regulatory event typically involves an inducer and a regulatory protein. When the regulatory event occurs, RNA polymerase initiates transcription at the promoter (usually upstream of the controlling site). The gene is "turned on," mRNA is made, and the protein coded for by the gene is produced. The controlling site itself does not code for any

~ FIGURE 15.1

General organization of an inducible gene.

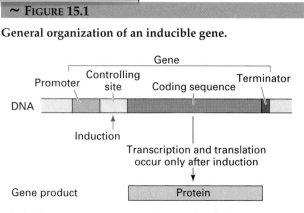

Inducible genes are expressed only when induced

product. As an example of such gene regulation, we now discuss regulation of the inducible *lac* genes of the *E. coli lac* operon.

Lactose as a Carbon Source for *E. coli*

E. coli can grow in a simple medium containing salts (including a nitrogen source) and a carbon source such as glucose. The energy for biochemical reactions in the cell comes from the metabolism of glucose. The enzymes required for glucose metabolism are coded for by constitutive genes. If lactose is provided to *E. coli* as a carbon source instead of glucose, a number of enzymes are rapidly synthesized that are needed for metabolizing lactose. (The same series of events, each involving a "sugar-specific" set of enzymes, is triggered by other sugars as well.) The enzymes are synthesized because the genes that code for them become actively transcribed in the presence of the sugar; the same genes are inactive if the sugar is absent. In other words, the genes are regulated genes whose products are needed only at certain times.

Lactose is a disaccharide consisting of the monosaccharides glucose and galactose. When lactose is present as the sole carbon source in the growth medium, three proteins are synthesized:

1. *β-galactosidase*. This enzyme breaks down lactose into glucose and galactose, and it catalyzes the isomerization ("conversion to a different form") of lactose to *allolactose*, a compound important in regulating expression of the *lac* operon (Figure 15.2).

2. *Lactose permease* (also called *M protein*). This protein, found in the *E. coli* cytoplasmic membrane, transports lactose into the cell.

3. *Transacetylase*. The function of this enzyme is poorly understood.

In wild-type *E. coli* growing in a medium containing glucose, only a low concentration of each of these three proteins is produced. In the presence of lactose but the absence of glucose, the amount of each enzyme increases coordinately (simultaneously) about a thousandfold. This occurs because the three essentially inactive genes are now being actively transcribed. This process is called **coordinate induction**. Allolactose, not lactose, is the inducer molecule directly responsible for the increased production of the three enzymes (see Figure 15.2). Further, the mRNAs for the enzymes have a relatively short half-life, so the transcripts must be made continually in order for the enzymes to be produced. When lactose is no longer present, transcription of the three genes is stopped and any mRNAs already present are broken down, so no more proteins are made. Existing proteins are diluted out by cell growth and division.

Experimental Evidence for the Regulation of the *lac* Genes

Our basic understanding of the organization of the genes, the controlling sites involved in lactose utilization, and the control of expression of the *lac* genes of *E. coli* came largely from the genetic experiments of François Jacob and Jacques Monod, for which they received the Nobel Prize. We now summarize their experiments.

MUTATIONS IN THE PROTEIN-CODING GENES. Mutations in the protein-coding genes were obtained following treatment of cells with mutagens (chemicals that induce mutations). The β-galactosidase gene was named *lacZ*, the permease gene *lacY*, and the transacetylase gene *lacA*. The three genes are tightly linked in the order *lacZ-lacY-lacA*. These three structural (= protein-coding) genes are transcribed onto a single polycistronic mRNA molecule rather than onto three separate mRNAs. That is, RNA polymerase initiates transcription at a single promoter, and a polycistronic mRNA is synthesized with the gene transcripts in the order 5'-*lacZ⁺*-*lacY⁺*-*lacA⁺*-3'. In translation, a ribosome loads onto the polycistronic mRNA at the 5' end, synthesizes β-galactosidase, then reinitiates translation at the permease sequence, synthesizes permease, then reinitiates at the transacetylase sequence, synthesizes transacetylase, and finally dissociates from the mRNA (Figure 15.3).

MUTATIONS AFFECTING THE REGULATION OF GENE EXPRESSION. In wild-type *E. coli* the three gene products are induced coordinately when lactose is present. Jacob and Monod isolated mutants in which all gene products of the operon were synthesized *con-*

~ FIGURE 15.2

Reactions catalyzed by the enzyme β-galactosidase. Lactose brought into the cell by the permease is either converted to glucose and galactose (top) or to allolactose (bottom), the true inducer for the lactose operon of *E. coli*.

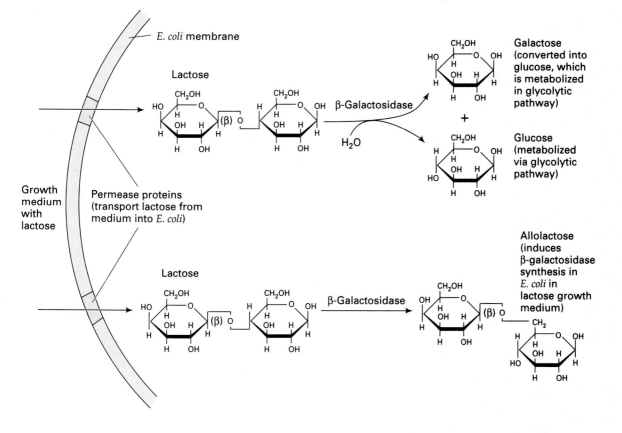

~ FIGURE 15.3

Translation of the polycistronic mRNA encoded by lactose utilization genes in wild-type *E. coli*.

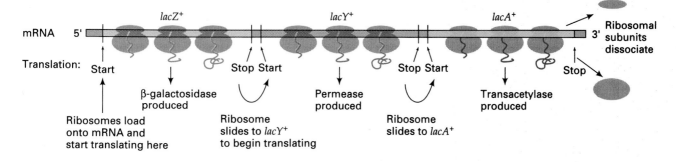

stitutively; that is, they were synthesized whether or not the inducer was present. Two classes of constitutive mutations were isolated. One class mapped to a small region next to the *lacZ* gene called the **operator** (*lacO*). The other class mapped to a gene a short distance away that they called the *lacI* gene or *lac* **repres-**

sor gene. Figure 15.4 depicts the organization of the *lac* structural gene cluster and the associated regulatory elements. This complex is the *lac* operon.

Operator Mutations. The mutations of the operator were called operator-constitutive, or *lacOᶜ*,

~ FIGURE 15.4

Organization of the *lac* genes of *E. coli* and the associated regulatory elements, the operator, promoter, and regulatory gene. The complex is called the *lac* operon.

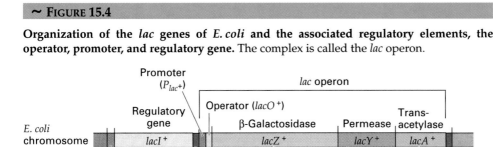

mutations. Through the use of partial diploid strains (*F'* strains; see Figure 6.7, p. 134), Jacob and Monod were able to define better the role of the operator in regulating expression of the *lac* genes. One such partial diploid was

$$\frac{F' \ \ lacO^+ \ lacZ^- \ lacY^+}{lacO^c \ lacZ^+ \ lacY^-}$$

(both gene sets have a normal promoter, and the *lacA* gene is omitted because it is not important to our discussions).

One *lac* region in the partial diploid has a normal operator (*lacO^+*), a mutated β-galactosidase gene (*lacZ^-*), and a normal permease gene (*lacY^+*). The other *lac* region has a constitutive operator mutation (*lacO^c*), a normal β-galactosidase gene (*lacZ^+*), and a mutated permease gene (*lacY^-*). This partial diploid was tested for production of β-galactosidase (from the *lacZ^+* gene) and of permease (from the *lacY^+* gene), both in the presence and the absence of the inducer.

Jacob and Monod found that active β-galactosidase is synthesized in the absence of inducer, and that permease is synthesized but is inactive because of the mutation. Only when lactose is added to the culture and the allolactose inducer is produced does synthesis of active permease occur. That is, the *lacZ^+* gene (which is on the same DNA molecule as *lacO^c*) is *constitutively expressed* (meaning the gene is active in the presence or absence of inducer), whereas the *lacY^+* gene is under normal inducible control—it is inactive in the absence of inducer and active in the presence of inducer. In other words, a *lacO^c* mutation *affects only those genes downstream from it on the same DNA molecule.* Similarly, the *lacO^+* region only controls *lac* structural genes adjacent to it and has

no effect on the genes on the other DNA molecule. This phenomenon of a gene or DNA sequence controlling only genes that are on the same, contiguous piece of DNA is called *cis*-**dominance**. The *lacO^c* mutation is cis-dominant since the defect affects the adjacent genes only and cannot be overcome by a normal *lacO^+* region elsewhere in the cell. That is, the operator must not encode a diffusible product. If it did, then in the *lacO^+/lacO^c* diploid state, one or the other of the regions would have controlled all the lactose utilization genes regardless of their location.

***lacI* Gene Regulatory Mutations.**   Again, the use of partial diploid strains illuminated the normal function of the *lacI* gene.

An example of the partial diploid used is

$$\frac{F' \ \ lacI^+ \ lacO^+ \ lacZ^- \ lacY^+}{lacI^- \ lacO^+ \ lacZ^+ \ lacY^-}$$

Both gene sets have normal operators and normal promoters. In the absence of the inducer, no β-galactosidase or permease was produced; both were synthesized in the presence of inducer. In other words, the expression of both operons was inducible. This means that the *lacI^+* gene in the cell can overcome the defect of the *lacI^-* mutation. Since the two *lacI* genes are located on different DNA molecules (that is, they are in a trans configuration), the *lacI^+* gene is said to be ***trans*-dominant** to the *lacI^-* gene.

Since the *lacI^+* gene controls the genes on the other DNA molecule, Jacob and Monod proposed that the *lacI^+* gene is a repressor gene that produces a **repressor molecule**. No functional repressor molecules are produced in *lacI^-* mutants. Thus, in a haploid *lacI^-* bacterial strain, the *lac* operon is constitutive. In a partial

~ **FIGURE 15.5**

Functional state of the *lac* operon in wild-type *E. coli* growing in the absence of lactose.

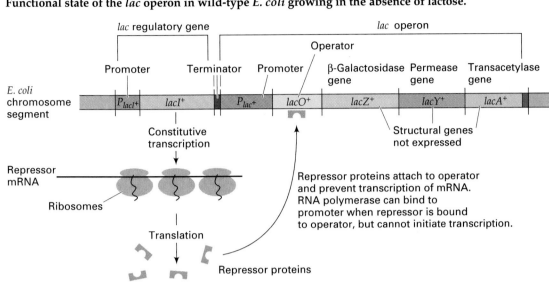

~ **FIGURE 15.6**

diploid with both a *lacI*⁺ and a *lacI*⁻, however, the functional repressor molecules produced by the *lacI*⁺ gene control the expression of both *lac* operons in the cell, making both operons inducible.

Jacob and Monod's Operon Model for the Regulation of the *lac* Genes

Based on the results of their studies, Jacob and Monod proposed their now-classical *operon model*. By definition, an **operon** is a *cluster of genes, the expressions of which are regulated together by operator-repressor protein interactions, plus the operator region itself and the promoter*. The promoter was not part of Jacob and Monod's original model; its existence was demonstrated by later studies. The order of the controlling elements and genes in the *lac* operon is promoter-operator-*lacZ*-*lacY*-*lacA*, and the regulatory gene *lacI* is located close to the structural genes, just upstream of the promoter (see Figure 15.4). The *lacI* gene has its own constitutive promoter and terminator.

The following description of the Jacob-Monod model for regulation of the *lac* operon has been embellished with up-to-date molecular information. Figure 15.5 depicts the state of the *lac* operon in wild-type *E. coli* growing in the absence of lactose. The repressor gene (*lacI*⁺) is transcribed constitutively, and translation of its mRNA produces a 360-amino-acid polypeptide. Four of these polypeptides asso-

ciate together to form a tetramer—the functional repressor protein (Figure 15.6).

The repressor binds to the operator (*lacO*⁺). When the repressor is bound to the operator, RNA polymerase can bind to the operon's promoter but is blocked from initiating transcription of the protein-coding genes—the *lac* operon is said to be under *negative control*. The low level of transcription of the genes that produces a few molecules of the enzymes, even

~ **FIGURE 15.6**

Molecular model of the *lac* repressor tetramer. The four monomers are colored green, violet, red, and yellow.

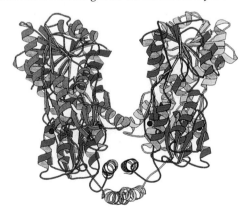

~ FIGURE 15.7

Functional state of the *lac* operon in wild-type *E. coli* growing in the presence of lactose as the sole carbon source.

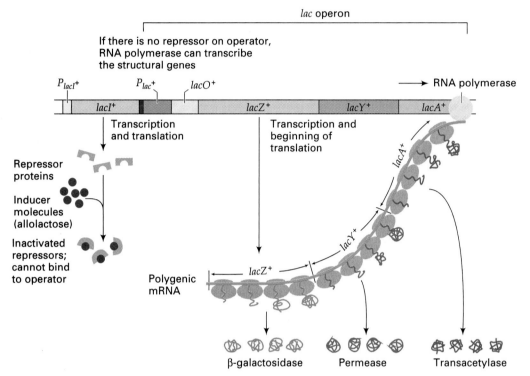

in the absence of inducer, occurs because repressors do not just bind and stay; they bind and unbind. In the split second after one repressor unbinds and before another binds, an RNA polymerase could initiate transcription of the operon, even in the absence of inducer.

When wild-type *E. coli* grows in the presence of lactose as the sole carbon source (Figure 15.7), some lactose is converted by β-galactosidase into allolactose (see Figure 15.2). Allolactose binds to the repressor and changes its shape—this is called an *allosteric shift*. As a result, the repressor loses its affinity for the *lac* operator, and it dissociates from the site. Free repressor proteins are also altered so that they cannot bind to the operator. Thus, the allolactose "induces" production of the *lac* operon enzymes.

With no repressor bound to the operator, RNA polymerase initiates synthesis of a single polycistronic mRNA molecule for the *lacZ⁺*, *lacY⁺*, and *lacA⁺* genes. The polycistronic mRNA for the *lac* operon is translated by a string of ribosomes to produce the three proteins specified by the operon. This efficient

mechanism ensures the coordinate production of proteins of related function.

EFFECT OF *lacOᶜ* MUTATIONS. The *lacOᶜ* mutations lead to constitutive expression of the *lac* operon genes and are cis-dominant to *lacO⁺* (Figure 15.8). The explanation for this is that base-pair alterations of the operator DNA sequence make it unrecognizable to the repressor protein. Since the repressor cannot bind, the structural genes physically linked to the *lacOᶜ* mutation become constitutively expressed.

EFFECTS OF *lacI⁻* GENE MUTATIONS. The *lacI⁻* mutations map within the repressor structural gene and result in amino acid changes in the repressor. The repressor's shape is changed, and it cannot now recognize and bind to the operator. As a consequence, in a haploid strain, transcription cannot be prevented, even in the absence of lactose, and constitutive expression of the *lac* operon results (Figure 15.9a).

The dominance of the *lacI⁺* (wild-type) gene over *lacI⁻* mutants is illustrated for the partial diploid

~ FIGURE 15.8

Cis-dominant effect of *lacO*ᶜ mutation in a partial-diploid strain of *E. coli*. (a) In the absence of the inducer, the *lacO*⁺ operon is turned off, while the *lacO*ᶜ operon produces functional β-galactosidase from the *lacZ*⁺ gene and nonfunctional permease molecules from the *lacY*⁻ gene. (b) In the presence of the inducer, the functional β-galactosidase and defective permease are produced from the *lacO*ᶜ operon, while the *lacO*⁺ operon produces nonfunctional β-galactosidase from the *lacZ*⁻ gene and functional permease from the *lacY*⁺ gene. Between the two operons in the cell, functional β-galactosidase and permease are produced.

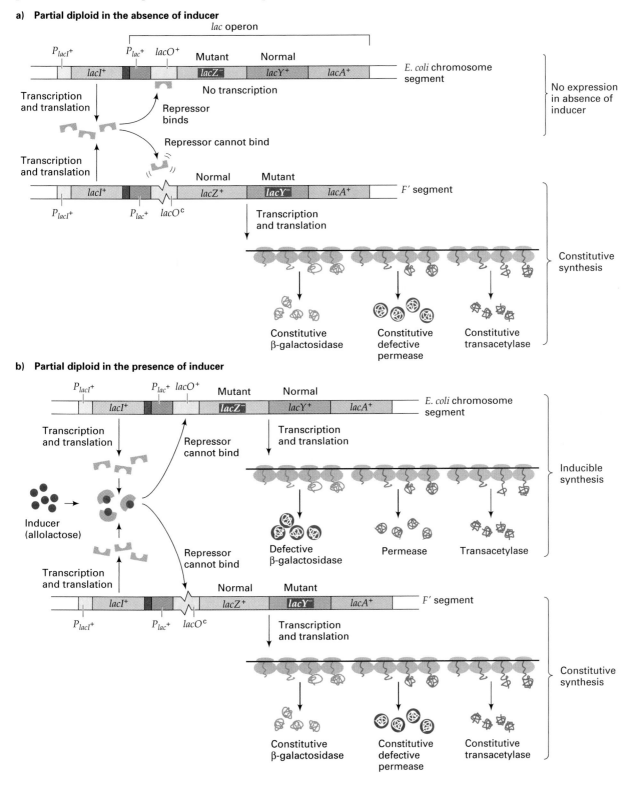

a) **Partial diploid in the absence of inducer**

b) **Partial diploid in the presence of inducer**

~ FIGURE 15.9

Effects of a *lacI⁻* mutation. (a) Effect on *lac* operon expression in a haploid cell, where mutant, inactive repressor molecules that cannot bind to the operator *lacO⁺* are produced: the structural genes are transcribed constitutively. Effect in a partial diploid strain *lacI⁺ lacO⁺ lacZ⁻ lacY⁺ / lacI⁻ lacO⁺ lacZ⁺ lacY⁻* in (b) the absence or (c) the presence of inducer.

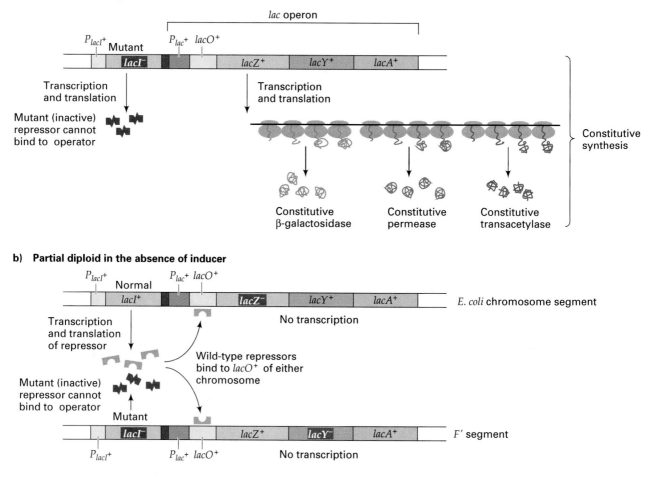

described earlier. In the absence of the inducer (Figure 15.9b), the defective *lacI⁻* repressor cannot bind to either normal operator (*lacO⁺*) in the cell. But sufficient normal repressors, produced from the *lacI⁺* gene, are present, and they bind to the two operators and block transcription of both operons. When the inducer is present (Figure 15.9c), the wild-type repressors are inactivated, so both operons are transcribed. One produces a defective β-galactosidase and a normal permease, while the other produces a normal β-galactosidase and a defective permease; between them, active β-galactosidase and permease are produced. Thus, in *lacI⁺/lac⁻* partial diploids, both operons present in the cell are under inducible control.

Other classes of *lacI* gene mutants have been identified since the time Jacob and Monod studied

the *lacI⁻* class of mutants. One of these classes, the *lacIˢ* (*superrepressor*) mutants, shows no production of *lac* enzymes in the presence or absence of lactose. In partial diploids with a *lacI⁺/lacIˢ* genotype, the *lacIˢ* allele is trans-dominant, affecting both operon copies (Figure 15.10). In this situation, the mutant repressor gene produces a superrepressor protein that can bind to the operator but cannot recognize the inducer allolactose. Therefore, the mutant superrepressors bind to the operators even in the presence of the inducer, and transcription of the operons can never occur. The presence of normal repressors in the cell has no effect since once a *lacIˢ* repressor is on the operator, the repressor cannot be induced to fall off. Low levels of transcription will occur since the superrepressor is not permanently (covalently) bound to the operator.

~ FIGURE 15.9 continued

c) Partial diploid in the presence of inducer

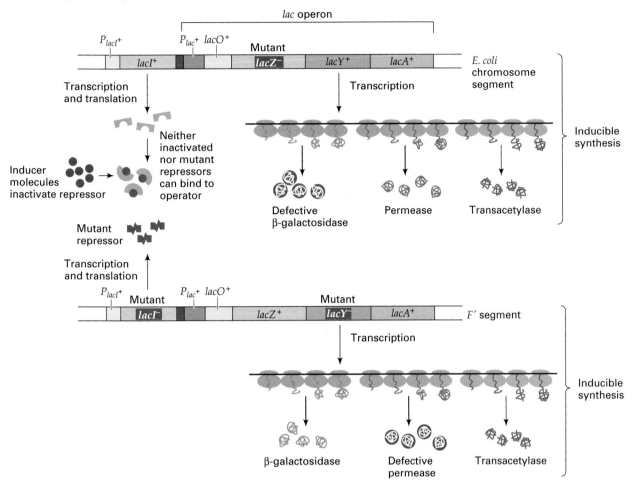

Cells with a *lacI*^{s} mutation cannot use lactose as a carbon source.

Positive Control of the *lac* Operon

In the *lac* operon the repressor protein exerts a negative effect on the expression of the *lac* operon by blocking RNA polymerase's action if the inducer is absent. Several years after Jacob and Monod proposed their operon model, researchers also found a positive control system that regulates the *lac* operon, a system that functions to turn on the expression of the operon. This system ensures that the *lac* operon will be expressed at high levels only if lactose is the sole carbon source, *and not if glucose is present as well.*

If only lactose is present, the positive regulation of the *lac* operon is as shown in Figure 15.11. First, a protein called *CAP* (catabolite activator protein) binds with *cAMP* (cyclic AMP, or cyclic adenosine 3′,5′-monophosphate) to form a CAP-cAMP complex. This complex is the positive-regulator molecule. The CAP protein is a dimer of two identical polypeptides. Next, the CAP-cAMP complex binds to the *CAP site*, which is located upstream of where RNA polymerase binds to the promoter. This binding facilitates binding of RNA polymerase and the initiation of transcription.

When glucose is in the medium, **catabolite repression (glucose effect)** occurs. In catabolite repression, the *lac* operon is expressed at low levels even if lactose is present in the medium. This occurs because glucose causes the amount of cAMP in the cell to be greatly reduced. As a result, insufficient CAP-cAMP complex is available to facilitate RNA polymerase binding to the *lac* promoter, and transcription is lowered significantly, even though repressors are removed from the operator by the presence of allolactose.

FIGURE 15.10

Dominant effect of *lacI*^{s} mutation over wild-type *lacI*$^+$ in a *lacI*$^+$ *lacO*$^+$ *lacZ*$^+$ *lacY*$^+$ *lacA*$^+$/ *lacI*^{s} *lacO*$^+$ *lacZ*$^+$ *lacY*$^+$ *lacA*$^+$ partial-diploid cell growing in the presence of lactose.

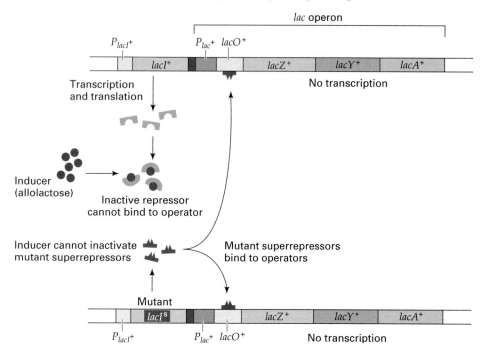

Catabolite repression occurs in the same way in a number of other bacterial operons related to catabolism of sugars other than glucose. These operons all have in common a CAP site in their promoters to which a specific CAP-cAMP complex binds to facilitate RNA polymerase binding.

KEYNOTE

In wild-type *E. coli*, the addition of lactose to the cells brings about a rapid synthesis of three enzymes. The genes for these enzymes are contiguous on the *E. coli* chromosome and are adjacent to a controlling site (an operator) and a single promoter. The promoter, operator, and genes constitute an operon; the genes are transcribed as a single unit. In the absence of lactose the operon is turned off by a repressor.

A positive control system also regulates the *lac* operon. That is, CAP-cAMP binds upstream of the promoter, and this facilitates binding of RNA polymerase to the promoter. If glucose is present, however, no CAP-cAMP is produced, so RNA polymerase binds less efficiently and the *lac* genes are not transcribed.

THE *trp* OPERON OF *E. COLI*

A bacterium has certain operons and other gene systems that enable it to manufacture any amino acid that is lacking in the medium so that it may grow and reproduce. When an amino acid is present in the growth medium, though, the genes encoding the enzymes for that amino acid's biosynthetic pathway are turned off. Unlike the *lac* operon, where gene activity is induced when a chemical (lactose) is added to the medium, in this case gene activity is repressed when a chemical (an amino acid) is added. We refer to amino acid biosynthesis operons controlled in this way as *repressible operons*. In general, operons for anabolic (biosynthetic) pathways are repressed (turned off) when the end product is readily available. One repressible operon in *E. coli* that has been extensively studied is the operon for the biosynthesis of the amino acid tryptophan (Trp).

Gene Organization of the Tryptophan Biosynthesis Genes

Figure 15.12 shows the organization of the controlling sites and of the genes that code for the tryptophan biosynthetic enzymes, and how they relate to

~ FIGURE 15.11

Role of cyclic AMP (cAMP) in the functioning of glucose-sensitive operons such as the *lac* operon of *E. coli*.

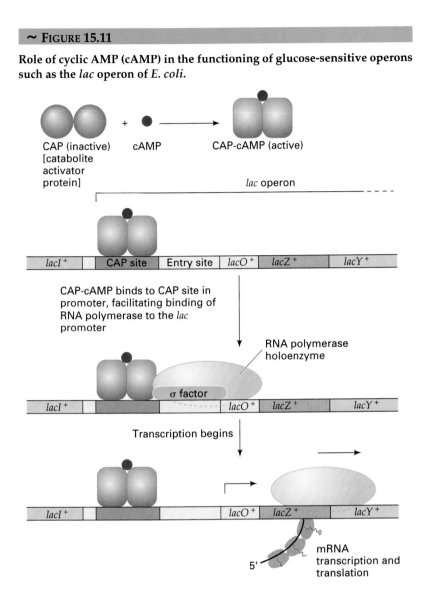

the biosynthetic steps. Much of the work that we will discuss is that of Charles Yanofsky and his collaborators.

Five structural genes (A–E) occur in the tryptophan operon. The promoter and operator regions are upstream from the *trpE* gene. Between the promoter-operator region and *trpE* is a short region called *trpL*, the leader region. Within *trpL*, relatively close to *trpE*, is an *attenuator site* (*att*) that plays an important role in the regulation of the tryptophan operon, as we will see later.

The entire tryptophan operon is approximately 7,000 base pairs long. Transcription of the operon results in the production of a polycistronic mRNA for the five structural genes.

Regulation of the *trp* Operon

Two regulatory mechanisms are involved in controlling the expression of the *trp* operon. One mechanism uses a repressor/operator interaction, and the other determines whether initiated transcripts include the structural genes or are terminated before those genes.

EXPRESSION OF THE *trp* OPERON IN THE PRESENCE OF TRYPTOPHAN. The regulatory gene for the *trp* operon is *trpR*; it is located some distance from the operon (and therefore does not appear in Figure 15.12). The product of *trpR* is an *aporepressor protein*, which alone cannot bind to the operator. When tryptophan is abundant within the cell, it binds to the

~ FIGURE 15.12

Organization of controlling sites and the structural genes of the *E. coli trp* operon. Also shown are the steps catalyzed by the products of the structural genes *trpA, trpB, trpC, trpD,* and *trpE.*

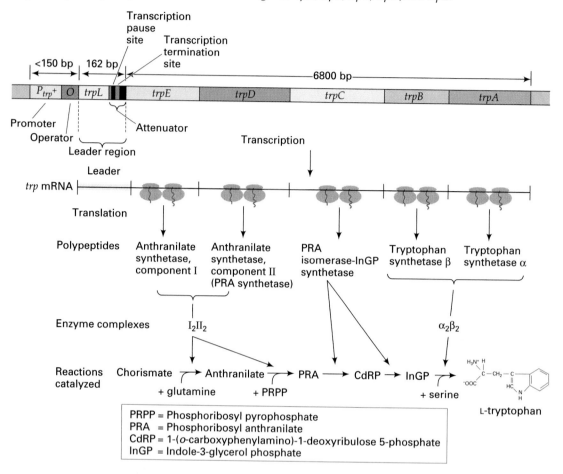

aporepressor and converts it to an active repressor. (Tryptophan is an example of an *effector* molecule, just as allolactose is the effector molecule for the *lac* operon.) The active repressor binds to the operator and prevents the initiation of transcription of the *trp* operon protein-coding genes by RNA polymerase. As a result, the tryptophan biosynthesis enzymes are not produced. By repression, transcription of the *trp* operon can be reduced about 70-fold.

EXPRESSION OF THE *trp* OPERON IN THE PRESENCE OF LOW CONCENTRATIONS OF TRYPTOPHAN. The second regulatory mechanism is involved in the expression of the *trp* operon under conditions of tryptophan starvation or tryptophan limitation. Under severe tryptophan starvation, the *trp* genes are expressed maximally, while under less severe starvation conditions, the *trp* genes are expressed at less than maximal levels. This is accomplished by a mech-

anism that controls the ratio of full-length transcripts that include the five *trp* structural genes to short transcripts that have terminated at the attenuator site within the *trpL* region (see Figure 15.12). The short transcripts have terminated by a process called **attenuation**. Attenuation can reduce transcription of the *trp* operon by eightfold to tenfold. Thus, repression and attenuation together can regulate the transcription of the *trp* operon over a range of about 560-fold to 700-fold.

MOLECULAR MODEL FOR ATTENUATION. The mRNA transcript of the leader region includes a sequence that can be translated to produce a short polypeptide. Just before the stop codon of the transcript are two adjacent codons for tryptophan that play an important role in attenuation.

There are four regions of the leader peptide mRNA that can fold and form secondary structures

~ FIGURE 15.13

Four regions of the *trp* operon leader mRNA and the alternate secondary structures they can form by complementary base-pairing.

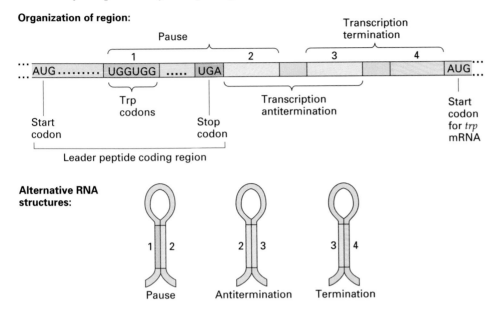

Organization of region:

Pause

Transcription termination

1   2   3   4

AUG .... UGGUGG ..... UGA   AUG

Trp codons

Start codon

Stop codon

Transcription antitermination

Start codon for *trp* mRNA

Leader peptide coding region

Alternative RNA structures:

1   2   2   3   3   4

Pause   Antitermination   Termination

by complementary base-pairing (Figure 15.13). Pairing of regions 1 and 2 results in a transcription *pause signal*, pairing of 3 and 4 is a *termination of transcription signal*, and pairing of 2 and 3 is an *antitermination signal* for transcription to continue.

Crucial to the attenuation model is the fact that transcription and translation are tightly coupled in prokaryotes, made possible by the absence of a nuclear envelope. In the *trp* regulatory system, this coupling of transcription and translation is brought about by a pause of the RNA polymerase caused by the pairing of RNA regions 1 and 2 just after they have been synthesized (see Figure 15.13). The pause lasts long enough for the ribosome to load onto the mRNA and to begin translating the leader peptide so that translation of the leader mRNA transcript occurs just behind the RNA polymerase.

As coupled transcription/translation continues, the position of the ribosome on the leader transcript plays an important role in the regulation of transcription termination at the attenuator. If the cells are starved for tryptophan, then the amount of Trp-tRNA molecules (charged tryptophan-tRNA) drops dramatically, because very few tryptophan molecules are available for the aminoacylation of the tRNA. A ribosome translating the leader transcript stalls at the tandem Trp codons in region 1 because the next specified amino acid in the peptide is in short supply; the leader peptide cannot be completed (Figure 15.14a).

Since the ribosome now "covers" region 1 of the attenuator region, the 1-2 pairing cannot happen because region 1 is no longer available. However, RNA region 2 will pair with RNA region 3 once region 3 is synthesized. Because region 3 is paired with region 2, region 3 cannot then pair with region 4 when it is synthesized. The 2:3 pairing is an antitermination signal because the termination signal of 3 paired with 4 does not form, and this allows RNA polymerase to continue past the attenuator and transcribe the structural genes.

If, instead, enough tryptophan is present that the ribosome can translate the Trp codons (Figure 15.14b), then the ribosome continues to the stop codon for the leader peptide. Since the ribosome is then covering part of RNA region 2, region 2 is unable to pair with region 3, and region 3 is then able to pair with region 4 when it is transcribed. The bonding of region 3 with region 4 is a termination signal. The 3:4 structure is referred to as the attenuator. The key signal for attenuation is the concentration of Trp-tRNA in the cell, since that determines how far the ribosome gets on the leader transcript, either to the Trp codons or to the stop codon.

Attenuation is involved in the genetic regulation of a number of other amino acid biosynthetic operons of *E. coli* and of *Salmonella typhimurium*. In every case there is a leader sequence with two or more codons for the particular amino acid, the synthesis of which

~ FIGURE 15.14

Models for attenuation in the *trp* operon of *E. coli*. The taupe structures are ribosomes that are translating the leader transcript. (a) Tryptophan-starved cells; (b) Non–tryptophan-starved cells.

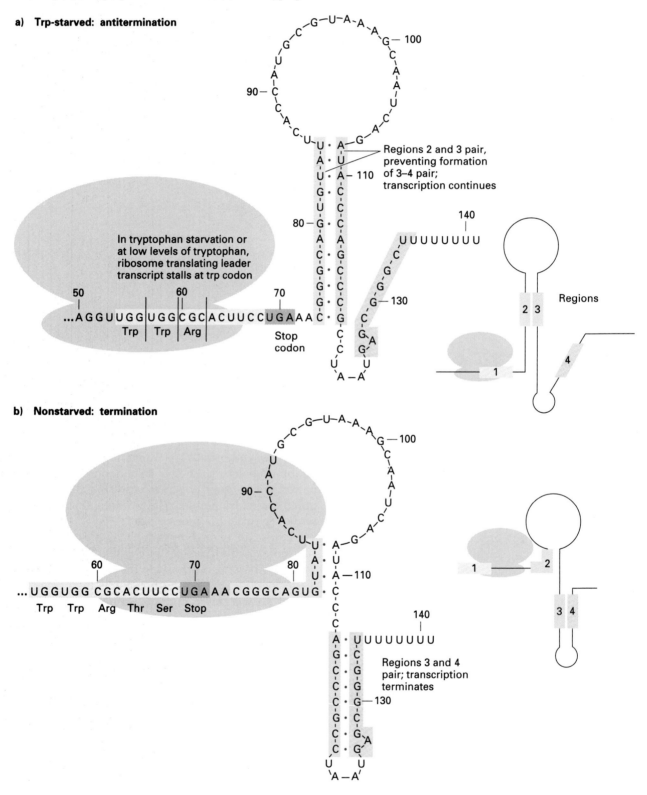

is controlled by the enzymes encoded by the operon. Attenuation has also been shown to regulate a number of genes not involved with amino acid biosynthesis—for example, the rRNA operons (*rrn*) and the *ampC* gene of *E. coli* (for ampicillin resistance).

KEYNOTE

Regulation of the *E. coli trp* operon is at the level of initiating or completing a transcript of the operon. This is accomplished through a repressor-operator system, which responds to free tryptophan levels, and through attenuation at a second controlling site called an attenuator, which responds to Trp-tRNA levels. The attenuator is located in the leader region between the operator region and the first *trp* structural gene. The attenuator acts to terminate transcription dependent on the concentration of tryptophan. In the presence of large amounts of tryptophan, attenuation is very effective; that is, enough Trp-tRNA is present so that the ribosome can continue and allow the leader transcript to form a secondary structure that causes transcription to be terminated. In the absence of or at low concentrations of tryptophan, the ribosomes stall at Trp codons, and the leader transcript forms a secondary structure that permits continued transcription activity.

REGULATION OF GENE EXPRESSION IN PHAGE LAMBDA

Bacteriophages exist by parasitizing bacteria, so many or all the essential components for phage reproduction are provided by the host cell, and the use of those components is controlled by the products of phage genes. Most genes of a phage, then, code for products that control the life cycle and the production of progeny phage particles. Much is known about gene regulation in a number of bacteriophages. In this section we discuss the regulation of gene expression as it relates to the lytic cycle and lysogeny in bacteriophage lambda (λ).

Early Transcription Events

Figure 15.15 shows the genetic map of λ. The mature λ chromosome is linear, and has complementary "sticky" ends. Once free in the host cell, the λ chromosome circularizes, so we show the genetic map in a circular form. Recall that λ is a temperate phage (see

Figure 6.14), so when it infects a bacterial cell, the phage has a choice of whether to enter the lytic pathway (when progeny phages are assembled and released from the cell) or the lysogenic pathway (when the λ chromosome integrates into the chromosome and no progeny phages are produced). The regulatory system involved in this choice is an excellent model for a genetic switch and as such has contributed to our thinking about how genetic switches operate in eukaryotic systems.

The choice between the lytic and lysogenic pathways occurs soon after λ infects the cell and its genome circularizes, and involves a sophisticated *genetic switch*. First, transcription begins at promoters P_L and P_R (Figure 15.16, part 1). Promoter P_L is for leftward transcription of the left early operon, and promoter P_R is for rightward transcription of the right early operon.

The first gene to be transcribed from P_R is *cro* (control of *r*epressor and *o*ther), the product of which is the Cro protein. This protein plays an important role in setting the genetic switch to the lytic pathway. The first gene to be transcribed from P_L is *N*. The resulting N protein is a transcription *antiterminator* that allows RNA synthesis to proceed past certain transcription terminators, in this case leftward of *N* and rightward of *cro*, thereby including all of the early genes (Figure 15.16, part 2). One of the genes transcribed as a result of the action of N protein is *cII*. The cII protein turns on genes *cI* (encodes the λ repressor), *O* and *P* (encode two DNA replication proteins), and *Q* (encodes a protein needed to turn on late genes for lysis and phage particle proteins). The Q protein is another antiterminator, permitting transcription to continue into the late genes involved in the lytic pathway. However, only when the switch is set to the lytic pathway and transcription continues from P_R for a sufficient time does enough Q protein accumulate to function effectively.

The Lysogenic Pathway

After the early transcription events, either the lysogenic or lytic pathway is followed (Figure 15.16, part 3). The switch is set for the lysogenic pathway as follows.

The establishment of lysogeny requires the protein products of the *cII* (right early operon) and *cIII* (left early operon; see Figure 15.15) genes. The cII protein (stabilized by cIII protein) activates transcription of the *cI* gene (located between the P_L and P_R promoters—see Figure 15.16, part 4a) leftwards from a promoter called P_{RE} (promoter for *r*epressor *e*stablishment). The product of the *cI* gene, the λ repressor, binds to two operator regions, O_L and O_R

~ FIGURE 15.15

A map of phage λ, showing the major genes. (Promoters discussed in text: P_L = promoter for leftward transcription of the left early operon; P_R = promoter for rightward transcription of the right early operon; P_{RE} = promoter for repressor establishment; and P_{RM} = promoter for repressor maintenance.)

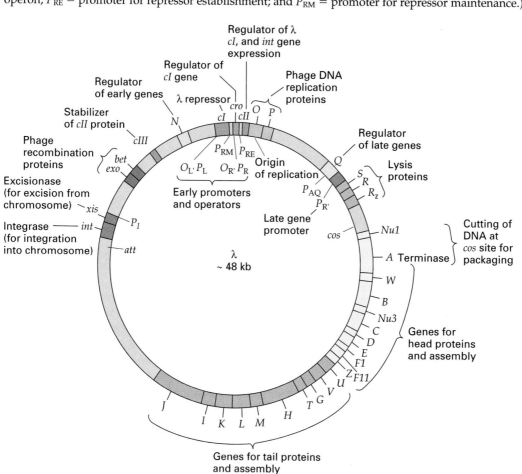

(see Figure 15.15 and Figure 15.16, part 5a), whose sequences overlap the P_L and P_R promoters, respectively. The binding of the λ repressor prevents the further transcription by RNA polymerase of the early operons controlled by P_L and P_R. As a result, transcription of the *N* and *cro* genes is blocked, and since the two proteins specified by these genes are unstable, the concentrations of these two proteins in the cell drop dramatically. Furthermore, a repressor bound to O_R stimulates synthesis of more repressor mRNA from a different promoter, P_{RM} (promoter for repressor *maintenance*), thereby maintaining repressor concentrations in the cell (see Figure 15.16, part 5a). Thus, if enough λ repressors are present, lysogeny is established by the binding of the repressor to operators O_L and O_R, followed by integration of λ DNA catalyzed by integrase. Integrase is the product of the cII-regulated promoter P_I. As the concen-

tration of cII drops, P_I transcription shuts off, leaving P_{RM} as the only active promoter.

The Lytic Pathway

The lysogenic pathway is favored when enough λ repressor is made so that early promoters are turned off, thereby repressing all the genes needed for the lytic pathway. One important lytic pathway gene that is repressed is *Q*; the Q protein is a positive regulatory protein required for the production of lysis proteins and phage coat proteins.

Let us consider the induction of the lytic pathway caused by ultraviolet light irradiation. Inducers such as ultraviolet light typically damage DNA, and this somehow causes a change in the function of the bacterial protein RecA (product of the *recA* gene). Normally RecA functions in DNA recombination,

~ FIGURE 15.16

Expression of λ genes after infecting *E. coli* and the transcriptional events that occur when either the lysogenic or lytic pathways are followed.

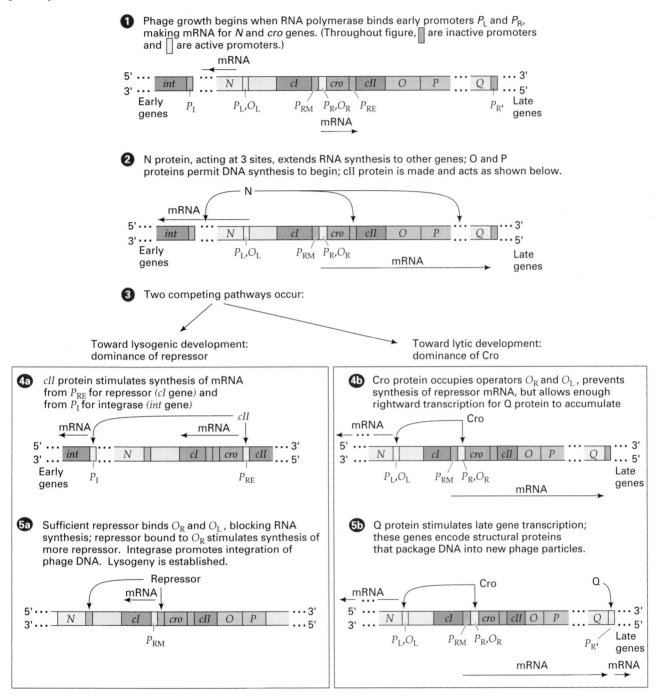

1 Phage growth begins when RNA polymerase binds early promoters P_L and P_R, making mRNA for *N* and *cro* genes. (Throughout figure, ▮ are inactive promoters and ▯ are active promoters.)

2 N protein, acting at 3 sites, extends RNA synthesis to other genes; O and P proteins permit DNA synthesis to begin; cII protein is made and acts as shown below.

3 Two competing pathways occur:

Toward lysogenic development: dominance of repressor

Toward lytic development: dominance of Cro

4a *cII* protein stimulates synthesis of mRNA from P_{RE} for repressor (*cI* gene) and from P_I for integrase (*int* gene)

5a Sufficient repressor binds O_R and O_L, blocking RNA synthesis; repressor bound to O_R stimulates synthesis of more repressor. Integrase promotes integration of phage DNA. Lysogeny is established.

4b Cro protein occupies operators O_R and O_L, prevents synthesis of repressor mRNA, but allows enough rightward transcription for Q protein to accumulate

5b Q protein stimulates late gene transcription; these genes encode structural proteins that package DNA into new phage particles.

but when DNA is damaged, RecA stimulates the λ repressor polypeptides to cleave themselves in two and therefore become inactivated. The resulting absence of repressor at O_R allows RNA polymerase to bind at P_R, and the *cro* gene is then further transcribed. The resulting Cro protein then acts to decrease RNA synthesis from P_L and P_R, and this reduces synthesis of cII protein, the regulator of λ repressor synthesis, and blocks synthesis of λ repressor mRNA from P_{RM} (Figure 15.16, part 4b). At the same time, transcription of the right early operon genes from P_R is decreased, but enough Q proteins are accumulated to set the genetic switch for transcription of the late genes for starting the lytic pathway (Figure 15.16, part 5b).

In summary, lambda uses complex regulatory systems to choose either the lytic or the lysogenic pathway. This decision depends on a sophisticated genetic switch that involves competition between the products of the *cI* ("c-one") gene (the repressor) and the *cro* gene (the Cro protein). If the repressor dominates, the lysogenic pathway is followed; if the Cro protein dominates, the lytic pathway is followed.

KEYNOTE

Bacteriophages are especially adapted for undergoing their reproduction within a bacterial host. Many of the genes related to the production of progeny phages, or to the establishment or reversal of lysogeny in temperate phages, are organized into operons. These operons (like bacterial operons) are controlled through the interaction of regulatory proteins with operators that are adjacent to clusters of structural genes.

SUMMARY

From studies of the regulation of expression of the three lactose-utilization genes in *E. coli*, Jacob and Monod developed a model that is the basis for the regulation of gene expression in a large number of bacterial and bacteriophage systems. The genes for the enzymes are contiguous in the chromosome and are adjacent to a controlling site (an operator) and a single promoter. This complex constitutes a transcriptional regulatory unit called an operon. A regulator gene, which may or may not be nearby, is associated with the operon. The addition of an appropriate substrate to the cell results in the coordinate induction of the operon's structural genes. Induction of the *lac* operon occurs as follows: lactose binds with a repressor protein that is encoded by the regulator gene, inactivating the repressor protein and preventing it from binding to the operator. As a result, RNA polymerase can transcribe the three genes from a single promoter onto a single polycistronic mRNA. As long as lactose is present, mRNA continues to be produced and the enzymes are made. When lactose is no longer present, the repressor pro-

tein is no longer inactivated, and it binds to the operator, thereby preventing RNA polymerase from transcribing the *lac* genes.

If both glucose and lactose are present in the medium, the lactose operon is not induced because glucose (which requires less energy to metabolize than does lactose) is the preferred energy source. This phenomenon is called catabolite repression and involves cellular levels of cyclic AMP. That is, in the presence of lactose and the absence of glucose, cAMP complexes with CAP to form a positive regulator needed for RNA polymerase to bind to the promoter. Addition of glucose results in a lowering of cAMP concentration, so no CAP-cAMP complex is produced and, therefore, RNA polymerase cannot bind efficiently to the promoter and transcribe the *lac* genes.

The genes for a number of bacterial amino acid biosynthesis pathways are also arranged in operons. Expression of these operons is accomplished by a repressor-operator system or, in some cases, through attenuation at a second controlling site called an attenuator, or by both mechanisms. The *trp* operon is an example of an operon with both types of transcription regulation systems. The *trp* repressor-operator system functions essentially like that of the *lac* operon, except that the addition of tryptophan to the cell activates the repressor, turning the operon off. The attenuator, located downstream from the operator in a leader region that is translated, is a transcription termination site that allows only a fraction of RNA polymerases that initiate at the *trp* promoter to transcribe the rest of the operon. Attenuation requires a coupling between transcription and translation and the formation of particular RNA secondary structures that signal whether or not transcription can continue.

Key to the attenuation phenomenon is the presence of multiple copies of codons for the amino acid synthesized by the enzymes encoded by the operon. When enough of the amino acid is present in the cell, enough charged tRNAs are produced so that the ribosome can quickly translate the key codons in the leader region, and this causes the RNA being made by the RNA polymerase ahead of the ribosome to assume a secondary structure that signals transcription to stop. However, when the cell is starved for that amino acid, there are insufficient charged tRNAs to be used at the key codons so that the ribosome stalls at that point. As a result the RNA ahead of that point assumes a secondary structure that permits continued transcription into the structural genes. The combination of repressor-operator regulation and attenuation control permits a fine degree of control of operon transcription.

Operons are also extensively used by bacteriophages to coordinate the synthesis of proteins with related functions. Many genes related to the production of progeny phages, or to the establishment or reversal of lysogeny of temperate phages, are organized into operons. These operons, like bacterial operons, are controlled through the interaction of regulatory proteins with operators that are adjacent to clusters of structural genes. Bacteriophage lambda has been an excellent model for studying the elaborate genetic switch that controls the choice between lytic and lysogenic pathways in a lysogenic phage. The switch involves two regulatory proteins, the lambda repressor and the Cro protein, which have opposite affinities for three binding sites within each of the two major operators that control the lambda life cycle.

In sum, operons are commonly encountered in bacteria and their viruses. They provide a simple way to coordinate the expression of genes with related function. Generally, there is an interaction between an effector molecule (such as lactose or tryptophan) and a repressor molecule. That interaction may inactivate (lactose) or activate (tryptophan) the repressor with respect to its ability to bind to the operator. If no repressor is bound to the operator, RNA polymerase can bind to the adjacent promoter and the genes can be transcribed.

ANALYTICAL APPROACHES FOR SOLVING GENETICS PROBLEMS

Q15.1 In the laboratory you are given 10 strains of *E. coli* with the following lactose operon genotypes, where $I = lacI$ (the repressor gene), $P = P_{lac}$ (the promoter), $O = lacO$ (the operator), and $Z = lacZ$ (the β-galactosidase gene).

For each strain, predict whether β-galactosidase will be produced (a) if lactose is absent from the growth medium and (b) if lactose is present in the growth medium.

1. $I^+ P^+ O^+ Z^+$

2. $I^- P^+ O^+ Z^+$

3. $I^+ P^+ O^c Z^+$

4. $I^- P^+ O^c Z^+$

5. $I^+ P^+ O^c Z^-$

6. $\dfrac{F'\ I^+ P^+ O^c Z^-}{I^+ P^+ O^+ Z^+}$

7. $\dfrac{F'\ I^+ P^+ O^+ Z^-}{I^+ P^+ O^c Z^+}$

8. $\dfrac{F'\ I^+ P^+ O^+ Z^+}{I^- P^+ O^+ Z^-}$

9. $\dfrac{F'\ I^+ P^+ O^c Z^-}{I^- P^+ O^+ Z^+}$

10. $\dfrac{F'\ I^- P^+ O^+ Z^-}{I^- P^+ O^c Z^+}$

(*Note:* In the partial-diploid strains (6–10), one copy of the *lac* operon is in the host chromosome and the other copy is in the F episome.)

A15.1 The answers are as follows, where "+" = β-galactosidase produced and "−" = it is not produced:

GENOTYPE	NONINDUCED: LACTOSE ABSENT	INDUCED: LACTOSE PRESENT
(1)	−	+
(2)	+	+
(3)	+	+
(4)	+	+
(5)	−	−
(6)	−	+
(7)	+	+
(8)	−	+
(9)	−	+
(10)	+	+

To answer this question completely, we need a good understanding of how the lactose operon is regulated in the wild type and of the consequences of particular mutations on the regulation of the operon.

Strain (1) is the standard wild-type operon. No enzyme is produced in the absence of lactose, since the repressor produced by the I^+ gene binds to the operator (O^+) and blocks the initiation of transcription. When lactose is added, it binds to the repressor, changing its conformation so that it no longer can bind to the O^+ region, thereby facilitating transcription of the structural genes for RNA polymerase.

Strain (2) is a haploid strain with a mutation in the *lacI* gene (I^-). The consequence is that the repressor protein cannot bind to the (normal) operator region, so there is no inhibition of transcription, even in the absence of lactose. This strain, then, is constitutive, meaning that β-galactosidase is produced by the $lacZ^+$ gene in the presence or absence of lactose.

Strain (3) is another constitutive mutant. In this case the repressor gene is a wild type and the β-galactosidase gene $lacZ^+$ is a wild type, but there is a mutation in the operator region (O^c). Therefore, repressor protein cannot bind to the operator, and transcription occurs in the presence or absence of lactose.

Strain (4) carries both regulatory mutations of the previous two strains. Functional repressor is not produced, but even if it were, the operator is changed so that it cannot bind. The consequence is the same: constitutive enzyme production.

Strain (5) produces functional repressor, but the operator (O^c) is mutated. Therefore, transcription cannot be blocked, and lac polycistronic mRNA is produced in the presence or absence of lactose. However, because there is also a mutation in the β-galactosidase gene (Z^-), no functional enzyme is generated.

In the partial-diploid strain (6), one lactose operon is completely wild type and the other carries a constitutive operator mutation and a mutant β-galactosidase gene. In the absence of lactose, no functional enzyme is produced. For the wild-type operon, the repressor binds to the operator and blocks transcription. For the operon with the two mutations, the operator region is mutated and cannot bind repressor, so the mRNA for the mutated operon is produced; however, the $lacZ$ gene is also mutated so that functional enzyme cannot be produced. In the presence of lactose, functional enzyme is produced, since repression of the wild-type operon is relieved so that the Z^+ gene can be transcribed. This type of strain provided one of the pieces of evidence that the operator region does not produce a diffusible substance.

In partial diploid (7), functional enzyme will be produced in the presence or absence of lactose, because one of the operons has an O^c mutation that does not respond to a repressor and that is linked to a wild-type Z^+ gene. That operon is transcribed constitutively. The other operon is inducible, but because there is a Z^- mutation, no functional enzyme is produced.

Partial diploid (8) has a wild-type operon and an operon with an I^- regulatory mutation and a Z^- mutation. The I^+ gene product is diffusible so that it can bind to the O^+ region of both operons, thereby putting both operons under inducer control. This strain demonstrates that the I^+ gene is trans-dominant to an I^- mutation. In this case the particular location of the one Z^- mutation is irrelevant: the same result would have been obtained had the Z^+ and Z^- been switched between the two operons. In this partial diploid, β-galactosidase is not produced unless lactose is present.

In strain (9), β-galactosidase is produced only when the lactose is present, because the O^c region controls only those genes that are adjacent to it on the same chromosome (cis dominance), and in this case one of the adjacent genes is Z^-, which codes for a nonfunctional enzyme. The partial diploid is heterozygous I^+/I^-, but I^+ is trans-dominant, as discussed for strain (8). Thus, the only normal Z^+ gene is under inducer control.

Partial diploid (10) has defective repressor protein as well as an O^c mutation adjacent to a Z^+ gene. On the latter ground alone, this partial diploid is constitutive. The other operon is also constitutively transcribed, but because there is a Z^- mutation, no functional enzyme is generated from it.

QUESTIONS AND PROBLEMS

15.1 How does lactose bring about the induction of synthesis of β-galactosidase, permease, and transacetylase? Why does this event not occur when glucose is also in the medium?

15.2 Operons produce polycistronic mRNA when they are active. What is a polycistronic mRNA? What advantages, if any, does it confer in terms of the cell's function?

*15.3 If an *E. coli* mutant strain synthesizes β-galactosidase whether or not the inducer is present, what genetic defect(s) might be responsible for this phenotype?

*15.4 Distinguish the effects you would expect from (a) a missense mutation and (b) a nonsense mutation in the *lacZ* (β-galactosidase) gene of the *lac* operon.

15.5 Elucidation of the regulatory mechanisms associated with the enzymes of lactose utilization in *E. coli* was a landmark in our understanding of regulatory processes in microorganisms. In formulating the operon hypothesis as applied to the lactose system, Jacob and Monod found that results from particular partial-diploid strains were invaluable. Specifically, in terms of the operon hypothesis, what information did the partial diploids provide that the haploids could not?

*15.6 For the *E. coli lac* operon, write the partial-diploid genotype for a strain that will produce β-galactosidase constitutively and permease by induction.

15.7 Mutants were instrumental in elaborating the model for regulation of the *lac* operon.
a. Discuss why *lacO^c* mutants are cis-dominant but not trans-dominant.
b. Explain why *lacI^s* mutants are trans-dominant to the wild-type *lacI^+* allele but *lacI^-* mutants are recessive.
c. Discuss the consequences of mutations in the repressor gene promoter as compared with mutations in the structural gene promoter.

*15.8 This question involves the *lac* operon of *E. coli*, where $I = lacI$ (the repressor gene), $P = P_{lac}$ (the promoter), $O = lacO$ (the operator), $Z = lacZ$ (the β-galactosidase gene), and $Y = lacY$ (the permease gene). Complete Table 15.A, using "+" to indicate that the enzyme in question will be synthesized and "−" to indicate that the enzyme will not be synthesized.

~ TABLE 15.A

	GENOTYPE	INDUCER ABSENT:		INDUCER PRESENT:	
		β-GALACTOSIDASE	PERMEASE	β-GALACTOSIDASE	PERMEASE
a.	$I^+ P^+ O^+ Z^+ Y^+$				
b.	$I^+ P^+ O^+ Z^- Y^+$				
c.	$I^+ P^+ O^+ Z^+ Y^-$				
d.	$I^- P^+ O^+ Z^+ Y^+$				
e.	$I^s P^+ O^+ Z^+ Y^+$				
f.	$I^+ P^+ O^c Z^+ Y^+$				
g.	$I^s P^+ O^c Z^+ Y^+$				
h.	$I^+ P^+ O^c Z^+ Y^-$				
i.	$I^- P^+ O^+ Z^+ Y^+$ $I^+ P^+ O^+ Z^- Y^-$				
j.	$I^- P^+ O^+ Z^+ Y^-$ $I^+ P^+ O^+ Z^- Y^+$				
k.	$I^s P^+ O^+ Z^+ Y^-$ $I^+ P^+ O^+ Z^- Y^+$				
l.	$I^+ P^+ O^c Z^- Y^+$ $I^+ P^+ O^+ Z^+ Y^-$				
m.	$I^- P^+ O^c Z^+ Y^-$ $I^+ P^+ O^+ Z^- Y^+$				
n.	$I^s P^+ O^+ Z^+ Y^+$ $I^+ P^+ O^c Z^+ Y^+$				
o.	$I^+ P^- O^c Z^+ Y^-$ $I^+ P^+ O^+ Z^- Y^+$				
p.	$I^+ P^- O^+ Z^+ Y^-$ $I^+ P^+ O^c Z^- Z^+$				
q.	$I^- P^- O^+ Z^+ Y^+$ $I^+ P^+ O^+ Z^- Y^-$				
r.	$I^- P^+ O^+ Z^+ Y^-$ $I^+ P^- O^+ Z^- Y^+$				

15.9 A new sugar, sugarose, induces synthesis of two enzymes from the *sug* operon of *E. coli*. Some properties of deletion mutations affecting the appearance of these enzymes are as follows (here, "+" = enzyme induced normally, i.e., synthesized only in the presence of the inducer; C = enzyme synthesized constitutively; 0 = enzyme cannot be detected):

MUTATION OF	ENZYME 1	ENZYME 2
Gene *A*	+	0
Gene *B*	0	+
Gene *C*	0	0
Gene *D*	C	C

a. The genes are adjacent, in the order *ABCD*. Which gene is most likely to be the structural gene for enzyme 1?

b. Complementation studies using partial-diploid (*F′*) strains were made. The episome (*F′*) and chromosome each carried one set of *sug* genes. The results were as follows:

GENOTYPE OF *F′*	CHROMOSOME	ENZYME 1	ENZYME 2
$A^+ B^- C^+ D^+$	$A^- B^+ C^+ D^+$	+	+
$A^+ B^- C^- D^+$	$A^- B^+ C^+ D^+$	+	0
$A^- B^+ C^- D^+$	$A^+ B^- C^+ D^+$	0	+
$A^- B^+ C^+ D^+$	$A^+ B^- C^+ D^-$	+	+

From all the evidence given, determine whether the following statements are true or false:

1. It is possible that gene *D* is a structural gene for one of the two enzymes.

2. It is possible that gene *D* produces a repressor.

3. It is possible that gene *D* produces a cytoplasmic product required to induce genes *A* and *B*.
4. It is possible that gene *D* is an operator locus for the *sug* operon.
5. The evidence is also consistent with the possibility that gene *C* could be a gene that produces a cytoplasmic product required to induce genes *A* and *B*.
6. The evidence is also consistent with the possibility that gene *C* could be the controlling end of the *sug* operon (the end from which mRNA synthesis presumably commences).

*15.10   What consequences would a mutation in the catabolite activator protein (CAP) gene of *E. coli* have for the expression of a wild-type *lac* operon?

15.11   The *lac* operon is an inducible operon, whereas the tryptophan operon is a repressible operon. Discuss the differences between these two types of operons.

*15.12   In the presence of high intracellular concentrations of tryptophan, only short transcripts of the *trp* operon are synthesized because of attenuation of transcription 5' to the structural genes. This is mediated by the recognition of two Trp codons in the leader sequence. If these codons were mutated to UAG stop codons, what effect would this have on the regulation of the operon in the presence or absence of tryptophan? Explain.

15.13   On infecting an *E. coli* cell, bacteriophage λ has a choice between the lytic and lysogenic pathways. Discuss the molecular events that determine which pathway is taken.

15.14   How do the lambda repressor protein and the Cro protein regulate their own synthesis?

*15.15   If a mutation in the phage lambda *cI* gene results in a nonfunctional *cI* gene product, what phenotype would you expect the phage to exhibit?

15.16   Bacteriophage λ can form a stable association with the bacterial chromosome because the virus manufactures a repressor. This repressor prevents the virus from replicating its DNA and making lysozyme and all the other tools used to destroy the bacterium. When you induce the virus with UV light, you destroy the repressor, and the virus goes through its normal lytic cycle. The repressor is the product of a gene called the *cI* gene and is a part of the wild-type viral genome. A bacterium that is lysogenic for λ⁺ is full of repressor protein, which confers immunity against any λ virus added to these bacteria. Added viruses can inject their DNA, but the repressor from the resident virus prevents replication, presumably by binding to an operator on the incoming virus. Thus, this system has many elements analogous to the *lac* operon. We could diagram a virus as shown in the figure. Several mutations of the *cI* gene are known. The c_i mutation results in an inactive repressor.

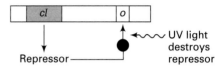

a. If you infect *E. coli* with λ containing a c_i mutation, can it lysogenize (form a stable association with the bacterial chromosome)? Why or why not?
b. If you infect a bacterium simultaneously with a wild-type c^+ and a c_i mutant of λ, can you obtain stable lysogeny? Why or why not?
c. Another class of mutants called c^{IN} makes a repressor that is insensitive to UV destruction. Will you be able to induce a bacterium lysogenic for c^{IN} with UV light? Why or why not?

CHAPTER *16*

EUKARYOTIC GENE REGULATION

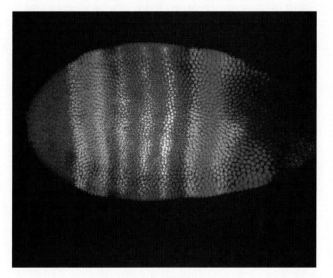

PRINCIPAL POINTS

~ There are no operons in eukaryotes. However, genes for related functions are often regulated coordinately.

~ In eukaryotes, gene expression is regulated at a number of distinct levels; there are regulatory systems for the control of DNA structure, transcription, precursor-RNA processing, transport of mature RNA out of the nucleus, translation of the mRNAs, degradation of the mature RNAs, and degradation of the protein products.

~ A eukaryotic protein-coding gene contains promoter elements and enhancers, and perhaps silencers. The promoter elements are required for transcription to begin. The regulation of transcription occurs mainly as a result of regulatory proteins binding to enhancers and silencers and transmitting their effect to the proteins bound to the promoter elements.

~ Transcriptionally active chromatin is more sensitive to DNase I than transcriptionally inactive chromatin. This sensitivity is a result of a loosened DNA-protein structure in the areas of transcriptional activity. For active genes, certain sites—called hypersensitive sites—are highly sensitive to DNase I and probably correspond to the binding sites for RNA polymerase and regulatory proteins.

~ Histones repress transcription by assembling nucleosomes on TATA boxes associated with genes. Gene activation occurs when proteins become bound to enhancers and disrupt the nucleosomes on the TATA boxes, so that the appropriate proteins can bind to the promoter elements to initiate transcription.

~ The DNA of many eukaryotes has been shown to be methylated at a certain proportion of the bases. Methyl groups are added mostly to cytosines. In those eukaryotes, transcriptionally active genes generally exhibit lower levels of DNA methylation than transcriptionally inactive genes.

~ In many eukaryotes, steroid hormones regulate the expression of particular sets of genes. To function in this short-term regulatory system, a steroid hormone diffuses into a cell, binds to a specific receptor, and activates it. The hormone-receptor complex then binds to hormone regulatory elements next to genes in the nucleus, thereby regulating the expression of those genes. The specificity of hormone action results from the presence of hormone receptors in only certain cell types and from interactions of hormone-receptor complexes with cell type-specific regulatory proteins.

~ Regulation at the level of RNA processing operates in the production of mature mRNA molecules from precursor-mRNA molecules. Two examples of such regulatory control are alternative polyadenylation (choice of poly(A) site) and alternative splicing (differential removal of introns). In both cases, different types of mRNAs are produced, and different proteins are synthesized.

~ Gene expression is regulated also by mRNA translation control and by mRNA degradation control. Structural features of individual mRNAs have been shown to be responsible for the range of mRNA degradation rates observed, although the precise roles of cellular factors and enzymes have yet to be determined.

~ Long-term regulation is involved in controlling events that activate and repress genes during development and differentiation. Those two processes result from differential gene activity in a genome that contains a constant amount of DNA, rather than from a programmed loss of genetic information.

~ Antibodies are specialized proteins called immunoglobulins, which bind specifically to antigens (*antibody gen*erators—chemicals recognized as foreign by the immune system of a host). Antibody molecules consist of two light chains and two heavy chains. On each type of chain, the amino acid sequence of one region is variable; this variation is responsible for the different antigen-binding sites on different antibody molecules. In germ-line DNA, the coding segments for immunoglobulin chains are scattered in tandem arrays of gene segments. During development, somatic recombination brings particular gene segments together to form functional antibody chain genes. A large number of different antibody chain genes results from the many possible ways in which the gene segments can recombine.

~ *Drosophila* has become a useful model system for studying the genetic control of development. *Drosophila* body structures result from specific gradients of molecules in the egg and the subsequent determination of embryo segments that directly correspond to adult body segments. Both processes are under genetic control as defined by mutations that disrupt the development events. Studies of mutants indicate that *Drosophila* development is directed by a regulatory cascade.

~ Once the basic segmentation pattern has been laid down in *Drosophila*, homeotic genes determine the developmental identity of the segments. Homeotic genes share DNA sequences called homeoboxes. Homeoboxes have been found in developmental genes in other organisms, and the homeodomains—the regions of the proteins the homeoboxes encode—probably play a role in regulating transcription by binding to specific DNA sequences.

*A*s we saw in Chapter 15, gene expression in prokaryotes is commonly regulated in a unit of protein-coding sequences and adjacent controlling sites (promoter and operator), collectively called an operon. Operons work in a simple way to coordinate the synthesis of proteins with related functions. In eukaryotes, gene expression is also regulated in a unit of protein-coding sequences and adjacent controlling sites. However, eukaryotic genes do not have operators, and no polycistronic mRNAs are transcribed, so eukaryotes do not have operons. Your goal in this chapter is to learn about some of the mechanisms for gene regulation in eukaryotes.

LEVELS OF CONTROL OF GENE EXPRESSION IN EUKARYOTES

In both unicellular and multicellular eukaryotes, the control of gene expression is more complicated than in prokaryotes. Some of the levels at which the expression of protein-coding genes can be regulated in eukaryotes are *transcriptional control, RNA processing control, transport control, mRNA translation control, mRNA degradation control*, and *protein degradation control* (Figure 16.1). These control processes help coordinate the generation of new proteins in different cells at different times. We now discuss some of these levels of regulation.

Transcriptional Control

GENERAL ASPECTS OF TRANSCRIPTIONAL CONTROL. **Transcriptional control** regulates whether or not a gene is to be transcribed and the rate at which transcripts are produced.

A eukaryotic protein-coding gene contains promoter elements and enhancer elements, and possibly silencers (see Chapter 12, p. 263, and Figure 12.5). Recall that promoters contain various combinations of basal promoter elements and promoter proximal elements. The initiation of transcription involves the assembly of RNA polymerase II and basal transcription factors (TFs) on the basal promoter elements. Other transcription factors bind to the proximal promoter elements and influence the rate of transcription of initiation. The regulation of transcription, though, occurs mainly as a result of the binding of activator proteins to enhancers and/or repressor proteins to silencers.

Activators and repressors are examples of regulatory proteins. An activator is a positive regulatory pro-

tein, and a repressor is a negative regulatory protein. Activators signal the basal transcription factors to stimulate transcription via bridging adapter proteins (see Figure 12.5b). The interaction of repressors with

~ **FIGURE 16.1**

Levels at which gene expression can be controlled in eukaryotes.

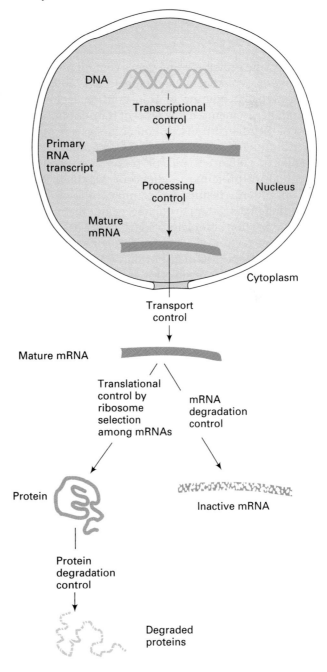

silencers causes a transcription inhibiting signal to be relayed to the basal transcription factors, perhaps involving adapter proteins. In short, the basal level of transcription resulting from proteins binding to the promoter elements may be regulated by the effects of regulatory proteins binding to enhancers and silencers. A gene may have both enhancers and silencers, so the rate of transcription of the gene will depend upon the various regulatory proteins bound to those sequences. Further, because activators and repressors are cell- and tissue-specific, the interactions of these proteins at enhancers and silencers are responsible for cell- and tissue-specific gene expression.

KEYNOTE

Eukaryotic protein-coding genes contain promoter elements, enhancers, and sometimes silencers. Through the binding of transcription factors to promoter elements, a low level of transcription can occur. Regulated transcription occurs as a result of the binding of activators (positive regulatory proteins) to enhancers and/or repressors (negative regulatory proteins) to silencers. Maximal transcription only results, for example, following the binding of activators to enhancers and the relay of a transcription stimulatory signal to the basal transcription factors via adapter proteins. The cell- and tissue-specific distribution of activators and repressors is responsible for the cell- and tissue-specific regulation of gene expression.

DNA BINDING PROTEINS AND THE REGULATION OF TRANSCRIPTION.

Many of the transcription factors and regulatory proteins involved in the process of transcription initiation in eukaryotes (and prokaryotes) are sequence-specific DNA-binding proteins. They interact with specific DNA sequences and exert an effect on transcription initiation. For a number of these proteins, but by no means all, certain well-defined structural motifs of the protein are responsible for binding of the protein to the DNA. Examples of some of the structural motifs—called *DNA-binding domains*—are zinc finger, leucine zipper, and the helix-turn-helix. For example, helix-turn-helix motifs in the bacteriophage λ repressor protein bind to the *lac* operator.

CHROMOSOME CHANGES AND TRANSCRIPTIONAL CONTROL.

As we saw in Chapter 10, the eukaryotic chromosome consists of DNA complexed with histones (to form nucleosomes) and nonhistone chromosomal proteins. In this section we examine the relationship between chromosome structure and transcriptional control.

DNase Sensitivity and Gene Expression. Compared with inactive genes, almost all transcriptionally active genes have an increased sensitivity to the DNA-degrading enzyme, DNase I. The interpretation is that transcriptionally active genes have a looser chromosome structure than transcriptionally inactive genes. Despite the increased DNase I, the DNA is still organized into nucleosomes. In fact, recent experiments show that the histone octamer can "step around" the transcribing RNA polymerase, indicating that there is no reason for the nucleosome to disassemble for transcription to occur.

DNase Hypersensitive Sites. Certain sites around transcriptionally active genes, called **hypersensitive sites** or **hypersensitive regions**, are highly sensitive to digestion by DNase I. These sites or regions are typically the first to be cut with DNase I. Most, but not all, DNase-hypersensitive sites are in the regions upstream from the start of transcription, probably corresponding to the DNA sequences where RNA polymerase and other gene regulatory proteins bind to initiate transcription.

KEYNOTE

Chromosome regions that are transcriptionally active have looser DNA-protein structures than chromosome regions that are transcriptionally inactive, resulting in sensitivity of the DNA to digestion by DNase. The promoter regions of active genes typically have an even looser chromosome structure, resulting in hypersensitivity to DNase.

Histones and Gene Regulation. In the chromosome, DNA is wrapped around histones to form nucleosomes. In the evolutionary sense, the histones are extremely conserved; their amino acid sequence, structure, and function are all very similar among all eukaryotic organisms. Furthermore, from cell to cell in an organism, the five different types of histones are arranged in an essentially constant way along the DNA. Clearly, this uniform distribution of histones along the DNA is not specific enough to control the expression of thousands of genes. Nonetheless, histones do become modified, primarily through acetylation and phosphorylation, thereby altering their ability to bind to the DNA. If the gene that they cover is to be expressed, the histones must be modified to loosen their grip on the DNA so that the DNA strands can interact with transcription factors and/or regulatory proteins. In essence, the histones act as general repressors of transcription. That is, when promoter elements (particularly the TATA box, which is important for transcription initiation; see Chapter 14) are associated with the histones of a nucleosome core, they cannot be found by the regulatory proteins and transcription factors that must bind to them to initiate transcription. (These regulatory proteins and transcription factors are in the nonhistone class of chromosomal proteins.)

So, how does gene activation occur in the cell? One model is as follows: histones repress transcription by forming nucleosomes on TATA boxes upstream of genes. Promoter-binding proteins are unable to disrupt the nucleosomes on the TATA boxes. However, enhancer-binding proteins bind to the enhancers, displacing any histones that are bound there, and interact with the histones in the TATA-bound nucleosomes. This causes the nucleosome core particle to break up, and promoter-binding proteins can now bind to the freed DNA.

Other research has established a correlation between the level of histone acetylation and transcriptional activity. That is, histones in transcriptionally active chromatin are hyperacetylated, while histones in transcriptionally inactive chromatin are hypoacetylated. When the histones become acetylated, nucleosome conformation is altered and higher-order chromatin structure is destabilized. As a result, the DNA becomes more accessible to transcription factors, and the generally repressive action of histones on transcription is overcome.

KEYNOTE

Histones repress transcription by assembling nucleosomes on TATA boxes associated with genes. Gene activation occurs when proteins become bound to enhancers and disrupt the nucleosomes on the TATA boxes, thereby allowing the appropriate proteins to bind to the promoter elements to initiate transcription.

DNA METHYLATION AND TRANSCRIPTIONAL CONTROL.

Once DNA has been replicated within the eukaryotic cell, a small proportion of bases becomes chemically modified by enzymes. A particular modified base has received significant attention since it has been correlated with gene activity in some cases. This base, 5-methylcytosine (5^mC), is produced from cytosine (C) shortly after replication by the enzyme DNA methylase (Figure 16.2). The percentage of cytosines present as 5^mC in eukaryotic DNA varies over a wide range, with about 3 percent in mammals and very little, if any, in less-complex eukaryotes, such as *Drosophila*, *Tetrahymena*, and yeast.

The relationship between DNA methylation and gene activity has been studied for a number of genes. For at least 30 genes examined, a negative correlation has been shown between DNA methylation and transcription; that is, there is a lower level of methylated DNA (hypomethylated DNA) in transcriptionally active genes compared with transcriptionally inactive genes. We must be cautious, though, because not all methylated C nucleotides in a gene region become demethylated when the gene is expressed. And, as stated earlier, some organisms do not have significant amounts of methylated C in their DNA. Further, we do not know whether methylation changes are necessary for the onset of transcriptional activity, or whether the changes are the result of transcriptional activity initiated by other means.

Nonetheless, there are now some observations indicating the importance of methylation in gene expression. For example, a gene encoding an enzyme that adds methyl groups to DNA is essential for development in mice; that is, mice homozygous for mutations in that gene die at an early stage.

KEYNOTE

The DNA of many eukaryotes has been shown to be methylated at a certain proportion of the bases, with most methylations occurring on cytosine residues. Some transcriptionally active genes exhibit lower levels of DNA methylation than transcriptionally inactive genes.

~ FIGURE 16.2

Production of 5-methylcytosine from cytosine in DNA by the action of the enzyme DNA methylase.

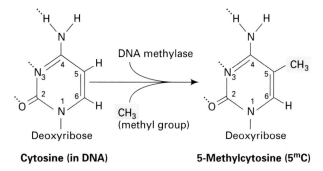

Cytosine (in DNA)   **5-Methylcytosine (5^mC)**

EXAMPLES OF TRANSCRIPTIONAL CONTROL OF GENE EXPRESSION IN EUKARYOTES. In this section we look at some examples of the short-term regulation of gene expression in eukaryotes. This regulation involves rapid responses to changes in environmental or physiological conditions. The responses occur at the transcriptional level and/or at the level of post-translational modulation of protein activity. We concentrate on the former here.

Galactose-Utilizing Genes of Yeast. In the galactose-utilizing system of yeast, three genes, *GAL1*, *GAL7*, and *GAL10*, encode enzymes that function in a pathway to metabolize the sugar galactose (Figure 16.3a). The enzymes are galactokinase (*GAL1* gene), galactose transferase (*GAL7* gene), and galactose epimerase (*GAL10* gene). (Note that genes are in italics, while the proteins they encode are not.) Once D-glucose 6-phosphate is made, it enters the glyco-

lytic pathway. In the absence of galactose, the *GAL* genes are not transcribed. When galactose is added, there is a rapid, coordinate induction of transcription of the *GAL* genes, and therefore a rapid production of the three galactose-utilizing enzymes.

The *GAL1*, *GAL7*, and *GAL10* genes are located near each other but do not constitute an operon (Figure 16.3b). An unlinked regulatory gene, *GAL4*, encodes the GAL4 protein, which has a zinc finger DNA-binding domain and a domain for activating transcription. GAL4 binds to a specific sequence between the *GAL1* and *GAL10* genes called *upstream activator sequence–galactose* (UAS$_G$); this sequence is a promoter element and consists of four similar, 17-bp sequences, each of which binds a dimer of GAL4. Transcription occurs in both directions from the UAS$_G$ (see Figure 16.3b).

A model for the regulation of transcription of the *GAL* genes is presented in Figure 16.4. In the absence of galactose, GAL4 dimers are bound to UAS$_G$. Another protein, GAL80 (encoded by the *GAL80* gene), is bound to the domain of GAL4 that is needed to turn on transcription. No transcription can occur under these circumstances. When galactose is added, a metabolite of galactose binds to GAL80, and this causes GAL4 to become phosphorylated (phosphate groups are added to certain amino acids). This changes the shape of the GAL4-GAL80 complex so that GAL4 is changed to the activating form needed to turn on transcription of the *GAL* enzyme-coding genes. In this system, then, the GAL4 protein acts as a positive regulator (activator), and galactose is an effector molecule.

Steroid Hormone Regulation of Gene Expression in Animals. The cells of animals are not exposed to rapid changes in environment, as are cells of bacteria and of microbial eukaryotes. The environment to which most cells of animals are exposed, the extracellular fluid, is relatively constant in the nutrients, ions, and other molecules it supplies. The constancy of the cell's environment is, in part, maintained through the action of chemicals called *hormones*, which are secreted by various cells in response to signals and which circulate in the blood until they stimulate their target cells. Elaborate feedback loops control the amount of hormone secreted, thereby controlling the response so that appropriate levels of chemicals in the blood and tissues are maintained.

A hormone, then, is an effector molecule that is produced in low concentrations by one cell and causes a physiological response in another cell. Polypeptide hormones interact with receptors on the cell surface, and a signal is produced that is transduced into the cell. One signal transduction pathway involves a kinase (phosphorylation) cascade, leading to activation of transcription factors. Another signal transduction

~ FIGURE 16.3

The galactose-metabolizing pathway of yeast. (a) The steps catalyzed by the enzymes encoded by the *GAL* genes; (b) The organization of the *GAL* protein-coding genes. The arrows indicate the direction of transcription. (UDP = uridine diphosphate)

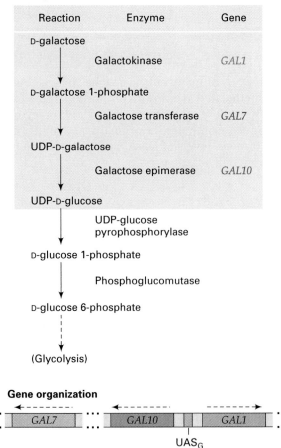

a) Pathway

Reaction	Enzyme	Gene
D-galactose		
	Galactokinase	*GAL1*
D-galactose 1-phosphate		
	Galactose transferase	*GAL7*
UDP-D-galactose		
	Galactose epimerase	*GAL10*
UDP-D-glucose		
	UDP-glucose pyrophosphorylase	
D-glucose 1-phosphate		
	Phosphoglucomutase	
D-glucose 6-phosphate		
(Glycolysis)		

b) Gene organization

GAL7   *GAL10*   *GAL1*

UAS$_G$

~ FIGURE 16.4

Model for the activation of the *GAL* genes of yeast.

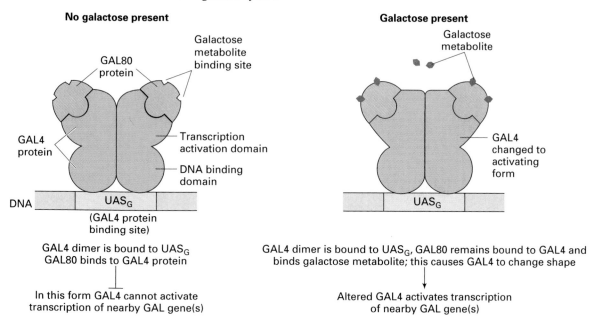

No galactose present

GAL80 protein

Galactose metabolite binding site

GAL4 protein

Transcription activation domain

DNA binding domain

DNA

UAS$_G$

(GAL4 protein binding site)

GAL4 dimer is bound to UAS$_G$
GAL80 binds to GAL4 protein

In this form GAL4 cannot activate
transcription of nearby GAL gene(s)

Galactose present

Galactose metabolite

GAL4 changed to activating form

UAS$_G$

GAL4 dimer is bound to UAS$_G$, GAL80 remains bound to GAL4 and
binds galactose metabolite; this causes GAL4 to change shape

Altered GAL4 activates transcription
of nearby GAL gene(s)

pathway commencing with a G protein leads to the activation of adenyl cyclase, which produces cyclic AMP (cAMP) from ATP. The cAMP acts as an intracellular signaling compound (called a *second messenger*) to activate enzymes controlled by the hormone.

A steroid hormone diffuses through the phospholipid bilayer of the plasma membrane and binds to a specific cytoplasmic receptor called a steroid hormone receptor (SHR), and then the complex binds directly to the target gene(s) to regulate gene expression. Steroid hormones influence the development and physiological regulation of organisms ranging from fungi to humans. Steroid hormones have tissue-specific effects. Estrogen, for example, induces the synthesis of the protein prolactin in the rat pituitary gland, and of the protein vitellogenin in the frog liver. Glucocorticoids induce the synthesis of growth hormone in the rat pituitary gland, and of the enzyme phosphoenolpyruvate carboxykinase in the rat kidney.

The specificity of the response to steroid hormones is controlled by the hormone receptors. That is, only target tissues that contain receptors for a particular steroid hormone can respond to that hormone, and this directly controls the sites of the hormone's action. With the exception of receptors for the steroid hormone glucocorticoid, which are widely distributed among tissue types, steroid receptors are found in a limited number of target tissues. While steroid hormones have well-substantiated effects on transcription, it also appears that steroids can affect the stability of mRNAs and possibly the processing of mRNA precursors.

Mammalian cells contain between 10,000 and 100,000 steroid hormone receptor (SHR) molecules. All steroid hormones work in the same general way. In the absence of a particular steroid hormone, the appropriate SHR is found in the cell associated with a large complex of proteins called *chaperones*, one of which is the 90-kDa heat shock protein, Hsp90. In this association, the SHR is inactive. When a steroid hormone such as glucocorticoid enters a cell, it binds to its specific SHR molecule, displacing Hsp90 and forming a glucocorticoid-SHR complex (Figure 16.5). The steroid-activated receptor complex now binds to specific DNA regulatory sequences, activating and/or repressing the transcription of the specific genes controlled by the hormone. For genes turned on by the hormone, new mRNAs appear within minutes after a steroid hormone encounters its target cell, enabling new proteins to be produced rapidly.

All genes regulated by a specific steroid hormone have in common a DNA sequence to which the steroid-receptor complex binds. The binding region is called a **steroid hormone response element** or **HRE**. The H in the acronym is replaced with another letter to indicate the specific steroid involved. Thus, GRE is the glucocorticoid response element, and ERE is the estrogen response element. HREs are located, often in multiple copies, in the enhancer regions of genes. GRE, for example, is located about 250 bp upstream from the transcription starting point and has the consensus sequence 5'-AGAACANNNTGTTCT-3', where N is any nucleotide.

~ **FIGURE 16.5**

Model for the action of the steroid hormone glucocorticoid in mammalian cells.

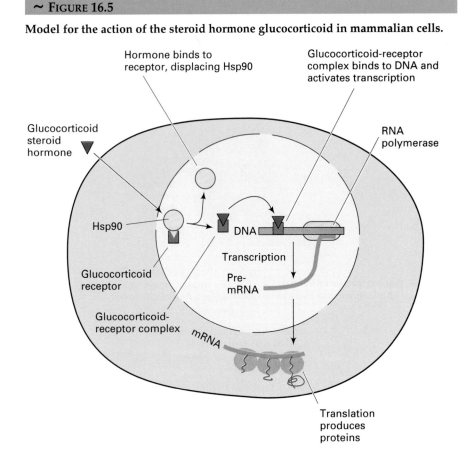

How the hormone-receptor complexes, once bound to the correct HREs, regulate transcriptional levels is not completely known. Potentially, the hormone-receptor complexes interact with transcription factors in the transcription initiation complex and thereby facilitate the initiation of transcription by RNA polymerase II. The unique action of each type of steroid results from the different receptor proteins and HREs involved.

In different types of cells the same steroid hormone may activate different sets of genes, even though the various cells have the same SHR. This is because a steroid-receptor complex can activate a gene only if the correct array of other regulatory proteins is present. Since the other regulatory proteins are specific for the cell type, different patterns of gene expression can result.

In sum, steroid hormones act as positive effector molecules, and SHRs act as regulatory molecules. When the two combine, the resulting complex binds to DNA and regulates gene transcription, resulting in a large (and specific) increase or decrease in cellular mRNA concentrations.

KEYNOTE

In higher eukaryotes, one of the well-studied systems of short-term gene regulation is the control of enzyme synthesis by steroid hormones. Steroid hormones exert their action by forming a complex with a specific receptor protein, thereby activating the receptor; the complex then binds directly to specific DNA regulatory sequences to control the expression of specific genes. The specificity of hormone action is caused by the presence of hormone receptors in only certain cell types and by interactions of steroid-receptor complexes with cell type-specific regulatory proteins.

RNA Processing Control

RNA processing control regulates the production of mature RNA molecules from precursor-RNA molecules. In Chapter 12 we discussed the synthesis of pre-mRNA and its processing to mature mRNA. When we examine living systems, we do not always see the "textbook" processing steps. There are many cases, for example, where *alternative polyadenylation* sites may be used to produce different pre-mRNA molecules, and *alternative splicing* may be used to produce different functional mRNAs. Which product is generated depends on regulatory signals. The outcome of alternative polyadenylation or alternative splicing are proteins that are encoded by the same gene but differ structurally and functionally. Such proteins are called *protein isoforms,* and their synthesis may be tissue specific. Alternative polyadenylation is independent of alternative splicing.

Figure 16.6 shows an example of how alternative polyadenylation *and* alternative splicing result in tissue-specific products of the human calcitonin gene (*CALC*). *CALC* consists of five exons and four introns. This gene is transcribed in certain cells of the thyroid gland and in certain neurons of the brain. Alternative polyadenylation occurs with the polyadenylation site next to exon 4, pA_1, used in thyroid cells, and the polyadenylation site next to exon 5, pA_2, used in the neuronal cells (Figure 16.6, part 1).

Alternative splicing occurs at the next stage of intron removal (Figure 16.6, part 2). The pre-mRNA in the thyroid is spliced to remove the three introns and bring together exons 1, 2, 3, and 4. The pre-mRNA in

~ **FIGURE 16.6**

Alternative polyadenylation and alternative splicing resulting in tissue-specific products of the human calcitonin gene, *CALC.* In the thyroid gland, calcitonin is produced, while in certain neurons CGRP (calcitonin-gene related peptide) is produced.

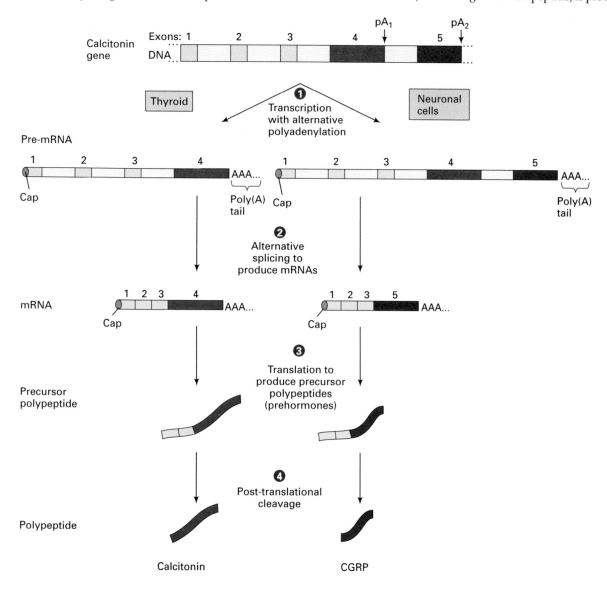

the neuronal cells is spliced to remove introns and to bring together introns 1, 2, 3, and 5; exon 4 is excised and discarded. The mRNAs produced are translated to produce precursor polypeptides (prehormones; Figure 16.6, part 3) from which the functional hormones are generated post-translationally by protease cleavage (Figure 16.6, part 4). The two products are calcitonin in the thyroid, with its amino acid sequence encoded by exon 4, and CGRP (calcitonin-gene related peptide), with its amino acid sequence encoded by part of exon 5. (The remainder of exon 5 is the 3' trailer part of the mRNA.) The mechanisms by which this alternative polyadenylation and alternative splicing occur are not known. The outcome is two different polypeptides encoded by the same gene synthesized in two different tissues. The thyroid hormone calcitonin is a circulating calcium-ion homeostatic hormone that aids the kidney in retaining calcium. CGRP is found in the hypothalamus and appears to have neuromodulatory and trophic (growth-promoting) activities.

KEYNOTE

Gene expression in eukaryotes can be regulated at the level of RNA processing. This type of regulation operates to determine the production of mature RNA molecules from precursor-RNA molecules. Two regulatory events that exemplify this level of control are alternative polyadenylation and alternative splicing. In both cases, different types of mRNAs are produced and different proteins result, typically in a tissue-specific manner.

mRNA Translation Control

Messenger RNA molecules are subject to **translational control** by ribosome selection among mRNAs. For example, mRNAs are stored in many vertebrate and invertebrate unfertilized eggs; in unfertilized eggs, the rate of protein synthesis is very slow but increases significantly after fertilization. This increase occurs without new mRNA synthesis, so translational control is responsible. The mechanisms involved are not completely understood, but typically, stored mRNAs are bound to proteins that both protect the mRNAs and inhibit their translation, and, in general, stored, inactive mRNAs have shorter poly(A) tails (15–90 As) than active mRNAs. Thus, mRNAs synthesized in growing oocytes that are destined for storage and later translation have short poly(A) tails.

In principle, an mRNA molecule can have a short poly(A) tail either because only a short string of A nucleotides was added at the time of polyadenylation or because a normal-length poly(A) tail was added and subsequently trimmed. At least for some mRNAs stored in growing oocytes of mice and frogs, the latter mechanism is involved. In one example, a pre-mRNA still in the process of intron removal has a long poly(A) tail (300–400 As), while the mature, stored mRNA has a short poly(A) tail (40–60 As). The decrease in length of the poly(A) tail for this mRNA occurs rapidly in the cytoplasm by a deadenylation enzyme. The rapid deadenylation is signaled by the *adenylate/uridylate (AU-rich) element* (ARE) in the 3' untranslated region (3' UTR) of the mRNA upstream of the AAUAAA polyadenylation site. Interestingly, to activate a stored mRNA in this class, a cytoplasmic polyadenylation enzyme recognizes the ARE and adds about 150 A nucleotides. Thus, the same sequence element is used to control poly(A) tail length, and therefore mRNA translatability, at different times and in opposite ways.

mRNA Degradation Control

In the cytoplasm, all RNA types are subjected to **degradation control**, in which the rate of RNA breakdown (also called RNA turnover) is regulated. Both rRNA (in ribosomes) and tRNA are very stable species, while mRNA molecules exhibit a range of stability from minutes to months. The stability of particular mRNA molecules may change in response to the presence of specific effector molecules (Table 16.1). This is accomplished by an increase in the rate of transcription of the gene(s) involved and/or an increase in the stability of the mRNAs produced.

mRNA degradation is a major control point in the regulation of gene expression in eukaryotes. Various sequences or structures have been shown to affect the half-lives of mRNAs, including the AU-rich elements (ARE) discussed earlier and various secondary structures. Two major mRNA decay pathways are the *deadenylation-dependent decay* and *deadenylation-independent decay pathways*. In the deadenylation-dependent decay pathway, the poly(A) tails are deadenylated until the tails are too short (5–15 As) to bind PAB (poly(A) binding protein). Once the tail is almost removed, the 5' cap structure is removed by an enzyme in a step called *decapping*. After an mRNA molecule is decapped, it is degraded from the 5' end by a 5'-to-3' exoribonuclease.

In yeast, the decapping enzyme is encoded by the *DCP1* gene. In *dcp1* mutants, mRNA degradation still occurs by a deadenylation-independent decay path-

~ TABLE 16.1

Examples of Tissues or Cells in Which Regulation of mRNA Stability Occurs in Response to Specific Effector Molecules[a]

mRNA	TISSUE OR CELL	REGULATORY Y SIGNAL (=EFFECTOR MOLECULE)	HALF-LIFE OF mRNA	
			WITH EFFECTOR	WITHOUT EFFECTOR
Vitellogenin	Liver (frog)	Estrogen	500 h	16 h
Vitellogenin	Liver (hen)	Estrogen	~24 h	<3 h
Apo-very low density lipoprotein (apoVLDL)	Liver (hen)	Estrogen	~20–24 h	<3 h
Ovalbumin, conalbumin	Oviduct (hen)	Estrogen, progesterone	>24 h	2–5 h
Casein	Mammary gland (rat)	Prolactin	92 h	5 h
Prostatic steroid-binding protein	Prostate (rat)	Androgen	Increases 30X	

[a]Note that the effector molecule in each case results in an increase in transcription as well as stabilization of the mRNA.

way. In one such pathway, mRNAs are decapped without being deadenylated, thereby exposing them to rapid degradation by 5'-to-3' exonucleases, and in another such pathway they are cleaved internally by endonucleases without being deadenylated, and then broken down further.

Protein Degradation Control

Regulatory mechanisms also exist at the post-translational level to determine the lifetime of a protein. There are many ways to regulate the amount of a particular protein in a cell. For example, a constitutively produced mRNA may be continuously translated, with the level of protein product controlled by the degradation rate of that protein. At another extreme, a short-lived mRNA may encode a protein that is very stable. Proteins in the lenses of higher vertebrate eyes, for example, fall into the latter situation. Their mRNAs have long since degraded, while the protein itself persists, usually for an individual's lifetime. On the other hand, steroid receptors and heat-shock proteins—the products themselves of relatively unstable mRNAs—have relatively short half-lives.

Degradation of proteins *(proteolysis)* in eukaryotes has been shown to require the protein cofactor *ubiquitin* (a protein apparently found ubiquitously). The binding of ubiquitin to a protein identifies it for degradation by proteolytic enzymes. Ubiquitin is released intact during the degradation process, enabling it to tag other proteins for degradation.

KEYNOTE

Gene expression is regulated also by mRNA translation control and by mRNA degradation control. The latter is a major control point in the regulation of gene expression. Structural features of individual mRNAs have been shown to be responsible for the range of mRNA degradation rates.

In sum, in eukaryotes, gene expression is regulated at transcriptional, post-transcriptional, and post-translational levels. There are regulatory systems for the control of transcription, precursor-RNA processing, transport out of the nucleus, degradation of mature RNA species, translation of the mRNA, and degradation of the protein product. The intertwining of regulatory events at these different levels leads to the fine tuning of the amount of the controlled protein in the cell.

GENE REGULATION IN DEVELOPMENT AND DIFFERENTIATION

During a eukaryotic organism's growth throughout its life cycle, cells that were genetically identical become physiologically and phenotypically differentiated in terms of their structure and function.

Two terms are used in the description of long-term gene regulation. **Development** refers to the process of regulated growth that results from the genome's interaction with cytoplasm and the external environment, and that involves a programmed sequence of phenotypic events that are typically irreversible. The total of the phenotypic changes constitutes the life cycle of an organism. **Differentiation**, the most spectacular aspect of development, involves the formation of different types of cells, tissues, and organs through the processes of specific regulation of gene expression; differentiated cells have characteristic structural and functional properties.

At a very general level, the processes involved in differentiation and development are the result of a highly programmed pattern of gene activation and gene repression. However, we are a long way from understanding the molecular bases for differentiation and development events. In the following sections we discuss selected aspects of gene regulation in development and differentiation.

Constancy of DNA in the Genome During Development

In early studies, a fundamental question was whether differentiation and development involve a *loss* of DNA, or whether these processes are the result of a programmed sequence of gene activation and repression events in a constant genome. In other words, is the adult genome identical to the zygote genome? Two studies addressing this question in plants and animals are described next.

REGENERATION OF CARROT PLANTS FROM MATURE SINGLE CELLS.   In the 1950s, Frederick Steward dissociated the tissue of a carrot to separate the cells, and then attempted to culture new carrot plants from those cells by using plant tissue culture techniques (Figure 16.7). Mature plants with edible carrots were successfully produced (cloned) from phloem cells. That the mature cells had the potential to act as zygotes and develop into complete plants indicated that the DNA content of a cell remains constant during development and differentiation. Regeneration of plants from tissues by somatic embryogenesis or organogenesis is now commonplace.

CLONING MAMMALS.   Researchers also wished to determine whether the DNA content of animal cells remained constant during development. That question was answered in the affirmative in 1997 when Ian Wilmut and his colleagues reported the cloning of a mammal (sheep) starting with an adult cell. This means that an adult nucleus has all the genetic information necessary to direct development again from the start—the nucleus is said to be totipotent, or to demonstrate **totipotency**.

Wilmut's experiment, illustrated in Figure 16.8, was as follows:

1. Mammary epithelial cells from a donor white-faced Finn Dorset ewe were grown in tissue culture, and then induced to enter the quiescent state (the G_0 phase of the cell cycle) by reducing the concentration of the growth serum.

2. The cells were fused with enucleated oocytes (egg cells), and the fusion cells were allowed to grow and divide for six days to produce embryos.

3. The embryos were implanted into black-faced recipient ewes.

One of 277 adult mammary epithelium-derived cells gave rise to a live lamb. That lamb, designated

---

~ FIGURE 16.7

Cloning of a mature carrot plant from a cell of a mature carrot.

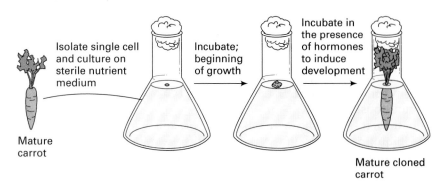

~ **FIGURE 16.8**

Representation of Wilmut's sheep cloning experiment, which showed the totipotency of the nucleus of a differentiated, adult cell.

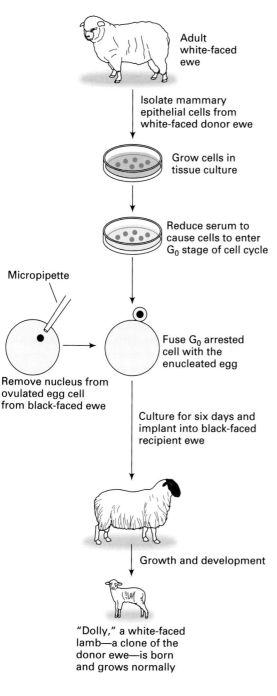

Adult white-faced ewe

Isolate mammary epithelial cells from white-faced donor ewe

Grow cells in tissue culture

Reduce serum to cause cells to enter G_0 stage of cell cycle

Micropipette

Fuse G_0 arrested cell with the enucleated egg

Remove nucleus from ovulated egg cell from black-faced ewe

Culture for six days and implant into black-faced recipient ewe

Growth and development

"Dolly," a white-faced lamb—a clone of the donor ewe—is born and grows normally

cell contained the nucleus from a white-faced Finn Dorset ewe and was implanted into a Scottish Blackface (black-faced) recipient ewe; Dolly is morphologically Finn Dorset. Second, and more definitive, DNA markers at four loci perfectly matched those of the donor mammary epithelial cells and not those of the recipient ewe.

Using a different procedure, Ryuzo Yanagimachi in 1998 cloned a large number of mice. In this procedure, the nuclei from ovary cells of brown mice were injected into enucleated eggs of black mice and then implanted into white surrogate mothers. The mice produced were brown, that is, clones of the nuclear donor. Yanagimachi and colleagues continued the cloning to produce a second generation, and then a third, for a total of 50 cloned mice. Also in 1998, Japanese researchers successfully cloned cows, so it seems possible to clone any mammal.

KEYNOTE

Long-term regulation is involved in controlling the events that activate and repress genes during development and differentiation. Development and differentiation result from differential gene activity in a genome that contains a constant amount of DNA, from the zygote stage to the mature organism stage. Thus, these events do not result from a loss of genetic information.

Differential Gene Activity Among Tissues and During Development

Specialized cell types have different cell morphologies. Since the amount of DNA typically remains constant during development, the simplest hypothesis is that the phenotypic differences between different cell types reflect differential gene activity. The following discussion presents some of the evidence supporting this hypothesis.

HEMOGLOBIN TYPES AND HUMAN DEVELOPMENT. Human adult hemoglobin Hb-A is made up of two α and two β polypeptides; each type of polypeptide is coded by a separate gene. The two genes appear to have arisen during evolution by duplication of a single ancestral gene, followed by alteration of the sequences in each gene.

Hemoglobin Hb-A is only one type of human hemoglobin. In fact, several distinct genes code for α- and β-like globin polypeptides that form different

6LL3 and named Dolly, is perfectly healthy; she progressed normally to sexual maturity and had a lamb herself in the summer of 1998.

Evidence that Dolly was truly the result of the cell fusion experiment is of two kinds. First, the fusion

~ FIGURE 16.9

Comparison of synthesis of different globin chains at given stages of embryonic, fetal, and postnatal development.

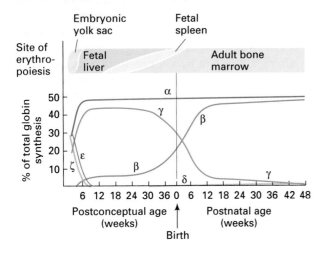

types of hemoglobin at different times during human development (Figure 16.9). In the human embryo the hemoglobin made in the yolk sac consists of two ζ (zeta) polypeptides and two ε (epsilon) polypeptides; ζ is an α-like polypeptide and ε is a β-like polypeptide. After about three months of development, synthesis of embryonic hemoglobin ceases, and the site of hemoglobin synthesis shifts to the fetal liver and spleen. Here, *fetal hemoglobin* (Hb-F) is made. Hb-F consists of two α polypeptides and two β-like γ (gamma) polypeptides, either two γA or two γG. γA and γG differ from each other by only one amino acid (out of 146 amino acids) and are each coded for by distinct genes.

Fetal hemoglobin is made until just before birth, when the site of hemoglobin synthesis switches to the bone marrow. In that tissue, α and β polypeptides are made, along with some β-like δ (delta) polypeptides. In the newborn through adult human, most of the hemoglobin is the $\alpha_2\beta_2$ tetramer (Hb-A), with about one in 40 molecules having the constitution $\alpha_2\delta_2$ (Hb-A2). Thus, globin gene expression switches during human development, and this switching must involve a sophisticated gene regulatory system that turns appropriate globin genes on and off over a long time period.

In the genome, the α-like genes (two α genes and one ζ gene) are all on chromosome 16, and the β-like genes (ε, γA, γG, δ, and β) are all on chromosome 11 (Figure 16.10). Between the ζ and α genes on chromosome 16 are three sequences, one of which has a nucleotide sequence that is very similar to the ζ sequence, and the other two of which closely resemble the α sequence. However, these sequences are inactive—they are example of *pseudogenes*. The β-like gene cluster has a pseudogene located between the γ genes and the δ gene.

Immunogenetics and Chromosome Rearrangements During Development

In this section we discuss how antibodies, the proteins that mediate the humoral immune response, are encoded in DNA. In this system, DNA rearrangements produce the genes from which antibody polypeptides are transcribed. These rearrangements are examples of *loss* of genetic information during development.

~ FIGURE 16.10

Linkage maps of human globin gene clusters. The function, if any, of the α-globin ϕ1 gene is not known. (Map is not to scale.)

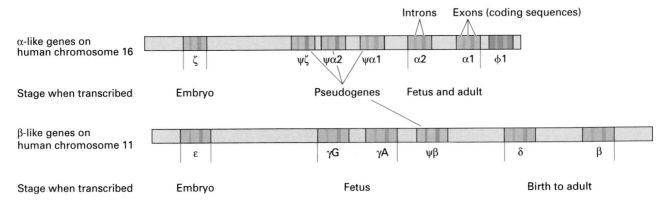

SYNOPSIS OF THE IMMUNE SYSTEM.

All vertebrates have an *immune system*. The immune system provides protection against infectious agents such as viruses, bacteria, fungi, and protozoans. Any substance that elicits an immune response in a foreign host is called an **antigen** (*anti*body *gen*erator). When activated by an antigen, B cells (a type of lymphocyte) develop into plasma cells that make **antibodies**; that is, specialized proteins called **immunoglobulins** (Igs). The antibodies bind specifically to the antigens that stimulated their production. Typically, the immune response involves the binding of many different antibodies to an array of antigens on such invaders as an infecting virus or bacterium. This binding mediates a variety of other mechanisms that inactivate the invading antigens.

ANTIBODY MOLECULES.

All antibody molecules made by a given plasma cell are identical; that is, they have the same protein chains and bind the same antigen. Because there are millions of different B cells in an organism, millions of different antibody types can be produced, each with a different amino acid sequence and a different antigen-binding specificity.

A stylized IgG antibody (immunoglobulin) molecule is shown in Figure 16.11a, and a molecular model of an antibody molecule is shown in Figure 16.11b. Both figures show the molecule's two short polypeptide chains, called *light (L) chains*, and two long polypeptide chains, called *heavy (H) chains*. The two H chains are held together by disulfide (–S–S–) bonds, and an L chain is bonded to each H chain by disulfide bonds. Other disulfide bonds within each L and H chain cause the chains to fold up into their characteristic shapes.

The overall structure resembles a Y, with the two arms containing the two antigen-binding sites. The two L chains in each Ig molecule are identical, as are the two H chains, so the two antigen-binding sites are identical. The hinge region (see Figure 16.11a) allows the two arms to move in space, making it easier for the antibody to bind an antigen. Also, one arm can then bind an antigen on, say, one virus, while the other arm binds the same antigen on a different virus. Such crosslinking of antibody molecules in solution helps inactivate infecting agents.

In mammals there are five major classes of antibodies: IgA, IgD, IgE, IgG, and IgM. They have different H chains—α (alpha), δ (delta), ϵ (epsilon), γ (gamma), and μ (mu), respectively. Two types of L chains are found: κ (kappa) and λ (lambda). Both L chain types are found in all Ig classes, but a given antibody molecule will have either two identical κ chains or two identical λ chains. A full discussion of the antibody types is beyond the scope of this text.

For our purposes, the most abundant class of immunoglobulin in the blood is IgG, and IgM plays an important role in the early stages of an antibody response to a previously unrecognized antigen.

Each polypeptide chain in an antibody is organized into domains of about 110 amino acids (see Figure 16.11a). Each L chain (κ or λ) has two domains, and the H chain of IgG (the γ chain) has four domains, while the IgM's H chain (the μ chain) has five domains. The N-terminal domains of the H and L chains have highly variable amino acid sequences that constitute the antigen-binding sites. These domains, representing in IgG the N-terminal half of the L chain and the N-terminal quarter of the H chain, are termed the *variable*, or *V, regions*. The V regions are symbolized generically as V_L (for the light chain) and V_H (for the heavy chain). The V_L and V_H regions together constitute the antigen-binding sites (see Figure 16.11a). The amino acid sequence of the rest of the L chain is constant for all antibodies (with the same L chain type, κ or λ) and is termed C_L. Similarly, the amino acid sequence of the rest of the H chain is constant and is termed C_H. For IgG, there are three approximately equal domains of C_H called C_H1, C_H2, and C_H3 (see Figure 16.11a). Thus, the production of antibody molecules involves synthesizing polypeptide chains, one part of which varies from molecule to molecule, and the other part of which is constant. How this occurs is discussed in the following section.

ASSEMBLY OF ANTIBODY GENES FROM GENE SEGMENTS DURING B CELL DEVELOPMENT.

A mammal produces 10^6 to 10^8 different antibodies. Since each antibody molecule consists of one kind of L chain and one kind of H chain, these antibodies theoretically would require 10^3 to 10^4 different L chains and 10^3 to 10^4 different H chains, and hence 10^6 to 10^8 genes, if L and H chains paired randomly. The dilemma is that the human genome is thought to contain perhaps a total of 10^5 genes. In fact, the variability in L and H chains results from particular DNA rearrangements that occur during B cell development. These rearrangements involve the joining of different gene segments to form a gene that is transcribed to produce an Ig chain; the process is called *somatic recombination*, and the many permutations of recombinations that are possible are largely responsible for the diversity of antibody structure. The process will now be illustrated for mouse immunoglobulin chains.

Light Chain Gene Recombination.   In mouse germline DNA, three types of gene segments encode the κ light chain (Figure 16.12):

~ FIGURE 16.11

IgG antibody molecule. (a) Diagram showing the two heavy and two light chains and the antigen-binding sites. The heavy and light chains are held together by disulfide bonds. (b) Model of IgG antibody molecule.

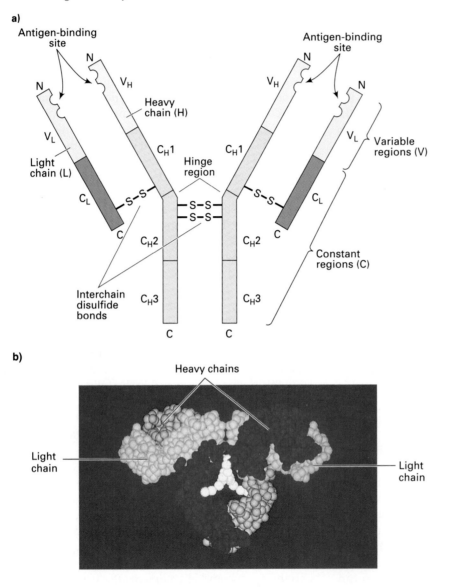

1. There are 350 or so L-V$_\kappa$ segments consisting of a leader sequence (L) and a sequence, V$_\kappa$, that varies from segment to segment. Each V$_\kappa$ segment encodes most of the amino acids of the light chain variable region. The leader sequence encodes a signal sequence (see Chapter 13) required for secretion of the Ig molecule.

2. There are four J$_\kappa$ joining segments used to join V$_\kappa$ and C$_\kappa$ segments in the production of a functional κ light chain gene.

3. There is one C$_\kappa$ segment that specifies the constant region of the κ light chain.

In the pre–B cell, the L-V$_\kappa$, J$_\kappa$, and C$_\kappa$ segments, in that order, are widely separated on the chromosome. As the B cell develops, a particular L-V$_\kappa$ segment becomes associated with one of the J$_\kappa$ segments and with the C$_\kappa$ segment (see Figure 16.12). In the example, L-V$_{\kappa 2}$ has recombined next to J$_{\kappa 3}$. Transcription of this new DNA arrangement produces the pre-mRNA molecule. Processing of the pre-mRNA, in this case to remove the sequence between the end of J$_{\kappa 3}$ and C$_\kappa$, produces the mature mRNA, which has the organization L-V$_{\kappa 2}$J$_{\kappa 3}$C$_\kappa$; translation and leader re-moval after secretion from the cell produces the κ light chain that the B cell is committed to make.

~ FIGURE 16.12

Production of the kappa (κ) light chain gene in mouse by recombination of V, J, and C gene segments during development. The rearrangement shown is only one of many possible recombinations.

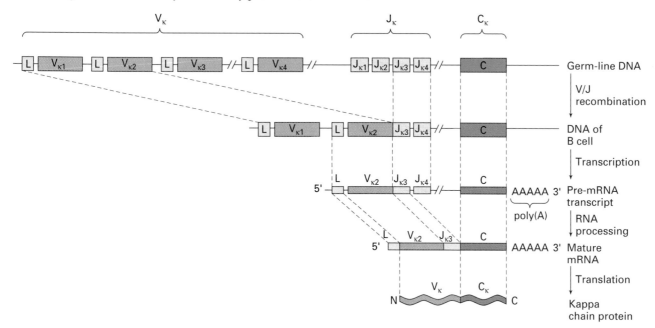

With 350 L-V_κ gene segments, four J_κ segments, and one C_κ gene segment, the number of possible κ chain variable regions that can be produced by this mechanism is $350 \times 4 \times 1 = 1{,}400$. Further diversity results from imprecise joining of the V_κ and J_κ gene segments. That is, during the joining process a few nucleotide pairs from V_κ and a few nucleotide pairs from J_κ are lost from the DNA at the $V_\kappa J_\kappa$ joint, generating significant diversity in sequence at that point.

A similar mechanism exists for mouse λ light chain gene assembly. In this case there are only two L-V_λ gene segments, and four J_λ gene segments each followed by its own C_λ gene segment. Thus, fewer λ variable regions can be produced than is the case for κ chains.

Heavy Chain Gene Recombination. The immunoglobulin heavy chain gene is also encoded by V_H, J_H, and C_H segments. In this case, additional diversity is provided by another gene segment, D (diversity), which is located between the V_H segments and the J_H segments. Heavy chain gene assembly is very similar to that described for the light chains. In the mouse, the C_H gene segments are arranged in the order μ, δ, γ, ε, and α for the H chain constant domains of IgM, IgD, IgG, IgE, and IgA, respectively. For γ there are four different sequences for the four different, but similar, IgG H chain constant regions. As in L chain gene rearrangements, further antibody diversity

results from imprecise joining of the gene segments that make up the chain's variable region.

KEYNOTE

Antibodies are specialized proteins called immunoglobulins, which bind specifically to antigens. Antibody molecules consist of two light (L) chains and two heavy (H) chains. The amino acid sequence of one region of each type of chain is variable; this variation is responsible for the different antigen-binding site on different antibody molecules. The other region(s) of each chain are constant in amino acid sequence. In germ-line DNA, the coding regions for immunoglobulin chains are scattered in tandem arrays of gene segments. Thus, for light chains, there are many variable region (V) gene segments, a few joining (J) gene segments, and one constant region (C) gene segment. During development, somatic recombination occurs to bring particular gene segments together into a functional L chain gene. A large number of different L chain genes results from the many possible ways in which the gene segments can recombine. Similar rearrangements occur for H chain genes, but with the addition of several D (diversity) segments that are between V and J, which increases the possible diversity of H chain genes.

GENETIC REGULATION OF DEVELOPMENT IN *DROSOPHILA*

For many years the fruit fly *Drosophila melanogaster* has been a prominent research organism for geneticists. In the several decades of classical genetics research with *Drosophila*, many mutations affecting development were identified. More recently, other organisms have become model systems for relating gene activity to developmental processes. In the nematode worm *Caenorhabditis elegans* (*C. elegans*) the genome is relatively small (8×10^7 bp), genetic crosses and selfings can be done readily, and the body of the worm is transparent (enabling all cells to be seen under the microscope). In addition, development of *C. elegans* is programmed in an invariant way—that is, each adult has a set number of cells, each of which is traceable back to the zygote along particular cell pathways or *cell lineages*. In the zebrafish, *Brachydanio rerio*, geneticists can easily observe several developmental steps in the transparent embryos. Genetic crosses can be made, large numbers of fish can be bred, and techniques have been developed to genetically screen the organism in an attempt to identify all genes that affect embryogenesis. In the plant *Arabidopsis thaliana*, individuals are small, making it possible to do genetic crosses and analyze large numbers of progeny. Genetic screens have isolated many mutations that affect the development, for example, of flowers and roots.

Nevertheless, *Drosophila* remains a system in which significant progress has been made. Many more developmental mutants have been isolated following extensive genetic screens so that virtually all areas of *Drosophila* development can be studied genetically and molecularly in detail. The discoveries made from such studies have become even more important as discoveries in other systems indicate that the genes discovered in *Drosophila* have counterparts in all highly complex organisms, including humans. This implies, of course, that the same mechanisms that control development in *Drosophila* are used in higher organisms as well. We focus on the genetics of *Drosophila* development in this section, providing a brief overview of what is known.

Drosophila Developmental Stages

About 24 hours after fertilization, a *Drosophila* egg hatches into a larva, which undergoes three molts, after which it is called a pupa (Figure 16.13). The pupa metamorphoses into an adult fly. The whole

~ **FIGURE 16.13**

Development of an adult *Drosophila* from a fertilized egg.

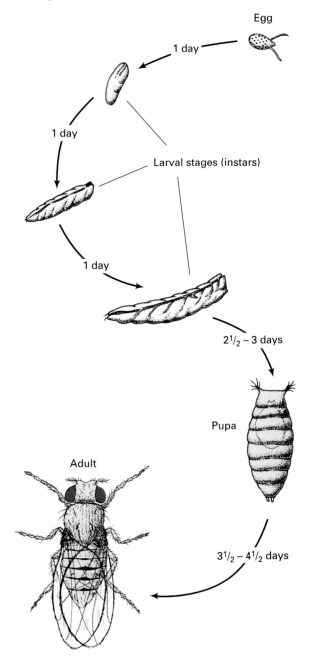

process from egg to adult fly takes about 10 to 12 days at 25°C.

Embryonic Development

Before a mature egg is fertilized, concentration gradients of particular proteins are established within it. The posterior end is indicated by the presence of a region called the *polar cytoplasm* (Figure 16.14a). At fertilization, the two parental nuclei are roughly centrally located in the egg. The two nuclei fuse to produce a 2N zygote nucleus (Figure 16.14b). For the first nine divisions, only the nuclei divide in a common cytoplasm—cytokinesis does not occur— to produce a multinucleate *syncytium* (Figure 16.14c). (After seven divisions, some nuclei migrate into the polar cytoplasm, where they become precursors to germ-line cells.) Next, the nuclei migrate and divide to form a layer at the surface of the egg, producing the *syncytial blastoderm* (Figure 16.14d). After four more divisions, membranes form around the nuclei to produce somatic cells, about 4,000 of which make up the *cellular blastoderm* (Figure 16.14e).

Subsequent development of body structures depends on two processes (Figure 16.15):

1. Concentration gradients of particular proteins are produced along the anterior-posterior axis and the dorsal-ventral axis of the egg. Somehow, a nucleus is aware of its position with reference to the molecular concentration in the two gradients.

2. Regions are determined in the embryo that correspond to adult body segments. In the cellular blastoderm stage, where the regions are not well defined, they are called *parasegments*. In subsequent stages, where they are visible, they are called *segments*, forming a striped pattern along the anterior-posterior axis of the embryo. The embryonic segments give rise to the segments of the adult fly.

Genes involved in regulating *Drosophila* development are defined by mutations that have a lethal phenotype early in development or that result in the development of abnormal structures (such as embryos with abnormal striping or two anterior ends, etc.).

How does the polarity of the *Drosophila* egg specify the segments of the adult fly body? Three major classes of developmental genes are involved: maternal effect genes, segmentation genes, and homeotic genes.

Maternal effect genes are expressed by the mother during oogenesis. The key genes are *bicoid*, *nanos*, and *torso*, which regulate the formation of the anterior, posterior, and terminal structures, respectively. The wild-type *bicoid* gene encodes a protein that is a morphogen; that is, it exerts an effect on the processes— collectively called *morphogenesis*—through which embryos change their form and give rise to an adult

~ **FIGURE 16.14**

Embryonic development in *Drosophila*.

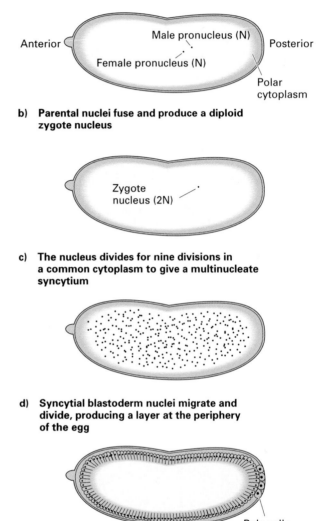

a) **Fertilized egg with two parental nuclei**

b) **Parental nuclei fuse and produce a diploid zygote nucleus**

c) **The nucleus divides for nine divisions in a common cytoplasm to give a multinucleate syncytium**

d) **Syncytial blastoderm nuclei migrate and divide, producing a layer at the periphery of the egg**

e) **Cellular blastoderm. Nuclei divide four times; membranes form around them and produce somatic cells**

organism. The *bicoid* gene is transcribed in the mother during oogenesis, and the resulting mRNA is deposited into the oocyte, where it becomes localized to the anterior pole. Translation of the mRNA occurs after

~ FIGURE 16.15

Drosophila development results from gradients in the egg that define parasegments in the cellular blastoderm, and segments in the embryo and adult. The adult segment organization directly reflects the segment pattern of the embryo.

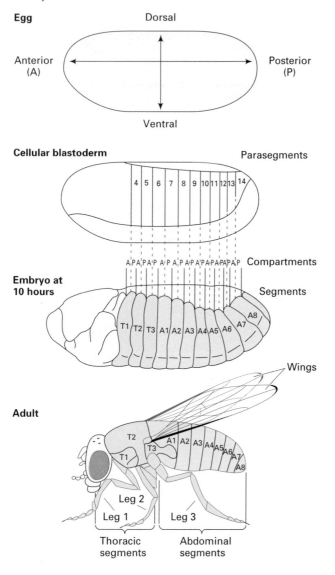

Similarly, *nanos* mRNAs are localized to the posterior end of the egg. Translation of those mRNAs produces the NANOS protein, which forms a posterior-to-anterior gradient and acts as a morphogen to direct abdomen formation. Mutations in *nanos* result in a no-abdomen phenotype.

The *torso* gene controls the formation of the terminal structures at the anterior and posterior ends of the embryo. The *torso* gene is transcribed and translated during oogenesis. The TORSO protein is distributed throughout the egg but becomes activated only at the two termini. Mutations in *torso* result in a no-terminals phenotype.

Next, about 25 *segmentation genes* act to subdivide the embryo into regions called segments, each of which is divided into anterior (A) and posterior (P) compartments (see Figure 16.15). Mutations in segmentation genes alter the number of segments or their internal organization but do not affect the overall organizational polarity of the egg. Three classes of segmentation genes are recognized based on their mutant phenotypes (Figure 16.16). Mutations in *gap genes* (such as *Krüppel* and *hunchback*) result in the deletion of several adjacent segments; mutations in *pair-rule genes* (such as *even-skipped* and *fushi tarazu*) result in the deletion of the same part of the pattern in every other segment; and mutations in *segment polarity genes* (such as *gooseberry* and *engrailed*) have portions of segments replaced by mirror images of adjacent half segments.

The three classes of segmentation genes have different roles in specifying regions of the embryo. Gap genes are activated or repressed by maternal effect genes. The gap gene products lead to an organization of the embryo into broad regions, each of which covers areas that will later develop into several distinct segments. Critical to this developmental step is expression of the *hunchback* gene, which is stimulated by the BICOID protein and inhibited by the NANOS protein. This generates a gradient of HUNCHBACK protein, with its highest concentration at the anterior end of the embryo. Next, through the transcription-regulating action of the gap genes, the pair-rule genes are expressed, leading to a division of the embryo into a number of regions (see Figure 16.15). The transcription factors encoded by the pair-rule genes regulate the expression of the segment polarity genes, which determines regions that will become the segments seen in larvae and adults.

Once the segmentation pattern has been determined, a major class of genes called the *homeotic* (structure-determining) *genes* specifies the identity of each segment with respect to the body part that will develop at metamorphosis (discussed later).

egg deposition. The BICOID protein produced forms a gradient with its highest concentration at the anterior end of the egg, fading to nothing in the posterior third of the egg. Mutations in the *bicoid* gene result in the conversion of head and thorax to abdomen, and terminal posterior structures are present at both ends.

~ FIGURE 16.16

Functions of segmentation genes as defined by mutations.

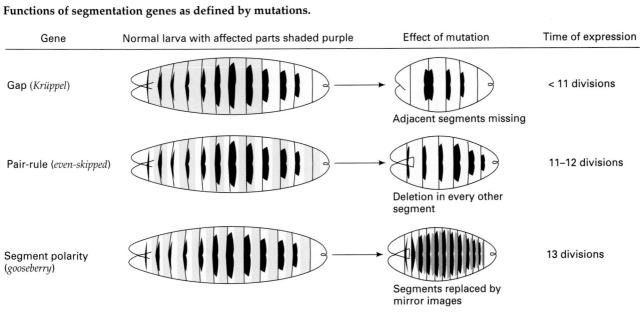

Gene	Normal larva with affected parts shaded purple	Effect of mutation	Time of expression
Gap (*Krüppel*)		Adjacent segments missing	< 11 divisions
Pair-rule (*even-skipped*)		Deletion in every other segment	11–12 divisions
Segment polarity (*gooseberry*)		Segments replaced by mirror images	13 divisions

KEYNOTE

Development of *Drosophila* body structures results from concentration gradients of particular proteins along the posterior-anterior and dorsal-ventral axes of the egg and from the subsequent determination of regions in the embryo that directly correspond to adult body segments. As defined by mutations, genes control *Drosophila* development in a regulatory cascade. First, maternal effect genes specify the gradients in the egg, then segmentation genes (gap genes, pair-rule genes, and segment polarity genes) determine the segments of the embryo and adult, and homeotic genes next specify the identity of the segments.

Imaginal Discs

During embryogenesis, certain groups of cells are defined that form larval structures called **imaginal discs** (*imago* means imitate). Imaginal discs are characteristic of metamorphic insects. When an imaginal disc consists of about 20 to 50 cells, it is already programmed to specify its given adult structure—its fate is determined. From then on the number of cells in each disc increases by mitotic division, until by the end of the larval stages there are many thousand cells per disc. Each imaginal disc differentiates into a specific part of the adult fly, including mouth parts, antennae, eyes, wings, halteres, legs, and the external genitalia. Other structures, such as the nervous system and gut, do not develop from imaginal discs. Figure 16.17 shows the positions of some imaginal discs in a mature larva and the adult structures that develop from them.

Homeotic Genes

Once the basic segmentation pattern has been laid down, homeotic genes give a specific developmental identity to each of the segments. The homeotic genes have been defined by mutations that affect the development of the fly. That is, **homeotic mutations** alter the identity of particular segments, transforming them into copies of other segments.

Pioneering studies by Edward Lewis were done on a cluster of homeotic genes called the *bithorax* complex (*BX-C*). *BX-C* determines the posterior identity of the fly, namely thoracic segment T3 and abdominal segments A1–A8. *BX-C* contains three genes called *Ultrabithorax* (*Ubx*), *abdominal-A* (*abd-A*), and *Abdominal-B* (*Abd-B*). Many, but not all, mutations in these homeotic genes are lethal, preventing the fly from developing past embryogenesis. The viable mutations have allowed the functions of this complex to be examined.

~ FIGURE 16.17

Locations of imaginal discs in a mature *Drosophila* larva and the adult structures derived from each disc.

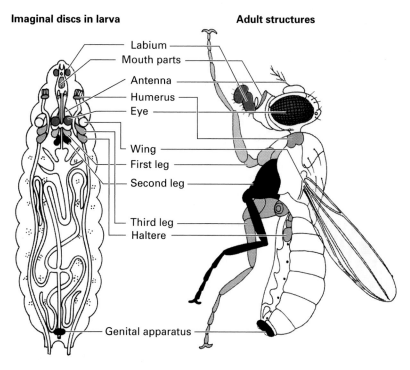

Imaginal discs in larva   **Adult structures**

Labium
Mouth parts
Antenna
Humerus
Eye
Wing
First leg
Second leg
Third leg
Haltere
Genital apparatus

Figure 16.18a shows the segments of a normal adult fly: the wings are located on segment T2, while the pair of halteres (rudimentary wings used as "balancers" in flight) are on segment T3. A photograph of normal adult fly is presented in Figure 16.18b. Figure 16.18c shows one type of developmental abnormality that can result from nonlethal homeotic mutations in *BX-C*: this fly is homozygous for three separate mutant alleles of the *Ubx* gene, designated *abx*, *bx3*, and *pbx*. Collectively these mutations transform segment T3 into an adult structure like that of T2, that is, a pair of wings. The fly lacks halteres, however, because no normal T3 segment is present.

Another well-studied group of mutations defines another large cluster of homeotic genes called the *Antennapedia* complex (*ANT-C*). *ANT-C* determines the anterior identity of the fly, namely, the head and segments T1 and T2. *ANT-C* contains at least four genes, called *Deformed* (*Dfd*), *fushi tarazu* (*ftz*), *Sex combs reduced* (*Scr*), and *Antennapedia* (*Antp*). Most *ANT-C* mutations are lethal. Among the nonlethal mutations is a group of mutations in *Antp* that result in a leg instead of an antenna growing out of the cells near the eye during the development of the eye disc.

The homeotic gene complexes *ANT-C* and *BX-C*, therefore, encode products that control the normal

development of the relevant adult fly structures. The *Antennapedia* complex (*ANT-C*) and the *bithorax* complex (*BX-C*) have been cloned. Both complexes are very large. In *ANT-C*, for example, the *Antp* gene is 103 kb long, with many introns; this gene encodes a mature mRNA of only a few kilobase pairs. *BX-C* covers more than 300 kb of DNA and contains only three protein-coding regions amounting to about 50 kb of that DNA: *Ubx*, *abdA*, and *AbdB* (Figure 16.19). The other 250 kb of DNA in the complex is not transcribed and is believed to consist of transcription regulatory regions of significant size and complexity. These regulatory regions control the expression of the protein-coding genes.

The *ANT-C* and *BX-C* protein-coding genes have similar functions but are located in different places in the genome. However, these genes all contain a similar sequence of about 180 bp that has been named the **homeobox**. The homeobox is part of the protein-coding sequence of each gene, and the corresponding 60-amino-acid part of each protein is called the **homeodomain**.

Homeoboxes have been found in over 20 *Drosophila* genes, most of which regulate development. The homeodomain-containing proteins they encode bind to an 8-bp consensus DNA sequence

~ **FIGURE 16.18**

Abnormal adult structures that can result from *bithorax* mutations. (a) Drawing of a normal fly. The haltere (rudimentary wing) is on T3 (see Figure 16.15). (b) Photograph of a normal fly with a single set of wings. (c) Photograph of a fly homozygous for three mutant alleles (*bx3*, *abx*, and *pbx*) that results in the transformation of segment T3 into a structure like T2: namely, a segment with a pair of wings. These flies therefore have two sets of wings but no halteres.

a)

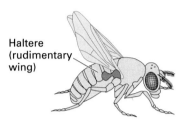

Haltere
(rudimentary
wing)

b)

c)

upstream of the genes regulated by the proteins, thereby regulating their transcription.

The complete set of homeotic genes and complexes in *Drosophila*—generically the *Hom* genes—consists of *iab*, *pb*, *Dfd*, *Scr*, *Antp*, *Ubx*, *abdA*, and *AbdB*. Most interestingly, these complexes are arranged in the same order along the chromosome as they are expressed along the anterior-posterior body axis: this is known as the *colinearity rule*. Homeotic gene complexes are found also in all major animal phyla with the exception of sponges and coelenterates, and in fungi. The homeobox sequences in the *Hom* genes are highly conserved, indicating common function in the wide range of organisms involved. As in *Drosophila*, the homeotic genes of vertebrates—the *Hox* genes—follow the colinearity rule. In mammals, for example, there are four clusters of homeotic genes designated *HoxA–D*. Each cluster is thought to have originated by duplication of a primordial gene cluster followed by subsequent evolutionary divergence. The patterns of *Hox* gene expression, the effects of mutations, and embryological analyses all indicate that the vertebrate genes have homeotic effects similar to *Drosophila* homeotic genes. Further, the studies indicate that the *Hox* genes specify the vertebrate body plan.

KEYNOTE

Homeotic genes in *Drosophila* determine the developmental identity of a segment, once the basic segmentation pattern has been laid down. Homeotic genes have in common similar DNA sequences called homeoboxes. The homeobox is part of the protein-coding sequence of each gene, and the corresponding approximately 60-amino-acid part of the protein is called the homeodomain. Homeoboxes have been found in developmental genes in other organisms, and homeodomains are thought to play a role in regulating transcription by binding to specific DNA sequences.

SUMMARY

In this chapter we have considered a number of examples of gene regulation in eukaryotes. The general picture is one of much greater complexity than in prokaryotes. Genes are not organized into operons, since genes of related function are often scattered around the genome; nonetheless, genes are regulated coordinately. The chromosome organization in eukaryotes also makes for greater complexity in regulating gene expression.

Gene expression in eukaryotes is regulated at a number of distinct levels. Regulatory systems have been found for: (1) the control of transcription; (2) the control of precursor-RNA processing; (3) transport of the mature RNA out of the nucleus; (4) translation of

~ FIGURE 16.19

Organization of the *bithorax* complex (*BX-C*). The DNA spanned by this complex is 300 kb long. The transcription units for *Ubx, abdA,* and *AbdB* are shown below the DNA: the exons are shown by colored blocks and the introns by bent, dotted lines. All three genes are transcribed from right to left. Shown above the DNA are regulatory mutants that affect the development of different fly segments.

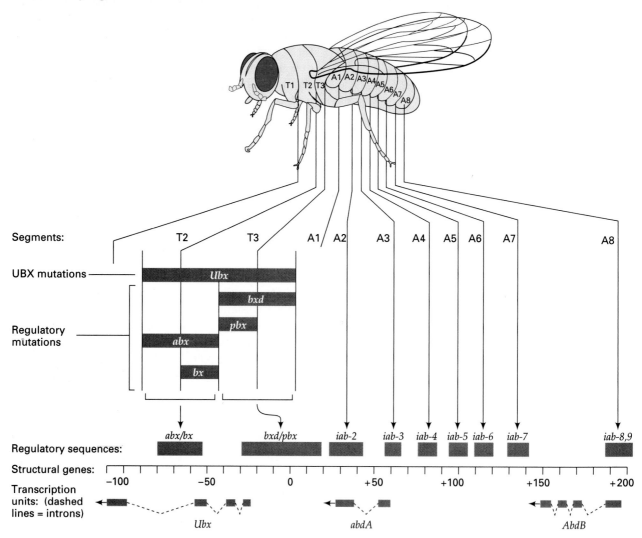

the mRNAs; (5) degradation of the mature RNAs; and (6) degradation of the protein products.

The control of transcription of protein-coding genes involves a number of elements. Initiation of transcription depends upon transcription factors binding to promoter elements upstream of the transcribed sequence. Regulation of the level of transcription then depends upon regulatory proteins binding to enhancers and/or silencers associated with the gene and transmitting their effects to the transcription factors on the promoter elements.

At the chromosome level, we can view the eukaryote chromosome as repressed for transcription by the interaction of histones and nonhistones with the DNA, so activation of transcription generally results from a loosening of the DNA-protein structure in the region of a gene being activated. Relevant to this, transcriptionally active chromatin is more sensitive to DNase I than is transcriptionally inactive chromatin. Moreover, upstream of active genes are DNase-hypersensitive sites, which are highly sensitive to DNase I digestion. These sites probably correspond to the binding sites for RNA polymerase and regulatory proteins.

Histones play an important role in gene repression by assembling nucleosomes in regions containing TATA boxes in promoter regions. Gene activation involves regulatory proteins binding to enhancers and then disrupting the nucleosomes in the TATA box region, thereby stimulating transcription.

Gene activation in a number of eukaryotes is accompanied in many instances by a decrease in the level of DNA methylation, although whether the methylation decrease is the cause or effect of the increase in transcription is not clear.

Gene expression can be regulated at the level of RNA processing. This type of regulation operates to determine the production of mature RNA molecules from precursor-RNA molecules. Two regulatory events that exemplify this level of control are alternative polyadenylation and alternative splicing. In both cases, different types of mRNAs are produced, depending on the choice made.

Gene expression is also regulated by mRNA translation control and by mRNA degradation control. The latter is believed to be a major control point in the regulation of gene expression, as evidenced by the wide range of mRNA stabilities found within organisms. Nucleases are ultimately responsible for degrading the RNAs, and particular structural features of the RNAs are signals for the degradation. For example, a group of AU-rich sequences in the 3'-untranslated regions of some short-lived mRNAs is responsible for their instability.

Lastly, we discussed the immune response, the structure of antibody molecules, and generation of antibody diversity in mammals. Here parts of the coding regions for the light and heavy polypeptide chains of antibody molecules are brought together by somatic recombination during cell development to produce the genes that are transcribed. The various permutations of segments that are possible in this process are the basis for the huge number of different types of antibodies that can be produced in mammals.

Gene regulation is a crucial aspect of development and differentiation. Classical and recent experiments have shown that development and differentiation result from differential gene activity of a genome that contains a constant amount of DNA, rather than from a programmed loss of genetic information. Development and differentiation, then, involve regulation of gene expression, using the levels of control we have just discussed. Adding to the complexity is the fact that we must consider the regulation of a large number of genes for each developmental process and communication between differentiating tissues, as well as systems for timing the activation and repression of genes during those events. For example, the structure and function of a cell are often determined early, even though the manifestations of this determination process are not seen until later in development. Such early determination events may involve some preprogramming of genes that will be turned on later. Neither the nature of these determination events nor the timing mechanism in developmental processes is well understood in vertebrates, although significant progress is being made in understanding the genetic regulation of development in organisms such as *Drosophila*.

In sum, a great deal of information has been learned in the past decade or so about gene regulation in eukaryotes. We have merely scratched the surface in this chapter. Thousands of researchers are currently working to elaborate the molecular details of gene regulation in model systems. Much of our advancing knowledge has been made possible by the application of recombinant DNA and related technologies, and we can look forward to substantial increases in our understanding of eukaryotic gene regulation in the near future.

ANALYTICAL APPROACHES FOR SOLVING GENETICS PROBLEMS

Q16.1 We learned in this chapter that, in humans, there are several distinct genes that code for α- and β-like globin polypeptides. These α, β, γ, δ, ε, and ζ globin genes are transcriptionally active at specific stages of development, resulting in the synthesis of polypeptides that are assembled in specific combinations to form different types of hemoglobin (see pp. 369–370 and Figure 16.9). Fill in the following table, indicating whether the globin gene in question is sensitive (S) or resistant (R) to DNase I digestion at each of the developmental stages listed.

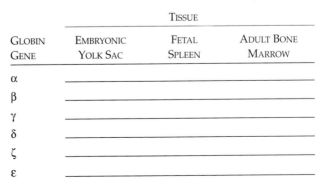

GLOBIN GENE	TISSUE		
	EMBRYONIC YOLK SAC	FETAL SPLEEN	ADULT BONE MARROW
α			
β			
γ			
δ			
ζ			
ε			

A16.1   The correctly filled-in table is as follows:

GLOBIN GENE	TISSUE		
	EMBRYONIC YOLK SAC	FETAL SPLEEN	ADULT BONE MARROW
α	R	S	S
β	R	R	S
γ	R	S	R
δ	R	R	S
ζ	S	R	R
ε	S	R	R

The explanation for the answers is as follows: DNase I typically digests regions of DNA that are transcriptionally active, while not digesting regions of DNA that are transcriptionally inactive. This is because transcriptionally inactive DNA is more highly coiled than transcriptionally active DNA. "R" means, then, that the gene was transcriptionally inactive, while "S" means that the gene was transcriptionally active.

To consider each globin gene type in turn, the α gene is transcriptionally inactive in the embryonic yolk sac, but active in spleen and adult bone marrow. That is, in the spleen, fetal hemoglobin (Hb-F) is made; Hb-F contains two α polypeptides, and two γ polypeptides. In the bone marrow, Hb-A is made, which contains two α and two β polypeptides.

The β-globin gene is inactive in yolk sac and spleen, and is active in bone marrow, making one of the two polypeptides found in Hb-A, the main adult form of hemoglobin. The β-like γ polypeptide is found in Hb-F, which is made only in the liver and the spleen; thus, the γ gene is active in spleen and inactive in yolk sac and bone marrow. The β-like δ polypeptide is found in $\alpha_2\delta_2$ hemoglobin, which is a minor class of hemoglobin found in adults; thus, the δ gene is active only in adult bone marrow.

The ζ gene makes an α-like polypeptide found only in the hemoglobin of the embryo, so the ζ gene is active in the yolk sac but inactive in spleen and bone marrow. Finally, ε gene encodes the β-like polypeptide of the embryo's hemoglobin, so this gene is also active in the yolk sac but inactive in spleen and bone marrow.

QUESTIONS AND PROBLEMS

16.1   Promoters, enhancers, transcription factors, and regulatory proteins that are active for one gene typically share structural similarities with these elements in other genes. Nonetheless, the transcriptional control of a gene can be exquisitely specific: it will be specifically transcribed in some tissues at very defined times. Explore how this specificity arises by addressing the following questions.
a.   Distinguish between the functions of promoters and enhancers in transcriptional regulation.
b.   What structural features are found in proteins that bind these DNA elements?
c.   Can an enhancer bound by a regulatory protein stimulate as well as suppress transcription? If so, how?
d.   Given that several different genes may contain the same types of promoter and enhancer elements, and a number of transcription factors contain the same structural features, how is transcriptional specificity generated?

16.2   A cloned DNA sequence was used to probe a Southern blot. There were two DNA samples on the blot, one from white blood cells and the other from a liver biopsy of the same individual. Both samples had been digested with *Hpa*II. The probe bound to a single 2.2-kb band in the white blood cell DNA, but bound to two bands (1.5 and 0.7 kb) in the liver DNA.
a.   Is this difference likely to be due to a somatic mutation in a *Hpa*II site? Explain.
b.   How would it affect your answer if you knew that white blood cell and liver DNA from this individual both showed the two-band pattern when digested with *Msp*I?

*16.3   Both fragile X syndrome and Huntington disease are caused by trinucleotide repeat expansion. Individuals with fragile X syndrome have at least 200 CGG repeats at the 5'-end of the *FMR-1* gene. Individuals with Huntington disease have at least 36 CAG repeats within the protein-coding region of the huntingtin gene.
a.   How is gene expression affected by these repeat expansions?
b.   Based on your answer to (a), why is the fragile X syndrome recessive, while Huntington disease is dominant?
c.   Why is the number of trinucleotide repeats needed to cause the phenotype different for each disease?

16.4   Steroid hormones can affect gene regulation of a targeted population of cells.
a.   What is a hormone?
b.   What role does each of the following have in a physiological response to a peptide or a steroid hormone?
   i.   steroid hormone receptor (SHR)
   ii.   steroid hormone response element (HRE)

*16.5   The following figure shows the effect of the hormone estrogen on ovalbumin synthesis in the oviduct of 4-day-old chicks. Chicks were given daily injections of estrogen ("Primary Stimulation") and then after 10 days the injections were stopped. Two weeks after withdrawal (25 days), the injections were resumed ("Secondary Stimulation").

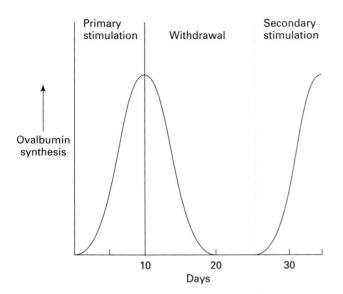

Provide possible explanations of these data.

16.6 Although the primary transcript of a gene may be identical in two different cell types, the translated mRNAs can be quite different. Consequently, in different tissues, distinct protein products can be produced from the same gene. Discuss two different mechanisms by which the production of mature mRNAs can be regulated to this end; give a specific example for each mechanism.

16.7 Distinguish between the terms *development* and *differentiation*.

16.8 What is totipotency? Give an example of the evidence for its existence.

*16.9 In Woody Allen's 1973 film *Sleeper*, the aging leader of a futuristic totalitarian society has been dismembered in a bomb attack. The government wishes to clone the leader from his only remaining intact body part, a nose. The characters Miles and Luna thwart the cloning by abducting the nose and flattening it under a steamroller.

a. In light of the 1996 cloning of the sheep Dolly, how should the cloning have proceeded, if Miles and Luna had not intervened?

b. If methods akin to those used for Dolly had been successful, in what *genetic* ways would the cloned leader be unlike the original?

c. Suppose that, instead of a nose, only mature B cells (B lymphocytes of the immune system) were available. What *genetic* deficits would you expect in the "new leader"?

d. In the set of experiments used to clone Dolly, six additional live lambs were obtained. Why is the production of Dolly more significant than the production of the other lambs?

e. If the cloning of the leader had succeeded, can you make any prediction about whether the "cloned leader" would be interested in perpetuating the totalitarian state?

*16.10 The enzyme lactate dehydrogenase (LDH) consists of four polypeptides (a tetramer). Two genes are known to specify two polypeptides, A and B, which combine in all possible ways (A_4, A_3B, A_2B_2, AB_3, and B_4) to produce five LDH isozymes. If, instead, LDH consisted of three polypeptides (it was a trimer), how many possible isozymes would be produced by various combinations of polypeptides A and B?

16.11 In humans, β-thalassemia is a disease caused by failure to produce sufficient β-globin chains. In many cases, the mutation causing the disease is a deletion of all or part of the β-globin structural gene. Individuals homozygous for certain of the β-thalassemia mutations are able to survive because their bone marrow cells produce γ-globin chains, which combine with α-globin chains to produce fetal hemoglobin. In these people, fetal hemoglobin is produced by the bone marrow cells throughout life, whereas normally it is produced in the fetal liver. Use your knowledge about gene regulation during development to suggest a mechanism by which this expression of γ-globin might occur in β-thalassemia.

*16.12 The following figure shows the percentage of ribosomes found in polysomes in unfertilized sea urchin oocytes (0 h) and at various times after fertilization:

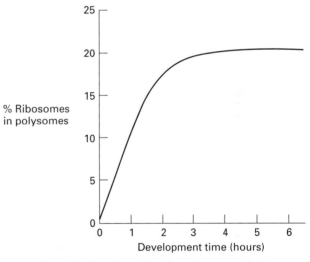

In the unfertilized egg, less than 1 percent of ribosomes are present in polysomes, while at 2 h post-fertilization, about 20 percent of ribosomes are present in polysomes. It is known that no new mRNA is made during the time period shown. How may the data be interpreted?

16.13 The mammalian genome contains about 10^5 genes. Mammals can produce about 10^6 to 10^8 different antibodies. Explain how it is possible for both of these sentences to be true.

*16.14 Ashley and Connie are identical twins. Ashley has blood type A, and therefore must have anti-B antibodies in her serum. (For the purpose of this problem, ignore the effects of possible new mutations.)

a. What is Connie's blood type?

b. If Ashley's blood type genotype is I^A/I^A, what is Connie's?

c. If Ashley has anti-B antibodies, what type of antibodies would Connie have?

d. Would Ashley's and Connie's anti-B antibodies have identical polypeptide sequences? Explain.

e. Would Ashley's and Connie's β-globin chain have the identical polypeptide sequence? Explain.

16.15 Antibody molecules (Ig) are composed of four polypeptide chains (two of one light chain type and two of one heavy chain type) held together by disulfide bonds.

a. If for the light chain there were 300 different V_κ and four J_κ segments, how many different light chain combinations would be possible?

b. If for the heavy chain there were 200 V_H segments, 12 D segments, and 4 J_H segments, how many heavy chain combinations would be possible?

c. Given the information in parts (a) and (b), what would be the number of possible types of IgG molecules (L + H chain combinations)?

16.16 Define *imaginal disc* and *homeotic mutant*.

*16.17 Imagine that you observed the following mutants (*a–e*) in *Drosophila*. Based on the characteristics given, assign each mutant to one of the following categories: maternal gene, segmentation gene, or homeotic gene.

a. Mutant *a*: In homozygotes the phenotype is normal, except wings are oriented backward.

b. Mutant *b*: Homozygous females are normal but produce larvae that have a head at each end and no distal ends. Homozygous males produce normal offspring (assuming the mate is not a homozygous female).

c. Mutant *c*: Homozygotes have very short abdomens, which are missing segments A2 through A4.

d. Mutant *d*: Affected flies have wings growing out of their heads in place of eyes.

e. Mutant *e*: Homozygotes have shortened thoracic regions and lack the second and third pairs of legs.

*16.18 If actinomycin D, an antibiotic that inhibits RNA synthesis, is added to newly fertilized frog eggs, there is no significant effect on protein synthesis in the eggs. Similar experiments have shown that actinomycin D has little effect on protein synthesis in embryos up until the gastrula stage. After the gastrula stage, however, protein synthesis is significantly inhibited by actinomycin D, and the embryo does not develop any further. Interpret these results.

*16.19 It is possible to excise small pieces of early embryos of a frog, transplant them to older embryos, and follow the course of development of the transplanted

material as the older embryo develops. A piece of tissue is excised from a region of the late blastula or early gastrula that would later develop into an eye and is transplanted to three different regions of an older embryo host (see part a of the following figure). If the tissue is transplanted to the head region of the host, it will form eye, brain, and other material characteristic of the head region. If the tissue is transplanted to other regions of the host, it will form organs and tissues characteristic of those regions in normal development (such as ear or kidney). In contrast, if tissue destined to be an eye is excised from a neurula (a stage which is later than blastula and gastrula) and transplanted into an older embryo host to exactly the same places as used for the blastula/gastrula transplants, in every case the transplanted tissue differentiates into an eye (see part b of the following figure). Explain these results.

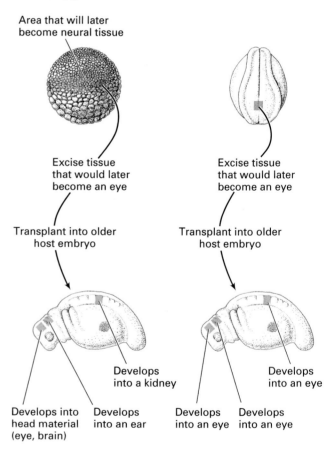

a) Tissue from late blastula or early gastrula

Area that will later become neural tissue

Excise tissue that would later become an eye

Transplant into older host embryo

Develops into head material (eye, brain)

Develops into an ear

Develops into a kidney

b) Tissue from neurula

Excise tissue that would later become an eye

Transplant into older host embryo

Develops into an eye

Develops into an eye

Develops into an eye

16.20 What features of the genes that control body segmentation have been conserved during evolution? How has this been shown?

CHAPTER *17*

GENETICS OF CANCER

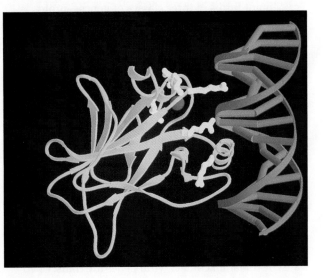

PRINCIPAL POINTS

~ Progression of a normal eukaryotic cell through the cell cycle is tightly controlled by a number of molecular factors. Healthy cells grow and divide only when the balance of stimulatory and inhibitory signals received from outside the cell favor cell proliferation. A cancerous cell does not respond to the usual signals and reproduces without constraints.

~ The two-hit mutation model for cancer states that two mutational events are critical for cancer to develop, one in each allele of a cancer-causing gene. In familial (hereditary) cancers, one mutation is inherited, predisposing the person to cancer; the other mutation occurs later in somatic cells. In sporadic (nonhereditary) cancers, both mutations occur in somatic cells.

~ Mutant forms of three classes of genes—proto-oncogenes, tumor suppressor genes, and mutator genes—have the potential to contribute to the transformation of a cell to a cancerous state. The products of proto-oncogenes normally stimulate cell proliferation, the products of tumor suppressor genes normally inhibit cell proliferation, and the products of mutator genes are involved in DNA replication and repair.

~ Retroviruses are RNA viruses that replicate via a DNA intermediate. All RNA tumor viruses are retroviruses, but not all retroviruses cause cancer. When a retrovirus infects a cell, the RNA genome is released from the viral particle, and through the action of reverse transcriptase a cDNA copy of the genome—called the proviral DNA—is synthesized. The proviral DNA integrates into the genome of the host cell. Then, using host transcriptional machinery, viral genes are transcribed,

and full-length viral RNAs are produced. Progeny viruses are assembled and exit the cell, where they can infect other cells.

~ When tumor induction occurs after retrovirus infection, it is because of the activity of a viral oncogene (v-*onc*) in that retroviral genome. Retroviruses carrying an oncogene are known as transducing retroviruses.

~ Normal animal cells contain genes with DNA sequences that are similar to those of the viral oncogenes. These cellular genes are proto-oncogenes. When a proto-oncogene is mutated to produce a cellular oncogene (c-*onc*), it induces tumor formation.

~ The normal products of tumor suppressor genes have inhibitory roles in cell growth and division. Therefore, when both alleles of a tumor suppressor gene are inactivated or lost, the inhibitory activity is lost, and unprogrammed cell proliferation can occur.

~ The development of most cancers involves the accumulation of mutations in a number of genes over a significant period of a person's life. This multistep path typically involves mutational events that activate oncogenes and inactivate tumor suppressor genes and mutator genes, thereby breaking down the multiple mechanisms that regulate growth and differentiation.

~ Various types of radiation and many chemicals increase the frequency with which cells become cancerous. These agents are known as carcinogens. Practically all carcinogens act by causing changes in the genome of the cell.

*I*n Chapter 16 we learned about some of the genetically controlled processes involved in development and differentiation. The picture we have is that, during development, specific tissues and organs arise by genetically programmed cell division and differentiation. Occasionally, dividing and differentiating deviate from their normal genetic program and give rise to tissue masses called *tumors*, or *neoplasms* ("new growth"). Figure 17.1 shows a mammogram indicating the presence of a tumor. The process by which a cell loses its ability to remain constrained in its growth properties is called **transformation** (not to be confused with transformation of a cell by uptake

of exogenous DNA). If the transformed cells stay together in a single mass, the tumor is said to be *benign*. Benign tumors are usually not life threatening, and their surgical removal generally results in a complete cure. Exceptions include many brain tumors, which are life threatening because they impinge on essential cells. If the cells of a tumor can invade and disrupt surrounding tissues, the tumor is said to be *malignant* and is identified as a **cancer**. Cells from malignant tumors can also break off and move through the blood system or lymphatic system, forming new tumors at other locations in the body. The spreading of malignant tumor cells throughout the

~ FIGURE 17.1

A mammogram showing a tumor.

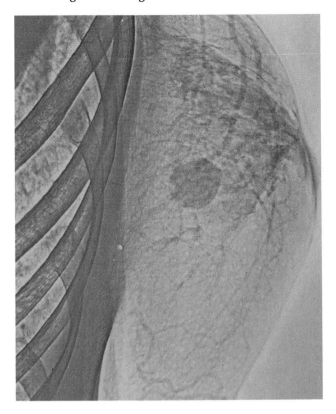

body is called **metastasis**. Malignancy can result in death because of damage to critical organs, starvation, secondary infection, metabolic problems, second malignancies, and/or hemorrhage.

The initiation of tumors in an organism is called **oncogenesis** (*onkos*, "mass" or "bulk"; *genesis*, "birth"). There are many genetic causes of cancer, such as spontaneous genetic changes (spontaneous gene mutations or chromosome mutations, for instance), exposure to mutagens or radiation, or the action of genes in cancer-inducing viruses *(tumor viruses)*. There is also hereditary predisposition to cancer. In this chapter we focus on the genetic basis of tumors and cancers.

RELATIONSHIP OF THE CELL CYCLE TO CANCER

During development a tissue is produced by cell proliferation. During a series of divisions, progeny cells begin to express genes that are specific for the tissue, a process called cell differentiation. Cell differentiation is also associated with the progressive loss of the ability of cells to proliferate: the most highly differentiated cell—the one that is fully functional in the tissue—can no longer divide. Such cells are known as *terminally differentiated cells*. They have a finite life span in the tissue and are replaced with younger cells produced by division of *stem cells*, a small fraction of cells in the tissue that are capable of *self-renewal*. However, in neoplastic diseases, both benign and malignant ones, growth and differentiation become unlinked.

Cells proliferate by going through the cell cycle, the cycle of cell growth, mitosis, and cell division in eukaryotes (see Chapter 1, pp. 10–11, and Chapter 11, pp. 244–246). Recall that the cell cycle consists of the mitotic phase (M) and an interphase between divisions consisting of three stages: G_1, S, and G_2. Several molecular factors control the progression of a normal cell through the cell cycle, the most important being the proteins that operate through interaction with receptors embedded in the plasma membrane. When these factors bind to the cell surface receptor, a signal transduction pathway is induced whereby the signal is transmitted into the cytoplasm, and the regulatory effect on cell division occurs. *Growth factors* turn on stimulatory pathways for cell division (Figure 17.2a), and *growth-inhibiting factors* turn on inhibitory pathways for cell division (Figure 17.2b). Growth factors cause genes that encode proteins needed for the cell division process to be turned on; growth-inhibiting factors cause genes that encode proteins with inhibitory effects on cell division to be turned on. Normal healthy cells give rise to progeny cells only when the balance of stimulatory and inhibitory signals from outside the cell favors cell division. A neoplastic cell, on the other hand, has lost control of cell division and reproduces without constraints (although not at a faster rate). This can occur when genes that encode inhibitory factors mutate, or when genes that encode stimulatory factors mutate.

KEYNOTE

Progression of a normal eukaryotic cell through the cell cycle is tightly controlled by a number of molecular factors. Healthy cells grow and divide only when the balance of stimulatory and inhibitory signals received from outside the cell favors cell proliferation. A cancerous cell does not respond to the usual signals and reproduces without constraints.

~ FIGURE 17.2

General events for regulation of cell division in normal cells. (a) When a growth factor binds to its cell membrane receptor, it acts as a signal to stimulate cell growth. To do that, the signal is transduced into the cell and relayed to the nucleus, activating the expression of a gene or genes that encode a protein or proteins required for the stimulation of cell division. (b) When a growth-inhibiting factor binds to its cell membrane receptor, it acts as a signal to inhibit cell growth. In this case the signal is transduced into the cell and relayed to the nucleus, activating the expression of a gene or genes that encode a protein or proteins required for the inhibition of cell division.

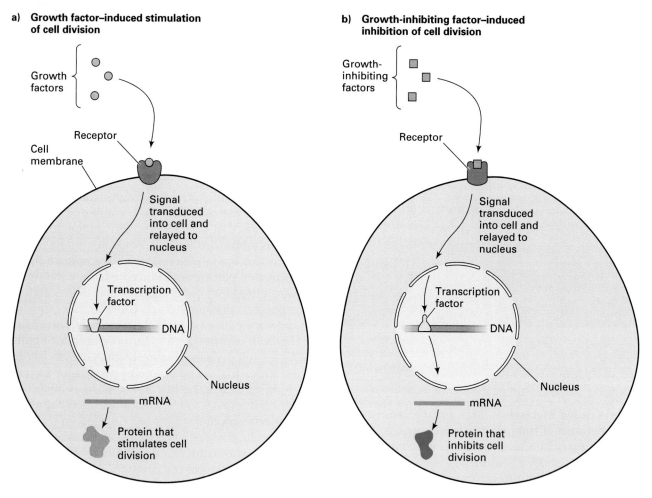

a) **Growth factor–induced stimulation of cell division**

Growth factors

Cell membrane

Receptor

Signal transduced into cell and relayed to nucleus

Transcription factor

DNA

Nucleus

mRNA

Protein that stimulates cell division

b) **Growth-inhibiting factor–induced inhibition of cell division**

Growth-inhibiting factors

Receptor

Signal transduced into cell and relayed to nucleus

Transcription factor

DNA

Nucleus

mRNA

Protein that inhibits cell division

THE TWO-HIT MUTATION MODEL FOR CANCER

All cancers are genetic disorders in that they are caused by changes in DNA that are stably inherited by progeny cells. That is, an accumulation of genetic mutations in particular classes of genes in a cell over a period of time causes cancer. The escalating genetic damage causes a progressive loss in the ability of the cell to respond properly to growth regulatory signals so that eventually the cell will divide uncon-trollably, thereby giving rise to a tumor. Research over the past 20 years or so has led to the identification of a number of the particular genes related to the onset of cancer.

Anecdotal evidence that genes have a role in cancer came from the observation that there was a high incidence of particular cancers in some human families. Cancers that "run" in families are known as *familial (hereditary) cancers*; cancers that do not appear to be inherited are known as *sporadic* (or *nonhereditary*) *cancers*. Sporadic cancers are more frequent than familial cancers.

~ **FIGURE 17.3**

An eye tumor in a patient with retinoblastoma.

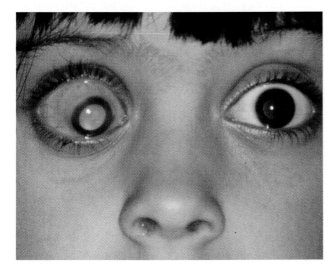

~ **FIGURE 17.4**

Knudson's two-hit mutation model. This model was proposed to explain (a) sporadic retinoblastoma by two independent mutations of the retinoblastoma (*RB*) gene and (b) hereditary retinoblastoma by a single mutation of the wild-type retinoblastoma gene in retinal cells in which a mutant *RB* was inherited through the germ line.

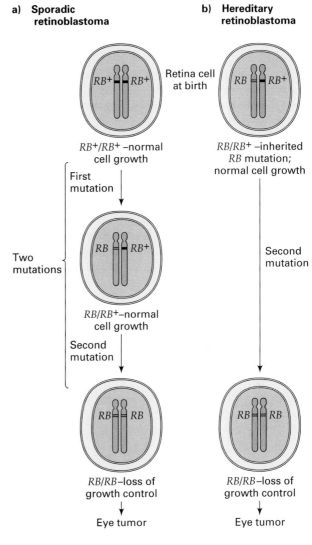

One model for the relation of mutations to cancer came from the study of the onset of retinoblastoma, a childhood cancer of the eye (Figure 17.3). Retinoblastoma occurs from birth to age four years and is the most common eye tumor in children. If discovered early enough, over 90 percent of the eye tumors can be permanently destroyed, usually by gamma radiation. There are two forms of retinoblastoma. In *sporadic retinoblastoma* (60 percent of cases) an eye tumor develops spontaneously in a patient from a family with no history of the disease. In these cases a *unilateral tumor* will develop—a tumor develops in one eye only. In *hereditary retinoblastoma* (40 percent of cases) the susceptibility to develop the eye tumors is inherited. Patients with this form of retinoblastoma typically develop multiple eye tumors involving both eyes (*bilateral tumors*), usually at an earlier age than is the case for unilateral tumor formation in sporadic retinoblastoma patients. A single gene is responsible for retinoblastoma.

In 1971, Alfred Knudson developed a two-hit mutational model for retinoblastoma (Figure 17.4). In sporadic retinoblastoma (Figure 17.4a), a child is born with two wild-type copies of the retinoblastoma gene (genotype RB^+/RB^+), and mutation of each to a mutant *RB* allele must then occur in the same eye cell. Since the chance of having two independent mutational events in the same cell is very low, sporadic retinoblastoma patients would be expected to develop mostly unilateral tumors, as is the case. Further, the rarity of the mutation event means that the two gene copies are mutated at different times,

the first mutation producing an RB/RB^+ cell, and the second mutation in that cell giving rise to the RB/RB genotype that results in eye tumor development. In hereditary retinoblastoma, patients inherit one copy of the mutated retinoblastoma gene through the germ line; that is, they are RB/RB^+ heterozygotes (Figure 17.4b). Only a single additional mutation of the retinoblastoma gene in an eye cell is needed to produce an RB/RB homozygote that would result in tumor formation. Given the number of cells in a developing retina and the rate of mutation per cell, *loss of heterozygosity* (LOH; here, a mutation in the RB^+ allele) is very likely for at least a few cells. Further, since only a single mutation is needed to produce homozygosity for *RB*, hereditary retinoblastoma is characterized on the average by earlier onset than sporadic retinoblastoma and by multiple bilateral tumors.

According to Knudson's model, the *retinoblastoma mutation is recessive*, because cancer develops only if both alleles are mutant. However, if one mutation is inherited through the germ line, tumor formation requires only a mutational event in the remaining wild-type allele in any one of the cells in that particular tissue. Owing to the high likelihood of such an event, the *disease appears dominant* in pedigrees. So for hereditary retinoblastoma, and in hereditary neoplasms in general, we say that inheritance of just one gene mutation predisposes a person to cancer, but does not cause it directly—a second mutation is required for loss of heterozygosity.

Support for Knudson's hypothesis came in the 1980s from the analysis of the chromosomes of tumor cells and normal tissues in retinoblastoma patients. Many patients carried deletions of a region of chromosome 13, and through genetic analysis the *RB* gene was mapped to chromosome location 13q14.1–q14.2 (see Figure 5.12). Retinoblastoma is among a very few cancers for which only one gene is critical for its development, in this case a mutation in a gene for a growth inhibitory factor, that is, a tumor suppressor gene (see pp. 397–401). In most cases, cancer develops as a multistep process involving mutations in several different key genes related to cell growth and division.

KEYNOTE

The two-hit mutation model for cancer explains the difference between familial (hereditary) cancers and sporadic (nonhereditary) cancers. In familial cancers, one mutation in a critical cancer-causing gene is inherited, thereby predisposing an individual to cancer. When the second mutation occurs later in a somatic cell, cancer may then develop. In sporadic cancers, both mutations occur in the somatic cells and, hence, such cancers typically occur later in life than familial cancers because the probability of two mutations is lower than the probability of one mutation.

GENES AND CANCER

Three classes of genes are mutated frequently in cancer. These are *proto-oncogenes*, *tumor suppressor genes*, and *mutator genes*. The products of proto-oncogenes normally stimulate cell proliferation. Mutant proto-oncogenes—now called oncogenes—either are more active than normal or are active at inappropriate times. The products of unmutated tumor suppressor genes normally inhibit cell proliferation. Mutant tumor suppressor genes have lost their inhibitory function. The products of wild-type mutator genes are needed to ensure fidelity of replication and maintenance of genome integrity. Mutant mutator genes have lost their normal function, and this makes the cell prone to accumulate mutational errors.

Oncogenes

Transformation of cells to the neoplastic state can result from infection with **tumor viruses,** which induce the cells they infect to proliferate in an uncontrolled fashion and produce a tumor. Tumor viruses, which may have RNA or DNA genomes, are widely found in animals. *RNA tumor viruses* and *DNA tumor viruses* cause cancer by entirely different mechanisms, as we will see. RNA tumor viruses transform a cell because of the property of a gene (or genes) in the viral genome called a *viral oncogene(s)*. By definition, an **oncogene** is a gene whose action stimulates unregulated cell proliferation.

RETROVIRUSES AND ONCOGENES.   RNA tumor viruses are all retroviruses, and the oncogenes carried by RNA tumor viruses are altered forms of normal host cell genes.

Structure of Retroviruses.   Examples of retroviruses are Rous sarcoma virus, feline leukemia virus, mouse mammary tumor virus, and human immunodeficiency virus (HIV-1, the causative agent of *acquired immunodeficiency syndrome*—AIDS). A retrovirus particle is shown in Figure 17.5. Within a protein core, which often is icosahedral in shape, are two copies of the 7-kb to 10-kb single-stranded RNA genome. The core is surrounded by an envelope derived from host membranes with viral-encoded glycoproteins inserted into it. When the virus infects a cell, the envelope glycoproteins interact with a host cell surface receptor to begin the process by which the virus enters the cell.

~ FIGURE 17.5

Stylized drawing of a retrovirus.

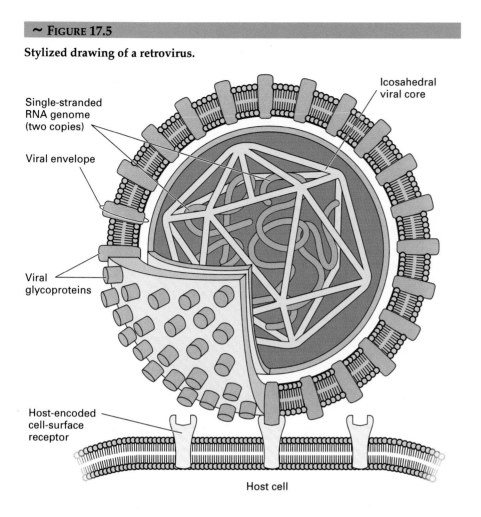

Icosahedral viral core

Single-stranded RNA genome (two copies)

Viral envelope

Viral glycoproteins

Host-encoded cell-surface receptor

Host cell

Life Cycle of Retroviruses.   A well-studied retrovirus is *Rous sarcoma virus* (RSV). RSV causes sarcomas—cancer of the connective tissue or muscle cells—in chickens. The RNA genome organization of RSV is shown in Figure 17.6a. When RSV infects a cell, the RNA genome is released from the viral particle, and a double-stranded DNA copy of the genome (*proviral DNA*) is made by reverse transcriptase,[1] an enzyme brought into the cell as part of the virus particle and encoded by the *pol* gene (Figure 17.6b). This RNA-to-DNA copying process is called reverse transcription. The proviral DNA next circularizes (Figure 17.6c) and

integrates into the host chromosome by a process that leads to a duplication of host DNA to give short direct repeats at the integration site (Figure 17.6d).

Once integrated, the proviral DNA is transcribed by the host RNA polymerase II to produce by alternative splicing (see Chapter 16) the various viral mRNAs that encode the individual viral proteins. Typical retroviruses have three protein-coding genes for the virus life cycle: *gag, pol,* and *env.* The *gag* gene encodes a precursor protein that, when cleaved, produces virus particle proteins. The *pol* gene encodes a precursor protein that is cleaved to produce reverse transcriptase and an enzyme needed for the integration of the proviral DNA into the host cell chromosome. The *env* gene encodes the precursor to the envelope glycoprotein. Progeny RNA genomes are produced by transcription of the entire, integrated viral DNA and packaged into new viral particles that exit the cell and can infect other cells.

---

[1]Reverse transcriptase, unlike most replication DNA polymerases, does not have 3'-to-5' exonuclease activity. Since 3'-to-5' exonuclease activity is involved in proofreading, reverse transcriptases have no capacity to correct errors introduced in the RNA-to-DNA synthesis process. Thus, significant numbers of mutations may be introduced in the DNA product by reverse transcriptase's normal polymerization activities.

~ **FIGURE 17.6**

The Rous sarcoma virus (RSV) RNA genome and the integration of the proviral DNA into the host (chicken) chromosome. (a) RSV genome RNA. (b) RSV proviral DNA produced by reverse transcriptase. (c) Circularization of the proviral DNA. (d) The integrated proviral DNA flanked by short direct repeats of host DNA.

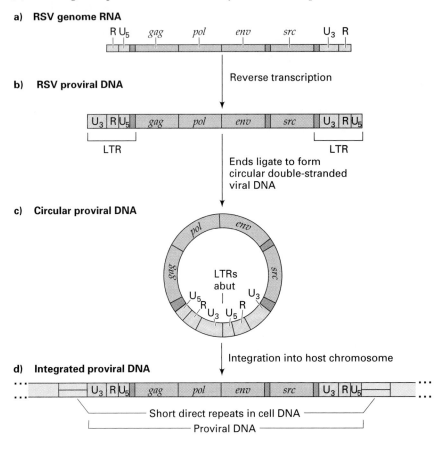

Some retroviruses also carry an oncogene that gives them the ability to transform the cells they infect; these are the *oncogenic retroviruses*. In the case of RSV the oncogene is called *src* (see Figure 17.6a), and, like other retroviral oncogenes, it is not involved in the viral life cycle. Different retroviruses carry different oncogenes. Most oncogenic retroviruses (RSV is an exception) cannot replicate because they do not have a full set of life-cycle genes. Retroviruses without oncogenes direct their own life cycle but do not change the growth properties of the cells they infect; these are *nononcogenic retroviruses*.

Concerning retroviruses, it is appropriate to discuss briefly HIV-1, the causative agent of AIDS, even though this virus does *not* cause cancer. The retrovirus HIV-1 has a bullet-shaped capsid and is surrounded by a viral envelope in which are embedded viral-encoded gp120 glycoproteins (Figure 17.7a). The HIV-1 genome contains complete *gag*, *pol*, and *env* genes, so HIV can self-propagate. In addition, HIV contains several other genes that are not oncogenes but that help control gene expression (Figure 17.7b). For example, *tat* encodes a protein that regulates transcription of the *gag* and *pol* genes and the translation of the resulting mRNA.

The gp120 glycoprotein of the HIV-1 envelope is the main basis for the infection of cells. This glycoprotein is recognized by the CD4 receptor found on the surfaces of immune system cells called *helper T cells* (a type of T lymphocyte). Certain other cell types without the CD4 receptor can also bind the HIV-1 virus, apparently because they have other receptors that recognize the gp120 glycoprotein. Once the virus has bound to a receptor on the cell surface, the viral parti-

~ FIGURE 17.7

HIV retrovirus. (a) Schematic drawing of a cross-section through an HIV particle. The capsid (protein coat) of this retrovirus is "bullet" shaped. (b) Organization of the HIV genome.

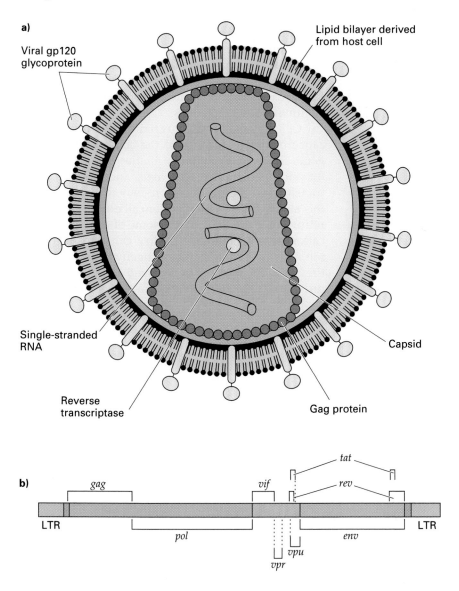

cles enter the cell. Next, the viral protein coat is lost, and the viral life cycle is initiated, starting with reverse transcription of the viral RNA into proviral DNA, which integrates into the genome.

Through normal viral replication, HIV-1 causes the death of the cell it infects. Thus, by repeated infection and viral replication a steady destruction of helper T cells and other infected cells by the virus takes place. The decrease in the population of helper T cells effectively disables the responses of mac-

rophages and B cells in the immune system, and the immune system becomes progressively less functional. As a result, a person infected with HIV-1 is unable to combat infections by pathogens such as bacteria, viruses, and fungi and also becomes susceptible to numerous types of cancers. AIDS patients die most frequently from infections. To date at least 19 million people worldwide have been infected with HIV-1. In Zimbabwe, for example, more than 25 percent of the population is infected.

Retroviruses are RNA viruses that replicate via a DNA intermediate. All RNA tumor viruses are retroviruses, but not all retroviruses cause cancer. When a retrovirus infects a cell, the RNA genome is released from the viral particle, and reverse transcriptase makes a cDNA copy of the genome called the proviral DNA. The proviral DNA integrates into the genome of the host cell. Then, using host transcriptional machinery, viral genes are transcribed and full-length viral RNAs are produced. Progeny viruses assembled within the cell exit the cell and can infect other cells.

Viral Oncogenes.   In the case of RSV, tumor induction is caused by a particular **viral oncogene** in the retroviral genome. Viral oncogenes (generically called v-*oncs*) are responsible for many different cancers. Only retroviruses that contain a v-*onc* gene are

tumor viruses. The v-*onc* genes are named for the tumor that the virus causes, with the prefix "v" to indicate that the gene is of viral origin. Thus, the v-*onc* gene of RSV is v-*src*. Bacteriophages that have picked up cellular genes are said to transduce the genes to other cells, so such retroviruses are called **transducing retroviruses** because they have picked up an oncogene from the genome of the cell. (We learn how this happens later.) Table 17.1 lists some transducing retroviruses and their viral oncogenes. Retroviruses that do not carry oncogenes are called *nontransducing retroviruses*.

Cells infected by RSV rapidly transform into the cancerous state because of the activity of the v-*src* gene. Since RSV contains all the genes necessary for viral replication (*gag*, *env*, and *pol*), an RSV-transformed cell produces progeny RSV particles. In this ability RSV is an exception; all other transducing retroviruses are defective in some of their viral replication genes (Figure 17.8): they can transform cells but are unable to produce progeny viruses because they lack one or more genes needed for virus reproduction. These defective retroviruses can produce

~ TABLE 17.1

Some Transducing Retroviruses and Their Viral Oncogenes

ONCOGENE	RETROVIRUS ISOLATE	V-ONC ORIGIN	V-ONC PROTEIN	TYPE OF CANCER
src	Rous sarcoma virus (RSV)	Chicken	$pp60^{src}$	Sarcoma
abl	Abelson murine leukemia virus (MLV)	Mouse	$P90\text{-}P160^{gag\text{-}abl}$	Pre–B cell leukemia
erbA	Avian erythroblastosis virus (AEV)	Chicken	$P75^{gag\text{-}erbA}$	Erythroblastosis and sarcoma
erbB	Avian erythroblastosis virus (AEV)	Chicken	$gp65^{erbB}$	Erythroblastosis and sarcoma
fms	McDonough (SM)-FeSV	Cat	$gp180^{gag\text{-}fms}$	Sarcoma
fos	FBJ (Finkel-Biskis-Jinkins)-MSV	Mouse	$pp55^{fos}$	Osteosarcoma
myc	MC29	Chicken	$P100^{gag\text{-}myc}$	Sarcoma, carcinoma, and myelocytoma
myb	Avian myeloblastosis virus (AMV) AMV-E26	Chicken Chicken	$p45^{myb}$ $P135^{gag\text{-}myb\text{-}ets}$	Myeloblastosis Myeloblastosis and erythroblastosis
raf	3611-MSV	Mouse	$P75^{gag\text{-}raf}$	Sarcoma
H-*ras*	Harvey MSV (Ha-MSV)	Rat	$pp21^{ras}$	Sarcoma and erythroleukemia
	RaSV	Rat	$P29^{gag\text{-}ras}$	Sarcoma?
K-*ras*	Kirsten MSV (Ki-MSV)	Rat	$pp21^{ras}$	Sarcoma and erythroleukemia

~ **FIGURE 17.8**

Structures of four defective transducing viruses (not to scale). (a) Avian myeloblastosis virus (AMV) contains the v-*myb* oncogene, which replaces the 3' end of *pol* and most of *env*. (b) Avian defective leukemia virus (DLV) contains the v-*myc* oncogene, which replaces the 3' end of *gag*, all of *pol*, and the 5' end of *env*. (c) Feline sarcoma virus (FeSV) contains the v-*fes* oncogene, which replaces the 3' end of *gag* and all of *pol* and *env*. (d) Abelson murine leukemia virus (AbMLV) contains the v-*abl* oncogene, which replaces the 3' end of *gag* and all of *pol* and *env*.

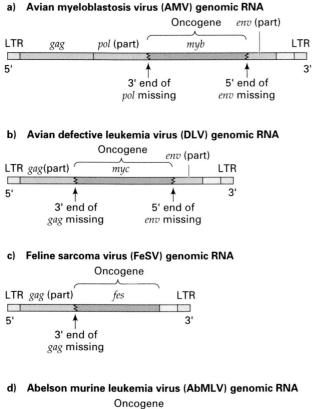

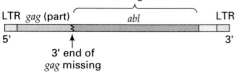

progeny viral particles if cells containing them are also infected with a normal virus (a *helper virus*) that can supply the missing gene products.

Cellular Proto-Oncogenes. In the mid-1970s, J. Michael Bishop and Harold Varmus, and others, demonstrated that normal animal cells contain genes with DNA sequences very closely related to the viral oncogenes. These genes are called proto-oncogenes.

(Bishop and Varmus received the Nobel Prize in 1989 for this research.) In the early 1980s, R. A. Weinberg and M. Wigler showed independently that a variety of human tumor cells contain oncogenes. These genes, when introduced into other cells growing in culture, transformed those cells into cancer cells. The human oncogenes were found to be very similar to viral oncogenes that had been characterized earlier, even though viruses did not induce the human cancers involved. These human oncogenes also were shown to be closely related to proto-oncogenes found in normally growing cells.

In short, most human and other animal oncogenes are mutant forms of normal cellular genes. Such genes in their normal state are called **proto-oncogenes**. Proto-oncogenes have important roles in regulating cell division and differentiation. When proto-oncogenes become mutated or translocated such that they induce tumor formation, they are called oncogenes (*onc*s). If they are carried by a virus, oncogenes are known as v-*onc*s. If they reside in the host chromosome, oncogenes are called **cellular oncogenes**, or c-*onc*s. A transducing retrovirus, then, carries a significantly altered form of a cellular proto-oncogene (now a v-*onc*). When the transducing retrovirus infects a normal cell, the hitchhiking oncogene transforms the cell into a cancer cell. As we will see, normal cells can also become transformed into cancer cells even if a tumor virus does not infect them if the proto-oncogene is converted into a cellular oncogene.

One significant difference between a cellular proto-oncogene and its viral oncogene counterpart is that most proto-oncogenes contain introns that are not present in the corresponding v-*onc*. This is the result of splicing that occurs in transcription of the genomic viral RNA from proviral DNA.

Formation of Transducing Retroviruses. The location at which retroviral DNA (the provirus) integrates into cellular DNA is totally random. Sometimes there occurs a genetic rearrangement by which the transcriptional unit of the provirus connects to nearby cellular genes, often by a deletion event involving the loss of some or all of the *gag*, *pol*, and *env* genes. In this way, viral RNA contains all or parts of a cellular gene. All viral progeny then carry the cellular gene and, under the influence of viral promoters, will express the cellular protein in infected cells. If the cellular gene picked up was an oncogene, the modified retrovirus will be oncogenic (see Figure 17.8). If the cellular gene picked up is a proto-oncogene, the modified retrovirus may still be oncogenic if the increased expression of the proto-oncogene causes oncogenesis.

PROTEIN PRODUCTS OF PROTO-ONCOGENES.

About 100 oncogenes have been identified. Based on similarities in DNA sequences and in amino acid sequences of the protein products, proto-oncogenes fall into several distinct classes, each with a characteristic type of protein product, as outlined in Table 17.2.

In the following sections we illustrate the stimulatory function of protein products of proto-oncogenes on cell growth and division by considering just two examples: growth factors and protein kinases.

~ TABLE 17.2

Classes of Oncogene Products

Growth factors	
sis	PDGF B-chain growth factor
int-2	FGF-related growth factor

Receptor and nonreceptor protein-tyrosine and protein-serine/threonine kinases

src	Membrane-associated nonreceptor protein-tyrosine kinase
fgr	Membrane-associated nonreceptor protein-tyrosine kinase
fps/fes	Nonreceptor protein-tyrosine kinase
kit	Truncated stem cell receptor protein-tyrosine kinase
pim-1	Cytoplasmic protein-serine kinase
mos	Cytoplasmic protein-serine kinase (cytostatic factor)

Receptors lacking protein kinase activity

mas	Angiotensin receptor

Membrane-associated G proteins activated by surface receptors

H-ras	Membrane-associated GTP-binding/GTPase
K-ras	Membrane-associated GTP-binding/GTPase
gsp	Mutant-activated form of G α

Cytoplasmic regulators

crk	SH-2/3 protein that binds to (and regulates?) phosphotyrosine-containing proteins

Nuclear transcription factors (gene regulators)

myc	Sequence-specific DNA-binding protein
fos	Combines with c-jun product to form AP-1 transcription factor
jun	Sequence-specific DNA-binding protein; part of AP-1
erbA	Dominant negative mutant thyroxine (T3) receptor
ski	Transcription factor?

Growth Factors. The effect of oncogenes on cell growth and division led to an early hypothesis that proto-oncogenes might be regulatory genes involved with the control of cell multiplication during differentiation. There is now a lot of evidence supporting that hypothesis.

We can generalize and say that some cancer cells can result from the excessive or untimely synthesis of growth factors in cells that do not normally produce the factors. Introduction of an altered growth factor gene such as a v-onc or mutation of a c-onc can cause tumor development.

Protein Kinases. Many proto-oncogenes encode protein kinases, enzymes that add phosphate groups to target proteins, thereby modifying the proteins' function. Protein kinases are integral members of signal transduction pathways. The src gene product, for example, is a nonreceptor protein kinase termed pp60src. The viral protein, pp60v-src, and the protein encoded by the cellular oncogene, pp60c-src, differ in only a few amino acids, and both proteins bind to the inner surface of the plasma membrane. A large class of proteins, including the receptors for growth factors, uses protein phosphorylation to transmit signals through the membrane. Thus, the action of protein kinases such as that encoded by src appears to be linked to growth factors and their activities through their role in signal transduction, and this explains how src can transform a normal cell into a metabolically different cancer cell.

CHANGING CELLULAR PROTO-ONCOGENES INTO ONCOGENES.

In normal cells, expression of proto-oncogenes is tightly controlled so that cell growth and division occur only as appropriate for the cell type involved. However, when proto-oncogenes are changed into oncogenes, the tight control can be lost, and unregulated cell proliferation can take place.

Four general types of changes have been found:

1. *Point mutations* (base-pair substitutions). Point mutations in the coding region of a gene or in the controlling sequences (promoter, regulatory elements, enhancers) can change a proto-oncogene into an oncogene by causing an increase in either the activity of the gene product or the expression of the gene, leading in turn to an increase in the amount of gene product.

2. *Deletions.* Deletions of part of the coding region or of part of the controlling sequences of a proto-oncogene have been found frequently in oncogenes. The deletions cause changes in the amount or activity of the encoded growth stimulatory protein, causing unprogrammed activation of some cell proliferation genes.

3. *Gene amplification* (increased number of copies of the gene). Some tumors have multiple (sometimes hundreds of) copies of proto-oncogenes. These have probably occurred by a random overreplication of small segments of the genomic DNA. In general, extra copies of the proto-oncogene in the cell result in an increased amount of gene product, thereby inducing or contributing to unscheduled cell proliferation.

4. *Chromosomal translocations.* Chromosomal translocations in human tumor cells are rather common, and some are specific for certain tumor types. We have already discussed a specific example of this and the oncogenes involved: chronic myelogenous leukemia (CML) and the Philadelphia chromosome produced by a reciprocal translocation involving chromosomes 9 and 22 (Chapter 7, Figure 7.11).

CANCER INDUCTION BY RETROVIRUSES.

Retroviruses are common causes of cancer in animals, although only one case is known for humans. A retrovirus can cause cancer if it is a transducing retrovirus and the v-*onc* it carries is expressed. In this case, transcription of the v-*onc* takes place under the control of retroviral promoters. Another way in which a retrovirus can cause cancer is if the proviral DNA integrates near a proto-oncogene. In this situation, expression of the proto-oncogene can come under control of retroviral promoter and enhancer sequences. These retroviral sequences do not respond to the environmental signals that normally regulate proto-oncogene expression, so overexpression of the proto-oncogene occurs, transforming the cell to the tumorous state. The process of proto-oncogene activation is called *insertional mutagenesis*. It occurs rarely in animals and is not known to occur in humans.

KEYNOTE

After retrovirus infection, tumor induction occurs as a result of the activity of a viral oncogene (v-*onc*) in the retroviral genome. Retroviruses carrying an oncogene are known as transducing retroviruses. Normal cellular genes, called proto-oncogenes, have DNA sequences that are similar to those of the viral oncogenes. Proto-oncogenes encode proteins that stimulate cell growth and division. In their mutated state, proto-oncogenes are called cellular oncogenes (c-*onc*s), and they may induce tumors. Retroviral oncogenes are in fact modified copies of cellular proto-oncogenes that have been picked up by the retrovirus.

DNA TUMOR VIRUSES.

DNA tumor viruses are oncogenic—they induce cell proliferation—but they do not carry oncogenes like those in RNA tumor viruses. Thus, as mentioned previously (p. 390), their mechanism for transforming cells is completely different. DNA tumor viruses transform cells to the cancerous state through the action of a gene or genes in the viral genome. Examples of DNA tumor viruses are found among five of six major families of DNA viruses—papovaviruses, hepatitis B viruses, herpes viruses, adenoviruses, and pox viruses.

DNA tumor viruses normally progress through their life cycles without transforming the cell to a cancerous state. Typically, the virus produces a viral protein that activates DNA replication in the host cell. Then, through the use of host proteins, the viral genome is replicated and transcribed, ultimately producing a large number of progeny viruses, which results in killing of the cell. The released viruses can then infect other cells. Rarely, the viral DNA is not replicated and becomes integrated into the host cell genome. If the viral protein that activates DNA replication of the host cell is now synthesized, this protein transforms the cell to the cancerous state by stimulating the quiescent host cell to proliferate; that is, it causes the cell to move from the G_0 phase to the S phase of the cell cycle.

The papovavirus family includes examples of DNA tumor viruses. In this family are the many known papillomaviruses, some of which cause benign tumors such as skin and venereal warts in humans. Other human papillomaviruses (*HPV-16*, *HPV-18*, or both) cause cervical cancer, which is a leading cause of cancer deaths among women worldwide. The key viral genes involved in this transformation are *E6* and *E7*. The protein products of these two genes cause a change in the levels of cellular proteins that are important in regulating cell growth and division.

Tumor Suppressor Genes

In the late 1960s, Henry Harris fused normal rodent cells with cancer cells and observed that some of the resultant hybrid cells did not form tumors, but established a normal growth pattern. Harris hypothesized that the normal cells contained gene products that had the ability to suppress the uncontrolled cell proliferation characteristic of cancer cells. The genes involved were called **tumor suppressor genes**. The normal products of tumor suppressor genes have an inhibitory role in cell growth and division. Thus, when tumor suppressor genes are inactivated, the inhibitory activity is lost, and unprogrammed cell

~ **TABLE 17.3**

Some Known or Candidate Tumor Suppressor Genes

GENE	CANCER TYPE	PRODUCT LOCATION	MODE OF ACTION	HEREDITARY SYNDROME	CHROMOSOME LOCATION
APC	Colon carcinoma	Cytoplasm?	Cell adhesion molecule	Hereditary adenomatous polyposis	5q21–q22
BRCA1	Breast cancer	Nucleus	Transcription factor	Breast cancer and ovarian cancer	17q21
BRCA2	Breast cancer	Nucleus	Transcription factor?	Breast cancer	13q12–q13
DCC	Colon carcinoma	Membrane	Cell adhesion molecule	Involved in colorectal cancer	18q21.3
NF1	Neurofibromas	Cytoplasm	GTPase-activator	Neurofibromatosis type 1	17q11.2
NF2	Schwannomas and meningiomas	Inner membrane?	Links membrane to skeleton?	Neurofibromatosis type 2	22q12.2
p16	Melanoma	Nucleus	Transcription factor	Melanoma	9p21
p53	Colon cancer; many others	Nucleus	Transcription factor	Li-Fraumeni syndrome	17p13.1
RB	Retinoblastoma	Nucleus	Transcription factor	Retinoblastoma	13q14.1–q14.2
VHL	Kidney carcinoma	Membrane?	Transcription elongation factor	von Hippel-Lindau disease	3p26–p25
WT1	Nephroblastoma	Nucleus	Transcription factor	Wilms tumor	11p13

Adapted with permission from J. Marx, *Science* 261 (1993): 1385–1387. Copyright © 1993 American Association for the Advancement of Science.

proliferation can begin. Inactivation of tumor suppressor genes has been linked to the development of a wide variety of human cancers, including breast, colon, and lung cancer. In essence, tumor suppressor genes are the opposites of proto-oncogenes. Two mutations are needed to inactivate a tumor suppressor gene and thereby cause a potential loss of cell growth and division control, while only one mutation is needed to change a proto-oncogene to an oncogene and thereby stimulate cell growth and division. Table 17.3 lists some of the known tumor suppressor genes in humans. The products of tumor suppressor genes are found throughout the cell.

THE RETINOBLASTOMA TUMOR SUPPRESSOR GENE, *RB*. Retinoblastoma was introduced earlier in this chapter in the context of Knudson's two-hit mutation model for cancer.

Genetics. The *RB* tumor suppressor gene has been mapped to 13q14.1–q14.2 (see Figure 5.12). The *RB* gene was cloned in 1986. It spans 180 kb of DNA and encodes a 110-kDa nuclear phosphoprotein (a phos-

phorylated protein)—pRB—that is involved in regulating cell growth.

Cell Biology. Recall that the cell cycle involves progression through the stages G_1-S-G_2-M (Chapter 1, pp. 10–11, and Figure 1.10). pRB plays a major role in the cell cycle by regulating the passage of cells from G_1 to S, a transition that commits the cell to progressing through the rest of the cell cycle. In G_1 in a normal cell or in a cell heterozygous for an *RB* mutation, unphosphorylated pRB binds to a complex of two transcription factors called E2F and DP1 (Figure 17.9a). As long as this status is maintained, the cells remain in G_1 or enter the quiescent state (the G_0 phase). If progression through the cell cycle is signaled, phosphorylation of pRB by a cyclin/cyclin dependent kinase (Cdk) complex (cyclins are proteins involved in control of the cell cycle) occurs, which makes pRB no longer able to bind to E2F. The released E2F molecules bind to genes with binding sites for this transcription factor, and those genes, whose activities are required for entry into S phase, are turned on. Progression of the cell into S is then

~ **FIGURE 17.9**

Role of pRB in regulating the passage of cells from G_1 to S. (a) In a normal cell, unphosphorylated pRB is in a complex with the E2F/DP1 transcription factors. When pRB becomes phosphorylated, that binding is blocked, and the transcription factors activate specific genes involved in the G_1-to-S transition. (b) A cell with two mutant *RB* alleles produces a shortened or unstable pRB that does not bind to E2F/DP1, allowing the transcription factors to activate genes for G_1-to-S transition. This leads to unprogrammed cell division.

a) **Normal cell**   b) **Cell with two mutant *RB* alleles**

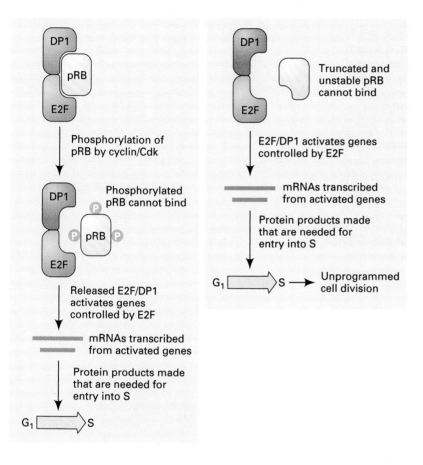

assured. Once a cell has completed mitosis, pRB is dephosphorylated.

In a cell with two mutant *RB* alleles, pRB is nonfunctional and does not bind to E2F/DP1, which then activates genes for transition to S phase (Figure 17.9b). As a result, unprogrammed cell division takes place.

Despite our present knowledge of the cellular activities of pRB, we do not yet know how children with *RB* mutations develop retinoblastoma. It may be through the role of pRB in development rather than its role in the cell cycle specifically. That is, cells programmed for terminal differentiation may depend critically on pRB in order to establish a state of permanent nonproliferation.

THE *P53* TUMOR SUPPRESSOR GENE. The tumor suppressor gene *p53* is so named because it encodes a protein of molecular weight 53 kDa called p53. When both alleles are mutated, *p53* may be involved in the development of perhaps 50 percent of all human cancers, including breast, brain, liver, lung, colorectal, bladder, and blood cancer. This does not mean that *p53* causes 50 percent of human cancers, but that mutations in *p53* are among the several genetic changes usually found in those cancers.

Genetics. The *p53* gene is at chromosome location 17p13.1. Individuals who inherit one mutant copy of *p53* develop Li-Fraumeni syndrome, a rare form of cancer that is an autosomal dominant trait because the cancer develops when the second copy of *p53* becomes mutated. Individuals with this syndrome develop cancers in a number of tissues, including breast and blood.

Cell Biology.   The p53 protein binds to DNA and acts as a transcription factor. p53 binds to several genes, one of which—*WAF1*—is specifically activated by wild-type p53. *WAF1* encodes a 21-kDa protein called p21, which, when its synthesis is activated by p53, causes cells to arrest in G_1. The arrest occurs because p21 binds to cyclin/cyclin dependent kinase (Cdk) complexes (see previous retinoblastoma discussion) and blocks the kinase activity required to activate the genes needed for the cell to make the transition from G_1 to S. In a normal cell, one of the most effective ways of causing p53 to initiate the cascade of events leading to arrest in G_1 is to damage the cellular DNA, for example, by irradiating the cell. In an unknown way, DNA damage results in stabilization of p53, and the cascade of events in Figure 17.10 occurs. The G_1 arrest gives the cell time to induce the necessary pathways to repair the DNA damage, after which the cell cycle can resume. If the extent of DNA damage is too great and all lesions cannot be repaired, the cell cycle will not resume, but the damaged cell will die by programmed cell death (*apoptosis*). The induction of apoptosis is an important function of p53, as the following paragraph shows.

If both alleles of *p53* are inactivated in the cell, active p53 is not present. Thus, *WAF1* cannot be activated, and no p21 is available to block cyclin dependent kinase activity, so the cell is unable to arrest in G_1. Therefore, the cell cycle may proceed to S. Cells with mutated *p53* alleles do not go into growth arrest following DNA damage because of the lack of functional p53 protein, and apoptosis does not occur. The progression of unrepaired cells through the cell cycle can cause the cells to accumulate yet further genetic damage and hence increase the probability of cancer.

The function of p53 is not as simple as we have just described, however. For example, at least 17 cellular or viral proteins have been shown to interact with p53. The interactions with viral proteins, for example, typically result in the inactivation of p53. This makes sense in the context of the propagation "needs" of DNA tumor viruses, which optimally multiply in dividing cells. Inactivation of p53, then, removes the inhibition of growth and allows the cell to proliferate, to the benefit of the virus.

BREAST CANCER TUMOR SUPPRESSOR GENES. In the United States over 185,000 new cases of breast cancer are diagnosed each year, representing more than 31 percent of all new cancers in women, and over 46,000 women die each year from this cancer. In devel-

~ **FIGURE 17.10**

Cascade of events by which DNA lesions cause an arrest in G_1. In an unknown fashion the lesions stabilize p53, which activates the *WAF1* gene giving rise to p21. The p21 protein binds to cyclin/Cdk, blocking the kinase activity needed for cells to progress from G_1 to S.

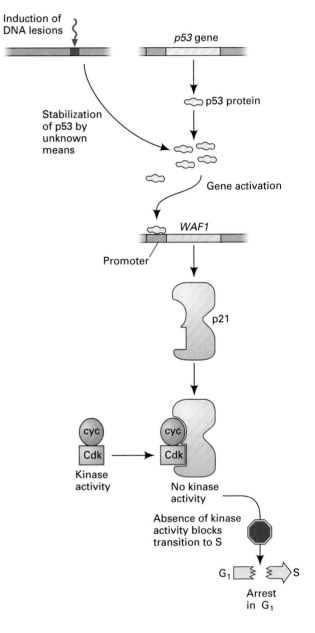

oped countries there is a one in ten chance a woman will be diagnosed with breast cancer in her lifetime. The average age of onset is 55. Approximately 5 percent of breast cancers are hereditary. As with hereditary retinoblastoma, this form of the cancer has an earlier age of onset than the sporadic form, and the cancer is often bilateral.

Among the several genes that play a role in familial breast cancer, two genes—*BRCA1* and *BRCA2*—have been hypothesized to be tumor suppressor genes. (Some studies have led to the alternative hypothesis that these two genes are mutator genes [see next section].) It is believed that most hereditary breast cancer in the United States results from mutations in either *BRCA1* or *BRCA2*, with most of the mutations occurring in *BRCA1*.

The *breast cancer* susceptibility gene *BRCA1* is at chromosome location 17q21 (see Figure 5.12). Mutations of the *BRCA1* gene also lead to susceptibility to ovarian cancer. The *BRCA1* gene encompasses over 100 kb of DNA; it is transcribed in numerous tissues, including breast and ovary, to produce a 7.8-kb mRNA that is translated to produce a 190-kDa protein with 1,863 amino acids. It is thought that the protein plays a role in regulating gene transcription.

BRCA2 is at chromosome location 13q12–q13. Unlike *BRCA1*, *BRCA2* does not have an associated high risk of ovarian cancer. The *BRCA2* encompasses approximately 70 kb of DNA and encodes a 3,418-amino-acid protein that has some similarity to the *BRCA1*-encoded protein but no similarity to other known proteins.

KEYNOTE

Tumor suppressor genes, like proto-oncogenes, are involved in the regulation of cell growth and division. Whereas the normal products of proto-oncogenes have a stimulatory role in those processes, the normal products of tumor suppressor genes have inhibitory roles. Therefore, when both alleles of a tumor suppressor gene are inactivated or lost, the inhibitory activity is lost, and unprogrammed cell proliferation can occur. Inactivation of tumor suppressor genes is involved in the development of a wide variety of human cancers, including breast, colon, and lung cancer.

Mutator Genes

A **mutator gene** is any gene that, when mutant, increases the spontaneous mutation frequencies of other genes. In a cell the normal (unmutated) forms of mutator genes are involved in such important activities as DNA replication and DNA repair. Mutations of these genes can significantly impair those processes and can make the cell error prone, so that it accumulates mutations. For an illustration of how a mutation in a mutator gene can result in cancer, we consider hereditary nonpolyposis colon cancer (HNPCC).

HNPCC is an autosomal dominant genetic disease in which there is an early onset of colorectal cancer. Unlike hereditary (or familial) adenomatous polyposis (FAP) (see next section), no adenomas (benign tumors or polyps) are seen in HNPCC, hence its name. HNPCC accounts for perhaps 5 to 15 percent of colorectal cancers.

Four human genes named *hMSH2*, *hMLH1*, *hPMS1*, and *hPMS2* have been identified, any one of which gives a phenotype of hereditary predisposition to HNPCC when it is mutated. Tumor formation requires only one mutational event to inactivate the remaining normal allele. Thus, owing to the high probability of such an event, HNPCC appears dominant in pedigrees. All four genes encode products that are involved in mismatch repair, a process for correcting mismatched base pairs left after improper DNA replication. (Mismatch repair is described in Chapter 18, p. 421.) In other words, these genes are mutator genes. Homologs of these genes are known in yeast, *E. coli*, and other organisms. Mu-tations in these genes make the DNA replication error prone, and mutation rates increase significantly compared with rates in normal cells.

THE MULTISTEP NATURE OF CANCER

The development of most cancers is a stepwise process involving an accumulation of mutations in a number of genes. It appears that perhaps six or seven independent mutations are needed over several decades of life in order for cancer to be induced. The multiple mutational events typically involve both activation of oncogenes and inactivation of tumor suppressor genes, with a resulting breakdown of the multiple cellular mechanisms that regulate growth and differentiation.

As an example, Figure 17.11 illustrates Bert Vogelstein's molecular model of multiple mutations leading to hereditary FAP, a form of colorectal cancer. Patients with FAP inherit the loss of a chromosome 5 tumor suppressor gene called *APC* (adenomatous polyposis coli). The same gene can be lost early in carcinogenesis in sporadic tumors. Once both alleles of *APC* are lost in a colon cell, increased cell growth will result. If hypomethylation (decreased methylation) of the DNA occurs, a benign tumor called an *adenoma class I* (a small polyp from the colon or rectum epithelium) can develop. Then, if a mutation converts the

~ **FIGURE 17.11**

A multistep molecular event model for the development of hereditary adenomatous polyposis (FAP), a colorectal cancer.

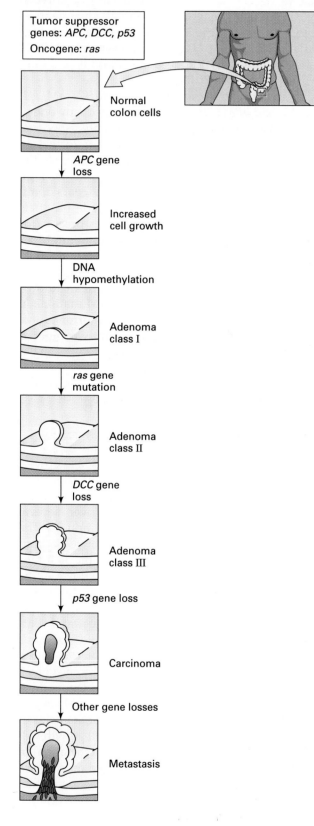

chromosome 12 *ras* proto-oncogene into an oncogene, the cells can progress to a larger benign tumor known as *adenoma class II* (a larger polyp). Next, if both copies of the chromosome 18 tumor suppressor gene *DCC* (deleted in colon cancer) are lost, the cells progress into *adenoma class III* (large benign polyps). Deletion of both copies of the chromosome 17 tumor suppressor gene *p53* results in the progression to a carcinoma (an epithelial cancer); with other gene losses the cancer metastasizes. Note that this is only one path whereby adenomatous polyposis can occur; others are possible. However, in all paths observed, deletions of *APC* and mutations of *ras* usually occur earlier in carcinogenesis than do deletions of *DCC* and *p53*. Progressive changes in function of oncogenes and tumor suppressor genes are also thought to occur for other cancers.

KEYNOTE

The development of most cancers involves an accumulation of mutations in a number of genes over a significant period of life. This multistep nature of cancer typically involves mutational events that activate oncogenes and inactivate tumor suppressor genes, thereby breaking down the multiple mechanisms that regulate growth and differentiation.

CHEMICALS AND RADIATION AS CARCINOGENS

Several natural and artificial agents increase the frequency with which cells become cancerous. These agents, mostly chemicals and types of radiation, are known as **carcinogens**. Because of the obvious human relevance, there is a vast amount of information about carcinogenesis spanning many areas of biology; only an overview is given here.

Although we have focussed much attention in this chapter on viruses as causes of cancer, chemicals are responsible for more human cancers than are viruses. Chemical carcinogenesis was discovered in the eighteenth century by Sir Percival Pott, an English surgeon who correlated the incidence of scrotal skin cancer in some of his patients to occupational exposure to coal soot when they had worked as chimney sweeps as children. From the beginning of industrial development in the eighteenth century to the present day, workers in many areas have been exposed to carcinogenic agents and have developed occupationally related cancers. For example, radiologists using X rays and radium (sources of ionizing radiation) and farmers exposed to the sun's ultraviolet (UV) light (nonionizing radiation) have developed skin cancer, asbestos and insulation workers exposed to asbestos have developed bronchial and lung cancers, and PVC workers exposed to vinyl chloride have developed liver cancer.

Chemical Carcinogens

Chemical carcinogens include both natural and synthetic chemicals. Two major classes of chemical carcinogens are recognized. *Direct-acting carcinogens* are chemicals that bind to DNA and act as mutagens. The second class, procarcinogens, must be converted metabolically to become active carcinogens; virtually all of these so-called *ultimate carcinogens* also bind to DNA and act as mutagens. In both cases the mutations typically are point mutations. (The mutageni-city of ultimate carcinogens can be demonstrated in a number of screening tests, including the Ames test described in Chapter 18, pp. 418–419.) Thus, direct-acting and most ultimate carcinogens bring about transformation of cells and the formation of tumors through binding to, and causing changes in, DNA. Direct-acting carcinogens include alkylating agents such as some anticancer therapeutic chemicals. Ex-amples of procarcinogens are polycyclic aromatic hydrocarbons (multi-ringed organic compounds found, for example, in the smoke from wood, coal, and cigarettes), azo dyes and natural metabolites (such as aflatoxin produced from fungal contamination of food), and nitrosamines (produced by nitrites in food). Most chemical carcinogens are procarcinogens.

The metabolic conversion of procarcinogens to ultimate carcinogens is carried out by normal cellular enzymes that function in a variety of pathways that involve, for example, hydrolysis, oxidation, and reduction. If a procarcinogen interacts with the active site of one of the enzymes, then it, too, can be modified by the enzyme activity—being hydrolyzed, oxidized, reduced, and so on—to give rise to the derivative ultimate carcinogen.

Chemical carcinogens are responsible for most cancer deaths in the United States, with the top two causes of cancer—tobacco smoke and diet—being responsible for 50 to 60 percent of those deaths. Since these two factors are environmental in nature, our risk of cancer can be significantly affected by our life habits.

Overall, smoking—mostly of cigarettes—is responsible for 30 percent of cancer deaths, making tobacco smoke (or rather the chemicals in the smoke) the most lethal carcinogen that exists. Tobacco smoke can cause a number of types of cancers, including lung, upper respiratory tract, esophagus, stomach, and liver cancer, and it can increase the risk of cancer in other organs, including breast, bladder, and vulva. The risk of developing cancer as a result of smoking tobacco is influenced by such factors as the amount of tobacco smoked, its tar content, and how long the person has been smoking. Thus, the younger a person is when he or she begins smoking, the greater the risk of developing cancer later in life. Moreover, there is good evidence that secondhand smoke increases the risk of cancer, making tobacco smoke an environmental concern for all individuals. One of the main types of carcinogens found in tobacco smoke is the polycyclic hydrocarbons. Once converted in the cell to their ultimate carcinogen derivatives, they react with negatively charged molecules, such as DNA, and this can result in mutations.

Radiation

We may be able to avoid or minimize environmental exposure to chemical carcinogens, but avoiding exposure to radiation in its various forms is more difficult. We are exposed to radiation, for example, from the sun, cellular telephones, radioactive radon gas, electric power lines, and some household appliances. Only about 2 percent of all cancer deaths are caused by radiation, and most of the cancers involved are the highly aggressive melanoma skin cancers that can be induced by exposure to the sun's UV light. Ionizing radiation, such as that emitted, for example, from X-ray machines, atomic bombs, decay of some radioactive materials, and radon gas, has the potential to be carcinogenic, although the risk to the general public is generally very low. Ionizing radiation most commonly causes leukemia and thyroid cancer.

Radiation carcinogens all act directly, causing mutations in DNA. The mutagenic effects of ultraviolet light and X rays are discussed in Chapter 18, p. 415. We discuss ultraviolet light as a carcinogen in more detail in the following paragraphs.

UV light is emitted by the sun, along with visible light and infrared radiation. The UV light that reaches Earth is classified into two types, based on its wavelength: ultraviolet A (UVA, spanning 320–400 nm) and ultraviolet B (UVB, spanning 290–320 nm). The intensity of UVA and UVB reaching an individual on Earth depends on a number of factors, including time of day, altitude, and materials in the atmosphere such as dust and other particles. Generally, the ambient level of UVA is one to three orders of magnitude higher than that of UVB.

UV light causes several forms of skin cancer, the most dangerous of which are directly related to long-term exposure to UV-light radiation. Both UVA and UVB play a role in carcinogenesis. Sunburn is mainly caused by UVB, which also induces skin cancer because the radiation in the wavelength range of UVB is mutagenic (see Chapter 18, p. 415, for a discussion of this). UVA plays a role in skin cancer by acting to increase the carcinogenic effects of UVB.

The risks of UV-light–induced skin cancers can be minimized by reducing one's exposure to the sun (or to UV light in any form) and by applying an effective sunscreen when out in the sun. The effectiveness of a sunscreen is indicated in terms of its sun protection factor (SPF). The SPF value tells how many times your natural sunburn protection against UVB the product will provide. However, one must be careful in interpreting what the different values mean. For example, a sunscreen with SPF 15 will block 93 percent of UVB, and a sunscreen with SPF 50 will block 98 percent of UVB. Most sunscreens provide limited or no protection against UVA, which can lead to harm if using a sunscreen with high SPF encourages long exposure to the sun.

Fortunately, many skin cancers are easy to detect and may be removed surgically.

KEYNOTE

Various types of radiation and many chemicals increase the frequency with which cells become cancerous. These agents are known as carcinogens. All carcinogens act by causing changes in the genome of the cell. A few chemical carcinogens act directly on the genome; the majority act indirectly. The latter are metabolically converted by cellular enzymes to ultimate carcinogens that bind to DNA and cause mutations. All carcinogenic forms of radiation act directly.

SUMMARY

To understand the development of cancer (neoplasia), it is necessary to understand how normal cell division is controlled. We know that a normal eukaryotic cell moves through the cell cycle in steps that are tightly controlled by a number of molecular factors. Healthy cells grow and divide only when the balance of stimulatory and inhibitory signals received from outside favors cell proliferation and the signals are appropriately transduced to the cytoplasm and nucleus. A cancerous cell does not respond to the usual signals and reproduces without constraints.

Three classes of genes have been shown to be mutated frequently in cancer: proto-oncogenes (the mutant forms are called oncogenes), tumor suppressor genes, and mutator genes. The products of proto-oncogenes stimulate cell proliferation, the products of wild-type tumor suppressor genes inhibit cell proliferation, and the products of wild-type mutator genes are involved in the replication and repair of DNA. Mutant forms of these three types of genes all have the potential to contribute to the transformation of a cell to a tumorous state.

Some forms of cancer are caused by tumor viruses. Both DNA and RNA tumor viruses are known. DNA tumor viruses transform cells to the cancerous state through the action of a gene or genes that are essential parts of the viral genome and that usually stimulate the transition from G_0 or G_1 to S. All RNA tumor viruses are retroviruses—RNA viruses that replicate via a DNA intermediate—but not all retroviruses cause cancer. When a retrovirus infects a cell, the RNA genome is released from the viral particle, and reverse transcriptase makes a cDNA copy of the genome—called the proviral DNA—that integrates into the host cell's genome. Expression of the proviral genes leads to the production of progeny viruses that exit the cell and infect other cells.

Tumor-causing RNA retroviruses contain cancer-inducing genes termed oncogenes. Tumor-causing retroviruses have picked up normal cellular genes—called proto-oncogenes—while simultaneously losing some of their genetic information. Proto-oncogenes in normal cells function in various ways to regulate cell proliferation and differentiation. However, in the retrovirus these genes have been modified or their expression regulated differently so that the oncogene protein product, now synthesized under viral control, is altered. The oncogene products, which include growth factors and protein kinases, are directly responsible for the transformation of cells to the cancerous state. Cellular proto-oncogenes may also mutate, resulting in a stimulatory effect on cell division. In their mutated state, proto-oncogenes are called cellular oncogenes. Since only one allele of a proto-oncogene needs to be mutated to cause changes in cell growth and division, the mutations are dominant mutations.

Tumor suppressor genes, like proto-oncogenes, are involved in the regulation of cell growth and division. In this case, the normal products of tumor suppressor genes have inhibitory roles. Therefore, when both alleles of a tumor suppressor gene are inactivated or lost, the inhibitory activity is lost, and unprogrammed cell proliferation can occur. In familial (hereditary) cancers, one mutation is inherited and the other mutation occurs later in somatic cells, leading to loss of heterozygosity. In other words, the inheritance of one gene mutation predisposes a person to cancer. In sporadic (nonhereditary) cancers, both mutations occur in the somatic cells.

The development of most cancers involves an accumulation of mutations in a number of genes over a significant period of life. This multistep nature of cancer typically involves mutational events that activate oncogenes and inactivate tumor suppressor genes, thereby breaking down the multiple mechanisms that regulate growth and differentiation. Mutations of mutator genes can also contribute to the development of cancer by adversely affecting the normal maintenance of the genome's integrity through accurate DNA replication and effi-

cient DNA repair, with the result that the cell accumulates mutations.

Various types of radiation and many chemicals—collectively known as carcinogens—increase the frequency with which cells become cancerous. Practically all carcinogens act by causing changes in the genome of the cell. A few chemical carcinogens act directly on the genome; the majority act indirectly by being converted by cellular enzymes to active derivatives called ultimate carcinogens. Through understanding what carcinogens exist in the environment, we can position ourselves to minimize their effects on us and thereby perhaps decrease our risk of cancer.

ANALYTICAL APPROACHES FOR SOLVING GENETICS PROBLEMS

Q17.1 An investigator has found a retrovirus capable of infecting human nerve cells. This is a complete virus, able to reproduce itself, and it contains no oncogenes. People who are infected suffer a debilitating encephalitis. The investigator has shown that when he infects nerve cells in culture with the complete virus, the nerve cells are killed as the virus reproduces. But if he infects cultured nerve cells with a virus in which he has created deletions in the *env* or *gag* genes, no cell death occurs. The investigator is interested in finding ways to bring about nerve cell growth or regeneration in people who have suffered nerve damage. For example, in a patient with a severed spinal cord, nerve regeneration might relieve paralysis. The investigator has cloned the human nerve growth factor gene and wants to insert it into the genome of the retrovirus from which he has deleted parts of the *env* and *gag* genes. He would then use the engineered retrovirus to infect cultured nerve cells. Adult nerve cells do not normally produce large amounts of nerve growth factor. If he is successful in inducing growth in them without causing any cell death, he would like to move on to clinical trials on injured patients. When the investigator applied for grant support to do this work, his application was denied on the grounds that there were inadequate safeguards in the plan. Why might this work be dangerous? What comparisons can you draw between the virus the investigator wants to create and, for example, avian myeloblastosis virus (see Figure 17.8)?

A17.1 In engineering the retrovirus in the way he plans, the investigator would probably be creating a new cancer virus in which the cloned nerve growth factor gene would be the oncogene. It is, of course, an advantage that the engineered virus would not be able to reproduce itself,

but we know that many "wild" cancer viruses are also defective and reproduce with the help of other viruses. If the engineered virus were to infect cells carrying other viruses (for example, wild-type versions of itself) that could supply the *env* and *gag* functions, the new virus could be reproduced and spread. Presumably, infection of normal nerve cells in vivo by the engineered retrovirus would sometimes result in abnormally high levels of nerve growth factor, and thus perhaps in the production of nervous system cancers.

In avian myeloblastosis virus the *pol* and *env* genes are partially deleted. Thus, like our investigator's virus, AMV needs a helper virus to reproduce. In AMV the *myb* oncogene has been inserted; it encodes a nuclear protein presumably involved in control of gene expression. In our new virus the oncogene would be the cloned nerve growth factor gene.[2]

QUESTIONS AND PROBLEMS

*17.1 What is the difference between a hereditary cancer and a sporadic cancer?

17.2 Distinguish between a transducing retrovirus and a nontransducing retrovirus.

*17.3 In what ways is the mechanism of cell transformation by transducing retroviruses fundamentally different from transformation by DNA tumor viruses? Even though the mechanisms are different, how are both able to cause neoplastic growth?

---

[2]Apropos this question, it is interesting to note that now that the safety of retroviral vectors has been established, they are being used successfully to deliver genes in the treatment of certain human diseases.

*17.4   Although there has been a substantial increase in our understanding of the genetic basis for cancer, the vast majority of cases of many types of cancer are not hereditary.

a. How might studying a hereditary form of a cancer provide insight into a similar, more frequent sporadic form?

b. The incidence of cancer in several members of an extended family might reasonably raise concern as to whether there is a genetic predisposition for cancer in the family. What does the term *genetic predisposition* mean? What might be the basis of a genetic predisposition to a cancer that appears as a dominant trait? What issues must be addressed before concluding that a genetic predisposition for a specific type of cancer exists in a particular family?

17.5   Cellular proto-oncogenes and viral oncogenes are related in sequence, but they are not identical. What fundamental difference is there between the two?

*17.6   Material that has been biopsied from tumors is useful for discerning both the type of tumor and the stage to which a tumor has progressed. It has been known for a long time that biopsied tissues with more differentiated cellular phenotypes are associated with less advanced tumors. Explain this finding in terms of the multistep nature of cancer.

*17.7   The sequences of proto-oncogenes are highly conserved among a large number of animal species. Based on this, what hypothesis can you make about the functions of the proto-oncogenes?

17.8   Explain why HIV-1, the causative agent of AIDS, is considered a nononcogenic retrovirus even though numerous types of cancers are frequently seen in AIDS patients.

17.9   Give two ways in which cancer can be induced by a retrovirus.

*17.10   Proto-oncogenes produce a diverse set of gene products.

a. What types of gene products are made by proto-oncogenes? Do these gene products share any feature(s)?

b. Which of the following mutations might result in an oncogene?

   i. A deletion of the entire coding region of a proto-oncogene

   ii. A deletion of a silencer that lies 5' to the coding region

   iii. A deletion of an enhancer that lies 3' to the coding region

   iv. A deletion of a 3' splice-site acceptor region

   v. The introduction of a premature stop codon

   vi. A point mutation

   vii. A translocation that places the coding region near a constitutively transcribed gene

   viii. A translocation that places the gene near constitutive heterochromatin

*17.11   You have a culture of normal cells and a culture of cells dividing uncontrollably (isolated from a tumor). Experimentally, how might you determine whether uncontrolled growth was the result of an oncogene or a mutated pair of tumor suppressor alleles?

17.12   What are the four main ways in which a proto-oncogene can be changed into an oncogene?

*17.13   After a retrovirus that does not carry an oncogene infects a particular cell, northern blots indicate that the amount of mRNAs transcribed from a particular proto-oncogene became elevated approximately 13-fold compared with uninfected control cells. Propose a hypothesis to explain this result.

17.14   Explain how progression through the cell cycle is regulated by the phosphorylation of the retinoblastoma protein pRB. What phenotype(s) might you expect in cells where

a. pRB was constitutively phosphorylated?

b. pRB was never phosphorylated?

c. a severely truncated pRB protein was produced that could not be phosphorylated?

d. a normal pRB protein was produced at higher than normal levels?

e. a normal pRB protein was produced at lower than normal levels?

17.15   Mutations in the *p53* gene appear to be a major factor in the development of human cancer.

a. Explain what the normal cellular functions of the *p53* gene product are and how alterations in these functions can lead to cancer.

b. Suppose cells in a cancerous growth are shown to have a genetic alteration that results in diminished *p53* gene function. Why can we not immediately conclude that the mutation has *caused* the cancer? How would the effect of the mutation be viewed in light of the current, multistep model of cancer?

*17.16   What is apoptosis? Why is the cell death associated with apoptosis desirable, and how is it regulated?

17.17   What mechanisms ensure that cells with heavily damaged DNA are unable to replicate?

17.18   How do radiation and chemical carcinogens induce cancers?

CHAPTER *18*

DNA MUTATION AND REPAIR

PRINCIPAL POINTS

~ Changes in heritable traits result from random mutation rather than by adaptation to environmental influences.

~ Mutation is the process by which the sequence of base pairs in a DNA molecule is altered. The alteration can be as simple as a single base-pair substitution, insertion, or deletion, or as complex as rearrangement, duplication, or deletion of whole sections of a chromosome. Mutations may occur spontaneously or may be induced experimentally by the application of mutagens.

~ Mutations at the level of the chromosome are called chromosomal mutations (see Chapter 7). Mutations in the sequences of genes at the level of the base pair are called gene mutations.

~ The consequences to an organism of a gene mutation depend on a number of factors, especially the extent to which the amino acid coding information for a protein is changed. For example, missense mutations cause the substitution of one amino acid for another, and nonsense mutations cause premature termination of polypeptide synthesis.

~ The effects of a gene mutation can be reversed either by reversion of the base-pair sequence to its original state, or by a mutation at a site distinct from that of the original mutation. The latter is called a suppressor mutation.

~ High-energy radiation may cause genetic damage by producing chemicals that interact with DNA or by causing unusual bonds between DNA bases. Mutations result if the genetic damage is not repaired. Ionizing radiation may also break chromosomes.

~ Gene mutations may also be caused by exposure to a variety of chemicals called chemical mutagens.

~ In prokaryotes and eukaryotes, a number of repair enzymes deal with different kinds of DNA damage. Not all DNA damage is repaired; hence, mutations do appear, but at relatively low frequencies. At high doses of mutagens, repair systems are unable to correct all of the damage, and cancer (in the case of eukaryotic, multicellular organisms) or cell death results.

~ Geneticists have made great progress in understanding how cellular processes take place by studying mutants that have defects in those processes. A number of screening procedures have been developed to help find mutants of interest after mutagenizing cells or organisms, and to detect mutations generated by specific molecular targeting methods.

---

*D*NA can be changed in a number of ways, such as by spontaneous changes, errors in the replication process, or the action of particular chemicals or radiation. There are two broad types of changes to the genetic material: *chromosomal mutations*, changes involving whole chromosomes or sections of them (the topic of Chapter 7), and *point mutations*, changes of one or a few base pairs.

A point mutation will have no phenotypic consequences to the organism unless it occurs within a gene or in the sequences regulating the gene. Thus, the point mutations that have been of particular interest to geneticists are *gene mutations*, that is, those that affect the function of genes.

In this chapter your goal is to learn about some of the mechanisms that cause point mutations, some of the repair systems that can repair genetic damage, and some of the methods used to select for genetic mutants. As we learn about the specifics of point mutations, we must be aware that mutations are a major source of genetic variation in a species and, hence, are important elements of the evolutionary process.

ADAPTATION VERSUS MUTATION

In the early part of the twentieth century, some geneticists believed that variation among organisms resulted from *adaptation* rather than mutation; that is, that the environment induced an inheritable change. Some observations with bacteria fueled the controversy. Wild-type *E. coli*, for example, is sensitive to the virulent bacteriophage T1. However, if a culture of wild-type *E. coli* started from a single cell is plated in the presence of an excess of phage T1, most of the bacteria are killed, but a very few survive and produce colonies because they are resistant to infection by T1.

The resistance trait is heritable. Supporters of the adaptation theory argued that the resistance trait arose as a result of the presence of the T1 phage in the environment. In the opposite camp, supporters of the mutation theory argued that mutations occur randomly, so that at any time in a large enough population of cells, some cells have undergone a mutation that makes them resistant to T1 (for this example) even though they have never been exposed to T1. When T1 is subsequently added, the T1-resistant bacteria are selected for.

The acquisition of resistance to T1 was used by Luria and Delbrück in 1943 to prove that the mutation mechanism was correct and the adaptation mechanism was incorrect. The test they used is known as the *fluctuation test*, which we now describe.

Consider a dividing population of wild-type *E. coli* that started with a single cell (Figure 18.1). Assume that phage T1 is added at generation 4, when there are 16 cells. If the adaptation theory is correct, a certain proportion of the generation 4 cells will be induced at that time to become resistant to T1. Most importantly, *that proportion will be the same for all identical cultures because adaptation would not commence until T1 was added.* However, if the mutation theory is correct, then the number of generation 4 cells that are resistant to T1 will depend on when in the culturing process the random mutational event occurred that confers resistance to T1. If the mutational event occurs in generation 3 in our example, then 2 of the 16 generation 4 cells will be T1-resistant (Figure 18.1a). However, if the mutational event occurs instead at generation 1, then 8 of the 16 generation 4 cells will be T1-resistant (Figure 18.1b). The key point is that if the mutation theory is correct, there should be a *fluctua-*

tion in the number of T1-resistant cells in generation 4 because the mutation to T1 resistance occurred randomly in the population and did not require the presence of T1.

Luria and Delbrück observed a large range in the number of resistant colonies among identical cultures. This high degree of fluctuation in the number of resistant bacteria was taken as proof that resistance was due to random mutation rather than adaptation.

KEYNOTE

Identical cultures of bacteria produced a large range in the number of phage-resistant colonies when phage was added. This was taken as evidence that heritable traits result from random mutation rather than by adaptation as a result of environmental influence.

MUTATIONS DEFINED

Mutation is the process by which the sequence of base pairs in a DNA molecule is altered. A mutation, then, is a DNA base-pair change or chromosome change.

A cell with a mutation is a mutant cell. If a mutant cell gives rise only to somatic cells (in multicellular organisms), the mutant characteristic only affects the individual in which the mutation occurs and is not passed on to the succeeding generation. This type of mutation is called a **somatic mutation**. However, mutations in the germ line of sexually

~ **FIGURE 18.1**

Representation of a dividing population of T1-phage-sensitive wild-type *E. coli*. At generation 4, T1 phage is added. (a) If one cell mutates to resistance to T1 phage infection at generation 3, then 2 of the 16 cells at generation 4 will be resistant to T1; (b) If one cell mutates to resistance to T1 phage infection at generation 1, then 8 of the 16 cells at generation 4 will be resistant to T1.

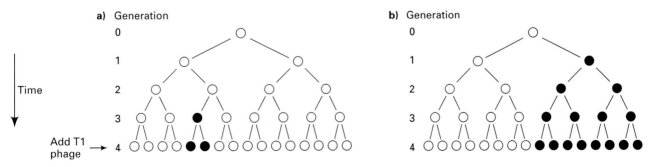

reproducing organisms—**germ-line mutations**—may be transmitted by the gametes to the next generation, producing an individual with the mutation in both its somatic and germ-line cells.

Types of Point Mutations

Mutations can occur spontaneously, but they can also be induced experimentally by the application of a **mutagen**, any physical or chemical agent that significantly increases the frequency of mutational events above the spontaneous mutation rate. Mutations that result from treatment with mutagens are called **induced mutations**; naturally occurring mutations are **spontaneous mutations**.

Figure 18.2 illustrates a number of terms used to describe different types of mutation. Each example is a type of point mutation (**base-pair substitution mutation**), that is, a change from one base pair to another base pair.

A **transition mutation** (Figure 18.2a) is a general term for a mutation from one purine-pyrimidine base pair to the other purine-pyrimidine base pair. The four types of transition mutations are AT to GC, GC to AT, TA to CG, and CG to TA.

A **transversion mutation** (Figure 18.2b) is a general term for a mutation from a purine-pyrimidine base pair to a pyrimidine-purine base pair. The four types of transversion mutations are AT to TA, GC to CG, AT to CG, and GC to TA.

When mutations occur in protein-coding genes, they are defined according to their effects on amino acid sequences in proteins, as follows:

A **missense mutation** (Figure 18.2c) is a gene mutation in which a base-pair change in the DNA causes a change in an mRNA codon so that a different amino acid is inserted into the polypeptide. A phenotypic change may or may not result depending on the amino acid change involved. In Figure 18.2c, an AT-to-GC transition mutation changes the DNA from $\frac{5'\text{-AAA-}3'}{3'\text{-TTT-}5'}$ to $\frac{5'\text{-GAA-}3'}{3'\text{-CTT-}5'}$, altering the mRNA codon from 5'-AAA-3' (lysine) to 5'-GAA-3' (glutamic acid).

A **nonsense mutation** (Figure 18.2d) is a gene mutation in which a base-pair change in the DNA results in the change of an mRNA codon from one for an amino acid to one for a stop (nonsense) codon (UAG, UAA, or UGA). For example, in Figure 18.2d, an AT-to-TA transversion mutation changes the DNA from $\frac{5'\text{-AAA-}3'}{3'\text{-TTT-}5'}$ to $\frac{5'\text{-TAA-}3'}{3'\text{-ATT-}5'}$ and this changes the mRNA codon from 5'-AAA-3' (lysine) to 5'-UAA-3', which is a nonsense codon. Because a nonsense mutation causes chain termination at an incorrect place

in a polypeptide, the mutation prematurely ends the polypeptide. Instead of complete polypeptides, polypeptide fragments (often nonfunctional) are released from the ribosomes.

A **neutral mutation** (Figure 18.2e) is a base-pair change in a gene that changes a codon in the mRNA such that the resulting amino acid substitution produces no detectable change in the function of the protein translated from that message. A neutral mutation is a subset of missense mutations where the new codon codes for a different amino acid that is chemically equivalent to the original and hence does not affect the protein's function. In Figure 18.2e, an AT-to-GC transition mutation changes the codon from 5'-AAA-3' to 5'-AGA-3', which substitutes the basic amino acid arginine for the basic amino acid lysine.

A **silent mutation** (Figure 18.2f) is also a subset of missense mutations and is when a base-pair change in a gene alters a codon in the mRNA such that the *same* amino acid is inserted in the protein. The protein in this case obviously has wild-type function. In Figure 18.2f, a silent mutation results from an AT-to-GC transition mutation that changes the codon from 5'-AAA-3' to 5'-AAG-3', both of which specify lysine.

A **frameshift mutation** (Figure 18.2g) changes the reading frame of an mRNA downstream of the mutation. An addition or deletion of one base pair, for example, shifts the mRNA's downstream reading frame by one base so that incorrect amino acids are added to the polypeptide chain after the mutation site. Often, frameshift mutations generate new codons resulting in a shortened protein, or they result in read-through of the normal stop codon, resulting in longer than normal proteins. In any case, a frameshift mutation usually results in a nonfunctional protein. In Figure 18.2g, an insertion of a GC base pair scrambles the message after the codon specifying glutamine. Since each codon consists of three bases, a frameshift mutation is produced by the insertion or deletion of any number of base pairs in the DNA that is not divisible by three.

KEYNOTE

Mutation is the process that alters the sequence of base pairs in a DNA molecule. The alteration can be as simple as a single base-pair substitution, insertion, or deletion, or as complex as rearrangement, duplication, or deletion of whole sections of a chromosome. Mutations in the sequences of genes are called gene mutations.

~ FIGURE 18.2

Types of base-pair substitution mutations. Transcription of the segment shown produces an mRNA with the sequence 5' . . . UCUCAAAAAUUUACG . . . 3', which encodes . . . -Ser-Gln-Lys-Phe-Thr- . . .

Sequence of part of a normal gene	Sequence of mutated gene

a) Transition mutation (AT to GC in this example)

5'···· TCTCAAAAATTTACG ···3'
3'···· AGAGTTTTTAAATGC ···5'

5'···· TCTCAAGAATTTACG ···3'
3'···· AGAGTTCTTAAATGC ···5'

b) Transversion mutation (CG to GC in this example)

5'···· TCTCAAAAATTTACG ···3'
3'···· AGAGTTTTTAAATGC ···5'

5'···· TCTGAAAAATTTACG ···3'
3'···· AGACTTTTTAAATGC ···5'

c) Missense mutation (change from one amino acid to another; here a transition mutation from AT to GC changes the codon from lysine to glutamic acid)

5'··· TCTCAAAAATTTACG ···3'
3'··· AGAGTTTTTAAATGC ···5'

5'··· TCTCAAGAATTTACG ···3'
3'··· AGAGTTCTTAAATGC ···5'

···· - Ser — Gln — Lys — Phe — Thr ····

···· - Ser — Gln — Glu — Phe — Thr ····

d) Nonsense mutation (change from an amino acid to a stop codon; here a transversion mutation from AT to TA changes the codon from lysine to UAA stop codon)

5'··· TCTCAAAAATTTACG ···3'
3'··· AGAGTTTTTAAATGC ···5'

5'··· TCTCAATAATTTACG ···3'
3'··· AGAGTTATTAAATGC ···5'

···· - Ser — Gln — Lys — Phe — Thr ····

···· - Ser — Gln — Stop

e) Neutral mutation (change from an amino acid to another amino acid with similar chemical properties; here an AT to GC transition mutation changes the codon from lysine to arginine)

5'··· TCTCAAAAATTTACG ···3'
3'··· AGAGTTTTTAAATGC ···5'

5'··· TCTCAAAGATTTACG ···3'
3'··· AGAGTTTCTAAATGC ···5'

···· - Ser — Gln — Lys — Phe — Thr ····

···· - Ser — Gln — Arg — Phe — Thr ····

f) Silent mutation (change in codon such that the same amino acid is specified; here an AT-to-GC transition in the third position of the codon gives a codon that still encodes lysine)

5'··· TCTCAAAAATTTACG ···3'
3'··· AGAGTTTTTAAATGC ···5'

5'··· TCTCAAAAGTTTACG ···3'
3'··· AGAGTTTTCAAATGC ···5'

···· - Ser — Gln — Lys — Phe — Thr ····

···· - Ser — Gln — Lys — Phe — Thr ····

g) Frameshift mutation (addition or deletion of one or a few base pairs leads to a change in reading frame; here the insertion of a GC base pair scrambles the message after glutamine)

5'··· TCTCAAAAATTTACG ···3'
3'··· AGAGTTTTTAAATGC ···5'

5'··· TCTCAAGAAATTTACG ···3'
3'··· AGAGTTCTTTAAATGC ···5'

···· - Ser — Gln — Lys — Phe — Thr ····

···· - Ser — Gln — Glu — Ile — Tyr ····

Reverse Mutations and Suppressor Mutations

There are two classes of point mutations in terms of their effects on the phenotype: (1) **forward mutations** change the genotype from wild type to mutant, and (2) **reverse mutations** (or **reversions** or **back mutations**) change the genotype from mutant to wild type or to partially wild type. Reversion of a nonsense mutation, for instance, occurs when a base-pair change results in a change of the mRNA nonsense codon to a codon for an amino acid. If this gene reversion is back to the wild-type amino acid, the mutation is a **true reversion**. If the reversion is to some other amino acid, the mutation is a *partial reversion,* and complete or partial function may be restored. Reversion of missense mutations occurs in the same way.

The effects of a mutation may be diminished or abolished by a **suppressor mutation**, that is, a mutation at a different site from the original mutation. *A suppressor mutation does not result in a reversal of the original mutation*; instead, it masks or compensates for the effects of the initial mutation.

Suppressor mutations may occur within the same gene as the original mutations but at a different site (called **intragenic** [*intra* = within] **suppressors**), or in a different gene (called **intergenic** [*inter* = between] **suppressors**). Intragenic suppressors act either by altering a different nucleotide in the same codon in which the original mutation occurred, or by altering a nucleotide in a different codon. An example of the latter is the suppression of a frameshift mutation by a base-pair addition near a base-pair deletion.

Intergenic suppression occurs as a result of a second mutation in another gene. Genes that cause suppression of mutations in other genes are called **suppressor genes**. Often, suppressor genes are tRNA genes. For example, in the case of nonsense suppressors, particular tRNA genes can mutate so that their anticodons now recognize a chain-terminating codon and put an amino acid into the chain. Thus, instead of polypeptide chain synthesis being stopped prematurely at a nonsense mutation, the altered (suppressor) tRNA inserts the amino acid it carries at that position, and a complete polypeptide results. How functional the polypeptide will be will depend on the effects of the inserted amino acid. If it is in an important part of the polypeptide, then the incorrect amino acid may not restore function to a significant degree. If it is in a less crucial area, the polypeptide may have some or complete function. Since there are three stop codons (UAG, UAA, and UGA), there are three classes of nonsense suppressors.

But we now have a dilemma. If we have changed a particular tRNA so that its anticodon can now read a nonsense codon, it cannot read the original codon that specifies the amino acid it carries. Thus, nonsense suppressor tRNAs are typically produced by mutation of tRNA genes that are redundant in the genome, that is, for which several different genes all specifying the same tRNA base sequence exist.

KEYNOTE

A suppressor mutation is a mutation at a second site that completely or partially restores a function lost because of a primary mutation at another site. Intragenic suppressors are suppressor mutations that occur within the same gene as the original mutation but at a different site. They act either by altering a different nucleotide in the same codon as that in which the original mutation occurred, or by altering a nucleotide in a different codon. Intergenic suppressors are suppressor mutations that occur in a different gene—a suppressor gene—from that with the original mutation. Intergenic suppressors typically involve a tRNA with an altered anticodon.

SPONTANEOUS AND INDUCED MUTATIONS

Spontaneous mutations occur naturally, without the use of experimentally applied chemical or physical mutagenic agents. Induced mutations occur as a result of treatment with known chemical or physical mutagens.

Spontaneous Mutations

Two different terms are used to give a quantitative measure of the occurrence of mutations. **Mutation rate** is the probability of a particular kind of mutation as a function of time, such as number per nucleotide pair per generation, or number per gene per generation. **Mutation frequency** is the number of occurrences of a particular kind of mutation expressed as the proportion of cells or individuals in a population, such as number per 100,000 organisms or number per 1 million gametes.

In humans, the spontaneous mutation rate for individual genes varies between 10^{-4} and 4×10^{-6} per gene per generation. For eukaryotes in general, the spontaneous mutation rate is 10^{-4} to 10^{-6} per gene per generation, and for bacteria and phages the rate is 10^{-5} to 10^{-7} per gene per generation. (The spontaneous mutation frequencies at specific loci for various organisms are presented in Table 21.4.) *Note that the rates and frequency values represent the mutations that become fixed in DNA.* Most spontaneous errors are corrected by cellular repair systems; only a few remain uncorrected as permanent changes.

All types of point mutations can occur spontaneously. Spontaneous mutations can occur during DNA replication as well as during the G_1 and G_2 phases of the cell cycle. Spontaneous mutations can also result from the movement of what are called transposable genetic elements, which we discuss in Chapter 19.

DNA REPLICATION ERRORS. Both point mutations and small additions and deletions can result from DNA replication errors. Point mutations, for example, can occur if mismatched base pairs occur during DNA replication. Mismatching may occur if the bases are involved in "wobble" pairing, in which the *normal* form of the base is paired with an incorrect partner because of a different spatial positioning of the atoms that can form hydrogen bonds. (Wobble in tRNA pairing with mRNA during translation is discussed in Chapter 13.) Figure 18.3a shows the normal AT and GC Watson-Crick base pairs, while Figure 18.3b shows wobble pyrimidine-purine pairing between T and G and between C and A, and Figure 18.3c shows wobble purine-purine pairing between A and G, and pyrimidine-pyrimidine pairing between T and C.

~ FIGURE 18.3

Normal and wobble base-pairing in DNA. (a) Normal Watson-Crick base-pairing between A and T and between G and C; (b) Wobble pyrimidine-purine base-pairing between T and G and between C and A; (c) Wobble purine-purine base-pairing between A and G and wobble pyrimidine-pyrimidine pairing between T and C.

a) Normal Watson-Crick base pairing

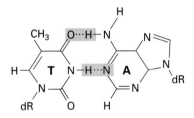

b) Wobble pyrimidine-purine base pairing

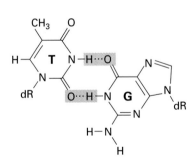
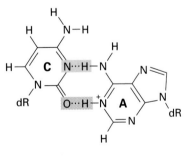

c) Wobble purine-purine and pyrimidine-pyrimidine base pairing

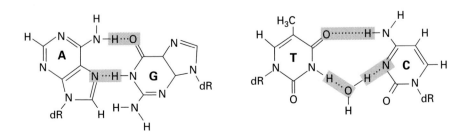

~ FIGURE 18.4

Production of a mutation as a result of a mismatch caused by wobble base-pairing. The details are explained in the text.

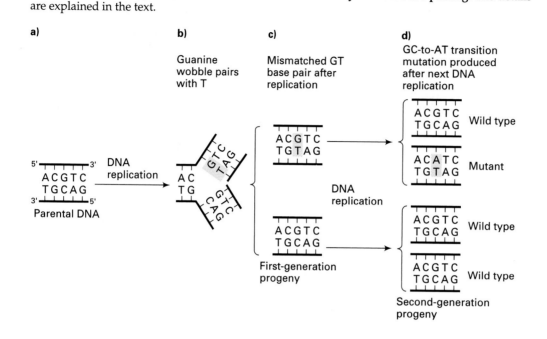

Figure 18.4 illustrates how a GC-to-AT transition mutation can be produced as a result of a mismatch caused by wobble pairing. When the parental DNA (Figure 18.4a) is replicating, a guanine could wobble pair with T (Figure 18.4b), giving a mismatched GT base pair after replication (Figure 18.4c). In the next round of DNA replication, the G likely will pair normally and give a GC pair in the progeny DNA, while the T specifies an AT in the other progeny DNA molecule (Figure 18.4d); this is the GC-to-AT transition mutation.

Note that the GT mismatched pair is a target for correction by proofreading during replication and by other repair systems. Only uncorrected mismatches lead to mutations in the next round of replication, as illustrated in Figure 18.4.

SPONTANEOUS CHEMICAL CHANGES. Two of the most common chemical events that occur to produce spontaneous mutations are *depurination* and *deamination* of particular bases. In depurination a purine, either adenine or guanine, is removed from the DNA when the bond breaks between the base and the deoxyribose sugar. If such lesions are not repaired, there is no base to specify a complementary base during DNA replication. Instead, a randomly chosen base is inserted, which can result in a mutation.

Deamination is the removal of an amino group from a base. For example, the deamination of cytosine

produces uracil (Figure 18.5). Uracil is not a normal base in DNA, although it is a normal base in RNA. A repair system described later in this chapter removes most of the uracils in DNA, thereby minimizing the mutational consequences of cytosine deamination. However, if the uracil is not repaired, an adenine will be incorporated in the new DNA strand opposite it during replication. Ultimately this will result in the conversion of a CG base pair to a TA base pair; that is, a transition mutation.

Induced Mutations

Since the rate of spontaneous mutation is so low, geneticists use mutagens to increase mutation frequency so that a significant number of organisms have

~ FIGURE 18.5

Deamination of cytosine to uracil.

Cytosine   Uracil

~ **FIGURE 18.6**

Production of thymine dimers by ultraviolet light irradiation. The two components of the dimer are covalently linked in such a way that the DNA double helix is distorted at that position.

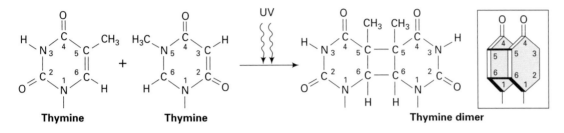

mutations in the gene being studied. Two classes of mutagens are used—radiation and chemical—both of which involve specific mechanisms of action.

RADIATION.   Both X rays and ultraviolet light (UV) are used to induce mutations. X rays penetrate tissue and collide with molecules, knocking electrons out of orbits and thus creating ions. The ions can break covalent bonds, including those in the sugar-phosphate backbone of DNA. In fact, ionizing radiation is the leading cause of gross chromosomal mutations in humans.

High doses of ionizing radiation kill cells, hence their use in treating some forms of cancer. At certain low levels of ionizing radiation, point mutations are commonly produced; at these levels, there is a linear relationship between the rate of point mutations and radiation dosage. Importantly, for many organisms, including humans, the effects of ionizing radiation doses are cumulative. That is, if a particular dose of radiation results in a certain number of point mutations, the same number of point mutations will be induced whether the radiation dose is received over a short period of time or over a long period of time. Clearly, then, people should be concerned about their exposure to ionizing radiation such as X rays.

Ultraviolet light (UV) rays have insufficient energy to induce ionizations. Nonetheless, ultraviolet light is a useful mutagen, and at high enough doses it can kill cells. For geneticists, other scientists, and medical personnel, this property is a useful one: UV light is used as a sterilizing agent in some applications. Ultraviolet light causes mutations because the purine and pyrimidine bases in DNA absorb light strongly in the ultraviolet range (254 to 260 nm). At this wavelength, UV light induces point mutations primarily by causing photochemical (light-induced chemical) changes in the DNA.

The sun is a very powerful source of UV radiation, but much of the UV light is screened out by the ozone in the atmosphere. Nonetheless, significant amounts of UV radiation are present in sunlight, so the potential for mutagenesis clearly exists.

One of the effects of UV radiation on DNA is the formation of abnormal chemical bonds between adjacent pyrimidine molecules in the same strand or between pyrimidines on the opposite strands of the double helix. This bonding is induced mostly between adjacent thymines, forming what are called *thymine dimers* (Figure 18.6), usually designated T^T. (C^C, C^T, and T^C pairs are also produced by UV radiation, but in much lower amounts.) This unusual pairing produces a bulge in the DNA strand and disrupts the normal pairing of Ts with corresponding As on the opposite strand. Many thymine dimers are repaired (see pp. 420–421). All kinds of pyrimidine dimers have the potential to cause problems during DNA replication, and if enough of them remain unrepaired in the cell, cell death may result.

KEYNOTE

Mutations may be induced by using radiation or chemical mutagens. Radiation may cause genetic damage by producing chemicals that affect the DNA (as in the case of X rays) or by causing the formation of unusual bonds between DNA bases, such as thymine dimers (as in the case of ultraviolet light). If radiation-induced genetic damage is not repaired, mutations or cell death may result. Ionizing radiation may also break chromosomes.

CHEMICAL MUTAGENS.   Chemical mutagens can be grouped into different classes based on their mechanism of action. Here we discuss base analogs, base-modifying agents, and intercalating agents and explain how they induce mutations.

Base Analogs. **Base analogs** are bases that are extremely similar to the bases normally found in DNA. Base analogs exist in alternate states, a normal state and a rare state. These alternate chemical states are called **tautomers**. In each of the two states the base analog pairs with a different base in DNA.

5-bromouracil (5BU) is a base analog mutation. 5BU has a bromine residue instead of the methyl group of thymine. In the normal state, 5BU resembles thymine and will pair only with adenine in DNA (Figure 18.7a). In its rare state it pairs only with guanine (Figure 18.7b). 5BU induces mutations by switching between the two forms once the base analog has been incorporated into the DNA (Figure 18.7c).

If, for example, 5BU is incorporated in its normal state, it pairs with adenine. If it changes into its rare

---

~ Figure 18.7

Mutagenic effects of the base analog 5-bromouracil (5BU). (a) In its normal state, 5BU pairs with adenine. (b) In its rare state, 5BU [indicated by white letters on magenta in part (c)] pairs with guanine. (c) The two possible mutation mechanisms. 5BU induces transition mutations when it incorporates into DNA in one state, then shifts to its alternate state during the next round of DNA replication.

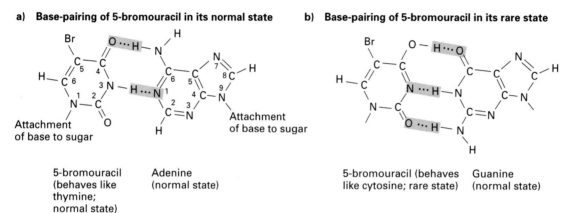

a) **Base-pairing of 5-bromouracil in its normal state**
b) **Base-pairing of 5-bromouracil in its rare state**

c) **Mutagenic action of 5BU**

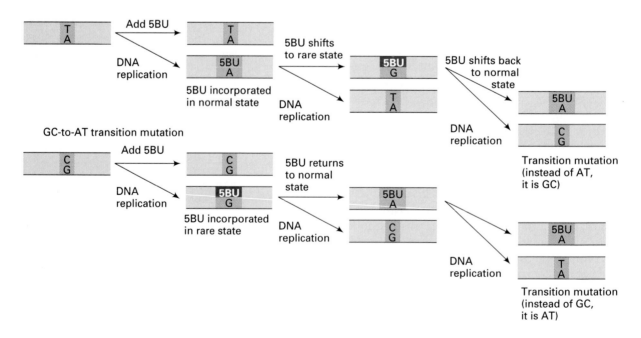

state during replication, it will pair instead with guanine. In the next round of replication the 5BU-G base pair will be resolved into a C-G base pair instead of the T-A base pair. By this process a transition mutation is produced, from TA to CG. 5BU can induce a mutation from CG to TA if it is first incorporated into DNA in its rare state and then switches to the normal state during replication (see Figure 18.7c). Thus, 5BU-induced mutations may be reverted by a second treatment of 5BU.

Not all base analogs are mutagens. One of the approved drugs given to AIDS patients, AZT (azidothymidine), is an analog of thymidine, but it is not a mutagen because it does not cause base-pair changes. The AIDS virus, HIV-1 (human immunodeficiency virus-1), is a retrovirus. That is, its genome is RNA, but when the virus enters a cell, the viral enzyme reverse transcriptase (see Chapters 14 and 17) makes a cDNA copy of the RNA. That cDNA is then incorporated into the cell's genomic DNA, from where it directs new viral synthesis. During the reverse transcriptase step, AZT (as a triphosphate derivative) is incorporated into the growing DNA chain as a thymidine analog. However, AZT has an azido group (N_3) on the 3' carbon of its deoxyribose sugar, and the missing OH group prevents further growth of the DNA chain. AZT is recognized as a substrate for reverse transcriptase, but not by DNA polymerase, so host DNA synthesis is not affected. Thus, AZT acts as a selective poison by inhibiting the production of the viral cDNA. This blocks new viral synthesis because a cDNA must be made and incorporated into the host genome in order to code for viral components.

Base-Modifying Agents.   A number of chemicals act as mutagens by modifying the chemical structure and properties of the bases. Alkylating agents, for example, introduce alkyl groups (such as $-CH_3$ and $-CH_2CH_3$)

onto the bases at a number of locations. Most mutations caused by alkylating agents result from the addition of an alkyl group to the 6-oxygen of guanine to produce O^6-alkylguanine. After treatment with methylmethane sulfonate (MMS), for example, some guanines are methylated to produce O^6-methylguanine (Figure 18.8). The methylated guanine pairs with thymine rather than cytosine, giving GC-to-AT substitutions.

Intercalating Agents.   Intercalating mutagens, including proflavin, acridine, and ethidium bromide (commonly used to stain DNA in gel electrophoresis experiments) insert themselves (*intercalate*) between adjacent bases in one or both strands of the DNA double helix (Figure 18.9). They are generally thin, platelike hydrophobic molecules. The chemical structures of proflavin and acridine orange are shown in Figure 18.9a.

If the intercalating agent inserts between adjacent base pairs of the DNA strand that is the template for new DNA synthesis (Figure 18.9b), an extra base (chosen at random: G in the figure) must be inserted in the new DNA strand opposite the intercalating agent. After one more round of replication, during which the intercalating agent is lost, the overall result is a base-pair addition mutation (CG is added in Figure 18.9b). If the intercalating agent inserts into the new DNA strand in place of a base (Figure 18.9c), then when that DNA double helix replicates after the intercalating agent is lost, the result is a base-pair deletion mutation (TA is lost in Figure 18.9c).

If a base-pair addition or base-pair deletion point mutation occurs in a protein-coding gene, the result is a frameshift mutation. Since intercalating agents can cause either additions or deletions, frameshift mutations induced by intercalating agents may be reverted by a second treatment with these same agents.

~ FIGURE 18.8

Example of the mutagenic action of a base-modifying agent. Methylmethane sulfonate (MMS) is an alkylating agent that alkylates guanine, which pairs with thymine to give a GC-to-AT transition. (Note: dr = deoxyribose.)

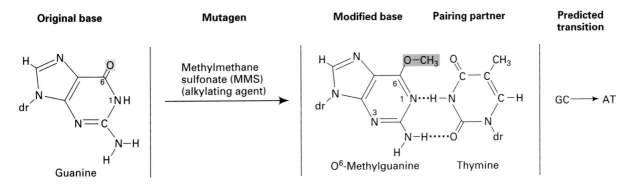

~ FIGURE 18.9

Intercalating mutations. (a) Structures of representative intercalating agents, proflavin and acridine orange; (b) Frameshift mutation by addition, when agent inserts into template strand; (c) Frameshift mutation by deletion, when agent inserts into newly synthesizing strand.

a) **Representative agents**

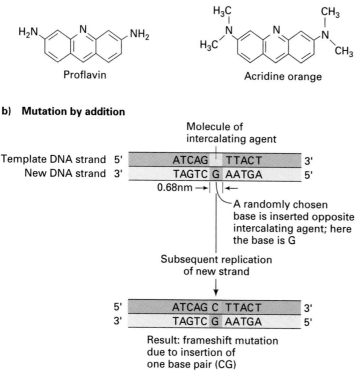

b) **Mutation by addition**

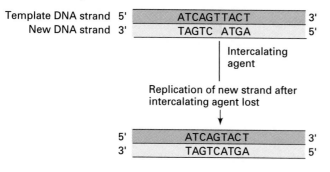

c) **Mutation by deletion**

Template DNA strand 5' ATCAGTTACT 3'
New DNA strand 3' TAGTC ATGA 5'

Intercalating agent

Replication of new strand after intercalating agent lost

5' ATCAGTACT 3'
3' TAGTCATGA 5'

KEYNOTE

Mutations may be produced by exposure to chemical mutagens. If the genetic damage caused by the mutagen is not repaired, mutations result. Chemical mutagens act in a variety of ways. Base analogs substitute for normal bases during DNA replication, then shift form so that their base-pairing properties change. Base-modifying agents cause chemical changes in existing bases, and this alters their base-pairing properties. Base analogs and base-modifying agents result in base-pair substitution mutations. Intercalating agents are inserted between adjacent bases during replication, causing single base-pair additions or deletions, depending on whether the insertion is in the parental or new strand. The result can be a frameshift mutation.

SITE-SPECIFIC IN VITRO MUTAGENESIS OF DNA.
Spontaneous and induced mutations do not occur in specific genes; rather they are scattered essentially randomly throughout the genome. However, most geneticists wish to study the effects of mutations in particular genes. After mutagenizing cells or organisms with radiation or chemical mutagens, they must screen the surviving population to identify those with the mutation(s) of interest. With recombinant DNA technology, we can clone genes and produce large amounts of that DNA for analysis and manipulation. That means it is now possible to mutate a gene at specific positions in the test tube by *site-specific in vitro mutagenesis*, introduce the mutated gene back into the cell by infection or transformation, and investigate the phenotypic changes produced by the mutation in vivo. Such techniques enable geneticists to study, for example, genes with unknown function and specific sequences involved in regulating a gene's expression.

There are many procedures for site-specific mutagenesis, a number of them using the polymerase chain reaction (PCR) (see Chapter 14, Figure 14.19 and pp. 317–320). In brief, PCR is an in vitro method to amplify a segment of target DNA using a specific pair of oligonucleotide primers. Figure 18.10 shows how a point mutation or small addition or deletion can be made in cloned DNA using PCR mutagenesis. Four primers are needed. Primer 1 is at the left end of the sequence to be amplified, and primer 2 is at the right end. Two other primers, 1M and 2M, match the target DNA sequence within its length, except where the mutation (M) is desired; 1M and 2M are complementary to each other. The mutation is symbolized in the figure as a "blip" in the primers. First, a PCR is done with primers 1 and 1M, and a second PCR is done with primers 2 and 2M. Then the primers are removed, and the two products A and B are mixed and the DNAs are denatured and allowed to reanneal. In some cases this will result in pairing of a molecule of single-stranded A with a molecule of single-stranded B. DNA polymerase can then extend the 3' ends of the strands in the central paired region, giving a full-length double-stranded DNA. This full-length molecule with the introduced mutation in the central region is then amplified using primers 1 and 2 and transformed into a cell to replace the wild-type sequence.

THE AMES TEST: A SCREEN FOR POTENTIAL MUTAGENS

Scientists are aware that many chemicals in our environment can have mutagenic effects. And since some chemicals cause mutations that result in cancerous

~ **FIGURE 18.10**

An example of site-specific mutagenesis using PCR.

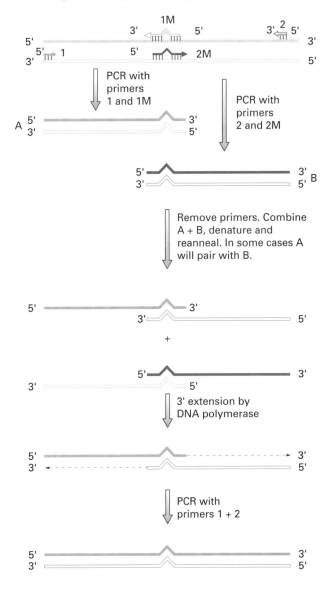

growth, there is a strong interest in testing new chemicals for their ability to induce mutations.

In the early 1970s a rapid test was developed by Bruce Ames. The **Ames test** uses the bacterium *Salmonella typhimurium* as the test organism. Approximately 10^9 cells of tester bacteria that are auxotrophic for histidine (*his* mutants) are spread on a culture plate that lacks the amino acid histidine, conditions under which the mutant bacteria cannot grow. Typically the tester bacteria are a mixture of several *his* strains that allow detection of both base-pair substitution mutations and frameshift mutations in the test. Further, because the strains are defective in a particular repair system, most kinds of DNA damage in the mutant strains are converted to mutations rather than being repaired. Finally, because in humans and other mammals many chemicals are converted to carcinogens in the liver and other tissues by enzymatic detoxification pathways, the culture medium contains a small amount of rat liver homogenate to simulate the effects of those mammalian enzymes. Control plates are also set up that lack the chemical being tested.

After incubation, his^+ colonies are present on both plates. The colonies on the control plate represent revertants that have arisen by spontaneous mutation. Thus, whether the potential mutagen is indeed a mutagen is demonstrated by a significantly higher number of his^+ revertants on the experimental plate, because those colonies represent both induced and spontaneous revertants.

The Ames test is so straightforward to perform that it is routinely used in many laboratories around the world. To date, the Ames test has identified a large number of mutagens, including many environmental chemicals (such as hair dye additives, vinyl chloride, and particular food colorings) that include both synthetic and natural compounds.

DNA REPAIR MECHANISMS

Spontaneous and induced mutations constitute damage to the DNA of a cell or an organism. Especially with high doses of mutagens, the mutational damage can be considerable. Both prokaryotic and eukaryotic cells have a number of enzyme-based repair systems to deal with damage to DNA. If the repair systems are unable to correct all of the lesions, the result is a mutant cell (or organism) or, if too many mutations remain, death of the cell. Life as we know it results from a delicate balance between perfect accuracy in the transmission of DNA to daughter cells and progeny organisms, and the occasional mutation that provides raw material for evolution. DNA repair systems provide that delicate balance.

Direct Correction of Mutational Lesions

REPAIR BY DNA POLYMERASE PROOFREADING.
The frequency of base-pair substitution mutations in bacterial genes varies from 10^{-7} to 10^{-11} errors per generation. However, DNA polymerase makes errors in inserting nucleotides while it is synthesizing the

new DNA strand perhaps at a frequency of 10^{-5}. Most of the difference between the two values is accounted for by 3'-to-5' exonuclease proofreading activity of the polymerase itself (see Chapter 11, p. 239).

The importance of the 3'-to-5' exonuclease activity of DNA polymerase for maintaining a low mutation rate is shown nicely by the existence of *mutator* mutations in *E. coli*. Strains carrying mutator mutations show a much higher-than-normal mutation frequency for all genes. These mutations have been shown to affect proteins whose normal functions are required for accurate DNA replication. For example, the *mutD* mutator gene of *E. coli* encodes a subunit of DNA polymerase III, the primary replication enzyme of *E. coli*. *mutD* mutants are defective in 3'-to-5' proofreading activity, so that many incorrectly inserted nucleotides are left unrepaired.

PHOTOREACTIVATION OF UV-INDUCED PYRIMIDINE DIMERS.
Using **photoreactivation** or **light repair**, UV-light-induced thymine (or other pyrimidine) dimers are reverted directly to the original form by exposure to near UV light in the wavelength range of 320 to 370 nm (see Figure 18.6). Photoreactivation occurs when an enzyme called *photolyase* is activated by a photon of light and splits the dimers apart. Strains with mutations in the *phr* gene are defective in light repair. Photolyase has been found in prokaryotes and in simple eukaryotes, but not in humans. Photolyases are apparently very effective since few thymine dimers (and, hence, mutations) are left after photoreactivation.

REPAIR OF ALKYLATION DAMAGE.
Alkylating agents transfer alkyl groups (usually methyl or ethyl groups) onto the bases at various locations, such as the oxygen of carbon 6 in guanine. The mutagen MMS methylates guanine, for example (see Figure 18.8). Alkylation damage like this can be removed by specific DNA repair enzymes. In this repair system, an enzyme encoded by the *ada* gene, called O^6-*methylguanine methyltransferase*, recognizes the O^6-methylguanine in the DNA and removes the methyl group, thereby changing the base back to its original form.

Repair Involving Excision of Base Pairs

EXCISION REPAIR.
Pyrimidine dimers induced by UV light can be repaired not only by photoreactivation but also by an **excision repair** (**dark repair**) system. As the name implies, this system works in the dark, while the photoreactivation system does

not. Excision repair was discovered when some UV-sensitive mutants of *E. coli* were found that exhibit a higher-than-normal rate of induced mutation in the dark.

The excision repair system in *E. coli* corrects not only pyrimidine dimers but also other serious damage-induced distortions of the DNA helix. The system works as diagrammed in Figure 18.11. UvrABC endonuclease, a multisubunit enzyme encoded by the three genes *uvrA*, *uvrB*, and *uvrC* (*uvr* means "UV repair"), recognizes the distorted DNA. This enzyme makes one cut in the damaged DNA strand eight nucleotides to the 5' side of the damage (for instance, the dimer) and four nucleotides to the 3' side of the damage, releasing a 12-nucleotide stretch of single-stranded DNA containing the damaged base(s). The gap is filled by the 5'-to-3' polymerizing activity of DNA polymerase I and sealed by DNA ligase. An excision repair system is found in most organisms that have been studied.

REPAIR BY MISMATCH REPAIR.   Despite proofreading by DNA polymerase, a significant number of errors still remain uncorrected after replication has been completed. In the next round of replication these errors can become fixed as mutations.

Many mismatched base pairs left after DNA replication may be corrected by another system of repair

called *mismatch repair.* In *E. coli,* the products of three genes, *mutS*, *mutL*, and *mutH*, are involved in the initial stages of mismatch repair (Figure 18.12). In this system, the *mutS* encoded protein, MutS, first binds to the mismatch. The system must now figure out which base is the correct one (the base on the parental DNA strand) and which is the erroneous one (the base on the new DNA strand). In *E. coli,* the two strands are distinguished by methylation of the A nucleotide in the sequence GATC. The GATC sequence is palindromic: that is, the same sequence is present 5'-to-3' on both DNA strands to give $\begin{array}{c}\text{5'-GATC-3'}\\\text{3'-CTAG-5'}\end{array}$. Both A nucleotides in the DNA segment are usually methylated. However, after replication, the parental DNA strand has a methylated A in the GATC sequence, while the A in the GATC of the *newly replicated DNA strand* is not methylated until a short time after its synthesis. In mismatch repair, the MutS protein bound to the mismatch forms a complex with the *mutL*- and *mutH*-encoded proteins, MutL and MutH, to bring the unmethylated GATC sequence close to the mismatch. The MutH protein then nicks the unmethylated DNA strand at the GATC site, the mismatch is removed by an exonuclease, and the gap is repaired by DNA polymerase III and ligase.

Mismatch repair also takes place in eukaryotes. It is not clear yet, however, how the new DNA strand is distinguished from the parental DNA strand. In humans, four genes named *hMSH2*, *hMLH1*, *hPMS1*, and *hPMS2* have been identified: *hMSH2* is homologous to *E. coli mutS*, and the other three genes have homologies to *E. coli mutL*. The genes are known as *mutator genes* because loss of function of such a gene results in increased accumulation of mutations in the genome. Mutations in any one of the four human mismatch repair genes confer a phenotype of hereditary predisposition to a form of colon cancer called hereditary nonpolyposis colon cancer (HNPCC). HNPCC and the role of mutator genes in cancer are described in Chapter 17.

SOS RESPONSE.   Some damage is particularly serious for DNA replication, because the replication enzymes have difficulty at bases in the template DNA strand that cannot specify a complementary base pair on the new strand, so that gaps are often left after the replication fork has passed by. Uncorrected, DNA damage of this kind can be lethal. In *E. coli,* DNA damage induces a complex system in what is called the *SOS response.* (The name *SOS* comes from the fact that the system is induced as an emergency response to mutational damage.) The SOS response operates to allow the cell to survive otherwise lethal events, although often at the expense of generating new mutations.

~ **FIGURE 18.11**

Excision repair of pyrimidine dimer and other damage-induced distortions of DNA initiated by the UvrABC endonuclease.

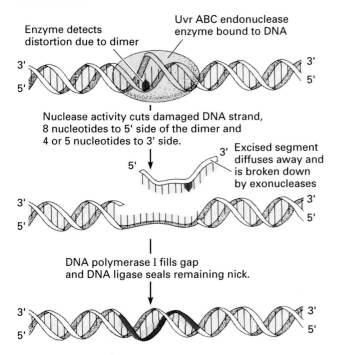

Enzyme detects distortion due to dimer

Uvr ABC endonuclease enzyme bound to DNA

Nuclease activity cuts damaged DNA strand, 8 nucleotides to 5' side of the dimer and 4 or 5 nucleotides to 3' side.

Excised segment diffuses away and is broken down by exonucleases

DNA polymerase I fills gap and DNA ligase seals remaining nick.

~ FIGURE 18.12

Mechanism of mismatch repair. The mismatch repair enzyme recognizes which strand the base mismatch is on by reading the methylation state of a nearby GATC sequence. If the sequence is unmethylated, a segment of that DNA strand containing the mismatch is excised and new DNA is inserted.

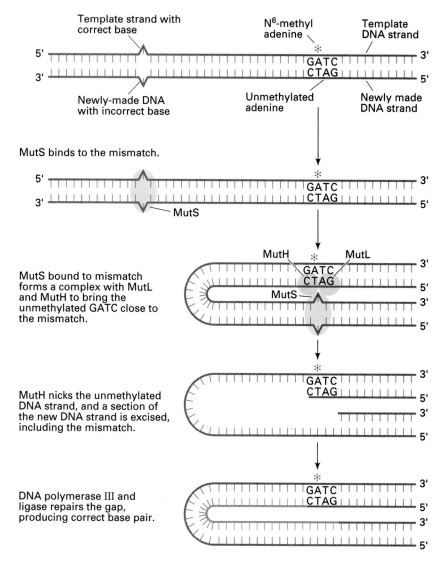

Two genes are key to controlling the SOS system in *E. coli*: *lexA* and *recA*. Cells with mutant *recA* and *lexA* genes have their SOS response permanently turned on. The SOS response works as follows. In the uninduced state—when there is no DNA damage—the *lexA*-encoded protein, LexA, represses the transcription of about 17 genes whose protein products are involved in the repair of various kinds of DNA damage, including excision repair and the repair of gaps (Figure 18.13a). When there has been sufficient DNA damage, the *recA*-encoded protein, RecA, is activated, and this stimulates the LexA protein to cleave itself, which in turn relieves the repression of the DNA repair genes (Figure 18.13b). As a result, the DNA repair genes are transcribed, and DNA repair proceeds. After the DNA damage is dealt with, RecA is inactivated, and newly synthesized LexA protein again represses the DNA repair genes.

The SOS response is, itself, a mutagenic system. For example, during repair by the SOS system, T^T dimers produced by UV irradiation are copied quickly and faithfully onto the new DNA segment to give

~ FIGURE 18.13

Outline of the SOS response. (a) The SOS system in the uninduced state; (b) The SOS system when induced by DNA damage. The details of the SOS response are given in the text.

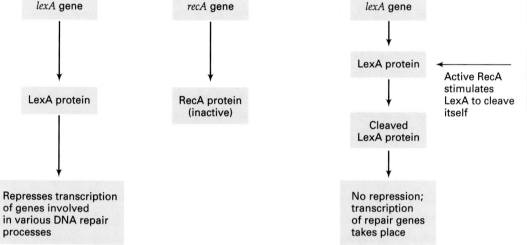

a) **Uninduced state**

b) **Induced state**

AA, the correct nucleotide sequence. However, for dimers involving C nucleotides, the SOS repair system becomes stalled at the dimer because of the mismatch. This delay is time enough for a C in a dimer to become chemically changed (deaminated) to a U (uracil), which remains dimerized to its partner. The U then specifies an A in the new DNA synthesized, and in the next generation the original C-G site will be a T-A site—this is a transition mutation. In fact, CG-to-TA transitions are the common result of UV mutagenesis. In essence, this is an *error-free bypass synthesis* since the template is faithfully copied into the new DNA strand.

KEYNOTE

Mutations constitute damage to the DNA. Both prokaryotes and eukaryotes have a number of repair systems that deal with different kinds of DNA damage. All of the systems use enzymes to make the correction. Without such repair systems, lesions would accumulate and be lethal to the cell or organism. Not all lesions are repaired; hence, mutations do appear, but at relatively low frequencies. At high doses of mutagens, repair systems are unable to correct all of the damage, and cell death may result.

Human Genetic Diseases Resulting from DNA Replication and Repair Errors

Some of the human genetic diseases that have been attributed to defects in DNA replication or repair are listed in Table 18.1. For example, *xeroderma pigmentosum* (Figure 18.14) is caused by homozygosity for a

~ FIGURE 18.14

An individual with xeroderma pigmentosum.

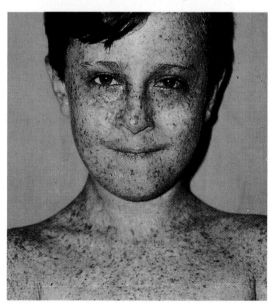

~ TABLE 18.1

Some Examples of Naturally Occurring Human Cell Mutants That Are Defective in DNA Replication or Repair

DISEASE AND MODE OF INHERITANCE	SYMPTOMS	FUNCTIONS AFFECTED	CHROMOSOME LOCATION[a]
Xeroderma pigmentosum (XP)—autosomal recessive	Sensitivity to sunlight, with skin freckling and cancerous growths on skin; lethal at relatively early age as a result of the malignancies	Repair of DNA damaged by UV irradiation or chemicals	9q34.1
Ataxia-telangiectasia (AT)—autosomal recessive	Muscle coordination defect; propensity for respiratory infection; progressive spinal muscular atrophy in significant proportion of patients in second or third decade of life; marked hypersensitivity to ionizing radiation; cancer prone; high frequency of chromosome breaks leading to translocations and inversions	Repair replication of DNA	11q22.3
Fanconi's anemia (FA)—autosomal recessive	Aplastic anemia[b]; pigmentary changes in skin; malformations of heart, kidney, and limbs; leukemia is a fatal complication; genital abnormalities common in males; spontaneous chromosome breakage	Repair replication of DNA, UV-induced pyrimidine dimers and chemical adducts not excised from DNA; a repair exonuclease, DNA ligase, and transport of DNA repair enzymes have been hypothesized to be defective in FA patients	16q24.3
Bloom's syndrome (BS)—autosomal recessive	Pre- and postnatal growth deficiency; sun-sensitive skin disorder; predisposition to malignancies; chromosome instability; diabetes mellitus frequently develops in second or third decade of life	Elongation of DNA chains intermediate in replication—candidate gene is homologous to *E. coli* helicase Q	15q26.1
Cockayne syndrome (CS)—autosomal recessive	Dwarfism; precociously senile appearance; optic atrophy; deafness; sensitivity to sunlight; mental retardation; disproportionately long limbs; knee contractures produce bowlegged appearance; early death	Precise molecular defect is unknown, but may involve transcription-coupled repair	5
Hereditary nonpolyposis colon cancer (HNPCC)—autosomal dominant	Inherited predisposition to non-polyp-forming colorectal cancer	Defect in mismatch repair develops when the remaining wild-type allele of the inherited mutant allele becomes mutated; homozygosity for mutations in any one of four genes (*hMSH2*, *hMLH1*, *hPMS1*, and *hPMS2*, known as mutator genes) has been shown to give rise to HNPCC	2p22–p21

[a]If multiple complementation groups exist, the location of the most common defect is given.

[b]Individuals with aplastic anemia make no, or very few, red blood cells.

recessive mutation in a repair gene. People with this lethal affliction are photosensitive, and portions of their skin that have been exposed to light show intense pigmentation, freckling, and warty growths that may become malignant. These people are deficient in excision repair of damage caused by ultraviolet light, X rays, or gamma radiation or by chemical treatment. Thus, individuals with xeroderma pigmentosum are unable to repair radiation damage to DNA and eventually die, often as a result of malignancies that arise from the damage.

SCREENING PROCEDURES FOR THE ISOLATION OF MUTANTS

Geneticists have made great progress over the years in understanding how normal processes take place by studying mutants that have defects in those processes. Researchers have used mutagens to induce mutations at a greater rate than the rate at which spontaneous mutations occur. However, mutagens change base pairs at random, without regard to the positions of the base pairs in the genetic material. Once mutations have been induced, then, they must be detected if they are to be studied. Mutations of haploid organisms are readily detectable because there is only one copy of the genome. In a diploid experimental organism such as *Drosophila*, dominant mutations are readily detectable. Sex-linked recessive mutations can be detected because they are expressed in half of the sons of a mutated, heterozygous female. Autosomal recessive mutations can only be detected if the mutation is homozygous.

The detection of mutations in humans is much more difficult than in *Drosophila* because geneticists cannot make controlled crosses. Dominant mutations can be readily detected, of course, but other types of mutations may be revealed only by pedigree analysis or by direct biochemical or molecular probing.

Fortunately, for some organisms of genetic interest, particularly microorganisms, selection and screening procedures have been developed to help geneticists isolate mutants of interest from a heterogeneous mixture in a mutagenized population. What follows are brief descriptions of some of these procedures.

Visible Mutations

Visible mutations affect the morphology or physical appearance of an organism (Figure 18.15). Examples of visible mutants are eye-color or wing-shape mutants of *Drosophila*, coat-color mutants of animals (such as albino organisms), colony-size mutants of yeast, and plaque morphology mutants of bacteriophages. Since visible mutations, by definition, are readily apparent, screening is done by inspection.

Nutritional Mutations

An *auxotrophic mutation* affects an organism's ability to make a particular molecule essential for growth. Auxotrophic mutations are most readily detected in microorganisms such as *E. coli* and *yeast* that grow on simple and defined growth media from which they synthesize the molecules essential to their growth. A number of selection and screening procedures are available to isolate auxotrophic mutants.

One simple procedure called **replica plating** can be used to screen for auxotrophic mutants of any microorganism that grows in discrete colonies on solid medium (Figure 18.16). In replica plating, samples from a culture of a mutagenized or unmutagenized colony-forming organism or cell type are plated onto a medium containing the nutrients appropriate for the mutants desired. For example, to isolate arginine auxotrophs, we would plate the culture on a master plate of minimal medium plus arginine (see Figure 18.16). On this medium, wild-type and arginine auxotrophs will grow, but no other auxotrophs will grow. After incubation, colonies will be seen on the plate. The pattern of the colonies is transferred onto sterile velveteen cloth, and replicas of the colony pattern on the cloth are then made by gently pressing new plates onto the velveteen. If the new plate contains minimal medium, the wild-type colonies will be able to grow, but the arginine auxotrophs will not. By comparing the patterns on the original minimal medium plus arginine master plate with those on the minimal medium replica plate, researchers can readily identify the potential arginine auxotrophs. They can then be picked from the original master plate and cultured for further study.

Conditional Mutations

The products of many genes—DNA polymerase and RNA polymerase, for example—are important for the growth and division of cells, and most mutations in these genes result in a lethal phenotype. The structure and function of such genes may be studied by inducing *conditional mutations* in the genes. A common type of conditional mutation to study is a heat-sensitive mutation that functions normally at the normal growth temperature and has no or severely impaired function at a higher temperature. In yeast, for instance, normal growth temperature is 23°C; heat-sensitive mutations

~ FIGURE 18.15

Examples of visible mutants and their wild-type counterparts. (a) White-eyed (mutant) (left) and red-eyed (wild-type) (right) *Drosophila*; (b) Albino (mutant) (left) and agouti (wild-type) (right) mice.

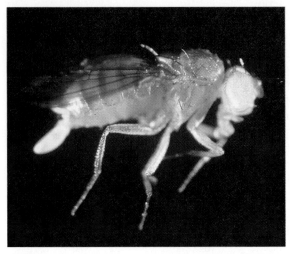

a)

b)

may be isolated at 36°C. Heat sensitivity typically results from a missense mutation causing a change in the amino acid sequence of a protein so that, at the higher temperature, the protein assumes a shape that is nonfunctional.

Essentially the same procedures are used to screen for heat-sensitive mutations of microorganisms as for auxotrophic mutations. For example, replica plating can be used to screen for temperature-sensitive mutants when the replica plate is incubated at a higher temperature than the master plate. That is, such mutants will grow on the master plate but not on the replica plate.

Modern Molecular Screens

Increasingly, the isolation of new mutants relies not on screening a population of randomly mutagenized organisms, but on using molecular targeting approaches to introduce the desired mutation. Screening is then crucial to detect those organisms in which the mutation has been achieved. Here we consider one mutational approach used to obtain information about gene function in humans. Because we cannot perform mutational studies with humans, researchers often attempt to mimic human mutations in mice. Such mouse models of human mutations are very valuable

~ FIGURE 18.16

Replica-plating technique to screen for mutant strains of a colony-forming microorganism.

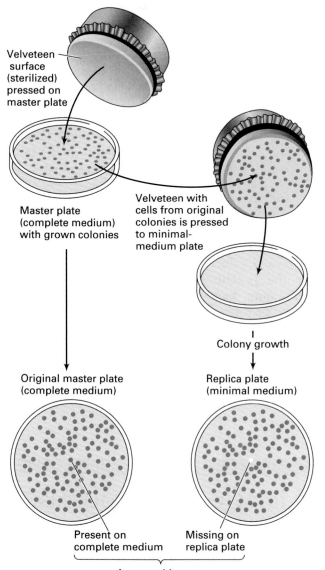

Velveteen surface (sterilized) pressed on master plate

Master plate (complete medium) with grown colonies

Velveteen with cells from original colonies is pressed to minimal-medium plate

Colony growth

Original master plate (complete medium)

Replica plate (minimal medium)

Present on complete medium

Missing on replica plate

Auxotrophic mutant

for furthering our understanding of the gene involved and may, in the case of disease genes, move us toward diagnosis and a cure.

If we have a cloned human gene, we can easily clone the equivalent mouse gene because the two genes show a high degree of homology. The cloned mouse gene can then be mutagenized in vitro—for example, by deleting part of it—to render its product completely nonfunctional. Using an appropriate vector, we can microinject the mutated gene into mouse embryos where, in some cells, it will replace one of the resident wild-type genes. The embryos are

implanted into a female mouse, and progeny mice are screened molecularly (through such techniques as PCR, restriction analysis, or DNA probing) to identify those carrying the mutation; such mice will be heterozygous $+/-$ for the mutation. Mice carrying the mutation are called *knockout mice* because the normal gene has been knocked out by the engineered gene. By interbreeding $+/-$ mice, knockout $-/-$ progeny can be produced with both alleles mutated. Molecular screens are used to identify these progeny (assuming they are viable), and they can then be used for study. Since the mutated genes used are completely nonfunctional, these mice will show a null phenotype for the gene—that is, a phenotype resulting from a complete absence of gene product.

Many types of knockout mice, both heterozygotes and homozygotes, have been produced to date. One example is knockout mice for the tumor suppressor gene *p53* (see Chapter 17, pp. 399–400). Recall that *p53* mutations are involved in a large proportion of human cancers. The *p53* ($-/-$) knockout mice show rapid development of spontaneous tumors and a number of other mutant phenotypes. Knockout mice have also been made for the cystic fibrosis (CF) gene (see Chapter 8, p. 190, and Chapter 14, pp. 321–324). The symptoms are very similar to those of human CF patients, notably major mucous membrane defects in the pharynx, lungs, intestine, and colon due to defects in chloride ion transport. CF is lethal in humans. In mice, 40 percent of CF gene knockouts die within one week of birth from intestinal obstruction.

The knockout approach for generating specific gene mutations is applicable for any organism that can be transformed by cloned DNA and whose resident genes can be replaced by the introduced DNA. In all applications, molecular screening is done to confirm the presence of the mutations.

KEYNOTE

Geneticists have made great progress in understanding how cellular processes take place by studying mutants that have defects in those processes. With microorganisms, a number of procedures are available to screen for mutants of interest from a heterogeneous mixture of cells in a mutagenized population of cells. More modern approaches involve making specific gene mutations called knockouts, using cloned genes that have been mutated in vitro. Molecular screens are used to detect heterozygous and homozygous mutant knockout organisms.

SUMMARY

In this chapter we have seen that genetic damage can occur to DNA spontaneously through replication errors or through treatment with radiation or chemical mutagens. If the genetic damage is not repaired, mutations will result, and if there has been too much damage, cell death may result.

Mutations occur spontaneously at a low rate. The mutation rate can be increased through the use of mutagens such as radiation and certain chemicals. Chemical mutagens work in a number of different ways, such as by acting as base analogs, by modifying bases, or by intercalating into the DNA. Intercalating results in frameshift mutations, while the others result in base-pair substitution mutations. Bruce Ames has devised a test—the Ames test—that can indicate whether chemicals (such as environmental or commercial chemicals) have the potential to cause mutations in humans. A large number of potential human carcinogens have been found in this way.

Cells possess a number of repair mechanisms that function to correct most damage to DNA. These repair mechanisms include: (1) repair by DNA polymerase proofreading, in which a base-pair mismatch in DNA being synthesized is immediately repaired by 3'-to-5' excision; (2) photoreactivation of pyrimidine dimers induced by UV light; (3) excision repair, in which pyrimidine dimers and other DNA damage that distorts the DNA helix are excised and replaced with new DNA; and (4) mismatch repair, in which the methylation state of a DNA sequence signals which DNA strand is newly synthesized so that the mismatched base on that strand is corrected. DNA damage that is not repaired is called a mutation. The collective array of repair enzymes serves to reduce mutation rates for spontaneous errors by several orders of magnitude. Such repair mechanisms cannot cope, however, with the extensive amount of DNA damage that arises from the experimental use of chemical mutagens or UV irradiation, so that many mutations typically result.

Lastly, we considered some examples of how a mutagenized population of cells can be screened for particular mutants of interest, such as auxotrophs and conditional mutants. Over the decades, spontaneous and induced mutants have proved invaluable for the genetic analysis of biological function. Techniques have been developed for mutating cloned genes in vitro and replacing resident normal genes to produce knockout mutant organisms. Molecular screens are used to detect heterozygous and homozygous knockouts.

ANALYTICAL APPROACHES FOR SOLVING GENETICS PROBLEMS

Q18.1 Five strains of *E. coli* containing base-substitution mutations that affect the tryptophan synthetase A polypeptide have been isolated. The figure below shows the changes produced in the protein itself in the indicated mutant strains. In addition, *A23* can be further mutated to insert Ile, Thr, Ser, or the wild-type Gly into position 210.

In the following questions, assume that only a single base change can occur at each step.

a. Using the genetic code (see Figure 13.5), explain how the two mutations *A23* and *A46* can result in two different amino acids being inserted at position 210. Give the nucleotide sequence of the wild-type gene at that position and the two mutants.

b. Can mutants *A23* and *A46* recombine? Why or why not? If recombination can occur, what would be the result?

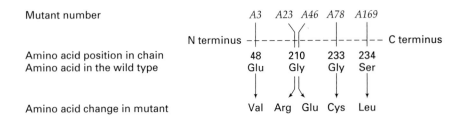

c. From what you can infer of the nucleotide sequence in the wild-type gene, indicate, for the codons specifying amino acids 48, 210, 233, and 234, whether or not a nonsense mutant could be generated by a single nucleotide substitution in the gene.

A18.1

a. There are no simple ways to answer questions like this one. The best approach is to scrutinize the genetic code dictionary and use a pencil and paper to try to define the codon changes that are compatible with all the data. The number of amino acid changes in position 210 of the polypeptide is helpful in this case. The wild-type amino acid is Gly, and the codons for Gly are GGU, GGC, GGA, and GGG. The *A23* mutant has Arg at position 210, and the arginine codons are AGA, AGG, CGU, CGC, CGA, and CGG. Clearly, any Arg codon could be generated by a single base change. Thus, we have to look at the amino acids at 210 generated by further mutations of *A23*. In the case of Ile, the codons are AUU, AUC, and AUA. The *only* way to get from Gly to Arg in one base change and then to Ile in a subsequent single base change is from GGA (Gly) → AGA (Arg) → AUA (Ile). Is this compatible with the other mutational changes from *A23*? There are four possible Thr codons, ACU, ACC, ACA, and ACG, so a mutation from AGA (Arg) to ACA (Thr) would fit. There are six possible Ser codons, UCU, UCC, UCA, UCG, AGU, and AGC, so a mutation from AGA to either AGU or AGC would fit.

Considering the *A46* mutant, the possible codons for Glu are GAA and GAG. Given that the wild-type codon is GGA (Glu), the only possible single base change to give Glu is if the Glu codon in the mutant is GAA. So, the answer to the question is that the wild-type sequence at position 210 is GGA, the sequence in the *A23* mutant is AGA, and the sequence in the *A46* mutant is GAA. In other words, the *A23* and *A46* mutations are in different bases of the codon.

b. The answer to this question follows from the answer deduced in part a. Mutants *A23* and *A46* can recombine since the mutations in the two mutant strains are in different base pairs. The results of a single recombination event (at the DNA level) between the first and second base of the codon in AGA × GAA are a wild-type GGA codon (Gly) and a double mutant AAA codon (Lys). Recombination can also occur between the second and third bases of the codon, but the products are AGA and GAA, that is, identical to the parents.

c. Amino acid 48 had a Glu-to-Val change. This change must have involved GAA to GUA or GAG to GUG. In either case the Glu codon can mutate with a single base-pair change to a nonsense codon, that is, UAA or UAG, respectively.

Amino acid 210 in the wild type has a GGA codon, as we have already discussed. This gene could mutate to the UGA nonsense codon with a single base-pair change.

Amino acid 233 had a Gly-to-Cys change. This change must have involved either GGU to UGU or GGC to UGC. In either case the Gly codon cannot mutate to a nonsense codon with one base-pair change.

Amino acid 234 had a Ser-to-Leu change. This change was either UCA to UUA or UCG to UUG. If the Ser codon was UCA, it could be changed to UGA in one step. But if the Ser codon was UCG, it cannot change to a nonsense codon in one step.

Q18.2 The chemically induced mutations *a*, *b*, and *c* show specific reversion patterns when subjected to treatment by the following mutagens: 2-aminopurine (AP), 5-bromouracil (BU), proflavin (pro), and hydroxylamine (HA). AP is a base-analog mutagen that induces mainly AT-to-GC changes and can cause GC-to-AT changes also. BU is a base-analog mutagen that induces mainly GC-to-AT changes and can cause AT-to-GC changes also. PRO is an intercalating agent that can cause a single base-pair addition or deletion with no specificity. HA is a base-modifying agent that modifies cytosine, causing one-way transitions from GC to AT. The reversion patterns are shown in the following table.

	MUTAGENS TESTED IN REVERSION STUDIES			
MUTATION	AP	BU	PRO	HA
a	−	−	+	−
b	+	+	−	+
c	+	+	−	−

(*Note:* "+" indicates many reversions to wild type were found; "−" indicates no reversions, or very few, to wild type were found.)

For each original mutation (*a*⁺ to *a*, *b*⁺ to *b*, etc.), indicate the probable base-pair change (AT to GC, deletion of GC, etc.) and the mutagen that was most probably used to induce the original change.

A18.2 This question tests knowledge of the base-pair changes that can be induced by the various mutagens used.

Mutagen AP induces mainly AT-to-GC changes and can cause GC-to-AT changes also. Thus, AP-induced mutations can be reverted by AP.

Base-analog mutagen BU induces mainly GC-to-AT changes and can cause AT-to-GC changes, so BU-induced mutations can be reverted by BU.

Proflavin causes single base-pair deletions or additions, so proflavin-induced changes can be reverted by a second treatment with proflavin.

Mutagen HA causes one-way transitions from GC to AT, so HA-induced mutations cannot be reverted by HA.

With these mutagen specificities in mind, we can answer the questions for each mutation in turn:

Mutation a^+ to a: The a mutation was reverted only by proflavin, indicating that it was a deletion or an addition (a frameshift mutation). Therefore, the original mutation was induced by an intercalating agent such as proflavin, since that is the only class of mutagen that can cause an addition or a deletion.

Mutation b^+ to b: The b mutation was reverted by AP, BU, or HA. A key here is that HA only causes GC-to-AT changes. Therefore, b must be GC, and the original b^+ must have been an AT. Thus, the mutational change of b^+ to b must have been caused by treatment with AP or BU, since these are the two mutagens in the list that can induce that change.

Mutation c^+ to c: The c mutation was reverted only by AP and BU. Since it could not be reverted by HA, c must be an AT, and c^+ a GC. The mutational change from c^+ to c therefore involved a GC-to-AT transition and could have resulted from treatment with AP, BU, or HA.

QUESTIONS AND PROBLEMS

*18.1 Mutations are (choose the correct answer):
a. Caused by genetic recombination.
b. Heritable changes in genetic information.
c. Caused by faulty transcription of the genetic code.
d. Usually but not always beneficial to the development of the individuals in which they occur.

18.2 Answer true or false: Mutations occur more frequently if there is a need for them.

*18.3 Which of the following is *not* a class of mutation?
a. Frameshift
b. Missense
c. Transition
d. Transversion
e. None of the above (meaning all are classes of mutation)

*18.4 Ultraviolet light usually causes mutations by a mechanism involving (choose the correct answer):
a. One-strand breakage in DNA.
b. Light-induced change of thymine to alkylated guanine.
c. Induction of thymine dimers and their persistence or imperfect repair.

d. Inversion of DNA segments.
e. Deletion of DNA segments.
f. All of the above.

18.5 For the middle region of a particular polypeptide chain, the wild-type amino acid sequence and the amino acid sequence of several mutants were determined, as shown below (. . . indicates additional, unspecified amino acids). For each mutant, say what DNA level change has occurred, whether the change is a base-pair substitution mutation (transversion or transition, missense or nonsense) or a frameshift mutation, and in which codon the mutation occurred. (Refer to the codon dictionary in Figure 13.5, p. 285.)

	CODON								
	1	2	3	4	5	6	7	8	9
a. Wild type: . . .	Phe	Leu	Pro	Thr	Val	Thr	Thr	Arg	Trp
b. Mutant 1: . . .	Phe	Leu	His	His	Gly	Asp	Asp	Thr	Val
c. Mutant 2: . . .	Phe	Leu	Pro	Thr	Met	Thr	Thr	Arg	Trp
d. Mutant 3: . . .	Phe	Leu	Pro	Thr	Val	Thr	Thr	Arg	
e. Mutant 4: . . .	Phe	Pro	Pro	Arg					
f. Mutant 5: . . .	Phe	Leu	Pro	Ser	Val	Thr	Thr	Arg	Trp

*18.6 In mutant strain X of *E. coli*, a leucine tRNA that recognizes the codon 5'-CUG-3' in normal cells has been altered so that it now recognizes the codon 5'-GUG-3'. A missense mutation, which affects amino acid 10 of a particular protein, is suppressed in mutant X cells.
a. What are the anticodons of the two Leu tRNAs, and what mutational event has occurred in mutant X cells?
b. What amino acid would normally be present at position 10 of the protein (without the missense mutation)?
c. What amino acid would be put in at position 10 if the missense mutation is not suppressed (that is, in normal cells)?
d. What amino acid is inserted at position 10 if the missense mutation is suppressed (that is, in mutant X cells)?

18.7 In any kind of chemotherapy, the object is to find a means to kill the invading pathogen or cancer cell without killing the cells of the host. To do this successfully, one must find and exploit biological differences between target organisms and host cells. Explain why AZT is effective in terminating viral cDNA synthesis but does not interfere with host DNA replication.

*18.8 The mutant *lacZ-1* was induced by treating *E. coli* cells with acridine, while *lacZ-2* was induced with 5BU. What kinds of mutants are these likely to be? Explain. How could you confirm your predictions by studying the structure of the β-galactosidase in these cells?

*18.9

a. The sequence of nucleotides in an mRNA is:

5'-AUGACCCAUUGGUCUCGUUAG-3'

Assuming that ribosomes could translate this mRNA, how many amino acids long would you expect the resulting polypeptide chain to be?

b. Hydroxylamine is a mutagen that results in the replacement of an AT base pair for a GC base pair in the DNA; that is, it induces a transition mutation. When applied to the organism that made the mRNA molecule shown in part a, a strain was isolated in which a mutation occurred at the 11th position of the DNA that coded for the mRNA. How many amino acids long would you expect the polypeptide made by this mutant to be? Why?

18.10 In a series of 94,075 babies born in a particular hospital in Copenhagen, 10 were achondroplastic dwarfs (this is an autosomal dominant condition). Two of these 10 had an achondroplastic parent. The other eight achondroplastic babies each had two normal parents. What is the apparent mutation rate at the achondroplasia locus?

*18.11 Three of the codons in the genetic code are chain-terminating codons for which no naturally occurring tRNAs exist. Just like any other codons in the DNA, though, these codons can change as a result of base-pair changes in the DNA. Confining yourself to single base-pair changes at a time, determine which amino acids could be inserted in a polypeptide by mutation of these chain-terminating codons: (a) UAG; (b) UAA; (c) UGA. (The genetic code is listed in Figure 13.5.)

18.12 The amino acid substitutions in the following figure occur in the α and β chains of human hemoglobin. Those amino acids connected by lines are related by single nucleotide changes. Propose the most likely codon or codons for each of the numbered amino acids. (Refer to the genetic code listed in Figure 13.5.)

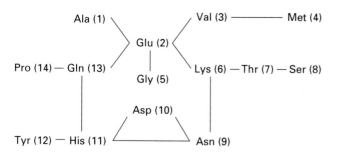

*18.13 Yanofsky studied the tryptophan synthetase of *E. coli* in an attempt to identify the base sequence specifying this protein. The wild type gave a protein with a glycine in position 38. Yanofsky isolated two *trp* mutants, *A23* and *A46*. Mutant *A23* had Arg instead of Gly at position 38, and mutant *A46* had Glu at position 38. Mutant

A23 was plated on minimal medium, and four spontaneous revertants to prototrophy were obtained. The tryptophan synthetase from each of the four revertants was isolated, and the amino acids at position 38 were identified. Revertant 1 had Ile, revertant 2 had Thr, revertant 3 had Ser, and revertant 4 had Gly. In a similar fashion, three revertants from *A46* were recovered, and the tryptophan synthetase from each was isolated and studied. At position 38, revertant 1 had Gly, revertant 2 had Ala, and revertant 3 had Val. A summary of these data is given in the figure. Using the genetic code in Figure 13.5, deduce the codons for the wild type, for the mutants *A23* and *A46*, and for the revertants, and place each designation in the space provided in the figure.

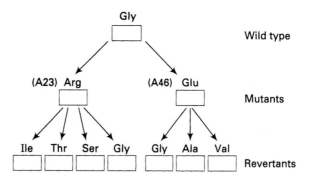

18.14 Consider an enzyme chewase from a theoretical microorganism. In the wild-type cell the chewase has the following sequence of amino acids at positions 39 to 47 (reading from the amino end) in the polypeptide chain:

-Met-Phe-Ala-Asn-His-Lys-Ser-Val-Gly-
　39　40　41　42　43　44　45　46　47

A mutant of the organism was obtained; it lacks chewase activity. The mutant was induced by a mutagen known to cause single base-pair insertions or deletions. Instead of making the complete chewase chain, the mutant makes a short polypeptide chain only 45 amino acids long. The first 38 amino acids are in the same sequence as the first 38 of the normal chewase, but the last seven amino acids are as follows:

-Met-Leu-Leu-Thr-Ile-Arg-Val
　39　40　41　42　43　44　45

A partial revertant of the mutant was induced by treating it with the same mutagen. The revertant makes a partly active chewase, which differs from the wild-type enzyme only in the following region:

-Met-Leu-Leu-Thr-Ile-Arg-Gly-Val-Gly-
　39　40　41　42　43　44　45　46　47

Using the genetic code given in Figure 13.5, deduce the nucleotide sequences for the mRNA molecules that specify this region of the protein in each of the three strains.

18.15   Two mechanisms in *E. coli* were described for the repair of DNA damage (thymine dimer formation) after exposure to ultraviolet light: photoreactivation and excision (dark) repair. Compare and contrast these mechanisms, indicating how each achieves repair.

18.16   DNA damage by mutagens has very serious consequences for DNA replication. Without specific base-pairing, the replication enzymes cannot specify a complementary strand, and gaps are left after the passing of a replication fork.
a.   What response has *E. coli* developed to large amounts of DNA damage by mutagens? How is this response coordinately controlled?
b.   Why is this response itself a mutagenic system?
c.   What effects would loss-of-function mutations in *recA* or *lexA* have on this response?

*18.17   A protein contains the amino acid proline (Pro) at one site. Treatment with nitrous acid, which deaminates C to make it U, produces two different mutants. One mutant has a substitution of serine (Ser) and the other has a substitution of leucine (Leu) at the site.

Treatment of the two mutants with nitrous acid produces new mutant strains, each with phenylalanine (Phe) at the site. Treatment of these new Phe-carrying mutants with nitrous acid produces no change. The results are summarized in the following figure:

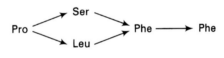

Using the appropriate codons, show how it is possible for nitrous acid to produce these changes, and why further treatment has no influence. (Assume only single nucleotide changes occur at each step.)

18.18   As genes have been cloned for a number of human diseases caused by defects in DNA repair and replication, striking evolutionary parallels have been found between human and bacterial DNA repair systems. Discuss the features of DNA repair systems that appear to be shared in these two types of organism.

*18.19   In the past few years, a considerable number of knockout mice that lack specific gene functions have been bred. In addition to the knockouts for *p53* and the *CFTR* genes described in this chapter, knockouts for many different *Hox* genes (see Chapter 16), lysosomal hexosaminidase (the enzyme missing in Tay-Sachs disease), and many other enzymes have been made.
a.   Why is the ability to make knockout mice important for the treatment and cure of human disease?
b.   Consider a situation in which a dominant mutation causes a human disease such as Huntington disease. A hypothetical researcher clones the disease gene and discovers that a novel, abnormal protein is responsible for the disease phenotype. Would you expect the phenotype of a knockout mouse to be similar to the disease phenotype? Why or why not?

CHAPTER *19*

TRANSPOSABLE ELEMENTS

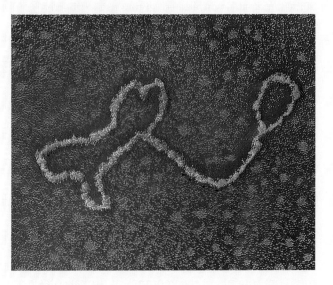

PRINCIPAL POINTS

~ Transposable elements are DNA segments that can insert themselves at one or more sites in a genome. The presence of transposable elements in a cell is usually detected by the changes they bring about in the expression and activities of the genes at or near the chromosomal sites into which they integrate.

~ In bacteria, two important types of transposable elements are insertion sequence (IS) elements and transposons (Tn). Insertion sequence elements and transposons have repeated sequences at their ends and encode proteins such as transposases that are responsible for their transposition. Transposons also have genes that encode other functions, such as drug resistance. When ISs and Tns integrate into the genome, a short host sequence at the target site is duplicated, giving rise to directly repeated sequences flanking the integrated element.

~ Transposable elements in eukaryotes resemble bacterial transposons in general structure and transposition properties. Transposons integrate at a target site by a precise mechanism so that the integrated elements are flanked at the insertion site by a short duplication of target site DNA. While most transposons move using a DNA-to-DNA mechanism, some eukaryotic transposons move via an RNA intermediate (using a transposon-encoded reverse transcriptase). Such transposons resemble retroviruses in genome organization and other properties.

~ Transposable elements have contributed to the evolution of genomes through the chromosome rearrangements they cause.

Certain genetic elements in the chromosomes of both prokaryotes and eukaryotes can move from one location to another in the genome. These mobile genetic elements are known **transposable elements**, since this term reflects the *transposition* ("change in position") events associated with them. Their discovery was a great surprise and altered our classical picture of genes and genomes.

Transposable elements are normal and ubiquitous components of the genomes of prokaryotes and eukaryotes. In structure and function, transposable elements are similar in both prokaryotes and eukaryotes. In prokaryotes, transposable elements can move to new positions on the same chromosome (since there is only one) or onto plasmids or phage chromosomes, while in eukaryotes, transposable elements may to move to new positions within the same chromosome or to a different chromosome. In both prokaryotes and eukaryotes, transposable elements insert into new chromosome locations with which they have no sequence homology; hence, transposition is a different process from homologous recombination and is referred to as *nonhomologous recombination*. Transposable elements are important because of the genetic changes they cause, and that is a reason they have been studied intensively. For example, they can produce mutations by inserting into genes, they can increase or decrease gene expression by inserting

into gene regulatory sequences (such as by disrupting promoter function or by stimulating a gene's expression through the activity of promoters on the element), and they can produce various kinds of chromosomal mutations because of the mechanics of transposition. In fact, transposable elements have made important contributions to the evolution of the genomes of both prokaryotes and eukaryotes through the chromosome rearrangements they cause. In this chapter we learn about the nature of transposable elements and about the genetic changes they cause.

TRANSPOSABLE ELEMENTS IN PROKARYOTES

We discuss two examples of transposable elements in prokaryotes: insertion sequence (IS) elements and transposons (Tn).

Insertion Sequences

An **insertion sequence (IS)**, or **IS element**, is the simplest transposable element found in prokaryotes. An IS element contains only genes required for mobilizing the element and inserting the element into a chromosome at a new location. IS elements are normal constituents of bacterial chromosomes and plasmids.

E. coli may have a number of IS elements, including IS1 (Figure 19.1), IS2, and IS10R, each present in 0 to 30 copies per genome and each with a characteristic length and unique nucleotide sequence. IS elements end with perfect or nearly perfect inverted terminal repeats (IRs) of between 9 and 41 bp. This means that essentially the same sequence is found at each end of an IS but in opposite orientations. The inverted repeats of IS1, for example, consist of 23 bp of not quite identical sequence (see Figure 19.1).

An IS element transposes as a result of the activity of an IS-element-encoded enzyme called **transposase**. This enzyme "reads" where the ends of the IS are by recognizing the IR sequences. The frequency of transposition is characteristic of each IS element, ranging between 10^{-5} and 10^{-7} events per generation. Typically, when an IS element transposes, a copy of the IS element inserts into a new chromosome location while the original IS element remains in place.

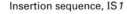

~ FIGURE 19.1

The insertion sequence (IS) transposable element, IS1. The 768-bp IS element has inverted repeat (IR) sequences at the ends. Shown below the element are the sequences for the 23-bp inverted terminal repeats (IR).

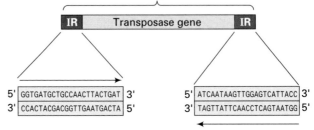

The chromosomal sites into which IS elements insert are called *target sites*. The process of IS insertion into a chromosome is shown in Figure 19.2. First, transposase makes a staggered cut in the target

~ FIGURE 19.2

Schematic of the integration of an IS element into chromosomal DNA. As a result of the integration event, the target site becomes duplicated to produce direct target repeats. Thus, the integrated IS element is characterized by its inverted repeat (IR) sequences flanked by direct target site duplications. Integration involves making staggered cuts in the host target site. After insertion of the IS, the gaps that result are filled in with DNA polymerase and DNA ligase. (Note: The base sequences given for the IR are for illustration only and are not the actual sequences found, either in length or sequence.)

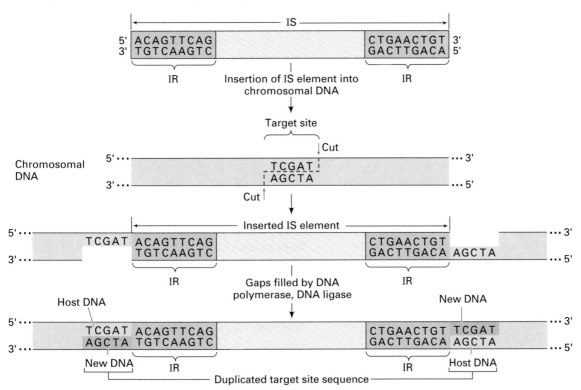

site. The IS element is then inserted, becoming joined to the single-stranded ends. The gaps are filled in by DNA polymerase and DNA ligase, producing an integrated IS element with two direct repeats of the target site sequence flanking the IS element called *target site duplications.* "Direct" in this case means that the two sequences are repeated in the same orientation (see Figure 19.2). The sizes of target site duplications vary with the IS element, but tend to be small (4 to 13 bp).

Transposons

Like an IS element, a **transposon (Tn)** contains genes for the insertion of the DNA segment into the chromosome and for the mobilization of the element to other locations on the chromosome. A transposon is more complex than an IS element, however, in that it contains additional genes.

There are two types of prokaryotic transposons: composite transposons and noncomposite transposons. *Composite transposons* are complex transposons with a central region containing genes (for example, antibiotic resistance genes) flanked on both sides by IS elements (also called *IS modules*). Composite transposons may be thousands of base pairs long. The IS elements are both of the same type and are called IS*L* (for "left") and IS*R* (for "right"). The Tn*10* transposon, for example, is 9,300 bp long and consists of 6,500 bp of central, nonrepeating DNA containing the tetracycline resistance gene flanked at each end with 1,400-bp IS elements called IS*10L* and IS*10R* (Figure 19.3) that are arranged in an inverted orientation. Cells containing Tn*10* are resistant to tetracycline because of the tetracycline resistance gene. Transposition of composite transposons occurs because of the function of the IS elements they contain. One or both IS elements encodes the transposase.

Noncomposite transposons, exemplified by the 4,957-bp Tn*3*, also contain genes such as those for drug resistance, but they do not terminate with IS elements. However, they do have the repeated sequences at their ends that are required for transposition. Enzymes for transposition are encoded by genes in the central region of the transposon.

As with the movement of IS elements, integration of a transposon into the chromosome causes a target site duplication. Transposition of transposons can also cause mutations. Insertion of a transposon into the reading frame of a gene will disrupt it, causing a loss of function of that gene. Insertion into a gene's controlling region can cause changes in the level of expression of the gene, depending on the promoter

~ FIGURE 19.3

Structure of the composite transposon Tn*10*. The general features of composite transposons are evident: a central region carrying a gene or genes, such as for drug resistance, flanked by either direct or inverted IS elements. The IS elements themselves each have terminal inverted repeats.

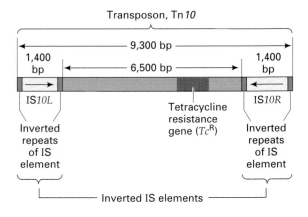

elements in the transposon and how they are oriented with respect to the gene. Deletion and insertion events also occur as a result of the activities of the transposons, and from crossing-over between duplicated transposons in the genome.

The transposition of transposons carrying antibiotic resistance genes is one way bacteria can acquire multiple antibiotic resistance. That is, transposons are found on plasmids that can be transmitted between some bacterial strains by conjugation (see Chapter 6, pp. 130–136). Through transposition, combinations of antibiotic resistance genes carried by the transposons can be accumulated on single plasmid types. This is an important issue in the effective treatment of human infectious diseases.

KEYNOTE

Transposable elements are unique DNA segments that can insert themselves at one or more sites in a genome. The presence of transposable elements in a cell is usually detected by the changes they bring about in the expression and activities of the genes at or near the chromosomal sites into which they integrate. In prokaryotes, two important types of transposable elements are insertion sequence (IS) elements and transposons (Tn). Tns often carry antibiotic resistance genes.

TRANSPOSABLE ELEMENTS IN EUKARYOTES

In the 1940s and 1950s, Barbara McClintock did a series of elegant genetic experiments with corn (*Zea mays*) that led her to hypothesize the existence of what she called "controlling elements," which modify or suppress gene activity in corn and which are mobile in the genome. Decades later, the controlling elements she studied were shown to be transposable elements. For her original work in deducing the presence of transposable elements, Barbara McClintock was awarded the Nobel Prize in 1983. A fascinating and moving biographical sketch of Barbara McClintock is given in Box 19.1.

Transposable elements have been identified in many eukaryotes, and they have been studied mostly in yeast, *Drosophila*, corn, and humans. In general, their structure and function are very similar to those of prokaryotic transposable elements. Eukaryotic transposable elements have genes that encode enzymes required for transposition, and they can integrate into chromosomes at a number of sites. Thus, such elements may affect the function of virtually any gene, turning it on or off, depending on the element involved and how it integrates into the gene. The integration events themselves, like those of prokaryotic transposable elements, involve nonhomologous recombination.

We discuss some examples of transposable elements in different eukaryotes next.

The *Ac-Ds* System in Corn

A number of different genes must function together for the synthesis of red anthocyanin pigment, which gives the corn kernel a purple color. Mutation of any one of these genes results in an unpigmented kernel. In her classic experiments, Barbara McClintock studied kernels that, rather than being solid purple or solid white, had spots of purple pigment on a white kernel (Figure 19.4).

Based on careful genetic and cytological studies, McClintock explained the spotted kernels as follows: if the corn plant carries a wild-type *C* gene, the kernel will be purple; *c* (colorless) mutations block purple pigment production, so the kernel is colorless. During kernel development, revertants of the mutation occur in certain cells, leading to spots of purple pigment. The original *c* (colorless) mutation resulted from a "mobile controlling element" (in modern terms, a transposon), called *Ds* for "dissociation," being inserted into the *C* gene (Figure 19.5a and b).

We now know that there are several families of transposons in corn and other plants. There are two forms of transposons in each family: *autonomous elements*, which can transpose by themselves; and *nonautonomous elements*, which cannot transpose by

---

~ FIGURE 19.4

Corn kernels, some of which show spots of pigment produced by cells in which a transposable genetic element had transposed out of a pigment-producing gene, thereby allowing function of that gene to be restored. The cells in the white areas of the kernel lack pigment because a pigment-producing gene continues to be inactivated by the presence of a transposable element within that gene.

themselves because they lack the gene for transposition—they require the presence of an autonomous element to supply the missing functions. Often the nonautonomous element is a deletion derivative of the autonomous element in the family. When either element is inserted into a gene, a mutant allele of that gene is produced. McClintock had discovered the *Ac-Ds* transposon family, where *Ac* stands for "activator" (an autonomous element) and *Ds* stands for "dissociation" (a nonautonomous element). *Ac* is required for *Ds* to transpose, for instance into the *C* gene to produce a *c* mutant (see Figure 19.5b). *Ac* can also result in *Ds* transposing (excising perfectly) out of the *c* gene, giving a wild-type revertant, a purple spot (Figure 19.5c). And, as McClintock's cytological studies showed, *Ds* may cause chromosome breakage near its insertion site.

The remarkable fact of McClintock's conclusion was that at the time there was no precedent for the existence of transposable genetic elements—indeed, the genome was thought to be very static with regard to gene locations. Only much more recently have transposable genetic elements been widely identified and studied, and only in 1983 was direct evidence obtained for the movable genetic elements proposed by McClintock.

KEYNOTE

Many plant transposons occur in families, the autonomous elements of which are able to direct their own transposition, and the nonautonomous elements of which are able to transpose only when activated by an autonomous element in the same genome. Most nonautonomous elements are derived from autonomous elements by internal deletions.

~ **FIGURE 19.5**

Kernel color in corn and transposon effects. (a) Purple kernels result from the active *C* gene. (b) Colorless kernels can result when the *Ac* transposon activates *Ds* transposition and *Ds* inserts into *C*, producing a mutation. (c) Spotted kernels result by reversion of the *c* mutation during kernel development when *Ac* activates *Ds* transposition out of the *C* gene.

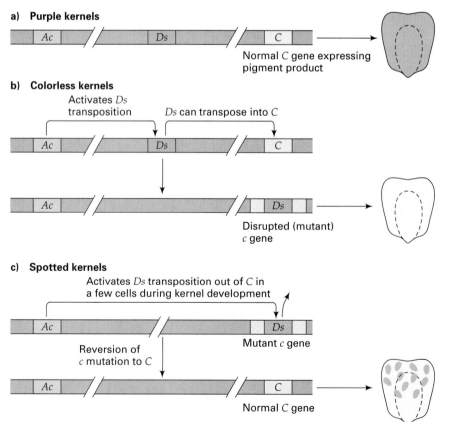

a) **Purple kernels**

Normal *C* gene expressing pigment product

b) **Colorless kernels**

Activates *Ds* transposition

Ds can transpose into *C*

Disrupted (mutant) *c* gene

c) **Spotted kernels**

Activates *Ds* transposition out of *C* in a few cells during kernel development

Mutant *c* gene

Reversion of *c* mutation to *C*

Normal *C* gene

BOX 19.1

BARBARA MCCLINTOCK
(1902–1992)

Barbara McClintock in 1947.

Barbara McClintock's remarkable life spanned the history of genetics in the twentieth century. Barbara McClintock was born in Hartford, Connecticut, to Sara Handy McClintock, an accomplished pianist as well as a poet and painter, and Thomas Henry McClintock, a physician. Both parents were quite unconventional in their attitudes toward child rearing: they were interested in what their children would and could be, rather than what they should be.

During her high school years Barbara discovered science, and she loved to learn and figure things out. After high school Barbara attended Cornell University, where she flourished both socially and intellectually. She enjoyed her social life, but her comfort with solitude and the tremendous joy she experienced in knowing, learning, and understanding were to be the defining themes of her life. The decisions she made during her university years were consistent with her adamant individuality and self-containment. In Barbara's junior year, after a particularly exciting undergraduate course in genetics, her professor invited her to take a graduate course in genetics. After that she was treated much like a graduate student; by the time she had finished her undergraduate course work, there was no question in her mind: she had to continue her studies of genetics.

At Cornell, genetics was taught in the plant-breeding department, which at the time did not take female graduate students. To circumvent this obstacle, McClintock registered in the botany department with a major in cytology and a minor in genetics and zoology. She began to work as a paid assistant to Lowell Randolph, a cytologist. McClintock and Randolph did not get along well and soon dissolved their working relationship, but as McClintock's colleague and lifelong friend Marcus Rhoades later wrote, "Their brief association was momentous because it led to the birth of maize cytogenetics." McClintock discovered that the metaphase or late prophase chromosomes in the first microspore mitosis were far better for cytological discrimination than were root tip chromosomes. In a few weeks she prepared

detailed drawings of the maize chromosomes, which she published in *Science.*

This was McClintock's first major contribution to maize genetics and laid the groundwork for a veritable explosion of discoveries that connected the behavior of chromosomes to the genetic properties of an organism, defining the new field of cytogenetics. McClintock was awarded a Ph.D. in 1927 and appointed an instructor at Cornell, where she continued to work with maize. The Cornell maize genetics group was small, including Professor R. A. Emerson, the founder of maize genetics, McClintock, George Beadle, C. R. Burnham, Marcus Rhoades, and Lowell Randolph, together with a few graduate students. By all accounts McClintock was the intellectual driving force of this talented group.

In 1929, a new graduate student, Harriet Creighton, joined the group and was guided by McClintock. Their work showed, for the first time, that genetic recombination is a reflection of the physical exchange of chromosome segments. A paper on the work by Creighton and McClintock, published in 1931, was perhaps McClintock's first seminal contribution to the science of genetics.

Although McClintock's fame was growing, she had no permanent position. Cornell had no female professors in fields other than home economics, so her prospects were dismal. She had already attained international recognition, but as a woman she had little hope of securing a permanent academic position at a major research university. R. A. Emerson obtained a grant from the Rockefeller Foundation to support her work for two years, allowing her to continue to work independently. McClintock was discouraged and resentful of the disparity between her prospects and those of her male counterparts. Her extraordinary talents and accomplishments were widely appreciated, but she was also seen as "difficult" by many of her colleagues, in large part because of her quick mind and intolerance of second-rate work and thinking.

In 1936, Lewis Stadler convinced the University of Missouri to offer McClintock an assistant professorship. She accepted the position and began to follow the behavior of maize chromosomes that had been broken by X-irradiation. However, soon after her arrival at Missouri she understood that hers was a special appointment. She found herself excluded from regular academic activities, including faculty meetings. In 1941, she took a leave of absence from Missouri and departed with no intention of returning. She wrote to her friend Marcus Rhoades, who was planning to go to Cold Spring Harbor for the summer to grow his corn. An invitation for McClintock was arranged through Milislav Demerec (member and later the director of the genetics department of the Carnegie Institution of Washington, then the dominant research laboratory at Cold Spring Harbor), who offered her a year's research appointment. Though hesitant to commit herself, McClintock accepted. When Demerec later offered her an appointment as a permanent member of the research staff, McClintock accepted, still unsure whether she would stay. Her dislike of making commitments was a given: she insisted that she would never have become a scientist in today's world of grants because she could not have committed herself to a written research plan. It was the unexpected that fascinated her, and she was always ready to pursue an observation that didn't fit. Nevertheless, McClintock did stay at Carnegie until 1967.

At Carnegie McClintock continued her studies on the behavior of broken chromosomes. In 1944, she was elected to the National Academy of Sciences and in 1945 to the presidency of the Genetics Society of America. In these same two years McClintock reported observing "an interesting type of chromosomal behavior" involving the repeated loss of one of the broken chromosomes from cells during development. What struck her as odd was that in this particular stock it was always chromosome 9 that broke, and it always broke at the same place. McClintock called the unstable chromosome site *Dissociation* or *Ds* because "the most readily recognizable consequence of its actions is this dissociation." She quickly established that the *Ds* locus would "undergo dissociation mutations only when a particular dominant factor is present." She named this factor *Activator* (*Ac*) because it activated chromosome breakage at *Ds*. She also reached the extraordinary conclusion that *Ac* was not only required for *Ds*-mediated chromosome breakage, but also could destabilize previously stable mutations. But more than that, and unprecedented, the chromosome-breaking *Ds* locus could "change its position in the chromosome," a phenomenon she called *transposition*. Moreover, she had evidence that the *Ac* locus was required for transposition of *Ds* and that, like the *Ds* locus, the *Ac* locus was mobile.

Within several years McClintock had established beyond a doubt that both the *Ac* and *Ds* loci were not only capable of changing their positions on the genetic map, but also of inserting into loci to cause unstable mutations. She presented a paper on her work at the Cold Spring Harbor Symposium of 1951. The reaction to her presentation ranged from perplexed to hostile. Later she published several papers in refereed journals, but from the paucity of reprint requests, she inferred on the part of the larger biological community an equally cool reaction to the astonishing news that genes could move.

McClintock's work had taken her far outside the scientific mainstream, and in a profound sense she had lost her ability to communicate with her colleagues. By her own admission McClintock had neither a gift for written exposition nor a talent for explaining complex phenomena in simple terms. But there are more important factors: the very notion that genes can move was in deep contradiction to the assumption of the regular relationships among genes that underlies the construction of linkage maps and the physical mapping of genes onto chromosomes. The concept that genetic elements can move would undoubtedly have met with resistance regardless of author and presentation.

McClintock was deeply frustrated by her failure to communicate, but her fascination with the unfolding story of transposition was sufficient to keep her working at the highest level of physical and mental intensity she could sustain. By the time of her formal retirement, she had accumulated a rich store of knowledge about the genetic behavior of two markedly different transposable element families. And beginning about the time her active fieldwork ended, transposable genetic elements began to surface in one experimental organism after another.

These later discoveries came in an altogether different age. In the two decades between McClintock's original genetic discovery of transposition and its rediscovery, genetics had undergone as profound a change as the cytogenetic revolution that had occurred in the second and third decades of the century. The genetic material had been identified as DNA, the manner in which information is encoded in the genes had been deciphered, and methods had been devised to isolate and study individual genes. Genes were no longer abstract entities known only by the consequences of their alteration or loss; they were real bits of nucleic acids that could be isolated, visualized, subtly altered, and reintroduced into living organisms.

By the time the maize transposable elements were cloned and their molecular analysis initiated, the importance of McClintock's discovery of transposition was widely recognized, and her public recognition was growing. For example, she received the National Medal of Science in 1970, she was named Prize Fellow laureate of the MacArthur Foundation and she received the Lasker Basic Medical Research Award in 1981, and in 1982 she shared the Horwitz Prize. Finally, in 1983, thirty-five years after publication of the first evidence for transposition, McClintock was awarded the Nobel Prize for physiology or medicine.

McClintock was sure she would die at ninety, and a few months after her ninetieth birthday she was gone, drifting away from life gently, as a leaf from an autumn tree. What Barbara McClintock was and what she left behind are eloquently expressed in a few short lines written many years earlier by her friend and champion, Marcus Rhoades, whose death preceded hers by a few months:

One of the remarkable things about Barbara McClintock's surpassingly beautiful investigations is that they came solely from her own labors. Without technical help of any kind she has by virtue of her boundless energy, her complete devotion to science, her originality and ingenuity, and her quick and high intelligence made a series of significant discoveries unparalleled in the history of cytogenetics. A skilled experimentalist, a master at interpreting cytological detail, a brilliant theoretician, she has had an illuminating and pervasive role in the development of cytology and genetics.

Adapted by permission of Nina Fedoroff and by courtesy of the National Academy of Sciences, Washington, DC.

Ty Elements in Yeast

Ty transposable elements of yeast structurally resemble bacterial transposons in that they have terminal repeated sequences, they integrate at sites with which they have no homology, and they generate a target site duplication (of 5 bp) upon insertion.

A *Ty* element is diagrammed in Figure 19.6. The element is about 5.9 kb long, includes two 334-bp-long, directly repeated sequences called long terminal repeats (LTR) or deltas (δ), and encodes two proteins. Each delta contains a promoter and sequences recognized by transposing enzymes. The *Ty* elements are transcribed into a single, 5,700-nucleotide mRNA that begins at the promoter in the delta at the 5' end of the element and is translated to give the two encoded proteins (see Figure 19.6). On average a yeast strain contains about 35 *Ty* elements.

Surprisingly, rather than transposing DNA to DNA, as is the case with bacterial transposons and most eukaryotic transposons, *Ty* elements transpose via an RNA intermediate, that is, by making an RNA

~ **FIGURE 19.6**

The *Ty* transposable element of yeast.

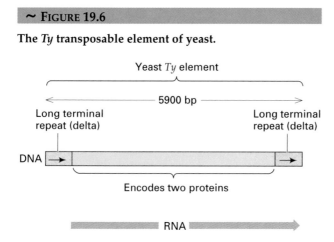

~ **FIGURE 19.7**

Illustration of the use of *P* elements to introduce genes into the *Drosophila* genome.

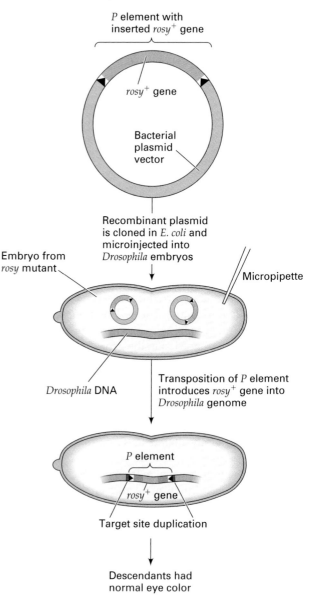

copy of the integrated DNA sequence and then by creating a new DNA *Ty* element by reverse transcription. Reverse transcription is the production of a DNA copy using RNA as a template in a reaction catalyzed by reverse transcriptase (see Chapter 17, p. 000); in this case, the reverse transcriptase is encoded by the *Ty* element. The new DNA element can then integrate at a new chromosome location. This transposition mechanism is similar to the mechanisms involved in the life cycle of retroviruses (see Chapter 17, pp. 391–394). Because of this similarity, *Ty* elements have been referred to as *retrotransposons*.

P Elements in *Drosophila*

A number of classes of transposons have been identified in *Drosophila*. In this organism it is estimated that about 15 percent of the genome is mobile—a remarkable percentage! One phenomenon caused by a transposon in *Drosophila* is **hybrid dysgenesis**, which involves the appearance of a series of defects, including mutations, chromosomal mutations, and eventually sterility, when certain strains of *Drosophila melanogaster* are crossed.

Hybrid dysgenesis primarily affects germ-line cells. That is, F_1 flies from a M♀ × P♂ cross have normal somatic tissues, but gonads do not develop, and the flies are sterile. Hybrid dysgenesis results in the germ line of the hybrid flies when *P* element transposition is induced at a high rate. The *P* elements integrate at various places in the genome, causing the characteristic defects of hybrid dysgenesis.

Apart from their role in hybrid dysgenesis, *P* elements are very important vectors for transferring genes into the germ line of *Drosophila* embryos, allowing genetic manipulation of the organism. Figure 19.7

illustrates an experiment in which the wild-type *rosy*⁺ (*ry*⁺) gene was introduced into a strain homozygous for a mutant *rosy* allele (flies have a rosy-red eye color). The wild-type *rosy* gene was introduced into the middle of a *P* element by recombinant DNA techniques and cloned in a plasmid. The recombinant plasmids were then microinjected into *rosy* embryos in the regions that would become the germ-line cells. *P*-element–encoded transposase then catalyzed the movement of the *P* element, along with the wild-type *rosy* gene it contained, to the *Drosophila* genome in some of the germ-line cells. When the flies resulting from these embryos produced gametes, they contained the wild-type *rosy* gene and, hence, descendants of these individuals had normal eye color.

KEYNOTE

Transposable genetic elements in eukaryotes are typically transposons. While most transposons move by using a DNA-to-DNA mechanism, some eukaryotic transposons such as yeast *Ty* elements transpose via an RNA intermediate (using a transposon-encoded reverse transcriptase), thereby resembling retroviruses.

Human Retrotransposons

We have learned that yeast *Ty* elements are retrotransposons, moving around the genome via RNA intermediates. There is good evidence to indicate that retrotransposons are present also in mammalian genomes.

In Chapter 10 we discussed the different repetitive classes of DNA sequences found in the genome. Of relevance here are the SINEs (short interspersed sequences) and LINEs (long interspersed sequences) found in the moderately repetitive class of sequences. SINEs are 100- to 300-bp repeated sequences interspersed between unique-sequence DNA 1,000 to 2,000 bp in length. LINEs are repeated sequences greater than 5,000 bp in length interspersed among unique-sequence DNA of up to approximately 35,000 bp in length. Both SINEs and LINEs occur in DNA families where family members are related by sequence.

In humans, a very abundant SINEs family is the *Alu family*. The repeated sequence in this family is about 300 bp long and is repeated between 300,000 and 500,000 times in the genome, amounting to up to 3 percent of the total genomic DNA. The name for the family comes from the fact that the sequence contains a restriction site for the enzyme *Alu*I ("Al-you-one"). Over evolutionary time, members of the family have diverged, so the sequences of the individuals in the family are related, but not identical.

Each Alu sequence is flanked by direct repeats of 7 to 20 bp, so they resemble transposable elements. Even though most of the moderately repetitive DNA in the genome is not transcribed, at least some members of the Alu family can be transcribed, leading to transposition within the genome. It is thought that these Alu sequences are actually retrotransposons that move via an RNA intermediate.

Members of one mammalian LINEs family, LINEs-1 (also called *L1 elements*), are also thought to be retrotransposons. In humans, there are 50,000 to 100,000 copies of the L1 element, making up about 5 percent of the genome. The maximum length of an L1 element is 6,500 bp, although only about 3,500 of them in the genome are full length, the rest having various length internal deletions (much like corn *Ds* elements). The full-length L1 elements contain a gene that probably encodes reverse transcriptase. Interestingly, in 1991 two unrelated cases of hemophilia in children were shown to result from insertions of an L1 element into the factor VIII gene, the product of which is required for normal blood clotting. Molecular analysis showed that the insertion was not present in either set of parents, indicating that the L1 element had newly transposed. More broadly, these results show that L1 elements in humans can indeed transpose, and that they can cause disease by insertional mutagenesis (see Chapter 17, p. 397).

SUMMARY

Bacteria and eukaryotic cells contain a variety of transposable elements that have the property of moving from one site to another in the genome. We discussed two important types of transposable elements in bacteria: insertion sequence (IS) elements and transposons (Tn). The simplest type of transposable element is an IS element. An IS element typically consists of inverted terminal repeat sequences flanking a coding region, the products of which provide transposition activity. Tn elements are more complex in that they contain other genes. There are two types of prokaryotic transposons. Composite transposons consist of a central region flanked on both sides by IS elements. The central region contains genes such as those for drug resistance. The IS elements contain the genes encoding the proteins required for transposition. Noncomposite transposons consist of a central region containing genes (such as for drug resistance), but they do not end with IS elements. Instead, short repeated sequences are found at their ends that are required for transposition. In these transposons, the transposition functions are encoded by genes in the central region.

Transposable elements in eukaryotes resemble bacterial transposons in general structure and transposition properties. Plant transposons often occur as families, each family containing an autonomous element (an element capable of transposing by itself) and one or more nonautonomous elements (elements that can only transpose if the autonomous element of the family is also present in the genome). Some eukaryotic transposons, such as yeast *Ty* elements and human (and maybe other mammalian) SINEs and LINEs family members, transpose via an RNA intermediate (using a transposon-encoded reverse transcriptase in the case of *Ty* and LINEs). These types of transposons resemble retroviruses in genome organization and other properties, and hence have been called retrotransposons.

The presence of transposable elements in a cell is usually detected by the changes they bring about in the expression and activities of the genes at or near the chromosomal sites into which they integrate. Gene expression may be increased or decreased if the element inserts into a promoter or other regulatory sequence, mutant alleles of a gene can be produced if an element inserts within the coding sequence of the gene, and chromosome rearrangements or chromosome breakage events can occur as a result of the transposition event.

ANALYTICAL APPROACHES FOR SOLVING GENETICS PROBLEMS

Q20.1 Imagine that you are a corn geneticist and are interested in a gene you call *zma*, which is involved in formation of the tiny hairlike structures on the upper surfaces of leaves. You have a cDNA clone of this gene. In a particular strain of corn that contains many copies of *Ac* and *Ds*, but no other transposable elements, you observe a mutation of the *zma* gene. You want to figure out whether this mutation does or does not involve the insertion of a transposable element into the *zma* gene. How would you proceed? Suggest at least two approaches, and say how your expectations for an inserted transposable element would differ from your expectations for an ordinary gene mutation.

A20.1 One approach would be to make a detailed examination of leaf surfaces in mutant plants. Since there are many copies of *Ac* in the strain, if a transposable element has inserted into *zma* it should be able to leave again, so that the mutation of *zma* would be unstable. The leaf surfaces should then show a patchy distribution of regions with and without the hairlike structures. A simple point mutation would be expected to be relatively stable.

A second approach would be to digest the DNA from mutant plants and the DNA from normal plants with a particular restriction endonuclease, run the digested DNAs on a gel and prepare a Southern blot, and probe the blot using the cDNA. If a transposable element has inserted into the *zma* gene in the mutant plants, then the probe should bind to different molecular weight fragments in mutant as compared to normal DNAs. This would not be the case if a simple point mutation had occurred.

QUESTIONS AND PROBLEMS

19.1 Distinguish between prokaryotic insertion elements and transposons. How do composite transposons differ from noncomposite transposons?

19.2 What properties do bacterial and eukaryotic transposable elements have in common?

*19.3 An understanding of the molecular structures of *Ac* and *Ds* elements, together with an understanding of the basis for their transposition, has allowed for a clearer, molecular-based interpretation of Barbara McClintock's observations. Ponder the significance of this on the acceptance of her work, and use your understanding of transposition of *Ac* and *Ds* elements in corn to propose an explanation for the following.

Two different true-breeding strains of corn with colorless kernels A and B are crossed with each of two different true-breeding strains with purple kernels C and D. The F_1 from each cross is selfed, with the following results:

P:	A × C	A × D
F_1:	all purple	all purple
F_2:	3 purple : 1 colorless	3 purple : 1 colorless
P:	B × C	B × D
F_1:	all purple	all purple
F_2:	3 purple : 1 spotted*	3 purple : 1 colorless

*spotted = kernels with purple spots in a colorless background

*19.4 Consider two theoretical yeast transposons, A and B. Each contains an intron. Each transposes to a new location in the yeast genome and then is examined for the presence of the intron. In the new locations, you find that A has no intron, but B does. What can you conclude about the mechanisms of transposon movement for A and B from these facts?

*19.5 When certain strains of *Drosophila melanogaster* are crossed, hybrid dysgenesis can result, producing mutations, chromosomal mutations, and sterility. A curious feature of hybrid dysgenesis is that it is seen when females from certain laboratory strains (M strains) are mated to males from wild populations (P strains), but not when males from the same laboratory strains are mated to females from wild populations.

a. Make an hypothesis for why hybrid dysgenesis occurs only in F_1 hybrids from M females $\times$ P males, and not in F_1 hybrids from P females $\times$ M males.

b. Make an hypothesis for why hybrid dysgenesis only affects the F_1 germ line, and not F_1 somatic cells.

19.6 In the experiment described in Figure 19.7, a wild-type ry^+ ($rosy^+$) gene was introduced into a strain homozygous for a mutant ry allele by using *P*-element–mediated gene transformation.

a. If a transformed male was mated to an M strain female (see Q19.5), would you expect the *P* element construct used in the experiment to transpose to another site in the genome of the F_1?

b. If a transformed male was mated to a P strain female (see Q19.5), would you expect the *P* element construct used in the experiment to transpose to another site in the genome of the F_1?

Chapter *20*

Extranuclear Genetics

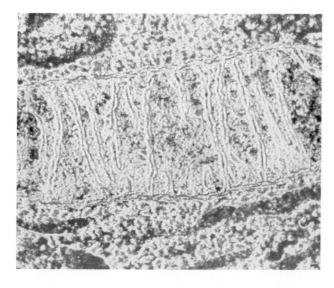

PRINCIPAL POINTS

~ Both mitochondria and chloroplasts contain their own DNA genomes. The DNA in most species' mitochondria and all chloroplasts is circular, double-stranded, and supercoiled. The mitochondrial DNA of some species is linear. Typically mitochondria and chloroplasts contain several nucleoid regions in which the DNA is located, and each nucleoid contains several copies of the DNA molecule.

~ The mitochondrial and chloroplast genomes contain genes for the rRNA components of the ribosomes of these organelles, for many (if not all) of the tRNAs used in organellar protein synthesis, and for a few proteins that remain in the organelles and perform functions specific to the organelles. All other proteins required by these organelles are nuclear-encoded, synthesized on cytoplasmic ribosomes, and transported into them. The mitochondrial genetic code of some organisms is different from the universal nuclear genetic code. Mitochondrial DNA analysis is used to study genetic relationships among individuals because of the maternal inheritance of mitochondria and polymorphisms of mitochondrial DNA.

~ The inheritance of mitochondrial and chloroplast genes follows rules different from those for nuclear genes: meiosis-based Mendelian segregation is not seen, uniparental (usually maternal) inheritance is typically exhibited, extranuclear genes cannot be mapped to the known nuclear linkage groups, and a phenotype resulting from an extranuclear mutation persists after nuclear substitution.

~ Examples of extranuclearly inherited gene mutations, involving genes in mitochondrial DNA or chloroplast DNA, include shoot variegation in four o'clock plants, certain slow-growing mutants in fungi, and certain antibiotic resistance or dependence traits in *Chlamydomonas.*

~ Not all cases of extranuclear inheritance result from genes on mitochondrial DNA or chloroplast DNA. Many other examples in eukaryotes result from infectious heredity, in which symbiotic, cytoplasmically located bacteria or viruses are transmitted when cytoplasms mix.

~ Maternal effect is defined as a phenotype in an individual that is established by the maternal nuclear genome, as the result of mRNA and/or proteins that are deposited in the oocyte prior to fertilization. These inclusions direct early development of the embryo. Maternal effect is different from extranuclear inheritance. The maternal inheritance pattern of extranuclear genes occurs because the zygote obtains most of its cellular organelles (containing the extranuclear genes) from the female parent.

~ While most genes are expressed in a way that is independent of parental origin, the expression of certain genes is determined by whether the gene is inherited from the female or male parent. This phenomenon is called genomic imprinting. In many cases, methylation patterns are the basis for imprinting.

*I*n our discussion of eukaryotic genetics up to now, we have analyzed the structure and expression of the genes located on chromosomes in the nucleus and have defined rules for the segregation of nuclear genes. Outside the nucleus, DNA is found in the mitochondrion (found in both animals and plants) and the chloroplast (found only in green plants and certain protists). The genes in these mitochondrial and chloroplast genomes are known as extrachromosomal genes, cytoplasmic genes, non-Mendelian genes, organellar genes, or extranuclear genes.

Although *extranuclear* is used in the discussions that follow, the term *non-Mendelian* is also informative because extranuclear genes do not follow the rules of Mendelian inheritance like nuclear genes. Recently, the application of molecular biology techniques has led to rapid advances in knowledge about the organization of extranuclear genomes. Your goal in this chapter is to learn about some of these advances and then obtain an understanding of the inheritance patterns of extranuclear genes.

ORGANIZATION OF EXTRANUCLEAR GENOMES

Mitochondrial Genome

Mitochondria, organelles found in the cytoplasm of all aerobic eukaryotic cells, are the principal producers of energy for the cell. They contain the enzymes of the Krebs cycle, carry out oxidative phosphorylation, and are involved in fatty acid biosynthesis. The complete DNA sequences of a number of mitochondrial (mt) DNA genomes are now known. Some of the properties of mitochondrial DNA, or mtDNA, will now be described.

STRUCTURE.   Many mitochondrial genomes are circular, double-stranded, supercoiled DNA molecules (Figure 20.1). Linear mitochondrial genomes are found in some protozoa and some fungi. No structural proteins (proteins required for chromosome folding) are associated with mtDNA. Multiple copies of the genomes are found within mitochondria located in multiple *nucleoid regions* (similar to those of bacterial cells).

The gene content is very similar in both number and function among mitochondrial genomes from different species. Despite that, the size of the genome varies tremendously from organism to organism. In animals the circular mitochondrial genome is less than 20 kb; for example, human mtDNA is 16,569 bp. In contrast, the mtDNA of yeast is about 80 kb (80,000 bp), and that of plants ranges from 100,000 to 2 million bp. The main difference between animal, plant,

and fungal mitochondria is that essentially the entire mitochondrial genomes of animals encode products, whereas the mitochondrial genomes of fungi and plants have extra DNA that does not code for products.

GENES OF MTDNA AND THEIR TRANSCRIPTION. Figure 20.2 is a map of the genes of human mtDNA. In general, mtDNA contains information for a number of mitochondrial components such as tRNAs, rRNAs, and some of the polypeptide subunits of the proteins cytochrome oxidase, NADH-dehydrogenase, and ATPase. Other components found in the mitochondria are encoded by nuclear genes and are imported into the mitochondria. These components include the DNA polymerase and other proteins for mtDNA replication, RNA polymerase and other proteins for transcription, ribosomal proteins for ribosome assembly, protein factors for translation, the aminoacyl tRNA synthetases, and the other polypeptide subunits for the three proteins mentioned above, cytochrome oxidase, NADH-dehydrogenase, and ATPase.

Transcription of mammalian mtDNA is unusual in that each strand is transcribed into a single RNA molecule that is then cut into smaller pieces. The origins for transcription of the two strands are near the DNA replication origin for the H strand (see Figure 20.2). How are the large RNA transcripts processed to produce the mature mRNAs, rRNAs, and tRNAs? Notice in Figure 20.2 that most of the genes encoding the rRNAs and the mRNAs are separated by tRNA genes. When the tRNAs are cut out of the transcript, essentially complete mRNAs and rRNAs are liberated. The processed transcripts are then modified to produce the mature RNAs. This involves adding a poly(A) tail to the 3' ends of the mRNAs and CCA to the 3' ends of the tRNAs. Mitochondrial mRNAs have no 5' cap.

In the much larger mitochondrial genomes of yeast and plants, tRNA genes do not separate the other genes, and the gaps between genes are large. In these systems, transcription termination is signaled by other, non-tRNA sequences in the DNA. Introns are never found in animal mitochondrial genomes, but they are found in yeast, plants, and some other organisms.

TRANSLATION IN THE MITOCHONDRIA.   Messenger RNAs synthesized within the mitochondria are translated by mitochondrial ribosomes assembled within the organelle. In humans, the 60S mitochondrial ribosomes consist of 45S and 35S subunits. There are only two rRNAs in the mitochondrial ribosome: 16S rRNA in the large subunit and 12S rRNA in the small subunit (see Figure 20.2). There is one gene for each rRNA in the mitochondrial genome. The ribosomal

~ FIGURE 20.1

Electron micrograph of mitochondrial DNA.

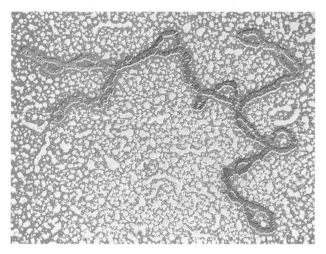

~ FIGURE 20.2

Map of the genes of human mitochondrial DNA. The outer circle shows the genes transcribed from the H (heavy) strand, and the inner circle shows the genes transcribed from the L (light) strand. The origins and directions of replication (*ori*), and the directions of transcription for the H and L strands are indicated. rRNA genes are shown in blue, tRNA genes in purple, and protein-coding genes in yellow. Code: ATPase 6 and 8: components of the mitochondrial ATPase complex. COI, COII, and COIII: cytochrome *c* oxidase subunits. cyt *b*: cytochrome *b*. ND1–6: NADH dehydrogenase components.

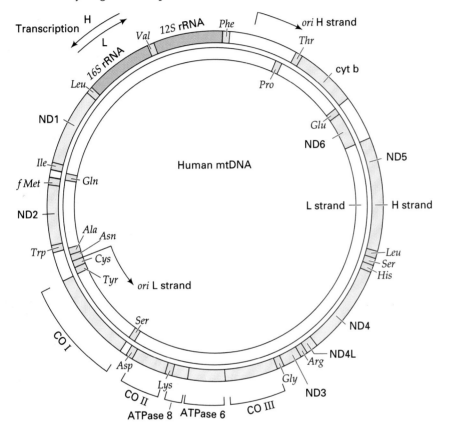

proteins found in mitochondrial ribosomes generally are encoded by nuclear genes and transported into the mitochondria from the cytoplasm.

In some ways protein synthesis in the mitochondrion is analogous to protein synthesis in bacteria. In all mitochondria, a special initiator tRNA—fMet-tRNA—is used in the initiation of protein synthesis, as is the case in *E. coli*. Special mitochondrial initiation factors (IFs), elongation factors (EFs), and release factors (RFs) distinct from cytoplasmic factors are used for translation.

Mitochondrial ribosomes are sensitive to most inhibitors of bacterial ribosome function, including streptomycin, neomycin, and chloramphenicol. Also, mt ribosomes are generally insensitive to antibiotics or other agents to which cytoplasmic ribosomes are

sensitive, such as cycloheximide. By the selective use of antibiotics, the site of synthesis of proteins found in mitochondria can be investigated. For example, mt proteins that are synthesized in the presence of cycloheximide, an inhibitor of cytoplasmic ribosomes, must be made on mt ribosomes. Conversely, mt proteins made in the presence of chloramphenicol, an inhibitor of mt ribosomes, must be made on cytoplasmic ribosomes. Using this approach, scientists have discovered that four of the seven subunits of the mitochondrial enzyme cytochrome oxidase in yeast are made in the cytoplasm, and the other three are made in the mitochondria (Figure 20.3).

For protein synthesis, only plant mitochondria use the universal nuclear genetic code. Mitochondria of other organisms have differences from that

~ FIGURE 20.3

Synthesis of the multisubunit protein cytochrome oxidase takes place on both cytoplasmic and mitochondrial ribosomes.

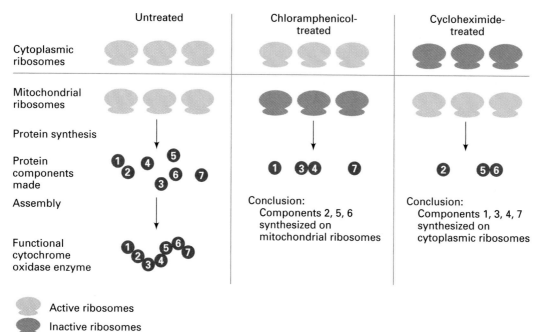

universal code, although there is no one pattern for the differences. For example, Table 20.1 shows the points of difference between human and yeast mitochondrial genetic codes.

There are also differences in mitochondria with respect to how tRNAs read the mRNAs. We learned in Chapter 13 about base-pairing wobble, in which one tRNA may read more than one codon (see Table 13.1). There is extended wobble in mitochondria, so that a minimum of 22 tRNAs is needed to read all sense codons. By contrast, 32 tRNAs theoretically are needed to read all sense codons using standard wobble.

INVESTIGATING GENETIC RELATIONSHIPS BY MTDNA ANALYSIS. In Chapter 14 we discussed some types of DNA analysis involving polymorphisms in DNA sequences. In human mtDNA, one 400-bp region is highly polymorphic. This polymorphism, along with the fact that the vast majority of mitochondria is maternally inherited, means that maternal lineages are practically unique. Thus, maternal-line relations between individuals can be investigated by using PCR to analyze mtDNA for polymorphisms.

An example of using mtDNA analysis involves the last tsar and tsarina of Russia and their children. During the Bolshevik Revolution of 1917, Tsar Nicholas Romanov II was overthrown and exiled,

and in 1918 the tsar and his family were executed by Bolshevik guards. Rumors persisted, however, that one of the tsar's daughters, Princess Anastasia, escaped the execution. In 1922 a woman came forward in Berlin claiming to be Anastasia. In 1928, using the name Anna Anderson, she came to the

~ TABLE 20.1

Differences Between Human and Yeast Mitochondrial Genetic Codes

| | | AMINO ACID | |
| | | MITOCHONDRIAL CODE | |
CODON[a]	NUCLEAR CODE	MAMMAL	YEAST
UGA	Termination	Tryptophan	Tryptophan
AUA	Isoleucine	Methionine	Isoleucine
CUN[b]	Leucine	Leucine	Threonine
AGG, AGA	Arginine	Termination	Arginine
CGN[b]	Arginine	Arginine	Termination?

[a]All sequences read 5' to 3'.
[b]N = any one of the four bases A, G, U, and C.

United States, where she lived until her death in 1984. Though she claimed until she died that she was Anastasia, there was insufficient information available to prove or disprove her claim during her lifetime. In 1993, mtDNA analysis was done on bones found two years earlier in a shallow grave in the Russian town to which the Romanovs had been exiled. The DNA samples were compared to a blood sample provided by Prince Philip, Duke of Edinburgh, who is the grand nephew of the Tsarina Alexandra. (Prince Philip's grandmother was Princess Victoria, Alexandra's sister.) The mtDNA patterns matched perfectly the mtDNA of Prince Philip, indicating they all belonged to the same maternal lineage and showing unequivocally that the bones were the remains of the tsarina and three of her five children. The bones of the tsar were identified in a similar way by matching mtDNA patterns with those of two living relatives. Soon afterward, mtDNA analysis proved that Anna Anderson was not Anastasia, since her mtDNA pattern did not match that of Prince Philip. It is not clear whether any of the three children was Anastasia, although a Russian government commission has stated that there is "definite proof" that one of the skeletons is that of Anastasia.

The case of the Romanovs is an example in which mtDNA analysis was a powerful tool for analyzing maternal lineages in humans. Mitochondrial DNA analysis is being used for studying genetic relationships in many other organisms as well (see Chapter 21, pp. 499–500). Mitochondrial DNA analysis is also being used in conservation biology studies to assess the extent of genetic variability in natural populations. One such study is analyzing the threatened grizzly bear in Yellowstone National Park as a model population for many endangered species of predators.

Chloroplast Genome

Chloroplasts are cellular organelles found only in green plants and photosynthetic protists, and are the site of photosynthesis in the cells containing them. Chloroplasts have a double membrane surrounding an internal, chlorophyll-containing lamellar structure embedded in a protein-rich stroma. Like mitochondria, chloroplasts contain their own genomes, although we do not know as much about the chloroplast (cp) genome as we do about the mitochondrial genome.

STRUCTURE AND REPLICATION.   The structure of the chloroplast genome is similar to the structure of mitochondrial genomes. In all cases the DNA is double-stranded, circular, devoid of structural proteins, and supercoiled.

~ **FIGURE 20.4**

The unicellular green alga, *Chlamydomonas*. This organism has a single chloroplast and a pair of flagella.

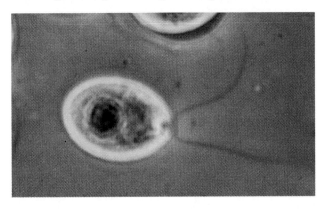

Chloroplast DNA is much larger than animal mtDNA, with a size between 80 kb and 600 kb. The DNA sequences of the chloroplast genomes of a few organisms have been completely determined. For example, the tobacco genome is 155,844 bp, and the rice genome is 134,525 bp. All chloroplast genomes contain a lot of noncoding DNA sequences.

The number of copies of cpDNA per chloroplast varies from species to species. In each case there are multiple copies per chloroplast, all found in multiple nucleoid regions that are also present in multiple copies. In the unicellular green alga *Chlamydomonas* (Figure 20.4), the one chloroplast in the cell contains between 500 and 1,500 cpDNA molecules.

GENE ORGANIZATION OF CPDNA.   The chloroplast genome contains two genes for each chloroplast rRNA (16S, 23S, 4.5S, and 5S). (By contrast, only one copy of each mitochondrial rRNA gene is found in mtDNA.) The genome also contains genes for the tRNAs, and genes for some, but not all, of the proteins required for transcription and translation of the cp-encoded genes (such as ribosomal proteins, RNA polymerase subunits, and translation factors) or for photosynthesis. Introns are found in some, but not all, of the protein-coding and tRNA genes in cpDNA. Other proteins found in the chloroplast are encoded by nuclear genes.

TRANSLATION IN CHLOROPLASTS.   Chloroplast protein synthesis uses organelle-specific 70S ribosomes that consist of 50S and 30S subunits. The 50S subunit contains one copy each of 23S, 5S, and 4.5S rRNAs, and the 30S subunit contains one copy of a 16S rRNA. The number of ribosomal proteins in each ribosomal subunit is less well defined; some are nuclear encoded and some are chloroplast encoded.

Protein synthesis in chloroplasts is similar to the process in prokaryotes. fMet-tRNA is used to initiate all proteins, and the chloroplast uses its own organelle-specific initiation factors (IFs), elongation factors (EFs), and release factors (RFs). The universal genetic code is used in chloroplast protein synthesis.

Like mitochondrial ribosomes, chloroplast ribosomes are resistant to cycloheximide, an inhibitor of cytoplasmic ribosomes, but are sensitive to virtually all inhibitors known to block prokaryotic protein synthesis. Using selective antibiotics in essentially the same way we described for the synthesis of multisubunit mitochondrial proteins (see Figure 20.3), the synthesis of chloroplast proteins has been examined. One important chloroplast protein is ribulose bisphosphate decarboxylase, the first enzyme used in the pathway for fixation of carbon dioxide in the photosynthetic process. This enzyme is a major protein and is found in the chloroplasts of all plant tissues. Since it constitutes about 50 percent of the protein found in green-plant tissue, *it is the most prevalent protein in the world.* Ribulose bisphosphate decarboxylase contains eight polypeptides, four identical small ones and four identical large ones. The small polypeptide is coded by a nuclear gene, while the large polypeptide is coded by a chloroplast gene.

KEYNOTE

Both mitochondria and chloroplasts contain their own DNA genomes. The DNA in most species' mitochondria and all chloroplasts is circular, double-stranded, and supercoiled. The mitochondrial DNA of some species is linear. The mitochondrial and chloroplast genomes contain genes for the rRNA components of the ribosomes of these organelles, for many (if not all) of the tRNAs used in organellar protein synthesis, and for a few proteins that remain in the organelles and perform functions specific to the organelles. All other mitochondrial and chloroplast proteins are nuclear-encoded, synthesized on cytoplasmic ribosomes, and imported into the organelle. In the mitochondria of some organisms, the genetic code is different from that found in nuclear protein-coding genes. Because humans and many other organisms receive most of their mitochondria from their mothers, maternal lineages are practically unique. Thus, maternal-line relations among individuals can be investigated by mtDNA analysis.

RNA Editing

In the mid-1980s a new phenomenon came to light in the production of certain mRNAs in the mitochondria of some protozoa. This phenomenon is called **RNA editing** and involves the post-transcriptional insertion and/or deletion of nucleotides in an mRNA molecule. For example, the sequences of the *COIII* gene for subunit III of cytochrome oxidase (COIII) and its mRNA transcripts were compared in the protozoans *Trypanosoma brucei* (*Tb*, for short), *Crithridia fasiculata* (*Cf*, for short), and *Leishmania tarentolae* (*Lt*, for short) (Figure 20.5). While the mRNA sequences are highly conserved among the three organisms, only the *Cf* and *Lt* mtDNA sequences are colinear with the mRNAs. Strikingly, the *Tb* gene has a sequence that cannot produce the mRNA it apparently encodes. The differences between the DNA and the mRNA sequences can be accounted for by U nucleotides in the mRNA not encoded in the DNA, and T nucleotides in the DNA not found in the transcript. The model is that the transcript of the *Tb COIII* gene is edited extensively once it is made by adding U nucleotides in the appropriate places and removing the U nucleotides encoded by the T nucleotides in the DNA (see Figure 20.5). Remarkably, when the whole sequence is examined, over 50 percent of the mature mRNA consists of U nucleotides that are added after transcription! This RNA editing must be accurate in order to reconstitute the appropriate reading frame for translation. A discussion of the models proposed to explain this type of RNA editing is beyond the scope of this text. A special RNA molecule, called a *guide RNA* (gRNA), is involved in the process. The gRNA pairs with the mRNA transcript and is thought to be responsible for cleaving the transcript, acting as a template for the missing U nucleotides, and ligating the transcript back together again.

Since the discovery of RNA editing in *T. brucei*, other types of RNA editing have been described. In the mitochondria of the slime mold *Physarum polycephalum*, single C nucleotides are added post-transcriptionally at many positions of several transcripts. In flowering plants, the sequences of most mitochondrial transcripts are edited by C-to-U changes, and to some extent by U-to-C changes. C-to-U editing is also involved in the production of an AUG initiation codon from an ACG codon in some chloroplast mRNAs in some flowering plants. Lastly, C-to-U editing occurs in the mRNA transcribed from the nuclear gene encoding the cytoplasmic apolipoprotein B in mammals. This editing results in tissue-specific generation of a stop codon, thereby leading to tissue-specific differences in the protein produced.

~ **FIGURE 20.5**

Comparison of the DNA sequences of the cytochrome oxidase subunit III gene (*COIII*) in the protozoans *Trypanosoma brucei* (*Tb*), *Crithridia fasiculata* (*Cf*), and *Leishmania tarentolae* (*Lt*), aligned with the conserved mRNA for *Tb*. The lowercase *u*'s are the U nucleotides added to the transcript by RNA editing. The template Ts in *Tb* DNA that are not in the RNA transcript are highlighted in blue.

Region of *COIII* gene transcript

Tb DNA	G GTTTTTGG AGG G GTTTTG G G A A GA GAG
Tb RNA	uuGuGUUUUUGGuuuAGGuuuuuuuuGuuG UUGuuGuuuuGuAuuAuGAuuGAGu
Cf DNA	TTTTTATTTTGATTTCGTTTTTTTTTATG TGTATTATTTGTGCTTTGATCCGCT
Lt DNA	TTTTTATTTTGATTTCGTTTTTTTTTATG TGTTTTATTTATGTTATGAGTAGGA
Tb Protein	Leu Cys Phe Trp Phe Arg Phe Phe Cys Cys Cys Cys Phe Val Leu Trp Leu Ser

Origin of Mitochondria and Chloroplasts

How might mitochondria and chloroplasts have arisen? The widely accepted **endosymbiont hypothesis** is that mitochondria and chloroplasts originated as free-living prokaryotes that invaded primitive eukaryotic cells and established a mutually beneficial (symbiotic) relationship. According to this hypothesis, eukaryotic cells started out as anaerobic organisms that lacked mitochondria and chloroplasts. At some point in their evolution, about a billion (10^9) years ago, a eukaryotic cell established a symbiotic relationship with a purple nonsulfur photosynthetic bacterium. Over time, the oxidative phosphorylation activities of the bacterium became used for the benefits of the eukaryotic cell, and, in the presence of atmospheric oxygen, photosynthetic activity (which is a generator of oxygen) was no longer needed and was lost. Eventually, the eukaryotic cell became dependent on the intracellular bacterium for survival, and the mitochondrion was formed. The chloroplasts of plants and algae are hypothesized to have occurred either simultaneously or later by the ingestion of an oxygen-producing photosynthetic bacterium (a cyanobacterium) by a eukaryotic cell.

As we have just discussed, many proteins found in mitochondria and chloroplasts are encoded by nuclear genes. Thus, further evolution of mitochondria and chloroplasts likely involved the extensive transfer of genes from the organelles to the nuclear DNA.

RULES OF EXTRANUCLEAR INHERITANCE

Since the pattern of inheritance shown by genes located in organelles differs strikingly from the pattern shown by nuclear genes, the term **non-Mendelian inheritance** is appropriate to use when we are discussing extranuclear genes. In fact, if the results obtained from genetic crosses do not conform to predictions based on the inheritance of nuclear genes, there is a good reason to suspect extranuclear inheritance.

Here are the four main characteristics of *extranuclear inheritance*:

1. Ratios typical of Mendelian segregation are not found because meiosis-based Mendelian segregation is not involved.

2. In complex eukaryotes, the results of reciprocal crosses involving extranuclear genes are not the same as reciprocal crosses involving nuclear genes because meiosis-based Mendelian segregation is not involved. (Recall from Chapters 2 and 3 that, in a reciprocal cross, the sexes of the parents are reversed in each case. For example, if A and B represent contrasting genotypes, A♀ × B♂ and A♂ × B♀ would represent a pair of reciprocal crosses.)

   Extranuclear genes usually show **uniparental inheritance** from generation to generation. In uniparental inheritance, all progeny (both males and females) have the phenotype of only one parent. Typically, for complex eukaryotes it is the mother's phenotype that is expressed exclusively, a phenomenon called **maternal inheritance**. Maternal inheritance occurs because the amount of cytoplasm in the female gamete usually greatly exceeds that in the male gamete. Therefore, the zygote receives most of its cytoplasm (containing the extranuclear genes in the mitochondria and, where applicable, the chloroplasts) from the female parent and a negligible amount from the male parent.

   In contrast, the results of reciprocal crosses between a wild-type and a mutant strain are identical if the genes are located on nuclear chromosomes. One exception occurs when X-linked genes are involved (see Chapter 3), but even then the results are distinct from those for extranuclear inheritance.

3. Extranuclear genes cannot be mapped to the chromosomes in the nucleus.

4. Extranuclear inheritance is not affected by substituting a nucleus with a different genotype.

KEYNOTE

The inheritance of extranuclear genes follows rules different from those for nuclear genes. In particular, meiosis-based Mendelian segregation is not involved, generally uniparental (and often maternal) inheritance is seen, extranuclear genes are not mappable to the known nuclear-linkage groups, and the phenotype persists even after nuclear substitution.

EXAMPLES OF EXTRANUCLEAR INHERITANCE

In this section we discuss the properties of a selected number of mutations in extranuclear chromosomes in order to illustrate the principles of extranuclear inheritance.

Shoot Variegation in the Four O'Clock

A variegated-shoot phenotype of the *albomaculata* strain of four o'clocks (*Mirabilis jalapa*, also called the marvel of Peru; Figure 20.6a) involves non-Mendelian inheritance. (A shoot of a plant consists of stem, leaves, and flowers.) This strain has mostly shoots with variegated leaves (leaves with yellowish-white patches) and occasional shoots that are wholly green or wholly yellow-white (Figure 20.6b).

Table 20.2 summarizes the results of crosses between flowers on variegated, green, and white shoots. The flowers on green shoots give only green progeny, regardless of whether pollen is from green, white, or variegated shoots. Flowers on white shoots give only white progeny, regardless of the pollen source. (However, because the whiteness indicates the absence of chlorophyll and hence an inability to carry out photosynthesis, white progeny die soon after seed germination.) Finally, flowers on variegated shoots all produce three types of progeny—completely green, completely white, and variegated—regardless of the type of pollen. In subsequent generations, maternal inheritance is always seen in these same patterns. In sum, the *progeny phenotype in each case was the same as that of the maternal parent* (the color of the progeny

~ FIGURE 20.6

Variegation in the four o'clock, *Mirabilis jalapa.* (a) Photograph of the four o'clock; (b) Drawing showing variegation. Shoots that are all green, all white, and variegated are found on the same plant, and flowers may form on any of these shoots.

a)

b)

Variegated shoot

All-white shoot

All-green shoot

Variegated main shoot

~ TABLE 20.2

Results of Crosses of Variegated Plants of *Mirabilis jalapa*

PHENOTYPE OF SHOOT-BEARING ♀ PARENT (EGG)	PHENOTYPE OF SHOOT-BEARING ♂ PARENT (POLLEN)	PHENOTYPE OF PROGENY
White	White	White
	Green	White
	Variegated	White
Green	White	Green
	Green	Green
	Variegated	Green
Variegated	White	Variegated, green, or white
	Green	Variegated, green, or white
	Variegated	Variegated, green, or white

shoots resembled the color of the parental flower), which is indicative of maternal inheritance.

Flowering plants are green because of the presence of the green pigment chlorophyll in large numbers of chloroplasts. Green shoots in four o'clocks have a normal complement of chloroplasts. White shoots have abnormal, colorless chloroplasts called leukoplasts. Leukoplasts lack chlorophyll and hence are incapable of carrying out photosynthesis. The explanation for the inheritance of shoot color in the *albomaculata* strain of four o'clocks is that the abnormal chloroplasts are defective as a result of a mutant gene in the cpDNA. During plant growth and development, chloroplasts and leukoplasts segregate so that a particular cell and its progeny cells may receive only chloroplasts (leading to green tissues), only leukoplasts (leading to white tissues), or a mixture of chloroplasts and leukoplasts (leading to variegation) (Figure 20.7). Let us take this model one step further. The flowers on a green shoot have only chloroplasts, and so through maternal inheritance these chloroplasts form the basis of the phenotype of the next generation. Similar arguments can be made for the flowers on white or variegated shoots.

KEYNOTE

The leaf color phenotypes in the variegated *albomaculata* strain of the four o'clock, *Mirabilis jalapa*, show maternal inheritance. The abnormal chloroplasts in white tissue are the result of a mutant gene in the cpDNA, and the observed inheritance patterns follow the segregation of cpDNA.

The [*poky*] Mutant of *Neurospora*

In *Neurospora crassa* the sexual phase of the life cycle is initiated following a fusion of nuclei from mating type *A* and *a* parents. Even though this fungus is a simple eukaryote, the two gametes are not equal in cytoplasmic content, so we can think of a female gamete and a male gamete in much the same way as we can for animals and plants and make reciprocal crosses.

Neurospora crassa is an obligate aerobe; that is, it requires oxygen to grow and survive, so mitochondrial functions are essential for its growth. The [*poky*] mutant is partially defective in aerobic respiration as a result of changes in the cytochromes of the mitochondria that affect the ability of the mitochondria to generate enough ATP to support rapid growth, so slow growth results. (The square brackets around the mutant symbols indicate an extranuclear gene.)

Reciprocal crosses between [*poky*] and the wild type give the following results:

[*poky*] ♀ × wild-type ♂ → all [*poky*] progeny

wild-type ♀ × [*poky*] ♂ → all wild-type progeny

As you can see, all the progeny show the same phenotype as the maternal parent, indicating maternal inheritance for the [*poky*] mutation.

KEYNOTE

The slow-growing [*poky*] mutant of *Neurospora crassa* shows maternal inheritance and deficiencies for some mitochondrial cytochromes.

~ FIGURE 20.7

Model for the inheritance of shoot color in the four o'clock, *Mirabilis jalapa*.

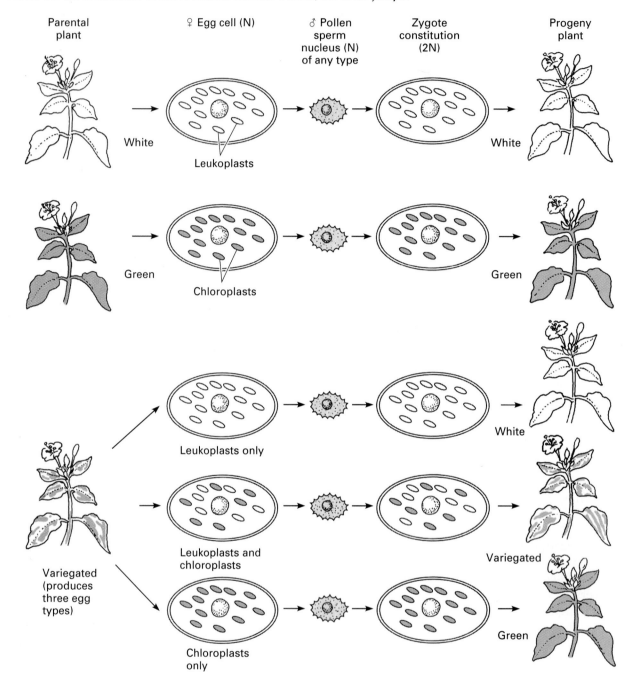

Yeast *petite* Mutants

Yeast grows as single cells. Thus, on solid media, yeast forms discrete colonies consisting of many thousands of individual cells clustered together. Yeast can grow with or without oxygen. In the absence of oxygen, yeast obtains energy for growth and metabolism through fermentation, in which the mitochondria are not involved. In the presence of oxygen, mitochondria carry out aerobic respiration and facilitate a faster growth rate than is the case in the absence of oxygen.

CHARACTERISTICS OF *PETITE* MUTANTS.
When yeast cells are spread onto solid medium, allowing growth to occur either by aerobic respiration or by fermentation, between 0.1 and 1 percent of the cells give rise to colonies that are much smaller than wild-type colonies. Since the discoverer of this phenomenon, Boris Ephrussi, was French, the small colonies are called *petites* (French for "small"), and the wild-type colonies are sometimes called *grandes* (French for "big"). *Petite* colonies are small because the growth rate of the mutant *petite* strain is significantly slower than that of the wild type, not because the cells are small. That is, there are fewer cells in the *petite* colonies.

Petites have cytochrome deficiencies and therefore are essentially incapable of carrying out aerobic respiration; they must obtain their energy primarily from fermentation, which is a relatively inefficient process. Thus, on a medium that supports only aerobic respiration, *petites* are unable to grow.

DIFFERENT TYPES OF *PETITE* MUTANTS.
Crosses between *petites* and wild type were made to determine how *petites* are inherited. Yeast are haploid, and crosses involve fusing two cells, one of mating type **a** and one of mating type α, to produce a diploid zygote (see p. 118 and Figure 5.13). That zygote can be grown into a colony to check its phenotype. When the zygote goes through meiosis (a process called sporulation), the four haploid meiotic products—the sexual spores (ascospores)—are contained within an ascus. Tetrad analysis can be done, that is, the analysis of all four products of meiosis (here, the four ascospores in an ascus) to determine segregation patterns.

Some *petites*, when crossed with wild type, give a 2:2 segregation of wild type : *petite* in tetrads (Figure 20.8a). This segregation pattern is that found for nuclear gene mutations, so these *petites* are *nuclear petites* (also called *segregational petites*) and symbolized *pet⁻*. The existence of *nuclear petites* is not surprising, since some subunits of some mitochondrial proteins are encoded by nuclear genes. Using genetic symbols, when a *pet⁻* mutant is crossed with the wild type (*pet⁺*), the diploid is *pet⁺/pet⁻*, which produces wild-type colonies. When this *pet⁺/pet⁻* cell goes through meiosis, each resulting tetrad shows a 2:2 segregation of wild-type (*pet⁺*) : *petite* (*pet⁻*) phenotype. *Nuclear petites* occur much less frequently than extranuclear *petites*.

Two other classes of *petites*, the *neutral petites* and the *suppressive petites*, show extranuclear inheritance.

Figure 20.8b shows the inheritance pattern of *neutral petites* (symbolized [*rho⁻N*]). When a *neutral petite* is crossed with normal wild-type cells ([*rho⁺N*]), the [*rho⁺N*]/[*rho⁻N*] diploids all produce wild-type colonies. When these diploids go through meiosis, all resulting tetrads show a 0:4 ratio of *petite* : wild type, while at the same time nuclear markers segregate 2:2. The name *neutral*, then, refers to that fact that this class of *petites* does not affect the wild type. This result is a classical example of uniparental inheritance, in which all progeny have the phenotype of only one parent. *This phenomenon is not maternal inheritance, however, since the two haploid cells that fuse to produce the diploid are the same size and contribute equal amounts of cytoplasm.*

An examination of the mitochondrial genetic material in the *neutral petites* reveals a remarkable characteristic: essentially 99 to 100 percent of the mtDNA is missing. Not surprisingly, then, the *neutral petites* are unable to perform mitochondrial functions. They survive because of cytoplasmically localized fermentation processes. In genetic crosses with the wild type, the normal mitochondria from the wild-type parent form a population from which new, normal mitochondria are produced in all progeny, and hence the *petite* trait is lost after one generation.

The second class of *petites* that shows extranuclear inheritance is the *suppressive petites* (symbolized [*rho⁻S*]). The *suppressive petites* are different from the *neutrals* because they *do* have an effect on the wild type. Most *petite* mutants are of the suppressive type.

The inheritance pattern of *suppressive petites* is different from those of *nuclear* and *neutral petites* (Figure 20.8c). When a [*rho⁺*]/[*rho⁻S*] diploid is formed, it has respiratory properties intermediate between those of normal and *petite* strains. If this diploid divides mitotically a number of times, the diploid population produced will be up to 99 percent *petites*. The name *suppressive*, then, refers to that fact that this class of *petites* overwhelms the wild type so that a respiratory-deficient phenotype results. Sporulation of any of the *petites* in that population produces tetrads with a 4:0 ratio of *petite* : wild type (see Figure 20.8c). Sporulation of any of the few wild-type diploids in the population produces tetrads with a 0:4 ratio of *petite* : wild type.

The *suppressive petites* have changes in the mtDNA. The changes start out as deletions of part of the mtDNA, and then, by some correction mechanism, sequences that are not deleted become duplicated until the normal amount of mtDNA is restored. During the correction events rearrangements of the mtDNA sometimes occur. Since the protein-coding genes in the mitochondrial genome are widely

~ FIGURE 20.8

Inheritance of yeast *petite* mutants. (a) Mendelian inheritance of *nuclear petites* (*pet⁻*). (b) Extrachromosomal inheritance of *neutral petites* ([*rho⁻N*]). (c) Extrachromosomal inheritance of *suppressive petites* ([*rho⁻S*]).

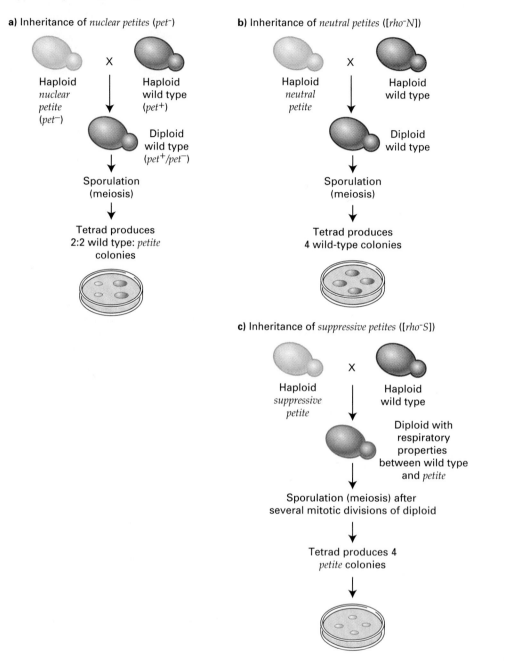

a) Inheritance of *nuclear petites* (*pet⁻*)

Haploid *nuclear petite* (*pet⁻*)   X   Haploid wild type (*pet⁺*)

Diploid wild type (*pet⁺/pet⁻*)

Sporulation (meiosis)

Tetrad produces 2:2 wild type: *petite* colonies

b) Inheritance of *neutral petites* ([*rho⁻N*])

Haploid *neutral petite*   X   Haploid wild type

Diploid wild type

Sporulation (meiosis)

Tetrad produces 4 wild-type colonies

c) Inheritance of *suppressive petites* ([*rho⁻S*])

Haploid *suppressive petite*   X   Haploid wild type

Diploid with respiratory properties between wild type and *petite*

Sporulation (meiosis) after several mitotic divisions of diploid

Tetrad produces 4 *petite* colonies

scattered, these deletions and rearrangements disrupt genes and lead to deficiencies in the enzymes involved in aerobic respiration, and a *petite* colony results. *Suppressive petites* are proposed to have a sup- pressive effect over normal mitochondria either: (1) by replicating faster than normal mitochondria and thereby outcompeting them, or (2) by fusion with normal mitochondria followed by recombination between *suppressive* mtDNA and normal mtDNA, thereby altering the latter's gene organization.

KEYNOTE

Yeast *petite* mutants grow slowly and have various deficiencies in mitochondrial functions as a result of alterations in mitochondrial DNA. Some *petite* mutants show Mendelian inheritance, while others show extranuclear inheritance: particular patterns of inheritance vary with the type of *petite*.

Extranuclear Genetics of *Chlamydomonas*

The motile, haploid unicellular alga *Chlamydomonas reinhardtii* has two flagella and a single chloroplast that contains many copies of cpDNA. There are two mating types, mt^+ and mt^-. In a process known as syngamous mating, a zygote is formed by fusion of two equal-sized cells (which, therefore, contribute an equal amount of cytoplasm), one of each mating type. A thick-walled cyst develops around the zygote. After meiosis, four haploid progeny cells are produced, and since mating type is determined by a nuclear gene, a 2:2 segregation of mt^+:mt^- mating types results.

Chlamydomonas has proved to be an excellent system for studying chloroplast and mitochondrial inheritance. One chloroplast trait that is inherited in an extranuclear manner is erythromycin resistance ([ery^r]). Wild-type *Chlamydomonas* cells are erythromycin sensitive ([ery^s]). From a cross of $mt^+[ery^r]$ × $mt^-[ery^s]$, the offspring are all erythromycin resistant (Figure 20.9a). This result is an example of uniparental inheritance. The reciprocal cross, $mt^-[ery^r]$ × $mt^+[ery^s]$ (Figure 20.9b), also shows uniparental inheritance, although here the erythromycin-sensitive phenotype is inherited. Thus, even though both parents contribute equal amounts of cytoplasm to the zygote, the progeny resemble the mt^+ parent with regard to chloroplast-controlled phenotypes. Somehow, preferential segregation of one chloroplast type occurs in this organism, or perhaps one parental type is inactivated preferentially.

A number of *Chlamydomonas* chloroplast mutants in addition to ery^r show uniparental inheritance, with progeny *always* resembling the phenotype of the mt^+ parent. The molecular explanation for this is that the cpDNA from the mt^- parent always disap-

pears from mt^+/mt^- zygotes. The reason is not known, however.

KEYNOTE

Chlamydomonas chloroplast genes are inherited in an extranuclear manner, with progeny always resembling the phenotype of the mt^+ parent.

Human Genetic Diseases and Mitochondrial DNA Defects

A number of human genetic diseases result from mtDNA gene mutations. These diseases show maternal inheritance. The following are some brief examples.

Leber's hereditary optic neuropathy (LHON): This disease affects mid-life adults and results in complete or partial blindness from optic nerve degeneration. Mutations in the mitochondrial genes for the proteins ND1, ND2, ND4, ND5, ND6, cyt *b*, COI, COIII, and ATPase 6 (see Figure 20.2) all lead to LHON. Those proteins are included in mitochondrial electron transport chain enzyme complexes. The electron transport chain drives cellular ATP production by oxidative phosphorylation. It appears that death of the optic nerve in LHON is a common result of oxidative phosphorylation defects, here brought about by inhibition of the electron transport chain.

Kearns-Sayre syndrome: People with this syndrome have encephalomyopathy, a brain disease. The cause of the syndrome is large deletions at various positions in the mtDNA. One model is that each deletion removes one or more tRNA genes, so mitochondrial protein synthesis is disrupted. In some unknown way, this leads to development of the syndrome.

Myoclonic epilepsy and ragged-red fiber (MERRF) disease: People with this disease have dementia, deafness, and seizures. Their mitochondria are abnormal in appearance. The disease is caused by a single nucleotide substitution in the gene for a lysine tRNA. The mutated tRNA adversely affects mitochondrial protein synthesis, and somehow this gives rise to the various phenotypes of the disease.

In most diseases resulting from mtDNA defects, the cells of affected individuals have a mixture of mutant and normal mitochondria. This condition is known as *heteroplasmy*. Characteristically the proportions of the two mitochondrial types vary from tissue to tissue and from individual to individual within a pedigree. The severity of the disease symptoms

~ FIGURE 20.9

Uniparental inheritance in *Chlamydomonas*. (a) From a cross of $mt^+[ery^r] \times mt^-[ery^s]$, 95 percent of the zygotes give tetrads that segregate 2:2 for the nuclear mating-type genes, and 4:0 for the extranuclear gene carried by the mt^+ parent (here, $[ery^r]$); (b) From the reciprocal cross of a $mt^-[ery^r] \times mt^+[ery^s]$, 95 percent of the zygotes give tetrads segregating $2\ mt^+ : 2\ mt^-$ and $0\ [ery^r] : 4\ [ery^s]$, again showing uniparental inheritance for the extranuclear trait of the mt^+ parent.

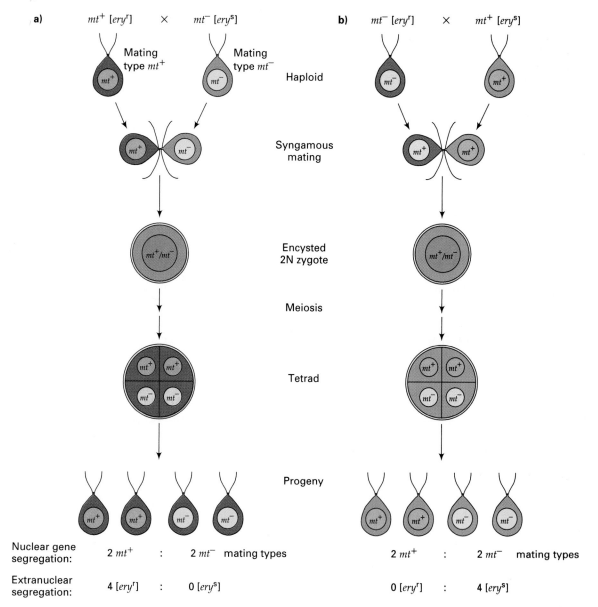

correlates approximately with the relative amount of mutant mitochondria.

Infectious Heredity—Killer Yeast

There are examples of eukaryotic extranuclear inheritance that are due to the presence of cytoplasmic bacteria or viruses symbiotically coexisting with the eukaryotes. One example is the killer phenomenon in yeast in which some strains secrete a toxin that kills sensitive yeast strains. (Killer strains are immune to their own toxin.) The killer phenomenon results from the presence in the cell's cytoplasm of two types of viruses, L and M (Figure 20.10a). Neither has adverse effects on the host cell.

~ FIGURE 20.10

The killer phenomenon in yeast. (a) Killer yeast contain two virus types, L and M, each of which contains a double-stranded RNA genome. L-dsRNA encodes both virus particles and the enzyme required for L and M virus replication. M-dsRNA encodes the killer toxin; (b) Sensitive yeast, which can be killed by killer toxin, either have L viruses but no M viruses, or have neither virus type.

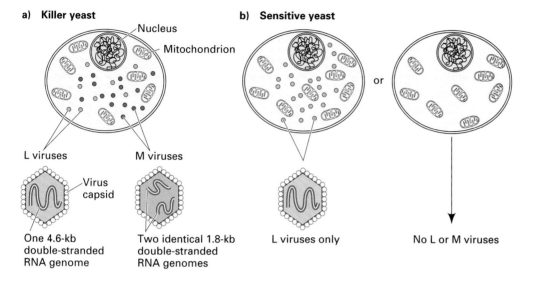

The L virus consists of a protein capsid surrounding a 4.6-kb double-stranded (ds) RNA genome called L-dsRNA. L-dsRNA encodes the capsid proteins for both L and M viruses and the viral polymerase required for viral RNA replication. So, since all viral particles are encoded by an L-dsRNA, M viruses are only found in cells if L viruses are also present. The M virus consists of a virus particle surrounding two identical copies of a 1.8-kb double-stranded RNA genome called M-dsRNA. M-dsRNA encodes the killer toxin protein, which is secreted from the cell. The same protein confers immunity on the killer cell.

Sensitive yeast cells can be killed by the M-encoded killer toxin; there are two types (Figure 20.10b). One type has only L viruses, and the other has neither L nor M viruses. In both types, no immunity is produced because killer toxin is not made.

Unlike most viruses, the yeast L and M viruses are not found outside the cell, so sensitive yeast cells cannot be infected by viruses that invade from outside. Rather, virus transmission from yeast to yeast occurs whenever yeast cells mate. All progeny of the mating will inherit copies of the viruses in the parental cells, illustrating an infectious mechanism of cytoplasmic inheritance.

KEYNOTE

Not all cases of extranuclear inheritance result from genes on mtDNA or cpDNA. Many other examples in eukaryotes result from infectious heredity in which symbiotic, cytoplasmically located bacteria or viruses are transmitted when cytoplasms mix. The killer phenomenon in yeast is such an example: it results from the infectious heredity of cytoplasmically located viruses.

CONTRASTS TO EXTRANUCLEAR INHERITANCE

Maternal Effect

The maternal inheritance pattern of extranuclear genes is distinct from the phenomenon of **maternal effect**, which is defined as the phenotype in an individual that is established by the maternal nuclear genome, as the result of mRNA and/or proteins that are deposited in the oocyte prior to fertilization. These inclusions direct early development of the embryo. That is, in maternal inheritance the progeny always have the maternal *phenotype*, whereas in maternal effect the progeny always have the phenotype specified by the maternal nuclear genotype. *Maternal effect does not involve any extranuclear genes and is discussed here to make the distinction from extranuclear inheritance clear.*

Maternal effect is seen, for instance, in the inheritance of the coiling direction in the shell of the snail *Limnaea peregra*. The shell-coiling trait is determined by a single pair of *nuclear* alleles, the dominant *D* allele for coiling to the right (dextral coiling) and the recessive *d* allele for coiling to the left (sinistral coiling). The shell-coiling phenotype is *always determined by the genotype of the mother*. The latter is shown by the results of reciprocal crosses between a true-breeding, dextral-coiling, and a sinistral-coiling snail (Figure 20.11). All the F_1s have the same genotype since a nuclear gene is involved, yet the *phenotype* is different for the reciprocal crosses.

In the cross of a dextral (D/D) female with a sinistral (d/d) male (Figure 20.11a), the F_1s are all D/d in genotype and dextral in phenotype. Selfing the F_1 produces F_2s with a 1:2:1 ratio of D/D, D/d, and d/d genotypes. *All* of the F_2s are dextral, even the d/d snails whose genotype would seem to indicate sinistral phenotype. Here is our first encounter with maternal effect; the d/d snails have a coiling phenotype not specified by the genotype they have, but one specified by the genotype of their mother (D/d). Selfing the F_2 snails gives F_3 progeny, 3/4 of which are dextral and 1/4 of which are sinistral. The latter are the d/d progeny of the F_2 d/d snails; these F_3 snails are sinistral because their phenotype reflects their mother's (the F_2) genotype.

Similar results are seen in the reciprocal cross of a sinistral (d/d) female with a dextral (D/D) male (Figure 20.11b). The F_1s are all D/d in genotype, yet they are sinistral in phenotype because the mother is genotypically d/d. Selfing the F_1 produces F_2s all of

---

~ FIGURE 20.11

Inheritance of the direction of shell coiling in the snail *Limnaea peregra* is an example of maternal effect. (a) Cross between true-breeding dextral-coiling female (D/D) and sinistral-coiling male (d/d); (b) Cross between sinistral-coiling female (d/d) and true-breeding dextral-coiling male (D/D).

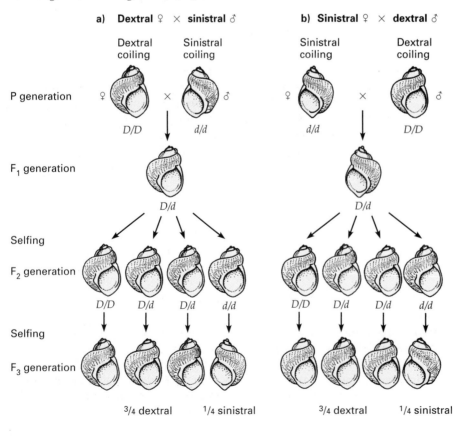

which are dextral for the same reason as the reciprocal cross already described. The genotypes and phenotypes of the F_2 and F_3 generations are the same as for the reciprocal cross (see Figure 20.11b), again for the same reasons.

These results do *not* fit our criteria for extranuclear inheritance. That is, if the coil-direction phenotype were controlled by an extranuclear gene, the progeny would always exhibit the *phenotype* of the mother, owing to maternal inheritance. Here, the coiling phenotype is governed directly by the nuclear *genotype* of the mother and is an example of maternal effect. But what is the basis for the coiling? The orientation of the mitotic spindle in the first mitotic division following fertilization controls the direction of coiling. The mother encodes products, deposited in the oocyte, that direct the orientation of the mitotic spindle and therefore the direction of cell cleavage. Thus, a mother of genotype $D/-$ deposits gene products that specify a dextral (right-handed) coiling, and a mother of genotype d/d deposits gene products that specify a sinistral (left-handed) coiling.

We encountered examples of maternal effect earlier in our discussion of the genetic control of *Drosophila* development (Chapter 16, pp. 375–376). The class of genes called *maternal effect genes* are nuclear genes that are expressed by the mother during oogenesis. The products of these genes are deposited in the egg and function to specify the gradients in the egg that control spatial organization in early development. These genes were identified by studying the properties of mutants that did not develop normally. For example, mothers homozygous for a mutated maternal effect gene, *bicoid* (*bcd*), produce mutant embryos that have no heads or thoraxes, only abdomens. On the basis of that result, the *bcd* gene was deduced to play an important role in determining normal anterior development. Relevant to our discussion here is that the *phenotype* (mutant) of the progeny embryos reflects the *genotype* (*bcd/bcd*) and not the phenotype (normal) of the mothers; that is, maternal effect is involved.

KEYNOTE

Maternal effect is different from extranuclear inheritance. The maternal inheritance pattern of extranuclear genes occurs because the zygote receives most of its organelles (containing the extranuclear genes) from the female parent, whereas in maternal effect the trait inherited is controlled by the maternal *nuclear* genotype before and after the fertilization of the egg and does not involve extranuclear genes.

Genomic Imprinting

An implicit assumption of Mendelian genetics is that the expression of a gene is totally independent of whether it is of maternal or paternal origin. Most genes align with this assumption. However, there are instances where the *expression of certain genes is determined by whether the gene is inherited from the female or male parent*, a phenomenon called **genomic imprinting** or **parental imprinting**. In other words, one or the other parent marks the offspring genetically, leading to functional differences between homologous alleles. As for maternal effect, genomic imprinting does not involve any extranuclear genes and is discussed here because of the maternal versus paternal influence on gene expression.

Some human genetic diseases appear to result from genomic imprinting, for example, the *Prader-Willi* and *Angelman syndromes*. Prader-Willi syndrome (PWS) occurs in about 1 in 25,000 births. PWS patients are typically small and weak at birth, and their symptoms include retardation and poor feeding due to diminished swallowing and sucking reflexes. The feeding difficulties improve by the age of 6 months and, from about 12 months onward, a pattern of uncontrollable eating develops, leading to obesity and associated psychological problems. Adolescents have poor motor skills and insatiable hunger. Adults are short compared to their family members and often develop a form of diabetes because of the eating disorder. PWS patients rarely live beyond 30 years unless they maintain strict weight-control programs to control the diabetes.

PWS is caused by deletion or disruption of a gene or several genes in region 15q11–q13 of chromosome 15. Pedigree analysis has shown that in 70 to 80 percent of cases examined, the deletion/disruption occurred in the father and that genomic imprinting plays a role. That is, in a PWS child, the activities of some genes in region 15q11–q13 on the maternal chromosome 15 are normally suppressed as a result of genomic imprinting. The suppression occurs by methylation of the genes. The paternally inherited alleles are necessary for normal development, but because of the gene deletion/disruption event in the father, those genes are also inactive and the PW phenotype results.

Angelman syndrome (AS) individuals have a number of symptoms, including severe motor and intellectual retardation, smaller-than-normal head size, jerky limb movements, hyperactivity, and frequent unprovoked laughter. In about 50 percent of AS patients, a deletion of region 15q11–q13 is seen. This is the same region affected in PWS patients. Indeed, it seems that AS can be caused in much the same way as

PWS, except that in AS maternally inherited alleles of the genes involved are needed for normal development. That is, the paternally inherited alleles are inactive because of methylation brought about by genomic imprinting, which causes AS to develop if the maternally inherited alleles are deleted or disrupted.

KEYNOTE

While most genes are expressed in a way that is totally independent of parental origin, the expression of certain genes is determined by whether the gene is inherited from the female or male parent. This phenomenon is called genomic imprinting. In many cases, methylation patterns are the basis for imprinting.

SUMMARY

Both mitochondria and chloroplasts contain DNA, the length of which varies from organism to organism. The genomes of both organelles are naked, usually circular, double-stranded DNAs. The two genomes contain genes that are not duplicated in the nuclear genome; hence, the organelle genes are contributing different information for the function of the cell. The organellar genes encode the rRNA components of the ribosomes (which are assembled and function in the organelles), and many (if not all) of the tRNAs used in organellar protein synthesis. Some of the organelle proteins are encoded by the organelle; the remainder are nuclear-encoded. Nuclear-encoded proteins found in organelles are synthesized on cytoplasmic ribosomes, then imported into the appropriate organelles.

The existence of polymorphic sequences in mtDNA and the fact that humans and many other organisms exhibit maternal inheritance of mitochondria mean that maternal lineages are practically unique. Thus, the genetic relationships between individuals can readily be investigated by mtDNA analysis. This method is used in many areas of biology, including conservation biology and paleoanthropology.

There are many examples of mitochondrial and chloroplast mutants. The inheritance of these extranuclear genes follows rules different from those for nuclear genes, and this is how extranuclear genes were originally identified. For extranuclear genes, meiosis-based Mendelian segregation is not involved, they show uniparental (and usually maternal) inheritance, the trait involved persists after nuclear substitution, and they cannot be mapped to nuclear chromosomes. Not all cases of extranuclear inheritance result from genes on mtDNA or cpDNA. Other examples result from infectious heredity, in which cytoplasmically located bacteria or viruses are transmitted when cytoplasms mix.

Maternal effect is the phenotype in an individual that is established by the maternal nuclear genome, as the result of mRNA and/or proteins that are deposited in the oocyte prior to fertilization. These inclusions direct early development of the embryo. Maternal effect is different from extranuclear inheritance. The maternal inheritance pattern of extranuclear genes occurs because the zygote receives most of its organelles (containing the extranuclear genes) from the female parent, whereas in maternal effect the trait inherited is controlled by the maternal *nuclear* genotype before fertilization of the egg and does *not* involve extranuclear genes.

Finally, we discussed genomic imprinting, the phenomenon in which the expression of a gene is determined by whether the gene is inherited from the female or male parent as opposed to being expressed independently of parental origin. This phenomenon does not involve extranuclear genes, but it does involve an influence on gene expression of one or the other parent. Methylation has been shown to be responsible for imprinting in some cases.

ANALYTICAL APPROACHES TO SOLVING GENETICS PROBLEMS

Q20.1 In *Neurospora*, strains of [*poky*] have been isolated that have reverted to nearly wild-type growth, although they still retain the abnormal metabolism and cytochrome pattern characteristic of [*poky*]. These strains are called *fast*-[*poky*]. A cross of a *fast*-[*poky*] female parent with a wild-type male parent gives a 1:1 segregation of

[*poky*]:*fast*-[*poky*] ascospores in all asci. Interpret these results, and predict the results you would expect from the reciprocal of the stated cross.

A20.1 The [*poky*] mutation shows maternal inheritance, so when it is used as a female parent, all the progeny ascospores will carry [*poky*]. We can designate the cross as [*poky*] × [*N*], where *N* signifies normal cytoplasm. All progeny are [*poky*]. The asci show a 1:1 segregation for the [*poky*] and *fast*-[*poky*] phenotypes. This ratio is characteristic of a nuclear gene segregating in the cross. Therefore, the simplest explanation is that the factor that causes [*poky*] strains to be *fast*-[*poky*] is a nuclear gene mutation; we can call it *F* (its actual designated name). The *F* gene segregates in meiosis, as do all nuclear genes. The cross can be rewritten as [*poky*]*F* ♀ × [*N*]+ ♂, which gives a 1:1 segregation of [*poky*]+ and [*poky*]*F*. The former is [*poky*] and the latter is *fast*-[*poky*].

With these results behind us, the reciprocal cross may be diagrammed as [*N*]+ ♀ × [*poky*]*F* ♂. All the progeny spores from this cross are [*N*], half of them being *F* and half of them being +. If *F* has no effect on normal cytoplasm, then these two classes of spores would be phenotypically indistinguishable, which is the case.

Q20.2 Four slow-growing mutant strains of *Neurospora crassa*, coded *a*, *b*, *c*, and *d*, were isolated. All have an abnormal system of respiratory mitochondrial enzymes. The inheritance patterns of these mutants were tested in controlled crosses with the wild type, with the following results:

FEMALE PARENT		MALE PARENT	PROGENY (ASCOSPORES)	
			WILD TYPE	SLOW GROWING
Wild type	×	*a*	847	0
a	×	Wild type	0	659
Wild type	×	*b*	1,113	0
b	×	Wild type	0	2,071
Wild type	×	*c*	596	590
Wild type	×	*d*	1,050	1,035

Give a genetic interpretation of these results.

A20.2 This question asks us to consider the expected transmission patterns for nuclear and extranuclear genes. The nuclear genes will have a 1:1 segregation in the offspring, since this organism is a haploid organism, and hence should exhibit no differences in the segregation patterns, whichever strain is the maternal parent. On the other hand, a distinguishing characteristic of extranuclear genes is a difference in the results of reciprocal crosses. In *Neurospora*, this characteristic is usually manifested by all progeny having the phenotype of the maternal parent. With these ideas in mind we can analyze each mutant in turn.

Mutant *a* shows a clear difference in its segregation in reciprocal crosses and is, in fact, a classic case of maternal inheritance. The interpretation here is that the gene is extranuclear; hence, the gene must be in the mitochondrion. The [*poky*] mutant described in this chapter shows this type of inheritance pattern.

By the same reasoning, the mutation in strain *b* must also be extranuclear.

Mutants *c* and *d* segregate 1:1, indicating that the mutations involved are in the nuclear genome. In these cases we need not consider the reciprocal cross, since there is no evidence for maternal inheritance. In fact, the actual mutations that are the basis for this question cause sterility, so the reciprocal cross cannot be done. We can confirm that the mutations are in the nuclear genome by doing mapping experiments, using known nuclear markers. Evidence of linkage to such markers would confirm that the mutations are not extranuclear.

QUESTIONS AND PROBLEMS

20.1 Compare and contrast the structure of the nuclear genome, the mitochondrial genome, and the chloroplast genome.

20.2 What genes are present in the human mitochondrial genome?

*20.3 A substantial body of evidence indicates that defects in mitochondrial energy production may contribute to the neuronal cell death seen in a number of late-onset neurodegenerative diseases, including Alzheimer's disease, Parkinson's disease, Huntington disease, and amyotrophic lateral sclerosis (ALS, Lou Gehrig disease). Some, but not all, of these diseases have been associated with mutations in the nuclear genome. One experimental system that has been developed to evaluate the contributions of the mitochondrial genome to these diseases uses a cytoplasmic hybrid known as a "cybid." Cybids are made by repopulating a tissue-culture cell line that has been made mitochondria-deficient with mitochondria from the cytoplasm of a human platelet cell. The cybids thus have nuclear DNA from the tissue-culture cell, and mitochondrial DNA from the human platelet cell.

The mitochondrial protein cytochrome oxidase has subunits encoded by both nuclear and mitochondrial genes. Patients with Alzheimer's disease have been reported to have lower levels of cytochrome oxidase than age-matched controls.

a. What is the evidence that cytochrome oxidase has subunits encoded by both nuclear and mitochondrial genes?

b. Given the means to assay cytochrome oxidase activity, how would you investigate whether the decreased levels of cytochrome oxidase activity in Alzheimer's patients could be ascribed to nuclear or mitochondrial genetic defects? What controls would you create?

c. If you are able to demonstrate that the mitochondrial contribution to cytochrome oxidase is responsible for lowered cytochrome oxidase activity, can you conclude that each mitochondrion of an affected individual has an identical defect?

*20.4 Discuss the differences between the universal genetic code of the nuclear genes of most eukaryotes and the code found in human mitochondria. Is there any advantage to the mitochondrial code?

*20.5 What features of extranuclear inheritance distinguish it from the inheritance of nuclear genes?

20.6 Distinguish between maternal effect and extranuclear inheritance.

20.7 Reciprocal crosses between two types of the evening primrose, *Oenothera hookeri* and *Oenothera muricata*, produce the following effects on the plastids:

O. hookeri female × *O. muricata* male → yellow plastids
O. muricata female × *O. hookeri* male → green plastids

Explain the difference between these results, noting that the chromosome constitution is the same in both types.

*20.8 A series of crosses are performed with a recessive mutation in *Drosophila* called *tudor*. Homozygous *tudor* animals appear normal and can be generated from the cross of two heterozygotes, but a true-breeding *tudor* strain cannot be maintained. When homozygous *tudor* males are crossed to homozygous *tudor* females, both of which appear to be phenotypically normal, a normal-appearing F_1 is produced. However, when F_1 males are crossed to wild-type females, or when F_1 females are crossed to wild-type males, no progeny are ever produced. The same results are seen in the F_1 progeny of homozygous *tudor* females crossed to wild-type males. The F_1 progeny of homozygous *tudor* males crossed to wild-type females appear normal, and they are capable of issuing progeny when mated either with each other or with wild-type animals.
a. How would you classify the *tudor* mutation? Why?
b. What might cause the *tudor* phenotype?

*20.9 A form of male sterility in corn is maternally inherited. Plants of a male-sterile line crossed with normal pollen give male-sterile plants. Some lines of corn carry a dominant, so-called restorer (*Rf*) gene that restores pollen fertility in male-sterile lines.
a. If a male-sterile plant is crossed with pollen from a plant homozygous for gene *Rf*, what will be the genotype and phenotype of the F_1?
b. If the F_1 plants of part a are used as females in a testcross with pollen from a normal plant (*rf/rf*), what would be the result? Give genotypes and phenotypes and designate the type of cytoplasm.

20.10 Distinguish between *nuclear* (segregational), *neutral*, and *suppressive petite* mutants of yeast.

*20.11 In yeast, a haploid *nuclear* (segregational) *petite* is crossed with a *neutral petite*. Assuming that both strains have no other abnormal phenotypes, what proportion of the progeny ascospores are expected to be *petite* in phenotype if the diploid zygote undergoes meiosis?

20.12 When grown on a medium containing acriflavin, a yeast culture produces a large number of very small (*tiny*) cells that grow very slowly. How would you determine whether the slow-growth phenotype was the result of a cytoplasmic factor or a nuclear gene?

20.13 *Drosophila melanogaster* has a sex-linked, recessive, mutant gene called *maroon-like* (*ma-l*). Homozygous *ma-l* females or hemizygous *ma-l* males have light-colored eyes owing to the absence of the active enzyme xanthine dehydrogenase, which is involved in the synthesis of eye pigments. When heterozygous *ma-l⁺/ma-l* females are crossed with *ma-l* males, all the offspring are phenotypically wild type. However, half the female offspring from this cross, when crossed back to *ma-l* males, give all *ma-l* progeny. The other half of the females, when crossed to *ma-l* males, give all phenotypically wild-type progeny. What is the explanation for these results?

*20.14 When females of a particular mutant strain of *Drosophila melanogaster* are crossed to wild-type males, all the viable progeny flies are females. Hypothetically, this result could be the consequence of either a sex-linked, male-specific lethal mutation or a maternally inherited factor that is lethal to males. What crosses would you perform in order to distinguish between these alternatives?

20.15 Reciprocal crosses between two *Drosophila* species, *D. melanogaster* and *D. simulans*, produce the following results:

melanogaster ♀ × *simulans* ♂ → females only
simulans ♀ × *melanogaster* ♂ → males, with few or no females

Propose a possible explanation for these results.

20.16 Some *Drosophila* are very sensitive to carbon dioxide—administering it to them anesthetizes them. The sensitive flies have a cytoplasmic particle called *sigma* that has many properties of a virus. Resistant flies lack *sigma*. The sensitivity to carbon dioxide shows strictly maternal inheritance. What would be the outcome of the following two crosses: (a) sensitive female × resistant male and (b) sensitive male × resistant female?

*20.17 A few years ago the political situation in Chile was such that very many young adults were kidnapped, tortured, and killed by government agents. When abducted young women had young children or were pregnant, those children were often taken and given to government supporters to raise as their own. Now that the political situation has changed, grandparents of stolen children are trying to locate and reclaim their grandchildren. Imagine that you are the judge in a trial centering on the custody of a child. Mr. and Mrs. Escobar believe Carlos Mendoza is the son of their abducted, murdered daughter. If this is true, then Mr. and Mrs. Sanchez are the paternal grandparents of the child, as their son (also abducted and murdered) was the husband of the Escobars' daughter. Mr. and Mrs. Mendoza claim Carlos is their natural child. The attorney for the Escobar and Sanchez couples informs you that scientists have discovered a series of RFLPs in human mitochondrial DNA. He tells you his clients are eager to be tested, and asks that you order that Mr. and Mrs. Mendoza and Carlos be tested also.

a. Can mitochondrial RFLP data be helpful in this case? In what way?

b. Do all seven parties need to be tested? If not, who actually needs to be tested in this case? Explain your choices.

c. Assume the critical people have been tested, and you have received the results. How would the results determine your decision?

20.18 The pedigree in the following figure shows a family in which an inherited disease called Leber's hereditary optic atrophy is segregating. This condition causes blindness in adulthood. Studies have recently shown that the mutant gene causing Leber's hereditary optic atrophy is located in the mitochondrial genome.

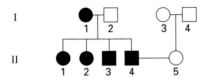

a. Assuming II-4 marries a normal person, what proportion of his offspring should inherit Leber's hereditary optic atrophy?

b. What proportion of the sons of II-2 should be affected?

c. What proportion of the daughters of II-2 should be affected?

*20.19 The inheritance of shell-coiling direction in the snail *Limnaea peregra* has been studied extensively. A snail produced by a cross between two individuals has a shell with a right-hand twist (dextral coiling). This snail produces only left-hand (sinistral) progeny on selfing. What are the genotypes of the F_1 snail and its parents?

*20.20 Beckwith-Wiedemann syndrome (BWS) is characterized by fetal malformation and cancer. BWS has been mapped to chromosome region 11p15.5 by its association with balanced chromosomal rearrangements. This region contains an embryonic tumor suppressor gene and a number of other genes, including *KVLQT1*. Mutations in *KVLQT1* cause a dominantly inherited heart defect that results in cardiac arrhythmia and sudden death. When a wide variety of childhood and adult tumors are examined for gene expression in region 11p15.5, a preferential loss of expression of the paternal alleles is seen. Most normal fetal tissues only express the maternal allele, but heart tissue typically expresses maternal and paternal alleles of *KVLQT1*.

a. What evidence would be necessary to show that genomic imprinting may play a role in Beckwith-Wiedemann syndrome? From the previous description, does such evidence exist?

b. What is the molecular basis for imprinting? How is it detected?

c. What kind of evidence would be necessary to show that imprinting is involved with the dominantly inherited heart defect associated with the *KVLQT1* gene?

d. What similarities, if any, do you see in the imprinting of genes in this region with imprinting for the Prader-Willi and Angelman syndromes discussed in this chapter (pp. 463–464)?

CHAPTER *21*

POPULATION GENETICS

PRINCIPAL POINTS

~ The genetic structure of a population is determined by the total of all alleles (the gene pool). In the case of diploid, sexually interbreeding individuals, the structure is also characterized by the distribution of alleles into genotypes. Thus, the genetic structure is described in terms of both allelic and genotypic frequencies. Except for rare mutations, individuals are born and die with the same set of alleles; what changes genetically over time (evolves) is the genetic structure of a group of individuals, reproductively connected in a Mendelian population.

~ There is usually a great deal of genetic variation among individuals within populations.

~ The genetic structure of a species can vary both geographically and over time.

~ Principles of population genetics are necessary to explain the evolution of species differences and adaptations. A change in the genetic structure of a population over time is equivalent to an evolutionary change.

~ Principles of population genetics can be applied to the management of rare and endangered species.

~ The Hardy-Weinberg law states that in a large, randomly mating population, free from evolutionary processes, the allelic frequencies do not change, and the genotypic frequencies stabilize after one generation in the proportions p^2, $2pq$, and q^2, where p and q equal the allelic frequencies of the population.

~ The classical, balance, and neutral mutation models generate testable hypotheses and are used to explain how much genetic variation should exist within natural populations and what processes are responsible for the observed variation.

~ Mutation, genetic drift, migration, and natural selection are processes that can alter allelic frequencies of a population.

~ Recurrent mutation changes allelic frequencies, and at equilibrium the relative rates of forward and reverse mutations determine allelic frequencies of a population in the absence of other processes.

~ Genetic drift is random change in allelic frequencies due to chance. Genetic drift produces genetic change within populations, genetic differentiation among populations, and loss of genetic variation within populations.

~ Migration, also termed gene flow, involves movement of alleles among populations. Migration can alter the allelic frequencies of a population, and it tends to reduce genetic divergence among populations.

~ Natural selection is differential reproduction of genotypes. It is measured by Darwinian fitness, which is the relative reproductive ability of genotypes. Natural selection can produce a number of different effects on the gene pool of a population.

~ Nonrandom mating affects the genotypic frequencies of a population. Inbreeding leads to an increase in homozygosity, and outbreeding leads to an increase in heterozygosity.

~ Rates of evolution can be measured by comparing DNA or RNA sequences. Different genes and different parts of the same gene tend to evolve at different rates.

~ In eukaryotic organisms, genes frequently occur in multiple copies with identical or similar sequences. A group of such genes is termed a multiple gene family.

~ The mitochondrial DNA of some organisms evolves at a faster rate than the nuclear DNA.

~ Concerted evolution is a process that maintains DNA sequence uniformity among multiple copies of the same sequence within a species.

~ Evolutionary relationships among organisms can be revealed by the study of DNA and RNA sequences.

Population genetics is the field of genetics that studies heredity in groups of individuals for traits that are determined by one or only a few genes. The focus of population genetics is on evolutionary issues. **Quantitative genetics**, the subject of the next chapter, also considers the heredity of traits in groups of individuals, but the traits of concern are determined by many genes. The focus of quantitative genetics, then, is more restricted to present populations. The impetus for the development of these areas came after the rediscovery of Mendel's work and the great implications it had for Darwinian theory. In fact, the fusion of Mendelian theory with Darwinian theory is called the neo-Darwinian synthesis and was championed by Sir Ronald Fisher, Sewall Wright, and J. B. S. Haldane (Figure 21.1).

~ FIGURE 21.1

Sir Ronald Fisher (a), Sewall Wright (b), and J. B. S. Haldane (c) are considered the major architects of neo-Darwinian theory.

a)

b)

c)

Population geneticists investigate the patterns of genetic variation among individuals within groups (the *genetic structure* of populations) and how these patterns vary geographically and evolve over time. In this discipline, our perspective shifts away from the individual and the cell and focuses instead on a **Mendelian population**, a group of *interbreeding* individuals that have a recent common ancestor and, therefore, share a common set of genes. All the genes contained by the individuals of a Mendelian population constitute the **gene pool**. To understand the genetics of the evolutionary process, we study the gene pool of a Mendelian population rather than the genotypes of its individual members.

Questions frequently studied by population geneticists include the following:

1. How much allelic variation is found in natural populations, and what processes control the amount of variation observed?

2. What evolutionary processes shape the genetic structures of populations?

3. What social or biological processes are responsible for producing genetic divergence among populations?

4. How do certain biological characteristics of a population, such as breeding system, fecundity, and age structure, influence the gene pool of the population?

Your goal in this chapter is to learn the basics of population genetics, including some mathematical models that have been developed to describe what happens to the gene pool of a population under various conditions.

GENETIC STRUCTURE OF POPULATIONS

To study the genetic structure of a Mendelian population, we must first describe the gene pool of the group quantitatively. This is done by calculating genotypic frequencies and allelic frequencies of particular alleles within the population. A frequency is a proportion and always ranges between 0 and 1. If 43 percent of the people in a group have red hair, the frequency of red hair in the group is 0.43.

Genotypic Frequencies

To calculate the **genotypic frequencies** at a specific locus, we count the number of individuals with one particular genotype and divide this number by the total number of individuals in the population. We do this for each genotype. The sum of the genotypic frequencies should be 1.

Consider a locus that determines the pattern of spots in the scarlet tiger moth, *Panaxia dominula*

~ FIGURE 21.2

Panaxia dominula, **the scarlet tiger moth.** The top two moths are wild-type homozygotes (*BB*), those in the middle two rows are heterozygotes (*Bb*), and the bottom moth is the rare homozygote (*bb*).

(Figure 21.2). Three genotypes are present in most populations, and each genotype produces a different phenotype. E. B. Ford collected moths at one locality in England and found the following numbers of genotypes: 452 *BB*, 43 *Bb*, and 2 *bb*, out of a total of 497 moths. The genotypic frequencies (where *f* = frequency of) are therefore:

$$f(BB) = 452/497 = 0.909$$
$$f(Bb) = 43/497 = 0.087$$
$$f(bb) = 2/497 = 0.004$$
$$\text{Total} \quad 1.000$$

Allelic Frequencies

Although genotypic frequencies at a single gene locus are useful for examining the effects of certain evolu-

tionary processes on a population, in most cases population geneticists use frequencies of alleles—**allelic frequencies**—to describe the gene pool. Allelic frequencies may be calculated in either of two ways: from the observed numbers of different genotypes at a particular locus, or from the genotypic frequencies. In the first method, we calculate the allelic frequencies directly from the *numbers* of genotypes. That is, we count the number of alleles of one type at a particular locus and divide it by the total number of alleles at that locus in the population:

$$\frac{\text{allelic}}{\text{frequency}} = \frac{\substack{\text{Number of copies of a given} \\ \text{allele in the population}}}{\text{Sum of all alleles in the population}}$$

So, when two alleles are present at a locus (such as the *A* locus), we can use the following formula for calculating the allelic frequency of *A* [*f(A)*], where *p* symbolizes that frequency:

$$p = f(A) = \frac{\substack{(2 \times \text{number of } AA \text{ homozygotes}) \\ + (\text{number of } Aa \text{ heterozygotes})}}{(2 \times \text{total number of individuals})}$$

As an example, imagine a population of 1,000 diploid individuals with 353 *AA*, 494 *Aa*, and 153 *aa* individuals. Each *AA* individual has two *A* alleles, while each *Aa* heterozygote has only a single *A* allele. From the previous equation, the number of *A* alleles in the population is (2 × 353) + 494 = 1,200. Because every diploid individual has two alleles, the total number of alleles in the population will be twice the number of individuals, or 2 × 1,000 = 2,000. Therefore, the allelic frequency of *A* is 1,200/2,000 = 0.60. The allelic frequency of *a* is then 1 − 0.60 = 0.40.

The second method of calculating allelic frequencies goes through the step of first calculating genotypic frequencies as demonstrated previously. In this example, *f(AA)* = 0.353; *f(Aa)* = 0.494; and *f(aa)* = 0.153. From these genotype frequencies we calculate the allele frequencies as follows:

$$p = f(A) = (\text{frequency of the } AA \text{ homozygote})$$
$$+ (\tfrac{1}{2} \times \text{frequency of the } Aa \text{ heterozygote})$$

$$q = f(a) = (\text{frequency of the } aa \text{ homozygote})$$
$$+ (\tfrac{1}{2} \times \text{frequency of the } Aa \text{ heterozygote})$$

(The allelic frequency of *a* is given as *f(a)*, and *q* symbolizes that frequency.) The allelic frequencies for a locus, like the genotypic frequencies, should always add up to 1. Therefore, once p is calculated, q is obtained by subtraction: $1 - p = q$.

Allelic Frequencies with Multiple Alleles.

Suppose we have three alleles—A^1, A^2, and A^3—at a locus, and we want to determine the allelic frequencies. As before, we add up the number of alleles of each type (in homozygotes and in relevant heterozygotes) and divide by the total number of alleles in the population:

$$p = f(A^1) = \frac{(2 \times A^1A^1) + (A^1A^2) + (A^1A^3)}{(2 \times \text{total number of individuals})}$$

$$q = f(A^2) = \frac{(2 \times A^2A^2) + (A^1A^2) + (A^2A^3)}{(2 \times \text{total number of individuals})}$$

$$r = f(A^3) = \frac{(2 \times A^3A^3) + (A^1A^3) + (A^2A^3)}{(2 \times \text{total number of individuals})}$$

As an example of such a calculation, consider data from a study of allelic frequencies at a locus that codes for the enzyme phosphoglucomutase (PGM). There are three alleles at this locus; each allele codes for a different molecular variant of the enzyme. In one population, the following numbers of genotypes were collected:

$$
\begin{array}{rcl}
AA & = & 4 \\
AB & = & 41 \\
BB & = & 84 \\
AC & = & 25 \\
BC & = & 88 \\
CC & = & \underline{32} \\
\text{Total} & = & 274
\end{array}
$$

The frequencies of the alleles are calculated as follows:

$$f(A) = p = \frac{(2 \times 4) + 41 + 25}{(2 \times 274)} = 0.135$$

$$f(B) = q = \frac{(2 \times 84) + 41 + 88}{(2 \times 274)} = 0.542$$

$$f(C) = r = \frac{(2 \times 32) + 88 + 25}{(2 \times 274)} = 0.323$$

The second method for calculating allelic frequencies from *genotypic frequencies* can also be used here. This calculation is quicker if we have already determined the frequencies of the genotypes. The frequency of the homozygote is added to half of the heterozygote frequency because half of the heterozygote's alleles are A and half are a. If three alleles (A^1, A^2, and A^3) are present in the population, the allelic frequencies are

$$p = f(A^1) = f(A^1A^1) + \tfrac{1}{2} f(A^1A^2) + \tfrac{1}{2} f(A^1A^3)$$

$$q = f(A^2) = f(A^2A^2) + \tfrac{1}{2} f(A^1A^2) + \tfrac{1}{2} f(A^2A^3)$$

$$r = f(A^3) = f(A^3A^3) + \tfrac{1}{2} f(A^1A^3) + \tfrac{1}{2} f(A^2A^3)$$

Allelic Frequencies at an X-linked Locus.

Calculating allelic frequencies at an X-linked locus is slightly more complicated, because males have only a single X-linked allele, while females have two. The frequencies of two alleles at an X-linked locus (X^A and X^a) are determined with the following equations:

$$p = f(X^A) = \frac{\begin{array}{c}(2 \times X^AX^A \text{ females}) + (X^AX^a \text{ females}) \\ + (X^AY \text{ males})\end{array}}{\begin{array}{c}(2 \times \text{number of females}) \\ + (\text{number of males})\end{array}}$$

$$q = f(X^a) = \frac{\begin{array}{c}(2 \times X^aX^a \text{ females}) + (X^AX^a \text{ females}) \\ + (X^aY \text{ males})\end{array}}{\begin{array}{c}(2 \times \text{number of females}) \\ + (\text{number of males})\end{array}}$$

That is, the frequency of an X-linked allele is given by twice the number of females homozygous for that allele plus the number of heterozygous females plus the number of males hemizygous for the allele. The total number of alleles in the population is given by the sum of twice the number of females (because each female has two X-linked alleles) and the number of males (who have a single allele at X-linked loci).

Allelic frequencies at an X-linked locus can also be determined from the genotypic frequencies by

$$p = f(X^A) = f(X^AX^A) + \tfrac{1}{2} f(X^AX^a) + f(X^AY)$$

$$q = f(X^a) = f(X^aX^a) + \tfrac{1}{2} f(X^AY) + f(X^aY)$$

Keynote

The genetic structure of a population reflects the total of all genes (the gene pool). In the case of diploid, sexually interbreeding individuals, the structure is also characterized by the distribution of alleles into genotypes. The genetic structure can be described in terms of allelic and genotypic frequencies. Except for rare mutations, individuals are born and die with the same set of alleles; what changes genetically over time (evolves) is the frequency of alternative alleles of a group of individuals, reproductively connected in a Mendelian population.

THE HARDY-WEINBERG LAW

The **Hardy-Weinberg law** is the most important principle in population genetics. In particular it offers a simple explanation for how the Mendelian principles that result from meiosis and sexual reproduction correlate with allelic and genotypic frequencies of a population. The Hardy-Weinberg law is named after the two individuals who independently discovered it in the early 1900s (Box 21.1).

The Hardy-Weinberg law is divided into three parts—a set of assumptions and two major results. A simple statement of the law follows:

Part 1: In an infinitely large, randomly mating population, free from mutation, migration, and natural selection (note that there are five assumptions);

Part 2: the frequencies of the alleles do not change over time; and

Part 3: as long as mating is random, the genotypic frequencies will remain in the proportions p^2 (frequency of *AA*), $2pq$ (frequency of *Aa*), and q^2 (frequency of *aa*), where p is the allelic frequency of *A* and q is the allelic frequency of *a*. The sum of the genotypic frequencies should be equal to 1 (that is, $p^2 + 2pq + q^2 = 1$).

In short, the Hardy-Weinberg law says what happens to the allelic and genotypic frequencies of a population as the alleles are passed from generation to generation in the absence of evolutionarily relevant processes. In other words, if the assumptions listed in part 1 are met, alleles would be expected to combine into genotypes based on simple laws of probability and therefore remain the same. Thus, genotype frequencies can be predicted from allele frequencies.

Assumptions of the Hardy-Weinberg Law

The Hardy-Weinberg law assumes that the population is infinitely large. If a population is limited in size, chance deviations from expected ratios can cause changes in allelic frequency, a phenomenon called *genetic drift* (discussed in more detail on pp. 484–486). The assumption of infinite size is unrealistic. But large populations will look very similar to populations that are infinitely large.

A second assumption of the Hardy-Weinberg law is that mating is random. **Random mating** refers to matings between genotypes occurring in proportion to the frequencies of the genotypes in the population. Humans mate preferentially for height, IQ, skin color, socioeconomic status, and some other traits, but most humans still mate randomly for many traits, blood

BOX 21.1

HARDY, WEINBERG, AND THE HISTORY OF THEIR CONTRIBUTION TO POPULATION GENETICS

Godfrey H. Hardy (1877–1947), a mathematician at Cambridge University, often met R. C. Punnett, the Mendelian geneticist, at the faculty club. One day in 1908 Punnett told Hardy of a problem in genetics that he attributed to a strong critic of Mendelism, G. U. Yule (Yule later denied having raised the problem). Supposedly, Yule said that if the allele for short fingers (brachydactyly) was dominant (which it is) and its allele for normal-length fingers was recessive, then short fingers ought to become more common with each generation. In time virtually everyone should have short fingers. Punnett believed the argument was incorrect, but he could not prove it.

Hardy was able to write a few equations showing that, given any particular frequency of

alleles for short fingers and alleles for normal fingers in a population, the relative number of people with short fingers and normal fingers will stay the same generation after generation, providing no natural selection is involved that favors one phenotype or the other in producing offspring. Hardy published a short paper describing the relationship between genotypes and phenotypes in populations, and within a few weeks a paper was published by Wilhelm Weinberg (1862–1937), a German physician of Stuttgart, that clearly stated the same relationship. The Hardy-Weinberg law signaled the beginning of modern population genetics.

To be complete, we should note that in 1903 the American geneticist W. E. Castle of Harvard University was the first to recognize the relationship between allelic and genotypic frequencies, but it was Hardy and Weinberg who clearly described the relationship in mathematical terms. Thus, the law is sometimes referred to as the Castle-Hardy-Weinberg law.

type being just one example. It is important to understand that the principles of the Hardy-Weinberg law apply to any locus for which random mating occurs, even if mating is nonrandom for other loci.

Finally, for the Hardy-Weinberg law to work, the population must be free from mutation, migration, and natural selection. In other words, the gene pool must be closed to the addition or subtraction of alleles. Later we will discuss these other evolutionary processes and their effect on the gene pool of a population. The assumption that no evolutionary processes act upon the population applies only to the locus in question—a population may be subject to evolutionary processes acting on some genes, while still meeting the Hardy-Weinberg assumptions at other loci.

Predictions of the Hardy-Weinberg Law

If the assumptions of the Hardy-Weinberg law are met, the population will be in genetic equilibrium, and two results are expected. First, the frequencies of alleles will not change from one generation to the next, and therefore the population is not evolving at this locus. Second, the genotypic frequencies will be in the proportions p^2, $2pq$, and q^2 after one generation of random mating and remain constant in these proportions in subsequent generations as long as all the assumptions required by the Hardy-Weinberg law continue to be met. When the genotypes are in these proportions, the population is said to be in Hardy-Weinberg equilibrium.

The Hardy-Weinberg law provides a mechanism for determining the genotypic frequencies from the allelic frequencies when the population is in equilibrium. This in turn provides a way of evaluating which assumptions are being violated when the expected theoretical distribution of genotypes does not match an empirically determined distribution.

Derivation of the Hardy-Weinberg Law

The Hardy-Weinberg law states that when a population is in equilibrium, the genotypic frequencies will be in the proportions p^2, $2pq$, and q^2. To understand the basis of these frequencies at equilibrium, consider a hypothetical population in which the frequency of allele A is p and the frequency of allele a is q. In producing gametes, each genotype passes on both alleles with equal frequency; therefore, the frequencies of A and a in the gametes are also p and q. If one thinks of the gametes as being in a gamete pool, the random formation of zygotes involves reaching into the pool and drawing two gametes at random. The genotypes that form after repeatedly drawing two gametes at a time

~ TABLE 21.1

Possible Combinations of A and a Gametes from Gametic Pools for a Population

	Gametes	♂	
		$p\ A$	$q\ a$
♀	$p\ A$	$p^2\ AA$	$pq\ Aa$
	$q\ a$	$pq\ Aa$	$q^2\ aa$

In sum, $p^2\ AA + 2pq\ Aa + q^2\ aa = 1.00$

will then be in frequencies that are predicted by the probabilities of drawing the particular allele-bearing gametes (the product rule; see Chapter 2). Table 21.1 shows the combinations of gametes when mating is random. This table illustrates the relationship between the allelic frequencies and the genotypic frequencies, which forms the basis of the Hardy-Weinberg law. When gametes pair randomly, the genotypes will occur in the proportions p^2 (AA), $2pq$ (Aa), and q^2 (aa). These genotypic proportions result from the expansion of the square of the allelic frequencies $(p + q)^2 = p^2 + 2pq + q^2$, and the genotypes reach these proportions after one generation of random mating.

The Hardy-Weinberg law also states that allelic and genotypic frequencies remain constant generation after generation if the population remains large, randomly mating, and free from mutation, migration, and natural selection (evolutionary processes). This result can also be understood by considering a hypothetical, randomly mating population, as illustrated in Table 21.2, where all possible matings are given. Random mating means that the frequency of mating between two genotypes will be equal to the product of the genotypic frequencies. For example, the frequency of an $AA \times AA$ mating will be equal to p^2 (the frequency of AA) $\times p^2$ (the frequency of AA) = p^4.

We see that the sum of the probabilities of $AA \times Aa$ ($2p^3q$) and $Aa \times AA$ ($2p^3q$) matings is $4p^3q$, and we know from Mendelian principles that these crosses produce 1/2 AA and 1/2 Aa offspring. Therefore, the probability of obtaining AA offspring from these matings is $4p^3q \times 1/2 = 2p^3q$. The frequencies of offspring produced by each type of mating are presented in the body of the table. At the bottom of the table, the total frequency for each genotype is obtained by addition. As we can see, after random mating the genotypic frequencies are still p^2, $2pq$, and q^2, and the allelic frequencies remain at p and q. The frequencies of the population can thus be represented in the zygotic and gametic stages as follows:

~ TABLE 21.2

Algebraic Proof of Genetic Equilibrium in a Randomly Mating Population for One Gene Locus with Two Alleles

TYPE OF MATING ♀ ♂	MATING FREQUENCY	OFFSPRING FREQUENCIES CONTRIBUTED TO THE NEXT GENERATION BY A PARTICULAR MATING		
		AA	Aa	aa
$p^2\,AA \times p^2\,AA$	p^4	p^4	—	—
$p^2\,AA \times 2\,pq\,Aa$]^a $2\,pq\,Aa \times p^2\,AA$]	$4\,p^3q$	$2\,p^3q$	$2\,p^3q$	—
$p^2\,AA \times q^2\,aa$] $q^2\,aa \times p^2\,AA$]	$2\,p^2q^2$	—	$2\,p^2q^2$	—
$2\,pq\,Aa \times 2\,pq\,Aa$	$4\,p^2q^2$	p^2q^2	$2\,p^2q^2$	p^2q^2
$2\,pq\,Aa \times q^2\,aa$] $q^2\,aa \times 2\,pq\,Aa$]	$4\,pq^3$	—	$2\,pq^3$	$2\,pq^3$
$q^2\,aa \times q^2\,aa$	q^4	—	—	q^4
Totals	$(p^2 + 2\,pq + q^2)^2 = 1$	$p^2(p^2 + 2\,pq + q^2) = p^2$	$2\,pq(p^2 + 2\,pq + q^2) = 2\,pq$	$q^2(p^2 + 2\,pq + q^2) = q^2$

Genotype frequencies = $(p + q)^2 = p^2 + 2\,pq + q^2 = 1$ in each generation afterward.

Gene (allele) frequencies = $p(A) + q(a) = 1$ in each generation afterward.

^{a}For example, matings between AA and Aa will occur at $p^2 \times 2\,pq = 2\,p^3q$ for $AA \times Aa$ and at $p^2 \times 2\,pq = 2\,p^3q$ for $Aa \times AA$ for a total of $4\,p^3q$. Two progeny types, AA and Aa, result in equal proportions from these matings. Therefore, offspring frequencies are $2\,p^3q$ ($1/2 \times 4\,p^3q$) for AA and for Aa.

Zygotes	Gametes
$p^2\,AA + 2pq\,Aa + q^2\,aa$	$p\,A + q\,a$

Each generation of zygotes produces A and a gametes in proportions p and q. The gametes unite to form AA, Aa, and aa zygotes in the proportions p^2, $2pq$, and q^2, and the cycle is repeated indefinitely as long as the assumptions of the Hardy-Weinberg law hold. This short proof gives the theoretical basis for the Hardy-Weinberg law.

The Hardy-Weinberg law indicates that at equilibrium, genotypic frequencies depend on the frequencies of alleles. This relationship between allelic frequencies and genotypic frequencies for a locus with two alleles is represented in Figure 21.3. Note that: (1) The maximum frequency of the heterozygote is 0.5, and this maximum value occurs only when the frequencies of A and a are both 0.5; (2) if allelic frequencies are between 0.33 and 0.66, the heterozygote is the most numerous genotype; and (3) when the frequency of one allele is low, the homozygote for that allele is the rarest of the genotypes.

This point is also illustrated by the distribution of genetic diseases, which are frequently rare and recessive, in humans. For a rare recessive trait, the frequency of the gene causing the trait will be much higher than

~ FIGURE 21.3

Relationship of the frequencies of the genotypes AA, Aa, and aa to the frequencies of alleles A and a (in values of p [top abscissa] and q [bottom abscissa], respectively) in populations that meet the assumptions of the Hardy-Weinberg law. Any single population will be defined by a single vertical line such as $p = 0.3$ and $q = 0.7$.

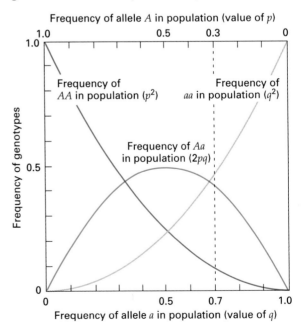

the frequency of the trait itself, because most of the rare alleles are in nonaffected heterozygotes (carriers). Albinism, for example, is a rare recessive condition in humans. One form of albinism is *tyrosinase-negative albinism* (see Chapter 8, p. 186). In this type of albinism, affected individuals have no tyrosinase activity, which is required for normal production of pigment. Among North American whites, the frequency of tyrosinase-negative albinism is roughly 1 in 40,000, or 0.000025. The genotype of affected individuals is *aa*. If the population meets the assumptions of the Hardy-Weinberg law, the frequency of the *aa* genotypes equals q^2. If $q^2 = 0.000025$, then $q = 0.005$ and $p = 1 - q = 0.995$. The heterozygote frequency is therefore $2pq = 2 \times 0.995 \times 0.005 = 0.00995$ (almost 1 percent). Thus, while the frequency of albinos is low (1 in 40,000), individuals heterozygous for albinism are much more common (almost 1 in 100). We have learned, therefore, that heterozygotes for recessive traits can be relatively common, even when the trait is rare.

KEYNOTE

The Hardy-Weinberg law describes what happens to allelic and genotypic frequencies of a large population when gametes fuse randomly and there is no mutation, migration, or natural selection. If these conditions are met, allelic frequencies do not change from generation to generation, and the genotypic frequencies stabilize after one generation in the proportions p^2, $2pq$, and q^2, where p and q equal the frequencies of the alleles in the population.

Extensions of the Hardy-Weinberg Law to Loci with More than Two Alleles

For two alleles at a locus, the Hardy-Weinberg law tells us that at equilibrium the frequencies of the genotypes will be p^2, $2pq$, and q^2, which is the square of the allelic frequencies $(p + q)^2$. If three alleles are present (for example, alleles *A*, *B*, and *C*) with frequencies equal to p, q, and r, the frequencies of the genotypes at equilibrium are also given by the square of the allelic frequencies:

$$(p + q + r)^2 = p^2 (AA) + 2pq (AB) + q^2 (BB) + 2pr (AC) + 2qr (BC) + r^2 (CC)$$

In the blue mussel found along the Atlantic coast of North America, three alleles are common at a locus coding for the enzyme leucine aminopeptidase (LAP). For a population of mussels inhabiting Long

Island Sound, the frequencies of the three alleles were as follows:

ALLELE	FREQUENCY
LAP^{98}	$p = 0.52$
LAP^{96}	$q = 0.31$
LAP^{94}	$r = 0.17$

If the population were in Hardy-Weinberg equilibrium, the expected genotypic frequencies would be

GENOTYPE	EXPECTED FREQUENCY		
LAP^{98}/LAP^{98}	p^2	$= (0.52)^2$	$= 0.27$
LAP^{98}/LAP^{96}	$2pq$	$= 2(0.52)(0.31)$	$= 0.32$
LAP^{96}/LAP^{96}	q^2	$= (0.31)^2$	$= 0.10$
LAP^{96}/LAP^{94}	$2qr$	$= 2(0.31)(0.17)$	$= 0.10$
LAP^{94}/LAP^{98}	$2pr$	$= 2(0.52)(0.17)$	$= 0.18$
LAP^{94}/LAP^{94}	r^2	$= (0.17)^2$	$= 0.03$

Extensions of the Hardy-Weinberg Law to Sex-Linked Alleles

If alleles are X-linked, females may be homozygous or heterozygous, but males carry only a single allele for each X-linked locus. For X-linked alleles in females, the Hardy-Weinberg frequencies are the same as those for autosomal loci: p^2 ($X^A X^A$), $2pq$ ($X^A X^B$), and q^2 ($X^B X^B$). In males, however, the frequencies of the genotypes will be p ($X^A Y$) and q ($X^B Y$), the same as the frequencies of the alleles in the population. For this reason, recessive X-linked traits are more frequent among males than among females. To illustrate this concept, consider the X-linked recessive trait red-green color blindness. The frequency of the color-blind allele varies among human ethnic groups; the frequency among African-Americans is 0.039. At equilibrium, the expected frequency of color-blind males in this group is $q = 0.039$, but the expected frequency of color-blind females is only $q^2 = (0.039)^2 = 0.0015$.

When random mating occurs within a population, the equilibrium genotypic frequencies are reached in one generation. However, if the alleles are X-linked and the sexes differ in allelic frequency, the equilibrium frequencies are approached over several generations. This is because males receive their X chromosome from their mothers only, while females receive an X chromosome from both the mother and the father. Consequently, the frequency of an X-linked allele in males will be the same as the frequency of that allele in their mothers, whereas the frequency in females will be the average of that in mothers and fathers. With random mating, the allelic frequencies in the two sexes oscillate back and forth each generation, and the difference in allelic frequency between the sexes is reduced by half each generation. Once the allelic frequencies of the males and females are equal, the frequencies of the genotypes will be in Hardy-Weinberg proportions after one more generation of random mating.

Testing for Hardy-Weinberg Proportions

To determine whether the genotypes of a population are in Hardy-Weinberg proportions, we first compute p and q from the observed frequencies of the genotypes (as described earlier). Once we have obtained these allelic frequencies, we calculate the expected genotypic frequencies (p^2, $2pq$, and q^2) and compare these frequencies with the actual observed frequencies of the genotypes using a chi-square test (see Chapter 2). The chi-square test gives us the probability that the difference between what we observed and what we expect under the Hardy-Weinberg law is due to chance. Note that while there are three classes, there is only one degree of freedom because the frequencies of alleles in a population have no theoretically expected values. Thus, p must be estimated from the observations themselves. So one degree of freedom is lost for every parameter (p in this case) that must be calculated from the data. Another degree of freedom is lost because, for the fixed number of individuals, once all but one of the classes have been determined, the last class has no degree of freedom and is set automatically. Therefore, with three classes of genotypes, two degrees of freedom are lost, leaving one degree of freedom.

Using the Hardy-Weinberg Law to Estimate Allelic Frequencies

An important application of the Hardy-Weinberg law is the calculation of allelic frequencies when one or more alleles is recessive. For example, in most human populations, the autosomal recessive trait of albinism is rare, but among the Hopi Indians of Arizona albinism is remarkably common (Figure 21.4). A 1969 study, for example, showed a frequency for the trait of 26 out of 6,000 individuals, or 0.0043, which is much higher than the frequency of albinism in most populations. Although we have calculated the frequency of the trait, we cannot directly determine the frequency of the allele for albinism because we cannot distinguish between heterozygous individuals and those individuals homozygous for the normal allele. Nevertheless, we can determine the allelic frequency from the Hardy-Weinberg law if we assume that the population is in equilibrium. At equilibrium, the frequency of the homozygous recessive genotype is q^2. For albinism among the Hopis, $q^2 = 0.0043$, and q can be obtained by taking the square root of the frequency of the trait. Therefore, $q = \sqrt{0.0043} = 0.066$, and $p = 1 - q = 0.934$.

~ FIGURE 21.4

Three Hopi girls, photographed about 1900. The middle child has albinism. Albinism, an autosomal recessive disorder, occurs in high frequency among the Hopi Indians of Arizona.

Following the Hardy-Weinberg law, the frequency of heterozygotes in the population will be $2pq = 2 \times 0.934 \times 0.066 = 0.123$. Thus, one out of eight Hopis, on the average, carries an allele for albinism!

Of course, if the assumptions of the Hardy-Weinberg law do not apply, then our estimate of allelic frequency will be inaccurate. Also, once we calculate allelic frequencies with these assumptions, we cannot then test the population to determine if the genotypic frequencies are in the Hardy-Weinberg expected proportions. To do so would involve circular reasoning, for we assumed Hardy-Weinberg proportions in the first place to calculate the allelic frequencies.

GENETIC VARIATION IN NATURAL POPULATIONS

One of the most significant questions addressed in population genetics is how much genetic variation exists within natural populations. The genetic variation within populations is important for several reasons. First, it determines the potential for evolutionary change and adaptation. The amount of variation also provides us with clues about the relative significance of various evolutionary processes, since some processes increase variation while others decrease it. The manner in which new species arise may depend on the amount of genetic variation harbored within populations. In addition, the ability of a population to persist over time can be influenced by how much genetic variation it has to draw on should environments change.

Models of Genetic Variation

During the 1940s and 1950s, population geneticists developed several opposing hypotheses concerning the amount of genetic variation within natural populations. These hypotheses, or models of genetic variation, have become known as the classical model, the balance model, and the neutral mutation model.

According to the **classical model**, within each population one allele functions best, and this allele is strongly favored by natural selection. As a consequence, almost all individuals in the population are homozygous for this "best" or "wild-type" allele. New alleles arise from time to time through mutation, but almost all are deleterious and are kept in low frequency by selection. Once in a long while, a new mutation arises that is better than the wild-type allele. This new allele increases the survival and reproduction of the individuals that carry it, so the frequency of the allele increases over time because of its selective advantage. Eventually the new allele reaches a high frequency and becomes the new wild-type allele. In this way a population evolves, but little genetic variation is found within the population at any one time. In other words, the classical model uses natural selection as the primary evolutionary force and predicts that little variation is present in any population.

The classical model is no longer tenable because the gene pool of a population has been found to consist of many alleles at each locus; therefore, individuals in populations are heterozygous at numerous loci. To account for the presence of this variation, the balance model was developed. The **balance model** proposes that natural selection actively maintains genetic variation within a population. Balancing selection is natural selection that prevents any single allele from reaching high frequency. One form of balancing selection is heterozygote superiority, in which the heterozygote has higher fitness than either homozygote.

An alternative explanation for the existence of large amounts of genetic variation in populations is provided by the proponents of the **neutral mutation model**. The neutral mutation model states that most molecular mutations do not affect the fitness of the organism and thus are not selected for or against.

Measuring Genetic Variation with Protein Electrophoresis

For many years it was difficult to determine which model for genetic variation was correct because pop-

ulation geneticists required data on genotypes at many loci of many individuals from multiple species, but no general procedure was available for obtaining such data. Then, in 1966, population geneticists began to apply the principle of protein electrophoresis to the study of natural populations. Electrophoresis is a technique that separates proteins with different molecular structures. (See Box 21.2 for a description of protein electrophoresis.) This procedure is now used to examine genetic variation in hundreds of plant and animal species.

Studies of electrophoretic variation in natural populations showed that the classical model was incorrect, because populations possess large amounts of genetic variation. However, this did not prove that the balance model was correct. The balance model predicts that much genetic variation is found within natural populations, and this turns out to be the case. But the balance model also proposes that large amounts of genetic variation are maintained by natural selection, and this fact was not conclusively demonstrated by electrophoretic studies. In other words, electrophoretic studies could not distinguish between the balance model and the neutral mutation model. Population geneticists still argue over the relative merits of the two models, and no clear consensus has emerged as to which model of genetic variation is correct. In reality, both models may be partly correct; at some loci genetic variation may be essentially neutral, while at others it may be acted upon by natural selection.

KEYNOTE

Three models were developed to explain the amount of variation in natural populations. The classical model predicted that alleles would rapidly become fixed or lost, and that variation would arise by new mutation. The balance model proposed that much genetic variation would be maintained in natural populations by natural selection. The neutral mutation model proposed that most molecular mutations do not affect the fitness of the organism and thus are not selected for or against. Through the use of protein electrophoresis, population geneticists eventually demonstrated that natural populations harbor much genetic variation, thus disproving the classical model. However, the results could not distinguish between the balance model and the neutral mutation model, and thus the processes responsible for keeping genetic variation within populations are still controversial.

BOX 21.2

ANALYSIS OF GENETIC VARIATION WITH PROTEIN ELECTROPHORESIS

Gel electrophoresis has been widely used to study genetic variation in natural populations. This process is a biochemical technique that separates large molecules on the basis of size, shape, and charge. To examine genetic variation in proteins with electrophoresis, separate tissue samples, taken from a number of individuals, are ground up, releasing the proteins into an aqueous solution. The individual solutions are then inserted into a gel made of starch, agar, polyacrylamide, or some other porous substance. Electrodes are placed at the ends of the gel, as shown in Box Figure 21.1a, and a direct electrical current is supplied, setting up an electrical field across the gel.

Because proteins are charged molecules, they will migrate within the electrical field. Molecules with different shape, size, and/or charge will migrate at different rates and will separate from one another within the gel over time. If two individuals have different genotypes at a locus coding for a protein, they will produce slightly different molecular forms of the protein, which can sometimes be separated and identified with electrophoresis.

After a current has been supplied to the gel for several hours and the proteins allowed to separate, the current is turned off. The gel now contains a large number of different enzymes and proteins; to identify the product of a single locus, one must stain for a specific enzyme or protein. This is frequently accomplished by having the enzyme in the gel carry out a specific biochemical reaction. The substrate for the reaction is added to the gel, along with a dye that changes color when the reaction takes place. A colored band will appear on the gel wherever the enzyme is located.

If two individuals have different molecular forms of the enzyme, indicating differences in genotypes, bands for those individuals will appear at different locations on the gel. By examining the pattern of bands produced, it is also possible to determine whether an individual is homozygous or heterozygous for a particular allele. A diagram of the banding pattern is shown in Box Figure 21.1b. Histochemical stains are available for dozens of enzymes, so the genotypes of many individuals may be quickly obtained for studying variation at a number of loci.

~ BOX FIGURE 21.1

The technique of protein electrophoresis used to measure genetic variation in natural populations. (a) Solutions of proteins from different individuals are inserted into slots in the gel, and a direct current is supplied to separate the proteins. (b) After electrophoresis, enzyme-specific staining reactions are used to reveal the proteins. The pattern of bands on the gel indicates the genotype of each individual.

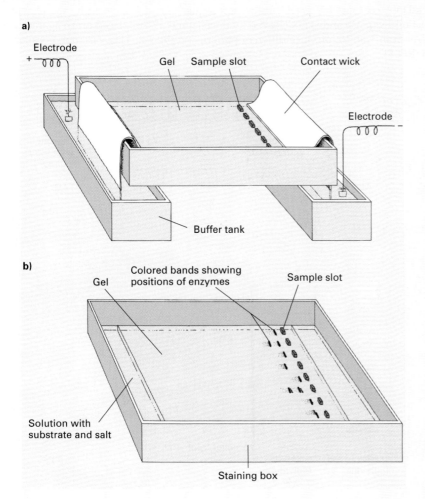

Measuring Genetic Variation with RFLPs and DNA Sequencing

Studies of nucleotide sequence differences in DNA can determine unambiguously the amount of genetic variation present in natural populations. For example, restriction enzymes, which were discussed in Chapter 14, can be useful for detecting genetic variation. Recall that restriction enzymes make double-stranded cuts in DNA at specific base sequences (see Table 14.1). The resulting fragments can be separated by agarose gel electrophoresis and can be observed by staining the DNA or by using probes for specific genes (see Chapter 14).

Suppose that two individuals differ in one or more nucleotides at a particular DNA sequence and that the differences occur at a site recognized by a restriction enzyme (Figure 21.5). One individual has a DNA molecule with the restriction site, but the other individual does not, because the sequences of DNA nucleotides differ. If DNA from these two individuals is cut with the restriction enzyme and the resulting fragments are separated on a gel, the two individuals produce different patterns of fragments, as shown in Figure 21.5. The different patterns are termed restriction fragment length polymorphisms, or RFLPs (see Chapter 14, p. 320). They indicate that the DNA sequences of the two individuals differ. RFLPs are inherited in the same way that alleles coding for other traits are inherited, only the RFLPs do not produce any outward phenotypes; their phenotypes are the fragment patterns produced on a gel when the DNA is cut by the restriction enzyme. Such differences involve only a small part of the DNA, specifically

~ **FIGURE 21.5**

DNA from individual 1 and individual 2 differs in one nucleotide found within the sequence recognized by the restriction enzyme _Bam_HI. Individual 1's DNA contains the _Bam_HI restriction sequence and is cleaved by the enzyme. Individual 2's DNA lacks the _Bam_HI restriction sequence and is not cleaved by the enzyme. When placed on an agarose gel and separated by electrophoresis, the DNAs from 1 and 2 produce different patterns on the gel. This variation is called a restriction fragment length polymorphism (RFLP).

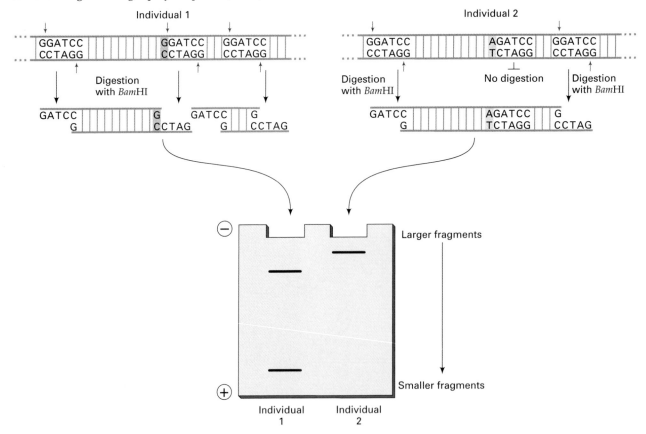

those few nucleotides recognized by the restriction enzyme. However, if we assume that restriction sites occur randomly in the DNA, which is not an unreasonable assumption, the presence or absence of restriction sites can be used to estimate the overall differences in DNA sequence.

To illustrate the use of RFLPs for estimating genetic variation, suppose we isolate DNA from five wild mice, cut the DNA with the restriction enzyme *Bam*HI, and separate the fragments with agarose gel electrophoresis. We then transfer the DNA to a membrane filter, using the Southern blot technique, and add a probe that will detect the gene for β-globin. A

set of restriction patterns that might be obtained is shown in Figure 21.6. A mouse could be $+/+$ (the restriction site is present on both chromosomes), $+/-$ (the restriction site is present on one chromosome and not on the other), or $-/-$ (the restriction site is absent on both chromosomes). For the 10 chromosomes present among these particular five mice, four have the restriction site and six do not.

To calculate the expected heterozygosity in nucleotide sequence, we use the formula:

$$H_{nuc} = \frac{n(\Sigma c_i) - \Sigma c_i^2}{j(\Sigma c_i)(n-1)}$$

~ FIGURE 21.6

Restriction patterns from five mice. The patterns differ in the presence $(+)$ or absence $(-)$ of a particular restriction site (the middle site of the three shown). Each mouse has two homologous chromosomes, each of which potentially carries the restriction site. Thus, a mouse may be $+/+$ (has the restriction site on both chromosomes), $+/-$ (has the restriction site on one chromosome), or $-/-$ (has the restriction site on neither chromosome). When the middle restriction site is present, the DNA is broken into two fragments after digestion with the restriction enzyme and separation with electrophoresis. When the middle restriction site is absent, only one DNA fragment is produced after restriction enzyme digestion.

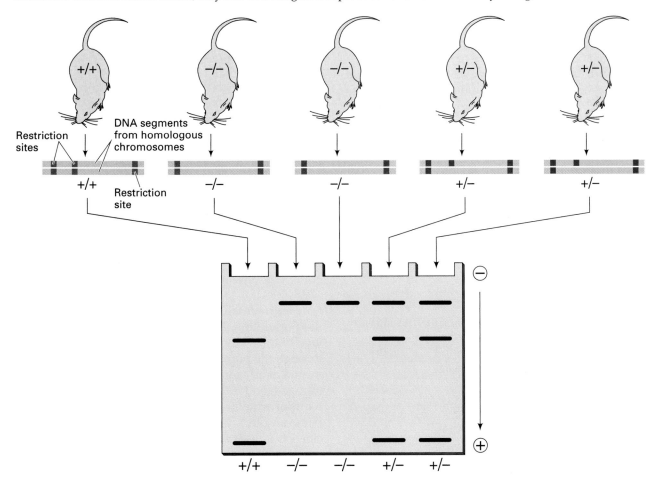

In this equation, *n* equals the number of homologous DNA molecules examined. In our example we looked at five mice, each with two homologous chromosomes, so $n = 10$. The quantity *j* equals the number of nucleotides in the restriction site; in our example $j = 6$, because *Bam*HI recognizes a six-base sequence. For each restriction site i, c_i represents the number of molecules in the sample that were cleaved at that restriction site. In our example, we examined only a single restriction site, so we have a single value of c_i, which equals 4. If additional restriction enzymes were used, we would have a value of c_i for each site recognized by an enzyme. The symbol Σc_i is the total number of cuts at all cleavage sites in all the chromosomes. Since we examined only a single restriction site, and that site was cut in four of the chromosomes, $\Sigma c_i = 4$. Thus, we obtain

$$H_{nuc} = \frac{10(4) - (4)^2}{6(4)(10-1)} = \frac{40-16}{24(9)} = \frac{24}{216} = 0.11$$

Nucleotide heterozygosity has been studied for a number of different organisms through the use of restriction enzymes. A few examples are shown in Table 21.3. Nucleotide heterozygosity typically varies from 0.002 to 0.02 in eukaryotic organisms. This means that an individual is heterozygous (contains different nucleotides on the two homologous chromosomes) at about one in every 50 to 500 nucleotides.

One disadvantage of using RFLP analysis is that we are taking a small sample of the nucleotides (those recognized by the restriction enzyme) and using this sample to estimate the overall level of variation. DNA sequencing (Chapter 14; pp. 315–317), by contrast, provides a method for detecting all nucleotide differences that exist among a set of DNA molecules. For example, Martin Krietman sequenced 11 copies (obtained from different fruit flies) of a 2,659-base-pair segment of the alcohol dehydrogenase gene in *Drosophila melanogaster*. Among the 11 copies, he found different nucleotides at 43 positions within the 2,659-base-pair segment. Only three of the 11 copies were identical at all nucleotides examined—thus, there were eight different alleles (at the nucleotide level) among the 11 copies of this gene! This suggests that populations harbor a tremendous amount of genetic variation in their DNA sequences.

CHANGES IN GENETIC STRUCTURE OF POPULATIONS

Meiosis and gamete fusion do not, by themselves, generate changes in the allelic frequencies of a population. When the population is large, randomly mating, and free from mutation, migration, and natural selection, no evolution occurs. The genetic structure of the population is in equilibrium. For many populations, however, the conditions required by the Hardy-Weinberg law do not hold. In the following sections we discuss the role of four evolutionary processes—mutation, genetic drift, migration, and natural selection—in changing the allele frequencies of a population. We also discuss the effects of nonrandom mating on genotype frequencies.

Mutation

Evolution is a two-step process: First, genetic variation arises; then, different alleles increase or decrease in frequency in response to evolutionary processes. Mutation affects evolution by being entirely responsible for the introduction of new variation and making a small contribution to changes in allelic frequency. In other words, mutation is the source of all new alleles. The rate at which mutations arise is generally low but varies among loci and among species (Table 21.4).

Detrimental mutations will be eliminated from the population. Mutations that confer some advantage to the individuals possessing them will spread through the population. Whether a mutation is detrimental or advantageous depends on the specific environment, and if the environment changes, previously harmful or neutral mutations may become beneficial. For example, after the widespread use of the insecticide DDT, insects with mutations that conferred resistance to DDT were capable of surviving and reproducing; because of this advantage, the mutations

~ TABLE 21.3

Estimates of Nucleotide Heterozygosity for DNA Sequences

DNA SEQUENCES	ORGANISM	H_{nuc}
β-Globin genes	Humans	0.002
Growth hormone gene	Humans	0.002
Alcohol dehydrogenase gene	Fruit fly	0.006
Mitochondrial DNA	Humans	0.004
H4 gene region	Sea urchin	0.019

~ TABLE 21.4

Spontaneous Mutation Frequencies at Specific Loci for Various Organisms[a]

ORGANISM	TRAIT	MUTATION PER 100,000 GAMETES[b]
T2 Bacteriophage (virus)	To rapid lysis ($r^+ \rightarrow r$)	7
	To new host range ($h^+ \rightarrow h$)	0.001
E. coli K12 (bacterium)	To streptomycin resistance	0.00004
	To phage T1 resistance	0.003
	To leucine independence	0.00007
	To arabinose dependence	0.2
Salmonella typhimurium (bacterium)	To threonine resistance	0.41
	To histidine dependence	0.2
Neurospora crassa	To adenine independence	0.0008–0.029
	To inositol independence	0.001–0.010
Drosophila melanogaster males	y^+ to yellow	12
	bw^+ to brown	3
Corn	Wx to waxy	0.00
	Sh to shrunken	0.12
	C to colorless	0.23
	R^r to r^r	49.20
Mouse	a^+ to nonagouti	2.97
	b^+ to brown	0.39
	c^+ to albino	1.02
Humans	Achondroplasia	0.6–1.3
	Huntington disease	0.5
	Neurofibromatosis	5–10
	Retinoblastoma	0.5–1.2

[a]Mutations to independence for nutritional substances are from the auxotrophic condition (e.g., *leu*) to the prototrophic condition (e.g., *leu*[+]).

[b]Mutation frequency estimates of viruses, bacteria, *Neurospora*, and Chinese hamster somatic cells are based on particle or cell counts rather than gametes.

Source: From *Genetics*, 3d ed. by Monroe W. Strickberger. Copyright © 1985. Adapted by permission of Prentice Hall, Inc., Upper Saddle River, NJ.

spread, and many insect populations quickly evolved resistance to DDT.

A mutation from *A* to *a* is referred to as a *forward mutation*. A mutation from *a* to *A* is a *reverse mutation,* and it typically occurs at a lower rate than a forward mutation. The forward mutation rate ($A \rightarrow a$) is symbolized with *u*; and the reverse mutation rate ($a \rightarrow A$) is symbolized with *v*. Consider a large hypothetical population in which the frequency of *A* is *p* and the

frequency of *a* is *q*, and in which no selection occurs. Each generation, a proportion *u* of all *A* alleles mutates to *a*. The actual number mutating depends both on *u* and the frequency of *A* alleles.

For example, suppose the population consists of 100,000 alleles. If *u* equals 10^{-4}, one out of every 10,000 *A* alleles mutates to *a*. When *p* = 1.00, all 100,000 alleles in the population are *A* and free to mutate to *a*, so $10^{-4} \times 100,000 = 10$ *A* alleles should mutate to a. However, if $p = 0.10$, only 10,000 alleles are A and free to mutate to a. Therefore, with a mutation rate of 10^{-4}, only 1 of the A alleles will undergo mutation. The decrease in the frequency of A resulting from mutation of $A \rightarrow a$ is equal to up; the increase in the frequency of A resulting from mutation of $a \rightarrow A$ is equal to vq. As a result of mutation, the amount that A decreases in one generation is equal to the increase in A alleles due to reverse mutations minus the decrease in A alleles due to forward mutations. Eventually the population achieves equilibrium, in which the number of alleles undergoing forward mutation is exactly equal to the number of alleles undergoing reverse mutation. At this point, no further change in allelic frequency occurs, in spite of the fact that forward and reverse mutations continue to take place. Using some simple algebra, population genetics theorists have shown that the equilibrium frequency for a, $\hat{q}$, is

$$\hat{q} = \frac{u}{u + v}$$

and the equilibrium value for A, $\hat{p}$, is

$$\hat{p} = \frac{v}{u + v}$$

Consider a population in which the initial allelic frequencies are $p = 0.9$ and $q = 0.1$ and the forward and reverse mutation rates are $u = 5 \times 10^{-5}$ and $v = 2 \times 10^{-5}$, respectively. (These values are similar to forward and reverse mutation rates observed for many genes.) In the first generation the change (symbolized as Δ) in allelic frequency is

$$\Delta p = vq - up$$
$$= (2 \times 10^{-5} \times 0.1) - (5 \times 10^{-5} \times 0.9)$$
$$\Delta p = -0.000043$$

The frequency of A decreases by only four-thousandths of 1 percent. At equilibrium, the frequency of the a allele, q, equals

$$\hat{q} = \frac{u}{u + v}$$

$$\hat{q} = \frac{5 \times 10^{-5}}{(5 \times 10^{-5}) + (2 \times 10^{-5})} = 0.714$$

If no other processes act on a population, after many generations the alleles will reach equilibrium. Therefore, mutation rates determine the allelic frequencies of the population in the absence of other evolutionary processes. However, because mutation rates are so low, the change in allelic frequency due to mutation pressure is exceedingly slow.

KEYNOTE

When we study what happens when we violate the assumption of the Hardy-Weinberg equilibrium of the absence of mutation, we see that mutation is the only way in which novel genetic material can come to exist within a species. The larger a population, the more potential there is for a novel mutation to arise. In addition, mutations that occur repeatedly can alter the allelic frequencies of a population over time, provided that other evolutionary processes are not active. Eventually, a mutational equilibrium is reached in which the allelic frequencies of the population remain constant in spite of continuing mutation; the allelic frequencies at this equilibrium are a function of the forward and reverse mutation rates.

Genetic Drift

Another major assumption of the Hardy-Weinberg law is that the population is infinitely large. Real populations are not infinite in size, but frequently they are large enough that expected ratios are realized and chance factors have small effects on allelic frequencies. Some populations are small, however, and in these groups chance factors may produce large changes in allelic frequencies. Random change in allelic frequency due to chance is called **genetic drift**, or simply *drift* for short. Sewall Wright (see Figure 21.1b), a brilliant population geneticist, championed the importance of genetic drift in the 1930s, so sometimes genetic drift is called the *Sewall Wright effect* in his honor.

CHANCE CHANGES IN ALLELIC FREQUENCY. Changes in allelic frequency resulting from random events can have important evolutionary implications in small populations. A natural disaster, for example, might kill a disproportionate number of members of a small population who have a particular genotype, thereby causing a significant change in allelic frequencies.

In essence, genetic drift occurs as a result of deviation from expected ratios of gametes and zygotes. Chance deviations from expected proportions arise from a general phenomenon called **sampling error**. Imagine that a population produces an infinitely large pool of gametes, with alleles in the proportions p and q. If random mating occurs and all the gametes unite to form zygotes, the proportions of the genotypes will be equal to p^2, $2pq$, and q^2, and the frequencies of the alleles in these zygotes will remain p and q. If the number of progeny is limited, however, the gametes that unite to form the progeny are a sample from the infinite pool of potential gametes. Just by chance, or by "error," this sample may deviate from the larger pool; the smaller the sample, the larger the potential deviation.

CAUSES OF GENETIC DRIFT.   All genetic drift arises from sampling error, but sampling error occurs in natural populations in several ways. First, genetic drift arises when population size remains continuously small over many generations. Undoubtedly, this situation is frequent, particularly where populations occupy marginal habitats or when competition for resources limits population growth. In such populations, genetic drift plays a major role in the evolution of allelic frequencies.

A second way in which genetic drift arises is through **founder effect**. Founder effect occurs when a population is initially established by a small number of breeding individuals. Although the population may subsequently grow in size and later consist of a large number of individuals, the gene pool of the population is derived from the genes present in the original founders. Chance may play a significant role in determining which genes were present among the founders, and this has a profound effect on the gene pool of subsequent generations.

Many excellent examples of founder effect come from the study of human populations. Consider the inhabitants of Tristan da Cunha, a small, isolated island in the South Atlantic. Three forms of genetic drift occurred in the evolution of the island's population. First, founder effect took place at the initial settlement. By 1855, the population of Tristan da Cunha consisted of about 100 individuals, but 26 percent of the genes of the population in 1855 were contributed by William Glass and his wife. Even in 1961, these original two settlers contributed 14 percent of all the genes in the 300 individuals of the population. The particular genes that Glass and other original founders carried heavily influenced the subsequent gene pool of the population. Second, population size remained small throughout the history of the settlement, and sampling error continually occurred.

A third form of sampling error, called **bottleneck effect**, also played a role in the population of Tristan da Cunha. Bottleneck effect is a form of genetic drift that occurs when a population is drastically reduced in size. Bottleneck effect is a type of founder effect, since the population is refounded by those few individuals that survive the reduction. Two bottlenecks occurred in the history of Tristan da Cunha. The first took place around 1856 and was precipitated by the death of William Glass and the arrival of a missionary who encouraged the inhabitants to leave the island. At this time the population dropped from 103 individuals to 33. A second bottleneck occurred in 1885, when 15 adult males drowned after their boat capsized, leaving only four adult males on the island, one of whom was insane and two of whom were old. Many of the widows and their families left the island during the next few years, and the population size then dropped from 106 to 59. Both bottlenecks had a major effect on the gene pool of the population. All the genes contributed by several settlers were lost, and the relative contributions of others were altered by these events.

EFFECTS OF GENETIC DRIFT.   Genetic drift produces changes in allelic frequencies, and these changes have several effects on the genetic structure of populations. First, genetic drift causes the allelic frequencies of a population to change over time (Figure 21.7). The different lines in the figure represent allelic frequencies in four populations over a number of generations. Although all populations

~ **FIGURE 21.7**

The effect of genetic drift on the frequency (q) of allele A^2 in four populations. Each population begins with q equal to 0.5, and the effective population size for each is 20. The mean frequency of allele A^2 for the four populations is indicated by the red line. These results were obtained by a computer simulation.

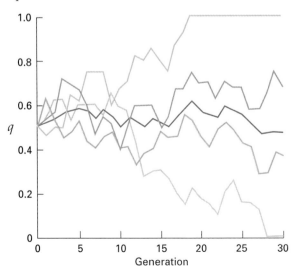

begin with an allelic frequency equal to 0.5, the frequencies in each population change over time as a result of sampling error. In each generation, the allelic frequency may increase or decrease, and over time the frequencies wander randomly or drift (hence the name "genetic drift"). As a result of genetic drift, then, different populations show genetic divergence. Sometimes, within 30 generations, just by chance, the allelic frequency reaches a value of 0.0 or 1.0. At this point, one allele is lost from the population, and the population is said to be *fixed* for the remaining allele in a one-locus, two-allele example. In fact, in the absence of other factors, genetic drift will always lead to the loss or fixation of an allele. The time it takes for fixation to occur depends on factors such as population size and the initial allele frequency.

Once an allele is fixed, no further change in allelic frequency occurs, unless the other allele is reintroduced through mutation or migration. If the initial allelic frequencies are equal, which allele becomes fixed is strictly random. On the other hand, if the initial allelic frequencies are not equal, the rare allele is more likely to be lost. During this process of genetic drift and fixation, the number of heterozygotes in the population also decreases, and after fixation, the population heterozygosity is zero. As heterozygosity decreases and alleles become fixed, populations lose genetic variation; thus, the second effect of genetic drift is a reduction in genetic variation within populations.

KEYNOTE

Genetic drift, or chance changes in allelic frequency due to sampling error, can have important evolutionary and survival implications for small populations. Genetic drift leads to loss of genetic variation within populations, genetic divergence among populations, and random fluctuation in the allelic frequencies of a population over time.

Migration

One of the assumptions of the Hardy-Weinberg law is that the population is closed and not influenced by other populations. Many populations are not completely isolated, however, and individuals migrating into a population may introduce new alleles to the gene pool and alter the frequencies of existing alleles. Thus, **migration** has the potential to disrupt Hardy-Weinberg equilibrium and may influence the evolution of allelic frequencies within populations. **Gene flow**

refers to the movement of alleles that takes place when organisms or gametes migrate, thereby contributing alleles to the gene pool of the recipient population.

Gene flow has two major effects on a population. First, it may introduce new alleles to the population if there are differences among populations. Such gene flow is a source of genetic variation for the population. Second, when the allelic frequencies of migrants and the recipient population differ, gene flow changes the allelic frequencies within the recipient population. Through exchange of genes, different populations remain similar, and thus, migration is a homogenizing force.

To illustrate the effect of migration on allelic frequencies, consider a simple model in which gene flow occurs in one direction, from population I to population II. Suppose that the frequency of allele A in population I (p_I) is 0.8 and the frequency of A in population II (p_{II}) is 0.5. Each generation, some individuals migrate from population I to population II, and these migrants are a random sample of the genotypes in population I. After migration, population II actually consists of two groups of individuals: the migrants with $p_I = 0.8$, and the residents with $p_{II} = 0.5$. The migrants now make up a proportion of population II, which we designate m. The frequency of A in population II after migration (p'_{II}) is

$$p'_{II} = mp_I + (1 - m)p_{II}$$

The frequency of A after migration is determined by the proportion of A alleles in the two groups that now make up population II. The first component, mp_I, represents the A alleles in the migrants—we multiply the proportion of the population that consists of migrants (m) by the allelic frequency of the migrants (p_I). The second component represents the A alleles in the residents, and equals the proportion of the population consisting of residents ($1 - m$) multiplied by the allelic frequency in the residents (p_{II}). Adding these two components gives us the allelic frequency of A in population II after migration.

The change in allelic frequency in population II as a result of migration (Δp) equals the original frequency of A subtracted from the frequency of A after migration:

$$\Delta p = p'_{II} - p_{II}$$

In the previous equation we found that p'_{II} equaled $mp_I + (1 - m)p_{II}$, so the change in allelic frequency can be written as

$$\Delta p = mp_I + (1 - m)p_{II} - p_{II}$$

Multiplying $(1 - m)$ by p_{II} in the above equation, we obtain

$$\Delta p = mp_{\mathrm{I}} + p_{\mathrm{II}} - mp_{\mathrm{II}} - p_{\mathrm{II}}$$

$$\Delta p = mp_{\mathrm{I}} - mp_{\mathrm{II}}$$

$$\Delta p = m(p_{\mathrm{I}} - p_{\mathrm{II}})$$

This final equation indicates that the change in allelic frequency from migration depends on two factors: the proportion of the migrants in the final population and the difference in allelic frequency between the migrants and the residents. If no differences exist in the allelic frequency of migrants and residents (p_{I} − p_{II} = 0), then we can see that the change in allelic frequency is zero. Populations must differ in their allelic frequencies in order for migration to affect the makeup of the gene pool. With continued migration, p_{I} and p_{II} become increasingly similar, and, as a result, the change in allelic frequency due to migration decreases. Eventually, allelic frequencies in the two populations will be equal, and no further change will occur. This is only true, however, when factors besides migration do not influence allelic frequencies.

Migration reduces divergence among populations, effectively increasing the size of the individual populations. Extensive gene flow has been shown to occur, for example, among populations of the monarch butterfly (Figure 21.8). Even a small amount of gene flow can reduce the effect of genetic drift. Calculations have shown that a single migrant moving between two populations every other generation will prevent the two populations from becoming fixed for different alleles.

KEYNOTE

Migration of individuals into a population may alter the makeup of the population gene pool, if the genes carried by the migrants differ from those of the resident population. Migration, also termed gene flow, tends to reduce genetic divergence among populations and increases the effective size of the population.

Natural Selection

Mutation, genetic drift, and migration alter the gene pool of a population, and they certainly influence the evolution of a species, but they do not result in adaptation. *Adaptation* is the process by which traits evolve that make organisms more suited to the environment. Adaptation is responsible for the many extraordinary traits seen in nature—wings that enable a hummingbird to fly backward, brains that allow humans to

~ **FIGURE 21.8**

Extensive gene flow occurs among populations of the monarch butterfly. The butterflies (a) overwinter in Mexico and then migrate (b) north during spring and summer to breeding grounds as far away as northern Canada. Extensive gene flow occurs during the migration period with the result that monarch populations display relatively little genetic divergence.

a)

b)

speak, read, and love. These biological features and countless other exquisite traits are the product of adaptation (Figure 21.9). Genetic drift, mutation, and migration all influence the pattern and process of adaptation, but adaptation arises chiefly from natural selection. **Natural selection** is the dominant force in the evolution of many traits and has shaped much of the phenotypic variation observed in nature. Charles Darwin and Alfred Russel Wallace (Figure 21.10) independently developed the concept of natural selection in the middle of the nineteenth century, with Darwin developing the concept much more than Wallace. For his innumerable contributions to our understanding of natural selection, Darwin is frequently regarded as the "father" of evolutionary theory.

~ FIGURE 21.9

Natural selection produces organisms that are finely adapted to their environment, such as this lizard with cryptic coloration, which allows it to blend in with its natural surroundings.

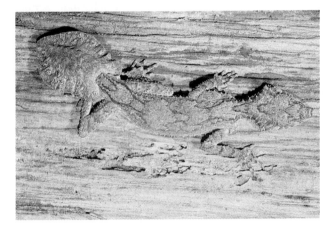

Natural selection can be defined as differential reproduction of genotypes. It simply means that individuals with certain genes produce more offspring than others; therefore, those genes increase in frequency in the next generation, as further discussed in Chapter 22. Through natural selection, traits that contribute to survival and reproduction increase over time. In this way, organisms adapt to their environments.

SELECTION IN NATURAL POPULATIONS. A classic example of selection in natural populations is the evolution of melanic (dark) forms of moths in association with industrial pollution, a phenomenon known as "industrial melanism." One of the best studied cases involves the peppered moth, *Biston betularia*. The common phenotype of this species, termed the *typical* form, is a greyish white color with black mottling over the body and wings (Figure 21.11).

Prior to 1848, all peppered moths collected in England possessed this *typical* phenotype, but in 1848 a single black moth was collected near Manchester, England. This new phenotype, called *carbonaria,* presumably arose by mutation and rapidly increased in frequency around Manchester and in other industrial regions. The *carbonaria* phenotype is dominant to the *typical* phenotype. By 1900, the *carbonaria* phenotype had reached a frequency of more than 90 percent in several populations. High frequencies of *carbonaria* appeared to be associated with industrial regions, whereas the *typical* phenotype remained common in more rural districts.

H. B. D. Kettlewell demonstrated that the increase in the *carbonaria* phenotype occurred as a result of strong selection against the *typical* form in polluted woods. Peppered moths are nocturnal; during the day they rest on the trunks of lichen-covered trees. Birds frequently prey on the moths during the day, but because the lichens that cover the trees are naturally grey, the *typical* form of the peppered moth is well camouflaged (see Figure 21.11a). In industrial areas,

~ FIGURE 21.10

Charles Darwin (a) and Alfred Russel Wallace (b), independent contributors to the development of the theory of evolution through natural selection.

a)   b)

~ FIGURE 21.11

Biston betularia, **the peppered moth, and its dark form** *carbonaria* **(a) on the trunk of a lichened tree in the unpolluted countryside, and (b) on the trunk of a tree with dark bark.** On the lichened tree, the dark form of the moth is readily seen, whereas the light form is well camouflaged. On the dark tree, the dark form of the moth is well camouflaged.

a)   b)

however, extensive pollution beginning with the industrial revolution in the mid-nineteenth century had killed most of the lichens and covered the tree trunks with black soot. Against this black background, the *typical* phenotype was quite conspicuous and was readily consumed by birds. In contrast, the *carbonaria* form was well camouflaged against the blackened trees and had a higher rate of survival in polluted areas (see Figure 21.11b). Because *carbonaria* survived better in polluted woods, more *carbonaria* alleles were transmitted to the next generation; thus, the *carbonaria* phenotype increased in frequency in industrial areas. In rural areas, where pollution was absent, the *carbonaria* phenotype was conspicuous and the *typical* form was camouflaged; in these regions the frequency of the *typical* form remained high.

FITNESS AND COEFFICIENT OF SELECTION.

Darwin described natural selection primarily in terms of survival. However, what is most important in the process of natural selection is the relative number of genes contributed to future generations. Therefore, we measure natural selection by assessing reproduction. Natural selection is measured in terms of **Darwinian fitness**, which is defined as the relative reproductive ability of a genotype.

Darwinian fitness is often symbolized as W and also called the adaptive value of a genotype. Since it is a measure of relative reproductive ability, population geneticists usually assign an adaptive value of 1 to a genotype that produces the most offspring. The fitnesses of the other genotypes are assigned relative

to this. For example, suppose that the genotype G^1G^1 on the average produces eight offspring, G^1G^2 produces an average of four offspring, and G^2G^2 produces an average of two offspring. The G^1G^1 genotype has the highest reproductive output, so its fitness is 1 ($W_{11} = 1.0$). Genotype G^1G^2 produces on the average four offspring for the eight produced by the most fit genotype, so the fitness of G^1G^2 (W_{12}) is $4/8 = 0.5$. Similarly, G^2G^2 produces two offspring for the eight produced by G^1G^1, so the fitness of G^2G^2 (W_{22}) is $2/8 = 0.25$. Table 21.5 illustrates the calculation of relative fitness values.

Related to Darwinian fitness is the **selection coefficient**, which is a measure of the relative intensity of selection against a genotype. The selection coefficient is symbolized by s and equals $1 - W$. In our example, the selection coefficients for G^1G^1 are $s = 0$; for G^1G^2, $s = 0.5$; and for G^2G^2, $s = 0.75$.

EFFECT OF SELECTION ON ALLELIC FREQUENCIES.

At times, natural selection eliminates genetic variation, and at other times it maintains variation; it can change allelic frequencies or prevent allelic frequencies from changing; it can produce genetic divergence among populations or maintain genetic uniformity. Which of these effects occurs depends primarily on the relative fitness of the genotypes and on the frequencies of the alleles in the population.

The change in allelic frequency that results from natural selection can be calculated by constructing a table such as Table 21.6. This "table method" can be used for any type of single-locus trait, whether

~ TABLE 21.5

Computation of Fitness Values and Selection Coefficients of Three Genotypes

	GENOTYPES		
	G^1G^1	G^1G^2	G^2G^2
Number of breeding adults in one generation	16	10	20
Number of offspring produced by all adults of the genotype in the next generation	128	40	40
Average number of offspring produced per breeding adult	128/16 = 8	40/10 = 4	40/20 = 2
Fitness W (relative number of offspring produced)	8/8 = 1	4/8 = 0.5	2/8 = 0.25
Selection coefficient ($s = 1 - W$)	1 − 1 = 0	1 − 0.5 = 0.5	1 − 0.25 = 0.75

~ TABLE 21.6

General Method of Determining Change in Allelic Frequency Due to Natural Selection When Initial Allelic Frequencies Are $p = 0.6$ and $q = 0.4$

	GENOTYPES		
	A^1A^1	A^1A^2	A^2A^2
Initial genotypic frequencies	p^2 $(0.6)^2= 0.36$	$2pq$ $2(0.6)(0.4) =$ 0.48	q^2 $(0.4)^2 = 0.16$
Fitness	$W_{11} = 0$	$W_{12} = 0.4$	$W_{22} = 1$
Frequency after selection	$p^2W_{11} =$ $(0.36)(0) = 0$	$2pqW_{12} =$ $(0.48)(0.4) =$ 0.19	$q^2W_{22} =$ $(0.16)(1) =$ 0.16
Relative genotypic frequency after selection	$P' = \dfrac{p^2W_{11}}{\overline{W}^a}$ $P' = 0/0.35 = 0$	$H' = \dfrac{2pqW_{12}}{\overline{W}}$ $H' = 0.19/0.35$ $= 0.54$	$Q' = \dfrac{q^2W_{22}}{\overline{W}}$ $Q' = 0.16/0.35$ $= 0.46$

Allelic frequency after selection $p' = P' + 1/2(H')$

$p' = 0 + 1/2(0.54) = 0.27$

$q' = 1 - p' = 1 - 0.27 = 0.73$

Change in allelic frequency due to selection $= \Delta p = p' - p$

$\Delta p = 0.27 - 0.6 = -0.33$

$^a\overline{W} = p^2W_{11} + 2pqW_{12} + q^2W_{22}$
$\overline{W} = 0 + 0.19 + 0.16$
$\overline{W} = 0.35$

dominant, codominant, recessive, or overdominant. To use the table method we begin by listing the genotypes (A^1A^1, A^1A^2, and A^2A^2) and their initial frequencies. If random mating has just taken place, the genotypes are in Hardy-Weinberg proportions and the initial frequencies are p^2, $2pq$, and q^2. We then list the fitnesses for each genotype, W_{11}, W_{12}, and W_{22}. Now, suppose that selection occurs and only some of the genotypes survive. The contribution of each genotype to the next generation will be equal to the initial frequency of the genotype multiplied by its fitness. For A^1A^1 this will be $p^2 \times W_{11}$. Notice that the contributions of the three genotypes do not add up to 1. We calculate the relative contributions of each genotype by dividing each by the mean fitness of the population. The *mean fitness of the population* equals $p^2W_{11} + 2pqW_{12} + q^2W_{22} = \overline{W}$. This gives us the relative frequencies of the genotypes after selection. We then calculate the new allelic frequency (p') from the genotypes after selection, using our familiar formula, $p' = $ (frequency of A^1A^1) + ($1/2 \times$ frequency of A^1A^2). Finally, the change in allelic frequency resulting from selection equals $p' - p$.

Natural selection results in a characteristic pattern of change in the genetic structure of a population depending on the starting genotypic frequencies. In many cases natural selection results in the elimination or at least great reduction in one of the alleles, a phenomenon called *directional selection*. The case of the peppered moth discussed earlier is an example of directional selection.

SELECTION AGAINST A RECESSIVE TRAIT.

When a trait is completely recessive, both the heterozygote and the dominant homozygote have a fitness of 1, while the recessive homozygote has reduced fitness, as shown below.

Genotype	Fitness (W)
AA	1
Aa	1
aa	$1 - s$

If the genotypes are initially in Hardy-Weinberg proportions, the contribution of each genotype to the next generation will be the frequency times the fitness.

AA	$p^2 \times 1 = p^2$
Aa	$2pq \times 1 = 2pq$
aa	$q^2 \times (1 - s) = q^2 - sq^2$

The mean fitness of the population is $p^2 + 2pq + q^2 - sq^2$. Since $p^2 + 2pq + q^2 = 1$, the mean fitness becomes

$1 - sq^2$, and the normalized genotypic frequencies after selection are

$$AA \qquad \frac{p^2}{1 - sq^2}$$

$$Aa \qquad \frac{2pq}{1 - sq^2}$$

$$aa \qquad \frac{q^2 - sq^2}{1 - sq^2}$$

To obtain q', the frequency after selection, we add the frequency of the aa homozygote and half the frequency of the heterozygote.

$$q' = \frac{q^2 - sq^2}{1 - sq^2} + \frac{1}{2} \times \frac{2pq}{1 - sq^2}$$

$$= \frac{q^2 - sq^2}{1 - sq^2} + \frac{pq}{1 - sq^2}$$

$$= \frac{q^2 - sq^2 + pq}{1 - sq^2}$$

$$= \frac{q^2 + pq - sq^2}{1 - sq^2}$$

$$= \frac{q(q + p) - sq^2}{1 - sq^2}$$

Since $(q + p) = 1$,

$$q' = \frac{q - sq^2}{1 - sq^2}$$

Therefore, the change in the frequency of a after one generation of selection is

$$\Delta q = q' - q$$

$$= \frac{q - sq^2}{1 - sq^2} - q$$

$$= \frac{q - sq^2}{1 - sq^2} - \frac{q(1 - sq^2)}{(1 - sq^2)}$$

$$= \frac{q - sq^2 - q(1 - sq^2)}{1 - sq^2}$$

$$= \frac{q - sq^2 - q + sq^3}{1 - sq^2}$$

$$= \frac{-sq^2 + sq^3}{1 - sq^2}$$

$$= \frac{-sq^2(1 - q)}{1 - sq^2}$$

So $\Delta q = -spq^2/(1 - sq^2)$ because $1 - q = p$. When $\Delta q = 0$, no further change occurs in allelic frequencies. Notice that there is a negative sign in the equation to the left of spq^2; because the values of s, p, and q are always positive or zero, Δq is negative or zero. Thus, the value of q will decrease with selection, exemplifying directional selection.

Selection against a recessive trait also depends on the actual frequencies of the allele in the population. Table 21.7 shows the magnitude of change in allelic frequency for each generation in three populations with different initial allelic frequencies. Population 1 begins with allelic frequency $q = 0.9$, population 2 begins with $q = 0.5$, and population 3 begins with $q = 0.1$. In this example (recessive lethal condition), the homozygous recessive genotype (aa) has a fitness of 0 and the other two genotypes (AA and Aa) have a fitness of 1. When the frequency of q is high, as in population 1, the change in allelic frequency is large; in the first generation, q drops from 0.9 to 0.474. However, when q is small, as in population 3, the change in q is much less; here q drops from 0.1 to 0.091 in the first generation. So, as q becomes smaller, the change in q becomes less. Because of this diminishing change in frequency, it is virtually impossible to eliminate a recessive trait from the population entirely. That is, the final recessive alleles in a population will almost always find themselves in the heterozygote condition.

HETEROZYGOTE SUPERIORITY. Natural selection does not always result in a directional change in allele frequency and a decrease in genetic variation. A type of natural selection called *balancing selection* results in the maintenance of genetic variation and forms the backbone of the balance model of genetic variation discussed earlier. The simplest type of balancing selection is called **heterosis, overdominance,** or **heterozygote superiority.** An equilibrium of allelic frequencies arises when the heterozygote has higher fitness than either of the homozygotes. In this case, both alleles are maintained in the population, because both are favored in the heterozygote genotype. Allelic frequencies will change as a result of selection until the equilibrium point is reached and then will remain stable. The allelic frequencies at which the population reaches equilibrium depend on the relative fitnesses of the two homozygotes. If the selection coefficient of AA is s and the selection coefficient of aa is t, it can be shown algebraically that at equilibrium

$$\hat{p} = f(A) = t/(s + t)$$

and

$$\hat{q} = f(a) = s/(s + t)$$

Notice that if selection against both homozygotes is the same ($s = t$), the equilibrium allele frequency is 0.5. As selection against the homozygotes becomes less symmetrical, the equilibrium allele

---

~ **TABLE 21.7**

Effectiveness of Selection Against a Recessive Lethal Genotype at Different Initial Frequencies

| | GENOTYPE (q^2 aa) AND ALLELE (q a) FREQUENCIES | | | | | |
| | POPULATION 1 | | POPULATION 2 | | POPULATION 3 | |
GENERATION	q^2	q	q^2	q	q^2	q
0	0.810	0.900	0.250	0.500	0.010	0.100
1	0.224	0.474	0.112	0.333	0.008	0.091
2	0.103	0.321	0.063	0.250	0.007	0.083
3	0.059	0.243	0.040	0.200	0.006	0.077
4	0.038	0.195	0.028	0.167	0.005	0.071
5	0.027	0.163	0.020	0.143	0.004	0.066
.						
.						
.						
10			0.007	0.08		

~ **FIGURE 21.12**

The distribution of malaria caused by the parasite *Plasmodium falciparum* **coincides with distribution of the** *Hb-S* **allele for sickle-cell anemia.** The frequency of *Hb-S* is high in areas where malaria is common, because *Hb-A/Hb-S* heterozygotes are resistant to malarial infection.

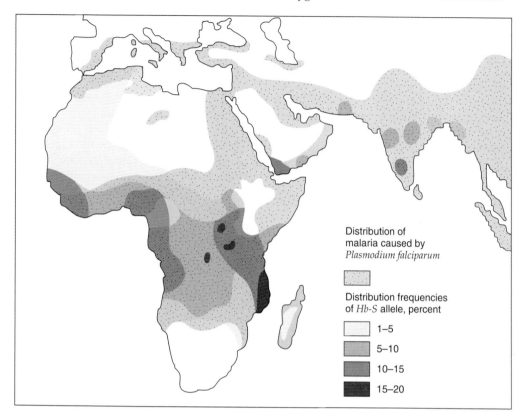

Distribution of
malaria caused by
Plasmodium falciparum

Distribution frequencies
of *Hb-S* allele, percent

	1–5
	5–10
	10–15
	15–20

frequency shifts in the direction of the most fit homozygote.

The most famous example of heterozygote superiority operating in nature is provided by human sickle-cell anemia. Sickle-cell anemia results from a mutation in the gene coding for beta hemoglobin. In some populations, there are three hemoglobin genotypes: *Hb-A/Hb-A*, *Hb-A/Hb-S*, and *Hb-S/Hb-S*. Individuals with the *Hb-A/Hb-A* genotype have completely normal red blood cells; *Hb-S/Hb-S* individuals have sickle-cell anemia; and *Hb-A/Hb-S* individuals have sickle-cell trait, a mild form of sickle-cell anemia. *Hb-A/Hb-S* heterozygotes have about a 10 to 15 percent better chance of survival—a higher fitness—than *Hb-A/Hb-A* individuals if infected with the malarial parasite. The *Hb-S/Hb-S* individuals are at a serious selective disadvantage because they have sickle-cell anemia. So, in malaria-infested areas in which the *Hb-S* gene is also found, an equilibrium state is established in which a significant number of *Hb-S* alleles are found in the heterozygotes because of the selective advantage of this genotype. The distributions of malaria and the *Hb-S* allele are illustrated in Figure 21.12. In sum, despite the seriousness of this genetic disease, natural selection cannot eliminate this detrimental allele from the population because the allele has beneficial effects in the heterozygote state.

KEYNOTE

Natural selection involves differential reproduction of genotypes and is measured in terms of Darwinian fitness, the relative reproductive ability of a genotype. The effects of selection depend on the relative fitnesses of the different genotypes. Directional selection results in the directional change in allele frequency, with the disfavored allele being eliminated from or significantly reduced in the population. Directional selection decreases the amount of genetic variation in a population. Balancing selection, exemplified here as heterozygote superiority, results in the maintenance of genetic variation in the population.

Simultaneous Effects of Mutation and Selection

Natural selection can do many things, including reducing the frequency of a deleterious recessive allele. As the frequency of the allele becomes low, the change in frequency diminishes with each generation. Opposing this reduction in the allele's frequency is mutation pressure, which will continually produce new alleles and tend to increase the frequency. Eventually an equilibrium is reached in which the input of new alleles by recurrent mutation is exactly counterbalanced by the loss of alleles through natural selection. When equilibrium is obtained, the allele's frequency remains stable, in spite of the fact that selection and mutation continue, unless the equilibrium is perturbed by some other process.

Consider a population in which selection occurs against a deleterious recessive allele, a. As we saw on pages 491–492, the amount a will change in one generation (Δq) as a result of selection is

$$\Delta q = -spq^2/(1 - sq^2)$$

For a rare recessive allele, q^2 will be near 0, and the denominator in this equation, $1 - sq^2$, will be approximately 1, so that the decrease in frequency caused by selection is given by

$$\Delta q = -spq^2$$

At the same time, the frequency of the a allele increases as a result of mutation from A to a. Provided the frequency of a is low, the reverse mutation of a to A essentially can be ignored. Equilibrium between selection and mutation occurs when the decrease in allelic frequency produced by selection is the same as the increase produced by mutation:

$$spq^2 = up$$

We can predict the frequency of a at equilibrium ($\hat{q}$) by rearranging this equation:

$$sq^2 = u$$
$$q^2 = u/s$$

and

$$\hat{q} = \sqrt{u/s}$$

If the recessive homozygote is lethal ($s = 1$), the equation becomes

$$\hat{q} = \sqrt{u}$$

As an example of the balance between mutation and selection, consider a recessive gene for which the mutation rate is 10^{-6} and s is 0.1. At equilibrium, the frequency of the gene will be $\hat{q} = \sqrt{10^{-6}/0.1}$. Most recessive deleterious traits remain within a population at low frequency because of equilibrium between mutation and selection.

For a deleterious dominant allele A, the frequency at equilibrium ($\hat{p}$) is

$$\hat{p} = u/s$$

If the mutation rate is 10^{-6} and s is 0.1, as in the above example, the frequency of the dominant gene at equilibrium will be $10^{-6}/0.1 = 0.00001$, which is considerably less than the equilibrium frequency for a recessive allele with the same fitness and mutation rate. This is because selection cannot act on a recessive allele in the heterozygote state, whereas both the homozygote and the heterozygote for a dominant allele have reduced fitness. For this reason, detrimental dominant alleles generally are less common than recessive ones.

Nonrandom Mating

A fundamental assumption of the Hardy-Weinberg law is that members of the population mate randomly. But individuals in many populations do not mate randomly for some traits, and when nonrandom mating occurs, the genotypes will not exist in the proportions predicted by the Hardy-Weinberg law.

One form of nonrandom mating is **positive assortative mating**, which occurs when individuals with similar phenotypes mate preferentially. Positive assortative mating is common in natural populations. For example, humans mate assortatively for height; tall men and tall women marry more frequently, and short men and short women marry more frequently, than would be expected on a random basis. **Negative assortative mating** occurs when phenotypically dissimilar individuals mate more often than randomly chosen individuals. If humans exhibited negative assortative mating for height, tall men and short women would marry preferentially, and short men and tall women would marry preferentially. Neither positive nor negative assortative mating affects the allelic frequencies of a population, but both may influence the genotypic frequencies if the phenotypes for which assortative mating occurs are genetically determined.

Two other departures from random mating are **inbreeding** and **outbreeding**. Inbreeding involves preferential mating between close relatives, and thus inbreeding is really positive assortative mating for relatedness. Outbreeding is preferential mating between nonrelated individuals and is a form of negative assortative mating for relatedness.

~ TABLE 21.8

Relative Genotype Distributions Resulting from Self-fertilization over Several Generations Starting with an *Aa* Individual

GENERATION	FREQUENCIES OF GENOTYPES		
	AA	*Aa*	*aa*
0	0	1	0
1	1/4	1/2	1/4
2	$1/4 + 1/8 = 3/8$	1/4	$1/4 + 1/8 = 3/8$
3	$3/8 + 1/16 = 7/16$	1/8	$3/8 + 1/16 = 7/16$
4	$7/16 + 1/32 = 15/32$	1/16	$7/16 + 1/32 = 15/32$
5	$15/32 + 1/64 = 31/64$	1/32	$15/32 + 1/64 = 31/64$
n	$[1 - (1/2)^n]/2$	$(1/2)^n$	$[1 - (1/2)^n]/2$
∞	1/2	0	1/2

Regular systems of inbreeding exist, such as self-fertilization, sib mating, and mating between first cousins. The most extreme case of inbreeding is self-fertilization, which occurs in many plants and a few animals, such as some snails. The effects of self-fertilization are illustrated in Table 21.8. Assume that we begin with a population consisting entirely of *Aa* heterozygotes, and that all individuals in this population reproduce by self-fertilization. After one generation of self-fertilization, the progeny will consist of 1/4 *AA*, 1/2 *Aa*, and 1/4 *aa*. Now only half of the population consists of heterozygotes. When this generation undergoes self-fertilization, the *AA* homozygotes will produce only *AA* progeny, and the *aa* homozygotes will produce only *aa* progeny. When the heterozygotes reproduce, however, only half of their progeny will be heterozygous like the parents, and the other half will be homozygous (1/4 *AA* and 1/4 *aa*). This means that in each generation of self-fertilization, the percentage of heterozygotes decreases by 50 percent. After a large number of generations, there will be no heterozygotes, and the population will be divided equally between the two homozygous genotypes.

The result of continued self-fertilization is to increase homozygosity at the expense of heterozygosity. The frequencies of alleles *A* and *a* remain constant, while the frequencies of the three genotypes change significantly. When less intensive inbreeding occurs, similar, but less pronounced, effects occur. On the other hand, outbreeding increases heterozygosity.

KEYNOTE

Inbreeding involves preferential mating between close relatives, and outbreeding is preferential mating between unrelated individuals. Continued inbreeding increases homozygosity within a population, whereas outbreeding tends to increase heterozygosity.

SUMMARY OF THE EFFECTS OF EVOLUTIONARY PROCESSES ON THE GENETIC STRUCTURE OF A POPULATION

Let us now review the major effects of the different evolutionary processes on (1) changes in allelic frequency within a population, (2) genetic divergence among populations, and (3) increases and decreases in genetic variation within populations.

Changes in Allelic Frequency Within a Population

Mutation, migration, genetic drift, and selection all have the potential to change the allelic frequencies of a population over time. Mutation, however, usually occurs at such a low rate that the change resulting from mutation pressure alone is frequently negligible.

Genetic drift will produce substantial changes in allelic frequency when population size is small. Furthermore, mutation, migration, and selection may lead to equilibria, where these processes continue to act, but the allelic frequencies no longer change. Nonrandom mating does not change allelic frequencies, but it does affect the genotypic frequencies of a population: inbreeding leads to increases in homozygosity and outbreeding produces an excess of heterozygotes.

Genetic Divergence Among Populations

Several evolutionary processes lead to genetic divergence among populations. Because genetic drift is a random process, allelic frequencies in different populations may drift in different directions, so genetic drift can produce genetic divergence among populations. Migration among populations has just the opposite effect, increasing effective population size and equalizing allelic frequency differences among populations. If the population size is small, different mutations may arise in different populations, and, therefore, mutation may contribute to population differentiation. Natural selection can increase genetic differences among populations by favoring different alleles in different populations, or it can prevent divergence by keeping allelic frequencies among populations uniform. Nonrandom mating will not, by itself, generate genetic differences among populations, although it may contribute to the effects of other processes by increasing or decreasing effective population size.

Increases and Decreases in Genetic Variation Within Populations

Migration and mutation tend to increase genetic variation within populations by introducing new alleles to the gene pool. Genetic drift produces the opposite effect, decreasing genetic variation within small populations through loss of alleles. Because inbreeding leads to increases in homozygosity, it also diminishes genetic variation within populations; outbreeding, on the other hand, increases genetic variation by increasing heterozygosity. Natural selection may increase or decrease genetic variation; if one particular allele is favored, other alleles decrease in frequency and may be eliminated from the population by selection. Alternatively, natural selection may increase genetic variation within populations through heterozygote superiority and other forms of balancing selection.

In reality, these evolutionary processes never act in isolation but combine and interact in complex ways. In most natural populations, the combined effects of these processes and their interaction determine the pattern of genetic variation observed in the gene pool over time.

SUMMARY OF THE EFFECTS OF EVOLUTIONARY PROCESSES ON THE CONSERVATION OF GENETIC RESOURCES

Over the past several decades considerable attention has been drawn to the problems of extinction and biodiversity. It is estimated that there are approximately 2 million known species and as many as 30 million that are yet to be described. Species are becoming extinct at a very rapid and increasing rate, with some not overly pessimistic estimates putting the value in the thousands each year. Many of the genetic principles and processes discussed in this chapter relate to this conservation problem. As we have seen, populations have genetic structure; conserving this structure may warrant special attention. For example, techniques of *population viability analysis* are concerned with estimating how large a population has to be in order to keep it from going extinct for a particular period of time with a certain degree of certainty.

If one wants to ensure that a population has the potential to evolve over long periods of time, an adequate gene pool must be maintained. The problem is particularly acute for rare and already endangered species. The effects of unintentional inbreeding of species in zoos and game management programs are diminishing as population genetic principles are being employed to manage genetic structures of populations more carefully. For example, we have seen that inbreeding, genetic drift, and selection can all decrease genetic variation, and populations may need to be maintained at certain genetic effective sizes to ensure that ample amounts of variation remain. In addition, migration between populations and how a population is subdivided geographically have been shown to have potentially large consequences on how much genetic variation is retained in a particular population or species.

MOLECULAR GENETIC TECHNIQUES AND EVOLUTION

Population geneticists now widely apply molecular genetic techniques to studies of genetic variation within populations and to questions about the molecular basis of evolution. By using restriction mapping and DNA sequencing methods (see also Chapter 14), biologists can now examine evolution at the most basic genetic level, the level of the DNA. These studies have not altered the basic principles of population genetics, but they have provided a more complete and detailed picture of how natural processes produce evolutionary change.

DNA Sequence Variation

A key question in the study of evolution is how patterns and rates of evolution differ among genes and among different parts of the same gene. Evolution rates of given DNA sequences can be measured by comparing sequences in two different organisms that diverged from a common ancestor at some known time in the past. We assume that the common ancestor had a single DNA sequence. After the two organisms diverged from this ancestor, their DNA sequences underwent independent evolutionary changes, producing the differences we see in their sequences today. For example, it is believed that most major groups of mammals diverged from a common ancestor some 65 million years ago. Suppose we examined a particular DNA sequence, perhaps the gene for growth hormone, in mice and humans; we might find that the sequences for this gene in mouse and human differ in 20 nucleotides. These 20 nucleotides must have changed during the 65 million years since the split between the mouse and human evolutionary lines.

To calculate the rate of evolutionary change in this gene, we first compute the number of nucleotide substitutions (changes) that occurred to produce the 20 nucleotide differences we observe today. There are a number of different mathematical methods for estimating this number, which is usually expressed as nucleotide substitutions per nucleotide site, so that the rate of evolution is independent of the length of the sequences compared. To obtain rates of change, we then divide our value of nucleotide substitutions per nucleotide site by the number of years that separate the two organisms. The rate of change obtained is then expressed as substitutions per nucleotide site per year. In our example with the growth hormone gene, we might find that the rate of change was 4×10^{-9} substitutions per site per year.

Different regions of the gene are apparently subject to different evolutionary processes and evolve at different rates. Table 21.9 shows relative rates of change in different parts of mammalian genes. Within functional genes, notice that the highest rate of change involves synonymous changes in the amino acid coding sequences. The rate of synonymous nucleotide change is about five times greater

~ TABLE 21.9

Relative Rates of Evolutionary Change in DNA Sequences of Mammalian Genes

SEQUENCE	NUCLEOTIDE SUBSTITUTIONS PER SITE PER YEAR ($\times 10^{-9}$)
Functional Genes	
5' Flanking region	2.36
Leader	1.74
Coding sequence—synonymous	4.65
Coding sequence—nonsynonymous	0.88
Intron	3.70
Trailer	1.88
3' Flanking region	4.46
Pseudogenes	4.85

than the observed rate of nonsynonymous changes. Synonymous changes do not alter the amino acid sequence of the protein. Thus, the high rate of evolutionary change seen there is not unexpected, because these changes do not affect a protein's functioning. Synonymous and nonsynonymous mutations probably arise with equal frequency. However, nonsynonymous changes that arise within coding sequences are usually detrimental to fitness and are eliminated by natural selection, whereas synonymous mutations are rarely detrimental, so they are tolerated.

Notice in Table 21.9 that high rates of evolutionary change also occur in the 3' flanking (noncoding) regions of functional genes. Like synonymous changes, sequences in the 3' flanking regions have no known effect on the amino acid sequence and usually have little effect on gene expression; most mutations that occur here will be tolerated by natural selection. Rates of change in introns are also high, but not as high as for synonymous changes and those in the 3' flanking regions. Although the sequences in the introns do not code for proteins, the intron must be properly spliced out for mRNA to be translated into a functional protein. A few sequences within the intron are critical for proper splicing (see Chapter 12, pp. 266–268). As a result, not all changes in introns will be tolerated, so the overall rate of change is a bit lower than that seen in synonymous coding sequences and in the 3' flanking region.

Lower rates of evolutionary change are seen in the 5' flanking region. Although this region is neither transcribed nor translated, it does contain the promoter for the gene; thus, sequences in the 5' flanking region are important for gene expression. Mutations in consensus sequences may prevent the gene from being transcribed and thus will have detrimental effects on the fitness of the organism; natural selection eliminates these mutations and keeps change in this region low.

Next in evolutionary rate are the leader and trailer regions (see Table 21.9), which have somewhat lower rates than the 5' flanking region. Leaders and trailers are transcribed and border the coding sequence, but they are not translated. They do, however, provide important signals for processing the mRNA and for attachment of the ribosome to the mRNA. Nucleotide substitutions in these regions are therefore limited. The lowest rate of evolution is seen in the nonsynonymous coding sequences. Alteration of these nucleotides changes the amino acid sequence of the protein, and most mutations that occur here are eliminated by natural selection.

One final thing to note in Table 21.9 is that the highest rate of evolution is seen in nonfunctional pseudogenes. Among human globin pseudogenes, for example, the rate of nucleotide change is approximately 10 times that observed in the coding sequences of functional globin genes. The high rate of evolution observed in these sequences occurs because pseudogenes no longer code for proteins; since changes in these genes do not affect an individual's fitness, changes are not eliminated by natural selection. In summary, we observe what seems to make intuitive sense: the highest rates of evolutionary change occur in sequences that have the least effect on function.

Keep in mind that while differences in evolutionary rate among parts of a gene exist and can often be related to the purported function of the particular region, when we average rates over an entire gene we find that different functional genes evolve at different rates (Table 21.10). For example, the rate of nonsynonymous nucleotide substitution in the prolactin gene of mammals is over 300 times higher than that found in the histone H4 gene of mammals, an observation paralleling that made on primary protein structure (amino acid sequence). Mitochondrial DNA, on the other hand, has been found to have a more uniform distribution and regular rate of accumulating nucleotide sequence differences. As a result, mitochondrial DNA is valuable as a "molecular clock" for evolutionary studies.

DNA Length Polymorphisms

In addition to evolution of nucleotide sequences through nucleotide substitution, variation frequently occurs in the number of nucleotides found within a gene. These variations, called DNA length polymorphisms, arise through deletions and insertions of relatively short stretches of nucleotides. For example, DNA length polymorphisms have been observed in the alcohol dehydrogenase gene of *Drosophila melanogaster*. In a study mentioned earlier (see p. 482), 11 copies of this gene were sequenced. In addition to extensive variation in nucleotide sequence, six insertions and deletions were found in the 11 copies. All were confined to introns and flanking regions of the DNA—none was found within exons. Insertions and deletions within exons usually alter the reading frame, so they will be selected against. As a result, insertions and deletions are most commonly found in noncoding regions of the DNA. Another class of DNA length polymorphisms involves variation in the number of copies of a particular gene. For example, among individual fruit flies, the number of copies of ribosomal RNA genes varies extensively.

Evolution of Multigene Families Through Gene Duplication

In eukaryotic organisms, we frequently find multiple copies of genes, all having identical or similar sequences. A group of such genes is termed a **multigene family**, which is a set of related genes that have evolved from some ancestral gene through the process of gene duplication. Members of a gene family may be clustered together or dispersed on different chromosomes.

~ **TABLE 21.10**

Relative Rates of Evolutionary Change in DNA Sequences of Different Mammalian Genes[a]

GENE	SYNONYMOUS RATE	NONSYNONYMOUS RATE
Histone H4	0.004	1.43
Insulin	0.16	5.41
Prolactin	1.29	5.59
α-globin	0.56	3.94
β-globin	0.87	2.96
Albumin	0.92	6.72
α-fetoprotein	1.21	4.90

[a]All rates are in nucleotide substitutions per site per year $\times 10^{-9}$.

An example is the globin gene family in vertebrates, which consists of genes that code for the polypeptide chains making up the hemoglobin molecule. The organization of this multigene family in humans is discussed in Chapter 16. The globin multigene family in humans is composed of seven α-like genes found on chromosome 16 and six β-like genes found on chromosome 11. Globin genes are also found in other animals, and globinlike genes are even found in plants, suggesting that this is a very ancient gene family. Almost all functional globin genes in animal species have the same general structure, consisting of three exons separated by two introns. However, the numbers of globin genes and their order vary among species, as is shown for the β-like genes in Figure 21.13. Because all globin genes have similarities in structure and sequence, it appears that an ancestral globin gene (perhaps most like the present day myoglobin gene) duplicated and diverged to produce an ancestral α-like gene and an ancestral β-like gene. These two genes then underwent repeated duplications, giving rise to the various α-like and β-like genes found in vertebrates today.

Repeated gene duplication, such as that giving rise to the globin gene family, appears to be a frequent evolutionary occurrence. Indeed, the number of copies of globin genes varies in some human populations. For example, most humans have two α-globin genes on chromosome 16. However, some individuals have a single α-globin gene on chromosome 16; others have three or even four copies of the α-globin gene on one of their chromosomes. These observations indicate that duplication and deletion of genes in a multigene family are constant, ongoing processes. In some cases, mutations arise that render a copy of the gene nonfunctional, creating a pseudogene. In other cases,

the change of nucleotide sequence may lead to different functions for the protein product of a gene.

Evolution in Mitochondrial DNA Sequences

In Chapter 20 we discussed the mitochondrial genome. We saw that animal mitochondrial DNA (mtDNA) is composed of approximately 15,000 base pairs and that this DNA encodes two rRNAs, 22 to 23 tRNAs, and 10 to 12 proteins. A number of recent studies have examined sequence variation in mtDNA. These studies reveal that mtDNA evolution differs from evolution typically observed in nuclear DNA. For example, for reasons that are not entirely clear, nucleotide sequences within animal mtDNA evolve at a rate that is 5 to 10 times faster than that typically observed for coding sequences in nuclear genes. However, the rapid rate of evolution observed in animal mtDNA is not universal for all eukaryotes; for unknown reasons, plant mtDNA evolves at a slower rate than plant nuclear DNA.

Whatever the cause of accelerated evolution of mtDNA, the fact that these sequences change rapidly makes them ideal for assessing evolutionary relationships among groups of closely related organisms. For example, analyses of mtDNA sequences have been used to investigate the relationships among humans, chimpanzees, gorillas, orangutans, and gibbons, a group whose precise evolutionary relationships have been controversial.

Mitochondrial DNA also differs from nuclear DNA in that all mtDNA is inherited clonally from the mother; mitochondria are located in the cytoplasm, and only the egg cell contributes cytoplasm to a zygote. As a consequence, mtDNA does not undergo meiosis, and all offspring should be identical to the

~ **FIGURE 21.13**

Organization of the globin gene families in several mammalian species.

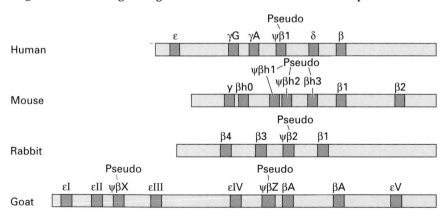

maternal genotype for mtDNA sequences (the off-spring are clones for mtDNA genes). This pattern of inheritance allows matriarchal lineages (descendants from one female) to be traced and provides a means for examining family structure in some populations. This, in conjunction with the rapid and regular rate of accumulation of nucleotide sequence differences, has allowed mtDNA to become a valuable tool for comparing closely related lineages.

Concerted Evolution

One of the surprising findings from studies of DNA sequence variation is the observation that some molecular force, or process, maintains uniformity of sequence in multiple copies of a gene. This phenomenon has been termed **concerted evolution** or **molecular drive**. Undoubtedly, multiple copies of genes arose through duplication. Following duplication, individual copies of a gene might be expected to acquire mutations and diverge. Selection might limit mutations in coding regions, but if many copies exist, we would expect some divergence to occur, especially in the noncoding sequences. Contrary to this expectation, numerous studies have revealed that nucleotide sequences in the different copies of a gene are frequently quite homogeneous. Furthermore, the noncoding sequences are also homogeneous, which suggests that natural selection is not responsible. When the same genes are examined in a second, closely related species, that group's sequences are also homogeneous, but are frequently different from the homogeneous sequence found in the first species.

These observations have led to the conclusion that some molecular process continually enforces uniformity among multiple copies of the same sequence within a species. At the same time, the process allows for rapid differentiation among species. The mechanism of concerted evolution is not fully understood, but concerted evolution has consequences for how genes evolve, and it represents an evolutionary force that was unknown before the application of modern molecular techniques to population genetics.

Evolutionary Relationships Revealed by RNA and Nuclear DNA Sequences

Within the past few years, molecular genetics has provided powerful tools for deciphering the evolutionary history of life. Because evolution is defined as genetic change, genetic relationships are of primary importance in constructing of evolutionary trees. In the past, evolutionary biologists were forced to rely entirely on comparison of phenotypes to infer genetic similarities and differences, and such comparisons of

structural, behavioral, ultrastructural, and biochemical characteristics were used to construct evolutionary trees for many groups of animals and plants and, indeed, are still the basis of most evolutionary studies today.

However, similar phenotypes can evolve in organisms that are distantly related. For example, if a naive biologist tried to construct an evolutionary tree on the basis of whether wings were present or absent, he might place birds, bats, and insects in the same evolutionary group, since all have wings.

In contrast, DNA sequences provide the most accurate and reliable information from which to infer evolutionary relationships. They allow direct comparison of the genetic differences among organisms, they are easily quantified, and all organisms have them (all organisms have at least some genes in common, such as tRNA genes, rRNA genes, and genes for a few proteins). Because of these considerable advantages, many evolutionary biologists have turned to DNA sequences for assessing evolutionary relationships and for constructing evolutionary trees.

One case where sequence data have provided new information about evolutionary relationships is in our understanding of the primary divisions of life. Many years ago, biologists divided all of life into two major groups, plants and animals. It was later recognized that organisms can be divided into prokaryotes and eukaryotes on the basis of cell structure. More recently, several primary divisions of life have been recognized, such as the five kingdoms: prokaryotes, protists, plants, fungi, and animals.

In recent years, RNA and DNA sequences have been used to uncover the primary lines of evolutionary history among all organisms. In one study, Norman Pace and his colleagues constructed an evolutionary tree of life based on the sequences found in 16S-like rRNA, which all organisms possess. As illustrated in Figure 21.14, their evolutionary tree revealed three major evolutionary groups: the Bacteria (the traditional prokaryotes), the Eukarya (eukaryotes), and the Archaea (a relatively little-known group of bacteria). Bacteria and Archaea, although both prokaryotic, were found to be as different genetically as Bacteria and Eukarya. The deep evolutionary differences that separate the Bacteria and the Archaea were not obvious on the basis of phenotype; this became clear only after their nucleotide sequences were compared. Sequences of other genes, including 5S rRNAs, large rRNAs, and genes for some basic proteins, support the idea that three major evolutionary groups exist among living organisms. Other researchers have proposed alternative evolutionary trees, and it will take more data in this very actively researched area to determine the most accurate one.

~ FIGURE 21.14

An evolutionary tree of life revealed by comparison of 16S (prokaryotes) and 18S rRNA (eukaryotes) sequences. (Reprinted with permission from N. Pace, "A Molecular View of Microbial Diversity in the Biosphere" in *Science* 276 (1997):735. Copyright © 1997 American Association for the Advancement of Science.)

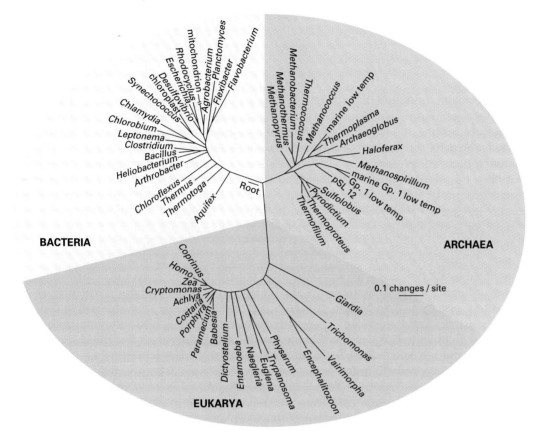

DNA sequences are being used also to study relationships in human evolution. In contrast to extensive variation in size, body shape, facial features, skin color, and so on, genetic differences among human populations are relatively small. For example, analysis of mtDNA sequences shows that the mean difference in sequence between two human populations is about 0.33 percent. Other primates exhibit much larger differences. For example, the two subspecies of orangutan differ by as much as 5 percent. This indicates that all human groups are closely related. Nevertheless, some genetic differences do occur among different human groups. Surprisingly, the greatest differences are not found between populations located on different continents, but are seen among human populations residing in Africa. All other human populations show fewer differences than we find among the African populations. Many experts interpret these findings to mean that humans

experienced their origin and early evolutionary divergence in Africa. After a number of genetically differentiated populations had evolved in Africa, it is hypothesized that a small group of humans may have migrated out of Africa and given rise to all other human populations. This hypothesis has been termed the "out-of-Africa" theory. Sequence data from both mitochondrial DNA and nuclear genes are consistent with this hypothesis. Although the out-of-Africa theory is not universally accepted, DNA sequence data are playing an increasingly important role in the study of human evolution, and indeed in the study of the evolution of many lineages.

SUMMARY

Population genetics is the study of the genetic structure of populations and species and how the structure changes or evolves over time. The gene pool of a population is the total of all genes within a Mendelian population, and it is described in terms of allelic and genotypic frequencies. The Hardy-Weinberg law describes what happens to allelic and genotypic frequencies of a large, randomly mating population free from evolutionary processes; when these conditions are met, allelic frequencies do not change, and genotypic frequencies stabilize after one generation in the proportions p^2, $2pq$, and q^2, where p and q equal the allelic frequencies of the population.

The classical, balance, and neutral mutation models have generated testable hypotheses that help explain how much genetic variation should exist within natural populations, and what processes are responsible for the variation observed. Protein electrophoresis showed that most populations of plants and animals contain large amounts of genetic variation, proving that the classical model was wrong. However, attempts to determine whether most genetic variation is maintained by natural selection (the balance model) or by the neutral processes of genetic drift and mutation (the neutral mutation model) have provided no clear answer, tending to indicate that the evolution of genetic structure proceeds as a result of many processes, sometimes acting simultaneously and sometimes in sequence.

Mutation, genetic drift, migration, and natural selection are processes that can alter allelic frequencies of a population. Recurrent mutation changes the allelic frequencies of a population; the relative rates of forward and reverse mutation will determine allelic frequencies of a population in the absence of other processes. Genetic drift, chance change in allelic frequencies due to small effective population size, leads to a loss of genetic variation within a population, genetic divergence among populations, and change of allelic frequency within a population. Migration tends to reduce genetic divergence among populations and increases effective population size. Natural selection is differential reproduction of genotypes. The relative reproductive contribution of genotypes is measured in terms of Darwinian fitness. The effects of natural selection depend on the fitnesses of the genotypes, the degree of dominance, and the frequencies of the alleles in the population. Nonrandom mating affects the effective population size and genotypic frequencies of a population; the allelic frequencies are unaffected. One type of nonrandom mating, inbreeding, leads to an increase in homozygosity.

New techniques of molecular genetics, including analysis of RFLPs and RNA and DNA sequences, have supported prior insights obtained from analyses of proteins with respect to evolutionary processes. Different parts of a gene are found to evolve at different rates; those parts of the gene that have the least effect on fitness appear to evolve at the highest rates. In addition to changes in nucleotide sequences, molecular evolution involves variation in DNA length polymorphisms. Multigene families evolve through repeated duplication of genes, followed by genetic divergence of their sequences. Mitochondrial DNA of animals appears to evolve at a faster and more uniform rate than nuclear DNA. From studies of DNA sequence variation it has been determined that a molecular force maintains uniformity of sequence in multiple copies of a gene, a phenomenon called concerted evolution. Finally, RNA and DNA sequences are being used for inferring evolutionary rates and relationships among organisms.

ANALYTICAL APPROACHES FOR SOLVING GENETICS PROBLEMS

Q21.1 In a population of 2,000 gaboon vipers, a genetic difference with respect to venom exists at a single locus. The alleles are incompletely dominant. The population shows 100 individuals homozygous for the t allele (genotype tt, nonpoisonous), 800 heterozygous (genotype Tt, mildly poisonous), and 1,100 homozygous for the T allele (genotype TT, deadly poisonous).

a. What is the frequency of the t allele in the population?
b. Are the genotypes in Hardy-Weinberg equilibrium?

A21.1 This question addresses the basics of calculating allelic frequencies and relating them to the genotype frequencies expected of a population in Hardy-Weinberg equilibrium.

a. The t frequency can be calculated from the information given, since the trait is an incompletely dominant one. There are 2,000 individuals in the population under study, meaning a total of 4,000 alleles at the T/t locus. The number of t alleles is given by

$$(2 \times tt \text{ homozygotes}) + (1 \times Tt \text{ heterozygotes})$$

$$= (2 \times 100) + (1 \times 800) = 1,000$$

This calculation is straightforward, since both alleles in the nonpoisonous snakes are *t*, while only one of the two alleles in the mildly poisonous snakes is *t*. Since the total number of alleles under study is 4,000, the frequency of *t* alleles is 1,000/4,000 = 0.25. This system is a two-allele system, so the frequency of *T* must be 0.75.

b. For the genotypes to be in Hardy-Weinberg equilibrium, the distribution must be p^2 *TT* + 2*pq* *Tt* + q^2 *tt* genotypes, where *p* is the frequency of the *T* allele and *q* is the frequency of the *t* allele. In part a we established that the frequency of *T* is 0.75 and the frequency of *t* is 0.25. Therefore, *p* = 0.75 and *q* = 0.25. Using these values, we can determine the expected genotype frequencies if this population is in Hardy-Weinberg equilibrium:

$$(0.75)^2 \ TT + 2(0.75)(0.25) \ Tt + (0.25)^2 \ tt$$

This expression gives 0.5625 *TT* + 0.3750 *Tt* + 0.0625 *tt*. Thus with 2,000 individuals in the population we would expect 1,125 *TT*, 750 *Tt*, and 125 *tt*. These values are close to the values given in the question, suggesting that the population is indeed in genetic equilibrium.

To check this result, we should perform a chi-square analysis (see Chapter 2), using the given numbers (not frequencies) of the three genotypes as the observed numbers and the calculated numbers as the expected numbers. The chi-square analysis is as follows, where *d* = (observed − expected):

Genotype	Observed	Expected	*d*	d^2	d^2/e
TT	1,100	1,125	−25	625	0.556
Tt	800	750	+50	2,500	3.334
tt	100	125	−25	625	5.000
Totals	2,000	2,000	0		8.890

The chi-square value (the sum of all the d^2/e values) is 8.89. For the reasons discussed in the text for a similar example (p. 477), there is only one degree of freedom. Looking up the chi-square value in the chi-square table (Table 2.5), we find a *P* value of approximately 0.0025. So about 25 times out of 10,000 we would expect chance deviations of the magnitude observed. In other words, our hypothesis that the population is in Hardy-Weinberg equilibrium is not substantiated. In this case, our guess that it was in equilibrium was inaccurate. Nonetheless, the population is not greatly removed from an equilibrium state.

Q21.2 About one normal allele in 30,000 mutates to the X-linked recessive allele for hemophilia in each human generation. Assume for the purposes of this problem that one *h* allele in 300,000 mutates to the normal alternative in each generation. (Note that in reality it is difficult to measure the reverse mutation of a human recessive allele that is essentially lethal, like the allele for hemophilia.) The mutation frequencies are indicated in the following equation:

$$h^+ \underset{v}{\overset{u}{\rightleftarrows}} h$$

where *u* = 10*v*. What allelic frequencies would prevail at equilibrium under mutation pressures alone in these circumstances?

A21.2 This question seeks to test our understanding of the effects of mutation on allelic frequencies. In the chapter we discussed the consequences of mutation pressure. The conclusion was that if *A* mutates to *a* at *n* times the frequency that *a* mutates back to *A*, then at equilibrium the value of *q* will be $\hat{q} = u/(u + v)$, or $\hat{q} = nv/(n + 1)v$. Applying that general derivation to this particular problem, we simply use the values given. We are told that the forward mutation rate is 10 times the reverse mutation rate, or *u* = 10*v*. At equilibrium the value of *q* will be $\hat{q} = u/(u + v)$. Since *u* = 10*v*, this equation becomes $\hat{q} = 10v/11v$, so *q* = 10/11, or 0.909. Therefore, at equilibrium brought about by mutation pressure the frequency of *h* (the hemophilia allele) will be 0.909 and the frequency of h^+ (the normal allele) will be $\hat{p}$, that is, $(1 - \hat{q}) = (1 - 0.909) = 0.091$.

QUESTIONS AND PROBLEMS

*21.1 In the European land snail, *Cepaea nemoralis*, multiple alleles at a single locus determine shell color. The allele for brown (C^B) is dominant to the allele for pink (C^P) and to the allele for yellow (C^Y). Pink is dominant to yellow, so yellow is recessive to pink and brown, and the dominance hierarchy among these alleles is $C^B > C^P > C^Y$. In one population of *Cepaea*, the following color phenotypes were recorded:

Brown	236
Pink	231
Yellow	33
Total	500

Assuming that this population is in Hardy-Weinberg equilibrium (large, randomly mating, and free from evolutionary processes), calculate the frequencies of the C^B, C^Y, and C^P alleles.

21.2 Three alleles are found at a locus coding for malate dehydrogenase (MDH) in the spotted chorus frog. Chorus frogs were collected from a breeding pond, and each frog's genotype at the MDH locus was determined with electrophoresis. The following numbers of genotypes were found:

M^1M^1	8
M^1M^2	35
M^2M^2	20
M^1M^3	53
M^2M^3	76
M^3M^3	62
Total	254

a. Calculate the frequencies of the M^1, M^2, and M^3 alleles in this population.
b. Using a chi-square test, determine whether the MDH genotypes in this population are in Hardy-Weinberg proportions.

*21.3 In a large interbreeding population, 81 percent of the individuals are homozygous for a recessive character. In the absence of mutation or selection, what percentage of the next generation would be homozygous recessives? Homozygous dominants? Heterozygotes?

*21.4 Let A and a represent dominant and recessive alleles whose respective frequencies are p and q in a given interbreeding population at equilibrium (with $p + q = 1$).
a. If 16 percent of the individuals in the population have recessive phenotypes, what percentage of the total number of recessive genes exist in the heterozygous condition?
b. If 1.0 percent of the individuals were homozygous recessive, what percentage of the recessive genes would occur in heterozygotes?

*21.5 A population has eight times as many heterozygotes as homozygous recessives. What is the frequency of the recessive gene?

21.6 In a large population of range cattle the following ratios are observed: 49 percent red (RR), 42 percent roan (Rr), and 9 percent white (rr).
a. What percentage of the gametes that give rise to the next generation of cattle in this population will contain allele R?
b. In another cattle population, only 1 percent of the animals are white and 99 percent are either red or roan. What is the percentage of r alleles in this case?

21.7 In a gene pool the alleles A and a have initial frequencies of p and q, respectively. Prove that the allelic frequencies and zygotic frequencies do not change from generation to generation as long as there is no selection,

mutation, or migration, the population is large, and the individuals mate at random.

*21.8 The S-s antigen system in humans is controlled by two codominant alleles, S and s. In a group of 3,146 individuals the following genotypic frequencies were found: 188 SS, 717 Ss, and 2,241 ss.
a. Calculate the frequency of the S and s alleles.
b. Determine whether the genotypic frequencies conform to the Hardy-Weinberg equilibrium by using the chi-square test.

21.9 Refer to Problem 21.8. A third allele is sometimes found at the S locus. This allele, S^u, is recessive to both the S and the s alleles and can only be detected in the homozygous state. If the frequencies of the alleles S, s, and S^u are p, q, and r, respectively, what would be the expected frequencies of the phenotypes $S-$, Ss, $s-$, and S^uS^u?

21.10 In a large interbreeding human population, 60 percent of individuals belong to blood group O (genotype i/i). Assuming negligible mutation and no selective advantage of one blood type over another, what percentage of the grandchildren of the present population will be type O?

*21.11 A selectively neutral, recessive character appears in 0.40 of the males and in 0.16 of the females in a randomly interbreeding population. What is the gene's frequency? How many females are heterozygous for it? How many males are heterozygous for it?

21.12 Suppose you found two distinguishable types of individuals in wild populations of some organism in the following frequencies:

	TYPE 1	TYPE 2
Females	99%	1%
Males	90%	10%

The difference is known to be inherited. What is its genetic basis?

*21.13 Red-green color blindness is due to a sex-linked recessive gene. About 64 women out of 10,000 are color-blind. What proportion of men would be expected to show the trait if mating is random?

*21.14 About 8 percent of the men in a population are red-green color-blind (owing to a sex-linked recessive gene). Answer the following questions, assuming random mating in the population, with respect to color blindness.
a. What percentage of women would be expected to be color-blind?
b. What percentage of women would be expected to be heterozygous?

c. What percentage of men would be expected to have normal vision two generations later?

21.15 List some of the basic differences in the classical, balance, and neutral mutation models of genetic variation.

*21.16 Two alleles of a locus, *A* and *a*, can be interconverted by mutation:

$$A \xrightarrow{\quad u \quad} a$$
$$A \xleftarrow{\quad v \quad} a$$

and *u* is a mutation rate of 6.0×10^{-7}, and *v* is a mutation rate of 6.0×10^{-8}. What will be the frequencies of *A* and *a* at mutational equilibrium, assuming no selective difference, no migration, and no random fluctuation caused by genetic drift?

21.17 The land snail *Cepaea nemoralis* is native to Europe but has been accidentally introduced into North America at several localities. These introductions occurred when a few snails were inadvertently transported on plants, building supplies, soil, or other cargo. The snails subsequently multiplied and established large, viable populations in North America.

Assume that today the average size of *Cepaea* populations found in North America is equal to the average size of *Cepaea* populations in Europe. What predictions can you make about the amounts of genetic variation present in European and North American populations of *Cepaea*? Explain your reasoning.

*21.18 A population of 80 adult squirrels resides on campus, and the frequency of the Est^1 allele among these squirrels is 0.70. Another population of squirrels is found in a nearby woods, and there, the frequency of the Est^1 allele is 0.5. During a severe winter, 20 of the squirrels from the woods population migrate to campus in search of food and join the campus population. What will be the allelic frequency of Est^1 in the campus population after migration?

*21.19 On sampling three populations and determining genotypes, you find the following three genotype distributions. What does each distribution imply with regard to selective advantages of population structure?

POPULATION	AA	Aa	aa
1	0.04	0.32	0.64
2	0.12	0.87	0.01
3	0.45	0.10	0.45

21.20 The frequency of two adaptively neutral alleles in a large population is 70 percent *A* : 30 percent *a*. The population is wiped out by an epidemic, leaving only four individuals, who produce many offspring. What is the probability that the population several years later will be 100 percent *AA*? (Assume no mutations.)

*21.21 A completely recessive gene, owing to changed environmental circumstances, becomes lethal in a certain population. It was previously neutral, and its frequency was 0.5.
a. What was the genotype distribution when the recessive genotype was not selected against?
b. What will be the allelic frequency after one generation in the altered environment?
c. What will be the allelic frequency after two generations?

21.22 Human individuals homozygous for a certain recessive autosomal gene die before reaching reproductive age. In spite of this removal of all affected individuals, there is no indication that homozygotes occur less frequently in succeeding generations. To what might you attribute the constant rate of appearance of recessives?

*21.23 As discussed in this chapter, the gene for sickle-cell anemia exhibits heterozygote superiority. An individual who is an *Hb-A/Hb-S* heterozygote has increased resistance to malaria and therefore has greater fitness than the *Hb-A/Hb-A* homozygote, who is susceptible to malaria, and the *Hb-S/Hb-S* homozygote, who has sickle-cell anemia. Suppose that the fitness values of the genotypes in Africa are:

$$Hb\text{-}A/Hb\text{-}A = 0.88$$
$$Hb\text{-}A/Hb\text{-}S = 1.00$$
$$Hb\text{-}S/Hb\text{-}S = 0.14$$

Give the expected equilibrium frequencies of the sickle-cell gene (*Hb-S*).

*21.24 Achondroplasia, a type of dwarfism in humans, is caused by an autosomal dominant gene. The mutation rate for achondroplasia is about 5×10^{-5}, and the fitness of achondroplastic dwarfs has been estimated to be about 0.2 compared with unaffected individuals. What is the equilibrium frequency of the achondroplasia gene based on this mutation rate and fitness value?

21.25 The frequencies of the L^M and L^N blood group alleles are the same in each of the populations I, II, and III, but the genotypes' frequencies are not the same, as shown below. Which of the populations is most likely to show each of the following characteristics: random mating, inbreeding, genetic drift? Explain your answers.

	M – M	M – N	N – N
I	0.50	0.40	0.10
II	0.49	0.42	0.09
III	0.45	0.50	0.05

*21.26 DNA was collected from 100 people randomly sampled from a population and was digested with the restriction enzyme *Bam*HI. The fragments were separated by electrophoresis and then transferred to a membrane filter using the Southern blot technique. The blots were probed with a particular cloned sequence. Three different patterns of hybridization were seen on the blots. Some DNA samples (56 of them) showed a single band of 6.3 kb, others (6) showed a single band at 4.1 kb, and yet others (38) showed both the 6.3 and the 4.1 kb bands.

a. Interpret these results in terms of *Bam*HI sites.

b. What are the frequencies of the restriction site alleles?

c. Does this population appear to be in Hardy-Weinberg equilibrium for the relevant restriction site(s)?

21.27 DNA was isolated from 10 nine-banded armadillos and cut with the restriction enzyme *Hind*III. *Hind*III recognizes the six-base sequence $\begin{smallmatrix} 5'- \text{ AAGCTT} - 3' \\ 3'- \text{ TTCGAA} - 5' \end{smallmatrix}$. The DNA fragments that resulted from the restriction reaction were separated with agarose electrophoresis and transferred to a membrane filter using Southern blotting. A labeled probe for the β-hemoglobin gene was added, which resulted in the following set of restriction patterns. Note: + / + indicates that the restriction site was present on both chromosomes of the individual, + / − indicates that the restriction site was present on one chromosome and absent on one chromosome of the individual, and − / − indicates that the restriction site was absent on both chromosomes of the individual:

+/+ −/− +/− +/− −/− +/− +/+ +/− −/− +/+

Calculate the expected heterozygosity in nucleotide sequence.

*21.28 Fifty tiger salamanders from one pond in west Texas were examined for genetic variation by using the technique of protein electrophoresis. The genotype of each salamander was determined for five loci (AmPep, ADH, PGM, MDH, and LDH-1). No variation was found at AmPep, ADH, and LDH-1; in other words, all individuals were homozygous for the same allele at these loci. The following numbers of genotypes were observed at the MDH and PGM loci:

MDH GENOTYPES	NUMBER OF INDIVIDUALS	PGM GENOTYPES	NUMBER OF INDIVIDUALS
AA	11	DD	35
AB	35	DE	10
BB	4	EE	5

Calculate the proportion of polymorphic loci and the heterozygosity for this population.

21.29 What factors cause genetic drift?

*21.30 What are the primary effects of the following evolutionary processes on the gene and genotypic frequencies of a population?

a. mutation

b. migration

c. genetic drift

d. inbreeding

21.31 Explain how heterozygote superiority leads to an increased frequency of sickle-cell anemia in areas where malaria is widespread.

*21.32 Suppose we examine the rates of nucleotide substitution in two 300-nucleotide sequences of DNA isolated from humans. In the first sequence (A), we find a nucleotide substitution rate of 4.88×10^{-9} substitutions per site per year. The rate is the same for synonymous and nonsynonymous substitutions. In the second sequence (B), we find a synonymous substitution rate of 4.66×10^{-9} substitutions per site per year and a nonsynonymous substitution rate of 0.70×10^{-9} substitutions per site per year. Referring to Table 21.9, what might you conclude about the possible functions of sequence A and sequence B?

21.33 What are some of the characteristics of mitochondrial DNA evolution in animals?

*21.34 What is concerted evolution?

21.35 What are some of the advantages of using DNA sequences to infer evolutionary relationships?

CHAPTER *22*

QUANTITATIVE GENETICS

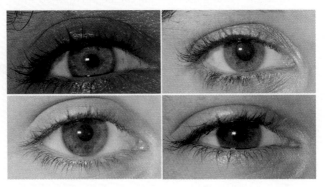

Principal Points

~ Discontinuous traits exhibit only a few distinct phenotypes. Continuous traits display a range of phenotypes.

~ Continuous traits have many phenotypes because they are encoded by many genotypes (are polygenic) and/or because environmental factors cause each genotype to produce a range of phenotypes.

~ Continuous traits are studied by using samples of populations and statistical concepts such as the mean, variance, and standard deviation.

~ The polygene or multiple-gene hypothesis of inheritance proposes that quantitative traits are determined by a number of genes whose effects add together to determine the phenotype.

~ Variation among individuals can be partitioned into genetic and environmental components. However, genotypes may behave differently in different environments, and thus caution must be exercised when designing and interpreting experiments that measure genetic and environmental contributions to phenotypic variation.

~ The broad-sense heritability of a trait is the proportion of the phenotypic variance that results from genetic differences among individuals. The narrow-sense heritability is the proportion of the phenotypic variance that results from additive genetic variance. Both of these measures are dependent on a particular population in a particular environment.

~ The amount that a trait changes in one generation as a result of selection for the trait is called the response to selection. The magnitude of the response to selection depends on the selection differential and the narrow-sense heritability.

---

*T*he mutations studied up to this point have only a few distinct phenotypes, and in many cases the phenotype could be used as a quick assay for the genotype. The seed coats of pea plants, for example, were either grey or white and the plants were tall or short. Traits such as these, with only a few distinct phenotypes, are called **discontinuous traits**.

For discontinuous traits, a simple relationship often is seen between the genotype and the phenotype. In most cases, each genotype produces only a single phenotype, and some phenotypes result from a single genotype. However, we saw in Chapter 4 that the relationship between the genotype and the phenotype may be affected by phenomena such as variable **penetrance** and **expressivity**, as well as **pleiotropy**. In addition, a single genotype can give rise to a range of phenotypes because the genotype interacts with variable environments during development to give rise to a **norm of reaction**. Indeed, many traits (probably most), such as birth weight and adult height in humans, and protein content in corn, exhibit a wide range of possible phenotypes. Traits such as these, with a continuous distribution of phenotypes, are called **continuous traits**. The distribution of a typical continuous trait—birth weight in humans—is illustrated in Figure 22.1. Since the phenotypes of continuous traits are not discrete, they must be described in quantitative terms. These traits are also known as quantitative traits, and the study of their inheritance makes up the field of **quantitative genetics**.

~ **Figure 22.1**

Distribution of birth weight of babies (males + females) born to teenagers in Portland, Oregon, in 1992.

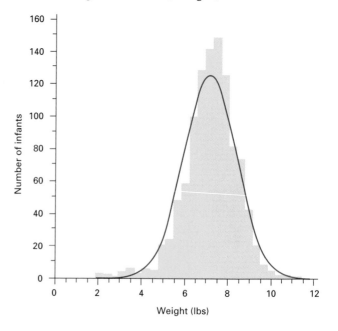

Quantitative genetics plays a role in our understanding of evolution, conservation, and many other areas of applied biology. In agriculture, for example, crop yield, rate of weight gain, milk production, and fat content are all continuous traits analyzed with the mathematical techniques of quantitative genetics. In psychology, methods of quantitative genetics are frequently employed to study complex behavioral traits, such as IQ, learning ability, and personality. Geneticists also use these methods to study other continuous traits found in humans, such as blood pressure, antibody titer, and fingerprint pattern.

THE NATURE OF CONTINUOUS TRAITS

Continuous traits by definition have a continuous range of phenotypes. To understand the inheritance of continuous traits, we must first determine why some traits have many phenotypes.

Why Some Traits Have Continuous Phenotypes

Multiple phenotypes of a trait are manifested in several ways. Frequently a range of phenotypes occurs because numerous genotypes exist among the individuals of a group—this happens when the trait is influenced by a large number of loci. For example, when a single locus with two alleles determines a trait, three genotypes are present: *AA*, *Aa*, and *aa*. With two loci, each with two alleles, the number of genotypes is $3^2 = 9$ (*AA BB, Aa BB, AA Bb, Aa Bb, AA bb, aa BB, aa Bb, Aa bb,* and *aa bb*). In general, the number of genotypes is 3^n, where *n* equals the number of loci with two alleles; if more than two alleles are present at a locus, the number of genotypes is even greater. As the number of loci influencing a trait increases, the number of genotypes quickly becomes large. Traits that are encoded by many loci are referred to as **polygenic traits**. If each genotype in a polygenic trait encodes a separate phenotype, many phenotypes will be present. And because many phenotypes are present and the differences between phenotypes are slight, the trait appears to be continuous.

More frequently, several genotypes of a polygenic trait produce the same phenotype. For example, with dominance, only two phenotypes are produced by the three genotypes at a locus (*AA* and *Aa* produce one phenotype and *aa* produces a different phenotype). If dominance occurs among the alleles at each of three loci, the number of genotypes is $3^3 = 27$, but the number of phenotypes is only $2^3 = 8$. When epistatic interactions occur among the alleles at different loci, as discussed in Chapter 4, several genotypes may code for the same phenotype. In many polygenic traits, multiple genotypes code for the same phenotype, and the relationship between genotype and phenotype is obscured.

A second reason a trait may have a range of phenotypes is that environmental factors also affect the trait. When environmental factors exert an influence on the phenotype, each genotype is capable of producing a range of phenotypes (the *norm of reaction*). Which phenotype is expressed depends both on the genotype and on the specific environment in which the genotype is found. For most continuous traits, both multiple genotypes and environmental factors influence the phenotype; such a trait is a **multifactorial trait**. A number of human traits are multifactorial, including intelligence, height, weight, obesity, and fingerprint pattern.

When multiple genes and environmental factors influence a trait, no simple relationship exists between genotype and phenotype in discontinuous traits. In order to understand the genetic basis of these traits and their inheritance, we must use special concepts and analytical procedures.

KEYNOTE

Discontinuous traits exhibit only a few distinct phenotypes and can be described in qualitative terms. Continuous traits, on the other hand, display a spectrum of phenotypes and must be described in quantitative terms. The relationship between the genotype and the phenotype is complex. One genotype may give rise to a range of phenotypes, and many genotypes can give rise to the same phenotype. Numerous phenotypes are present in a continuous trait because the trait may be encoded by many loci, producing many genotypes, and because environmental factors may cause each genotype to produce a range of phenotypes.

Questions Studied in Quantitative Genetics

Continuous traits require special attention. What follows is a set of questions frequently studied by quantitative geneticists.

1. To what degree does the observed variation in phenotype result from differences in genotype, and to what degree does this variation reflect the

influence of different environments? (In our study of discontinuous traits, this question assumed little importance, because the differences in phenotype reflected differences in genes.)

2. How many genes determine the phenotype of a trait? (When only a few loci are involved and the trait is discontinuous, the number of loci involved can often be determined by examining the phenotypic ratios in genetic crosses. With complex, continuous traits, determining the number of loci involved is difficult.)

3. Are the contributions of the determining genes equal? Or do a few genes have major effects on the trait, and other genes only modify the phenotype slightly?

4. To what degree do alleles at the different loci interact with one another? Are the effects of alleles additive?

5. When selection occurs for a particular phenotype, how rapidly does the trait change? Do other traits change at the same time?

6. What is the best method for selecting and mating individuals to produce desired phenotypes in the progeny?

STATISTICAL TOOLS

One of the fundamental questions of quantitative traits is how much of the variation that exists among individuals in populations is genetically determined and how much is environmentally induced. Thus, the perennial question of **nature versus nurture** or genes versus environment is at the heart of the field of quantitative genetics. We can phrase the problem as follows: how much of the variation in some aspect of the phenotype (V_P) is due to genetic variation (V_G), and how much is due to environmental variation (V_E), or

$$V_P = V_G + V_E$$

In order to use this equation we have to learn how to measure variation in phenotype and how to divide the variation into genetic and environmental components. For this we need to investigate some statistical methodology.

Samples and Populations

Suppose we want to describe some aspect of a trait in a large group of individuals. For example, we might be interested in the average birth weight of infants born in New York City during the year 1987. Since thousands of babies are born in New York City every year, collecting information on each baby's weight

might not be practical. An alternative method would be to collect information on a subset of the group, say birth weights of 100 infants born in New York City during 1987, and then use the average obtained on this subset as an estimate of the average for the entire city. Biologists and other scientists commonly employ this sampling procedure in data collection, and statistics are necessary for analyzing such data. The group of interest (in our example, all infants born in New York City during 1987) is called the **population**, and the subset used to give us information about the population (our set of 100 babies) is called a **sample**. For a sample to give us confidence in the information about the population, it must be large enough so that chance differences between the sample and the population are not misleading. If our sample consisted of only a single baby and that infant was unusually large, then our estimate of the average birth weight of all babies would not be very accurate. The sample must also be a random subset of the population. If all the babies in our sample came from Hope Hospital for Premature Infants, then we would grossly underestimate the true birth weight of the population. While these points might seem obvious, a great many errors in judgment are made because data were not collected randomly.

KEYNOTE

To describe and study a large group of individuals, scientists frequently examine a subset of the group. This subset is called a sample, and the sample provides information about the larger group, which is termed the population. The sample must be of reasonable size, and it must be a random subset of the larger group to provide accurate information about the population.

Distributions

For discontinuous traits, we were able to describe the phenotypes found among a group of individuals by stating the proportion of individuals falling into each phenotypic class. Continuous traits exhibit a range of phenotypes, however, and describing the phenotypes found within a group of individuals is more complicated. One means of summarizing the phenotypes of a continuous trait is with a **frequency distribution**, which is a description of the population in terms of the proportion of individuals that fall within a certain range of phenotypes.

   A frequency distribution is made by constructing classes that consist of individuals falling within a

Weight of 5,494 F$_2$ Beans (Seeds of *Phaseolus vulgaris*) Observed by Johannsen in 1903								
Weight (mg)	50–150	150–250	250–350	350–450	450–550	550–650	650–750	750–850	850–950
(Midpoint of range)	(100)	(200)	(300)	(400)	(500)	(600)	(700)	(800)	(900)
Number of beans	5	38	370	1,676	2,255	928	187	33	2

specified range of the phenotype, and then by counting the number of individuals in each class. Table 22.1 presents a frequency distribution constructed from data in Johannsen's study of weight inheritance in the dwarf bean, *Phaseolus vulgaris.* Johannsen weighed 5,494 beans from the F$_2$ progeny of a cross and classified them into nine classes, each of which covered a 100-milligram (mg) range of weight. A frequency distribution such as this can be visualized by plotting the phenotypes in a *frequency histogram*, as shown in Figure 22.2 for Johannsen's beans. In the histogram, the phenotypic classes are indicated along the horizontal axis (*x* axis), and the number in each class is plotted on the vertical axis (*y* axis). If a curve is drawn tracing the outline of the histogram, the curve has a shape that is characteristic of the frequency distribution. Several different types of distributions are illustrated in Figure 22.3.

Many continuous phenotypes show a symmetrical, bell-shaped distribution similar to the one shown in Figure 22.3a and superimposed over the data in Figure 22.2. This type of distribution is called a **normal distribution**. The normal distribution has specific properties and is produced when a large number of independent factors influence the measurement. Since many continuous traits are multifactorial (influenced by multiple genes and multiple environmental factors), observing a nearly normal distribution for these traits is not surprising. Two

~ **FIGURE 22.2**

Frequency histogram for bean weight in *Phaseolus vulgaris* plotted from the data in Table 22.1. A normal curve has been fitted to the data and is superimposed on the frequency histogram.

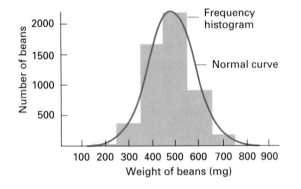

~ **FIGURE 22.3**

Three different types of distributions. (a) A normal distribution of percent sucrose in sugar beets; (b) A skewed distribution representing coat color in guinea pigs; (c) A bimodal distribution of the size of female rattlesnakes.

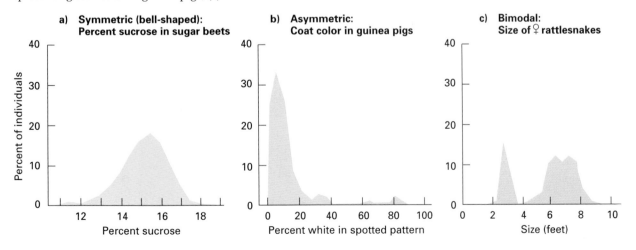

other types of distribution are illustrated in Figures 22.3b and 22.3c.

The Mean

A frequency distribution of a phenotypic trait can be summarized with two convenient statistics, the mean and the variance. The **mean**—also known as the average—gives us information about where the center of the distribution of the phenotypes in a sample is located along a continuous range of possibilities. The mean of a sample, symbolized $\bar{x}$, is calculated by simply adding up all the individual measurements (x_1, x_2, x_3, etc.) and dividing by the number of measurements we added (n). We can represent this calculation with the following formula:

$$\text{Mean} = \bar{x} = \frac{x_1 + x_2 + x_3 + \ldots x_n}{n}$$

This equation can be abbreviated by using the symbol Σ, which means the summation, and x_i, which means the ith (or individual) value of x:

$$\text{Mean} = \bar{x} = \frac{\Sigma x_i}{n}$$

Table 22.2 includes a sample calculation of the mean (and other statistics; see next section) for body lengths of 10 spotted salamanders from Penobscot County, Maine.

The mean is frequently used in quantitative genetics to characterize the phenotypes of a group of individuals. For example, Edward M. East examined the inheritance of flower length in several strains of the tobacco plant. He crossed a short-flowered strain of tobacco with a long-flowered strain. Within each strain, flower length varied somewhat, so East reported that the mean phenotype of the short strain was 40.4 mm and the mean phenotype of the long strain was 93.1 mm. The F_1 progeny had a mean flower length of 63.5 mm. In this situation, the mean provides a convenient way to characterize quickly the phenotypes of parents and offspring.

The Variance and the Standard Deviation

The **variance** is a measure of how much the individual measurements spread out around the mean—how variable the measurements are. Two distributions may have the same mean but very different variances,

～ TABLE 22.2

Sample Calculations of the Mean, Variance, and Standard Deviation for Body Length of 10 Spotted Salamanders from Penobscot County, Maine

BODY LENGTH (x_i) (mm)	($x_i - \bar{x}$)	($x_i - \bar{x}$)^2
65	(65 − 57.1) = 7.9	7.9^2 = 62.41
54	(54 − 57.1) = −3.1	-3.1^2 = 9.61
56	(56 − 57.1) = −1.1	-1.1^2 = 1.21
60	(60 − 57.1) = 2.9	2.9^2 = 8.41
56	(56 − 57.1) = −1.1	1.1^2 = 1.21
55	(55 − 57.1) = −2.1	2.1^2 = 4.41
53	(53 − 57.1) = −4.1	-4.1^2 = 16.81
55	(55 − 57.1) = −2.1	-2.1^2 = 4.41
58	(58 − 57.1) = 0.9	0.9^2 = 0.81
59	(59 − 57.1) = 1.9	1.9^2 = 3.61
Σx_i = 571		$\Sigma(x_i - \bar{x})^2$ = 112.90

$$\text{Mean} = \bar{x} = \frac{\Sigma x_i}{n} = \frac{571}{10} = 57.1$$

$$\text{Variance} = s^2 = \frac{\Sigma(x_i - \bar{x})^2}{n-1} = \frac{112.9}{9} = 12.54$$

$$\text{Standard deviation} = s_x = \sqrt{12.54} = 3.54$$

~ FIGURE 22.4

Graphs showing three distributions with the same mean but different variances.

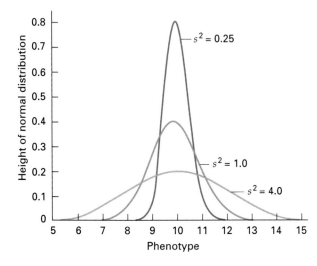

~ FIGURE 22.5

Normal distribution curve showing the proportions of the data in the distribution that are included within certain multiples of the standard deviation.

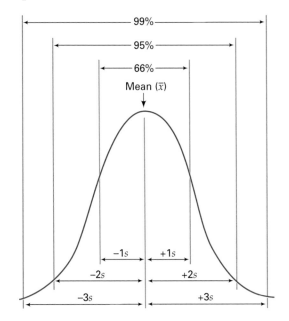

as shown in Figure 22.4. The variance, symbolized as s^2, is defined as the average deviation from the mean:

$$\text{Variance} = s^2 = \frac{\Sigma(x_i - \overline{x})^2}{n - 1}$$

Variance is calculated by first subtracting the mean from each individual measurement, squaring the value obtained, adding all the squared values, and finally dividing this sum by the number of original measurements minus one. (For mathematical reasons, which we will not discuss here, the variance is obtained by dividing by $n - 1$ instead of n.)

A related statistic is the **standard deviation**, which is simply the square root of the variance:

$$\text{Standard deviation} = s = \sqrt{s^2}$$

The standard deviation is often preferred to the variance, because the standard deviation is in the same units as the original measurements, while the variance is in the units squared. Sample calculations for the variance and the standard deviation are presented in Table 22.2. Look again at Figure 22.4. A broad curve implies considerable variability in the quantity measured and a correspondingly large standard deviation. A narrow curve, in contrast, indicates relatively little variability in the quantity measured and a correspondingly small standard deviation.

A key property of the normal distribution can now become clear. Once we know the mean and the standard deviation, a theoretical normal distribution has the shape indicated in Figure 22.5, where 66 per-

cent of the individual observations have values within one standard deviation above or below ($\pm 1s$) the mean of the distribution, about 95 percent of the values fall within two standard deviations ($\pm 2s$) of the mean, and over 99 percent fall within three standard deviations ($\pm 3s$). We can infer many things about our data and experiments using these objectively determined percentages.

The variance and the standard deviation provide us with valuable information about the phenotypes of a group of individuals. In our discussion of the mean, we saw how East used the mean to describe flower length of parents and offspring in crosses of the tobacco plant. When East crossed a strain of tobacco with short flowers to a strain with long flowers, the F_1 offspring had a mean flower length of 63.5 mm, which was intermediate to the phenotypes of the parents. When he intercrossed the F_1, the mean flower length of the F_2 offspring was 68.8 mm, approximately the same as the mean phenotype of the F_1. However, the F_2 progeny differed from the F_1 in an important attribute that is not apparent if we only examine the means of the phenotype—the F_2 were more variable in phenotype than the F_1. The variance in the flower length of the F_2 was 42.4 mm^2, whereas the variance in the F_1 was only 8.6 mm^2. This finding indicated that more genotypes were present among the F_2 progeny than the F_1. Thus, mean and variance are both necessary for fully describing the distribution of phenotypes within a group of individuals.

POLYGENIC INHERITANCE

In the examples of polygenic inheritance described in the following sections, geneticists did not know at first how these traits were inherited, although it was apparent that their pattern of inheritance differed from that of discontinuous traits.

Inheritance of Ear Length in Corn

In a study of ear length in corn (*Zea mays*) reported in 1913, Rollins Emerson and Edward East crossed the true-breeding Black Mexican sweet corn variety, which had short ears of mean length 6.63 cm, with the true-breeding Tom Thumb popcorn variety, which had long ears of mean length 16.80 cm, and then they interbred the F_1 plants. Figure 22.6 presents the results in both photographs and histograms. The F_1s have a mean ear length of 12.12 cm, which is approxi-

mately intermediate between the mean ear lengths of the two parental lines. Since the parental plants are true breeding and therefore are homozygous for whatever genes control the lengths of their ears, the F_1 plants will be genetically identical with respect to those genes; that is, they will be heterozygous. Therefore, the range of ear length phenotypes seen in the F_1 plants must be due to environmental rather than genetic differences.

In the F_2 the mean ear length of 12.89 cm is about the same as the mean for the F_1 population, but the F_2 population has a much larger variation around the mean than the F_1 population has. This variation is easily seen in Figure 22.6b; it can also be shown by calculating the standard deviation s. The standard deviation of the long-eared parent is 1.887 cm, and that of the short-eared parent is 0.816 cm. In the F_1, $s = 1.519$ cm, and in the F_2, $s = 2.252$ cm. These numbers confirm that the F_2 has greater variability.

If the environment was responsible for variation in the parental and the F_1 generations, it would likely have a similar effect on the F_2. Thus, there must be another explanation for the greater variation in ear

---

~ **FIGURE 22.6**

Inheritance of ear length in corn. (a) Representative corn ears from the parental, F_1, and F_2 generations from an experiment in which two pure-breeding corn strains that differ in ear length were crossed and then the F_1s interbred. (b) Histograms of the distributions of ear length (in centimeters) of ears of corn from the experiment represented in part (a); the vertical axes represent the percentages of the different populations found at each ear length.

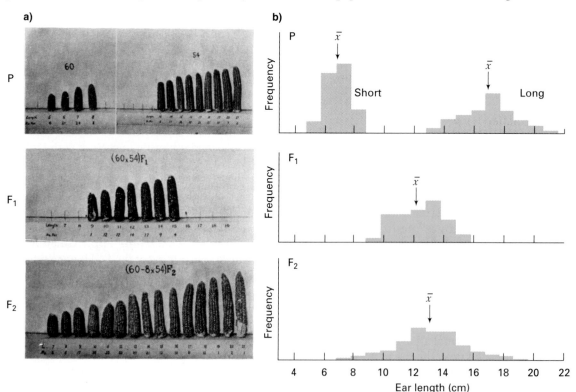

length in the F_2 generation. A more reasonable hypothesis is that the increased variability of the F_2 results from greater genetic variation in the F_2.

Setting aside the environmental influence for the moment, the data reveal four observations that apply generally to quantitative-inheritance studies similar to this one:

1. The mean value of the quantitative trait in the F_1 is approximately intermediate between the means of the two true-breeding parental lines.

2. The mean value for the trait in the F_2 is approximately equal to the mean for the F_1 population.

3. The F_2 shows more variability around the mean than the F_1 does.

4. The extreme values for the quantitative trait in the F_2 extend farther into the distribution of the two parental values than do the extreme values of the F_1.

The data presented cannot be explained in terms of standard, single-gene-locus Mendelian genetic principles that govern the inheritance of discontinuous traits. That is, if a single gene were responsible for the two phenotypes of the original parents (AA = homozygous, long; aa = homozygous, short), then the F_1 data could be explained if we assume incomplete dominance. However, crossing the F_1 heterozygote (Aa) should produce a 1:2:1 ratio of AA, Aa, and aa, or long, intermediate, and short phenotypes. Clearly, the data do not fall into such discrete classes.

KEYNOTE

For a quantitative trait, the F_1 progeny of a cross between two phenotypically distinct, pure-breeding parents has a phenotype intermediate between the parental phenotypes. The F_2 shows more variability than the F_1, with a mean phenotype close to that of the F_1. The F_1 variation is due to the environment, while the F_2 variation is due to the environment and genetics. The extreme phenotypes of the F_2 extend well beyond the range of the F_1 and into the ranges of the two parental values.

Polygene Hypothesis for Quantitative Inheritance

The simplest explanation for the data obtained from Emerson and East's experiments on corn ear length and from other experiments with quantitative traits is that such traits are controlled by many genes. This explanation, called the **polygene** or **multiple-gene hypothesis for quantitative inheritance**, is regarded as one of the landmarks of genetic thought.

The polygene hypothesis can be traced back to 1909 and the classic work of Hermann Nilsson-Ehle, who studied the color of wheat kernels. Nilsson-Ehle crossed true-breeding lines of plants with red kernels and plants with white kernels. The F_1 had grains that were all the same shade of an intermediate color between red and white. At this point he could not rule out incomplete dominance as the basis for the F_1 results. However, when he intercrossed the F_1s, a number of the F_2 progeny showed a ratio of approximately 15 red (all shades) : 1 white kernels, clearly a deviation from a 3:1 ratio expected for a monohybrid cross. He recognized four discrete shades of red, in addition to white, among the progeny. He counted the relative number of each class and found a 1:4:6:4:1 phenotypic ratio of wheat with dark red, medium red, intermediate red, light red, and white kernels. Note that 1/16 of the F_2 has a kernel phenotype as extreme as the original red parent, and 1/16 has a kernel phenotype as extreme as the original white parent.

How can the data be interpreted in genetic terms? In Chapter 4 (Analytical Approaches for Solving Genetics Problems, question 4.3c) the explanation for a 15:1 ratio was that two allelic pairs are involved in determining the phenotypes segregating in the cross. Since several of the F_2 populations from the wheat crosses exhibited a 15 red : 1 white ratio, we can apply that explanation to the kernel trait.

Let us hypothesize that two pairs of independently segregating alleles control the production of red pigment: alleles R (red) and C (crimson) result in red pigment, and alleles r and c result in the lack of pigment. Nilsson-Ehle's parental cross and the F_1 genotypes are then as follows:

$$P \qquad RR\ CC \quad \times \quad rr\ cc$$
$$\text{(dark red)} \qquad \text{(white)}$$
$$F_1 \qquad\qquad Rr\ Cc$$
$$\text{(intermediate red)}$$

When the F_1 is interbred, the distribution of genotypes in the F_2 is that typical of dihybrid inheritance: 1/16 $RR\ CC$ + 2/16 $Rr\ CC$ + 1/16 $rr\ CC$ + 2/16 $RR\ Cc$ + 4/16 $Rr\ Cc$ + 2/16 $rr\ Cc$ + 1/16 $RR\ cc$ + 2/16 $Rr\ cc$ + 1/16 $rr\ cc$. If R and C are dominant to r and c, the 9:3:3:1 phenotypic ratio characteristic of dihybrid inheritance will result. For the wheat kernel color phenotype, then, dominance is not the simple answer, since the observed phenotypic ratio approximates 1:4:6:4:1. These numbers in the phenotypic ratio are the same as the coefficients in the *binomial expansion* of $(a + b)^4$. The following calculation demonstrates how we arrive at the coefficients and their associated terms of this expansion. In essence, we multiply $(a + b)$ by $(a + b)$, then multiply the product by $(a + b)$, and so on:

$$
\begin{array}{r}
a + b \\
\times\ a + b \\
\hline
= a^2 + ab \\
+\ ab + b^2 \\
\hline
= a^2 + 2ab + b^2 \qquad \text{—this is } (a + b)^2 \\
\times\ a + b \\
\hline
= a^3 + 2a^2b + ab^2 \\
+\ a^2b + 2ab^2 + b^3 \\
\hline
= a^3 + 3a^2b + 3ab^2 + b^3 \qquad \text{—this is } (a + b)^3 \\
\times\ a + b \\
\hline
= a^4 + 3a^3b + 3a^2b^2 + ab^3 \\
+\ a^3b + 3a^2b^2 + 3ab^3 + b^4 \\
\hline
= a^4 + 4a^3b + 6a^2b^2 + 4ab^3 + b^4 \qquad \text{—this is } (a + b)^4 \\
\hline
\end{array}
$$

A simple explanation for the wheat kernel color phenotypic distribution, then, is that each dose of a gene controlling pigment production allows the synthesis of a certain amount of pigment. Therefore, the intensity of red coloration depends on the number of R or C alleles in the genotype; $RR\ CC$ (term a^4 in the expansion) would be dark red, and $rr\ cc$ (b^4 in the expansion) would be white. Table 22.3 summarizes this situation with regard to the five phenotypic classes observed by Nilsson-Ehle. In other words, each capital-letter allele causes more red pigment to be added to the wheat kernel color phenotype. Alleles that contribute to the phenotype are called **contributing alleles**. The alleles that do not have any effect on the phenotypes of the quantitative trait, such as the r and c alleles in the wheat kernel color trait, are called **noncontributing alleles**. The inheritance of red kernel color in wheat is an example of a multiple gene or polygene series of as many as four contributing alleles.

We must be cautious in interpreting the genetic basis of this particular quantitative trait, though. Some F_2 populations show only three phenotypic classes with a 3:1 ratio of red to white, while other F_2 populations show a 63:1 ratio of red to white, with discrete classes of color between dark red and white. These results indicate that the genetic basis for the quantitative trait can vary with the strain of wheat involved. The 3:1 case could be explained by a single-gene system with two contributing alleles, while the 63:1 case could indicate a polygene series with six contributing alleles. The number of discrete classes in the latter case would be seven, with the proportion of each class following the coefficients in the binomial expansion of $(a + b)^6$, that is, 1:6:15:20:15:6:1.

~ TABLE 22.3

Genetic Explanation for the Number and Proportions of F_2 Phenotypes for the Quantitative Trait Red Kernel Color in Wheat

GENOTYPE	NUMBER OF CONTRIBUTING ALLELES FOR RED	PHENOTYPE	FRACTION OF F_2
$RR\ CC$	4	Dark red	1/16
$RR\ Cc$ or $Rr\ CC$	3	Medium red	4/16
$RR\ cc$ or $rr\ CC$ or $Rr\ Cc$	2	Intermediate red	6/16
$rr\ Cc$ or $Rr\ cc$	1	Light red	4/16
$rr\ cc$	0	White	1/16

The multiple-gene hypothesis has been applied to other examples of quantitative inheritance, including ear length in corn. In its basic form, the multiple-gene hypothesis proposes that a number of the attributes of quantitative inheritance can be explained on the basis of the action and segregation of a number of allelic pairs that each have a small but additive effect on the phenotype. These allelic pairs with small effects are called *polygenes*. For the most part the multiple-gene hypothesis is satisfactory as a working hypothesis for interpreting many quantitative traits. The whole picture of quantitative traits is very complicated, though, and there are still many gaps in our understanding of quantitative inheritance. In addition, we have a lot to learn about the molecular aspects of quantitative traits.

KEYNOTE

Quantitative traits are based on genes in a multiple-gene series, or polygenes. The multiple-gene hypothesis assumes that contributing and noncontributing alleles in the series operate so that as the number of contributing alleles increases, there is an additive effect on the phenotype.

Determining the Number and Location of Polygenes for a Quantitative Trait

How does the genetic structure of the population interact with the environment to determine the phenotypic structure of the population? One way of getting at this question is to try to determine how many genes generally influence the variation in quantitative characters.

It is not easy to estimate the number of genes controlling a quantitative trait. If a large number of genes are involved, many different phenotypes may be present, and the progeny are not easily placed into distinct phenotypic classes. The fewer the genes, the easier the task. Artificial breeding programs, where the changes in variance among inbred, hybrid, and backcrossed lines are calculated, can be used to make these estimates. In this way it has been estimated that skin color in humans is determined by a minimum of 4 to 6 genes; fruit weight in tomatoes by 7 to 11 genes; and oil content in corn by a minimum of 17 genes.

Where discrete phenotypic classes can be identified, we can estimate the number of genes involved. In general, the relationship between the number of independently segregating allelic pairs in a polygene series and the number of quantitative-trait phenotypes in the F_2 is as shown in Table 22.4. When one pair of alleles controls a trait, then 1/4 of the F_2 should show one phenotypic extreme (such as the darkest red) and 1/4 should show the other phenotypic extreme (such as white). The general formula $(1/2)^{2n} (= (1/4)^n)$ describes the predicted probability for an extreme phenotype when n is the number of pairs of segregating alleles involved. For two genes, and therefore two pairs of alleles, the fraction of the F_2 population with an extreme character is $(1/2)^4 = 1/16$. For three genes (three pairs of alleles) the fraction is 1/64, as in the wheat kernel example. As the number of genes increases, the fraction of the F_2 with an extreme phenotype decreases, and it does so very rapidly. For five genes the fraction is 1/1,024, and for 10 genes it is 1/1,048,576.

The number of genotypic classes also increases rapidly as the number of pairs of alleles increases. The beginnings of this increase are shown in Table 22.4. Three pairs of alleles yield 27 possible genotypic classes in the F_2; five pairs of alleles yield 243 genotypes; and 10 pairs of alleles yield 59,049 genotypes. In general, the number of genotypic classes in the F_2 is $(3)^n$, where n is the number of pairs of alleles involved.

As the number of pairs of alleles in a multiple-gene series increases, the F_2 population rapidly assumes a continuum of phenotypic variation for which distinctions between classes are impossible. In fact, when the number of pairs of alleles is five or more, it is difficult to be sure of the number of pairs of alleles involved in controlling a quantitative effect.

Traditional and molecular methods for finding genes that influence traits can also be used in the case of quantitative traits. As discussed in Chapter 5, we can use traditional methods of linkage analysis. We can also use molecular linkage methods as discussed in Chapter 6.

HERITABILITY

Heritability is the proportion of a population's phenotypic variation that is attributable to genetic factors. The term has frequently been misused. For example, when individuals in a family resemble each other in some aspect of the phenotype, be it stature or intelligence, a genetic basis is often assumed to be responsible for the similarity. But all the resemblance among family members could just as easily be due to their shared environment. The concept of heritability,

~ TABLE 22.4

Probability Information for a Quantitative Character Controlled by a Polygenic Series in Which Independently Segregating Allelic Pairs Have Equivalent, Additive Effects

NUMBER OF ALLELIC PAIRS	NUMBER OF SEGREGATING ALLELES	FRACTION OF POPULATION SHOWING AN EXTREME EXPRESSION OF THE CHARACTER	NUMBER OF GENOTYPES IN F_2	NUMBER OF PHENOTYPES IN F_2	F_2 PHENOTYPIC RATIOS ARE THE COEFFICIENTS IN THE BINOMIAL EXPANSION
1	2	$(1/2)^2 = 1/4$	$(3)^1 = 3$	3	$(a+b)^2$
2	4	$(1/2)^4 = 1/16$	$(3)^2 = 9$	5	$(a+b)^4$
3	6	$(1/2)^6 = 1/64$	$(3)^3 = 27$	7	$(a+b)^6$
4	8	$(1/2)^8 = 1/256$	$(3)^4 = 81$	9	$(a+b)^8$
n	$2n$	$(1/2)^{2n} = (1/4)^n$	$(3)^n$	$2n + 1$	$(a+b)^{2n}$

analyzed through carefully designed quantitative genetics experiments, is used to examine the relative contributions of genes and environment to variation in a specific trait. For example, the extent to which genes affect human behaviors, such as alcoholism and criminality, may be useful to know for establishing social policy, but utmost care needs to be taken because of how easy it is to misuse supposedly scientific information.

To assess heritability, we must first measure the variation in the trait, and then we must partition that variance into components attributable to different causes.

Components of the Phenotypic Variance

As noted earlier, the **phenotypic variance** (V_P) is a measure of the variability of a trait. It is calculated by computing the variance of the trait for a group of individuals (see p. 000). Differences among individuals arise from several factors, so we can partition the phenotypic variance into several components. First, some of the phenotypic variation may arise because of genetic differences among individuals (different genotypes within the group)—this is **genetic variance** (V_G). Additional variation often results from environmental differences among the individuals—this is **environmental variance** (V_E). Temperature, nutrition, and parental care are examples of obvious environmental factors that may cause differences during development among individuals. Thus, we have the basic nature-nurture equation discussed earlier:

$$V_P = V_G + V_E$$

One hundred percent of the variation among individuals is accounted for by genetic and environmental influences; however, the partitioning of phenotypic variance is more complicated than this. The sum of the genetically caused variance and the environmentally caused variance may not add up to the total phenotypic variance. This is because the genetically caused variance and environmentally caused variance may covary, and another term ($COV_{G \times E}$) is needed. $COV_{G \times E}$ is often minimal and therefore ignored by geneticists. For example, let's say milk production in cows is influenced by genes, but it is also influenced by the amount of feed a farmer provides. The farmer knows the cows and provides the offspring of good milking cows more feed and those of poor milking cows less feed. In this way the variance in milk production is increased beyond what we would expect on the basis of genes and environment operating independently.

In addition, we may want to know if we can expect offspring to resemble their parents. Just knowing that there is a genetic component to variation does not answer this question, because another source of phenotypic variance is genetic-environmental interaction ($V_{G \times E}$). Genetic-environmental interaction occurs when the relative effects of the genotypes differ among environments. For example, in a cold environment, genotype *AA* of a plant may be 40 cm in height, and genotype *Aa* may be 35 cm in height. However, when the genotypes are moved to a warm climate, genotype *Aa* is now 60 cm and *AA* is only 50 cm in height. In this example, both genotypes grow taller in the warm environment, so there is an environmental effect on variance. There is also a genetic effect, but it is not the kind that results in consistent resemblance among relatives. The genetic effect depends on the environment. The relative performance of the genotypes switches in the two environments. Therefore, both environmental differences (temperature) and genetic differences (genotypes) contribute to the phenotypic variance. However, the effects of genotype and environment cannot simply be added together. $V_{G \times E}$ must also be considered.

The phenotypic variance, composed of differences arising from genetic variation, environmental variation, genetic-environmental covariation, and genetic-environmental interaction, can be represented by the following equation:

$$V_P = V_G + V_E + COV_{G \times E} + V_{G \times E}$$

The relative contributions of these four factors to the phenotypic variance depend on the genetic composition of the population, the specific environment, and the manner in which the genes interact with the environment.

KEYNOTE

Variation among individuals can be partitioned into genetic and environmental components. The fact that genotypes might not be distributed randomly across environments, and that genotypes may behave differently in different environments, provides a caution: the results of an experiment determining the relative importance of genetic and environmental factors may depend, in nonobvious ways, on the environment in which the experiment is performed.

The complications continue in that the genetic variance V_G can be further subdivided into components arising from different types of interactions among genes. Some genetic variance occurs as a result of the average effects of the different alleles on the phenotype. For example, an allele *g* may, on the average, contribute 2 centimeters in height to a plant, and the allele G may contribute 4 centimeters. In this case, the gg homozygote would contribute 2 + 2 = 4 centimeters in height, the Gg heterozygote would contribute 2 + 4 = 6 centimeters in height, and the GG homozygote would contribute 4 + 4 = 8 centimeters in height. To determine the genetic contribution to height, we would then add the effects of alleles at this locus to the effects of alleles at other loci that might influence the phenotype. Genes such as these are said to have additive effects, and this type of genetic variation is called **additive genetic variance**. The genes that determine kernel color in wheat (see p. 515) are additive in this way. The phenotypic variance arising from the additive effects of genes is the additive genetic variance and is symbolized by V_A.

Some genes may exhibit dominance, and this constitutes another source of genetic variance, the **dominance variance** (V_D). If dominance is present, the individual effects of the alleles are not strictly additive; we must also consider how alleles at a locus interact. In the presence of dominance, the heterozygote Gg would contribute 8 centimeters in height to the phenotype, the same amount as the GG homozygote. Thus, a population consisting of both genotypes would have genetic variation, but there would be no corresponding phenotypic variation. As the degree of dominance diminishes, the genotypic differences more clearly become phenotypic differences as the dominance variance turns into additive genetic variance.

The presence of epistasis adds another source of genetic variation, called epistatic or **interaction variance** (V_I). So we can partition the genetic variance as follows:

$$V_G = V_A + V_D + V_I$$

and the total phenotypic variance can then be summarized as

$$V_P = V_A + V_D + V_I + V_E + \text{COV}_{G\times E} + V_{G\times E}$$

It is very difficult to design experiments that can analyze all of these components simultaneously, and assumptions about some usually have to be made. For example, it is often assumed that there is no genetic-environmental interaction ($V_{G\times E}$), but the well-trained geneticist would always remember that the results of such an experiment must be presented with appropriate caution.

Broad-Sense and Narrow-Sense Heritability

Geneticists are interested in how much of the phenotypic variance V_P can be attributed to genetic variance V_G. This quantity, the proportion of the phenotypic variance that consists of genetic variance, is called the **broad-sense heritability** and is expressed as follows:

$$\text{Broad-sense heritability} = h_B^2 = \frac{V_G}{V_P}$$

where h^2 symbolizes heritability.

The heritability of a trait can range from 0 to 1. A broad-sense heritability of 0 indicates that none of the variation in phenotype among individuals results from genetic differences. A heritability of 0.5 means that 50 percent of the phenotypic variation arises from genetic differences among individuals, and a heritability of 1 would suggest that all the phenotypic variance is genetically based. Broad-sense heritability includes genetic variation from all types of genes. It ignores the fact that the genetic variance may be of the additive, dominance, or interactive sort. It also assumes that genetic-environmental interaction ($V_{G\times E}$) is not important. Thus, its usefulness is questionable.

Because additive genetic variance allows one to make accurate predictions about the resemblance between offspring and parents, quantitative geneticists frequently determine the proportion of the phenotypic variance that results from additive genetic variance, a quantity referred to as the **narrow-sense heritability**. The additive genetic variance is also that variation that responds to selection in a predictable way, and thus the narrow-sense heritability provides information about how a trait will evolve. The narrow-sense heritability is expressed as follows:

$$\text{Narrow-sense heritability} = h_N^2 = \frac{V_A}{V_P}$$

Understanding Heritability

Despite their utility, heritability estimates have a number of significant limitations. Unfortunately, these limitations are often ignored. As a result, heritability is one of the most misunderstood and widely abused concepts in all of genetics. The following are some of the important qualifications and limitations of heritability.

1. Heritability does not indicate the extent to which a trait is genetic. What it does measure is the *proportion of the phenotypic variance* among individuals in a population that results from genetic differences. The "proportion of the phenotypic variance among individuals in a population that results from genetic differences" may seem like the same thing as "the extent to which a trait is genetic," but these two statements actually mean quite different things. Genes often influence the development of a trait, and thus the trait may be said to be genetic.

However, the differences in phenotype among individuals may not be genetic at all.

2. Heritability is based on the variance, which can only be calculated for a group of individuals. Thus, heritability is characteristic of a *population* and not an individual.

3. Heritability is not fixed for a trait. Thus, there is no universal heritability for a trait like human stature. Rather, the heritability value for a trait depends on the genetic makeup and the specific environment of the population.

4. Even if heritability is high in each of two populations and the populations differ markedly in a particular trait, one cannot assume that the populations are genetically different. Therefore, heritability cannot be used to draw conclusions about the nature of differences between populations.

5. Traits shared by members of the same family do not necessarily have high heritability. When members of the same family share a trait, the trait is said to be a **familial trait**. Familial traits (as mentioned earlier) may arise because family members share genes or because they are exposed to the same environmental factors. Thus, familiality is not the same as heritability.

Heritability values for a number of traits in different species are given in Table 22.5. These heritability values are based upon various populations and have been determined using a variety of methods. Heritability of human traits, for example, is often studied by comparing identical and fraternal twins. Note that estimates of heritability are rarely precise, and most measured heritability values have large standard errors. Therefore, heritability values calculated for human traits must be viewed with special caution given the difficulties of separating genetic and environmental influences.

KEYNOTE

The broad-sense heritability of a trait represents the proportion of the phenotypic variance that results from genetic differences among individuals. The narrow-sense heritability is more restricted—it measures the proportion of the phenotypic variance that results from additive genetic variance. Narrow-sense heritability allows quantitative geneticists to make predictions about the resemblance between parents and offspring, and it represents that part of the phenotypic variance that responds to natural or artificial selection in a predictable manner.

RESPONSE TO SELECTION

Quantitative genetics has played a key role in plant and animal breeding, and in evolutionary biology. Both fields are concerned with genetic change within groups of organisms. In the case of plant and animal breeding, genetic change can lead to improvement in yield, hardiness, size, and other agriculturally important qualities; in the case of evolutionary biology, genetic change occurs in natural populations as a result of the processes discussed in Chapter 21. **Evolution** can be defined as genetic change that takes place over time within a group of organisms. Therefore, both evolutionary biologists and plant and animal breeders are interested in the process of evolution, and both use the methods of quantitative genetics to predict the rate and magnitude of genetic change.

The essential element of natural selection is that individuals with certain genotypes leave more offspring than others. In this way, groups of individuals change or evolve over time and become better adapted to their particular environment. Humans bring about evolution in domestic plants and animals through the similar process of **artificial selection**. In artificial selection, humans, not nature, select the individuals that are to survive and reproduce. If the selected traits have a genetic basis, then the genetic structure of the selected population will change over time and evolve. Artificial selection can be a powerful tool in bringing about rapid evolutionary change, as evidenced by the extensive variation observed among domesticated plants and animals. For example, all breeds of domestic dogs are derived from one species that was domesticated some 10,000 years ago. The large number of breeds that exist today, encompassing a tremendous variety of sizes, shapes, colors, and even behaviors, has been produced by artificial selection and selective breeding.

Both natural selection, as described by Charles Darwin, and artificial selection, practiced by plant and animal breeders, depend on the presence of genetic variation. Only if genetic variation exists within a population can that population change genetically and evolve. Furthermore, the amount and type of genetic variation is extremely important in determining how fast evolution will occur. Both evolutionary biologists and breeders are interested in the question of how much genetic variation for a particular trait exists within a population. As we have seen, quantitative genetics is often employed to answer this question.

How can we estimate the response to selection? When natural or artificial selection is imposed on a phenotype, the phenotype will change from one generation to the next, provided that there is genetic

~ TABLE 22.5

Narrow-Sense Heritability Values for Some Traits in Humans, Domesticated Animals, and Natural Populations[a]

ORGANISM	TRAIT	HERITABILITY
Humans	Stature	0.65
	Serum immunoglobulin (IgG) level	0.45
	General intelligence	0.52
	Scholastic achievement	0.38
	Good memory	0.22
	Verbal reasoning ability	0.50
	Spelling ability	0.53
	Aptitude for science	0.34
	Aptitude for history and literature	0.45
	Foot tapping speed	0.50
	Spina bifida (spinal cord not enclosed)	0.60
	Cleft lip and palate	0.76
Cattle	Milk yield	0.35
	Butterfat content	0.40
	Body weight	0.65
Pigs	Back-fat thickness	0.70
	Litter size	0.05
Poultry	Egg weight	0.50
	Egg production (to 72 weeks)	0.10
	Body weight (at 32 weeks)	0.55
Mice	Body weight	0.35
Drosophila	Abdominal bristle number	0.50
Jewelweed	Germination time	0.29
Milkweed bugs	Wing length (females)	0.87
	Fecundity (females)	0.50
Spring peeper (frog)	Size at metamorphosis	0.69
Wood frog	Development rate (mountain population)	0.31
	Size at metamorphosis (mountain population)	0.62

[a]The estimates given in this table apply to particular populations in particular environments; heritability values for other populations may differ.

variation for the trait in the population. The amount the phenotype changes in one generation is termed the **selection response**. To illustrate the concept of selection response, suppose a geneticist wishes to produce a strain of *Drosophila melanogaster* with large body size. The geneticist would first examine flies from a genetically diverse population and would measure the body size of these *unselected flies*. Suppose that our geneticist found the mean body weight of the unselected flies to be 1.3 mg. After determining the mean body weight of this population, the geneticist would select flies with large bodies (assume that the mean body weight of these selected flies was 3.0 mg). She would then place the large, selected flies in a separate culture vial and allow them to interbreed. After the F_1 offspring of these selected parents emerged, the geneticist would measure the body weights of the F_1 flies and compare them with the body weights of the original, unselected population.

What our geneticist has done in this procedure is to apply selection for large body size to the population of fruit flies. If genetic variation underlies the variation in body size of the original population, the offspring of the selected flies will resemble their parents, and the mean body size of the F_1 generation will be greater than the mean body size of the original population. If the F_1 flies have a mean body weight of 2.0 mg, which is considerably larger than the mean body weight of 1.3 mg observed in the original, unselected population, a response to selection has occurred.

The amount of change that occurs in one generation, or the selection response, depends on two things: the narrow-sense heritability and the **selection differential**. The selection differential is the difference between the mean phenotype of the selected parents and the mean phenotype of the unselected population. In our example of body size in fruit flies, the original population had a mean weight of 1.3 mg, and the mean weight of the selected parents was 3.0 mg, so the selection differential is 3.0 mg − 1.3 mg = 1.7 mg. The selection response is related to the selection differential and the heritability by the following formula:

$$\text{Selection response} = \text{narrow-sense heritability} \times \text{selection differential}$$

When the geneticist applied artificial selection to body size in fruit flies, the difference between mean body weights of the F_1 flies and the original population was 2.0 mg − 1.3 mg = 0.7 mg, which is the response to selection. We now have values for two of the three parameters in the above equation: the selection response (0.7 mg) and the selection differential (1.7 mg). By rearranging the formula for the selection response, we can solve for the narrow-sense heritability:

$$\text{Narrow-sense heritability} = h_N^2 = \text{selection response}/\text{selection differential}$$

$$h_N^2 = 0.7 \text{ mg}/1.7 \text{ mg} = 0.41$$

Measuring the response to selection provides another means for determining the narrow-sense heritability, and heritabilities for many traits are determined in this way.

~ **FIGURE 22.7**

Selection for phototaxis in *Drosophila pseudoobscura*. The upper graph is the line selected for avoidance of light. The lower graph is the line selected for attraction to light. The phototactic score is the number of times the fly moved toward the light out of a total of 15 light-dark choices.

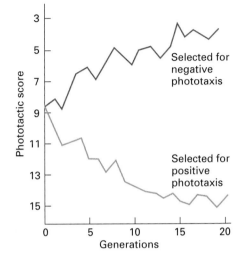

A trait will continue to respond to selection as long as genetic variation for the trait remains within the population. The results from an actual, long-term selection experiment on phototaxis in *Drosophila pseudoobscura* are presented in Figure 22.7. Phototaxis is a behavioral response to light. In this study, flies were scored for the number of times the fly moved toward light in a total of 15 light-dark choices. Two different response-to-selection experiments were carried out. In one, attraction to light was selected, and in the other, avoidance of light was selected. As can be seen in Figure 22.7, the fruit flies responded to selection for positive and negative phototactic behavior for a number of generations. Eventually the response to selection tapered off, and finally no further directional change in phototactic behavior occurred. One possible reason for this lack of response in later generations is that no more genetic variation for phototactic behavior existed within the population. In other words, all flies at this point were homozygous for all the alleles affecting the behavior. If this were the case, phototactic behavior could not undergo further evolution in this population without input of additional genetic variation. More often, some variation still exists for the trait, even after the selection response levels off, but the population fails to respond to selection because the genes for the selected trait have detrimental effects on other traits. These detrimental effects occur because the phenotypes of two or more traits may be associated; this means that the traits do not vary independently. (As an example, fair skin, blond hair, and blue eyes are often found together in the same individual.)

KEYNOTE

The amount a trait changes in one generation as a result of selection for the trait is called the selection response or the response to selection. The magnitude of selection response depends on both the intensity of selection, what is termed the selection differential, and the narrow-sense heritability.

SUMMARY

Quantitative genetics is the field that studies the inheritance of continuous, or quantitative traits—those traits with a range of phenotypes. Continuous traits usually result from the influence of multiple genes and environmental factors. Statistics such as the mean, variance, and standard deviation can be used to describe continuous traits.

Polygenic traits are caused by genes at multiple loci, each of which follows the principles of Mendelian inheritance. The multiple gene hypothesis assumes that the effects of the alleles at a locus are small and additive, but at present this is at best a working model and much more needs to be known.

The broad-sense heritability is the proportion of the phenotypic variance in a population that is due to genetic differences. The narrow-sense heritability indicates the proportion of the phenotypic variance that results from additive genetic variance.

The response to selection is the amount a trait changes in one generation as a result of selection; it depends on the narrow-sense heritability and the selection differential. Evolutionary responses that result from the effects of natural selection depend on heritability in the narrow sense.

The field of quantitative genetics strives to clarify the relationship between a complex genetic architecture and a complex phenotypic architecture. Are most continuous traits controlled by many genes with small effects, or are they controlled by relatively few genes with major effects? How important are epistatic interactions and pleiotropic effects? These are not easy questions to answer. But in the answers will lie findings that will affect the way we view the relationship between genotype and phenotype.

ANALYTICAL APPROACHES FOR SOLVING GENETICS PROBLEMS

Q22.1 Assume that genes *A*, *B*, *C*, and *D* are members of a multiple-gene series that controls a quantitative trait. Each gene has a duplicate, cumulative effect in that each contributes 3 cm of height to the organism when it is present. Each gene assorts independently. In addition, gene *L* is always present in the homozygous state, and the *LL* genotype contributes a constant 40 cm of height. The alleles *a*, *b*, *c*, and *d* do not contribute anything to the height of the organism. If we ignore height variation caused by environmental factors, an organism with genotype *AA BB CC DD LL* would be 64 cm high, and one with genotype *aa bb cc dd LL* would be 40 cm high. A cross is made of *AA bb CC DD LL* × *aa BB cc DD LL* and is carried into the F_2 by selfing of the F_1.

a. How does the size of the F_1 individuals compare with the size of each parent?

b. Compare the mean of the F_1 with the mean of the F_2, and comment on your findings.

c. What proportion of the F_2 population would show the same height as the *AA bb CC DD LL* parent?

d. What proportion of the F_2 population would show the same height as the *aa BB cc DD LL* parent?

e. What proportion of the F_2 population would breed true for the height shown by the *aa BB cc DD LL* parent?

f. What proportion of the F_2 population would breed true for the height characteristic of F_1 individuals?

A22.1 This question explores our understanding of the basic genetics involved in a multiple-gene series that in this case controls a quantitative trait. The approach we take is essentially the same as the approach used with a series of independently assorting genes that control distinctly different traits. That is, we make predictions on the basis of genotypes and relate the results to phenotypes, or we make predictions on the basis of phenotypes and relate the results to genotypes.

a. Each allele represented by a capital letter contributes 3 cm of height to the base height of 40 cm, which is controlled by the ever-present *LL* homozygosity. Therefore, the *AA bb CC DD LL* parent, which has six capital-letter alleles from the *A*-through-*D*, multiple-gene series, is 40 + (6 × 3) = 58 cm high. Similarly, the *aa BB cc DD LL* parent has four capital-letter alleles and therefore is 40 + 12 = 52 cm high. The F_1

from a cross between these two individuals would be heterozygous for the *A*, *B*, and *C* loci and homozygous for *D* and *L*, that is, *Aa Bb Cc DD LL*. This progeny has five capital-letter alleles apart from *LL* and hence is 40 + 15 = 55 cm high.

b. The F_2 is derived from a self of the *Aa Bb Cc DD LL* F_1. All the F_2 individuals will be *DD LL*, making them at least 40 + 6 = 46 cm high. Now we must deal with the heterozygosity at the other three loci. What we need to calculate is the relative proportions of individuals with all the various possible numbers of capital-letter alleles. This calculation is equivalent to determining the relative distribution of three independently assorting traits, each showing incomplete dominance. In other words, we must calculate directly the relative frequencies of all possible genotypes for the three loci, and collect those with no, one, two, three, four, five, and six capital-letter alleles. Thus, the probability of getting an individual with two capital-letter alleles for each locus is 1/4, the probability of getting an individual with one capital-letter allele for each locus is 1/2, and the probability of getting an individual with no capital-letter alleles for each locus is 1/4. So the probability of getting an F_2 individual with six capital-letter alleles for the *A*, *B*, and *C* loci is $(1/4)^3 = 1/64$, and the same probability is obtained for an individual with no capital-letter alleles. This analysis gives us a clue about how we should consider all the possible combinations of genotypes that have the other numbers of capital-letter alleles. That is, the simplest approach is to compute the coefficients in the binomial expansion of $(a + b)^6$, as explained on pp. 515–516. The expansion gives a 1:6:15:20:15:6:1 distribution of zero, one, two, three, four, five, and six capital-letter alleles, respectively. Now, since each capital-letter allele in the *A*, *B*, and *C* set contributes 3 cm of height over the 46-cm height given by the *DD LL* genotype common to all, the F_2 individuals would fall into the following distribution:

NUMBER OF CAPITAL-LETTER ALLELES	HEIGHT ADDED TO BASIC HEIGHT OF 46 CM FOR COMMON *DD LL* GENOTYPE (cm)	HEIGHT OF INDIVIDUALS (cm)	FREQUENCY
6	18	64	1
5	15	61	6
4	12	58	15
3	9	55	20
2	6	52	15
1	3	49	6
0	0	46	1

The distribution is clearly symmetrical, giving an average of 55 cm, the same height shown in F_1 individuals.

c. The *AA bb CC DD LL* parent was 58 cm, so we can read the proportion of F_2 individuals that show this same height directly from the table in part b. The answer is 15/64.

d. The *aa BB cc DD LL* parent was 52 cm, and from the table in part b the proportion of F_2 individuals that show this same height is 15/64.

e. We are asked to determine the proportion of the F_2 population that would breed true for the height shown by the *aa BB cc DD LL* parent, which was 52 cm. To breed true, the organism must be homozygous. We have also established that *DD LL* is a constant genotype for the F_2 individuals, giving a basic height of 46 cm. Therefore, for a height of 52 cm, two additional, active, capital-letter alleles must be present apart from those at the *D* and *L* loci. With the requirement for homozygosity, there are only three genotypes that give a 52-cm height; they are *AA bb cc DD LL*, *aa BB cc DD LL*, and *aa bb CC DD LL*. The probability of each combination occurring in the F_2 is 1/64, so the answer to the problem is 1/64 + 1/64 + 1/64 = 3/64. (Note that the individual probability for each genotype may be calculated. That is, probability of *AA* = 1/4, probability of *bb* = 1/4, probability of *cc* = 1/4, and probability of *DD LL* = 1, giving an overall probability for *AA bb cc DD LL* of 1/64.)

f. We are asked to determine the proportion of the F_2 population that would breed true for the height characteristic of F_1 individuals. Again, the basic height given by *DD LL* is 46 cm. The F_1 height is 55 cm, so three capital-letter alleles must be present in addition to *DD LL* to give that height, since (3×3) cm = 9 cm, and 9 cm + 46 cm = 55 cm. However, since an individual must be homozygous to be true breeding, the answer to this question is none, since 3 is an odd number, meaning that at least one locus must be heterozygous in order to get the 55-cm height.

Q22.2 Five field mice collected in Texas had weights of 15.5 g, 10.3 g, 11.7 g, 17.9 g, and 14.1 g. Five mice collected in Michigan had weights of 20.2 g, 21.2 g, 20.4 g, 22.0 g, and 19.7 g. Calculate the mean weight and the variance in weight for mice from Texas and for mice from Michigan.

A22.2 To answer this question, we use the formulas given in the section on statistics. The formula for the mean is

$$\bar{x} = \frac{\Sigma x_i}{n}$$

The symbol Σ means to add, and the x_i represents all the individual values. We begin by summing up all the weights of the mice from Texas:

$$\Sigma x_i = 15.5 + 10.3 + 11.7 + 17.9 + 14.1 = 69.5$$

Next, we divide this summation by n, which represents the number of values added together. In this case, we added together five weights, so $n = 5$. The mean for the Texas mice is therefore

$$\frac{\Sigma x_i}{n} = \frac{69.5}{5} = 13.9$$

To calculate the variance in weight among the Texas mice, we utilize the formula

$$s^2 = \frac{\Sigma(x_i - \bar{x})^2}{n - 1}$$

We must take each individual weight and subtract it from the mean weight of the group. Each value obtained from this subtraction is then squared, and all squared values are added up, as shown below.

$15.5 - 13.9 =$	1.6	$(1.6)^2 =$	2.56
$10.3 - 13.9 =$	-3.6	$(-3.6)^2 =$	12.96
$11.7 - 13.9 =$	-2.2	$(-2.2)^2 =$	4.84
$17.9 - 13.9 =$	4.0	$(4.0)^2 =$	16.0
$14.1 - 13.9 =$	0.2	$(0.2)^2 =$	0.04
			36.4

The sum of all the squared values is 36.4. All that remains for us to do is to divide this sum by $n - 1$, which is $5 - 1 = 4$.

$$s^2 = \frac{\Sigma(x_i - \bar{x})^2}{n - 1} = \frac{36.4}{4} = 9.1$$

The mean and the variance for the Texas mice are therefore 13.9 and 9.1, respectively.

We now repeat these steps for the mice from Michigan.

$$\Sigma x_i = 20.2 + 21.2 + 20.4 + 22.0 + 19.7 = 103.5$$

$$\frac{\Sigma x_i}{n} = \frac{103.5}{5} = 20.7$$

$$s^2 = \frac{\Sigma(x_i - \bar{x})^2}{n - 1}$$

$20.2 - 20.7 =$	-0.5	$(-0.5)^2 =$	0.25
$21.2 - 20.7 =$	0.5	$(0.5)^2 =$	0.25
$20.4 - 20.7 =$	-0.3	$(-0.3)^2 =$	0.9
$22.0 - 20.7 =$	1.3	$(1.3)^2 =$	1.69
$19.7 - 20.7 =$	-1.0	$(-1.0)^2 =$	1.0
			4.09

$$s^2 = \frac{\Sigma(x_i - \bar{x})^2}{n - 1} = \frac{4.09}{4} = 1.023$$

The mean and the variance for the Michigan mice are 20.7 and 1.023, respectively.

We conclude that the Michigan mice are considerably heavier than the Texas mice and that the Michigan mice also exhibit less variance in weight.

QUESTIONS AND PROBLEMS

*22.1 The following measurements of head width and wing length were made on a series of steamer-ducks:

SPECIMEN	HEAD WIDTH (cm)	WING LENGTH (cm)
1	2.75	30.3
2	3.20	36.2
3	2.86	31.4
4	3.24	35.7
5	3.16	33.4
6	3.32	34.8
7	2.52	27.2
8	4.16	52.7

Calculate the mean and the standard deviation of head width and of wing length for these eight birds.

22.2 Answer the following questions:
a. In a family of six children, what is the probability that three will be girls and three will be boys?
b. In a family of five children, what is the probability that one will be a boy and four will be girls?
c. What is the probability that in a family of six children, all will be boys?

*22.3 In flipping a coin, there is a 50 percent chance of obtaining heads and a 50 percent chance of obtaining tails on each flip. If you flip a coin 10 times, what is the probability of obtaining exactly five heads and five tails?

*22.4 Two pure-breeding parents that differ in a size character are crossed. The F_1 generation is no more variable in size than the parents. Explain.

22.5 If two pure-breeding strains, differing in a size trait, are crossed, is it possible for F_2 individuals to have phenotypes that are more extreme than either grandparent (that is, be larger than the largest or smaller than the smallest in the parental generation)? Explain.

22.6 Two pairs of genes with two alleles each, A/a and B/b, determine plant height additively in a population. The homozygote $AA\ BB$ is 50 cm tall, the homozygote $aa\ bb$ is 30 cm tall.

a. What is the F_1 height in a cross between the two homozygous stocks?

b. What genotypes in the F_2 will show a height of 40 cm after an $F_1 \times F_1$ cross?

c. What will be the F_2 frequency of the 40-cm plants?

*22.7 Three independently segregating genes (*A*, *B*, *C*), each with two alleles, determine height in a plant. Each capital-letter allele adds 2 cm to a base height of 2 cm. The environment does not affect this trait.

a. What are the heights expected in the F_1 progeny of a cross between homozygous strains *AA BB CC* (14 cm) $\times$ *aa bb cc* (2 cm)?

b. What distribution of heights (frequency and phenotype) is expected in an $F_1 \times F_1$ cross?

c. What proportion of F_2 plants will have heights equal to the heights of the original two parental strains?

d. What proportion of the F_2 will breed true for the height shown by the F_1?

22.8 Repeat Problem 22.7, but assume that each capital-letter allele acts to double the existing height; for example, *Aa bb cc* = 4 cm, *AA bb cc* = 8 cm, *AA Bb cc* = 16 cm, and so on.

22.9 Assume that three equally and additively contributing pairs of alleles control flower length in nasturtiums. A completely homozygous plant with 10-mm flowers is crossed to a completely homozygous plant with 30-mm flowers. F_1 plants all have flowers about 20 mm long. F_2 plants show a range of lengths from 10 to 30 mm, with about 1/64 of the F_2 having 10-mm flowers and 1/64 having 30-mm flowers. What distribution of flower length would you expect to see in the offspring of a cross between an F_1 plant and the 30-mm parent?

*22.10 In an experiment, the mean internode length in spikes of the barley variety *asplund* was found to be 2.12 mm. In the variety *abed binder* the mean internode length was found to be 3.17 mm. The mean of the F_1 of a cross between the two varieties was approximately 2.7 mm. The F_2 gave a continuous range of variation from one parental extreme to the other. Analysis of the F_3 generation showed that in the F_2, 8 out of the total 125 individuals were of the *asplund* type, giving a mean of 2.19 mm. Eight other individuals were similar to the parent *abed binder*, giving a mean internode length of 3.24 mm. Is the internode length in spikes of barley a discontinuous or a quantitative (continuous) trait? Why?

22.11 From the information given in Problem 22.10, determine how many gene pairs involved in determination of internode length are segregating in the F_2.

22.12 Assume that the difference between a type of oats yielding about 4 g per plant and a type yielding 10 g is the result of three equal and cumulative multiple-gene pairs *AA BB CC*. If you cross the type yielding 4 g with the type yielding 10 g, what will be the phenotypes of the F_1 and the F_2? What will be their distributions?

*22.13 Assume that in squashes the difference in fruit weight between a 3-lb type and a 6-lb type is due to three allelic pairs, *A/a*, *B/b*, and *C/c*. Each capital-letter allele contributes a half pound to the weight of the squash. From a cross of a 3-lb plant (*aa bb cc*) with a 6-lb plant (*AA BB CC*), what will be the phenotypes of the F_1 and the F_2? What will be their distributions?

22.14 Refer to the assumptions stated in Problem 22.13. Determine the range in fruit weight of the offspring in the following squash crosses: (a) *AA Bb CC* $\times$ *aa Bb Cc*; (b) *AA bb Cc* $\times$ *Aa BB cc*; (c) *aa BB cc* $\times$ *AA BB cc*.

*22.15 Assume that the difference between a corn plant 10 dm (decimeters) high and one 26 dm high is due to four pairs of equal and cumulative multiple alleles, with the 26-dm plants being *AA BB CC DD* and the 10-dm plants being *aa bb cc dd*.

a. What will be the size and genotype of an F_1 from a cross between these two true-breeding types?

b. Determine the limits of height variation in the offspring from the following crosses:

(1) *Aa BB cc dd* $\times$ *Aa bb Cc dd*;

(2) *aa BB cc dd* $\times$ *Aa Bb Cc dd*;

(3) *AA BB Cc DD* $\times$ *aa BB cc Dd*;

(4) *Aa Bb Cc Dd* $\times$ *Aa bb Cc Dd*.

22.16 Refer to the assumptions given in Problem 22.15. But for this problem, two 14-dm corn plants, when crossed, give nothing but 14-dm offspring (case A). Two other 14-dm plants give one 18-dm, four 16-dm, six 14-dm, four 12-dm, and one 10-dm offspring (case B). Two other 14-dm plants, when crossed, give one 16-dm, two 14-dm, and one 12-dm offspring (case C). What genotypes for each of these 14-dm parents (cases A, B, and C) would explain these results? Would it be possible to get a plant taller than 48 dm by selection in any of these families?

*22.17 Pigmentation in the imaginary river-bottom dweller *Mucus yuccas* is a quantitative character controlled by a set of five independently segregating polygenes with two alleles each: *A/a*, *B/b*, *C/c*, *D/d*, and *E/e*. Pigment is deposited at three different levels, dependent on the threshold of gene products produced by the capital-lettered alleles. That is, greyish-brown pigmentation is seen if at least four capital alleles are present, light-tan pigmentation is seen if two or three capital alleles are present, and whitish-blue pigmentation is seen if these thresholds are not met. If an *AA BB CC DD EE* animal is crossed to a true-breeding *aa bb cc dd ee* animal and the progeny are selfed, what kinds of phenotypes are expected in the F_1 and F_2?

*22.18 Since monozygotic twins share all of their genetic material and dizygotic twins share, on average, half of their genetic material, twin studies can sometimes be useful for evaluating the genetic contribution to a trait. Consider the following two instances.

An intelligence quotient (IQ) assesses intellectual performance on a standardized test that involves reasoning ability, memory, and knowledge of an individual's language and culture. IQ scores are transformed so that the population mean score is 100, and 95 percent of the individuals have scores in the range between 70 and 130. Observations in the United States and England found that monozygotic twins had an average difference of 6 IQ points, dizygotic twins had an average difference of 11 points, and random pairs of individuals had an average difference of 21 points.

In a large sample of pairs of twins in the United States, where one twin was a smoker, 83 percent of monozygotic twins both smoked, while 62 percent of dizygotic twins both smoked.

From these data, can you infer the genetic determination of IQ or smoking?

22.19 A quantitative geneticist determines the following variance components for leaf width in a population of wildflowers growing along a roadside in Kentucky:

Additive genetic variance (V_A)	= 4.2
Dominance variance (V_D)	= 1.6
Interaction variance (V_I)	= 0.3
Environmental variance (V_E)	= 2.7
Genetic-environmental interaction variance ($V_{G \times E}$)	= 0.0

a. Calculate the broad-sense heritability and the narrow-sense heritability for leaf width in this population of wildflowers.
b. What do the heritabilities obtained in part a indicate about the genetic nature of leaf width variation in this plant?

*22.20 Assume that all variance affecting seed weight in beans is genetically determined and is additive. From a population where the mean seed weight was 0.88 g, a farmer selected two seeds, each weighing 1.02 g. The farmer planted these and crossed the resulting plants to each other, then collected and weighed their seeds. The mean weight of their seeds was 0.96 g. What is the narrow-sense heritability of seed weight?

*22.21 Members of the inbred rat strain SHR are salt sensitive: they respond to a high salt environment by developing hypertension. Members of a different inbred rat strain, TIS, are not salt sensitive. Imagine you placed a population consisting only of SHR rats in an environment that was variable in regard to distribution of salt,

so that some rats would be exposed to more salt than others. Discuss the heritability of blood pressure in this population.

*22.22 In Kansas a farmer is growing a variety of wheat called TK138. He calculates the narrow-sense heritability for yield (the amount of wheat produced per acre) and finds that the heritability of yield for TK138 is 0.95. The next year he visits a farm in Poland and observes that a Russian variety of wheat growing there, UG334, has only about 40 percent as much yield as TK138 grown on his farm in Kansas. Since he found the heritability of yield in his wheat to be very high, he concludes that the American variety of wheat (TK138) is genetically superior to the Russian variety (UG334), and he tells the Polish farmers that they can increase their yield by using TK138. What is wrong with his conclusion?

*22.23 We wish to determine the narrow-sense heritability of tail length in mice. We measure tail length among the mice of a population and find a mean tail length of 9.7 cm. We then select the 10 mice in the population with the longest tails; mean tail length in these selected mice is 14.3 cm. We interbreed the mice with the long tails and examine tail length in their progeny. The mean tail length in the F_1 progeny of the selected mice is 13 cm.

Calculate the selection differential, the response to selection, and the narrow-sense heritability for tail length in these mice.

*22.24 The narrow-sense heritability of egg weight in a particular flock of chickens is 0.60. A farmer selects for increased egg weight in this flock. The difference in the mean egg weight of the unselected chickens and the selected chickens is 10 g. How much should egg weight increase in the offspring of the selected chickens?

*22.25 The variances tabulated below were determined for measurements of body length, antenna bristle number, and egg production in a moth species. Which of these characters would be most rapidly changed by natural selection? Which character would be least affected by natural selection?

VARIANCE	BODY LENGTH	ANTENNA BRISTLE NUMBER	EGG PRODUCTION
Phenotypic (V_P)	798	342	145
Additive (V_A)	132	21	21
Dominance (V_D)	122	126	24
Interaction (V_I)	118	136	34
Genetic-environmental ($V_{G \times E}$)	81	23	21
Maternal effects (V_{Em})	345	36	45

GLOSSARY

−10 box (Pribnow box) A part of the promoter sequence in prokaryotic genomes that is located at about 10 base pairs upstream from the transcription starting point. The consensus sequence for the Pribnow box is TATAAT. The Pribnow box is often referred to as the TATA box.

acrocentric chromosome A chromosome with a centromere near the end such that it has one long arm plus a stalk and a satellite.

additive genetic variance (V_A) Genetic variance that arises from the additive effects of genes on the phenotype.

adenine (A) A purine base found in RNA and DNA; in double-stranded DNA, adenine pairs with the pyrimidine thymine.

allele One of two or more alternative forms of a single gene locus. Different alleles of a gene each have a unique nucleotide sequence, and their activities are all concerned with the same biochemical and developmental process, although their individual phenotypes may differ.

allelic frequencies The frequencies of alleles at a locus occurring among individuals in a population.

allelomorph (allele) A term coined by William Bateson; literally means "alternative form"; later shortened by others to *allele*.

allopolyploidy Polyploidy involving two or more genetically distinct sets of chromosomes.

alternation of generations The two distinct reproductive phases of green plants in which stages alternate between haploid cells and diploid cells (gametophyte cells and sporophyte cells).

Ames test A test developed by Bruce Ames in the early 1970s that investigates new or old environmental chemicals for carcinogenic effects. It uses the bacterium *Salmonella typhimurium* as a test organism for mutagenicity of compounds.

amino acids The building blocks of polypeptides. There are 20 different amino acids.

aminoacyl-tRNA A tRNA molecule covalently bound to an amino acid. This complex brings the amino acid to the ribosome so that it can be used in polypeptide synthesis.

aminoacyl-tRNA synthetase An enzyme that catalyzes the addition of a specific amino acid to a tRNA molecule. Since there are 20 amino acids, there are also 20 synthetases.

amniocentesis A procedure in which a sample of amniotic fluid is withdrawn from the amniotic sac of a developing fetus and cells are cultured and examined for chromosomal abnormalities.

anaphase The stage in mitosis or meiosis during which the sister chromatids (mitosis) or homologous chromosomes (meiosis) separate and migrate toward the opposite poles of the cell.

anaphase II The second stage of meiosis during which the centromeres (and therefore the chromatids) are pulled to the opposite poles of the spindle. The separated chromatids are now referred to as chromosomes in their own right.

aneuploidy The abnormal condition in which one or more whole chromosomes of a normal set of chromosomes either are missing or are present in more than the usual number of copies.

antibody A protein molecule that recognizes and binds to a foreign substance introduced into the organism.

anticodon A three-nucleotide sequence that pairs with a codon in mRNA by complementary base-pairing.

antigen Any large molecule that stimulates the production of specific antibodies or that binds specifically to an antibody.

applied research Research done with an eye toward making products that can be commercialized, or at least made available to humankind for practical benefit.

artificial selection Human determination as to which individuals will survive and reproduce. If the selected traits have a genetic basis, they will change and evolve.

asexual (vegetative) reproduction Reproduction in which a new individual develops either from a single cell or from a group of cells in the absence of any sexual process.

attenuation A regulatory mechanism in certain bacterial biosynthetic operons that controls gene expression by causing RNA polymerase to terminate transcription.

autonomously replicating sequences (ARSs) Specific sequences (e.g., in baker's yeast, *Saccharomyces cerevisiae*) that, when included as part of an extrachromosomal, circular DNA molecule, confer on that molecule the ability to replicate autonomously.

autopolyploidy Polyploidy involving more than two chromosome sets of the same species.

autosome A chromosome other than a sex chromosome.

auxotroph A mutant strain of a given organism that is unable to synthesize a molecule required for growth and therefore must have the molecule supplied in the growth medium in order for it to grow.

bacteria Spherical, rod-shaped, or spiral-shaped, single-cellular or multicellular, filamentous prokaryotic organisms.

bacterial artificial chromosome (BAC) Vectors for cloning large DNA fragments up to about 300 kb in *E. coli*. BACs are modified versions of the *E. coli* F factor that is responsible for conjugation.

balance model A hypothesis of genetic variation proposing that balancing selection maintains large amounts of genetic variation within populations.

Barr body A highly condensed mass of chromatin found in the nuclei of normal female cells, but not in the nuclei of normal male cells. It represents a cytologically condensed and inactivated X chromosome.

basal transcription factor Protein required for the initiation of transcription by a eukaryotic RNA polymerase.

base analog A chemical whose molecular structure is extremely similar to the bases normally found in DNA.

base-pair substitution mutation A change in a gene such that one base pair is replaced by another base pair; for instance, an AT pair is replaced by a GC pair.

basic research Research done to further knowledge for knowledge's sake.

bivalent A pair of homologous, synapsed chromosomes during the first meiotic division.

bottleneck effect A form of genetic drift that occurs when a population is drastically reduced in size. Some genes may be lost from the gene pool as a result of chance.

branch-point sequence The consensus sequence in mammalian cells, YNCURA, (where Y is a pyrimidine, R is a purine, and N is any base), to which the free 5' end of the intron loops and binds to the A nucleotide in the sequence during intron splicing.

broad-sense heritability A quantity representing the proportion of the phenotypic variance that consists of genetic variance.

C value The amount of DNA found in the haploid set of chromosomes.

CAAT box One of the eukaryotic promoter elements found in approximately 80 base pairs upstream of the initiation site, but it can function at a number of other locations and in either orientation with respect to the start point. The consensus sequence is 5'-GGCCAATCT-3'.

cancer Diseases characterized by the uncontrolled and abnormal division of eukaryotic cells and by the spread of the disease (metastasis) to disparate sites in the organism.

5'-capping The addition of a methylated guanine nucleotide (a "cap") to the 5' end of a premessenger RNA molecule; the cap is retained on the mature mRNA molecule.

catabolite repression (glucose effect) The inactivation of an inducible bacterial operon in the presence of glucose even though the operon's inducer is present.

cDNA DNA copies made from an RNA template catalyzed by the enzyme reverse transcriptase.

cDNA library The collection of molecular clones that contains cDNA copies of the entire mRNA population of a cell. *See also* **cDNA**.

cell cycle The cyclical process of growth and cellular reproduction in unicellular and multicellular eukaryotes. The cycle includes nuclear division, or mitosis, and cell division, or cytokinesis.

cell division A process whereby one cell divides to produce two cells.

cell-free, protein-synthesizing system A system, isolated from cells, that contains ribosomes, mRNA, tRNAs with amino acids attached, and all the necessary protein factors for the in vitro synthesis of polypeptides.

cellular oncogene (c-*onc*) The genes, present in a functional state in cancerous cells, that are responsible for the cancerous state.

centimorgan (cM) The map unit: named in honor of T. H. Morgan.

centromere (kinetochore) A specialized region of a chromosome seen as a constriction under the microscope. This region is important in the activities of the chromosomes during cellular division.

chain-terminating codon One of three codons for which no normal tRNA molecule exists with an appropriate anticodon. A nonsense codon in an mRNA specifies the termination of polypeptide synthesis.

character An observable phenotypic feature of the developing or fully developed organism that is the result of gene action.

charged tRNA The product of an amino acid added to a tRNA.

charging The act of adding an amino acid to the tRNA.

chiasma A cross-shaped structure formed during crossing-over and visible during the diplonema stage of meiosis.

chiasma interference (chromosomal interference) The physical interference caused by the breaking and rejoining of chromatids that reduces the probability of more than one crossing-over event occurring near another one in a portion of the meiotic tetrad.

chi-square (χ^2) test A statistical procedure that determines what constitutes a significant difference between observed results and results expected on the basis of a particular hypothesis; a goodness-of-fit test.

chloroplast The cellular organelle found only in green plants that is the site of photosynthesis in the cells containing it.

chorionic villus sampling A procedure in which a sample of chorionic villus tissue of a developing fetus is examined for chromosomal abnormalities.

chromatid One of the two visibly distinct longitudinal subunits of all replicated chromosomes that becomes visible between early prophase and metaphase of mitosis.

chromatin The piece of DNA-protein complex that is studied and analyzed. Each chromatin fragment reflects the general features of chromosomes but not the specifics of any individual chromosome.

chromosomal interference *See* **chiasma interference**.

chromosomal mutation The variation from the wild-type condition in either chromosome number or chromosome structure.

chromosome The physical structures in which the genetic material of the cell is organized. In bacteria, chromosomes may be linear or circular, and in archaea the chromosomes are circular. A few proteins are associated with these two types of chromosome. In eukaryotes, the genomic DNA is complexed with proteins and distributed among a number of linear chromosomes.

chromosome aberration *See* **chromosomal mutation**.

chromosome theory of inheritance The theory that the chromosomes are the carriers of the genes. The first clear formulation of the theory was made by both Sutton and Boveri, who independently recognized that the transmission of chromosomes from one generation to the next closely paralleled the pattern of transmission of genes from one generation to the next.

chromosome walking A process to identify adjacent clones in a genomic library. In chromosome walking, a piece of DNA is used to probe a genomic library to find an overlapping clone; then a piece of that clone is used as a probe to screen the library again for an overlapping clone; and so on.

cis-dominance The phenomenon of a gene or DNA sequence controlling only genes that are on the same contiguous piece of DNA.

cis-trans (complementation) test A test developed by E. Lewis, used to determine whether two different mutations are within the same cistron (gene).

classical model A hypothesis of genetic variation proposing that natural populations contain little genetic variation as a result of strong selection for one allele.

cloning The generation of many copies of a DNA molecule (e.g., a recombinant DNA molecule) by replication in a suitable host.

cloning vector (cloning vehicle) A double-stranded DNA molecule that is able to replicate autonomously in a host cell and with which a DNA fragment (or fragments) can be bonded to form a recombinant DNA molecule for cloning.

coding sequence The part of an mRNA molecule that specifies the amino acid sequence of a polypeptide during translation.

codominance The situation in which the heterozygote exhibits the phenotypes of both homozygotes.

codon A group of three adjacent nucleotides in an mRNA molecule that specifies either one amino acid in a polypeptide chain or the termination of polypeptide synthesis.

coefficient of coincidence A number that expresses the extent of chiasma interference throughout a genetic map; the ratio of observed double-crossover frequency to expected double-crossover frequency. Interference is equal to 1 minus the coefficient of coincidence.

complementary base-pairing The hydrogen bonding between a particular purine and a particular pyrimidine in double-stranded nucleic acid molecules (DNA–DNA, DNA–RNA, or RNA–RNA). The major specific pairings are guanine with cytosine and adenine with thymine or uracil.

complementary DNA *See* **cDNA.**

complementation test *See* **cis-trans test.**

complete dominance The case in which one allele is dominant to the other so that at the phenotypic level the heterozygote is essentially indistinguishable from the homozygous dominant.

complete recessiveness The situation in which an allele is phenotypically expressed only when it is homozygous.

concerted evolution (molecular drive) A poorly understood evolutionary process that produces uniformity of sequence in multiple copies of a gene.

conjugation A process having a unidirectional transfer of genetic information through direct cellular contact between a donor ("male") and a recipient ("female") bacterial cell.

consensus sequence The sequence indicating which nucleotide is found most frequently at each position.

conservative model A DNA replication scheme in which the two parental strands of DNA remain together and serve as a template for the synthesis of a new daughter double helix.

constitutive heterochromatin Condensed chromatin that is always genetically inactive and is found at homologous sites on chromosome pairs.

continuous (quantitative) traits Traits that show a continuous variation in phenotype over a range.

contributing allele An allele that contributes to the phenotype.

controlling site A specific sequence of nucleotide pairs adjacent to the gene where the transcription of a gene occurs in response to a particular molecular event.

coordinate induction The simultaneous transcription and translation of two or more genes brought about by the presence of an inducer.

core enzyme The portion of the *E. coli* RNA polymerase that is the active enzyme and can be written as $\alpha_2\beta\beta'\omega$.

cotransduction The simultaneous transduction of two or more bacterial genes; a good indication that the bacterial genes are closely linked.

cotranslational transport The movement of a protein into the ER simultaneously with its synthesis.

coupling An arrangement in which the two wild-type alleles are on one homologous chromosome and the two recessive mutant alleles are on the other.

crisscross inheritance A type of gene transmission passed from a male parent to a female child to a male grandchild.

cross *See* **cross-fertilization.**

cross-fertilization (cross) A term used for the fusion of male and female gametes from different individuals; the bringing together of genetic material from different individuals for the purpose of genetic recombination.

crossing-over A term introduced by Morgan and E. Cattell, in 1912, to describe the process of reciprocal chromosomal interchange by which recombinants arise.

cytosine (C) A pyrimidine base found in RNA and DNA. In double-stranded DNA, cytosine pairs with the purine guanine.

dark repair *See* **excision repair.**

Darwinian fitness The relative reproductive ability of a genotype.

degeneracy A multiple coding; more than one codon per amino acid.

degradation control Regulation of the RNA breakdown rate in the cytoplasm.

deletion (deficiency) A chromosomal mutation resulting in the loss from a chromosome of a segment of the genetic material and the genetic information contained therein.

deoxyribonuclease (DNase) An enzyme that catalyzes the degradation of DNA to nucleotides.

deoxyribonucleic acid (DNA) A polymeric molecule consisting of deoxyribonucleotide building blocks that in a double-stranded, double-helical form is the genetic material of most organisms.

deoxyribonucleotide The basic building block of DNA, consisting of a sugar (deoxyribose), a base, and a phosphate.

deoxyribose The pentose (5-carbon) sugar found in DNA.

development The process of regulated growth that results from the interaction of the genome with cytoplasm and the environment. It involves a programmed sequence of phenotypic events that are typically irreversible.

diakinesis The stage that follows diplonema and during which the four chromatids of each tetrad are most condensed.

dicentric bridge *See* **dicentric chromosome.**

dicentric chromosome A chromosome with two centromeres. For example, as a result of the crossover between genes B and C in the inversion loop, one recombinant chromatid becomes stretched across the cell as the two centromeres begin to migrate, forming a dicentric bridge.

dideoxynucleotide A modified nucleotide that has a 3'-H on the deoxyribose sugar rather than a 3'-OH. If a dideoxynucleoside triphosphate (ddNTP) is used in a DNA synthesis reaction, the ddNTP can be incorporated into the growing chain. However, no further DNA synthesis can then occur because no phosphodiester bond can be formed with an incoming DNA precursor.

dideoxy sequencing A method of rapid sequencing of DNA molecules developed by Fred Sanger. This technique incorporates the use of dideoxy nucleotides in a DNA polymerase-catalyzed DNA synthesis reaction.

differentiation An aspect of development that involves the formation of different types of cells, tissues, and organs through the processes of specific regulation of gene expression.

dihybrid cross A cross between two dihybrids of the same type. Individuals that are heterozygous for two pairs of alleles at two different loci are called dihybrid.

dioecious A term referring to plant species that have both male and female sex organs on different individuals.

diploid (2N) A eukaryotic cell with two sets of chromosomes.

diplonema The second stage of prophase I in which the chromosomes begin to repel one another and tend to move apart.

discontinuous trait A heritable trait in which the mutant phenotype is sharply distinct from the alternative, wild-type phenotype.

disjunction The process in anaphase during which sister chromatid pairs undergo separation.

dispersive model A DNA replication scheme in which the parental double helix is cleaved into double-stranded DNA segments that act as templates for the synthesis of new double-stranded DNA segments. Somehow, the segments reassemble into complete DNA double helices, with parental and progeny DNA segments interspersed.

DNA *See* **deoxyribonucleic acid.**

DNA fingerprinting *See* **DNA typing.**

DNA helicase An enzyme that catalyzes the unwinding of the DNA double helix during replication in *E. coli*; product of the *rep* gene.

DNA ligase (polynucleotide ligase) An enzyme that catalyzes the formation of a covalent bond between free single-stranded ends of DNA molecules during DNA replication and DNA repair.

DNA polymerase An enzyme that catalyzes the synthesis of DNA.

DNA polymerase I An *E. coli* enzyme that catalyzes DNA synthesis, originally called the Kornberg enzyme.

DNA primase The enzyme in DNA replication that catalyzes the synthesis of a short nucleic acid primer.

DNA typing The use of restriction fragment length polymorphisms DNA analysis to identify an individual.

docking protein An integral protein membrane of the endoplasmic reticulum (ER) to which the nascent polypeptide-signal recognition particle (SRP)-ribosome complex binds to facilitate the binding of the polypeptide's signal sequence and associated ribosome to the ER.

dominance variance (represented by V_D) Genetic variance that arises from the dominance effects of genes.

dominant An allele or phenotype that is expressed in either the homozygous or the heterozygous state.

dominant lethal allele Allele that will exhibit a lethal phenotype when present in the heterozygous condition.

dosage compensation A mechanism in mammals that compensates for X chromosomes in excess of the normal complement. *See also* **Barr body**.

double crossover Two crossovers occurring in a particular region of a chromosome in meiosis.

Down syndrome *See* **trisomy-21**.

duplication A chromosomal mutation that results in the doubling of a segment of a chromosome.

effector A small molecule involved in the control of expression of many regulated genes.

endosymbiont hypothesis The hypothesis that mitochondria and chloroplasts originated as free-living prokaryotes that invaded primitive eukaryotic cells and established a mutually beneficial (symbiotic) relationship.

enhancer sequence (enhancer element) In eukaryotes, a type of DNA sequence element having a strong, positive effect on transcription by RNA polymerase II.

environmental sex determination The process by which the environment plays a major role in determining the sex of an organism.

environmental variance (represented by V_E) Any nongenetic source of phenotypic variation among individuals.

episome An autonomously replicating plasmid (a circular, double-stranded DNA molecule) that is capable of integrating into the host cell's chromosome.

epistasis A form of gene interaction in which one gene masks the phenotypic expression of another.

essential genes Genes that when mutated can result in a lethal phenotype.

euchromatin Chromatin that is condensed during division but becomes uncoiled during interphase.

eukaryote A term that literally means "true nucleus." Eukaryotes are organisms that have cells in which the genetic material is located in a membrane-bound nucleus. Eukaryotes can be unicellular or multicellular.

euploid The condition in which an organism or cell has one complete set of chromosomes, or an exact multiple of complete sets.

excision repair (dark repair) An enzyme-catalyzed, light-independent process of repair of ultraviolet-light–induced thymine dimers in DNA that involves removal of the dimers and synthesis of a new piece of DNA complementary to the undamaged strand.

exon The sequences present in mRNA molecule that are separated by introns. Exons include 5' and 3' untranslated regions and the amino-acid-coding sequences.

expressivity The degree to which a particular genotype is expressed in the phenotype.

F$_1$ generation The first filial generation produced by crossing two parental strains.

F$_2$ generation The second filial generation produced by selfing the F$_1$.

facultative heterochromatin Chromatin that may become condensed throughout the cell cycle and may contain genes that are inactivated when the chromatin becomes condensed.

familial trait A trait shared by members of a family.

fine-structure mapping A high-resolution mapping of allelic sites within a gene.

first filial generation *See also* **F$_1$ generation** The offspring that result from the first experimental crossing of animals or plants.

first law *See* **principle of segregation**.

fitness *See* **Darwinian fitness**.

formylmethionine (fMet) A specially modified amino acid involving the addition of a formyl group to the amino group of methionine. It is the first amino acid incorporated into a polypeptide chain in prokaryotes and in mitochondria and chloroplasts of eukaryotes.

forward mutation A mutational change from a wild-type allele to a mutant allele.

founder effect A phenomenon that occurs when the isolate effect is exhibited by a small breeding unit that has formed by migration of a small number of individuals from a large population.

F-pili (sex pili) Hairlike cell surface components produced by cells containing the F factor, which allow the physical union of F^+ and F^- cells or Hfr and F^- cells to take place.

frameshift mutation A mutational addition or deletion of a base pair in a gene that disrupts the normal reading frame of an mRNA, which is read in groups of three bases.

frequency distribution A means of summarizing the phenotypes of a continuous trait whereby the population is described in terms of the proportion of individuals that have each phenotype.

gametes Mature reproductive cells that are specialized for sexual fusion. Each gamete is haploid and fuses with a cell of similar origin but of opposite sex to produce a diploid zygote.

gametogenesis The formation of male and female gametes by meiosis.

gametophyte The haploid sexual generation in the life cycle of plants that produces the gametes.

GC box A eukaryotic promoter element with the consensus sequence 5'-GGGCGGG-3' that can be found in either orientation upstream of the transcription initiation site. The GC boxes appear to help the RNA polymerase near the transcription start point.

gene (Mendelian factor) The determinant of a characteristic of an organism. Genetic information is coded in the DNA, which is responsible for species and individual variation. A gene's nucleotide sequence specifies a polypeptide or RNA and is subject to mutational alteration.

gene flow The movement of genes that takes place when organisms migrate and then reproduce, contributing their genes to the gene pool of the recipient population.

gene locus *See* **locus**.

gene pool The total genetic information encoded in the total genes in a breeding population existing at a given time.

generalized transduction A type of transduction in which any gene may be transferred between bacteria.

gene regulatory elements Base-pair sequences associated with a gene, which are involved in the regulation of gene expression.

gene segregation *See* **principle of segregation (first law)**.

genetic code The base-pair information that specifies the amino acid sequence of a polypeptide.

genetic counseling The procedures whereby the risks of prospective parents having a child who expresses a genetic disease are evaluated and explained to them. The genetic counselor typically makes predictions about the probabilities of particular traits (deleterious or not) occurring among children of a couple.

genetic drift Any change in gene frequency due to chance in a population.

genetic map (linkage map) A representation of the genetic distance separating nonallelic gene loci in a linkage structure.

genetic mapping The use of genetic crosses to locate genes on chromosomes relative to one another.

genetic recombination A process by which parents with different genetic characters give rise to progeny so that genes in which the parents differed are associated in new combinations. For example, from *A B* and *a b* the recombinants *A b* and *a B* are produced.

genetics The science of heredity that involves the structure and function of genes and the way genes are passed from one generation to the next.

genetic variance (represented by V_G) Genetic sources of phenotypic variation among individuals of a population; includes dominance genetic variance, additive genetic variance, and epistatic genetic variance.

genome The total amount of genetic material in a chromosome set; in eukaryotes, this is the amount of genetic material in the haploid set of chromosomes of the organism.

genomic imprinting Phenomenon in which the expression of certain genes is determined by whether the gene is inherited from the female or male parent.

genomic library The collection of molecular clones that contains at least one copy of every DNA sequence in the genome.

genotype The complete genetic makeup of an organism.

genotypic frequencies The frequencies or percentages of different genotypes found within a population.

genotypic sex determination The process by which the sex chromosomes play a decisive role in the inheritance and determination of sex.

germ-line mutations Mutations in the germ line of sexually reproducing organisms may be transmitted by the gametes to the next generation, giving rise to an individual with the mutant state in both its somatic and germ-line cells.

glucose effect *See* **catabolite repression.**

Goldberg-Hogness box (TATA box, or TATA element) Found approximately at position -30 from the transcription initiation site. The Goldberg-Hogness sequence is considered to be the likely eukaryotic promoter sequence. The consensus sequence for the Goldberg-Hogness box is TATAAAAA.

group I-intron self-splicing *See* **self-splicing.**

guanine (G) Purine base found in RNA and DNA. In double-stranded DNA, guanine pairs with the pyrimidine cytosine.

haploid (N) A cell or an individual with one copy of each nuclear chromosome.

Hardy-Weinberg law (Hardy-Weinberg equilibrium, Hardy-Weinberg law of genetic equilibrium) An extension of Mendel's laws of inheritance that describes the expected relationship between gene frequencies in natural populations and the frequencies of individuals of various genotypes in the same populations.

harlequin chromosomes 5-bromodeoxyuridine (5-BUdR), a thymidine analog, is incorporated into DNA during replication. When both DNA strands contain 5-BUdR, the chromatid stains less intensely than when only one DNA strand contains the analog. When cells are grown in the presence of 5-BUdR for two replication cycles, the two sister chromatids stained differentially are called harlequin chromosomes.

hemizygous The condition of X-linked genes in males. Males that have an X chromosome with an allele for a particular gene but do not have another allele of that gene in the gene complement are hemizygous.

hereditary trait A characteristic under control of the genes that is transmitted from one generation to another.

heritability The proportion of phenotypic variation in a population attributable to genetic factors.

hermaphroditic For animals (e.g., a nematode), refers to species in which each individual has both testes and ovaries; in plants, refers to species that have both stamens and pistils on the same flower.

heterochromatin Chromatin that remains condensed throughout the cell cycle and is genetically inactive.

heterogametic sex The sex that has sex chromosomes of different types (e.g., XY) and therefore produces two kinds of gametes with respect to the sex chromosomes.

heterogeneous nuclear RNA (hnRNA) The RNA molecules of various sizes that exist in a large population in the nucleus. Some of the RNA molecules are precursors to mature mRNAs.

heterosis The phenomenon in which the heterozygous genotypes with respect to one or more characters are superior in comparison with the corresponding homozygous genotypes in terms of growth, survival, phenotypic expression, and fertility.

heterozygote superiority *See* **overdominance.**

heterozygous A term describing a diploid organism having different alleles of one or more genes and therefore producing gametes of different genotypes.

***Hfr* (high-frequency recombination)** A male cell in *E. coli* with the *F* factor integrated into the bacterial chromosome. When the *F* factor promotes conjugation with a female (F^-) cell, bacterial genes are transferred to the female cell with high frequency.

highly repetitive sequence A DNA sequence that is repeated between 10^5 and 10^7 times in the genome.

histone One of a class of basic proteins that are complexed with DNA in chromosomes and that play a major role in determining the structure of eukaryotic nuclear chromosomes.

homeobox A 180-bp consensus sequence found in the protein-coding sequences of genes that regulate development.

homeodomain The 60-amino-acid part of proteins that corresponds to the homeobox sequence of genes. All homeodomain-containing proteins appear to be located in the nucleus.

homeotic mutations Mutations that alter the identity of particular segments, transforming them into copies of other segments.

homogametic sex The sex in a species, most often the female, that produces only the X sex chromosome.

homolog Each individual member of a pair of homologous chromosomes.

homologous chromosomes The members of a chromosome pair that are identical in the arrangement of genes they contain and in their visible structure.

homozygous A term describing a diploid organism having the same alleles at one or more genes and therefore producing gametes of identical genotypes.

Human Genome Project A project to obtain the sequence of the complete 3 billion (3×10^9) nucleotide pairs of the human genome, and to map all of the estimated 50,000 to 100,000 human genes.

hybrid dysgenesis The appearance of a series of defects, including mutations, chromosomal aberrations, and sterility, when certain strains of *Drosophila melanogaster* are crossed.

hypersensitive sites (hypersensitive regions) Sites in the regions of DNA around transcriptionally active genes that are highly sensitive to digestion by DNase I.

hypothetico-deductive method of investigation Research method involving making observations, forming hypotheses to explain the observations, making experimental predictions based on the hypotheses, and, finally, testing the predictions. The last step produces new observations, and so a cycle is set up leading to a refinement of the hypotheses and perhaps eventually to the establishment of a law or an accepted principle.

imaginal discs In the *Drosophila* blastoderm, undifferentiated cells that will develop into adult tissue and organs.

immunoglobulins Specialized proteins (antibodies) secreted by B cells that circulate in the blood and lymph and that are responsible for humoral immune responses.

inbreeding Preferential mating between close relatives.

incomplete (partial) dominance The condition resulting when one allele is not completely dominant to another allele so that the heterozygote has a phenotype between that shown in individuals homozygous for either individual allele involved. An example of partial dominance is the Andalusian blue chicken.

induced mutation A mutation that results from treatment with mutagens.

inducer A chemical or environmental agent for bacterial operons that brings about the transcription of an operon.

induction The synthesis of a gene product (or products) in response to the action of an inducer, that is, a chemical or environmental agent.

insertion sequence (IS) element The simplest transposable genetic element found in prokaryotes. It is a mobile segment of DNA that contains genes required for the process of insertion of the DNA segment into a chromosome and for the mobilization of the element to different locations.

interaction variance (represented by V_I) Genetic variance that arises from epistatic interactions among genes.

intergenic suppressor A mutation whose effect is to suppress the phenotypic consequences of another mutation in a gene distinct from the gene in which the suppressor mutation is located.

internal control region (ICR) Promoter sequence, recognized by RNA polymerase III, that is located within the gene sequence, for instance, in tRNA genes and 5S rRNA genes of eukaryotes.

intervening sequence (ivs) *See* **intron**.

intragenic suppressors A mutation whose effect is to suppress the phenotypic consequences of another mutation within the same gene in which the suppressor mutation is located.

intron A nucleotide sequence in eukaryotes that must be excised from a structural gene transcript in order to convert the transcript into a mature messenger RNA molecule containing only coding sequences that can be translated into the amino acid sequence of a polypeptide.

inversion A chromosomal mutation that results when a segment of a chromosome is excised and then reintegrated in an orientation 180° from the original orientation.

IS element *See* **insertion sequence (IS) element**.

karyotype A complete set of all the metaphase chromatid pairs in a cell (literally, "nucleus type").

kinetochore *See* **centromere**.

Klinefelter syndrome A human clinical syndrome that results from disomy for the X chromosome in a male, which results in a 47,XXY male. Many of the affected males are mentally deficient, have underdeveloped testes, and are taller than average.

lagging strand In DNA replication, the DNA strand that is synthesized discontinuously in the 5' to 3' direction away from the replication fork.

leader sequence One of three main parts of the mRNA molecule. The leader sequence is located at the 5' end of the mRNA molecule and contains the coded information that the ribosome and special proteins read to tell it where to begin the synthesis of the polypeptide.

leading strand In DNA replication, the DNA strand synthesized continuously in the 5' to 3' direction toward the replication fork.

leptonema The stage during meiosis in prophase I at which the chromosomes have begun to coil and are visible.

lethal allele An allele that results in the death of an organism.

light repair *See* **photoreactivation**.

LINEs (long interspersed repeated sequences) The dispersed families of repeated sequences in mammals that are several thousand base pairs in length and occur more than 20,000 times in the genome.

linkage A term describing genes that do not assort independently.

linkage map *See* **genetic map**.

linked genes Genes that do not assort independently.

linker *See* **restriction site linker**.

locus (*plural*, **loci**) The position of a gene on a genetic map; the specific place on a chromosome where a gene is located.

lod score method The lod (logarithm of **od**ds) score method is a statistical analysis, usually performed by computer programs, based on data from pedigrees. It is used to test for linkage between two loci in humans.

lyonization A mechanism in mammals that allows them to compensate for X chromosomes in excess of the normal complement. The excess X chromosomes are cytologically condensed and inactivated, and they do not play a role in much of the development of the individual. The name derives from the discoverer of the phenomenon, Mary Lyon.

lysogenic A term describing a bacterium that contains a temperate phage in the prophage state. The bacterium is said to be lysogenic for that phage. On induction, phage reproduction is initiated, progeny phages are produced, and the bacterial cell lyses.

lysogenic pathway A path, besides the lytic cycle, that a phage can follow. The chromosome does not replicate; instead, it inserts itself physically into a specific region of the host cell's chromosome in a way that is essentially the same as *F* factor integration.

lysogeny The phenomenon of the insertion of a temperate phage chromosome into a bacterial chromosome, where it replicates when the bacterial chromosome replicates. In this state the phage genome is repressed and is said to be in the prophage state.

lytic cycle A type of phage life cycle in which the phage takes over the bacterium and directs its growth and reproductive activities to express the phage's genes and to produce progeny phages.

macromolecule A large molecule (such as DNA, RNA, and proteins) that has a molecular weight of at least a few thousand daltons.

mapping functions Mathematical formulas that are used to correct the observed recombination values for the incidence of multiple crossovers.

map unit (mu) A unit of measurement used for the distance between two gene pairs on a genetic map. A crossover frequency of 1 percent between two genes equals 1 map unit. *See also* **centimorgan**.

maternal effect The phenotype in an individual that is established by the maternal nuclear genome, as the result of mRNA and/or proteins that are deposited in the oocyte prior to fertilization. These inclusions direct early development in the embryo.

maternal inheritance A phenomenon in which the mother's phenotype is expressed exclusively.

mating types A genic system in which two sexes are morphologically indistinguishable but carry different alleles and will mate.

mean The average of a set of numbers, calculated by adding all the values represented and dividing by the number of values.

meiosis Two successive nuclear divisions of a diploid nucleus that result in the formation of haploid gametes or of meiospores having one-half the genetic material of the original cell.

meiosis I The first meiotic division that results in the reduction of the number of chromosomes. This division consists of four stages: prophase I, metaphase I, anaphase I, and telophase I.

meiosis II The second meiotic division, resulting in the separation of the chromatids.

Mendelian factor *See* **gene**.

Mendelian population An interbreeding group of individuals sharing a common gene pool; the basic unit of study in population genetics.

messenger RNA (mRNA) The RNA molecule that contains the coded information for the amino acid sequence of a protein.

metacentric chromosome A chromosome that has the centromere approximately in the center of the chromosome.

metaphase A stage in mitosis or meiosis in which chromosomes become aligned along the equatorial plane of the spindle.

metaphase II The second stage of meiosis during which the centromeres line up on the equator of the second-division spindles (in each of two daughter cells formed from meiosis I).

metaphase plate The plane where the chromosomes become aligned during metaphase.

metastasis The spreading of malignant tumor cells throughout the body so that tumors develop at new sites.

migration Movement of organisms from one location to another.

missense mutation A gene mutation in which a base-pair change in the DNA causes a change in an mRNA codon, with the result that a different amino acid is inserted into the polypeptide in place of one specified by the wild-type codon.

mitochondria Organelles found in the cytoplasm of all aerobic animal and plant cells; the principal sources of energy in the cell.

mitosis The process of nuclear division in haploid or diploid cells producing daughter nuclei that contain identical chromosome complements and that are genetically identical to one another and to the parent nucleus from which they arose.

moderately repetitive sequence A DNA sequence that is reiterated from a few to as many as 10^5 times in the genome.

molecular cloning *See* **cloning**.

molecular drive *See* **concerted evolution**.

molecular genetics A subdivision of the science of genetics involving how genetic information is encoded within the DNA and how biochemical processes of the cell translate the genetic information into the phenotype.

monoecious A term referring to plants in which male and female gametes are produced in the same individual.

monohybrid cross A cross between two individuals that are both heterozygous for the same pair of alleles (e.g., $Aa \times Aa$). By extension, the term also refers to crosses involving pure-breeding parents that differ with respect to the alleles of one locus (e.g., $AA \times aa$).

monoploidy An aberrant, aneuploid state in a normally diploid cell or organism in which one chromosome is missing, leaving one chromosome with no homolog.

monosomy An aberrant, aneuploid state in a normally diploid cell or organism in which one chromosome is missing, leaving one chromosome with no homolog.

mRNA splicing A process whereby an intervening sequence between two coding sequences in an RNA molecule is excised and the coding sequences ligated (spliced) together.

multifactorial trait A trait influenced by multiple genes and environmental factors.

multigene family A set of related genes that have evolved from some ancestral gene through the process of gene duplication.

multiple alleles Many alternative forms of a single gene.

multiple cloning site *See* **polylinker**.

multiple crossovers More than one crossover occurring in a particular region of a chromosome in meiosis.

mutagen Any physical or chemical agent that significantly increases the frequency of mutational events above a spontaneous mutation rate.

mutant allele Any alternative to the wild-type allele of a gene. Mutant alleles may be dominant or recessive to wild-type alleles.

mutation Any detectable and heritable change in the genetic material not caused by genetic recombination.

mutation frequency The number of occurrences of a particular kind of mutation in a population of cells or individuals.

mutation rate The probability of a particular kind of mutation as a function of time.

mutator gene A mutant gene that increases the spontaneous mutation frequencies of other genes.

narrow-sense heritability The proportion of the phenotypic variance that results from additive genetic variance.

natural selection Differential reproduction of genotypes.

negative assortative mating A mating that occurs between dissimilar individuals more often than it does between randomly chosen individuals.

neutral mutation A base-pair change in a gene that changes a codon in the mRNA so that there is no change in the function of the protein translated from that message.

neutral mutation model A hypothesis that replaced the classical model by acknowledging the presence of extensive genetic variation in proteins, but proposing that this variation is neutral with regard to natural selection.

nitrogenous base A nitrogen-containing base that, along with a pentose sugar and a phosphate, is one of the three parts of a nucleotide, the building block of RNA and DNA.

noncontributing alleles The alleles that do not have any effect on the phenotype of the quantitative trait.

nondisjunction (primary nondisjunction) A failure of homologous chromosomes or sister chromatids to separate at anaphase.

nonhistone A type of acidic or neutral protein found in chromatin.

nonhomologous chromosomes The chromosomes containing dissimilar genes that do not pair during meiosis.

non-Mendelian inheritance (cytoplasmic inheritance) The inheritance of characters determined by genes not located on the nuclear chromosomes but on mitochondrial or chloroplast chromosomes. Such genes show inheritance patterns distinctly different from those of nuclear genes.

nonparental-ditype (NPD) One of three types of tetrads possible when two genes are segregating in a cross. The NPD tetrad contains four nuclei, all of which have recombinant (nonparental) genotypes, that is, two of each possible type.

nonsense codon *See* **chain-terminating codon**.

nonsense mutation A gene mutation in which a base-pair change in the DNA causes a change in an mRNA codon from an amino-acid-coding codon to a chain-terminating (nonsense) codon. As a result, polypeptide chain synthesis is terminated prematurely and is therefore either nonfunctional or, at best, partially functional.

nontranscribed spacer (NTS) sequence Sequences, which are not transcribed, found between transcription units in rDNA. Important sequences that control transcription of the rDNA are within the NTS.

normal distribution A probability distribution in statistics, graphically displayed as a bell-shaped curve.

norm of reaction The range of potential phenotypes that a single genotype could develop if exposed to a range of environmental conditions.

northern blot analysis A similar technique to Southern blotting except that RNA rather than DNA is separated and transferred to a filter for hybridization with a probe.

nuclease An enzyme that catalyzes the degradation of a nucleic acid by breaking phosphodiester bonds. Nucleases specific for DNA are termed deoxyribonucleases (DNases), and nucleases specific for RNA are termed ribonucleases (RNases).

nucleofilament A fiber seen in chromatin. It is approximately 10 nm in diameter and consists of DNA wrapped around nucleosome cores.

nucleoid Central region in a bacterial cell in which the chromosome is compacted.

nucleoside phosphate *See* **nucleotide**.

nucleosome The basic structural unit of eukaryotic nuclear chromosomes, consisting of two molecules each of the four core histones (H2A, H2B, H3, and H4, the histone octamer), a single molecule of the linker histone H1, and about 180 bp of DNA.

nucleotide A monomeric molecule of RNA and DNA that consists of three distinct parts: a pentose sugar (ribose in RNA, deoxyribose in DNA), a nitrogenous base, and a phosphate group.

nucleus A discrete structure within the cell that is bounded by a nuclear membrane. It contains most of the genetic material of the cell.

nullisomy The aberrant, aneuploid state in a normally diploid cell or organism in which there is a loss of one pair of homologous chromosomes.

Okazaki fragments The relatively short, single-stranded DNA fragments in discontinuous DNA replication that are synthesized during DNA replication and that are subsequently covalently joined to make a continuous strand.

oligomers (*oligo* = few) Short DNA molecules.

oncogene A gene whose action stimulates unregulated cell proliferation. Cellular oncogenes are altered forms of cellular proto-oncogenes.

oncogenesis Tumor (cancer) initiation in an organism.

one gene–one enzyme hypothesis The hypothesis, based on Beadle and Tatum's studies in biochemical genetics, that each gene controls the synthesis of one enzyme.

one gene–one polypeptide hypothesis Updated version of the one gene–one enzyme hypothesis, which states that each gene controls the synthesis of a polypeptide chain.

oogenesis The development in the gonad of the female germ cell (egg cell) of animals.

open reading frame In a segment of DNA, a potential protein-coding sequence identified by an initiator codon in frame with a chain-terminating codon.

operator The controlling site that is adjacent to a promoter, and that is responsible for controlling the transcription of genes that are contiguous to the promoter.

operon A cluster of genes whose expressions are regulated together by operator-regulator protein interactions, plus the operator region itself and the promoter.

origin A specific site on the chromosome at which the double helix denatures into single strands and continues to unwind as the replication fork(s) migrates.

origin of replication A specific DNA sequence that is required for the initiation of DNA replication in prokaryotes.

outbreeding Preferential mating between nonrelated individuals.

overdominance (heterozygote superiority) Condition in which the heterozygote has higher fitness than either of the homozygotes.

ovum A mature egg cell. In the second meiotic division, the secondary oocyte produces two haploid cells; the large cell rapidly matures into the ovum.

pachynema The stage in meiosis (mid-prophase I) during which the homologous pairs of chromosomes exchange chromosome regions.

paracentric inversion An inversion in which the inverted segment occurs on one chromosome arm and does not include the centromere.

parental-ditype (PD) One of three types of tetrads possible when two genes are segregating in a cross. The PD tetrad contains four nuclei, all of which are parental genotypes, with two of one parent and two of the other parent.

parental genotypes (parental classes, parentals) Individuals among progeny of crosses that have combinations of genetic markers like one or other of the parents in the parental generation.

parental imprinting *See* **genomic imprinting**.

partial dominance *See* **incomplete (partial) dominance**.

particulate factors The term Mendel used to describe the factors that carried hereditary information and were transmitted from parents to progeny through the gametes. We now know these factors by the name *genes*.

pedigree analysis A family tree investigation that involves the careful compilation of phenotypic records of the family over several generations.

penetrance The frequency with which a dominant or homozygous recessive gene manifests itself in the phenotype of an individual.

pentose sugar A 5-carbon sugar that, along with a nitrogenous base and a phosphate group, is one of the three parts of a nucleotide, the building block of RNA and DNA.

peptide bond A covalent bond in a polypeptide chain that joins the a-carboxyl group of one amino acid to the a-amino group of the adjacent amino acid.

peptidyl transferase The enzyme that catalyzes the formation of the peptide bond in protein synthesis.

pericentric inversion An inversion in which the inverted segment includes the parts of both chromosome arms and therefore includes the centromere.

P generation The parental generation, that is, the immediate parents of an F_1.

phage lysate The progeny phages released following lysis of phage-infected bacteria.

phage vector A phage that carries pieces of bacterial DNA between bacterial strains in the process of transduction.

phenocopy An abnormal individual resulting from special environmental conditions. It mimics a similar phenotype caused by gene mutation.

phenotype The observable properties (structural and functional) of an organism that are produced by the interaction between its genotype and the environment.

phenotypic variance (represented by V_P) A measure of a trait's variability.

phosphate group A component, along with a pentose sugar and a nitrogenous base, of a nucleotide, the building block of RNA and DNA. Because phosphate groups are acidic in nature, DNA and RNA are called nucleic acids.

phosphodiester bond A covalent bond in RNA and DNA between a sugar and a phosphate. Phosphodiester bonds form the repeating sugar-phosphate array of the backbone of DNA and RNA.

photoreactivation (light repair) One way by which thymine dimers can be repaired. The dimers are reverted directly to the original form by exposure to visible light in the wavelength range 320–370 nm.

physical map Map of physically identifiable regions or markers on genomic DNA, constructed without genetic recombination analysis.

pistil The female reproductive organ in a flowering plant that typically consists of the stigma, the style, and the ovary.

plaque A round, clear area in a lawn of bacteria on solid medium that results from the lysis of cells by repeated cycles of phage lytic growth.

plasmid An extrachromosomal genetic element consisting of double-stranded DNA that replicates autonomously from the host chromosome.

pleiotropy Term referring to multiple phenotypic effects resulting from a single mutant gene.

point mutation A mutation caused by a substitution of one base pair for another.

poly(A) tail A sequence of 50 to 250 adenine nucleotides that is added as a post-transcriptional modification at the 3' ends of most eukaryotic mRNAs.

polygene (multiple-gene) hypothesis for quantitative inheritance The hypothesis that quantitative traits are controlled by many genes.

polygenic traits Traits encoded by many loci.

polylinker (multiple cloning site) A region of clustered unique restriction sites in a cloning vector.

polymerase chain reaction (PCR) A method used to replicate defined DNA sequences selectively and repeatedly from a DNA mixture.

polynucleotide A linear sequence of nucleotides in DNA or RNA.

polypeptide A polymeric, covalently bonded linear arrangement of amino acids joined by peptide bonds.

polyploidy The condition of a cell or organism that has more than its normal number of sets of chromosomes.

polyribosome (polysome) The complex between an mRNA molecule and all the ribosomes that are translating it simultaneously.

polytene chromosome A special type of chromosome representing a bundle of numerous chromatids that have arisen by repeated cycles of replication of single chromatids without nuclear division. This type of chromosome is characteristic of various tissues of Diptera.

population A group of interbreeding individuals that share a set of genes.

population genetics A branch of genetics that describes in mathematical terms the consequences of Mendelian inheritance on the population level.

positional cloning The isolation of a gene associated with a genetic disease on the basis of its approximate chromosomal position.

positive assortative mating A mating that occurs more frequently between individuals who are phenotypically similar than it does among randomly chosen individuals.

precursor-mRNA (pre-mRNA) molecule The initial transcript of a gene that is modified and/or processed to produce the mature, functional mRNA molecule. In eukaryotes, for example, the transcript is modified at both the 5' and the 3' ends, and in a number of cases RNA sequences that do not code for amino acids are present and must be excised.

precursor rRNA (pre-rRNA) A primary transcript of adjacent rRNA genes (16S, 23S, and 5S rRNA genes in prokaryotes; 18S, 5.8S, and 28S rRNA genes in eukaryotes) plus flanking and spacer DNA that must be processed to release the mature rRNA molecules.

pre-tRNA (precursor tRNA) A primary transcript of a tRNA gene whose bases must be extensively modified and that must be processed to remove extra RNA sequences in order to produce the mature tRNA molecule. In some cases, the primary transcript may contain the sequences of two or more tRNA molecules.

Pribnow box A part of the promoter sequence in prokaryotic genomes that is located at about 10 base pairs upstream from the transcription starting point. The consensus sequence for the Pribnow box is TATAAT. The Pribnow box is often referred to as the TATA box.

primer *See* **RNA primer.**

primosome A complex of *E. coli* primase, helicase, and perhaps other polypeptides that together become functional in catalyzing the initiation of DNA synthesis.

principle of independent assortment (second law) The law that the factors (genes) for different traits assort independently of one another. In other words, genes on different chromosomes behave independently in the production of gametes.

principle of segregation (first law) The law that two members of a gene pair (alleles) segregate (separate) from each other during the formation of gametes. As a result, one-half the gametes carry one allele and the other half carry the other allele.

proband In human genetics, an affected person with whom the study of a character in a family begins. (*See also* **propositus; proposita.**)

product rule The rule that the probability of two independent events occurring simultaneously is the product of each of their probabilities.

prokaryote A cellular organism whose genetic material is not located within a membrane-bound nucleus (*cf.* eukaryote).

promoter site (promoter sequence, promoter) A specific regulatory nucleotide sequence in the DNA to which RNA polymerase binds for the initiation of transcription.

proofreading mechanism In DNA synthesis, the process of recognizing a base-pair error during the polymerization events and correcting it. Proofreading is a property of the DNA polymerase in prokaryotic cells.

prophage A temperate bacteriophage integrated into the chromosome of a lysogenic bacterium. It replicates with the replication of the host cell's chromosome.

prophase The first stage in mitosis or meiosis during which the chromosomes (already replicated) condense and become visible under the microscope.

prophase I The first stage of meiosis. There are several stages of prophase I, including leptonema, zygonema, pachynema, diplonema, and diakinesis.

prophase II The second stage of meiosis during which there is chromosome contraction.

proposita In human genetics, an affected female person with whom the study of a character in a family begins. (*See also* **proband.**)

propositus In human genetics, an affected male person with whom the study of a character in a family begins. (*See also* **proband.**)

protein One of a group of high-molecular-weight, nitrogen-containing organic compounds of complex shape and composition.

proto-oncogenes A gene that in normal cells functions to control the normal proliferation of cells, and that when mutated or changed in any other way becomes an oncogene.

prototroph A strain that is a wild type for all nutritional requirement genes and thus requires no supplements in its growth medium.

Punnett square A matrix that describes all the possible gametic fusions that will give rise to the zygotes that will produce the next generation.

pure-breeding *See* **true-breeding (pure-breeding) strain.**

purine A type of nitrogenous base. In DNA and RNA the purines are adenine and guanine.

pyrimidine A type of nitrogenous base. Cytosine is a pyrimidine in DNA and RNA; thymine is a pyrimidine in DNA; and uracil is a pyrimidine in RNA.

quantitative genetics Study of the inheritance of quantitative traits.

random mating Matings between genotypes occurring in proportion to the frequencies of the genotypes in the population.

rDNA repeat units The tandem arrays of rRNA genes, 18S-5.8S-28S, repeated many times along the chromosome.

recessive An allele or phenotype that is expressed only in the homozygous state.

recessive lethal allele An allele that causes lethality when it is homozygous.

reciprocal cross A cross of males and females of one trait with males and females of another trait. In the garden pea example, a reciprocal cross for smooth and wrinkled seeds is smooth female × wrinkled male and wrinkled female × smooth male.

recombinant chromosome A chromosome that emerges from meiosis with a combination of genes different from a parental combination of genes.

recombinant DNA molecule A new type of DNA sequence that has been constructed or engineered in the test tube from two or more distinct DNA sequences.

recombinants The individuals or cells that have nonparental combinations of genes as a result of the processes of genetic recombination.

recombination *See* **genetic recombination.**

release factors *See* **termination factors.**

replica plating The procedure for transferring the pattern of colonies from a master plate to a new plate. In this procedure, a velveteen pad on a cylinder is pressed lightly onto the surface of the master plate, thereby picking up a few cells from each colony to inoculate onto the new plate.

replication bubble Opposing replication forks found with the local denaturing of DNA during replication.

replication fork A Y-shaped structure formed when a double-stranded DNA molecule unwinds to expose the two single-stranded template strands for DNA replication.

replicon (replication unit) The stretch of DNA in eukaryotes from the origin of replication to the two termini of replication on each side of the origin.

repressor *See* **repressor gene.**

repressor gene A regulatory gene whose product is a protein that controls the transcriptional activity of a particular operon.

repressor molecule The protein product of a repressor gene.

repulsion An arrangement in which each homologous chromosome carries the wild-type allele of one gene and the mutant allele of the other one.

restriction endonucleases (restriction enzymes) Enzymes important for analyzing DNA and for constructing recombinant DNA molecules because of their ability to cleave double-stranded DNA molecules at specific nucleotide-pair sequences.

restriction enzymes *See* **restriction endonucleases.**

restriction fragment length polymorphisms (RFLPs) The different restriction maps that result from different patterns of distribution of restriction sites. They are detected by the presence of restriction fragments of different lengths on gels.

restriction map A genetic map of DNA showing the relative positions of restriction enzyme cleavage sites.

restriction site linker (linker) A relatively short, double-stranded oligodeoxyribonucleotide about 8 to 12 nucleotide pairs long that is synthesized by chemical means and that contains the cleavage site for a specific restriction enzyme within its sequence.

reverse genetics *See* **positional cloning.**

reverse mutation (reversion) A mutational change from a mutant allele back to a wild-type allele.

reverse transcriptase An enzyme (an RNA-dependent DNA polymerase) that makes a complementary DNA copy of an mRNA strand.

ribonuclease (RNase) An enzyme that catalyzes the degradation of RNA to nucleotides.

ribonucleic acid (RNA) A usually single-stranded polymeric molecule consisting of ribonucleotide building blocks. RNA is chemically very similar to DNA. The three major types of RNA in cells are ribosomal RNA (rRNA), transfer RNA (tRNA), and messenger RNA (mRNA), each of which performs an essential role in protein synthesis (translation). In some viruses, RNA is the genetic material.

ribonucleotide The basic building block of RNA consisting of a pentose sugar (ribose), a base, and a phosphate.

ribose The pentose sugar component of the nucleotide building block of RNA.

ribosomal DNA (rDNA) The regions of the DNA that contain the genes for the rRNAs in prokaryotes and eukaryotes.

ribosomal proteins The proteins that along with rRNA molecules make up the ribosomes of prokaryotes and eukaryotes.

ribosomal RNA (rRNA) The RNA molecules of discrete sizes that along with ribosomal proteins make up the ribosomes of prokaryotes and eukaryotes.

ribosome A complex cellular particle composed of ribosomal protein and rRNA molecules that is the site of amino acid polymerization during protein synthesis.

RNA *See* **ribonucleic acid.**

RNA editing Post-transcriptional insertion and/or deletion of nucleotides in an mRNA molecule.

RNA polymerase An enzyme that catalyzes the synthesis of RNA molecules from a DNA template in a process called transcription.

RNA polymerase I An enzyme in eukaryotes located in the nucleolus that catalyzes the transcription of the 18S, 5.8S, and 28S rRNA genes.

RNA polymerase II An enzyme in eukaryotes found only in the nucleoplasm of the nucleus. It catalyzes the transcription of mRNA-coding genes.

RNA polymerase III An enzyme in eukaryotes found only in the nucleoplasm. It catalyzes the transcription of the tRNA and 5S rRNA genes.

RNA primer A preexisting polynucleotide chain in DNA replication to which new nucleotides can be added.

RNA processing control The second level of control of gene expression in eukaryotes. This level involves regulating the production of mature RNA molecules from precursor-RNA molecules.

RNA synthesis *See* **transcription.**

Robertsonian translocation A type of nonreciprocal translocation in which the long arms of two nonhomologous acrocentric chromosomes become attached to a single centromere.

sample The subset used to give information about a population. It must be of reasonable size and it must be a random subset of the larger group in order to provide accurate information about the population.

sampling error The phenomenon in which chance deviations from expected proportions arise in small samples.

secondary oocyte A large cell produced by the primary oocyte. In the ovaries of female animals, the diploid primary oocyte goes through meiosis I and unequal cytokinesis to produce two cells; the large cell is called the secondary oocyte.

second law *See* **principle of independent assortment.**

selection coefficient A measure of the relative intensity of selection against a genotype.

selection differential In natural and artificial selection, the difference between the mean phenotype of the selected parents and the mean phenotype of the unselected population.

selection response The amount that a phenotype changes in one generation when selection is applied to a group of individuals.

self-fertilization (selfing) The union of male and female gametes from the same individual.

self-splicing The excision of introns from some precursor RNA molecules that occurs by a protein-independent reaction in some organisms.

semiconservative model A DNA replication scheme in which each daughter molecule retains one of the parental strands.

semidiscontinuous Concerning DNA replication, when one new strand is synthesized continuously and the other discontinuously. *See also* **discontinuous DNA replication**.

sense codon A codon in an mRNA molecule that specifies an amino acid in the corresponding polypeptide.

sequence-tagged site (STS) A short segment of DNA that defines a unique position in the human genome; an STS is usually detected by the *polymerase chain reaction (PCR)*.

sex chromosome A chromosome in eukaryotic organisms that is represented differently in the two sexes. In many organisms, one sex possesses a pair of visibly different chromosomes. One is an X chromosome, and the other is a Y chromosome. Commonly, the XX sex is female and the XY sex is male.

sex-influenced traits The traits that appear in both sexes, but either the frequency of occurrence in the two sexes is different or there is a different relationship between genotype and phenotype.

sex-limited trait A genetically controlled character that is phenotypically exhibited in only one of the two sexes.

sex-linked *See* **X-linked**.

sexual reproduction The reproduction involving the fusion of haploid gametes produced by meiosis.

shuttle vector A cloning vector that can replicate in two or more host organisms. Shuttle vectors are used for experiments in which recombinant DNA is to be introduced into organisms other than *E. coli*.

signal hypothesis The hypothesis that the secretion of proteins from a cell occurs through the binding of a hydrophobic amino terminal extension to the membrane and the subsequent removal and degradation of the extension in the cisternal space of the endoplasmic reticulum.

signal peptidase An enzyme in the cisternal space of the ER that catalyzes removal of the signal sequence from the polypeptide.

signal recognition particle (SRP) In eukaryotes, a complex of a small RNA molecule with six proteins, which can temporarily halt protein synthesis by recognizing the signal sequence of a nascent polypeptide destined to be translocated through the ER, binding to it, and thereby blocking further translation of the mRNA.

signal sequence The hydrophobic, amino terminal extension found on proteins that are secreted from a cell. The amino terminus (extension) is removed and degraded in the cisternal space of the endoplasmic reticulum.

silencer *See* **silencer element**.

silencer element In eukaryotes, a transcriptional regulatory element that decreases RNA transcription rather than stimulating it like other enhancer elements.

silent mutation A mutational change resulting in a protein with a wild-type function because of an unchanged amino acid sequence.

simple telomeric sequences Simple, tandemly repeated DNA sequences at, or very close to, the extreme ends of the chromosomal DNA molecules.

SINEs (short interspersed repeated sequences) One class of interspersed and highly repeated sequences that consists of dispersed families with unit lengths of fewer than 500 base pairs and repeated for as many as hundreds of thousands of copies in the genome.

single-strand DNA binding (SSB) proteins (helix-destabilizing proteins) Proteins that help the DNA unwinding process by stabilizing the single-stranded DNA.

sister chromatid A chromatid derived from replication of one chromosome during interphase of the cell cycle.

small nuclear ribonucleoprotein particles (snRNPs) The complexes formed by small nuclear RNAs and proteins in which the processing of pre-mRNA molecules occurs.

small nuclear RNA (snRNA) Found only in eukaryotes, one of four major classes of RNA molecules produced by transcription. snRNAs are used in the processing of pre-mRNA molecules.

somatic cell hybridization The fusion of two genetically different somatic cells of the same or different species to generate a somatic hybrid for genetic analysis.

somatic mutation A mutation in a cell that produces a mutant spot or area, but the mutant characteristic is not passed on to the succeeding generation.

Southern blot technique A technique invented by E. M. Southern and used in analyzing genes and gene transcripts, in which DNA fragments are transferred from a gel to a nitrocellulose filter.

spacer sequences Transcribed sequences that are found between, and flanking, coding RNA sequences. Spacer sequences are removed during processing of pre-rRNA and pre-tRNA to produce mature molecules.

specialized transduction A type of transduction in which only specific genes are transferred.

spermatogenesis Development of the male animal germ cell within the male gonad.

sperm cells (spermatozoa) The male gametes; the spermatozoa produced by the testes in male animals.

spliceosomes The splicing complexes formed by the association of several snRNPs bound to the pre-mRNA.

spontaneous mutations The mutations that occur without the use of chemical or physical mutagenic agents.

sporophyte The haploid, asexual generation in the life cycle of plants that produces haploid spores by meiosis.

stamen The male reproductive organ in a flowering plant that usually consists of a stalk, called a filament, bearing a pollen-producing anther.

standard deviation The square root of the variance. It measures the extent to which each measurement in the data set differs from the mean value and is used as a measure of the extent of variability in a population.

steroid response element (RE) The DNA sequence to which steroid hormones will bind to activate a gene.

stop codon *See* **chain-terminating codon**.

structural gene A gene that codes for an mRNA molecule and hence for a polypeptide chain.

submetacentric chromosome A chromosome that has the centromere nearer one end than the other. Such chromosomes appear J-shaped at anaphase.

sum rule The rule that the probability of either one of two mutually exclusive events occurring is the sum of their individual probabilities.

suppressor gene A gene that causes suppression of mutations in other genes.

suppressor mutation A mutation at a second site that totally or partially restores a function lost because of a primary mutation at another site.

synapsis The intimate association of homologous chromosomes brought about by the formation of a zipperlike structure along the length of the chromatids called the *synaptonemal complex*.

synaptonemal complex A complex structure spanning the region between meiotically paired (synapsed) chromosomes that is concerned with crossing-over rather than with chromosome pairing.

syntenic The genes that are localized to a particular chromosome by using an experimental approach (literally "together thread"; the term is similar to *linked*).

TATA element *See* **Goldberg-Hogness box.**

tautomers Alternate chemical forms in which DNA (or RNA) bases are able to exist.

telocentric chromosome A chromosome that has the centromere more or less at one end.

telomere-associated sequences Repeated, complex DNA sequences extending from the molecular gene of chromosomal DNA, suspected to mediate many of the telomere-specific interactions.

telophase A stage during which the migration of the daughter chromosomes to the two poles is completed.

telophase II The last stage of meiosis II during which a nuclear membrane forms around each set of chromosomes, and cytokinesis takes place.

template strand The unwound single strand of DNA on which new strands are made (following complementary base-pairing rules).

termination factors (release factors; RF) The specific proteins in polypeptide synthesis (translation) that read the chain termination codons and then initiate a series of specific events to terminate polypeptide synthesis.

terminator *See* **transcription terminator sequence.**

testcross A cross of an individual of unknown genotype, usually expressing the dominant phenotype, with a homozygous recessive individual in order to determine the genotype of the individual.

testis-determining factor Gene product in placental mammals that sets the switch toward male sexual differentiation.

tetrad analysis Genetic analysis of all the products of a single meiotic event. Tetrad analysis is possible in those organisms in which the four products of a single nucleus that has undergone meiosis are grouped together in a single structure.

tetrasomy The aberrant, aneuploid state in a normally diploid cell or organism in which an extra chromosome pair results in the presence of four copies of one chromosome type and two copies of every other chromosome type.

tetratype (T) One of the three types of tetrads possible when two genes are segregating in a cross. The T tetrad contains two parental and two recombinant nuclei, one of each parental type and one of each recombinant type.

three-point testcross A test involving three genes within a relatively short section of the chromosome. It is used to map genes for their order in the chromosome and for the distance between them.

thymine (T) A pyrimidine base found in DNA but not in RNA. In double-stranded DNA, thymine pairs with adenine.

topoisomerases A class of enzymes that catalyze the supercoiling of DNA.

totipotency The capacity of a nucleus to direct events through all the stages in development and therefore produce a normal adult.

trailer sequence The sequence of the mRNA molecule beginning at the end of the amino-acid-coding sequence and ending at the 3' end of the mRNA. The trailer sequence is not translated and varies in length from molecule to molecule.

transconjugants In bacteria, the recipients inheriting donor DNA in the process of conjugation.

transcription The transfer of information from a double-stranded DNA molecule to a single-stranded RNA molecule. It is also called *RNA synthesis.*

transcriptional control The first level of control of gene expression in eukaryotes. This level involves regulating whether or not a gene is to be transcribed and the rate at which transcripts are produced.

transcription factors (TFs) *See* **basal transcription factor.**

transcription terminator sequence (terminator) A transcription regulatory sequence located at the distal end of a gene that signals the termination of transcription.

trans-dominant The phenomenon of a gene or DNA sequence controlling genes that are on a different piece (strand) of DNA.

transducing phage The phage that is the vehicle by which genetic material is shuttled between bacteria.

transducing retroviruses Retroviruses that have picked up an oncogene from the cellular genome.

transductants In bacteria, the recombinant recipients resulting from the process of transduction.

transduction A process by which bacteriophages mediate the transfer of bacterial genetic information from one bacterium (the donor) to another (the recipient); a process whereby pieces of bacterial DNA are carried between bacterial strains by a phage.

transfer RNA (tRNA) One of the four classes of RNA molecules produced by transcription and involved in protein synthesis; molecules that bring amino acids to the ribosome, where they are matched to the transcribed message on the mRNA.

transformant The genetic recombinant generated by the transformation process.

transformation (a) The unidirectional transfer of extracellular DNA into cells resulting in a phenotypic change in the recipient. (b) The failure of cells to remain constrained in their growth properties that gives rise to tumors.

transgene A gene introduced into the genome of an organism by genetic manipulation in order to alter its genotype.

transgenic organism An organism that has had its genotype altered by the introduction of a gene into its genome by genetic manipulation.

transition mutation A specific type of base-pair substitution mutation that involves a change in the DNA from one purine-pyrimidine base pair to the other purine-pyrimidine base pair at a particular site (e.g., AT to GC).

translation (protein synthesis) The conversion in the cell of the mRNA base-sequence information into an amino acid sequence of a polypeptide.

translational control The regulation of protein synthesis by ribosome synthesis among mRNAs.

translocation (transposition) (a) A chromosomal mutation involving a change in position of a chromosome segment (or segments) and the gene sequences it contains. (b) In polypeptide synthesis, the movement of the ribosome, one codon at a time, along the mRNA toward the 3' end.

transmission genetics (classical genetics) A subdivision of the science of genetics primarily dealing with how genes are passed from one individual to another.

transposable element A genetic element of chromosomes of both prokaryotes and eukaryotes that has the capacity to mobilize itself and move from one location to another in the genome.

transposase An enzyme encoded by the IS element of a transposon that catalyzes transposition activity of a transposable element.

transposon (Tn) A mobile DNA segment that contains genes for the insertion of the DNA segment into the chromosome and for mobilization of the element to other locations on the chromosomes.

transversion mutation A specific type of base-pair substitution mutation that involves a change in the DNA from a purine-pyrimidine base pair to a pyrimidine-purine base pair at the same site (e.g., AT to TA or GC to TA).

trihybrid cross A cross between individuals of the same type that are heterozygous for three pairs of alleles at three different loci.

trisomy An aberrant, aneuploid state in a normally diploid cell or organism in which there are three copies of a particular chromosome instead of two copies.

trisomy-21 A human clinical condition characterized by various abnormalities. It is caused by the presence of an extra copy of chromosome 21.

true-breeding (pure-breeding) strain A strain allowed to self-fertilize for many generations to ensure that the traits to be studied are inherited and unchanging.

true reversion A point mutation from mutant back to wild type in which the change codes for the original amino acid of the wild type.

tumor viruses Viruses that induce cells to dedifferentiate and to divide to produce a tumor.

Turner syndrome A human clinical syndrome that results from monosomy for the X chromosome in the female, which gives a 45,X female. These females fail to develop secondary sexual characteristics, tend to be short, have weblike necks, have poorly developed breasts, are usually infertile, and exhibit mental deficiencies.

uniparental inheritance A phenomenon, usually exhibited by extranuclear genes, in which all progeny have the phenotype of only one parent.

unique (single-copy) sequence A class of DNA sequences that has one to a few copies per genome.

uracil (U) A pyrimidine base found in RNA but not in DNA.

variance A statistical measure of how values vary from the mean.

vegetative reproduction *See* **asexual reproduction.**

viral oncogene A viral gene that transforms a cell it infects to a cancerous state. *See also* **cellular oncogene; oncogenesis.**

virulent phage A phage (such as T4) that always follows the lytic cycle when it infects bacteria.

visible mutation A mutation that affects the morphology or physical appearance of an organism.

wild type A strain, organism, or gene of the type that is designated as the standard for the organism with respect to genotype and phenotype.

wild-type allele The allele designated as the standard ("normal") for a strain of organism.

wobble hypothesis A theory proposed by Francis Crick that proposes that the base at the 5' end of the anticodon (3' end of the codon) is not as constrained as the other two bases. This feature allows for less exact base-pairing, so that the 5' end of the anticodon can potentially pair with one of three different bases at the 3' end of the codon.

X chromosome A sex chromosome present in two copies in the homogametic sex and in one copy in the heterogametic sex.

X chromosome-autosome balance system A genotypic sex determination system. The main factor in sex determination is the ratio between the numbers of X chromosomes and autosomes. Sex is determined at the time of fertilization, and sex differences are assumed to be due to the action during development of two sets of genes located in the X chromosomes and in the autosomes.

X chromosome nondisjunction An event occurring when the two X chromosomes fail to separate in meiosis so that eggs are produced either with two X chromosomes or with no X chromosomes, instead of the usual one X chromosome.

X-linked Referring to genes located on the X chromosome.

X-linked dominant trait A trait due to a dominant mutant gene carried on the X chromosome.

X-linked recessive trait A trait due to a recessive mutant gene carried on the X chromosome.

Y chromosome A sex chromosome that when present is found in one copy in the heterogametic sex, along with an X chromosome, and is not present in the homogametic sex. Not all organisms with sex chromosomes have a Y chromosome.

yeast artificial chromosome (YAC) A cloning vector in which DNA fragments several hundred kilobase pairs long can be cloned in yeast. A YAC is a linear vector with a yeast telomere at each end, a centromere, a sequence for autonomous replication in yeast, a selectable marker for yeast, and a polylinker.

Y-linked (holandric ["wholly male"]) trait A trait due to a mutant gene carried on the Y chromosome but with no counterpart on the X.

zygonema The stage during meiosis in prophase I at which homologous chromosomes begin to pair in a highly specific way (like a zipper).

zygote The cell produced by the fusion of the male and female gametes.

SUGGESTED READING

CHAPTER 1: GENETICS: AN INTRODUCTION

Egel, R. 1995. The synaptonemal complex and the distribution of meiotic recombination events. *Trends Genet.* 11:206–208.

Hawley, R. S., and Arbel, T. 1993. Yeast genetics and the fall of the classical view of meiosis. *Cell* 72:301–303.

Staehelin, L. A., and Hepler, P. K. 1996. Cytokinesis in higher plants. *Cell* 84:821–824.

Sturtevant, A. H. 1965. *A history of genetics.* New York: Harper & Row.

CHAPTER 2: MENDELIAN GENETICS

Bateson, W. 1909. *Mendel's principles of heredity.* Cambridge: Cambridge University Press.

Dice, L. R. 1946. Symbols for human pedigree charts. *J. Hered.* 37:11–15.

Mendel, G. 1866. Experiments in plant hybridization (translation). In *Classic papers in genetics,* J. A. Peters, ed. 1959. Englewood Cliffs, NJ: Prentice-Hall.

Peters, J. A., ed. 1959. *Classic papers in genetics.* Englewood Cliffs, NJ: Prentice-Hall.

Sandler, I., and Sandler, L. 1985. A conceptual ambiguity that contributed to the neglect of Mendel's paper. *Hist. Phil. Life Sci.* 7:3–70.

Tschermak-Seysenegg, E. von. 1951. The rediscovery of Mendel's work. *J. Hered.* 42:163–171.

CHAPTER 3: CHROMOSOMAL BASIS OF INHERITANCE, SEX LINKAGE, AND SEX DETERMINATION

Barr, M. L. 1960. Sexual dimorphism in interphase nuclei. *Am. J. Hum. Genet.* 12:118–127.

Bogan, J. S., and Page, D. C. 1994. Ovary? Testis?—A mammalian dilemma. *Cell* 76:603–607.

Bridges, C. B. 1916. Nondisjunction as a proof of the chromosome theory of heredity. *Genetics* 1:1–52, 107–163.

———. 1925. Sex in relation to chromosomes and genes. *Am. Natur.* 59:127–137.

Ellis, N., and Goodfellow, P. N. 1989. The mammalian pseudoautosomal region. *Trends Genet.* 5:406–410.

Farabee, W. C. 1905. Inheritance of digital malformations in man. *Papers Peabody Museum Amer. Arch. Ethnol. (Harvard Univ.)* 3:65–78.

Haqq, C. M., King, C.-Y., Ukiyama, E., Falsafi, S., Haqq, T. N., Donahoe, P. K., and Weiss, M. A. 1994. Molecular basis of mammalian sexual determination: Activation of Müllerian inhibiting substance gene expression by SRY. *Science* 266:1494–1500.

Jiménez, R., Sánchez, A., Burgos, M., and Díaz de la Guardia, R. 1996. Puzzling out the genetics of mammalian sex determination. *Trends Genet.* 12:164–166.

Kay, G. F., Barton, S. C., Surani, M. A., and Rastan, S. 1994. Imprinting and X chromosome counting mechanisms determine *Xist* expression in early mouse development. *Cell* 77:639–650.

Koopman, P., Gubbay, J., Vivian, N., Goodfellow, P., and Lovell-Badge, R. 1991. Male development of chromosomally female mice transgenic for *Sry. Nature* 351:117–121.

Lee, J. T., Strauss, W. M., Dausman, J. A., and Jaenisch, R. 1996. A 450 kb transgene displays properties of the mammalian X-inactivation center. *Cell* 86:83–94.

Lyon, M. F. 1962. Sex chromatin and gene action in the mammalian X-chromosome. *Am. J. Hum. Genet.* 14:135–148.

McClung, C. E. 1902. The accessory chromosome—sex determinant? *Biol. Bull.* 3:43–84.

McElreavy, K., Vilain, E., Abbas, N., Costa, J.-M., Souleyreau, N., Kucheria, K., Boucekkine, C., Thibaud, E., Brauner, R., Flamant, F., and Fellous, M. 1992. XY sex reversal associated with a deletion 5′ to the *SRY* "HMG box" in the testis-determining region. *Proc. Natl. Acad. Sci. USA* 89:11016–11020.

McKusick, V. A. 1965. The royal hemophilia. *Sci. Am.* 213:88–95.

Migeon, B. R. 1994. X-chromosome inactivation: Molecular mechanisms and genetic consequences. *Trends Genet.* 10:230–235.

Morgan, L. V. 1922. Non criss-cross inheritance in *Drosophila melanogaster. Biol. Bull.* 42:267–274.

Morgan, T. H. 1910. Sex-limited inheritance in *Drosophila. Science* 32:120–122.

———. 1911. An attempt to analyze the constitution of the chromosomes on the basis of sex-limited inheritance in *Drosophila. J. Exp. Zool.* 11:365–414.

Page, D. C., de la Chapelle, A., and Weissenbach, J. 1985. Chromosome Y-specific DNA in related human XX males. *Nature* 315:224–226.

Penny, G. D., Kay, G. F., Sheardown, S. A., Rastan, S., and Brockdorff, N. 1996. Requirement for *Xist* in X chromosome inactivation. *Nature* 379:131–137.

Stern, C., Centerwall, W. P., and Sarkar, Q. S. 1964. New data on the problem of Y-linkage of hairy pinnae. *Am. J. Hum. Genet.* 16:455–471.

Sutton, W. S. 1903. The chromosomes in heredity. *Biol. Bull.* 4:231–251.

Willard, H. F. 1996. X chromosome inactivation, *XIST,* and pursuit of the X-inactivation center. *Cell* 86:5–7.

Wilson, E. B. 1905. The chromosomes in relation to the determination of sex in insects. *Science* 22:500–502.

CHAPTER 4: EXTENSIONS OF MENDELIAN GENETIC ANALYSIS

Bultman, S. J., Michaud, E. J., and Woychik, R. P. 1992. Molecular characterization of the mouse *agouti* locus. *Cell* 71:1195–1204.

Ginsburg, V. 1972. Enzymatic basis for blood groups. *Methods Enzymol.* 36:131–149.

Huntington's Disease Collaborative Research Group. 1993. A novel gene containing a trinucleotide repeat that is expanded and unstable on Huntington's disease chromosomes. *Cell* 72:971–983.

Landauer, W. 1948. Hereditary abnormalities and their chemically induced phenocopies. *Growth Symposium* 12:171–200.

Landsteiner, K., and Levine, P. 1927. Further observations on individual differences of human blood. *Proc. Soc. Exp. Biol. Med.* 24:941–942.

Siracusa, L. D. 1994. The *agouti* gene: turned on to yellow. *Trends Genet.* 10:423–428.

CHAPTER 5: GENETIC MAPPING IN EUKARYOTES

Bateson, W., Saunders, E. R., and Punnett, R. G. 1905. Experimental studies in the physiology of heredity. *Rep. Evol. Committee R. Soc.* II:1–55, 80–99.

Blixt, S. 1975. Why didn't Mendel find linkage? *Nature* 256:206.

Creighton, H. S., and McClintock, B. 1931. A correlation of cytological and genetical crossing-over in *Zea mays*. *Proc. Natl. Acad. Sci. USA.* 17:492–497.

Dib, C., and many other authors. 1996. A comprehensive genetic map of the human genome based on 5,264 microsatellites. *Nature* 380:152–154.

Ephrussi, B., and Weiss, M. C. 1969. Hybrid somatic cells. *Sci. Am.* 220:26–35.

Kao, F., Jones, C., and Puck, T. T. 1976. Genetics of somatic mammalian cells: Genetic, immunologic, and biochemical analysis with Chinese hamster cell hybrids containing selected human chromosomes. *Proc. Natl. Acad. Sci. USA* 73:193–197.

McKusick, V. A. 1971. The mapping of human chromosomes. *Sci. Am.* 224:104–113.

Morgan, T. H. 1910. Sex-limited inheritance in *Drosophila*. *Science* 32:120–122.

———. 1910. The method of inheritance of two sex-limited characters in the same animal. *Proc. Soc. Exp. Biol. Med.* 8:17.

———. 1911. An attempt to analyze the constitution of the chromosomes on the basis of sex-limited inheritance in *Drosophila*. *J. Exp. Zool.* 11:365–414.

———. 1911. Random segregation versus coupling in Mendelian inheritance. *Science* 34:384.

Morgan, T. H., Sturtevant, A. H., Müller, H. J., and Bridges, C. B. 1915. *The mechanism of Mendelian heredity*. New York: Henry Holt.

Ried, T., Baldini, A., Rand, T. C., and Ward, D. C. 1992. Simultaneous visualization of seven different DNA probes by *in situ* hybridization using combinatorial fluorescence and digital imaging microscopy. *Proc. Natl. Acad. Sci. USA* 89:1388-1392.

Sturtevant, A. H. 1913. The linear arrangement of six sex-linked factors in *Drosophila* as shown by their mode of association. *J. Exp. Zool.* 14:43–59.

Sutton, W. S. 1903. The chromosomes in heredity. *Biol. Bull.* 4:231–251.

CHAPTER 6: GENETIC ANALYSIS IN BACTERIA AND BACTERIOPHAGES

Benzer, S. 1959. On the topology of the genetic fine structure. *Proc. Natl. Acad. Sci. USA* 45:1607–1620.

———. 1961. On the topography of the genetic fine structure. *Proc. Natl. Acad. Sci. USA* 47:403–415.

———. 1962. The fine structure of the gene. *Sci. Am.* 206:70–84.

Curtiss, R. 1969. Bacterial conjugation. *Annu. Rev. Microbiol.* 23:69–136.

Ellis, E. L., and Delbruck, M. 1939. The growth of bacteriophage. *J. Gen. Physiol.* 22:365–384.

Hayes, W. 1968. *The genetics of bacteria and their viruses,* 2nd ed. New York: Wiley.

Hershey, A. D., and Rotman, R. 1949. Genetic recombination between host-range and plaque-type mutants of bacteriophage in single bacterial cells. *Genetics* 34:44–71.

Hotchkiss, R. D., and Gabor, M. 1970. Bacterial transformation with special reference to recombination processes. *Annu. Rev. Genet.* 4:193–224.

Jacob, F., and Wollman, E. L. 1951. *Sexuality and the genetics of bacteria*. New York: Academic Press.

Susman, M. 1970. General bacterial genetics. *Annu. Rev. Genet.* 4:135–176.

Wollman, E. L., Jacob, F., and Hayes, W. 1962. Conjugation and genetic recombination in *E. coli* K-12. *Cold Spring Harbor Symp. Quant. Biol.* 21:141–162.

Zinder, N., and Lederberg, J. L. 1952. Genetic exchange in *Salmonella*. *J. Bacteriol.* 64:679–699.

CHAPTER 7: CHROMOSOMAL MUTATIONS

Barr, M. L., and Bertram, E. G. 1949. A morphological distinction between neurones of the male and female, and the behavior of the nucleolar satellite during accelerated nucleoprotein synthesis. *Nature* 163:676–677.

Blackwell, T. K., and Alt, F. W. 1989. Mechanism and developmental program of immunoglobulin gene rearrangements in mammals. *Annu. Rev. Genet.* 23:605–636.

Borst, P., and Greaves, D. R. 1987. Programmed gene rearrangements altering gene expression. *Science* 235:658–667.

Caskey, C. T., Pizzuti, A., Fu, Y.-H., Fenwick, R. G., and Nelson, D. L. 1992. Triplet repeat mutations in human disease. *Science* 256:784–789.

Dalla-Favera, R., Martinotti, S., Gallo, R., Erickson, J., and Croce, C. 1983. Translocation and rearrangements of the *c-myc* oncogene locus in human undifferentiated B-cell lymphomas. *Science* 219:963–997.

DeKlein, A., van Kessel, A. G., Grosveld, G., Bartram, C. R., Hagemeijer, A., Bootsma, D., Spurr, N. K., Heisterkamp, N., Groffen, J., and Stephenson, J. R. 1982. A cellular oncogene is translocated to the Philadelphia chromosome in chronic myelocytic leukemia. *Nature* 300:765–767.

Huntington's Disease Collaborative Research Group. 1993. A novel gene containing a trinucleotide repeat that is expanded and unstable on Huntington's disease chromosome. *Cell* 72:971–983.

Kremer, E., Pritchard, M., Lynch, M., Yu, S., Holman, K., Baker, E., Warren, S. T., Schlessinger, D., Sutherland, G. R., and Richards, R. I. 1991. Mapping of DNA instability at the fragile X to a trinucleotide repeat sequence p(CGG)*n*. *Science* 252:1711–1714.

Lyon, M. F. 1961. Gene action in the X-chromosomes of the mouse (*Mus. musculus L*). *Nature* 190:372–373.

Penrose, L. S., and Smith, G. F. 1966. *Down's anomaly*. Boston: Little, Brown.

Richards, R. I., and Sutherland, G. R. 1992. Dynamic mutations: A new class of mutations causing human disease. *Cell* 70:709–712.

———. 1992. Fragile X syndrome: The molecular picture comes into focus. *Trends Genet.* 8:249–255.

Ried, T., Baldini, A., Rand, T. C., and Ward, D. C. 1992. Simultaneous visualization of seven different DNA probes by *in situ* hybridization using combinatorial fluorescence and digital imaging microscopy. *Proc. Natl. Acad. Sci. USA* 89:1388–1392.

Rowley, J. D. 1973. A new consistent chromosomal abnormality in chronic myelogenous leukemia identified by quinacrine fluorescence and Giemsa staining. *Nature* 243:290–293.

Shaw, M. W. 1962. Familial mongolism. *Cytogenetics* 1:141–179.

Siomi, H., Siomi, M. C., Nussbaum, R. L., and Dreyfuss, G. 1993. The protein product of the fragile X gene, *FMR1*, has characteristics of an RNA-binding protein. *Cell* 74:291–298.

Tarleton, J. C., and Saul, R. A. 1993. Molecular genetic advances in fragile X syndrome. *J. Pediatrics* 122:169–185.

Verkerk, A. J. M. H., Piertti, M., Sutcliff, J. S., Fu, Y.-H., Kuhl, D. P. A., Pizzuti, A., Reiner, O., Richards, S., Victoria, M. F., Zhang, F., Eussen, B. E., van Ommen, G.-J. B., Blonden, L. A. J., Riggins, G. J., Chastain, J. L., Kunst, C. B., Galjaard, H., Caskey, C. T., Nelson, D. L., Oostra, B. A., and Warrent, S. T. 1991. Identification of a gene (*FMR–1*) containing a CGG repeat coincident with a breakpoint cluster region exhibiting length variation in fragile X syndrome. *Cell* 65: 905–914.

CHAPTER 8: GENE CONTROL OF PROTEINS

Beadle, G. W., and Tatum, E. L. 1942. Genetic control of biochemical reactions in *Neurospora*. *Proc. Natl. Acad. Sci. USA* 27:499–506.

Collins, F. 1992. Cystic fibrosis: Molecular biology and therapeutic implications. *Science* 256:774–779.

Doggett, N. A., Cheng, J.-F., Smith, C. L., and Cantor, C. R. 1989. The Huntington's disease locus is most likely within 325 kilobases of the chromosome 4p telomere. *Proc. Natl. Acad. Sci. USA* 86:10011–10014.

Galjaard, H. 1986. Biochemical diagnosis of genetic diseases. *Experientia* 42:1075–1085.

Garrod, A. E. 1909. *Inborn errors of metabolism.* New York: Oxford University Press.

Gusella, J. F., Wexler, N. S., Conneally, P. M., Naylor, S. L., Anderson, M. A., Tanzi, R. E., Watkins, P. C., Ottina, K., Wallace, M. R., Sakaguchi, A. Y., Young, A. B., Shoulson, I., Bonilla, E., and Martin, J. B. 1993. A polymorphic DNA marker genetically linked to Huntington's disease. *Nature* 306:234–238.

Guttler, F., and Woo, S. L. C. 1986. Molecular genetics of PKU. *J. Inherited Metab. Dis.* 9 Suppl.1:58–68.

Ingram, V. M. 1963. *The hemoglobins in genetics and evolution.* New York: Columbia University Press.

Maniatis, T., Fritsch, E. F., Lauer, J., and Lawn, R. M. 1980. The molecular genetics of human hemoglobins. *Annu. Rev. Genet.* 14:145–178.

McIntosh, I., and Cutting, G. R. 1992. Cystic fibrosis transmembrane conductance regulator and the etiology and pathogenesis of cystic fibrosis. *FASEB J.* 6:2775–2782.

Motulsky, A. G. 1964. Hereditary red cell traits and malaria. *Am. J. Trop. Med. Hyg.* 13:147–158.

———. 1973. Frequency of sickling disorders in U.S. blacks. *N. Engl. J. Med.* 288:31–33.

Neel, J. V. 1949. The inheritance of sickle-cell anemia. *Science* 110:64–66.

Pauling, L., Itano, H. A., Singer, S. J., and Wells, J. C. 1949. Sickle cell anemia, a molecular disease. *Science* 110:543–548.

Scriver, C. R., and Clow, C. L. 1980. Phenylketonuria and other phenylalanine hydroxylation mutants in man. *Annu. Rev. Genet.* 14:179–202.

Srb, A. M., and Horowitz, N. H. 1944. The ornithine cycle in *Neurospora* and its genetic control. *J. Biol. Chem.* 154:129–139.

Stout, J. T., and Caskey, C. T. 1988. The Lesch-Nyhan syndrome: Clinical, molecular and genetic aspects. *Trends Genet.* 4:175–178.

Woo, S. L. C., Lidsky, A. S., Guttler, F., Chandra, T., and Robson, K. J. H. 1983. Cloned human phenylalanine hydroxylase gene allows prenatal diagnosis and carrier detection of classical phenylketonuria. *Nature* 300:151–155.

CHAPTER 9: DNA: THE GENETIC MATERIAL

Avery, O. T., MacLeod, C. M., and McCarty, M. 1944. Studies on the chemical nature of the substance inducing transformation of pneumococcal types. Induction of transformation by a deoxyribonucleic acid fraction isolated from pneumococcus type III. *J. Exp. Med.* 79:137–158.

Chargaff, E. 1951. Structure and function of nucleic acids as cell constituents. *Fed. Proc.* 10:654–659.

Dickerson, R. E. 1983. The DNA helix and how it is read. *Sci. Am.* 249 (December):94–111.

Fraenkel-Conrat, H., and Singer, B. 1957. Virus reconstitution: Combination of protein and nucleic acid from different strains. *Biochim. Biophys. Acta* 24:540–548.

Franklin, R. E., and Gosling, R. 1953. Molecular configuration of sodium thymonucleate. *Nature* 171:740–741.

Geis, I. 1983. Visualizing the anatomy of A, B, and Z-DNAs. *J. Biomol. Struct. Dynam.* 1:581–591.

Gierer, A., and Schramm, G. 1956. Infectivity of ribonucleic acid from tobacco mosaic virus. *Nature* 177:702–703.

Griffith, F. 1928. The significance of pneumococcal types. *J. Hyg.* (Lond.) 27:113–159.

Hershey, A. D., and Chase, M. 1952. Independent functions of viral protein and nucleic acid in growth and bacteriophage. *J. Gen. Physiol.* 36:39–56.

Jaworski, A., Hsieh, W.-T., Blaho, J. A., Larson, J. E., and Wells, R. D. 1988. Left-handed DNA in vivo. *Science* 238:773–777.

Krishna, P., Kennedy, B. P., van de Sande, J. H., and McGhee, J. D. 1988. Yolk proteins from nematodes, chickens, and frogs bind strongly and preferentially to left-handed Z-DNA. *J. Biol. Chem.* 263:19066–19070.

Pauling, L., and Corey, R. B. 1956. Specific hydrogen-bond formation between pyrimidines and purines in deoxyribonucleic acids. *Arch. Biochem. Biophys.* 65:164–181.

Rich, A., Nordheim, A., and Wang, A. H.-J. 1984. The chemistry and biology of left-handed Z-DNA. *Annu. Rev. Biochem.* 53:791–846.

Wang, A. H.-J., Quigley, G. J., Kolpak, F. J., Crawford, J. L., van Boom, J. H., van der Marel, G., and Rich, A. 1979. Molecular structure of a left-handed double helical DNA fragment at atomic resolution. *Nature* 282:680–686.

Wang, J. C. 1982. DNA topoisomerases. *Sci. Am.* 247:94–109.

Watson, J. D. 1968. *The double helix.* New York: Atheneum.

Watson, J. D., and Crick, F. H. C. 1953. Genetical implications of the structure of deoxyribonucleic acid. *Nature* 171:964–969.

———. 1953. Molecular structure of nucleic acids. A structure for deoxyribose nucleic acid. *Nature* 171:737–738.

Wilkins, M. H. F., Stokes, A. R., and Wilson, H. R. 1953. Molecular structure of deoxypentose nucleic acids. *Nature* 171:738–740.

CHAPTER 10: DNA: ORGANIZATION IN CHROMOSOMES

Amati, B. B., and Gasser, S. M. 1988. Chromosomal ARS and CEN elements bind specifically to the yeast nuclear scaffold. *Cell* 54:967–978.

Blackburn, E. H. 1994. Telomeres: No end in sight. *Cell* 77:621–623.

Blackburn, E. H., and Szostak, J. W. 1984. The molecular structure of centromeres and telomeres. *Annu. Rev. Biochem.* 53:163–194.

Bloom, K. S., Amaya, E., Carbon, J., Clarke, L., Hill, A., and Yeh, E. 1984. Chromatin conformation of yeast centromeres. *J. Cell Biol.* 99:1559–1568.

Britten, R. J., and Kohne, D. E. 1968. Repeated sequences in DNA. *Science* 161:529–540.

Burlingame, R. W., Love, W. E., Wang, B.-C., Hamlin, R., Xuang, N.-H., and Moudranakis, E. N. 1985. Crystallographic structure of the octameric histone core of the nucleosome at a resolution of 33 Å. *Science* 228:546–553.

Cai, M., and Davis, R. W. 1990. Yeast centromere binding protein CBF1 of the helix-loop-helix protein family is required for chromosome stability and methionine prototrophy. *Cell* 61:437–446.

Carbon, J. 1984. Yeast centromeres: Structure and function. *Cell* 37:351–353.

Clarke, L. 1990. Centromeres of budding and fission yeasts. *Trends Genet.* 6:150–154.

Comings, D. 1978. Mechanisms of chromosome banding and implications for chromosome structure. *Annu. Rev. Genet.* 12:25–46.

D'Ambrosio, E., Waitzikin, S. D., Whitney, F. R., Salemme, A., and Furano, A. V. 1985. Structure of the highly repeated, long interspersed DNA family (LINE or L1Rn) of the rat. *Mol. Cell. Biol.* 6:411–424.

Freifelder, D. 1978. *The DNA molecule. Structure and properties.* San Francisco: Freeman.

Grosschedl, R., Giese, K., and Pagel, J. 1994. HMG domain proteins: Architectural elements in the assembly of nucleoprotein structures. *Trends Genet.* 10:94–100.

Jelinek, W. R., and Schmid, C. W. 1982. Repetitive sequences in eukaryotic DNA and their expression. *Annu. Rev. Biochem.* 51:813–844.

Korenberg, J. R., and Rykowski, M. C. 1988. Human genome organization: Alu, LINES, and the molecular structure of metaphase chromosome bands. *Cell* 53:391–400.

Kornberg, R. D. 1977. Structure of chromatin. *Annu. Rev. Biochem.* 46:931–954.

Kornberg, R. D., and Klug, A. 1981. The nucleosome. *Sci. Am.* 244 (2):52–64.

Levis, R. W., Ganesan, R., Houtchens, K., Tolar, L. A., and Sheen, F. 1993. Transposons in place of telomeric repeats at a Drosophila telomere. *Cell* 75:1083–1093.

Lewin, B. 1980. *Gene expression,* 2nd ed., vol. 2, *Eucaryotic chromosomes.* New York: Wiley.

Long, E. O., and Dawid, I. B. 1980. Repeated genes in eukaryotes. *Annu. Rev. Biochem.* 49:727–764.

Ludérus, M. E. E., den Blaauwen, J. L., de Smit, O. J. B., Compton, D. A., and van Driel, R. 1994. Binding of matrix attachment regions to lamin polymers involves single-stranded regions and the minor groove. *Mol. Cell. Biol.* 14:6297–6305.

Mason, J. M., and Biessmann, A. 1995. The unusual telomeres of *Drosophila. Trends Genet.* 11:58–62.

Mirkovich, J., Mirault, M.-E., and Laemmli, U. K. 1984. Organization of the higher-order chromatin loop: Specific DNA attachment sites on nuclear scaffold. *Cell* 39:223–232.

Morse, R. H., and Simpson, R. T. 1988. DNA in the nucleosome. *Cell* 54:285–287.

Moyzis, R. K. 1991. The human telomere. *Sci. Am.* 265 (Aug):48–55.

Olins, A. L., Carlson, R. D., and Olins, D. E. 1975. Visualization of chromatin substructure: nu-bodies. *J. Cell Biol.* 64:528–537.

Pettijohn, D. E. 1988. Histone-like proteins and bacterial chromosome structure. *J. Biol. Chem.* 263:12793–12796.

Pluta, A. F., Mackay, A. M., Ainsztein, A. M., Goldberg, I. G., and Earnshaw, W. C. 1995. The centromere: Hub of chromosomal activities. *Science* 270:1591–1594.

Pruss, D., Bartholomew, B., Persinger, J., Hayes, J., Arents, G., Moudrianakis, E. N., and Wolfe, A. P. 1996. An asymmetric model for the nucleosome: A binding site for linker histones inside the DNA gyres. *Science* 274:614–617.

Roth, S. Y., and Allis, C. D. 1996. Histone acetylation and chromatin assembly: A single escort, multiple dances? *Cell* 87:5–8.

Singer, M. F. 1982. Highly repeated sequences in mammalian genomes. *Int. Rev. Cytol.* 76:67–112.

———. 1982. SINEs and LINEs: Highly repeated short and long interspersed sequences in mammalian genomes. *Cell* 28:133–134.

Sinsheimer, R. L. 1959. A single-stranded deoxyribonucleic acid from bacteriophage ΦX174. *J. Mol. Biol.* 1:43–53.

van Holde, K., and Zlatanova, J. 1996. What determines the folding of the chromatin fiber? *Proc. Natl. Acad. Sci. USA.* 93:10548–10555.

Williamson, J. R., Raghuraman, M. K., and Cech, T. R. 1989. Monovalent cation-induced structure of telomeric DNA: The G-quartet model. *Cell* 59:871–880.

Woodcock, C. L. F., Frado, L.-L. Y., and Rattner, J. B. 1984. The higher-order structure of chromatin: Evidence for a helical ribbon arrangement. *J. Cell Biol.* 99:42–52.

Worcel, A. 1978. Molecular architecture of the chromatin fiber. *Cold Spring Harbor Symp. Quant. Biol.* 42:313–324.

Worcel, A., and Benyajati, C. 1977. Higher order coiling of DNA in chromatin. *Cell* 12:83–100.

Worcel, A., and Burgi, E. 1972. On the structure of the folded chromosome of *Escherichia coli. J. Mol. Biol.* 71:127–147.

Zimmerman, S. B. 1982. The three-dimensional structure of DNA. *Annu. Rev. Biochem.* 51:395–427.

CHAPTER 11: DNA REPLICATION

Baker, T. A., and Wickner, S. H. 1992. Genetics and enzymology of DNA replication in *Escherichia coli. Annu. Rev. Genet.* 26:447–477.

Beese, L. S., Derbyshire, V., and Steitz, T. A. 1993. Structure of DNA polymerase I Klenow fragment bound to duplex DNA. *Science* 260:352–355.

Bell, S. P., and Stillman, B. 1992. ATP-dependent recognition of eukaryotic origins of DNA replication by a multiprotein complex. *Nature* 357:128–134.

Benbow, R. M., Zhao, J., and Larson, D. D. 1992. On the nature of origins of DNA replication in eukaryotes. *BioEssays* 14:661–670.

Blackburn, E. H. 1991. Structure and function of telomeres. *Nature* 350:569–573.

———. 1992. Telomerases. *Annu. Rev. Biochem.* 61:113–129.

Coverley, D., and Laskey, R. A. 1994. Regulation of eukaryotic DNA replication. *Annu. Rev. Biochem.* 63:745–776.

DeLucia, P., and Cairns, J. 1969. Isolation of an *E. coli* strain with a mutation affecting DNA polymerase. *Nature* 224:1164–1166.

Diller, J. D., and Raghuraman, M. K. 1994. Eukaryotic replication origins: Control in space and time. *Trends Biochem.* 19:320–325.

Gilbert, W., and Dressler, D. 1968. DNA replication: The rolling circle model. *Cold Spring Harbor Symp. Quant. Biol.* 33:473–484.

Gilley, D., and Blackburn, E. H. 1996. Specific RNA residue interactions required for enzymatic functions of Tetrahymena telomerase. *Mol. Cell. Biol.* 16:66–75.

Hamlin, J. L. 1992. Mammalian origins of replication. *BioEssays* 14:651–659.

Huberman, J. A. 1995. Prokaryotic and eukaryotic replicons. *Cell* 82:535–542.

Huberman, J. A., and Riggs, A. D. 1968. On the mechanism of DNA replication in mammalian chromosomes. *J. Mol. Biol.* 32:327–341.

Kornberg, A. 1960. Biologic synthesis of deoxyribonucleic acid. *Science* 131:1503–1508.

Kornberg, A., and Baker, T. A. 1992. *DNA replication,* 2nd ed. New York: Freeman.

Lendvay, T. S., Morris, D. K., Sah, J., Balasubramanian, B., and Lundblad, V. 1996. Senescence mutants of *Saccharomyces cerevisiae* with a defect in telomere replication identify three additional *EST* genes. *Genetics* 144:1399–1412.

Marians, K. J. 1992. Prokaryotic DNA replication. *Annu. Rev. Biochem.* 61:673–719.

Mason, J. M., and Biessmann, A. 1995. The unusual telomeres of *Drosophila*. *Trends Genet.* 11:58–62.

Meselson, M., and Radding, C. M. 1975. A general model for genetic recombination. *Proc. Natl. Acad. Sci. USA* 72:358–361.

Meselson, M., and Stahl, F. W. 1958. The replication of DNA in *Escherichia coli*. *Proc. Natl. Acad. Sci. USA* 44:671–682.

Modrich, P. 1987. DNA mismatch correction. *Annu. Rev. Biochem.* 56:435–466.

Nasmyth, K. 1996. At the heart of the budding yeast cell cycle. *Trends Genet.* 12:405–412.

Ogawa, T., Baker, T. A., van der Ende, A., and Kornberg, A. 1985. Initiation of enzymatic replication at the origin of the *Escherichia coli* chromosome: Contributions of RNA polymerase and primase. *Proc. Natl. Acad. Sci. USA* 82:3562–3566.

Ogawa, T., and Okazaki, T. 1980. Discontinuous DNA replication. *Annu. Rev. Biochem.* 49:424–457.

Okazaki, R. T., Okazaki, K., Sakobe, K., Sugimoto, K., and Sugino, A. 1968. Mechanism of DNA chain growth. I. Possible discontinuity and unusual secondary structure of newly synthesized chains. *Proc. Natl. Acad. Sci. USA* 59:598–605.

Runge, K. W., and Zakian, V. A. 1996. *TEL2*, an essential gene required for telomere length regulation and telomere position effect in *Saccharomyces cerevisiae*. *Mol. Cell. Biol.* 16:3094–3105.

Shippen-Lentz, D., and Blackburn, E. H. 1990. Functional evidence for an RNA template in telomerase. *Science* 247:546–552.

Steiner, B. R., Hidaka, K., and Futcher, B. 1996. Association of the Est1 protein with telomerase activity in yeast. *Proc. Natl. Acad. Sci. USA* 93:2817–2821.

Stillman, B. 1994. Smart machines at the DNA replication fork. *Cell* 78:725–728.

Stukenberg, P. T., Turner, J., and O'Donnell, M. 1994. An explanation for lagging strand replication: Polymerase hopping among DNA sliding clamps. *Cell* 78:877–887.

Szostak, J., Orr-Weaver, T., Rothstein, R., and Stahl, F. 1983. The double-strand break repair model for recombination. *Cell* 33:25–35.

Wellinger, R. J., Ethier, K., Labrecque, P., and Zakian, V. A. 1996. Evidence for a new step in telomere maintenance. *Cell* 85:423–433.

West, S. C. 1992. Enzymes and molecular mechanisms of genetic recombination. *Annu. Rev. Biochem.* 61:603–640.

CHAPTER 12: GENE EXPRESSION: TRANSCRIPTION

Atchison, M. L. 1988. Enhancers: Mechanisms of action and cell specificity. *Annu. Rev. Cell Biol.* 4:127–153.

Belfort, M. 1989. Bacteriophage introns: Parasites within parasites? *Trends Genet.* 5:209–213.

Bogenhagen, D. F., Sakonju, S., and Brown, D. D. 1980. A control region in the center of the 5S RNA gene directs specific initiation of transcription II: The 3' border of the region. *Cell* 19:27–35.

Brand, A. H., Breeden, L., Abraham, J., Sternglanz, R., and Nasmyth, K. 1987. Characterization of a "silencer" in yeast: A DNA sequence with properties opposite to those of a transcriptional enhancer. *Cell* 41:41–48.

Breathnach, R., Mandel, J. L., and Chambon, P. 1977. Ovalbumin gene is split in chicken DNA. *Nature* 270:314–318.

Breitbart, R. E., Andreadis, A., and Nadal-Ginard, B. 1987. Alternative splicing: A ubiquitous mechanism for the generation of multiple protein isoforms from single genes. *Annu. Rev. Biochem.* 56:467–495.

Brody, E., and Abelson, J. 1985. The "spliceosome": Yeast premessenger RNA associates with a 40S complex in a splicing-dependent reaction. *Science* 228:963–967.

Buratowski, S. 1994. The basics of basal transcription by RNA polymerase II. *Cell* 77:1–3.

Busby, S., and Ebright, R. H. 1994. Promoter structure, promoter recognition, and transcription activation in prokaryotes. *Cell* 79:743–746.

Cech, T. R. 1983. RNA splicing: Three themes with variations. *Cell* 34:713–716.

———. 1985. Self-splicing RNA: Implications for evolution. *Int. Rev. Cytol.* 93:3–22.

———. 1986. The generality of self-splicing RNA: Relationship to nuclear mRNA splicing. *Cell* 44:207–210.

Choi, Y. D., Grabowski, P. J., Sharp, P. A., and Dreyfuss, G. 1986. Heterogeneous nuclear ribonucleoproteins: Role in RNA splicing. *Science* 231:1534–1539.

Chu, F. K., Maley, G. F., West, D. K., Belfort, M., and Maley, F. 1986. Characterization of the intron in the phage T4 thymidylate synthase gene and evidence for its self-excision from the primary transcript. *Cell* 45:157–166.

Clark, D. J., and Felsenfeld, G. 1992. A nucleosome core is transferred out of the path of a transcribing polymerase. *Cell* 71:11–22.

Crick, F. H. C. 1979. Split genes and RNA splicing. *Science* 204:264–271.

Eick, D., Wedel, A., and Heumann, H. 1994. From initiation to elongation: Comparison of transcription by prokaryotic and eukaryotic RNA polymerases. *Trends Genet.* 10:292–296.

Geiduschek, E. P., and Tocchini-Valentini, G. P. 1988. Transcription by RNA polymerase III. *Annu. Rev. Biochem.* 57:873–914.

Gesteland, R. F., and Atkins, J. F., eds. 1993. *The RNA world.* Cold Spring Harbor, NY: Cold Spring Harbor Laboratory Press.

Goodrich, J. A., Cutler, G., and Tjian, R. 1996. Contacts in context: Promoter specificity and macromolecular interactions in transcription. *Cell* 84:825–830.

Green, M. R., 1986. Pre-mRNA splicing. *Annu. Rev. Genet.* 20:671–708.

———. 1991. Biochemical mechanisms of constitutive and regulated pre-mRNA splicing. *Annu. Rev. Cell Biol.* 7:559–599.

Greenblatt, J. 1991. RNA polymerase-associated transcription factors. *Trends Biochem. Sci.* 11:408–411.

Guarente, L., and Birmingham-McDonogh, O. 1992. Conservation and evolution of transcriptional mechanisms in eukaryotes. *Trends Genet.* 8:27–32.

Guthrie, C. 1992. Messenger RNA splicing in yeast: Clues to why the spliceosome is a ribonucleoprotein. *Science* 253:157–163.

Halle, J.-P., and Meisterernst, M. 1996. Gene expression: Increasing evidence for a transcriptosome. *Trends Genet.* 12:161–163.

Helman, J. D., and Chamberlin, M. J. 1988. Structure and function of bacterial sigma factors. *Annu. Rev. Biochem.* 57:839–872.

Horowitz, D. S., and Krainer, A. R. 1994. Mechanisms for selecting 5' splice sites in mammalian pre-mRNA splicing. *Trends Genet.* 10:100–105.

Jeffreys, A. J., and Flavell, R. A. 1977. The rabbit beta-globin gene contains a large insert in the coding sequence. *Cell* 12:1097–1108.

Lang, W. H., Morrow, B. E., Ju, Q., Warner, J. R., and Reeder, R. H. 1994. A model for transcription termination by RNA polymerase I. *Cell* 79:527–534.

Marmur, J., Greenspan, C. M., Palecek, E., Kahan, F. M., Levine, J., and Mandel, M. 1963. Specificity of the complementary RNA formed by *Bacillus subtilis* infected with bacteriophage SP8. *Cold Spring Harbor Symp. Quant. Biol.* 28:191–199.

Nilsen, T. W. 1994. RNA-RNA interactions in the spliceosome: Unraveling the ties that bind. *Cell* 78:1–4.

Nomura, M. 1973. Assembly of bacterial ribosomes. *Science* 179:864–873.

O'Hare, K. 1995. mRNA 3' ends in focus. *Trends Genet.* 11:253–257.

Pabo, C. O., and Sauer, R. T. 1992. Transcription factors: Structural families and principles of DNA recognition. *Annu. Rev. Biochem.* 61:1053–1093.

Roeder, R. G. 1991. The complexities of eukaryotic transcription initiation: Regulation of preinitiation complex assembly. *Trends Biochem. Sci.* 16:402–408.

Rogers, J. H. 1989. How were introns inserted into nuclear genes? *Trends Genet.* 5:213–216.

Sharp, P. A. 1985. On the origin of RNA splicing and introns. *Cell* 42:397–400.

Sharp, P. A. 1994. Split genes and RNA splicing. Nobel lecture. *Cell* 77:805–815.

Sollner-Webb, B. 1988. Surprises in polymerase III transcription. *Cell* 52:153–154.

Srivastava, A. K., and Schlessinger, D. 1990. Mechanism and regulation of bacterial ribosomal RNA processing. *Annu. Rev. Microbiol.* 44:105–129.

Stragier, P. 1991. Dances with sigmas. *EMBO J.* 10:3559–3566.

Struhl, K. 1996. Chromatin structure and RNA polymerase II connection: Implications for transcription. *Cell* 84:179–182.

Symons, R. H. 1992. Small catalytic RNAs. *Annu. Rev. Biochem.* 61:641–671.

Thompson, C. C., and McKnight, S. L. 1992. Anatomy of an enhancer. *Trends Genet.* 8:232–236.

Tilghman, S. M., Curis, P. J., Tiemeier, D. C., Leder, P., and Weissman, C. 1978. The intervening sequence of a mouse β-globin gene is transcribed within the 15S β-globin mRNA precursor. *Proc. Natl. Acad. Sci. USA* 75:1309–1313.

Tilghman, S. M., Tiemeier, D. C., Seidman, J. G., Peterlin, B. M., Sullivan, M., Maizel, J. V., and Leder, P. 1978. Intervening sequence of DNA identified in the structural portion of a mouse beta-globin gene. *Proc. Natl. Acad. Sci. USA* 78:725–729.

Wahle, E., and Keller, W. 1992. The biochemistry of 3′-end cleavage and polyadenylation of messenger RNA precursors. *Annu. Rev. Biochem.* 61:419–440.

Weinstock, R., Sweet, R., Weiss, M., Cedar, H., and Axel, R. 1978. Intragenic DNA spacers interrupt the ovalbumin gene. *Proc. Natl. Acad. Sci. USA* 75:1299–1303.

Weis, L., and Reinberg, D. 1992. Transcription by RNA polymerase II: Initiator-directed formation of transcription-competent complexes. *FASEB J.* 6:3300–3309.

White, R. J., and Jackson, S. P. 1992. The TATA-binding protein: A central role in transcription by RNA polymerases I, II, and III. *Trends Genet.* 8:284–288.

Wolffe, 1994. Transcription: In tune with the histones. *Cell* 77:13–16.

Zaug, A. J., and Cech, T. R. 1986. The intervening sequence RNA of *Tetrahymena* is an enzyme. *Science* 231:470–475.

CHAPTER 13: GENE EXPRESSION: TRANSLATION

Blobel, G., and Dobberstein, B. 1975. Transfer of proteins across membranes. I. Presence of proteolytically processed and unprocessed nascent immunoglobulin light chains on membrane-bound ribosomes of murine myeloma. *J. Cell Biol.* 67:835–851.

Brenner, S., Jacob, F., and Meselson, M. 1961. An unstable intermediate carrying information from genes to ribosomes for protein synthesis. *Nature* 190:576–581.

Cold Spring Harbor Symposia for Quantitative Biology. 1966: *The genetic code*, vol. 31. Cold Spring Harbor, NY: Cold Spring Harbor Laboratory.

Crick, F. H. C. 1966. Codon-anticodon pairing: The wobble hypothesis. *J. Mol. Biol.* 19:548–555.

Crick, F. H. C., Barnett, L., Brenner, S., and Watts-Tobin, R. J. 1961. General nature of the genetic code for proteins. *Nature* 192:1227–1232.

Dingwall, C., and Laskey, R. A. 1986. Protein import into the cell nucleus. *Annu. Rev. Cell Biol.* 2:367–390.

Garen, A. 1968. Sense and nonsense in the genetic code. *Science* 160:149–159.

Griffiths, G., and Simons, K. 1986. The trans Golgi network: Sorting at the exit site of the Golgi complex. *Science* 234:438–442.

Horowitz, S., and Gorovsky, M. A. 1985. An unusual genetic code in nuclear genes of *Tetrahymena*. *Proc. Natl. Acad. Sci. USA* 82:2452–2455.

Jackson, R. J., and Standart, N. 1990. Do the poly(A) tail and 3′ untranslated region control mRNA translation? *Cell* 62:15–24.

Khorana, H. G. 1966–67. Polynucleotide synthesis and the genetic code. *Harvey Lectures* 62:79–105.

Kozak, M. 1983. Comparison of initiation of protein synthesis in procaryotes, eucaryotes, and organelles. *Microbiol. Rev.* 47:145.

———. 1989. Context effects and inefficient initiation at non-AUG codons in eucaryotic cell-free translation systems. *Mol. Cell. Biol.* 9:5073–5080.

McCarthy, J. E. G., and R. Brimacombe. 1994. Prokaryotic translation: The interactive pathway leading to initiation. *Trends Genet.* 10:402–407.

Meyer, D. I. 1982. The signal hypothesis—A working model. *Trends Biochem. Sci.* 7:320–321.

Morgan, A. R., Wells, R. D., and Khorana, H. G. 1966. Studies on polynucleotides. LIX. Further codon assignments from amino acid incorporation directed by ribopolynucleotides containing repeating trinucleotide sequences. *Proc. Natl. Acad. Sci. USA* 56:1899–1906.

Nierhaus, K. H. 1990. The allosteric three-site model for the ribosomal elongation cycle: Features and future. *Biochemistry* 29:4997–5008.

Nirenberg, M., and Leder, P. 1964. RNA code words and protein synthesis. *Science* 145:1399–1407.

Nirenberg, M., and Matthaei, J. H. 1961. The dependence of cell-free protein synthesis in *E. coli* upon naturally occurring or synthetic polyribonucleotides. *Proc. Natl. Acad. Sci. USA* 47:1588–1602.

Noller, H. F., Hoffarth, V., and Zimniak, L. 1992. Unusual resistance of peptidyl transferase to protein extraction procedures. *Science* 256:1416–1419.

Pfeffer, S. R., and Rothman, J. E. 1987. Biosynthetic protein transport and sorting by the endoplasmic reticulum and Golgi. *Annu. Rev. Biochem.* 56:829–852.

Rogers, S., Wells, R., and Rechsteiner, M. 1986. Amino acid sequences common to rapidly degraded proteins: The PEST hypothesis. *Science* 234:364–368.

Ryan, K. R., and Jensen, R. E. 1995. Protein translocation across mitochondrial membranes: What a long, strange trip it is. *Cell* 83:517–519.

Schekman, R. 1985. Protein localization and membrane traffic in yeast. *Annu. Rev. Cell Biol.* 1:115–143.

Schnell, D. J. 1995. Shedding light on the chloroplast protein import machinery. *Cell* 83:521–524.

Shine, J., and Dalgarno, L. 1974. The 3′-terminal sequence of *Escherichia coli* 16S ribosomal RNA: Complementarity to nonsense triplet and ribosome binding sites. *Proc. Natl. Acad. Sci. USA* 71:1342–1346.

Silver, P. A. 1991. How proteins enter the nucleus. *Cell* 64:489–497.

Verner, K., and Schatz, G. 1988. Protein translocation across membranes. *Science* 241:1307–1313.

Watson, J. D. 1963. The involvement of RNA in the synthesis of proteins. *Science* 140:17–26.

Zheng, N., and Gierasch, L. M. 1996. Signal sequences: The same yet different. *Cell* 86:849–852.

CHAPTER 14: CLONING AND MANIPULATION OF DNA

Arber, W. 1965. Host-controlled modification of bacteriophage. *Annu. Rev. Microbiol.* 19:365–378.

Arber, W., and Dussoix, D. 1962. Host specificity of DNA produced by *Escherichia coli* I. Host controlled modification of bacteriophage lambda. *J. Mol. Biol.* 5:18–36.

Arnheim, N., and Erlich, H. 1992. Polymerase chain reaction strategy. *Annu. Rev. Biochem.* 61:131–156.

Boyer, H. W. 1971. DNA restriction and modification mechanisms in bacteria. *Annu. Rev. Microbiol.* 25:153–176.

Bult, C. J., and 39 other authors. 1996. Complete genome sequence of the methanogenic archaeon, *Methanococcus jannaschii. Science* 273:1058–1073.

Collins, F. 1992. Cystic fibrosis: Molecular biology and therapeutic implications. *Science* 256:774–779.

Culver, K. V., and Blaese, R. M. 1994. Gene therapy for cancer. *Trends Genet.* 10:174–178.

Danna, K., and Nathans, D. 1971. Specific cleavage of simian virus 40 DNA by restriction endonuclease of *Haemophilus influenzae. Proc. Natl. Acad. Sci. USA* 68:2913–2917.

Dib, C., Fauré, S., Fizames, C., Samson, D., Drouot, N., Vignal, A., Millasseau, P., Marc, S., Hazan, J., Seboun, E., Lathrop, M., Gyapay, G., Morissette, J., and Weissenbach, J. 1996. A comprehensive genetic map of the human genome based on 5,264 microsatellites. *Nature* 380:152–154.

Erlich, H. A., and Arnheim, N. 1992. Genetic analysis using the polymerase chain reaction. *Annu. Rev. Genet.* 26:479–506.

Green, E. D., and Olson, M. V. 1990. Chromosomal region of the cystic fibrosis gene in yeast artificial chromosomes: A model for human genome mapping. *Science* 250:94–98.

Harris, J. D., and Lemoine, N. R. 1996. Strategies for targeted gene therapy. *Trends Genet.* 12:400–405.

Kay, M. A., and Woo, S. L. C. 1994. Gene therapy for metabolic disorders. *Trends Genet.* 10:253–257.

Kerem, B.-S., Rommens, J. M., Buchanan, J. A., Markiewicz, D., Cox, T. K., Chakravarti, A., Buchwald, M., and Tsui, L.-C. 1989. Identification of the cystic fibrosis gene: Genetic analysis. *Science* 245:1073–1080.

Klee, H., Horsch, R., and Rogers, S. 1987. *Agrobacterium*-mediated plant transformation and its further applications to plant biology. *Annu. Rev. Plant Physiol.* 38:467–486.

Knowlton, R. G., Cohen-Haguenauer, O., Van Cong, N., Frézal, J., Brown, V. A., Barker, D., Braman, J. C., Schumm, J. W., Tsui, L.-C., Buchwald, M., and Donis-Keller, H. 1985. A polymorphic DNA marker linked to cystic fibrosis is located on chromosome 7. *Nature* 318:380–385.

Koenig, M., Hoffman, E. P., Bertelson, C. J., Monaco, A. P., Feener, C., and Kunkel, L. M. 1987. Complete cloning of the Duchenne muscular dystrophy (DMD) cDNA and preliminary genomic organization of the DMD gene in normal and affected individuals. *Cell* 50:509–517.

Luria, S. E. 1953. Host-induced modification of viruses. *Cold Spring Harbor Symp. Quant. Biol.* 18:237–244.

Mandel, J.-L., Monaco, A. P., Nelson, D. L., Schlessinger, D., and Willard, H. 1992. Genome analysis and the human X chromosome. *Science* 258:103–109.

Maxam, A. M., and Gilbert, W. 1977. A new method for sequencing DNA. *Proc. Natl. Acad. Sci. USA* 74:560–564.

Morgan, R. A., and Anderson, W. F. 1993. Human gene therapy. *Annu. Rev. Biochem.* 62:191–217.

Mulligan, R. C. 1993. The basic science of gene therapy. *Science* 260:926–932.

Mullis, K. B. 1990. The unusual origin of the polymerase chain reaction. *Sci. Am.* (April):56–65.

Murray, J. M., Davies, K. E., Harper, P. S., Meredith, L., Mueller, C. R., and Williamson, R. 1982. Linkage relationship of a cloned DNA sequence on the short arm of the X chromosome to Duchenne muscular dystrophy. *Nature* 300:69–71.

Pääbo, S. 1993. Ancient DNA. *Sci. Am.* (November):86–92.

Riordan, J. R., Rommens, J. M., Kerem, B., Alon, N., Rozmahel, R., Grzelczak, Z., Zielenski, J., Lok, S., Plavsic, N., Chou, J. L., Drumm, M. L., Ianuzzi, M. C., Collins, F. S., and Tsui, L.-C. 1989. Identification of the cystic fibrosis gene: Cloning and characterization of complementary DNA. *Science* 245:1066–1073.

Rommens, J. M., Ianuzzi, M. C., Kerem, B., Drumm, M. L., Melmer, G., Dean, M., Rozmahel, R., Cole, J. L., Kennedy, D., Hidaka, N., Zsiga, M., Buchwald, M., Riordan, J. R., Tsui, L.-C., and Collins, F. S. 1989. Identification of the cystic fibrosis gene: Chromosome walking and jumping. *Science* 245:1059–1065.

Sambrook, J., Fritsch, E. F., and Maniatis, T. 1989. *Molecular cloning: A laboratory manual,* 2nd ed. Cold Spring Harbor, NY: Cold Spring Harbor Laboratory.

Sanger, F., and Coulson, A. R. 1975. A rapid method for determining sequences in DNA by primed synthesis with DNA polymerase. *J. Mol. Biol.* 94:441–448.

Schlessinger, D. 1990. Yeast artificial chromosomes: Tools for mapping and analysis of complex genomes. *Trends Genet.* 6:248–258.

Southern, E. M. 1975. Detection of specific sequences among DNA fragments separated by gel electrophoresis. *J. Mol. Biol.* 98:503–517.

Vollrath, D., Foote, S., Hilton, A., Brown, L. G., Beer-Romero, P., Bogan, J. S., and Page, D. C. 1992. The human Y chromosome: A 43-interval map based on naturally occurring deletions. *Science* 258:52–59.

Wainwright, B. J. 1993. The isolation of disease genes by positional cloning. *Medical J. Australia* 159:170–174.

Watson, J. D., Gilman, M., Witkowski, J., and Zoller, M. 1992. *Recombinant DNA,* 2nd ed. New York: Scientific American Books, Freeman.

White, R., and Lalouel, J.-M. 1988. Chromosome mapping with DNA markers. *Sci. Am.* 258(February):40–48.

Wicking, C., and Williamson, B. 1991. From linked marker to gene. *Trends Genet.* 7:288–293.

CHAPTER 15: REGULATION OF GENE EXPRESSION IN BACTERIA AND BACTERIOPHAGES

Bertrand, K., Korn, L., Lee, F., Platt, T., Squires, C. L., Squires, C., and Yanofsky, C. 1975. New features of the structure and regulation of the tryptophan operon of *Escherichia coli. Science* 189:22–26.

Bertrand, K., and Yanofsky, C. 1976. Regulation of transcription termination in the leader region of the tryptophan operon of *Escherichia coli* involves tryptophan as its metabolic product. *J. Mol. Biol.* 103:339–349.

Fisher, R. F., Das, A., Kolter, R., Winkler, M. E., and Yanofsky, C. 1985. Analysis of the requirements for transcription pausing in the tryptophan operon. *J. Mol. Biol.* 182:397–409.

Gilbert, W., Maizels, N., and Maxam, A. 1974. Sequences of controlling regions of the lactose operon. *Cold Spring Harbor Symp. Quant. Biol.* 38:845–855.

Gilbert, W., and Muller-Hill, B. 1966. Isolation of the *lac* repressor. *Proc. Natl. Acad. Sci. USA* 56:1891–1898.

Jacob, F. 1965. Genetic mapping of the elements of the lactose region of *Escherichia coli. Biochem. Biophys. Res. Commun.* 18:693–701.

Jacob, F., and Monod, J. 1961. Genetic regulatory mechanisms in the synthesis of proteins. *J. Mol. Biol.* 3:318–356.

Lee, F., and Yanofsky, C. 1977. Transcription termination at the *trp* operon attenuators of *Escherichia coli* and *Salmonella typhimurium:* RNA secondary structure and regulation of termination. *Proc. Natl. Acad. Sci. USA* 74:4365–4369.

Lewis, M., Chang, G., Horton, N. C., Kercher, M. A., Pace, H. C., Schumacher, M. A., Brennan, R. G., and Lu, P. 1996. Crystal structure of the lactose operon repressor and its complexes with DNA and inducer. *Science* 271:1247–1254.

Maizels, N. 1974. *E. coli* lactose operon ribosome binding site. *Nature (New Biol.)* 249:647–649.

Matthews, K. S. 1996. The whole lactose repressor. *Science* 271:1245–1246.

Niu, W., Kim, Y., Tau, G., Heyduk, T., and Ebright, R. H. 1996. Transcription activation at class II CAP-dependent promoters: Two interactions between CAP and RNA polymerase. *Cell* 87:1123–1134.

Pabo, C. O., Sauer, R. T., Sturtevant, J. M., and Ptashne, M. 1979. The λ repressor contains two domains. *Proc. Natl. Acad. Sci. USA* 76:1608–1612.

Ptashne, M. 1967. Isolation of the λ phage repressor. *Proc. Natl. Acad. Sci. USA* 57:306–313.

———. 1992. *A genetic switch,* 2nd ed. Oxford: Cell Press and Blackwell Scientific Publications.

Ptashne, M., and Gilbert, W. 1970. Genetic repressors. *Sci. Am.* 222 (June):36–44.

Yanofsky, C. 1981. Attenuation in the control of expression of bacterial operons. *Nature* 289:751–758.

———. 1987. Operon-specific control by transcription attenuation. *Trends Genet.* 3:356–360.

Yanofsky, C., and Kolter, R. 1982. Attenuation in amino acid biosynthetic operons. *Annu. Rev. Genet.* 16:113–134.

CHAPTER 16: EUKARYOTIC GENE REGULATION

Ashburner, M. 1990. Puffs, genes, and hormones revisited. *Cell* 61:1–3.

Atwater, J. A., Wisdom, R., and Verma, I. M. 1990. Regulated mRNA stability. *Annu. Rev. Genet.* 24:519–541.

Bachvarova, R. F. 1992. A maternal tail of poly(A): The long and the short of it. *Cell* 69:895–897.

Beachy, P. A. 1990. A molecular view of the *Ultrabithorax* homeotic gene of *Drosophila. Trends Genet.* 6:46–51.

Beato, M., Herrlich, P., and Schützm, G. 1995. Steroid hormone receptors: Many actors in search of a plot. *Cell* 83:851–857.

Beelman, C. A., and Parker, R. 1995. Degradation of mRNA in eukaryotes. *Cell* 81:179–182.

Beelman, C. A., Stevens, A., Caponigro, G., LaGrandeur, T. E., Hatfield, L., Fortner, D. M., and Parker, R. 1996. An essential component of the decapping enzyme required for normal rates of mRNA turnover. *Nature* 382:642–646.

Beerman, W., and Clever, U. 1964. Chromosome puffs. *Sci. Am.* 210:50–58.

Blackwell, T. K. and F. W. 1988. Immunoglobin genes. In *Molecular immunology,* B. D. Homes and D. M. Glover, eds., pp. 1–60. Washington, DC: IRL Press.

Bonner, J. J., and Pardue, M. L. 1976. Ecdysone-stimulated RNA synthesis in imaginal discs of *Drosophila melanogaster. Chromosoma* 58:87–99.

Cedar, H. 1988. DNA methylation and gene activity. *Cell* 53:3–4.

Chen, C.-Y. A., and Shyu, A.-B. 1995. AU-rich elements: Characterization and importance of mRNA degradation. *Trends Biochem. Sci.* 20:465–470.

Davis, M. M., Calame, K., Early, P. W., Livant, D. L., Joho, R., Weissman, I. L., and Hood, L. 1980. An immunoglobulin heavy chain gene is formed by at least two recombinational events. *Nature* 283:733–739.

Davis, M. M., Kim, S. K., and Hood, L. 1980. Immunoglobulin class switching: Developmentally regulated DNA rearrangements during differentiation. *Cell* 22:1–2.

De Robertis, E. M., and Gurdon, J. B. 1977. Gene activation in somatic nuclei after injection into amphibian oocytes. *Proc. Natl. Acad. Sci. USA* 74:2470–2474.

Efstratiadis, A., Posakony, J. W., Maniatis, T., Lawn, R. M., O'Connell, C., Spritz, R. A., DeRiel, J. K., Forget, B. G., Weissman, S. M., Slighton, J. L., Blechtl, A. E., Smithies, O., Baralle, F. E., Shoulders, C. C., and Proudfoot, N. J. 1980. The structure and evolution of the human β-globin gene family. *Cell* 21:653–668.

Gellert, M. 1992. V(D)J recombination gets a break. *Trends Genet.* 8:408–412.

Green, M. R. 1989. Pre-mRNA processing and mRNA nuclear export. *Curr. Opinion Cell Biol.* 1:519–525.

Gross, D. S., and Garrard, W. T. 1987. Poising chromatin for transcription. *Trends Biochem. Sci.* (August):293–297.

———. 1988. Nuclease hypersensitive sites in chromatin. *Annu. Rev. Biochem.* 57:159–197.

Grunstein, M. 1992. Histones as regulators of genes. *Sci. Am.* 267 (October):68–74B.

Gurdon, J. B. 1968. Transplanted nuclei and cell differentiation. *Sci. Am.* 219:24–35.

Gurdon, J. B., Laskey, R. A., and Reeves, R. 1975. The developmental capacity of nuclei transplanted from keratinized skin cells of adult frogs. *J. Embryol. Exp. Morph.* 34:93–112.

Hanna-Rose, W., and Hansen, U. 1996. Active repression mechanisms of eukaryotic transcription repressors. *Trends Genet.* 12:229–234.

Hochstrasser, M. 1996. Protein degradation or regulation: Ub the judge. *Cell* 84:813–815.

Johnston, M., Flick, J. S, and Pexton, T. 1994. Multiple mechanisms provide rapid and stringent repression of *GAL* gene expression in *Saccharomyces cerevisiae. Mol. Cell. Biol.* 14:3834–3841.

Karlsson, S., and Nienhuis, A. W. 1985. Development regulation of human globin genes. *Annu. Rev. Biochem.* 54:1071–1078.

Landschulz, W. H., Johnson, P. F., and McKnight, S. L. 1988. The leucine zipper: A hypothetical structure common to a new class of DNA binding protein. *Science* 240:1759–1763.

Lucas, P. C., and Granner, D. K. 1992. Hormone response domains in gene transcription. *Annu. Rev. Biochem.* 61:1131–1173.

O'Malley, B. W., and Schrader, W. T. 1976. The receptors of steroid hormones. *Sci. Am.* 234:32–43.

Pabo, C. O., and Sauer, R. T. 1992. Transcription factors: Structural families and principles of DNA recognition. *Annu. Rev. Biochem.* 61:1053–1093.

Paranjape, S. M., Kamakaka, R. T., and Kadonaga, J. T. 1994. Role of chromatin structure in the regulation of transcription by RNA polymerase II. *Annu. Rev. Biochem.* 63:265–297.

Parthun, M. R., and Jaehning, J. A. 1992. A transcriptionally active form of GAL4 is phosphorylated and associated with GAL80. *Mol. Cell. Biol.* 12:4981–4987.

Rhodes, D., and Klug, A. 1993. Zinc fingers. *Sci. Am.* 259 (February):56–65.

Rivera-Pomar, R., and Jäckle, H. 1996. From gradients to stripes in *Drosophila* embryogenesis: Filling in the gaps. *Trends Genet.* 12:478–483.

Rogers, J. O., Early, H., Carter, C., Calame, K., Bond., M., Hood, L., and Wall, R. 1980. Two mRNAs with different 3' ends encode membrane-bound and secreted forms of immunoglobin chain. *Cell* 20:303–312.

Ross, J. 1996. Control of messenger RNA stability in higher eukaryotes. *Trends Genet.* 12:171–175.

Scott, M. P., Tamkun, J. W., and Hartzell III, G. W. 1989. The structure and function of the homeodomain. *Biochim. Biophys. Acta* 989:25–48.

Studitsky, V. M., Clark, D. J., and Felsenfeld, G. 1994. A histone octamer can step around a transcribing polymerase without leaving the template. *Cell* 76:371–382.

Takagaki, Y., Seipelt, R. L., Peterson, M. L., and Manley, J. L. 1996. The polyadenylation factor CstF–64 regulates alternative processing of IgM heavy chain pre-mRNA during B cell differentiation. *Cell* 87:941–952.

Tobler, H., Etter, A., and Müller, F. 1992. Chromatin diminution in nematode development. *Trends Genet.* 8:427–432.

Tsai, M.-J., and O'Malley, B. W. 1994. Molecular mechanisms of action of steroid/thyroid receptor superfamily members. *Annu. Rev. Biochem.* 63:451–486.

Varshavsky, A. 1996. The N-end rule: Functions, mysteries, uses. *Proc. Natl. Acad. Sci. USA* 93:12142–12149.

Verdine, G. L. 1994. The flip side of DNA methylation. *Cell* 76:197–200.

Wilmut, I., Schnieke, A. E., McWhir, J., Kind, A. J., and Campbell, K. H. S. 1997. Viable offspring derived from fetal and adult mammalian cells. *Nature* 385:810–813.

Wolffe, A. P. 1994. Transcription: In tune with the histones. *Cell* 77:13–16.

Wolffe, A. P., and Pruss, D. 1996. Targeting chromatin disruption: Transcription regulators that acetylate histones. *Cell* 84:817–819.

CHAPTER 17: GENETICS OF CANCER

Baltimore, D. 1985. Retroviruses and retrotransposons: The role of reverse transcription in shaping the eukaryotic genome. *Cell* 40:481–482.

Bishop, J. M. 1987. The molecular genetics of cancer. *Science* 235:305–311.

Cavenee, W. K., and White, R. L. 1995. The genetic basis of cancer. *Sci. Am.* (March): 72–79.

Cleaver, J. E. 1994. It was a very good year for DNA repair. *Cell* 76:1–4.

Fishel, R., Lescoe, M. K., Rao, M. R. S., Copeland, N. G., Jenkins, N. A., Garber, J., Kane, M., and Kolodner, R. 1994. The human mutator gene homolog MSH2 and its association with hereditary nonpolyposis colon cancer. *Cell* 75:1027–1038.

Hartwell, L. H., and Kastan, M. B. 1994. Cell cycle control and cancer. *Science* 266:1821–1828.

Jiricny, J. 1994. Colon cancer and DNA repair: Have mismatches met their match? *Trends Genet.* 10:164–168.

Kamb, A. 1995. Cell-cycle regulators and cancer. *Trends Genet.* 11:136–140.

Kingston, R. E., Baldwin, A. S., and Sharp, P. A. 1985. Transcription control by oncogenes. *Cell* 41:3–5.

Leach, F. S., and many other authors. 1993. Mutations of a *mutS* homolog in hereditary nonpolyposis colorectal cancer. *Cell* 75:1215–1225.

Mancini, M. A., Shan, B., Nickerson, J. A., Penman, S., and Lee, W.-H. 1994. The retinoblastoma gene product is a cell cycle-dependent, nuclear matrix-associated protein. *Proc. Natl. Acad. Sci. USA* 91:418–422.

Rabbitts, T. H. 1994. Chromosomal translocations in human cancer. *Nature* 372:143–149.

Ratner, L., Gallo, R. C., and Wong-Staal, F. 1985. Cloning of human oncogenes. In *Recombinant DNA research and virus*, edited by Y. Becker. Boston: Martinus Nijhoff.

Ratner, L., Josephs, S. F., and Wong-Staal, F. 1985. Oncogenes: Their role in neoplastic transformation. *Annu. Rev. Microbiol.* 39:419–449.

Rebbeck, T. R., Couch, F. J., Kant, J., Calzone, K., DeShano, M., Peng, Y., Chen, K., Garber, J. E., and Weber, B. L. 1996. Genetic heterogeneity in hereditary breast cancer: Role of *BRCA1* and *BRCA2*. *Am. J. Hum. Genet.* 59:547–553.

Weinberg, R. A., 1995. The retinoblastoma protein and cell cycle protein. *Cell* 81:323–330.

Wooster, R., and many other authors. 1994. Localization of a breast cancer susceptibility gene, *BRCA2*, to chromosome 13q12–13. *Science* 265:2088–2090.

———. 1995. Identification of the breast cancer susceptibility gene *BRCA2*. *Nature* 378:789–792.

Wooster, R., and Stratton, M. R. 1995. Breast cancer susceptibility: A complex disease unravels. *Trends Genet.* 11:3–5.

CHAPTER 18: DNA MUTATION AND REPAIR

Ames, B. N., Durston, W. E., Yamasaki, E., and Lee, F. 1973. Carcinogens are mutagens: A simple test system combining liver homogenates for activation and bacteria for detection. *Proc. Natl. Acad. Sci. USA* 70:2281–2285.

Boyce, R. P., and Howard-Flanders, P. 1964. Release of ultraviolet light-induced thymine dimers from DNA in *E. coli* K12. *Proc. Natl. Acad. Sci. USA* 51:293–300.

Cleaver, J. E. 1994. It was a very good year for DNA repair. *Cell* 76:1–4.

Devoret, R. 1979. Bacterial tests for potential carcinogens. *Sci. Am.* 241:40–49.

Fishel, R., Lescoe, M. K., Rao, M. R. S., Copeland, N. G., Jenkins, N. A., Garber, J., Kane, M., and Kolodner, R. 1993. The human mutator gene homolog MSH2 and its association with hereditary nonpolyposis colon cancer. *Cell* 75:1027–1038.

Lederberg, J., and Lederberg, E. M. 1952. Replica plating and indirect selection of bacterial mutants. *J. Bacteriol.* 63:399–406.

Luria, S. E., and Delbrück, M. 1943. Mutations of bacteria from virus sensitivity to virus resistance. *Genetics* 28:491–511.

Modrich, P. 1987. DNA mismatch correction. *Annu. Rev. Biochem.* 56:435–466.

Morgan, A. R. 1993. Base mismatches and mutagenesis: How important is tautomerism? *Trends Biochem. Sci.* 18:160–163.

Setlow, R. B., and Carrier, W. L. 1964. The disappearance of thymine dimers from DNA: An error-correcting mechanism. *Proc. Natl. Acad. Sci. USA* 51:226–231.

Tessman, I., Liu, S.-K., and Kennedy, A. 1992. Mechanism of SOS mutagenesis of UV-irradiated DNA: Mostly error-free processing of deaminated cytosine. *Proc. Natl. Acad. Sci. USA* 89:1159–1163.

CHAPTER 19: TRANSPOSABLE ELEMENTS

Bhattacharyya, M. K., Smith, A. M., Ellis, T. H. N., Hedley, C., and Martin, C. 1990. The wrinkled-seed character of pea described by Mendel is caused by a transposon-like insertion in a gene encoding starch-branching enzyme. *Cell* 60:115–122.

Boeke, J. D., and Corces, V. G. 1989. Transcription and reverse transcription of retrotransposons. *Annu. Rev. Microbiol.* 43:403–434.

Boeke, J. D., Garfinkel, D. J., Styles, C. A., and Fink, G. R. 1985. Ty elements transpose through an RNA intermediate. *Cell* 40:491–500.

Bucheton, A. 1990. I transposable elements and I-R hybrid dysgenesis in *Drosophila*. *Trends Genet.* 6:16–21.

Cohen, S. N., and Shapiro, J. A. 1980. Transposable genetic elements. *Sci. Am.* 242:40–49.

Engles, W. R. 1983. The P family of transposable elements in *Drosophila*. *Annu. Rev. Genet.* 17:315–344.

Federoff, N. V. 1989. About maize transposable elements and development. *Cell* 56:181–191.

Foster, T. J., Davis, M. A., Roberts, D. E., Takashita, K., and Kleckner, N. 1981. Genetic organization of transposon Tn10. *Cell* 23:201–213.

Kingsman, A. J., and Kingsman, S. M. 1988. Ty: A retroelement moving forward. *Cell* 53:333–335.

Kleckner, N. 1981. Transposable elements in prokaryotes. *Annu. Rev. Genet.* 15:341–404.

McClintock, B. 1939. The behavior in successive nuclear divisions of a chromosome broken at meiosis. *Proc. Natl. Acad. Sci. USA* 25:405–416.

———. 1950. The origin and behavior of mutable loci in maize. *Proc. Natl. Acad. Sci. USA* 36:344–355.

———. 1951. Chromosome organization and genic expression. *Cold Spring Harbor Symp. Quant. Biol.* 16:13–47.

———. 1953. Induction of instability at selected loci in maize. *Genetics* 38:579–599.

———. 1956. Controlling elements and the gene. *Cold Spring Harbor Symp. Quant. Biol.* 21:197–216.

———. 1961. Some parallels between gene control systems in maize and in bacteria. *Am. Naturalist* 95:265–277.

———. 1965. The control of gene action in maize. *Brookhaven Symp. Biol.* 18:162 ff.

———. 1984. The significance of responses of the genome to challenge. Nobel lecture. *Science* 226:792–801.

Varmus, H. E. 1982. Form and function of retroviral proviruses. *Science* 216:812–821.

CHAPTER 20: EXTRANUCLEAR GENETICS

Birky, C. W. 1978. Transmission genetics of mitochondria and chloroplasts. *Annu. Rev. Genet.* 12:471–512.

Brown, M. D., Voljavec, A. S., Lott, M. T., MacDonald, I., and Wallace, D. C. 1992. Leber's hereditary optic neuropathy: A model for mitochondrial neurodegenerative diseases. *FASEB J.* 6:2791–2799.

Cattaneo, R. 1991. Different types of messenger RNA editing. *Annu. Rev. Genet.* 25:71–88.

Chiu, W.-L., and Sears, B. B. 1993. Plastome-genome interactions affect plastid transmission in Oenothera. *Genetics* 133:989–997.

Clayton, D. A. 1982. Replication of animal mitochondrial DNA. *Cell* 28:693–705.

Ephrussi, B. 1953. *Nucleo-cytoplasmic relations in microorganisms.* New York: Oxford University Press.

Freeman, G., and Lundelius, J. W. 1982. The developmental genetics of dextrality and sinistrality in the gastropod *Limnaea peregra. Wilhelm Roux Arch. Dev. Biol.* 191:69–83.

Grivell, L. 1983. Mitochondrial DNA. *Sci. Am.* 225(March):78–89.

Gyllensten, U., Wharton, D., Josefsson, A., and Wilson, A. C. 1991. Paternal inheritance of mitochondrial DNA in mice. *Nature* 352:255–257.

Herbert, A. 1996. RNA editing, introns and evolution. *Trends Genet.* 12:6–9.

Hoch, B., Maier, R. M., Appel, K., Igloi, G. L., and Kössel, H. 1991. Editing of chloroplast mRNA by creation of an initiation codon. *Nature* 353:178–180.

Kirk, J. T. O. 1971. Chloroplast structure and biogenesis. *Annu. Rev. Biochem.* 40:161–196.

Lander, E. S., and Lodish, H. 1990. Mitochondrial diseases: Gene mapping and gene therapy. *Cell* 61:925–926.

Lee, S. B., and Taylor, J. W. 1992. Uniparental inheritance and replacement of mitochondrial DNA in *Neurospora tetrasperma. Genetics* 134:1063–1075.

Levings, C. S. 1983. The plant mitochondrial genome and its mutants. *Cell* 32:659–661.

Lu, B., and Hanson, M. R. 1992. A single nuclear gene specifies the abundance and extent of RNA editing of a plant mitochondrial transcript. *Nucl. Acids Res.* 20:5699–5703.

Matzke, M., and Matzke, A. J. M. 1993. Genomic imprinting in plants: Parental effects and *trans*-inactivation phenomena. *Annu. Rev. Plant. Physiol. Plant Mol. Biol.* 44:53–76.

Reik, W. 1989. Genomic imprinting and genetic disorders in man. *Trends Genet.* 5:331–336.

Schuster, W., and Brennicke, A. 1994. The plant mitochondrial genome: Physical structure, information content, RNA editing, and gene migration to the nucleus. *Annu. Rev. Plant. Physiol. Plant Mol. Biol.* 45:61–78.

Scott, J. 1995. A place in the world for RNA editing. *Cell* 81:833–836.

Simpson, L., and Thiemann, O. H. 1995. Sense from nonsense: RNA editing in mitochondria of kinetoplastid protozoa and slime molds. *Cell* 81:837–840.

Swain, J. L., Stewart, T. A., and Leder, P. 1987. Parental legacy determines methylation and expression of an autosomal transgene: A molecular mechanism for parental imprinting. *Cell* 50:719–727.

Turmel, M., Lemieux, B., and Lemieux, C. 1988. The chloroplast genome of the green alga *Chlamydomonas moewusii:* Localization of protein-coding genes and transcriptionally active regions. *Mol. Gen. Genet.* 214:412–419.

Umesono, K., and Ozeki, H. 1987. Chloroplast gene organization in plants. *Trends Genet.* 3:281–287.

Van Winkle-Swift, K. P., and Birky, C. W. 1978. The nonreciprocality of organelle gene recombination in *Chlamydomonas reinhardi* and *Saccharomyces cerevisiae. Mol. Gen. Genet.* 166:193–209.

Wallace, D. C. 1992. Diseases of the mitochondrial DNA. *Annu. Rev. Biochem.* 61:1175–1212.

Yang, X., and Griffiths, A. J. F. 1992. Male transmission of linear plasmids and mitochondrial DNA in the fungus *Neurospora. Genetics* 134:1055–1062.

CHAPTER 21: POPULATION GENETICS

Avise, J. C. 1986. Mitochondrial DNA and the evolutionary genetics of higher animals. *Phil. Trans. Roy. Soc. Lond.,* Ser. B 321:325–342.

Avise, J., Giblin-Davidson, C. C., Laerm, J., Patton, J. C., and Lansman, R. A. 1979. Mitochondrial DNA clones and matriarchal phylogeny within and among geographic populations of the pocket gopher *Geomys pinetis. Proc. Natl. Acad. Sci. USA* 76:6694–6698.

Ayala, F. J., ed. 1976. *Molecular evolution.* Sunderland, MA: Sinauer.

Buri, P. 1956. Gene frequency in small populations of mutant *Drosophila. Evolution* 10:367–402.

Clarke, C. A., and Sheppard, P. M. 1966. A local survey of the distribution of industrial melanic forms in the moth *Biston betularia* and estimates of the selective values of these in an industrial environment. *Proc. R. Soc. Lond. [Biol.]* 165:424–439.

Crow, J. F. 1986. *Basic concepts in population, quantitative, and evolutionary genetics.* New York: Freeman.

Darwin, C. 1860. *On the origin of species by means of natural selection, or the preservation of favoured races in the struggle for life.* New York: Appleton.

Dobzhansky, T. 1951. *Genetics and the origin of species,* 3rd ed. New York: Columbia University Press.

Fisher, R. A. 1930. *The genetical theory of natural selection.* Oxford: Clarendon Press.

Ford, E. B. 1971. *Ecological genetics,* 3rd ed. London: Chapman and Hall.

Glass, B., Sacks, M. S., Jahn, E. F., and Hess, C. 1952. Genetic drift in a religious isolate: An analysis of the causes of variation in blood group and other gene frequencies in a small population. *Am. Naturalist* 86:145–159.

Gray, M. W. 1989. Origin and evolution of mitochondrial DNA. *Annu. Rev. Cell Biol.* 5:25–50.

Hardy, G. H. 1908. Mendelian proportions in a mixed population. *Science* 28:49–50.

Hartl, D. L., and Clark, A. G. 1988. *Principles of population genetics,* 2nd ed. Sunderland, MA: Sinauer.

Hedrick, P. H. 1983. *Genetics of populations.* Boston: Science Books International.

Hillis, D. M., and Moritz, C. 1990. *Molecular systematics.* Sunderland, MA: Sinauer.

Hillis, D. M., Moritz, C., Porter, C. A., and Baker, R. J. 1991. Evidence for biased gene conversion in concerted evolution of ribosomal DNA. *Science* 251:308–309.

Kettlewell, H. B. D. 1961. The phenomenon of industrial melanism in the Lepidoptera. *Annu. Rev. Entomol.* 6:245–262.

Kreitman, M. 1983. Nucleotide polymorphism at the alcohol dehydrogenase locus of *Drosophila melanogaster.* *Nature* 304:412–417.

Lewontin, R. C. 1974. *The genetic basis of evolutionary change.* New York: Columbia University Press.

Lewontin, R. C. 1985. Population genetics. *Annu. Rev. Genet.* 19:81–102.

Lewontin, R. C., Moore, J. A., Provine, W. B., and Wallace, B. 1981. *Dobzhansky's genetics of natural populations I-XLIII.* New York: Columbia University Press.

Li, W-H. 1991. *Fundamentals of molecular evolution.* Sunderland, MA: Sinauer.

Li, W-H., Luo, C. C., and Wu, C. I. 1985. Evolution of DNA sequences. *Molecular evolutionary genetics,* R. J. MacIntyre, ed., pp. 1–94. New York: Plenum.

MacIntyre, R. J., ed. 1985. *Molecular evolutionary genetics.* New York: Plenum.

Maniatis, T., Fritsch, E. F., Lauer, L., and Lawn, R. M. 1980. The molecular genetics of human hemoglobin. *Annu. Rev. Genet.* 14:145–178.

Maynard Smith, J. 1989. *Evolutionary genetics.* Oxford: Oxford University Press.

Nei, M. 1987. *Molecular evolutionary genetics.* New York: Columbia University Press.

Nei, M., and Koehn, R. K. 1983. *Evolution of genes and proteins.* Sunderland, MA: Sinauer.

Pace, N. R., Olsen, G. J., and Woese, C. R. 1986. Ribosomal RNA phylogeny and the primary lines of evolutionary descent. *Cell* 45:325–326.

Schonewald-Cox, C. M., Chambers, S. M., MacBryde, B., and Thomas, L. 1983. *Genetics and conservation: A reference for managing wild animal and plant populations.* Menlo Park, CA: Benjamin/Cummings.

Selander, R. K., and Kaufman, D. W. 1975. Self-fertilization and genetic population structure in a colonizing land snail. *Proc. Natl. Acad. Sci. USA* 70:1186–1190.

Soulé, M. E., ed. 1986. *Conservation biology: The science of scarcity and diversity.* Sunderland, MA: Sinauer.

Speiss, E. 1977. *Genes in populations.* New York: Wiley.

Stringer, C. B. 1990. The emergence of modern humans. *Sci. Am.* (December):98–104.

Stringer, C. B., and Andrews, P. 1988. Genetic and fossil evidence for the origin of modern humans. *Science* 239:1263–1268.

Woese, C. R. 1981. Archaebacteria. *Sci. Am.* 244:98–122.

CHAPTER 22: QUANTITATIVE GENETICS

Blumer, M. G. 1980. *The mathematical theory of quantitative genetics.* Oxford: Clarendon Press.

Darwin, C. 1860. *On the origin of species by means of natural selection, or the preservation of favoured races in the struggle for life.* New York: Appleton.

Dobzhansky, T., and Pavlovsky, O. 1969. Artificial and natural selection for two behavioral traits in *Drosophila pseudoobscura.* *Proc. Natl. Acad. Sci. USA.* 62:75–80.

East, E. M. 1910. A Mendelian interpretation of variation that is apparently continuous. *Am. Natur.* 44:65–82.

———. 1916. Studies on size inheritance in *Nicotiana.* *Genetics* 1:164–176.

East, E. M., and Jones, D. F. 1919. *Inbreeding and outbreeding.* Philadelphia: Lippincott.

Falconer, D. S. 1989. *Introduction to quantitative genetics.* New York: Wiley.

Hill, W. G., ed. 1984. *Quantitative genetics,* Parts I and II. New York: Van Nostrand Reinhold.

Lander, E. S., and Botstein, D. 1989. Mapping Mendelian factors underlying quantitative traits using RFLP linkage maps. *Genetics* 121:185–199.

Mather, K. 1943. Polygenic inheritance and natural selection. *Biol. Rev.* 18:32–64.

Nilsson-Ehle, H. 1909. Kreuzungsuntersuchungen an Hafer und Weizen. *Lunds Univ. Aarskr. N. F. Atd.,* Ser. 2, 5 (2):1–122.

Paterson, A. H., Lander, E. S., Hewitt, J. D., Person, S., Lincoln, S. E., and Tanksley, S. D. 1988. Resolution of quantitative traits into Mendelian factors by using a complete RFLP linkage map. *Nature* 335:721–726.

Selander, R. K., and Kaufman, D. W. 1975. Self-fertilization and genetic population structure in a colonizing land snail. *Proc. Natl. Acad. Sci. USA* 70:1186–1190.

Sokal, R. R., and Rohlf, F. J. 1981. *Biometry,* 2nd ed. San Francisco: Freeman.

Sokal, R. R., and Thoday, J. M., eds. 1979. *Quantitative genetic variation.* New York: Academic Press.

Thoday, J. M. 1961. Location of polygenes. *Nature* 191:368–370.

Thompson, J. N., Jr., and Thompson, J. M., eds. 1979. *Quantitative genetic variation.* New York: Academic Press.

Weir, B. S., Eisen, E. J., Goodman, M. M., and Namkoong, G., eds. 1988. *Proceedings of the second international conference on quantitative genetics.* Sunderland, MA: Sinauer.

SOLUTIONS TO SELECTED QUESTIONS AND PROBLEMS

CHAPTER 1

1.1 c

1.3 c

1.6 a. Yes, if a sexual mating system exists in that species. In that case, two haploid cells can fuse to produce a diploid cell, which can then go through meiosis to produce haploid progeny. The fungi *Neurospora crassa* and *Saccharomyces cerevisiae* exemplify this positioning of meiosis in the life cycle.

b. No, because a diploid cell cannot be formed and meiosis occurs only starting with a diploid cell.

1.8 c

1.10 a. metaphase

b. anaphase

1.14 a. The chance that a gamete would have a particular maternal chromosome is $\frac{1}{2}$. Therefore, the chance of obtaining a gamete with all three maternal chromosomes is $(\frac{1}{2})^3 = \frac{1}{8}$.

b. The set of gametes with some maternal and paternal chromosomes is composed of all gametes *except* those that have only maternal or only paternal chromosomes. That is, p(gamete with both maternal and paternal chromosomes) = 1 − p(gamete with only maternal or only paternal chromosomes). From part (a), the chance of a gamete having chromosomes from only one parent is $\frac{1}{8}$. Using the sum rule, p(gamete with both maternal and paternal chromosomes) = 1 − ($\frac{1}{8} + \frac{1}{8}$) = $\frac{3}{4}$.

1.15 One of the long chromosomes and the short chromosome might be members of a heteromorphic pair—that is, X and Y chromosomes, respectively.

1.17 False. Owing to the randomness of independent assortment and to crossing-over, both of which characterize meiosis, the probability of any two sperm cells being genetically identical is extremely remote.

1.19 a. 17 + 26 = 43 chromosomes.

b. Similar chromosomes pair in meiosis. The pairing pattern seen in the hybrid indicates that some of the chromosomes in Arctic and red foxes share evolutionary similarity, but others do not. Unpaired chromosomes will not segregate in an orderly manner, giving rise to unbalanced meiotic products with either extra or missing chromosomes. This can lead to sterility for two reasons. First, meiotic products that are missing chromosomes may not have genes necessary to form gametes. Second, even if gametes are able to form, a zygote generated from them will not have the chromosome set from the hybrid, the red, or the Arctic fox. The zygote will have missing or extra genes, causing it to be inviable.

1.20 The probability of a given homolog going to one particular pole is $\frac{1}{2}$. The probability of all five paternal chromosomes going to the same pole is $(\frac{1}{2})^5 = \frac{1}{32}$. The same answer applies for all maternal chromosomes going to one pole.

CHAPTER 2

2.1 a. red

b. 3 red, 1 yellow

c. all red

d. $\frac{1}{2}$ red, $\frac{1}{2}$ yellow

2.3 The F_2 genotypic ratio (if *C* is colored, *c* is colorless) is $\frac{1}{4}$ *CC* : $\frac{1}{2}$ *Cc* : $\frac{1}{4}$ *cc*. If we consider just the colored plants, there is a 1:2 ratio of *CC* homozygotes to *Cc* heterozygotes. Therefore, if a colored plant is picked at random, the probability that it is *CC* is $\frac{1}{3}$ (i.e., $\frac{1}{3}$ of the colored plants are homozygous), and the probability that it is *Cc* is $\frac{2}{3}$. Only if a *Cc* plant is selfed will more than one phenotypic class be found among its progeny; therefore, the answer is $\frac{2}{3}$.

2.4 a. Parents are *Rr* (rough) and *rr* (smooth); F_1s are *Rr* (rough) and *rr* (smooth).

b. *Rr* × *Rr* → $\frac{1}{4}$ rough, $\frac{1}{4}$ smooth

2.6 Progeny ratio approximates 3:1, and so the parent is heterozygous. Of the dominant progeny, there is a 1:2 ratio of homozygous to heterozygote, so $\frac{1}{3}$ will breed true.

2.7 Black is dominant to brown. If *B* is the allele for black and *b* for brown, then female X is *Bb* and female Y is *BB*. The male is *bb*.

2.10

	PARENTS		PROGENY		FEMALE PARENT
	female ×	male	grey	white	GENOTYPE
a.	grey ×	white	81	82	*Gg*
b.	grey ×	grey	118	39	*Gg*
c.	grey ×	white	74	0	*GG*
d.	grey ×	grey	90	0	*GG* or *Gg*

2.11 a. To obtain a white babbit in a cross between F_1 babbits, both babbits must be *Bb* in genotype, and a *bb* offspring must be produced by these parents. Among the black F_1 progeny, there is a $\frac{1}{3}$ chance of picking a *BB* individual and a $\frac{2}{3}$ chance of picking a *Bb* individual. The chance of two *Bb* parents giving a *bb* offspring is $\frac{1}{4}$. Thus:

P(white offspring) = P(both F_1 babbits are *Bb* and a *bb* offspring is produced)

= P(both F_1 babbits are *Bb*) × P(*bb* offspring)

= ($\frac{2}{3} × \frac{2}{3}$) × ($\frac{1}{4}$)

= $\frac{1}{9}$

b. If he crosses an F_1 male (*Bb* or *BB*) to the parental female (*Bb*), two types of crosses are possible. The crosses and probabilities are (1) *Bb* (F_1 male) × *Bb* (parental female); P = $\frac{2}{3} × 1 = \frac{2}{3}$ and (2) *BB* (F_1 male) × *Bb* (parental female); P = $\frac{1}{3} × 1 = \frac{1}{3}$. Only the first cross can produce white progeny, $\frac{1}{4}$ of the time. Thus, the chance that this strategy will yield white progeny is:

P = $\frac{2}{3}$ (chance of *Bb* × *Bb* cross) × $\frac{1}{4}$ (chance of *bb* offspring)

= $\frac{1}{6}$

c. The best strategy is as follows. Remate the initial two black babbits (both are *Bb*) to obtain a white male offspring (P = $\frac{1}{4}$ [white *bb*]

$\times \frac{1}{2}$ [male] = $\frac{1}{8}$). Retain this male and breed it back to its mother. This cross would be $Bb \times bb$ and give $\frac{1}{2}$ white (*bb*) and $\frac{1}{2}$ black (*Bb*) offspring. The progeny of this cross could be used to develop a "breeding colony" consisting of black (*Bb*) females and white (*bb*) males. These would consistently produce half white and half black offspring.

2.13 Try fitting the data to a model in which catnip sensitivity/insensitivity is controlled by a pair of alleles at one gene. Hypothesize that, because sensitivity is seen in all of the progeny of the initial mating between catnip-sensitive Cleopatra and catnip-insensitive Antony, sensitivity is dominant. If *S* is the sensitive allele and *s* the insensitive allele, then the initial cross was $S- \times ss$, and the progeny are *Ss*. If two of the *Ss* kittens mate, one would expect 3 *Ss* (sensitive) : 1 *ss* (insensitive) kittens. In the mating with Augustus, the cross would be $Ss \times ss$ and should give a 1 *Ss* (sensitive) : 1 *ss* (insensitive) progeny ratio. The observed progeny ratios are not far off from these expectations.

An alternative hypothesis is that sensitivity (*s*) is recessive, and insensitivity (*S*) is dominant. For Antony and Cleopatra to have sensitive (*ss*) offspring, they would need to be *Ss* and *ss*, respectively. When two of their *ss* progeny mate, only sensitive, *ss* offspring should be produced. This is not observed, so this hypothesis does not explain the data.

2.15 a. *WW Dd* × *ww dd*

b. *Ww dd* × *Ww dd*

c. *ww DD* × *WW dd*

d. *Ww Dd* × *Ww dd*

e. *Ww Dd* × *Ww dd*

2.16 The cross is *Aa Bb Cc* × *Aa Bb Cc*.

a. Considering the *A* gene alone, the probability of an offspring showing the *A* trait from $Aa \times Aa$ is $\frac{3}{4}$. Similarly, the probability of showing the *B* trait from $Bb \times Bb$ is $\frac{3}{4}$ and the *C* trait from $Cc \times Cc$ is $\frac{3}{4}$. Therefore, the probability of a given progeny being phenotypically *ABC* (using the product rule) is $\frac{3}{4} \times \frac{3}{4} \times \frac{3}{4} = \frac{27}{64}$.

b. Considering the *A* gene, the probability of an *AA* offspring from $Aa \times Aa$ is $\frac{1}{4}$. The same probability is the case for a *BB* offspring and for a *CC* offspring. Therefore, the probability of an *AA BB CC* offspring (from the product rule) is $\frac{1}{4} \times \frac{1}{4} \times \frac{1}{4} = \frac{1}{64}$.

2.20 a. The F_1 has the genotype *Aa Bb Cc^h* and is all agouti, black. The F_2 is $\frac{27}{64}$ agouti, black; $\frac{9}{64}$ agouti, black, Himalayan; $\frac{9}{64}$ agouti, brown; $\frac{9}{64}$ black; $\frac{3}{64}$ agouti, brown, Himalayan; $\frac{3}{64}$ black, Himalayan; $\frac{3}{64}$ brown; $\frac{1}{64}$ brown, Himalayan.

b. F_2 animals that are black and agouti (and either Himalayan or have full-body pigmentation) have the genotypes *A– B– C–* and *A– B– c^{h}c^h*. Among the *A–* animals, $\frac{2}{3}$ are *Aa*. Among the *B–* animals, $\frac{1}{3}$ are *BB*. Among the *C–* and *c^{h}c^h* animals, $\frac{1}{2}$ are *Cc^h*. Thus, among all of the black, agouti F_2, $\frac{2}{3} \times \frac{1}{3} \times \frac{1}{2} = \frac{1}{9}$ are *Aa BB Cc^h*. If one limits the animals under consideration to those that are black, agouti, *and not Himalayan* (i.e., *A– B– C–*), then $\frac{2}{3}$ of the *C–* animals are *Cc^h*, and the proportion of *Aa BB Cc^h* animals is $\frac{2}{3} \times \frac{1}{3} \times \frac{2}{3} = \frac{4}{27}$.

c. From the cross *Aa Bb Cc^h* × *Aa Bb Cc^h*, $\frac{1}{4}$ of the progeny will be *bb* and show brown pigment. This will be the case regardless of whether the animals are pigmented over their entire bodies or are Himalayan. Thus, $\frac{1}{4}$ of the Himalayan mice will show brown pigment.

d. From the cross *Aa Bb Cc^h* × *Aa Bb Cc^h*, $\frac{3}{4}$ of the progeny will be *B–* and show black pigment. This will be the case regardless of whether the animals are agouti or nonagouti. Thus, $\frac{3}{4}$ of the agouti mice will show black pigment.

2.24 Mating type C is determined only by the genotype *aa bb*. Thus, "C" must be genotype *aa bb*. Crosses of the other strains to "C," then, are testcrosses and can tell us the genotypes of the strains. Therefore, "A" is *Aa Bb*, "B" is *aa Bb*, and "D" is *Aa bb*.

2.25 a. The cross is $W/w\ R/r \times W\ r$. Workers are females, and in the progeny of the cross they are: $\frac{1}{4}$ *W/W R/r* (black, use wax) : $\frac{1}{4}$ *W/w R/r* (black, use wax) : $\frac{1}{4}$ *W/W r/r* (black, use resin) : $\frac{1}{4}$ *W/w r/r*

(black, use resin). In sum, all workers are black-eyed, while $\frac{1}{2}$ use wax and $\frac{1}{2}$ use resin.

b. $\frac{1}{4}$ each *W R, W r, w R,* and *w r*.

c. Here the point is that the female in question is heterozygous, not that her progeny will be wingless. Madonna is *C/c*, so there is a $\frac{1}{4}$ chance that her granddaughter will be heterozygous and that wingless males will be found in the hive (i.e., $\frac{1}{2}$ that her daughter is heterozygous × $\frac{1}{2}$ that the daughter of the daughter is heterozygous).

d. The chance that the F_4 generation great-great-granddaughter is heterozygous is $\frac{1}{2} \times \frac{1}{2} \times \frac{1}{2} \times \frac{1}{2} = \frac{1}{16}$.

2.26 a. The mother must be heterozygous *Aa* in order to have children who have the trait.

b. The father is homozygous *aa* because he expresses the trait.

c. The cross is $Aa \times aa$, so children will be *aa* if they have the trait (II-2 and II-5) and *Aa* if they do not have the trait (II-1, II-3, and II-4).

d. From the cross $Aa \times aa$, the prediction is that $\frac{1}{2}$ of the progeny will be *Aa* (normal) and $\frac{1}{2}$ will be *aa* (expressing the trait). There are five children, two of whom have the trait and three of whom are normal. Thus, the ratio fits as well as it could for five children.

CHAPTER 3

3.1

	MOTHER'S		FATHER'S	
	MOTHER	FATHER	MOTHER	FATHER
X chromosome	Yes	Yes	No	No
Y chromosome	No	No	No	Yes

3.3 Fathers always give their X chromosome to their daughters, so the woman must be heterozygous for the color-blindness trait and is c^+c. Her husband received his X chromosome from his mother and has normal color vision, so he is c^+Y. The cross is therefore $c^+c \times c^+Y$. All daughters will receive the paternal X bearing the c^+ allele and have normal color vision. Sons will receive the maternal X, so half will be *cY* and be color-blind, and half will be c^+Y and have normal color vision.

3.5 If *c* is the red-green color-blindness allele and *a* is the albino allele, the parents' genotypes are $c^+c^+\ aa$ ♀ and $cY\ a^+a^+$ ♂. All children will be normal visioned (c^+c ♀ and c^+Y ♂) and normally pigmented (a^+a, both sexes).

3.6 The parentals are $ww\ vg^+vg^+$ ♀ and $w^+Y\ vgvg$ ♂.

a. The F_1 males are all $wY\ vg^+vg$, white eyes, long wings. The F_1 females are all $w^+w\ vg^+vg$, red eyes, long wings.

b. The F_2 females are $\frac{3}{8}$ red, long; $\frac{3}{8}$ white, long; $\frac{1}{8}$ red, vestigial; $\frac{1}{8}$ white, vestigial. The same ratios of the respective phenotypes apply for the males.

c. The cross of F_1 male with the parental female is $wY\ vg^+vg \times ww\ vg^+vg^+$. All progeny have white eyes and long wings. The cross of an F_1 female with the parental male is $w^+w\ vg^+vg \times w^+Y\ vgvg$. Female progeny: All have red eyes, half have long wings, and half have vestigial wings. Male progeny: $\frac{1}{4}$ red, long; $\frac{1}{4}$ red, vestigial; $\frac{1}{4}$ white, long; $\frac{1}{4}$ white, vestigial.

3.9 The simplest hypothesis is that brown-colored teeth are determined by a sex-linked dominant mutant allele. Man A was *BY* and his wife was *bb*. All sons will be *bY* normals and cannot pass on the trait. All the daughters receive the X chromosome from their father, and so they are *Bb* with brown enamel. Half their sons will have brown teeth because half their sons receive the X chromosome with the *B* mutant allele.

3.11 a. To have produced a child with cystic fibrosis, both parents must have been heterozygous for the autosomal recessive mutant gene. Therefore, the probability of their next child having cystic fibrosis is $\frac{1}{4}$ (i.e., $\frac{1}{4}$ of the progeny from $Aa \times Aa$ will be *aa*).

b. Nonaffected children can be either AA or Aa. From a cross $Aa \times Aa$, there will be a progeny ratio of 1 AA : 2 Aa : 1 aa. Therefore, $\frac{2}{3}$ of the nonaffected children will be Aa heterozygotes.

3.16

	PEDIGREE A	PEDIGREE B	PEDIGREE C
Autosomal recessive	Yes	Yes	Yes
Autosomal dominant	Yes	Yes	No
X-linked recessive	Yes	Yes	No
X-linked dominant	No	No	No

3.18 a. Y-linked inheritance can be excluded because females are affected. X-linked recessive inheritance can also be excluded because an affected mother (I.2) has a normal son (II.5). Autosomal recessive inheritance can also be excluded because in such a case, two affected parents such as II.1 and II.2 could only have affected offspring, which they do not.

b. The two remaining mechanisms of inheritance are X-linked dominant and autosomal dominant. Genotypes can be written to satisfy both mechanisms of inheritance. Of these two, X-linked dominant inheritance may be more likely because II.6 and II.7 have only affected daughters, indicating crisscross inheritance. If the trait were autosomal dominant, one would expect half of the daughters and half of the sons to be affected.

3.21 a is false. The father passes on his X to his daughters, so *none* of the sons will be affected.

b is false. Because the disease is rare, the mother would be heterozygous, and only 50% of her daughters would receive the X with the dominant mutant allele.

c is true. All daughters receive the father's X chromosome.

d is false. Only 50% of her sons would receive the X with the dominant mutant allele.

3.23 a. True. Two affected individuals will always have affected children ($aa \times aa$ can give only aa offspring).

b. False. An autosomal trait is inherited independent of sex type.

c. Need not be true. The trait could be masked by normal dominant alleles through many generations before two heterozygotes marry and produce affected, homozygous offspring.

d. Could be true. If the trait is rare, an unaffected individual marrying into the pedigree is likely to be homozygous for a normal allele. The trait is recessive, and the children receive the dominant, normal allele from the unaffected parent, so the children will be normal. This answer would not be true if the unaffected individual was heterozygous. In this case, half of the children would be affected.

3.25 Because hemophilia is an X-linked trait, the most likely explanation is that random inactivation of X chromosomes (lyonization) produces individuals with different proportions of cells with the normal allele. Thus, if some women had only 40% of their cells with an active h^+ allele and 60% with the h (hemophilia) allele, but other women had 60% of their cells with an active h^+ allele, and 40% with the h allele, then the amount of clotting factor these women would make would differ significantly.

CHAPTER 4

4.2 Six possible genotypes: w/w, $w/w1$, $w/w2$, $w1/w1$, $w1/w2$, and $w2/w2$.

4.6 The woman's genotype is I^A/I^B, and the man's genotype is I^A/i.

a. $\frac{1}{2} \times \frac{1}{2} = \frac{1}{4}$.

b. Zero. A blood group O baby is not possible.

c. $\frac{1}{2}$ (probability of male) $\times$ $\frac{1}{4}$ (probability of AB) $\times$ $\frac{1}{2}$ (probability of male) $\times$ $\frac{1}{4}$ (probability of B) = $\frac{1}{64}$ probability that all four conditions will be fulfilled.

4.8 Blood type O, because genotype is i/i.

4.10 Half will be C^R/C^W and hence will resemble the parents.

4.12 a. The cross is $F/F\ G^N/G^N \times f/f\ G^O/G^O$. The F_1 is $F/f\ G^O/G^N$, which is fuzzy with round leaf glands.

b. Interbreeding the F_1 gives $\frac{3}{16}$ fuzzy, oval-glanded; $\frac{6}{16}$ fuzzy, round-glanded; $\frac{3}{16}$ fuzzy, no-glanded; $\frac{1}{16}$ smooth, oval-glanded; $\frac{2}{16}$ smooth, round-glanded; $\frac{1}{16}$ smooth, no-glanded.

c. Cross is $F/f\ G^O/G^N \times f/f\ G^O/G^O$. The progeny are $\frac{1}{4}$ fuzzy, oval-glanded; $\frac{1}{4}$ fuzzy, round-glanded; $\frac{1}{4}$ smooth, oval-glanded; $\frac{1}{4}$ smooth, round-glanded.

4.18 To show no segregation among the progeny, the chosen plant must be homozygous. The genotypes comprising the $\frac{9}{16}$ colored plants are: 1 $A/A\ B/B$: 2 $A/A\ B/b$: 2 $A/a\ B/B$: 4 $A/a\ B/b$. Only one of these genotypes is homozygous; the answer is $\frac{1}{9}$.

4.19 Two independently assorting genes are involved. Because of epistasis, only two phenotypes are seen in the F_2; that is, $A/-\ B/-$ genotypes are runner, and the other three genotypes are bunch. Thus, the original cross was $A/A\ b/b \times a/a\ B/B$, giving an F_1 of $A/a\ B/b$.

4.20 a. The cross is $A/a\ B/b \times A/a\ B/b$, which gives $\frac{9}{16}\ A/-\ B/-$: $\frac{3}{16}\ A/-\ b/b$: $\frac{3}{16}\ a/a\ B/-$: $\frac{1}{16}\ a/a\ b/b$. The $A/-\ B/-$ rabbits are not deaf because they produce both substances needed for hearing. The other three genotypic classes result in deafness because one or the other or both of the enzymes needed for hearing are not produced. Thus, the phenotypic ratio is 9 hearing rabbits : 7 deaf rabbits.

b. Epistasis. In this case, it is duplicate recessive epistasis.

c. The cross is $a/a\ B/b \times A/a\ B/b$, which gives $\frac{3}{8}\ A/a\ B/-$: $\frac{1}{8}\ A/a\ b/b$: $\frac{3}{8}\ a/a\ B/-$: $\frac{1}{8}\ a/a\ b/b$. Only the $A/a\ B/-$ rabbits can hear; the rest are deaf. Thus, the phenotypic ratio is 3 hearing rabbits : 5 deaf rabbits.

4.23 a. If A^Y governs yellow and a^+ governs nonyellow (agouti), then A^Y/A^Y are lethal, A^Y/a^+ are yellow, and a^+/a^+ are agouti. Let c^+ determine colored coat and c determine albino. The parental genotypes, then, are $A^Y/a^+\ c^+/c$ (yellow) and $A^Y/a^+\ c/c$ (white).

b. The proportion is 2 yellow : 1 agouti : 1 albino. None of the yellows breed true because they are all heterozygous, with homozygous A^Y/A^Y individuals being lethal.

4.25 a. $Y/Y\ R/R$ (crimson) $\times$ $y/y\ r/r$ (white) gives $Y/y\ R/r$ F_1 plants, which have magenta-rose flowers. Selfing the F_1 gives an F_2 as follows: $\frac{1}{16}$ crimson ($Y/Y\ R/R$), $\frac{2}{16}$ orange-red ($Y/Y\ R/r$), $\frac{1}{16}$ yellow ($Y/Y\ r/r$), $\frac{2}{16}$ magenta ($Y/y\ R/R$), $\frac{4}{16}$ magenta-rose ($Y/y\ R/r$), $\frac{2}{16}$ pale yellow ($Y/y\ r/r$), and $\frac{4}{16}$ white ($y/y\ R/R$, $y/y\ R/r$, and $y/y\ r/r$). Progeny of the F_1 backcrossed to the crimson parent are $\frac{1}{4}$ crimson, $\frac{1}{4}$ orange-red, $\frac{1}{4}$ magenta, and $\frac{1}{4}$ magenta-rose.

b. $Y/Y\ R/r \times Y/y\ r/r$ gives $\frac{1}{4}$ orange-red ($Y/Y\ R/r$), $\frac{1}{4}$ yellow ($Y/Y\ r/r$), $\frac{1}{4}$ magenta-rose ($Y/r\ R/r$), $\frac{1}{4}$ pale yellow ($Y/y\ r/r$).

c. $Y/Y\ r/r$ (yellow) $\times$ $y/y\ R/r$ (white) gives $\frac{1}{2}$ magenta-rose ($Y/y\ R/r$) and $\frac{1}{2}$ pale yellow ($Y/y\ r/r$).

4.27 a. The simplest approach is to calculate the proportion of progeny that will be black and then subtract that answer from 1. The black progeny have the genotype $A/-\ B/-\ C/-$, and the proportion of these progeny is $(\frac{3}{4})^3$. Therefore, the proportion of colorless progeny is $1 - (\frac{3}{4})^3 = 1 - \frac{27}{64} = \frac{37}{64}$.

b. Black is produced only when $c/c\ A/-\ B/-$ results, and this offspring occurs with the frequency $(\frac{1}{4})(\frac{3}{4})(\frac{3}{4}) = \frac{9}{64}$ black, which gives $\frac{55}{64}$ colorless.

4.29 In males, H/H and H/h are horned, and h/h is hornless; in females, H/H is hornless. The cross is an $H/H\ W/W$ male $\times$ $h/h\ w/w$ female. The F_1 is $H/h\ W/w$, which gives horned white males and hornless white females. Interbreeding the F_1 gives the following F_2:

	MALE	FEMALE
$\frac{3}{16}\ H/H\ W/-$	horned, white	horned, white
$\frac{6}{16}\ H/h\ W/-$	horned, white	hornless, white
$\frac{3}{16}\ h/h\ W/-$	hornless, white	hornless, white
$\frac{1}{16}\ H/H\ w/w$	horned, black	horned, black
$\frac{2}{16}\ H/h\ w/w$	horned, black	hornless, black
$\frac{1}{16}\ h/h\ w/w$	hornless, black	hornless, black

In sum, the ratios are $\frac{9}{16}$ horned white : $\frac{3}{16}$ hornless white : $\frac{3}{16}$ horned black : $\frac{1}{16}$ hornless black males, and $\frac{3}{16}$ horned white : $\frac{9}{16}$ hornless white : $\frac{1}{16}$ horned black : $\frac{3}{16}$ hornless black females.

CHAPTER 5

5.2 From the χ^2 test, $\chi^2 = 16.10$; P is less than 0.01 at 3 degrees of freedom. This test reveals that the two genes do not fit a 1:1:1:1 ratio. It does not say why. Linkage might seem reasonable until it is realized that the minority classes are not reciprocal classes (both carry the aa phenotype). If the segregation at each locus is considered, however, the $B/- : b/b$ ratio is about 1:1 (203:197), but the $A/- : a/a$ ratio is not (240:160). The departure, then, specifically results from a deficiency of a/a individuals. This departure should be confirmed in other crosses that test the segregation at locus A. In corn, further evidence would be that the a/a deficiency might show up as a class of ungerminated seeds or of seedlings that die early.

5.4

Note that the map distances are not strictly additive—e.g., a-$d = 38$ and a-$c = 8$, but d-$e = 45$—because of the effects of multiple crossovers, as described in the chapter.

5.6 45% $a\ b^+$, 45% $a^+\ b$, 5% $a^+\ b^+$, 5% $a\ b$

5.7 a. $0.035 + 0.465 = 0.50$

    b. All the daughters have $a^+\ b^+$ phenotype.

5.9 Each chromosome pair segregates independently. We can compute the relative proportions of gametes produced for each homologous pair of chromosomes separately from the known map distances (P = parental; R = recombinant):

P		R		P		R		P		R	
AB	0.4	Ab	0.1	CD	0.45	Cd	0.05	EF	0.35	Ef	0.15
ab	0.4	aB	0.1	cd	0.45	cD	0.05	ef	0.35	eF	0.15

To answer the question, simply multiply the probabilities of getting the particular gamete from the F_1 multiple heterozygote.

    a. $A\ B\ C\ D\ E\ F = 0.4 \times 0.45 \times 0.35 = 0.063$ (6.3%)

    b. $A\ B\ C\ d\ e\ f = 0.4 \times 0.05 \times 0.35 = 0.007$ (0.7%)

    c. $A\ B\ c\ D\ E\ f = 0.1 \times 0.05 \times 0.15 = 0.00075$ (0.075%)

    d. $a\ B\ C\ d\ e\ f = 0.1 \times 0.05 \times 0.35 = 0.00175$ (0.175%)

    e. $a\ b\ c\ D\ e\ F = 0.4 \times 0.05 \times 0.15 = 0.003$ (0.3%)

5.10 a. 47.5% each of $DP\ h$ and $dp\ h$; 2.5% each of $Dp\ h$ and $dP\ h$.

    b. 23.75% each of $Dp\ H$, $Dp\ h$, $dP\ H$, and $dP\ h$; 1.25% each of $DP\ H$, $DP\ h$, $dp\ H$, and $dp\ h$.

5.14 a. By doing a three-point mapping analysis as described in the chapter, we find that the order of genes in the chromosome is dp-b-hk, with 35.5 mu between dp and b, and 5.4 mu between b and hk.

    b. (1) The frequency of observed double crossover is 1.4%, and the frequency of expected double crossovers is 1.9%. The coefficient of coincidence, therefore, is $1.4/1.9 = 0.74$. (2) The interference value is given by $1 -$ coefficient of coincidence $= 0.26$.

5.15 a. $a^+\ c\ b/a\ c^+\ b^+$

    b. $a^+\ c^+\ b^+/Y$

    c.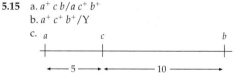

5.18 The two X chromosomes in the female have the genotypes $a +$ and $+ b$. The lethal must be on the $+ b$ chromosomes of the female parent because there are far fewer parental $+ b$ chromosomes in the progeny males than $a +$ chromosomes. Therefore, we can diagram the cross as:

All female progeny of the cross will be wild type. At this point, however, we do not know the order of the three genes on the chromosome.

You would expect 1,000 males normally, but you get only 499. The remainder can be considered the lethal (l) progeny. This fact allows you to predict the entire set of eight genotypes expected among the zygotes of the cross. This is done by knowing that $a +$ is really $+ +$, where the last $+$ sign is the wild-type allele of the lethal. Thus, each genotype seen is accompanied in this cross by one carrying the l allele. Writing them out and recognizing the equality of reciprocal classes, we have:

405 $a + +$	and	405 $+ b\ l$
44 $+ b +$	and	44 $a + +$
48 $+ + +$	and	48 $a\ b\ l$
2 $a\ b +$	and	2 $+ + l$

This is:

810 parentals ($a + +$ and $+ b\ l$)
88 single crossovers, one region ($+ b +$ and $a + +$)
96 single crossovers, other region ($+ + +$ and $a\ b\ l$)
4 double crossovers ($a\ b +$ and $+ + l$)

Traditional methods lead us to the map:

5.25 From applying the tetrad analysis formula,

$$\text{map distance} = \frac{\frac{1}{2}T + NPD}{Total} \times 100\%$$

the a-b distance is 19.6 map units, the b-c distance is 11 map units, and the a-c distance is 14 map units. The gene order is a-c-b.

CHAPTER 6

6.1 The whole chromosome would have to be transferred in order for the recipient to become a donor in an $Hfr \times F^-$ cross; that is, the F factor in the Hfr strain is transferred to the F^- cell last. This transfer takes approximately 100 min, and usually the conjugal unions break apart before then.

6.2

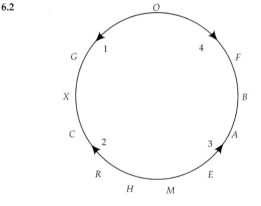

6.4 Strain A is $thy\ leu^+$ and B is $thy^+\ leu$. Transformed B should be $thy^+\ leu^+$, and its presence should be detected on a medium containing neither thymine nor leucine.

6.6 0.07 mu. The plaques produced on $K12(\lambda)$ are wild-type r^+ phages, and in the undiluted lysate there are 470×5 per milliliter (because 0.2 mL was plated), or 2,350/mL. The r^+ phages were generated by recombination between the two *rII* mutations. The other product of the recombination event is the double mutant, and it does not grow on $K12(\lambda)$. Therefore, the true number of recombinants in the population is actually equivalent to twice the number of r^+ phages because for every wild-type phage produced, there ought to be a doubly mutant recombinant produced. Thus, there are 4,700 recombinants/mL. The total number of phages in the lysate is 672 × (dilution factor) × (1 mL divided by the sample size plated) per milliliter, or 672 × 1,000 × 10 = 6,720,000/mL. The map distance between the mutations is $\left(\frac{4700}{6,720,000}\right)$ × 100% = 0.07% = 0.07 mu.

6.8   There are two answers that are compatible with the data:

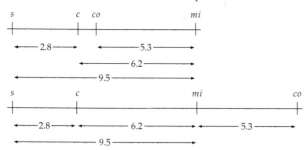

6.10   0.5% recombination (the numbers of plaques counted are so small that this value is a rather rough approximation).

6.13   a. 3 genes

b. *A, D,* and *F* are in one; *B* and *G* are in the second; and *C* and *E* are in the third.

CHAPTER 7

7.1   a. pericentric inversion (D∘E F inverted)

b. nonreciprocal translocation (BC moved to other arm)

c. tandem duplication (EF duplicated)

d. reverse tandem duplication (EF duplicated)

e. deletion (C deleted)

7.2   See text, pp. 159–161.

7.4   a.

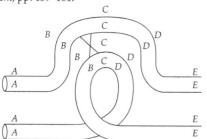

b. From part a, the crossover between B and C results in the following:

A B C D E

A B C D A

E B C D E

A D C B E

c. Paracentric inversion, because the centromere is not included in the inverted DNA segment.

7.6   a. Paracentric inversion.

b. Single crossover within the inversion loop:

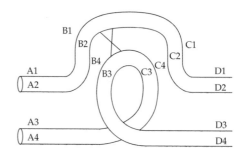

7.7   a. Mr. Lambert is heterozygous for a pericentric inversion of chromosome 6. One of the breakpoints is within the fourth light band up from the centromere, and the other is in the sixth dark band below the centromere. Mrs. Lambert's chromosomes are normal.

b. When Mr. Lambert's chromosome 6s paired, they formed an inversion loop that included the centromere. Crossing-over occurred within the loop and gave rise to the partially duplicated, partially deficient 6, which the child received.

c. The child's abnormalities stem from having three copies of some, and only one copy of other, chromosome 6 regions. The top part of the short arm is duplicated, and there is a deficiency of the distal part of the long arm in this case.

d. The inversion appears to cover more than half of the length of chromosome 6, so crossing-over will occur in this region in the majority of meioses. In the minority of meioses where crossing-over has occurred outside the loop or where it has occurred within the loop, but the child receives an uncrossed-over chromatid, the child can be normal. There is significant risk for abnormality, so monitoring of fetal chromosomes should be done.

7.9

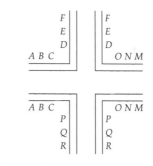

7.11   a. Mr. Denton has normal chromosomes. Mrs. Denton is heterozygous for a balanced reciprocal translocation between chromosomes 6 and 12. Most of the short arm of chromosome 6 has been reciprocally translocated onto the long arm of chromosome 12. The breakpoints appear to be in the thick dark band just above the centromere of 6 and in the third dark band below the centromere of 12.

b. The child received a normal 6 and a normal 12 from his father. In meiosis in Mrs. Denton, the 6s and 12s paired and formed a cross-like figure in prophase I, segregation of adjacent nonhomologous centromeres to the same pole occurred, and the child received a gamete containing a normal 6 and one of the translocation chromosomes. (See Figure 7.10.)

c. The child received a normal chromosome 6 and a normal chromosome 12 from Mr. Denton. The child is not phenotypically normal because of partial trisomy (it has three copies of part of the short arm of chromosome 6) and partial monosomy (it has only one copy of most of the long arm of chromosome 12).

d. Given that segregation of adjacent homologous centromeres to the same pole will be relatively rare in this case, there should be about a 50% chance that a given conception will be chromosomally unbalanced. Of the 50% balanced ones, 50% would be translocation heterozygotes.

e. Prenatal monitoring of fetal chromosomes could be done, followed by therapeutic abortion of chromosomally unbalanced fetuses.

7.15 a. 45
   b. 47
   c. 23
   d. 69
   e. 48

7.16 b. An individual with three instead of two chromosomes is said to be *trisomic*.

7.17 a. Mother. The color-blind Turner syndrome child must have received her X chromosome from her father because the father carries the only color-blind allele. Therefore, the mother must have produced an egg with no X chromosomes.

b. Father. Because the Turner syndrome child does not have color blindness, she must have received her X chromosome from the homozygous normal mother. Therefore, the father must have produced a sperm with no X or Y chromosome.

7.21 $(7N + 10N) \times 2 = 34$ (4N)

CHAPTER 8

8.2 The double homozygote should have PKU but not AKU. The PKU block should prevent most homogentisic acid from being formed, so it could not accumulate to high levels.

8.4 a. The simplest approach is to calculate the proportion of the F_2 that are colored and to subtract that answer from 1. In this case, the noncolorless are the brown or black progeny. The proportion of progeny that make at least the brown pigment is given by the probability of having the following genotype: $a^+/- b^+/- c^+/- (d^+/d^+, d^+/d,$ or $d/d)$. The answer is $\frac{3}{4} \times \frac{3}{4} \times \frac{3}{4} = \frac{27}{64}$. Therefore, the proportion of colorless is $1 - \frac{27}{64} = \frac{37}{64}$.

b. The brown progeny have the following genotype: $a^+/- b^+/- c^+/- d/d$. The probability of getting individuals with this genotype is $\frac{3}{4} \times \frac{3}{4} \times \frac{3}{4} \times \frac{1}{4} = \frac{27}{256}$.

8.6 a. Since any of the normal alleles $a^+, b^+,$ or c^+ is sufficient to catalyze the reaction leading to color, in order for color to fail to develop, all three normal alleles must be missing. That is, the colorless F_2 must be $a/a\ b/b\ c/c$. The chance of obtaining such an individual is $\frac{1}{4} \times \frac{1}{4} \times \frac{1}{4} = \frac{1}{64}$.

b. Now, colorless F_2 are obtained if *either* one or both steps of the pathway are blocked. That is, colorless F_2 are obtained in either of the following genotypes: $d/d\ -/-\ -/-\ -/-$ (the first or both steps blocked) or $d^+/-\ a/a\ b/b\ c/c$ (second step blocked). The chance of obtaining such individuals is
$$\left(\tfrac{1}{4} \times 1 \times 1 \times 1\right) + \left(\tfrac{3}{4} \times \tfrac{1}{4} \times \tfrac{1}{4} \times \tfrac{1}{4}\right) = \frac{(64+3)}{256} = \frac{67}{256}.$$

8.8 a. Half have white eyes (the sons), and half have fire red eyes (the daughters).

b. All have fire red eyes.

c. All have brown eyes.

d. All are w^+; $\frac{1}{4}$ are $bw^+/- st^+/-$, red; $\frac{1}{4}$ are $bw^+/- st/st$, scarlet; $\frac{1}{4}$ are $bw/bw\ st^+/-$, brown; $\frac{1}{4}$ are $bw/bw\ st/st$, the color of 3-hydroxykynurenine plus the color of the precursor to biopterin, or colorless.

8.9 Wild-type T4 will produce progeny phages at all three temperatures. Let us suppose that model (1) is correct. If cells infected with the double mutant are first incubated at 17°C and then shifted to 42°C, progeny phages will be produced, and the cells will lyse. The explanation is as follows: the first step, A to B, is controlled by a gene whose product is heat-sensitive. At 17°C the enzyme works, and A is converted to B, but B cannot then be converted to mature phages because that step is cold-sensitive. When the temperature is raised to 42°C, the A-to-B step is blocked, but the accumulated B can be con-

verted to mature phages because the enzyme involved with that step is cold-sensitive and functional at the high temperature. If model (2) is the correct pathway, then progeny phages should be produced in a 42°-to-17°C temperature shift, but not vice versa. In general, two gene product functions can be ordered by this method whenever one temperature shift allows phage production and the reciprocal shift does not, according to the following rules: (1) if a low-to-high temperature shift results in phages but a high-to-low temperature shift does not, the *hs* step precedes the *cs* step (model 1); (2) if a high-to-low temperature shift results in phages but a low-to-high temperature does not, the *cs* step precedes the *hs* step (model 2).

8.11 One approach to this problem is to try to fit the data sequentially to each pathway, as if each were correct. Check where each mutant *could* be blocked (remember, each mutant carries only one mutation), whether the mutant would be able to grow if supplemented with the *single* nutrient that is listed, and whether the mutant would *not* be able to grow if supplemented with the "no growth" intermediate. It will not be possible to fit the data for mutant *4* to pathway b, or data for mutants *3* and *4* to pathway c. The data for all mutants can be fit only to pathway d. Thus, d must be the correct pathway.

A second approach to this problem is to realize that in any linear segment of a biochemical pathway (a segment without a branch), a block early in the segment can be circumvented by *any* metabolites that normally appear later in the same segment. Consequently, if two (or more) intermediates can support growth of a mutant, they normally are made after the blocked step in the same linear segment of a pathway. From the data given, compounds D and E both circumvent the single block in mutant *4*. This means that compounds D and E lie after the block in mutant *4* on a linear segment of the metabolic pathway. The only pathway where D and E lie in an unbranched linear segment is pathway d. Mutant *4* could be blocked between A and E in this pathway. Mutant *4* cannot be fit to a *single* block in any of the other pathways that are shown, so the correct pathway is d.

8.12 a. Three distinct mutational sites exist based on the recombination analysis: one in the *b* gene, and two in the *c* gene.

b. Two polypeptide chains are affected based on the complementation analysis: one encoded by the *b* gene, and one encoded by the *c* gene.

c. The genotypes are:
   $1 = b^1\ c^+$
   $2 = b^+\ c^2$
   $3 = b^+\ c^3$
   $4 = b^4\ c^4$

d. There are 0.004 mu between the two mutant sites in the *c* gene identified by c^2 and c^3. The *b* and *c* genes are unlinked.

e. The *b* and *c* genes assort independently of one another. Thus, strains 1, 2, or 3, when mated with the wild type, will give 50 percent prototrophs and 50 percent auxotrophs. As strain 4 is a double mutant, it will only give 25 percent prototrophs. Four equally frequent genotypes would be expected from the cross $b^4\ c^4 \times + +$. Only one of the four will be $+ +$.

8.14 Baby Joan, whose blood type is O, must belong to the Smith family. Mrs. Jones, being of blood type AB, could not have an O baby. Therefore, Baby Jane must be hers.

8.15 a. Normal parents have affected offspring, so the disease appears to be recessive. However, patients with 50 percent of GSS activity have a mild form of the disease; therefore, an individual who was heterozygous (mutant/+) for a complete loss-of-function GSS mutation might show mild symptoms. Thus, in a population, the disease may show variable penetrance. The expressivity of the disease in individuals will depend on the nature of their GSS mutation. The alleles discussed here appear to be recessive, but a complete loss-of-function allele may show partial dominance.

b. Patient 1, with 9 percent of normal GSS activity, has a more severe form of the disease. Patient 2, with 50 percent of normal GSS activity, has a less severe form of the disease. Thus, increased disease severity is associated with less GSS enzyme activity.

c. The two amino acid substitutions may disrupt different regions of the enzyme's structure (consider the effect of various amino acid substitutions on hemoglobin function that are discussed in the text, p. 189). Amino acids vary in their polarity and charge, so different amino acid substitutions within the same structural region could have varying chemical effects on protein structure. This too could lead to disparate levels of enzymatic function. (For a discussion of the chemical dissimilarities between amino acids, see Chapter 13).

d. By analogy with the disease PKU discussed in the text, 5-oxoproline is produced only when a precursor to glutathione accumulates in large amounts due to a block in a biosynthetic pathway. When GSS levels are 9 percent of normal, this occurs. When GSS levels are 50 percent of normal, sufficient GSS enzyme is present to complete the pathway sufficiently to prevent high levels of 5-oxoproline.

e. The mutations are allelic, because both the severe and the mild forms of the disease are associated with alterations in the same polypeptide that is a component of the GSS enzyme. (Although the data in this problem suggest that the GSS enzyme is composed of a single polypeptide, they do not exclude the possibility that GSS has multiple polypeptide subunits.)

f. If GSS is normally found in fetal fibroblasts, one could, in principle, measure GSS activity in fibroblasts obtained via amniocentesis. The GSS enzyme level in cells from at-risk fetuses could be compared to that in normal control samples to predict disease due to inadequate GSS levels. Some variation in GSS level might be seen, depending on the allele(s) present. More than one mutation is present in the population, so it is important to devise a functional test that assesses GSS activity, rather than a test that identifies a single mutant allele based on the nature of its DNA mutation.

CHAPTER 9

9.4 b

9.5 a, b and c. Phage ghosts (supernatant) and progeny would have isotopes. Both proteins and nucleic acids have carbon, nitrogen, and hydrogen. If a parental phage was labeled with isotopes of C, N, or H, one would expect the phage to have a labeled protein coat as well as labeled DNA. As a consequence, these isotopes would be recovered in the DNA of the progeny phage as well as in the phage ghosts left behind in the supernatant after phage infection. This experiment would not be very helpful to determine whether DNA or protein was the genetic material, as neither material would be labeled selectively.

9.6 3' and 5' carbons are connected by a phosphodiester bond.

9.11 The evidence for only two base-pair combinations in DNA, that is, A-T and G-C, comes from quantitative measurements of the four bases in double-stranded DNA isolated from a wide variety of organisms. In all cases, the amount of A was shown to equal the amount of T, and the amount of G was shown to equal the amount of C. Moreover, different DNAs exhibit different base ratios (stated as the %GC), so although A = T and G = C, in most organisms (A + T) does not equal (G + C). The simplest hypothesis, therefore, is that there are two base pairs in DNA, A-T and G-C, and that the proportion of the two base pairs varies from organism to organism.

9.13 a. 3' TCAATGGACTAGCAT 5'

b. 3'AAGAGTTCTTAAGGT 5'

9.14 The adenine-thymine base pair is held together by two hydrogen bonds, but the guanine-cytosine base pair is held together by three hydrogen bonds. Thus, the guanine-cytosine base pair requires more energy to break it and is the harder pair to break apart.

9.16 b, c, and d

9.18 C = 17%; therefore, G = 17% and the % GC = 34. Therefore, the % AT = 66 and the % A = $^{66}/_2$ = 33.

9.25 a. 200,000

b. 10,000

c. 3.4×10^4 nm

CHAPTER 10

10.2 a. $^9/_{24}$ = 0.375. Since S is 37.5% of the cycle, 37.5% of cells should be in S at any instant, and these are the ones that would incorporate label.

b. A little over 4 h. Such cells would have to take up some ^{3}H (a few minutes), pass through G_2 (4 h), and pass through mitotic prophase (several minutes).

c. Both. Each chromatid is a double-stranded DNA molecule containing one old and one newly synthesized DNA strand.

d. A little more than 13 h. Such cells would have to pass through nearly all of S, all of G_2, and through prophase.

10.3 a. 16 hours is longer than any single stage of the cell cycle. In particular, a cell that just entered G_2 at the beginning of the labeling period would, at the end of a 16-hour labeling period, be in early S phase. Therefore, it would be labeled. Indeed, every cell would be labeled.

b. Labeled metaphase chromosomes would already be present by the time the ^{3}H-thymidine was removed, as some of the cells in G_1 at the beginning of the labeling period would go through S and G_2, and be in M after 16 hours.

c. Both chromatids would be labeled, as in a 5-min pulse of ^{3}H-thymidine.

d. Consider what would happen if metaphase chromosomes were examined in cells collected periodically after the radioactive medium was washed out. Start with cells sampled immediately after the radioactive medium is washed out. Since these cells were left in the radioactive medium for 16 hours, and there is only a little over 13 hours between the beginning of the S period and metaphase of mitosis, the previous S phase of these cells was spent entirely in the presence of radioactive thymidine. Any metaphase chromosomes seen at this time point would be labeled in their entirety, so these chromosomes would not be useful to identify regions replicated at a particular time during the S period. Suppose instead that cells were sampled after about 6 hours in nonradioactive medium. These cells would have been in the first part of G_1 when the radioactive medium was added and spent about the first 8 hours of their S period in the radioactive medium. All of their chromosomal regions would be labeled except for the regions that replicated late in the S period. By sampling cells over increasing lengths of time and comparing the results of different samplings, chromosomal regions that replicated at successively earlier stages could be identified. Finally, consider the cell sample taken after 13 hours in nonradioactive medium. These cells would have just entered the S period when the nonradioactive medium had been washed out. Since they have only spent the beginning of their S period in radioactive medium, chromosomal regions that replicated early in the S period would be labeled. However, since the culture was left in radioactive medium for 16 hours, the parents of these cells spent about the last hour of their S period in the presence of radioactive thymidine. Thus, these cells would have chromosomes that were labeled both at the beginning and at the end of the S period. Only by comparison of this sample to samples obtained at earlier time points might these regions be distinguished.

10.4 Chromosomes are placed in groups according to their size (see Figure 10.5). The following table summarizes members of each group in a normal and the abnormal individual.

GROUP	NORMAL CHROMOSOME NUMBER	NORMAL CHROMOSOME TYPE	ABNORMAL CHROMOSOME NUMBER
A	6	1, 2, 3	6
B	4	4, 5	4
C	15 or 16	6, 7, 8, 9, 10, 11, 12, X (XX in females)	16
D	6	13, 14, 15	6
E	6	16, 17, 18	6
F	4	18, 20	4
G	4 or 5	21, 22, (Y in males)	5

Both groups C and G could have a potentially unusual chromosome number. The most likely explanations are that either the individual is XXY, or female and trisomy 21 (XX, 21, 21, 21). Other explanations are possible (e.g., XY, with trisomy for any of 6, 7, 8, 9, 10, 11, 12, or XX with trisomy for 22) but less likely, as such aneuploids typically are spontaneously aborted before the stage (12 weeks) at which amniocentesis would be possible. To distinguish between these possibilities, use a staining technique that affords more resolution of the chromosomal banding patterns, such as G banding (Figure 10.6), and/or use chromosome-specific probes and perform FISH (p. 115) to determine which chromosomes are present.

10.6 c.

10.8 a. D, G, J

b. A, E, F, H

c. I

10.11 a. The belt forms a right-handed helix. Although you wrapped the belt around the can axis in a counterclockwise direction from your orientation (looking down at the can), the belt was winding up and around the side of the can in a clockwise direction from its orientation. While the belt is wrapped around the can, curve the fingers of your right hand over the belt and use your index finger to trace the direction of the belt's spiral. Your right index finger will trace the spiral upward, the same direction your thumb points when you wrap your hand around the can. Therefore, the belt has formed a right-handed helix.

b. Three

c. Three. The number of helical turns is unchanged, although the twist in the belt is.

d. The belt appears more twisted because the pitch of the helix was altered, and the edges of the belt (positioned much like the complementary base pairs of a double helix) are twisted more tightly.

e. When it is twisted around the can, the length of the belt decreases by about 70–80 percent, depending on the initial length of the belt and the diameter of the can.

f. Yes. As the DNA of linear chromosomes is wrapped around histones to form the 10-nm nucleofilament, it becomes supercoiled. In much the same manner as you must add twists to the belt in order for it to lie flat on the surface of the can, supercoils must be introduced into the DNA for it to wrap around the histones.

g. Topoisomerases increase or reduce the level of negative supercoiling in DNA. In order for linear DNA to be packaged, negative supercoils must be added to it.

10.13 A telomere is the region of DNA at the end of a linear chromosome. Telomeres are characteristically heterochromatic, and in a given species they share a common sequence. For most organisms, the telomeric sequences found at the extreme end of a chromosome are simple, highly repeated, and species specific. Nearby, but not at the end of a chromosome, are telomere-associated sequences. These are repeated, complex sequences that extend for thousands of bases from the simple telomeric sequences. Telomere organization is quite different in some organisms such as *Drosophila*. In these organisms, telomeres consist of transposons that belong to the LINE family of repeated sequences.

10.16 a. Histones are relatively small, basic, positively charged proteins that are rich in the amino acids arginine and lysine. There are five types of histones (H1, H2A, H2B, H3, and H4). Histones H2A, H2B, H3, and H4 are among the most highly conserved of all proteins. In contrast to the small number of histones, there are many types of nonhistones. They are quite variable in size, and are usually negatively charged, acidic proteins.

b. The five histones (H1, H2A, H2B, H3, and H4) are present in all cells of all eukaryotic organisms. Relative to the total amount of DNA, they are present in fairly constant amounts. In contrast, nonhistones differ in number and type from cell type to cell type within an organism, at different times in the same cell type, and from organism to organism.

c. Histones are positively charged and mostly interact with the negatively charged sugar-phosphate backbone of DNA. Nonhistones are typically negatively charged and can interact with the positively charged histones or with DNA.

d. The histones H2A, H2B, H3, and H4 are fundamental to packaging DNA into nucleosome core particles and the formation of the 10-nm nucleofilament. When histone H1 is added, this nucleofilament can be compacted to produce a 30-nm chromatin fiber. Nonhistones are important in higher-order packaging of DNA. They form the protein scaffold that anchors the looped domains of the 30-nm chromatin fiber, and they control the degree of chromosome condensation.

10.17 a. Moderately repetitive sequences appear in a few to as many as 10^5 copies in the genome and can be tandemly repeated or dispersed. Some tandemly repeated sequences are genes that are present in multiple copies, such as the genes for ribosomal RNA, transfer RNA (tRNA), histones, and globin. Some dispersed moderately repetitive sequences are LINEs, long, interspersed repeats of about 5,000 bp or more, and SINEs, short interspersed repeats of about 100–500 bp. Some LINEs are transposons and are mobile within the genome. In *Drosophila*, such transposons are found at telomeres. An example of SINEs are Alu repeats, short interspersed repeats that are about 300 bp. Although these can be transcribed, it is unclear exactly what function they provide.

b. A gene that is present in a greater copy number can express its product at a higher level. Thus, for genes whose products are required in large amounts, this may be an effective production strategy. Two examples of such gene products are (1) the RNAs that are components of ribosomes that are encoded by tandemly repeated rRNA genes and needed in large numbers to produce ribosomes for protein synthesis, and (2) the histone proteins, which are encoded by tandemly repeated genes and required in large numbers in order to package DNA.

CHAPTER 11

11.3 Key: ^{15}N-^{15}N DNA = HH; ^{15}N-^{14}N DNA = HL; ^{14}N-^{14}N DNA = LL

a. Generation 1: all HL; 2: ½ HL, ½ LL; 3: ¼ HL, ¾ LL; 4: ⅛ HL, ⅞ LL; 6: ¹⁄₃₂ HL, ³¹⁄₃₂ LL; 8: ¹⁄₁₂₈ HL, ¹²⁷⁄₁₂₈ LL

b. Generation 1: ½ HH, ½ LL; 2: ¼ HH, ¾ LL; 3: ⅛ HH, ⅞ LL; 4: ¹⁄₁₆ HH, ¹⁵⁄₁₆ LL; 6: ¹⁄₆₄ HH, ⁶³⁄₆₄ LL; 8: ¹⁄₂₅₆ HH, ²⁵⁵⁄₂₅₆ LL

11.6 a. That DNA replication is semiconservative does not a priori require DNA replication to be semidiscontinuous. For example, if both of the two old strands were completely unwound and replication were initiated from the 3' end of each, it could proceed continuously in a 5'-to-3' direction along both strands. Alternatively, if DNA polymerase were able to synthesize DNA in both 3'-to-5' and 5'-to-3' directions, DNA replication could proceed continuously on both DNA strands.

b. That DNA replication is semidiscontinuous does ensure that it is semiconservative. In the semidiscontinuous model, each old, separated strand serves as a template for a new strand. This is the essence of the semiconservative model.

c. That DNA polymerase synthesizes just one new strand from each "old" single-stranded template and can synthesize in only one direction (5'-to-3') ensures that replication is semiconservative.

11.7 See text, pp. 238–239. DNA polymerase I; a fragment of DNA; dATP, dGTP, dTTP, and dCTP; and magnesium ions are needed.

11.8 a. When DNA polymerase III reaches an RNA primer and dissociates from the DNA, DNA polymerase I continues to synthesize the DNA in a 5'-to-3' direction. It simultaneously removes the RNA primer using its 5'-to-3' exonuclease activity, and replaces the RNA with DNA nucleotides. See text pp. 242–244 and Figure 11.7.

11.10 None is an analog for adenine, B and D are analogs of thymine, C is an analog of cytosine, and A is an analog of guanine.

11.12 A primer strand is a nucleic acid sequence that is extended by DNA synthesis activities. A template strand directs the base sequence of the DNA strand being made; for example, an A on the template causes a T to be inserted on the new chain, and so on.

11.14 Since a replication fork moves at a rate of 10^4 bp/minute and each replicon has two replication forks moving in opposite directions, in one replicon, replication would be occurring at a rate of 2×10^4 bp/minute. Since all of the organism's DNA is replicated in 3 minutes, the number of replicons in the diploid genome is

$$\frac{4.5 \times 10^8 \text{ bp}}{3 \text{ minutes}} \times \frac{1 \text{ replicon}}{2 \times 10^4 \text{ bp / minute}} = 7,500.$$

11.17 See text, pp. 241–244.

11.19 Assume the amount of a gene's product is directly proportional to the number of copies of the gene present in the *E. coli* cell. Assay the enzymatic activity of genes at various positions in the *E. coli* chromosome during the replication period. Then, some genes (those immediately adjacent to the origin) will double their activity very shortly after replication begins. Relate the map position of genes having doubled activity to the amount of time that has transpired since replication was initiated. If replication is bidirectional, there should be a doubling of the gene products both clockwise and counterclockwise from the origin.

11.21 Clearly, DNA replication in the Jovian bug does not occur as it does in *E. coli*. Assuming that the double-stranded DNA is antiparallel as it is in *E. coli*, the Jovian DNA polymerases must be able to synthesize DNA in the 5'-to-3' direction (on the leading strand) as well as in the 3'-to-5' direction (on the lagging strand). This is unlike any DNA polymerase on Earth.

11.23 Assuming these cells have the typical 4 h G_2 period, you would add ³H thymidine to the medium, wait 4.5 h, and prepare a slide of metaphase chromosomes. Autoradiography would then be done on this chromosome preparation. Regions of chromosomes displaying silver grains are then the late-replicating regions. Cells at earlier stages in the S period, when they began to take up ³H, will not have had sufficient time to reach metaphase.

11.25 Evidence that the length of simple-sequence telomeric repeats is under genetic control comes from analysis of yeast mutants that affect telomere length. In yeast, mutants at the *TLC1* gene or the *EST1* (*Ever Shorter Telomeres*) gene cause telomeres to shorten continuously until the cells die. Mutations at the *TEL1* and *TEL2* genes cause cells to maintain their telomeres at a new, shorter-than-wild-type length.

Judging from the analysis of these yeast mutants, telomere length is important to cell viability. Telomerase activity has been found in immortal cells such as tumor-cells, but not in normal somatic cells. This suggests that when telomerase activity is abnormally present, it may play a role in tumor-cell proliferation, and that its absence in normal somatic cells may be important for their aging or senescence.

CHAPTER 12

12.1 DNA contains deoxyribose and thymine, whereas RNA contains ribose and uracil, respectively. Also, DNA is usually double-stranded, but RNA is usually single-stranded.

12.4 See text, pp. 261–267, 269–272, and 274–275.

12.5 RNA polymerase I transcribes the major rRNA genes that code for 18S, 5.8S, and 28S rRNAs; RNA polymerase II transcribes the protein-coding genes to produce mRNA molecules; and RNA polymerase III transcribes the 5S rRNA genes, the tRNA genes, and the genes for small nuclear RNA (snRNA) molecules.

In the cell, the 18S, 5.8S, 28S, and 5S rRNAs are structural components of ribosomes. The mRNAs are translated to produce proteins, the tRNAs bring amino acids to the ribosome to donate to the growing polypeptide chain during protein synthesis, and at least some of the snRNAs are involved in RNA processing events.

12.7 An enhancer element is defined as a DNA sequence that somehow, without regard to its position relative to the gene or its orientation in the DNA, increases the amount of DNA synthesized from the gene it controls (discussed in the chapter on p. 263).

12.9 The three classes of RNA are mRNA (pp. 264–266), tRNA (pp. 274–275), and rRNA (pp. 268–269).

12.12 For mRNA: 5' capping and 3' polyadenylation in eukaryotes. For rRNA in prokaryotes, a pre-rRNA molecule is processed to the mature 16S, 23S, and 5S rRNAs. In eukaryotes, a precursor molecule is processed to the 18S, 5.8S, and 28S rRNA molecules; 5S rRNA is made elsewhere. The large rRNAs in both types of organisms are methylated. Additionally, the same rRNAs in eukaryotes are pseudouridylated.

12.14 a. This would prevent splicing out of any sequences corresponding to introns and would result in abnormal sequences of most proteins. This should be lethal.

b. If splicing is not affected, this should have no phenotypic consequence.

c. This would prevent the splicing out of intron 2 material. Since the 5' cut could still be made, it is likely that this mutation would lead to the absence of functional mRNA and the absence of β-globin. (This should produce a phenotype similar to that seen when the β-globin gene is deleted. In that case, there is anemia compensated by increased production of fetal hemoglobin. The condition is called β-thalassemia.)

12.16 The first two bases of an intron are typically 5'-GU-3', which are essential for base-pairing with the U1 snRNA during spliceosome assembly. A GC to TA mutation at the initial base pair of the first intron impairs base-pairing with the U1 snRNA, so that the 5' splice site of the first intron is not identified. This causes the retention of the first intron in the *tub* mRNA and a longer mRNA transcript in *tub/tub* mutants. When the mutant *tub* mRNA is translated, the retention of the first intron could result either in the introduction of amino acids not present in the *tub*⁺ protein, or, if the intron contained a chain termination (stop) codon, premature translation termination and the production of a truncated protein. In either case, a nonfunctional gene product is produced.

The *tub* mutation is recessive because the single *tub*⁺ allele in a *tub/tub*⁺ heterozygote produces mRNAs that are processed normally, and when these are translated, enough normal (*tub*⁺) product is produced to obtain a wild-type phenotype. Only the *tub* allele produces abnormal transcripts. When both copies of the gene are mutated in *tub/tub* homozygotes, no functional product is made, and a mutant, obese phenotype results.

12.17 a. This would be possible because the 5S genes occur in clusters apart from the other rRNA genes.

b. This would not be possible because the 18S gene copies are interspersed among 5.8S and 28S genes.

c. This would be possible. These three occur in tandem arrays.

d. This would not be possible because the 5S genes are located separately within the chromosomes.

12.19 a. 3

b. 1, 2, 3, 4

c. 3

d. 1, 2

e. 1 (note that some rRNAs have introns)

f. 4

g. 1

CHAPTER 13

13.1 b. mRNA

13.2 c. Linear, folded chains of amino acids

13.5 Consider Figure 13.5 in answering this question.

AMINO ACID	tRNAs NEEDED	RATIONALE
Ile	1	3 codons can use 1 tRNA (wobble)
Phe	1	2 codons can use 1 tRNA (wobble)
Tyr	1	" " " " "
His	1	" " " " "
Gln	1	" " " " "
Asn	1	" " " " "
Lys	1	" " " " "
Asp	1	" " " " "
Glu	1	" " " " "
Cys	1	" " " " "
Trp	1	1 codon
Met	2	Single codon, but need one tRNA for initiation and one tRNA for elongation
Val	2	4 codons: 2 can use 1 tRNA (wobble)
Pro	2	" " " " "
Thr	2	" " " " "
Ala	2	" " " " "
Gly	2	" " " " "
Leu	3	6 codons: 2 can use 1 tRNA (wobble)
Arg	3	" " " " "
Ser	3	" " " " "
Total	**32**	**61 codons**

13.9 a. 4 A : 6 C

$AAA = (^4/_{10})(^4/_{10})(^4/_{10}) = 0.064$, or 6.4% Lys

$AAC = (^4/_{10})(^4/_{10})(^6/_{10}) = 0.096$, or 9.6% Asn

$ACA = (^4/_{10})(^6/_{10})(^4/_{10}) = 0.096$, or 9.6% Thr

$CAA = (^6/_{10})(^4/_{10})(^4/_{10}) = 0.096$, or 9.6% Gln

$CCC = (^6/_{10})(^6/_{10})(^6/_{10}) = 0.216$, or 21.6% Pro

$CCA = (^6/_{10})(^6/_{10})(^4/_{10}) = 0.144$, or 14.4% Pro

$CAC = (^6/_{10})(^4/_{10})(^6/_{10}) = 0.144$, or 14.4% His

$ACC = (^4/_{10})(^6/_{10})(^6/_{10}) = 0.144$, or 14.4% Thr

In sum, 6.4% Lys, 9.6% Asn, 9.6% Gln, 36.0% Pro, 24.0% Thr, and 14.4% His.

b. 1 A : 3 U : 1 C; the same logic is followed here, using $^1/_5$ as the fraction for A

$AAA = 0.008$, or 0.8% Lys

$AAU = 0.024$, or 2.4% Asn

$AUA = 0.024$, or 2.4% Ile

$UAA = 0.024$, or 2.4% Chain terminating

$AUU = 0.072$, or 7.2% Ile

$UAU = 0.072$, or 7.2% Tyr

$UUA = 0.072$, or 7.2% Leu

$UUU = 0.216$, or 21.6% Phe

$AAC = 0.008$, or 0.8% Asn

$ACA = 0.008$, or 0.8% Thr

$CAA = 0.008$, or 0.8% Gln

$ACC = 0.008$, or 0.8% Thr

$CAC = 0.008$, or 0.8% His

$CCA = 0.008$, or 0.8% Pro

$CCC = 0.008$, or 0.8% Pro

$UUC = 0.072$, or 7.2% Phe

$UCU = 0.072$, or 7.2% Ser

$CUU = 0.072$, or 7.2% Leu

$UCC = 0.024$, or 2.4% Ser

$CUC = 0.024$, or 2.4% Leu

$CCU = 0.024$, or 2.4% Pro

$UCA = 0.024$, or 2.4% Ser

$UAC = 0.024$, or 2.4% Tyr

$CUA = 0.024$, or 2.4% Leu

$CAU = 0.024$, or 2.4% His

$AUC = 0.024$, or 2.4% Ile

$ACU = 0.024$, or 2.4% Thr

In sum, 0.8% Lys, 3.2% Asn, 12.0% Ile, 2.4% Chain terminating, 9.6% Tyr, 19.2% Leu, 28.8% Phe, 4.0% Thr, 0.8% Gln, 3.2% His, 4.0% Pro, and 12.0% Ser. The likelihood is that the chain would not be long because of the chance of the chain-terminating codon.

13.10

WORD SIZE	NUMBER OF COMBINATIONS
a. 5	$2^5 = 32$
b. 3	$3^3 = 27$
c. 2	$5^2 = 25$

(The minimum word sizes must uniquely designate 20 amino acids.)

13.11 Stage A: $4^2 = 16$ different meaningful triplets

Stage B: $4^2 \times 2 = 32$ different meaningful triplets

13.12 a. AAA ATA AAA ATA etc.

b. TTT TAT TTT TAT etc.

c. AAA for Phe and AUA for Tyr

13.13 No chain-terminating codons can be produced from only As and Gs. But the stop codon UAA can be made from As and Us. Therefore, the population A proteins will be longer than those from population B. Most of the population B proteins will be free in solution rather than attached to ribosomes.

13.15 Met-Val-Ser-Ser-Pro-Ile-Gly-Ala-Ala-Ile-Ser ... (In fact, either of the Ile residues might be replaced by Met in a particular molecule.) The normal tRNA recognizes the AUC codon and therefore carries Ile. The mutant tRNA will recognize the codon AUG and insert Ile there (although the normal tRNA-Met will compete for these sites). The N-terminal Met will not be replaced by Ile because it requires the special tRNA-fMet for initiation to occur.

13.17 Normal: Met-Phe-Ser-Asn-Tyr-(. . .)-Met-Gly-Trp-Val.

Mutant *a*: Met-Phe-Ser-Asn

Mutant *b*: Starts at later AUG to give: Met-Gly-Trp-Val

Mutant *c*: Met-Phe-Ser-Asn-Tyr-(. . .)-Met-Gly-Trp-Val

Mutant *d*: Met-Phe-Ser-Lys-Tyr-(. . .)-Met-Gly-Trp-Val

Mutant *e*: Met-Phe-Ser-Asn-Ser-(. . .)-Trp- Gly-Gly-Trp (. . .) (no stop codon; protein continues)

Mutant *f*: Met-Phe-Ser-Asn-Tyr-(. . .)-Met-Gly-Trp-Val-Trp (. . .)(no stop codon; protein continues)

13.18 One approach to this problem is to infer the possible coding sequence(s) that could be used for the normal protein (using N = any nucleotide, R = purine, Y = pyrimidine), and then examine this sequence to deduce what possible mutations could have resulted in the mutant proteins.

Based on the normal amino acid sequence, the mRNA sequence is:

amino acid sequence: Met- Gly- Glu- Thr- Lys- Val- Val- ...-Pro

potential mRNA: 5'-AUG GGN GAR ACN AARGUN GUN ... CCN–3'

In mutant 1, premature chain termination has occurred. This could have occurred if, in the DNA transcribed into the third (GAR

(Glu)) codon, a GC base pair was changed to a TA base pair. This would lead to a UAR (stop) codon.

normal sequence: Met- Gly- Glu- Thr- Lys- Val- V a l -...-Pro
normal mRNA: 5'– AUG GGN GAR ACN AAR GUN GUN ... CCN–3'
mutant mRNA: 5'– AUG GGN UAR ACN AAR GUN GUN ... CCN–3'
mutant sequence: Met- Gly- s t o p

It could also have occurred if a TA base-pair insertion mutation occurred in the DNA, so that a U was transcribed in between the normal second and third codons, resulting in a UGA (stop) third codon.

amino acid sequence:Met- Gly- Glu- Thr- Lys- Val- V a l -...-Pro
potential mRNA: 5'– AUG GGN GAR ACN AAR GUN GUN ...CCN–3'
mutant mRNA: 5'– AUG GGN UGA RAC NAA RGU NGU N.. . CC–3'
mutant sequence: Met- Gly- stop

In mutant 2, premature chain termination has occurred after a wrong amino acid has been inserted. To explain both of these results as a consequence of a single mutational event, try either insertion or deletion mutations that would alter the reading frame. One possible explanation is that a GC base-pair insertional mutation in the DNA resulted in a G being inserted after the third codon. If the N of the fourth codon were a U, then such a frameshifting insertion would change the Thr to Asp, and also introduce a chain termination codon into the fifth codon position. If the R in the GAR coding for Glu (the third amino acid) is a G, this mutation could also be caused by a GC or AT base-pair insertional mutation after the first two nucleotides coding for Glu. As before, assume the N of the fourth codon is a U.

amino acid sequence:Met- Gly- Glu- Thr- Lys- Val- V a l -...-Pro
potential mRNA: 5'– AUG GGN GAR ACN AAR GUN GUN ...CCN–3'
mutant mRNA: 5'– AUG GGN GAR GAC UAA RGU NGU N.. .CC–3'
mutant sequence: Met- Gly- Glu- Asp- stop

In mutant 3, the situation is similar to that with mutant 2. This time, however, several wrong amino acids are inserted before chain termination. To explain all of these consequences as the result of a single mutational event, check for the consequences of deletions or insertions in the region of the second and third codons. One possible explanation is that a deletion mutation in the DNA resulted in the N of the second codon being deleted. If, as in mutant 2, the N of the fourth codon is a U, and the R of the third codon is a G, the R of the fifth codon is an A, and the N of the sixth codon is an A, the mutant sequence would be obtained.

normal sequence: Met- Gly- Glu- Thr- Lys- Val- V a l -...-Pro
normal mRNA: 5'–AUG GGN GAR ACN AAR GUN GUN ... CCN–3'
normal mRNA: 5'–AUG GGN GAG ACU AAA GUA GUN ... CCN–3'
mutant mRNA: 5'–AUG GGG AGA CUA AAG UAG UN . .C CN–3'
mutant sequence: Met- Gly- Arg- Leu- Lys- Stop

In mutant 4, the normal second amino acid (Gly) has been replaced with Arg. Arg is encoded by AGR or CGN, while Gly is encoded by GGN. If a GC base pair were substituted for a CG base pair in the DNA so that the first G of the second codon were replaced by a C, Arg would be inserted as the second amino acid.

normal sequence: Met- Gly- Glu- Thr- Lys- Val- V a l -...-Pro
normal mRNA: 5'–AUG GGN GAR ACN AAR GUN GUN ...CCN–3'
mutant mRNA: 5'–AUG CGN GAR ACN AAR GUN GUN ...CCN–3'
mutant sequence: Met- Arg- Glu- Thr- Lys- Val- V a l -...-Pro

13.21 a. The pre-mRNA must be substantially processed by RNA splicing (removal of introns) and polyadenylation to a smaller mature mRNA. (As is discussed in Chapter 15, this gene has 24 exons!)

b. A 1,480-amino-acid protein requires $1,480 \times 3 = 4,440$ bases of protein-coding sequence. This leaves $6,500 - 4,440 = 2,060$ bases of 5' untranslated leader and 3' untranslated trailer sequence in the mature mRNA, about 32 percent.

c. The ΔF508 mutation could be caused by a DNA deletion for the three base pairs encoding the mRNA codon for phenylalanine. This codon is UUY (Y = U or C), and the DNA sequence of the non-template strand is 5'-TTY-3'. The segment of DNA containing these bases could be deleted in the appropriate region of the gene.

d. If positioned at random and solely within the coding region of a gene (i.e., not in 3' or 5' untranslated sequences or in intronic sequences), a deletion of three base pairs results either in an mRNA missing a single codon or an mRNA missing bases from two adjacent codons. If three of the six bases from two adjacent codons were deleted, the remaining three bases would form a single codon. In this case, an incorrect amino acid might be inserted into the polypeptide at the site of the left codon, and the amino acid encoded by the right codon would be deleted. If the 3' base of the left codon were deleted, it would be replaced by the 3' base of the right codon. The code is degenerate and wobble occurs in the 3' base, so this type of deletion might not alter the amino acid specified by the left codon. The adjacent amino acid would still be deleted, however.

CHAPTER 14

14.3 Genomic libraries are discussed on pp. 305–306, and cloning in a lambda vector by replacing a central section of the λ chromosome with foreign DNA on pp. 303–304. The steps to make a yeast genomic library are:

1. Isolate high-molecular-weight DNA from yeast nuclei.

2. Perform a partial digest of the DNA with a restriction enzyme appropriate for the λ cloning vector (e.g., *Eco*RI), and isolate DNA fragments of the correct size range for cloning in the λ vector by sucrose gradient centrifugation.

3. Remove the central section of an appropriate λ vector by digestion with the same enzyme, *Eco*RI in this case.

4. Ligate the left and right λ arms to the yeast DNA fragments.

5. Package the recombinant DNA molecules in vitro into λ phage particles.

6. Infect *E. coli* cells with the λ phage population, and collect progeny phages produced by cell lysis. These phages represent the yeast genomic library.

14.6 Use cDNA. Human genomic DNA contains introns. mRNA transcribed off genomic DNA needs to be processed before it can be translated. Bacteria do not contain these processing enzymes. So in the bacterium, even if the human mRNA were translated, the protein it would code for would not be insulin. cDNA is the complementary copy of a functional mRNA molecule. So when this is the template, its mRNA transcript will be functional and when translated, human pro-insulin will be synthesized.

14.10 The restriction map is:

The map is built by considering the large fragments first. For example, the 2,500-bp B fragment is cut with A to produce 1,900-bp and 600-bp fragments. The 2,100-bp A fragment is cut with B to produce the same 1,900-bp fragment and a 200-bp fragment. Thus, the 200-bp and 600-bp fragments must be on opposite sides of the 1,900-bp fragment. The map is extended in a step-by-step fashion by next considering other cuts that produced 200-bp fragments, 600-bp fragments, and so on.

14.11 The sequencing gel banding pattern is shown in the following figure:

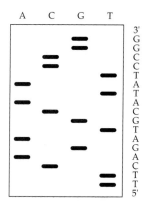

14.14 Chromosomes bearing CF mutations have a shorter fragment than chromosomes bearing wild-type alleles. Both parents (M and P lanes) have two bands, indicating that each has a normal and mutant chromosome. The fetus (F lane) also has two bands, indicating that it, too, is heterozygous. It has a normal gene, so it will be normal.

CHAPTER 15

15.3 A constitutive phenotype can be the result of either an *lacI⁻* or an *lacOᶜ* mutation.

15.4 a. A missense mutation results in partial or complete loss of β-galactosidase activity, but there would be no loss of permease and transacetylase activities.

b. A nonsense mutation is likely to have polar effects unless the mutation is very close to the normal chain-terminating codon for β-galactosidase. Thus, permease and transacetylase activities

would be lost in addition to the loss of β-galactosidase activity if nonsense mutation occurred near the 5' end of the *lacZ* gene.

15.6 *lacI⁺ lacOᶜ lacZ⁺ lacY⁻/lacI⁺ lacO⁺ lacZ⁻ lacY⁺*
(It cannot be ruled out that one of the repressor genes is *lacI⁻*.)

15.8 Answer is in Table 15.A.

15.10 The CAP, in a complex with cAMP, is required to facilitate RNA polymerase binding to the *lac* promoter. RNA polymerase binding occurs only in the absence of glucose and only if the operator is not occupied by a repressor (i.e., if lactose is absent). A mutation in the CAP gene, then, would render the *lac* operon incapable of expression because RNA polymerase would not be able to recognize the promoter.

15.12 See pp. 346–369. For a wild-type *trp* operon, the absence of tryptophan results in antitermination; that is, the structural genes are transcribed, and the tryptophan biosynthetic enzymes are then made. This occurs because the absence of tryptophan results in the absence of, or at least a very low level of, Trp-tRNA, which, in turn, causes the ribosome translating the leader sequence to stall at the Trp codons. With the ribosome stalled at that location, the RNA being synthesized just ahead of the ribosome by RNA polymerase assumes a secondary structure that favors continued transcription of the structural genes by the polymerase. If the two Trp codons were mutated to stop codons, then the mutant operon would function in the same way as the wild-type operon in the absence of tryptophan because the ribosome would stall in the same place, and the antitermination would be produced, resulting in transcription of the structural genes.

For a wild-type *trp* operon, the presence of tryptophan turns the transcription of the structural genes off. This happens because the presence of tryptophan leads to the accumulation of Trp-tRNA, which allows the ribosome to read the two Trp codons and stall at the normal stop codon for the leader sequence. When stalled in that position, the antitermination signal cannot form in the RNA being synthesized; instead, a termination signal is formed, resulting in

TABLE 15.A

| GENOTYPE | INDUCER ABSENT | | INDUCER PRESENT | |
	β-GALACTOSIDASE	PERMEASE	β-GALACTOSIDASE	PERMEASE
a. *I⁺ P⁺ O⁺ Z⁺ Y⁺*	−	−	+	+
b. *I⁺ P⁺ O⁺ Z⁻ Y⁺*	−	−	−	+
c. *I⁺ P⁺ O⁺ Z⁺ Y⁻*	−	−	+	−
d. *I⁻ P⁺ O⁺ Z⁺ Y⁺*	+	+	+	+
e. *Iˢ P⁺ O⁺ Z⁺ Y⁺*	−	−	−	−
f. *I⁺ P⁺ Oᶜ Z⁺ Y⁺*	+	+	+	+
g. *Iˢ P⁺ Oᶜ Z⁺ Y⁺*	+	+	+	+
h. *I⁺ P⁺ Oᶜ Z⁺ Y⁻*	+	−	+	−
i. *I⁻ P⁺ O⁺ Z⁺ Y⁺* / *I⁺ P⁺ O⁺ Z⁻ Y⁻*	−	−	+	+
j. *I⁻ P⁺ O⁺ Z⁺ Y⁻* / *I⁺ P⁺ O⁺ Z⁻ Y⁺*	−	−	+	+
k. *Iˢ P⁺ O⁺ Z⁺ Y⁻* / *I⁺ P⁺ O⁺ Z⁻ Y⁺*	−	−	−	−
l. *I⁺ P⁺ Oᶜ Z⁺ Y⁺* / *I⁺ P⁺ O⁺ Z⁻ Y⁻*	−	+	+	+
m. *I⁻ᵈ P⁺ Oᶜ Z⁺ Y⁻* / *I⁺ P⁺ O⁺ Z⁻ Y⁺*	+	−	+	+
n. *Iˢ P⁺ O⁺ Z⁺ Y⁺* / *I⁺ P⁺ Oᶜ Z⁺ Y⁺*	+	+	+	+
o. *I⁻ᵈ P⁺ O⁺ Z⁺ Y⁻* / *I⁺ P⁺ O⁺ Z⁻ Y⁺*	+	+	+	+
p. *I⁺ P⁻ Oᶜ Z⁺ Y⁻* / *I⁺ P⁺ O⁺ Z⁺ Y⁻*	−	−	+	+
q. *I⁺ P⁻ O⁺ Z⁺ Y⁻* / *I⁺ P⁺ Oᶜ Z⁻ Z⁺*	−	+	−	+
r. *I⁻ P⁻ O⁺ Z⁺ Y⁺* / *I⁺ P⁺ O⁺ Z⁻ Y⁻*	−	−	−	−

the termination of transcription. In a mutant *trp* operon with two stop codons instead of the Trp codons, the ribosome will still stall at the premature stop codons even in the presence of tryptophan (and therefore Trp-tRNA). And, as discussed previously, this results in an antitermination signal so that the structural genes will be transcribed.

In sum, in both the presence and the absence of tryptophan, the mutant *trp* operon will show no attenuation; that is, the structural genes will be transcribed in both cases, and the tryptophan biosynthetic enzymes will be synthesized.

15.15 The *cI* gene codes for a repressor protein that functions to keep the lytic functions of the phage repressed when lambda is in the lysogenic state. A *cI* mutant strain would be unable to establish lysogeny, and thus it would also follow the lytic pathway.

Chapter 16

16.3 a. In the fragile X syndrome, the expanded CGG repeat results in hypermethylation and transcriptional silencing. In Huntington disease, the expanded CAG repeat results in the inclusion of a polyglutamine stretch within the huntingtin protein, which causes it to have a novel, abnormal function.

b. A heterozygote with a CGG repeat expansion near one copy of the *FMR-1* gene will still have one normal copy of the *FMR-1* gene. The normal gene can produce a normal product, even if the other is silenced, so fragile X syndrome is recessive. In contrast, a novel, abnormal protein is produced by the CAG expansion in the disease allele in Huntington disease. The disease phenotype is due to the presence of the abnormal protein, so the disease trait is dominant.

c. Transcriptional silencing may require significant amounts of hypermethylation and so require more CGG repeats for an effect to be seen. In contrast, protein function may be altered by a relatively smaller-sized polyglutamine expansion.

16.5 The data show clearly that the synthesis of ovalbumin is dependent upon the presence of the hormone estrogen. The data do not indicate at what level estrogen works. Theoretically, it could act to increase transcription of the ovalbumin gene, to stabilize the ovalbumin mRNA, to stabilize the ovalbumin protein, or to stimulate transport of the ovalbumin mRNA out of the nucleus. Experiments in which the levels of the ovalbumin mRNA were measured have shown that the production of ovalbumin is primarily regulated at the level of transcription.

16.9 a. Based on the work of Wilmut and his colleagues, the nose cells would first be dissociated and grown in tissue culture. The cells would be induced into a quiescent state (the G_0 phase of the cell cycle) by reducing the concentration of growth serum in the medium. Then they would be fused with enucleated oocytes from a donor female and allowed to grow and divide by mitosis to produce embryos. The embryos would be implanted into a surrogate female.

b. Although the nuclear genome would generally be identical to that in the original nose cell, cytoplasmic organelles would presumably derive from those in the enucleated oocyte. Hence, the mitochondrial DNA would not derive from the original leader. In addition, telomeres in an older individual are "shorter," so one might expect the telomeres in the cloned leader to be those of an older individual.

c. In mature B cells, DNA rearrangements at the heavy- and light-chain immunoglobulin genes have occurred. One would expect the cloned leader to be immunocompromised because he would be unable to make the wide spectrum of antibodies present in a normal individual.

d. The production of Dolly was significant because it demonstrated the apparent complete totipotency of a nucleus from a mature mammary epithelium cell. The other six lambs developed from the transplanted nuclei of embryonic or fetal cells, which one

might expect to be less determined and have a higher degree of totipotency due to their younger developmental age.

e. There is no way to predict the psychological profile of the cloned leader based on his genetic identity. Even identical twins, who are genetically more identical than such a clone, do not always share behavioral traits.

16.10 four: A_3, $A_2 B$, AB_2, and B_3

16.12 Preexisting mRNAs (stored in the oocyte) are recruited into polysomes as development begins following fertilization.

16.14 a. Connie must also have blood type A.

b. Connie must also have the blood type genotype I^A/I^A.

c. Connie has anti-B antibodies.

d. No, Ashley's and Connie's anti-B antibodies would not have identical polypeptide sequences. During development, their IgG DNAs were rearranged independently, so their IgG mRNAs would be different. (In addition, the splicing process would further diversify their functional mRNAs.) Thus, their IgG antibodies would be different; that is, they would have different polypeptide sequences.

e. Yes. Their genes for β-globin would be identical, so the sequences of their β-globin chains would be identical (as would be those of all other sequences except for the immunoglobulins).

16.17

Mutant	Class
a	segmentation gene (segment polarity)
b	maternal gene (anterior-posterior gradient)
c	segmentation gene (gap)
d	homeotic (eye to wing transdetermination)
e	segmentation gene (gap)

16.18 Preexisting mRNA that was produced by the mother and deposited in the egg prior to fertilization is translated up until the gastrula stage. After gastrulation, new mRNA synthesis is necessary for production of proteins needed for subsequent embryo development.

16.19 The tissue taken from the blastula/gastrula has not yet been committed to its final differentiated state in terms of its genetic programming; that is, it is not yet determined. Thus, when it is transplanted into the host, the determined tissues surrounding the transplant in the host communicate with the transplanted tissue and cause it to be determined in the same way as they are; for example, tissue transplanted to a future head area will differentiate into head material, and so on. In contrast, tissues in the neurula stage are now determined as to their final tissue state once development is complete. Thus, tissue transplanted from a neurula to an older embryo cannot be influenced by the determined, surrounding tissues and will develop into the tissue type for which it is determined, in this case, an eye.

Chapter 17

17.1 Hereditary cancer is associated with the inheritance of a germ-line mutation; sporadic cancer is not. Consequently, hereditary cancer "runs" in families. For some cancers, both hereditary and sporadic forms exist, with the hereditary form being much less frequent. For example, retinoblastoma occurs when both normal alleles of the tumor suppressor gene *RB* are inactivated. In hereditary retinoblastoma, a mutated, inactive allele is transmitted via the germ line. Retinoblastoma occurs in cells of an *RB*/+ heterozygote when an additional somatic mutation occurs. In the sporadic form of the disease, retinoblastoma occurs when both alleles are inactivated somatically.

17.3 Tumor growth induced by transforming retroviruses results either from the activity of a single viral oncogene or the activation of a proto-oncogene due to the nearby integration of the proviral DNA. The oncogene can cause abnormal cellular proliferation via the variety of mechanisms discussed in the text. The expression of a proto-oncogene, normally tightly regulated during cell growth and development, can be altered if it comes under the control of the promoter and enhancer sequences in the retroviral LTR.

DNA tumor viruses do not carry oncogenes. They transform cells through the action of a gene or genes within their genomes. For example, in a rare event, a DNA virus can be integrated into the host genome, and the DNA replication of the host cell may be stimulated by a viral protein that activates viral DNA replication. This would cause the cell to move from the G_0 to the S phase of the cell cycle.

For both transducing retroviruses and DNA tumor viruses, an abnormally expressed protein(s) leads to the activation of the cell from G_0 to S, and abnormal cell growth occurs.

17.4 a. Studies of hereditary forms of cancer have led to insights into the fundamental cellular processes affected by cancer. For example, substantial insights into the important role of DNA repair and the relationship between the control of the cell cycle and DNA repair have come from analyses of the genes responsible for hereditary forms of human colorectal cancer. For breast cancer, studying the normal functions of the *BRCA1* and *BRCA2* genes promises to provide substantial insights into breast and ovarian cancer.

b. Genetic predisposition for cancer refers to the presence of an inherited mutation that, with additional somatic mutations during the lifespan of an individual, can lead to cancer. For diseases such as retinoblastoma, a genetic predisposition has been associated with the inheritance of a recessive allele of the *RB* tumor suppressor gene. Retinoblastoma occurs in *RB*/+ individuals when the normal allele is mutated in somatic cells and the pRB protein no longer functions. Because somatic mutation is likely, the disease appears dominant in pedigrees.

Although there is a substantial understanding of the genetic basis for cancer and the genetic abnormalities present in somatic cancerous cells, there are also substantial environmental risk factors for specific cancers. Environmental risk factors must be investigated thoroughly when a pedigree is evaluated for the presence of a genetic predisposition for cancer.

17.6 Tumors result from multiple mutational events that typically involve both the activation of oncogenes and the inactivation of tumor suppressor genes. The analysis of hereditary adenomatous polyposis, an inherited form of colorectal cancer, has shown that the more differentiated cells found in benign, early-stage tumors are associated with fewer mutational events, while the less differentiated cells found in malignant and metastatic tumors are associated with more mutational events. Although the path by which mutations accumulate varies among tumors, additional mutations that activate oncogenes and inactivate tumor suppressor genes generally result in dedifferentiation and increased cellular proliferation.

17.7 The high degree of conservation of proto-oncogenes suggests that they function in normal, essential, conserved cellular processes. Given the relationship between oncogenes and proto-oncogenes, it also suggests that cancer occurs when these processes are not correctly regulated.

17.10 a. Proto-oncogenes encode a diverse set of gene products that include growth factors, receptor and nonreceptor protein kinases, receptors lacking protein kinase activity, membrane-associated GTP-binding proteins, cytoplasmic regulators involved in intracellular signaling, and nuclear transcription factors. These gene products all function in intercellular and intracellular circuits that regulate cell division and differentiation.

b. In general, mutations that activate a proto-oncogene convert it into an oncogene. Since (i), (iii), and (viii) cause a decrease in gene expression, they are unlikely to result in an oncogene. Since (ii) and (vii) could activate gene expression, they could result in an oncogene. Mutations (iv), (v), and (vi) cannot be predicted with certainty. The deletion of a 3' splice-site acceptor would alter the mature mRNA and possibly the protein produced and may or may not affect the protein's function and regulation. Similarly, it is difficult to predict the effect of a nonspecific point mutation or a premature stop codon. The

text presents examples where these types of mutations have caused the activation of a proto-oncogene and resulted in an oncogene.

17.11 Experimentally fuse cells from the two cell lines and then test the resultant hybrids for their ability to form tumors. If the uncontrolled growth of the tumor cell line was due to a mutated pair of tumor suppressor alleles, the normal alleles present in the normal cell line would "rescue" the tumor cell line defect. The hybrid line would grow normally and be unable to form a tumor. If the uncontrolled growth of the tumor cell line was due to an oncogene, the oncogene would also be present in the hybrid cell line. The hybrid line would grow uncontrollably and form a tumor.

17.13 One hypothesis is that the proviral DNA has integrated near the proto-oncogene, and the expression of the proto-oncogene has come under the control of promoter and enhancer sequences in the retroviral LTR. This could be assessed by performing a whole-genome Southern blot analysis to determine whether the organization of the genomic DNA sequences near the proto-oncogene has been altered.

17.16 Apoptosis is programmed, or suicidal, cell death. Cells targeted for apoptosis are those that have large amounts of DNA damage and so are at a greater risk for neoplastic transformation. Apoptosis is regulated by p53, among other proteins. In cells with large amounts of DNA damage, p53 is stabilized. This leads to the activation of *WAF1*, whose product, p21, causes cells to arrest in G_1 by binding to cyclin/Cdk complexes and blocking kinase activity needed for progression from G_1 to S. If the DNA damage is unable to be repaired, apoptosis will be induced by p53.

CHAPTER 18

18.1 b. Mutations are heritable changes in genetic information.

18.3 e. None of these (i.e., all are classes of mutations).

18.4 c. Ultraviolet light usually causes the induction of thymine dimers and their persistence or imperfect repair.

18.6 a. The normal anticodon was 5'-CAG-3', and the mutant one is 5'-CAC-3'. The mutational event was a CG-to-GC transversion.

b. Presumably Leu

c. Val

d. Leu

18.8 Acridine is an intercalating agent, and so it can be expected to induce frameshift mutations. 5BU is incorporated in place of T. During DNA replication, it is likely to be read as C by DNA polymerase because of its rare chemical state. This results in point mutations, usually TA-to-CG transitions. Considering these expectations, *lacZ-1* would probably result in a completely altered amino acid sequence after some point, although it might be truncated (due to the introduction of an out-of-frame nonsense codon). *lacZ-2* is likely to contain a single amino acid difference due to a missense mutation, although it too could contain a nonsense codon.

18.9 a. Six.

b. Three. The mutation results in replacement of the UGG codon in the mRNA by UAG, which is a chain-termination codon.

18.11 a. UAG: CAG Gln; AAG Lys; GAA Glu; UUG Leu; UCG Ser; UGG Trp; UAU Tyr; UAC Tyr; UAA chain terminating.

b. UAA: CAA Gln; AAA Lys; GAA Glu; UUA Leu; UCA Ser; UGA chain terminating; UAU Tyr; UAC Tyr; UAG chain terminating.

c. UGA: CGA Arg; AGA Arg; GGA Gly; UUA Leu; UCA Ser; UAA chain terminating; UGU Cys; UGC Cys; UGG Trp.

18.13

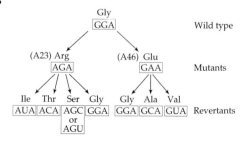

18.17 At the DNA level, nitrous acid deaminates C to make it U. As a result, CG base pairs are changed to TA base pairs. At the RNA level, this means that C nucleotides can be changed to U nucleotides.

According to the genetic code dictionary, Pro is specified by CCX (where X is U, C, G, or A); Ser is specified by UCX or AGU or AGC; Leu is specified by CUX or UUA or UUG; and Phe is specified by UUU or UUC.

To go from Pro to Ser to Phe requires changing the first two C nucleotides of the CCX codon to two U nucleotides of the UUU or UUC codon. This rules out the AGU and AGC codons for Ser. Since the Phe is unaffected by further nitrous acid treatment, the Phe codon must not contain a C (a CG at the DNA level); thus, the Phe codon must be UUU, and this means the Ser codon must be UCU and the Pro codon must be CCU.

Then, to go from Pro to Leu to Phe starting with CCU requires that the Leu codon must be CUU.

In sum, the changes that are compatible with the given data are:

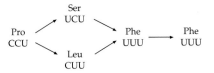

18.19 a. The production of knockout mice allows for the development of an animal model system to investigate human disease. It allows for greater investigation into the consequences of not having a particular gene function and also allows for the development of therapeutic approaches in a model system that closely parallels the human disease.

b. Knockout mice lack a particular gene function, so the phenotype of a knockout mouse will not necessarily mimic the phenotype of a dominant mutation that has an additional, abnormal function. To design a mouse model in this instance, one must add a copy of a dominant allele to the normal genetic background or replace a normal gene with the dominant allele. Only in this way would the aberrant protein product of the gene be expressed and would the phenotype it causes be observed.

CHAPTER 19

19.3 a. As is described in Figure 19.5, the spotted kernel phenotype arises from the transposition of Ds elements out of the C gene during kernel development. This means that the mutation in colorless strain B is caused by an insertion of a Ds element in the C gene. It is able to transpose because the purple strain C has Ac elements that can activate Ds transposition.

Spotted kernels are not seen in any of the other crosses, so transposition of Ds elements is not occurring in them. Since spotted kernels are not seen in the A × C cross, the colorless phenotype in strain A is not caused by the insertion of a Ds element. Spotted kernels are not seen in the B × D cross, so the D strain lacks Ac elements. For these reasons, the crosses A × C, A × D, and B × D all show only expected Mendelian patterns of inheritance.

b. The degree of spotting will depend on the number of intact Ac elements in the true-breeding purple strain. A greater number of intact Ac elements will result in a greater amount of transposase and so cause both an increase in the excision of the Ds element from the C gene and a greater degree of chromosome breakage at the C gene. The timing and frequency of transposition of the Ac and Ds elements, as well as the genetic rearrangements associated with the elements, are developmentally controlled. Transposition later in the development of the kernel will produce smaller spots.

19.4 Since a transposition results in the loss of the intron, it may be hypothesized that transposition in A occurs via an RNA intermediate. That is, intron removal occurs only at the RNA level.

The lack of intron removal during B transposition suggests that in B there is a DNA-DNA transposition mechanism, without any RNA intermediate.

19.5 a. In a cross of M females × P males, P males contribute chromosomes bearing P elements to an M cytoplasm. Since the M cytotype lacks a repressor, the P elements mobilize, causing hybrid dysgenesis. In the reciprocal cross, P elements remain in a P cytoplasm (contributed by the P female), where repressor exists, and therefore mobilization of P elements is repressed.

b. The simplest hypothesis is that P element transposition is restricted to the germ line, and does not occur in somatic cells. (In fact, this occurs because transposase is produced only in germ-line cells due to the alternative splicing of the transposase pre-mRNA in germ-line cells.)

CHAPTER 20

20.3 a. Selective use of antibiotics that inhibit translation by mitochondrial vs. cytoplasmic ribosomes has shown that components 2, 5, and 6 of cytochrome oxidase are synthesized on mitochondrial ribosomes, and that components 1, 3, 4, and 7 are synthesized on cytoplasmic ribosomes. See Figure 20.3 and the accompanying text on pp. 449–450.

b. Compare cytochrome oxidase activity in cybids made with platelets from diseased individuals and in cybids made with platelets from age-matched control individuals. It is important to assess a number of different enzyme activities associated with mitochondrial proteins to ensure that the deficits in cytochrome oxidase were specific.

c. As discussed in the text (pp. 459–460), the cells of individuals with diseases resulting from mitochondrial DNA defects have a mixture of mutant and normal mitochondria. That is, they show heteroplasmy. Thus, assays in cybids are measurements of the enzyme activity present in a population of mitochondria in a cell. It would be unlikely that each of the mitochondria of an affected individual has an identical defect.

20.4 The two codes are different in codon designations. The mitochondrial code has more extensive wobble so that fewer tRNAs are needed to read all possible sense codons. As a consequence, fewer mitochondrial genes are necessary. The advantage is that fewer tRNAs are needed, and hence fewer tRNA genes need be present than for cytoplasmic tRNAs.

20.5 The features of extranuclear inheritance are differences in reciprocal-cross results (not related to sex), nonmappability to known nuclear chromosomes, Mendelian segregation not followed, and the indifference to nuclear substitution.

20.8 a. The *tudor* mutation is a maternal effect mutation. Homozygous *tudor* mothers give rise to sterile progeny, regardless of their mate.

b. The "grandchildless" phenotype results from the absence of some maternally packaged component in the egg needed for the development of the F₁'s germ line.

20.9 a. If normal cytoplasm is [N] and male-sterile cytoplasm is [Ms], then the F₁ genotype is [Ms] Rf / rf, and the phenotype is male-fertile.

b. The cross is [Ms] Rf / rf ♀ × [N] rf / rf ♂, giving 50 percent [Ms] Rf / rf and 50 percent [Ms] rf / rf progeny. Thus, the phenotypes are ½ male-fertile and ½ male-sterile.

20.11 ½ petite, ½ wild-type (*grande*)

20.14 The first possibility is that the results are the consequences of a sex-linked lethal gene. The females would be homozygous for a dominant gene L that is lethal in males but not in females. In this case, mating the F₁ females of an L / L × +/Y cross to +/Y males should give a sex ratio of 2 females : 1 male in the progeny flies.

The second possibility is that the trait is cytoplasmically transmitted via the egg and is lethal to males. In this case, the same F₁ females should continue to have only female progeny when mated with +/Y males.

20.17 a. Carlos Mendoza will have inherited his mitochondrial DNA from his mother, and she will have inherited it from her mother. If Mrs. Mendoza and Mrs. Escobar have different mitochondrial RFLPs, then it can be determined which of them contributed mitochondria to Carlos.

b. None of the potential grandfathers need to be tested, because they will not have given any mitochondria to Carlos. In addition, there is no point in testing Mrs. Sanchez. She may have given mitochondria to Carlos's father, but the father did not pass them on to Carlos.

c. If Mrs. Mendoza and Mrs. Escobar do not differ in RFLP, the data will not be helpful. If they do differ, and if Carlos matches Mrs. Mendoza, the case should be dismissed. If Carlos matches Mrs. Escobar, then the Escobar and Sanchez couples are indeed the grandparents, and the Mendozas have claimed a stolen child.

20.19 The parental snails were D/d female and $d/–$ male. The F_1 snail is d/d. (Given the F_1 genotype, the male can be either homozygous d or heterozygous, but the determination cannot be made from the data given.)

20.20 a. It would be necessary to show that individuals with Beckwith-Wiedemann syndrome have the loss of expression of a parental allele that contributes to the onset of BWS. From what is described in the question, there is a tumor suppressor gene that, if not expressed correctly, could result in tumor formation. In addition, the region appears to be subject to imprinting because of the expression pattern of genes in the region in tumors and the expression pattern of *KVLQT1* in some tissues of normal individuals. The loss of expression of one tumor suppressor allele could give rise to tumor formation due to the sole expression of a mutant allele.

b. The molecular basis for imprinting is a decrease in the expression of a specific allele due to the heritable methylation of that allele in one parent. To demonstrate imprinting, one must identify which of a paternally or maternally inherited allele is transcribed. This can be done if the parental alleles are polymorphic and the differences between them can be detected by analysis of their mRNA or protein products.

c. It would be necessary to demonstrate that only one of the two parental alleles is expressed in the heart of diseased individuals. For example, if imprinting occurred abnormally, the levels of mRNA for the *KVLQT1* gene could be lowered, and this might lead to the disease state.

d. In both situations, a region of a chromosome appears to be subject to imprinting, and multiple genes are involved. This suggests that there may be signals for methylation in certain regions, and if these signals are disrupted, abnormal imprinting can result in disease.

CHAPTER 21

21.1 Brown = C^BC^B, C^BC^P, $C^BC^Y = p^2 + 2pq + 2pr = 236/500 = 0.472$
Pink = C^PC^P, $C^PC^Y = q^2 + 2qr = 231/500 = 0.462$
Yellow = $C^YC^Y = r^2 = 33/500 = 0.066$

$f(C^YC^Y) = r^2$

$\sqrt{f(C^YC^Y)} = qr^2$

$\sqrt{33/500} = r = \sqrt{0.066} = 0.26$

$f(C^PC^P + C^PC^Y + C^YC^Y) = q^2 + 2qr + r^2$

$f(C^PC^P + C^PC^Y + C^YC^Y) = (q + r)^2$

$\dfrac{231 + 33}{500} = (q + r)^2$

$0.528 = (q + r)^2$

$\sqrt{0.528} = q + r$

$0.727 = q + r$

$r = 0.26$

$0.727 - r = q$

$q = 0.467$

$p = 1 - q - r$

$= 1 - 0.467 - 0.26 = 0.273$

So, $f(C^B) = p = 0.273$

$f(C^Y) = r = 0.26$

$f(C^P) = q = 0.467$

21.3 The conditions of this problem meet the requirements for a population in Hardy-Weinberg equilibrium. In such a population, if p equals the frequency of A, and q equals the frequency of a, one expects p^2 AA, $2pq$ Aa, and q^2 aa genotypes after random mating. Here, $q^2 = 0.81$, so $q = 0.9$. Since $p + q = 1$, $p = 1 - 0.9 = 0.1$. In the next generation, one would expect $p^2 = (0.1)^2 = 0.01$ (or 1%) AA genotypes, $2pq = 2(0.1)(0.9) = 0.18$ (or 18%) Aa genotypes, and $q^2 = (0.9)^2 = 0.81$ (or 81%) aa genotypes.

21.4 a. $\sqrt{0.16} = 0.4 = 40\%$ = frequency of recessive alleles; $1 - 0.4 = 0.6 = 60\%$ = frequency of dominant alleles; $2pq = (2)(0.4)(0.6) = 0.48$ = probability of heterozygous diploids. Then $(0.48)/[(2 \times 0.16) + 0.48] = 0.48/0.80 = 60\%$ of recessive alleles are heterozygotes.

b. If $q^2 = 1\% = 0.01$, then $q = 0.1$, $p = 0.9$, and $2pq = 0.18$ heterozygous diploids. Therefore, $(0.18)/[(2 \times 0.01) + 0.18] = 0.18/0.20 = 90\%$ of recessive alleles in heterozygotes.

21.5 $2pq/q^2 = 8$, and so $2p = 8q$; then $2(1 - q) = 8q$, and $2 = 10q$, or $q = 0.2$

21.8 a. Let p = the frequency of S and q = the frequency of s. Then

$$p = \frac{2(188)\ SS + 717\ Ss}{2(3,146)} = \frac{1,093}{6,292} = 0.1737$$

$$q = \frac{717\ Ss + 2\ (2,241)\ ss}{2(3146)} = \frac{5\ 199}{6,292} = 0.8263$$

b.

CLASS	OBSERVED	EXPECTED	d	d^2/e
SS	188	95	+93	91.0
Ss	717	903	−186	38.3
ss	2,241	2,148	+93	4.0
	3,146	3,146	0	133.3

There is only one degree of freedom because the three genotypic classes are completely specified by two gene frequencies, namely, p and q. Thus, the number of degrees of freedom = number of genes − 1. The $\chi 2$ value for this example is 133.3, which, for one degree of freedom, gives a P value less than 0.0001. Therefore, the distribution of genotypes differs significantly from the Hardy-Weinberg equilibrium.

21.11 Since the frequency of the trait is different in males and females, the character might be caused by a sex-linked recessive gene. If the frequency of this gene is q, females would occur with the character at a frequency of q^2, and males with the character would occur at a frequency of q. The frequency of males is $q = 0.4$, and thus we may predict that the frequency of females would be $(0.4)^2 = 0.16$ if this is a sex-linked gene. This result fits the observed data. Therefore, the frequency of heterozygous females is $2pq = 2(0.6)(0.4) = 0.48$. For sex-linked genes, no heterozygous males exist.

21.13 $^{64}/_{10,000}$ are color-blind, i.e., $0.0064 = q^2$, and so $q = 0.08$ = probability of color-blind male.

21.14 a. Let q equal the frequency of the recessive allele, and p equal the frequency of the dominant allele. The frequency of males with the trait will be q, and the frequency of females will be q^2. Here, $q = 0.08$, so $q^2 = 0.0064$, or 0.64 percent.

b. Since $q = 0.08$, $p = 0.92$. The frequency of heterozygotes is $2pq = 0.1472$, or 14.72 percent. Only women can be heterozygotes, so the frequency of heterozygous women is 14.72 percent.

21.16

$$q = \frac{u}{u + v} = \frac{6 \times 10^{-7}}{(6 \times 10^{-7}) + (6 \times 10^{-8})}$$

$$= \frac{6 \times 10^{-7}}{(6 \times 10^{-7}) + (0.6 + 10^{-7})} = \frac{6}{6.6} = 0.91$$

$$p = 1 - q = 1 - 0.91 = 0.09$$

Thus, the frequencies are 0.008 AA, 0.164 Aa, and 0.828 aa.

21.18 $p'_{II} = mp_I + (1 - m)p_{II}$
$p'_{II} = (0.20)(0.5) + (1 - 0.20)(0.70) = 0.66$

21.19 For each population, determine the frequency of the *A* and *a* alleles, and compare how the population structure compares to that expected if no selection was occurring. To do this calculation, suppose there were 100 individuals in each population containing 200 alleles.

In population 1, there would be $(0.04)(100)(2) + (0.32)(100)(1) = 40$ *A* alleles, and $(0.64)(100)(2) + (0.32)(100)(1) = 160$ *a* alleles. Thus, $f(A) = p = {}^{40}/_{200} = 0.20$, and $f(a) = q = {}^{160}/_{200} = 0.80$. If the population was in Hardy-Weinberg equilibrium, one would expect $p^2 = 0.04$ *AA*, $2pq = 0.32$ *Aa*, and 0.64 *aa* individuals, as is observed. *If selection is acting at all*, it may be acting to maintain the observed frequencies of the *A* and *a* alleles in equilibrium.

In population 2, there would be $(0.12)(100)(2) + (0.87)(100)(1) = 111$ *A* alleles, and $(0.01)(100)(2) + (0.87)(100)(1) = 89$ *a* alleles. Here, $p = 0.555$ and $q = 0.455$. If the population was in Hardy-Weinberg equilibrium, one would expect $p^2 = 0.308$ *AA*, $2pq = 0.505$ *Aa*, and 0.207 *aa* individuals. The population is not in equilibrium, as there are far more heterozygotes, and far fewer homozygotes, than would be expected. This suggests that the heterozygote is being selected for in this population.

In population 3, there would be $(0.45)(100)(2) + (0.10)(100)(1) = 100$ *A* alleles, and $(0.10)(100)(1) + (0.45)(100)(2) = 100$ *a* alleles. Here, $p = q = 0.5$. If the population was in Hardy-Weinberg equilibrium, one would expect $f(AA) = f(aa) = p^2 = q^2 = 0.25$ and $f(Aa) = 2pq = 0.50$. The population is not in equilibrium, as there are far more homozygotes (of either type) and far fewer heterozygotes than expected. This suggests that the heterozygote is being selected against in this population.

21.21 a. When selectively neutral, the genes distribute themselves according to the Hardy-Weinberg law, so 0.25 are *AA*, 0.05 are *Aa*, and 0.25 are *aa*.

b. $q = 0.33$

c. $q = 0.25$

21.23 From the text discussion on pp. 000–000, at equilibrium, $p = f(Hb\text{-}A) = t/(t + s)$, and $q = f(Hb\text{-}S) = s/(s + t)$, where t equals the selection coefficient of *Hb-S/Hb-S*, and s equals the selection coefficient of *Hb-A/Hb-A*. Since fitness = 1 − selection coefficient, $f(Hb\text{-}S) = 0.12 / (0.12 + 0.86) = 0.122$.

21.24 $q = u/s = (5 \times 10^{-5})/0.8 = 0.0000625$

21.26 a. The data fit the idea that a single *Bam*HI site varies. The probe is homologous to a region wholly within the 4.1-kb piece bounded on one end by the variable *Bam*HI site and on the other end by a constant site. When the variable site is present, the hybridized fragment is 4.1 kb. When the variable site is absent, the fragment extends to the next constant *Bam*HI site and is 6.7 kb long. People with only 4.1- or only 6.7-kb bands are homozygotes; people with both are heterozygotes.

b. The "+" allele of the variable site is present in $2(6) + 38 = 50$ chromosomes, and the "−" allele is present in $2(56) + 38 = 130$ chromosomes. Thus, $f(+)$ is 0.25 and $f(−)$ is 0.75.

c. If the population is in Hardy-Weinberg equilibrium, we would expect $(0.25)^2$ or 0.0625 of the sample to show only the 4.1-kb band. This would be 6.25 individuals. We observed 6. We expect $(0.75)^2$ or 0.5625 to be homozygous for the 6.7-kb band, which is 56.25 individuals. We saw 56. Finally, we would expect $2(0.25)(0.75)$ or 0.375 to be heterozygotes, or 37.5 individuals. We observed 38. The observed numbers are so close to the expected that a χ^2 test is unnecessary.

21.28 To calculate the percent polymorphic loci, we divide the number of loci with more than one allele (two) by the total number of loci examined (five).

$$\frac{2}{5} = 0.40$$

Heterozygosity is calculated by averaging the frequency of heterozygotes for each locus. The frequency of heterozygotes for the *AmPep*, *ADH*, and *LDH-1* loci is zero. Thirty-five out of 50 individuals are heterozygous at the *MDH* locus, and 10 out of 50 individuals are heterozygous at the *PGM* locus. Thus,

$$\frac{0 + 0 + 0 + \dfrac{35}{50} + \dfrac{10}{50}}{5} = \frac{0.9}{5} = 0.18$$

21.30 a. Mutation leads to change in gene frequencies within a population if no other forces are acting and introduce new genetic variation. If population size is small, mutation may lead to genetic differentiation among populations.

b. Migration increases population size and genetic variation within populations and equalizes gene frequencies among populations.

c. Genetic drift reduces genetic variation within populations, leads to genetic change over time, increases genetic differences among populations, and increases homozygosity within populations.

d. Inbreeding increases homozygosity within populations and decreases genetic variation.

21.32 Comparison with the rates of nucleotide substitution in Table 21.9 indicates that sequence A has as high a rate as that typically observed in mammalian pseudogenes. In addition, the rates of synonymous and nonsynonymous substitutions are the same. These observations suggest that sequence A is either a pseudogene or a sequence that provides no function, because high rates of substitution are observed when sequences are functionless.

Sequence B has a relatively low rate of nonsynonymous substitution but a relatively high rate of synonymous substitution. This is the pattern we expect when a sequence codes for a protein; thus, sequence B probably encodes a protein.

21.34 Concerted evolution is maintenance of sequence uniformity in multiple copies of a gene. It leads to similar sequences in multiple copies of the same gene within a species but different sequences among different species. It can be found in both coding and noncoding sequences (e.g., ribosomal RNA genes). It is intriguing that the process(es) involved in concerted evolution to continually enforce uniformity among multiple copies of a sequence must also allow for rapid differentiation among species.

CHAPTER 22

22.1 The mean is obtained by summing the individual values and dividing by the total number of values. The mean head width is ${}^{25.21}/_8 = 3.15$ cm, and the mean wing length is ${}^{281.7}/_8 = 35.21$ cm.

The standard deviation equals the square root of the variance (the square root of s^2). The variance is computed by squaring the sum of the difference between each measurement and the mean value, and dividing this sum by the number of measurements minus one. One has:

$$s_{headwidth} = \sqrt{s^2} = \sqrt{\frac{\sum (x_i - \bar{x})^2}{n-1}} = \sqrt{\frac{1.70}{7}} = \sqrt{0.24} = 0.49 \text{ cm}$$

$$s_{winglength} = \sqrt{s^2} = \sqrt{\frac{\sum (x_i - \bar{x})^2}{n-1}} = \sqrt{\frac{413.35}{7}} = \sqrt{59.05} = 7.68 \text{ cm}$$

22.3 ${}^{252}/_{1,024}$

22.4 Each pure-breeding parent is homozygous for the genes (however many there are) controlling the size character, and hence each parent is homogeneous in type. A cross of two pure-breeding strains will generate an F_1 heterozygous for those loci controlling the size trait. Since the F_1 is genetically homogeneous (all heterozygotes), it shows no greater variability than the parents.

22.7 a. 8 cm

b. 1 (2 cm) : 6 (4 cm) : 15 (6 cm) : 20 (8 cm) : 15 (10 cm) : 6 (12 cm) : 1 (14 cm)

c. Proportion of F_2 of *AA BB CC* is $(\frac{1}{4})^3$; proportion of F_2 of *aa bb cc* is $(\frac{1}{4})^3$; proportion of F_2 with heights equal to one or other is $(\frac{1}{4})^3 + (\frac{1}{4})^3 = \frac{2}{64}$.

d. The F_1 height of 8 cm can be produced only by maintaining heterozygosity at no less than one gene pair (e.g., *AA BB CC*). Thus, although 20 of 64 F_2 plants are expected to be 8 cm tall, none of these will breed true for this height.

22.10 It is a quantitative trait. Variation appears to be continuous over a range rather than falling into three discrete classes. Also, the pattern in which the F_1 mean falls between the parental means and in which the F_2 individuals show a continuous range of variation from one parental extreme to the other is typical of quantitative inheritance.

22.13 The cross is *aa bb cc* (3 lb) $\times$ *AA BB CC* (6 lb), which gives an F_1 that is *Aa Bb Cc* and which weighs 4.5 lb. The distribution of phenotypes in the F_2 is 1 (3 lb) : 6 (3.5 lb) : 15 (4 lb) : 20 (4.5 lb) : 15 (5 lb) : 6 (5.5 lb) : 1 (6 lb).

22.15 a. The progeny are *Aa Bb Cc Dd*, which are 18 dm high.

b. (1) The minimum number of capital-letter alleles is one, and the maximum number is four, giving a height range of 12 to 18 dm. (2) The minimum number is one, and the maximum number is four, giving a height range of 12 to 18 dm. (3) The minimum number is four, and the maximum number is six, giving a height range of 18 to 22 dm. (4) The minimum number is zero, and the maximum number is seven, giving a height range of 10 to 24 dm.

22.17 Let n = total number of alleles, s = number of capital alleles, t = number of lowercase alleles, a = chance of obtaining a capital allele, b = chance of obtaining a lowercase allele, and $x! = (x)(x-1)(x-2)...(1)$, with $0! = 1$. Then the chance p of obtaining progeny with a specified number of each type of allele is given by:

$$p(s,t) = \frac{n!}{s!t!} a^s b^t$$

$$p(0,10) = \frac{10!}{0!10!} \left(\frac{1}{2}\right)^0 \left(\frac{1}{2}\right)^{10} = \frac{1}{1,024}$$

$$p(1,9) = \frac{10!}{1!9!} \left(\frac{1}{2}\right)^1 \left(\frac{1}{2}\right)^9 = \frac{10}{1,024}$$

$\left.\right\} \frac{11}{1,024}$ bluish-white

$$p(2,8) = \frac{10!}{2!8!} \left(\frac{1}{2}\right)^2 \left(\frac{1}{2}\right)^8 = \frac{45}{1,024}$$

$$p(3,7) = \frac{10!}{3!7!} \left(\frac{1}{2}\right)^3 \left(\frac{1}{2}\right)^7 = \frac{120}{1,024}$$

$\left.\right\} \frac{165}{1,024}$ light tan

$$1 - \frac{11 + 165}{1,024} = \frac{848}{1,024}$$ greyish-brown

22.18 In neither of these cases is any information provided about the environments in which the twins were raised. While it would initially appear that there is some genetic component in each case, there is no way to evaluate the role of environment. It could be that the identical twins sampled in each case were raised in an identical environment, and this could entirely account for the smoking behavior as well as the IQ results.

22.20 The selection differential is 0.14 g. The selection response is 0.08 g. $h^2 = \frac{0.08}{0.14} = 0.57$.

22.21 The heritability of blood pressure is zero because there would be no genetic variability in the population to affect blood pressure. Blood pressure would vary only because of salt exposure.

22.22 Heritability is specific to a particular population and to a specific environment and cannot be used to draw conclusions about the basis of populational differences. Because the environments of the farms in Kansas and in Poland differ and because the two wheat varieties differ in their genetic makeup, the heritability of yield calculated in Kansas cannot be applied to the wheat grown in Poland. Furthermore, the yield of TK138 would most likely be different in Poland and might even be less than the yield of the Russian variety when grown in Poland.

22.23 Selection differential = $14.3 - 9.7 = 4.6$

Selection response = $13 - 9.7 = 3.3$

Narrow-sense heritability = selection response/selection differential = $\frac{3.3}{4.6} = 0.72$

22.24 Selection response = narrow-sense heritability $\times$ selection differential. Selection response = 0.60×10 g = 6 g, so egg weight should increase 6 g in the offspring of the selected chickens.

22.25 The variance that responds to selection is the additive genetic variance. The narrow-sense heritability for each of the traits is V_A/V_P and can be used to predict how a trait will respond to selection. This is 0.165 for body length, 0.061 for antenna bristle number, and 0.144 for egg production. Thus, body length will respond most to selection, and antenna bristle number will respond least to selection.

CREDITS

INDEX

Note: Page numbers in *italics* indicate material in figures and tables.